John Hill

A General Natural History

Or, New and Accurate Descriptions of the Animals, Vegetables, and Minerals of the Various Parts of the World

John Hill

A General Natural History
Or, New and Accurate Descriptions of the Animals, Vegetables, and Minerals of the Various Parts of the World

ISBN/EAN: 9783742801258

Manufactured in Europe, USA, Canada, Australia, Japa

Cover: Foto ©Klaus-Uwe Gerhardt /pixelio.de

Manufactured and distributed by brebook publishing software (www.brebook.com)

John Hill

A General Natural History

A GENERAL

NATURAL HISTORY:

OR,

NEW and ACCURATE

DESCRIPTIONS

OF THE

Animals, Vegetables, and Minerals

Of the VARIOUS

PARTS of the WORLD;

WITH

Their VIRTUES, and USES, in MEDICINE and MECHANICS:

ILLUSTRATED

BY a GENERAL REVIEW of the Knowledge of the Ancients; and the Improvements and Discoveries of later Ages in these Studies.

WITH

A GREAT NUMBER OF FIGURES, ELEGANTLY ENGRAVED.

By JOHN HILL, M.D.

MEMBER of the IMPERIAL ACADEMY, of the ACADEMY of FENNES and BOURDEAUX, &c. &c.

ENLARGED WITH ADDITIONAL PLATES.

LONDON:

Printed for the AUTHOR in St. James's Street.

MDCCLXXIII.

THE
PREFACE.

A LARGE field was before the Author of these volumes, when he first laid down the plan of the Work. Part of it had been more, part less cultivated, and a part, not to reproach the Naturalists too much, left almost in it's native wildness. When I mention the part most cultivated, I shall readily enough be understood to mean that which relates to vegetables; the animal world has, of late years, began also to be examined, but what has been written concerning it, is not by any means to be compared with the labours of the late and present age, in regard to Botany. The Fossils, although of equal number, and of equal value with even the Animals, or with the Plants in use; most valuable of all in point of property and intrinsic value; had been scarce at all regarded as the objects of a regular disquisition.

FROM the earliest time indeed men had written of them. Theophrastus, whose works on this subject I some time since published, although one of the earliest of those whose works have travelled down to us, mentions the study as not new, and even quotes Diocles, and some others who went before him. But although the Greeks, and the Latins after them, have treated of minerals, what they have written serves but to shew how little they were acquainted with them, nor has the industry of the Germans served to much better effect, than the genius of those earlier investigators. The one have described a few, the others have understood a few. The far greater part lay untouched, unnoticed, and even unnamed: And, of the small number that either had considered, none were yet arranged into any kind of order.

IF we except our Dr Woodward, it will be hard to mention any author, who has made but a tolerable attempt toward distributing the Fossils into method, and forming the study of them into a science. Linnaeus has followed him, but, instead of order, he has introduced confusion: neither did the arrangement of Woodward, in his catalogues, by any means approach towards a perfect system. The real characters of those bodies he was wholly unacquainted with, and it was therefore impossible he should arrive at any sound method of disposing them.

IN

The PREFACE.

IN this situation was the study of this part of Natural History, at the time when these volumes were undertaken. The Author thought it therefore eligible, in that part, to proceed solely upon the plan of his own thoughts, and that volume is written accordingly, in a manner different from all that has appeared upon the subject, and does not refer to any other work, or adopt the opinions of any other author.

I have been accused of establishing a new method, and of introducing new words on this occasion: but it was necessary, and therefore it was excusable. In regard to the subjects of the two succeeding volumes, others had written of them scientifically, and they had been already arranged into, if not perfect, at least very rational methods. It had therefore been indeed culpable to have formed new systems, and introduced new words in these, because it was not necessary. Some good might have arisen from it; but that good would not have been proportioned to the trouble, which it would have occasioned to such as had already studied in those systems.

TO avoid this, I have, both in the History of Plants, and that of Animals, adopted so much of the method of Linnæus, as is consistent with observation: and have, so far as my examinations of the bodies would enable me, filled up the deficiencies, and corrected the errors of that plan. The reader is not to understand this, as regarding any part of the work, beside the characters, which are prefixed in a few lines to each genus: the descriptions are, in general, formed from the objects themselves; those which are not, that is, those which had for their subjects plants or animals, of which neither my own private collection, nor that of any of our naturalists, afforded specimens, are the result of a careful examination of what authors have said, compared with one another.

THAT I chose the method of Linnæus for the arrangement of these bodies, was, in part, because it is nearer a natural system than that of any other, and, in part, because it has been most read: those who study these subjects, had acquainted themselves with his new names, much better than the old, and, not to load them with unnecessary words, I have, in as many parts as I could, retained these with their original sense.

LINNÆUS, though full of merit, is full of faults, as well as of imperfections. He has written no History either of plants or animals, so that it was but very little in which I could be obliged to him, if I had inclined to be so; and in that the inquiry has generally pointed out errors, which it was necessary to amend. Where I depart from him in the characters of any species, or of a genus, it is not in carelessness or wantonness, but from observation. I do not lay down his characters as perfect; but characters of which his are the foundation, and which an examination of the bodies they concern, has altered or approved.

A very

The PREFACE.

A very considerable part of botany lay untouched, even by this author. All that concerned the lesser, commonly called the imperfect plants, was to be begun de novo; *in this I have been obliged to form a new method, and to introduce new words as in the mineralogy; but, as there were none before for the same objects, these must be excused: they have a sufficient plea for this, as they lay down, in a regular and systematic method, a large portion of the History of Plants, never disposed in any degree of order before.*

AS in regard to the Plants, so with the Animals, the lesser have been disregarded. It required a peculiar series of experiments to discover the characters of that minute part of the insect creation, which was too minute for the unassisted sight; and authors, either through unacquaintance with the only means by which these could be discovered, the assistance of glasses, or from mere neglect, have passed them over in silence. Certainly, their want of magnitude does not exclude them from their rank among animated beings; yet this is the only History of Animals, in which they have been honoured with a name.

I have been at the pains of viewing all these by the microscope, and not according to the erroneous accounts of those superficial writers, who have treated of some of them, or of those who only pretend to have seen others; but, from what appeared on that examination, I have arranged them into a regular method, and given them denominations. The reader who is addicted to cavilling, must not find fault with me for this. I have not changed their names, the greatest part of them had none before. For the other parts of this volume, they are finished on the same plan with those of the second: and have just as much obligation as they to the writers who have gone before.

IT is with great pleasure that I see a spirit of this part of knowledge raising itself among us. We have the foundation in our hands, and there requires little more than application for the superstructure. Mr da Costa, a person of great knowledge in mineralogy, promises us an account of many things in his excellent collection; I don't know whether he does not mean to call it a History of Fossils. Mr Baker is about to publish figures of many objects seen through microscopes: though this collection will hardly be without error, yet, doubtless, there will be much curiosity in it; and Captain Armstrong, in a history which he has just published of Minorca, has shewn an uncommon knowledge in these things, for one who could have but occasionally studied them. I should say more of this Author, had he said less of me. The Royal Society of London is a name also, from which much might be expected. I am sorry to confess, that their labours have, to this time, by no means answered the expectation of the world, but they have now

entered

entered on new regulations, under the influence of which it is scarce to be doubted, but they will do honour to themselves and to their country. I have fallen under their displeasure for giving cause to these; but that is of little consequence, in comparison with the advantages that will accrue from those censures which occasioned the reformation.

UPON the whole, it is a singular satisfaction to me to see a study, which I so much love, rising at this time to a new height; and it is not the least part of that satisfaction that I have in these volumes, and some other Works on the same subject, contributed somewhat towards it. With what advantages I have delivered my sentiments on these subjects, others are to judge; it satisfies me, that I have done it with truth.

THE

THE HISTORY OF ANIMALS.

PART I.

Of the lesser Animals, called, ANIMALCULES and INSECTS.

BOOK I.

Of ANIMALCULES.

ANIMALCULES are animals so minute in their size, as not to be the immediate objects of our senses, and are only seen by the assistance of microscopes. These are vastly more numerous than any other part of the animal creation; but the species, on a close examination, are found to be extreamly few, in proportion to the number of the individuals. If we were to take all the accounts of the people who have written on the subject of microscopical discoveries as authentic, we should indeed be led to imagine nature as fruitful in the diversity of genera, as well as species, among these her minuter, as in her larger, works; but an impartial and attentive examination excludes the far greater number of what are described and figured by those writers, from the rank of distinct existences. Many of them have had their origin from the mere imperfect observations of other known species; and not a few from the absolute want of candour and ingenuity in the writers, who have described and figured things they never saw.

We have delineations of little men and women in the figures of the Animalcula in semine of some authors; and other extravagancies of a little less ridiculous kind in those of the Animalcules, found in the infusions of vegetables as given by others. These would need no more than a repetition to condemn both them and their authors; but the far greater number of the additional existences in this part of natural history has been owing to error, not to disingenuity. The same creature, in it's different positions, will appear of very various shapes and figures; and among these creatures, as among the insect tribe, many have a power of enlarging and contracting their bodies occasionally, in a surprising manner. It is difficult to guard against the being deceived by the figures given with the greatest truth and accuracy from these, in their several states, as of so many different creatures: indeed there is no way of avoiding it, but by making our own observation the basis of our accounts, and paying a very limited credit to those which we receive on the testimony of others. In consequence of a close investigation, with an apparatus more calculated for making discoveries than any of the common kinds, I have added some unknown species to the Animalcule kingdom; but, notwithstanding this, the discoveries made of the errors and fictions in the accounts of others, in improperly multiplying the known species, have so greatly reduced the list of the supposed Animalcules, that the number will appear much smaller here, than in any work that has been published on the subject.

The most obvious distinctions among these minute creatures are, that some have, and others have not, tails; and that some have, and others have not visible limbs. According to these characters, they may be conveniently arranged under three classes: the first, containing those which have no visible limbs, nor any tail; the second, those which have tails; the third, those which have visible limbs; distinguished by the names of, 1. Gymnia; 2. Cercaria; 3. Arthronia.

ANIMALCULES.

Class the First.

GYMNIA.

Those which have no tails, nor any visible limbs.

ENCHELIDES. — Capillary Eels.

ANIMALCULES, which have no visible limbs or tails, and are of a cylindric or subcylindric figure.

These have been mentioned by all the microscopic writers, under the general name of Capillary Eels, but their figures have not been sufficiently attended to, and consequently the several species have never been distinguished.

Enchelis corpore utrinque attenuato graciliori.
The Enchelis, with a slender body smaller at the ends. — Pepper Water Eel.

This is an Animalcule of the smaller kind, and is not visible without considerably powerful glasses. It's whole body is of the same thickness, except that it becomes gradually smaller at each end: it's surface appears uniform and smooth; it's colour is whitish: it is very thin, in proportion to it's length, and very nimble in it's motions; it's whole body is frequently convoluted and drawn into different figures; and, but for the observing it's progressive motion, which are always made with the head forwards, it would not be easy to distinguish at which extremity of the body that were placed. It is considerably transparent, but less so than some of the others of this genus.

It is frequent in the thin skin that appears on the surface of the cold infusions of vegetable substances, after they have stood three or four days; an infusion of pepper usually furnishes great numbers of it, but it is in nothing so certain and numerous, as in water in which cubebs bruised have been some time infused.

Enchelis corpore crassiori subaequali.
The Enchelis, with a thicker and nearly equal body.

This is not of the number of the very smallest Animalcules. It is oblong and thin, but considerably thicker than the precedent species: it is so perfectly transparent, that there is scarce any colour visible in any part of it, but what there does appear is whitish: it's surface is perfectly smooth and even, and the head and tail scarce at all smaller than the middle: it is but languid in it's motions, and rarely alters the figure of it's body.

It is found in infusions of elder flowers, and many other vegetable substances: a third magnifier shews it very distinctly.

Enchelis

Enchelis capite attenuato, cauda truncata.
The Enchelis, with the head small, and the tail truncated.

This is one of the larger Animalcules: it is considerably long, in proportion to it's thickness, and is very transparent: it's surface is smooth; it's skin appears extremely thin, and scarce at all coloured: it is smallest at the head; it's whole body, except at this part, is of the same thickness, and the end, where the tail should be, is truncated. It is but languid in it's motions.

It is found in fermented liquors, that have stood till vapid, particularly in cyder. A third magnifier shews it very distinctly.

Enchelis corpore utrinque attenuato, annulato. **Vinegar Eel.**
The Enchelis, with an annulated body, small at each end.

This is a very minute Animalcule. It is very slender, in proportion to it's length, and is transparent; but it's skin seems thicker than in either of the former, and of a brownish colour. It's extremities are both small, but the head runs more to a point than in the other; and the surface of the body, when strictly examined, appears not to be smooth as in the others, but annulated or composed of a number of rings or joints, like that of an earthworm; and it's motions are performed, by extending and contracting it's body in the same manner.

It is found, like the former, in fermented liquors, that have stood till they begin to decay, but particularly in dead vinegar. The other species are also sometimes found in that fluid, but more rarely, and never any of them till it becomes quite dead and vapid; hence the absurdity of the opinion, that vinegar owes it's sharpness to these Animalcules, is sufficiently evident.

These are the distinct species of this genus, which I have hitherto observed. Authors in general seem to understand the capillary eels as of one species, and even add to them the larger animals of this form, found in dead small beer, and very much decayed vinegar, under the same name, though these are in reality the offspring of a common small black fly, to be described in it's place.

CYCLIDIA.

ANIMALCULES which have no visible limbs or tail, and are of a roundish or elliptic figure. Some of these have been figured and described by the microscopical writers, but in a very vague and indeterminate manner, and without names.

Cyclidium corpore orbiculari pellucidissimo.
The Cyclidium, with a round and very pellucid body.

This is of the number of the smallest Animalcules. It is of an exactly and perfectly round shape, and so pellucid, that in a full light nothing is distinguishable of it but a thin line marking it's circumference: in the best adapted light it appears no other than an extremely thin and smooth membrane, filled with clear water; it is colourless as the fluid it swims in; it's motions are very quick, and it is a continual prey to the larger inhabitants of the same fluid.

It is common in ditch water in summer; scarce a drop can be any where taken up, that does not shew millions of it. It requires a second magnifier, or even a first, to see it distinctly.

Cyclidium corpore orbiculari subfusco.
The Cyclidium, with a brownish, orbicular body.

This is of the number of the smaller Animalcules, but it is considerably larger than the former. It is less exactly round also in it's shape, though it never much recedes from that figure. It is pellucid, but less so than the former; it's skin has somewhat of a brownish tinge, and is smooth; but there are seen within the body some spots and lines which mark out where the intestines lie. It is very nimble in it's motions, and

and frequently will turn round with great rapidity, moving progressively forward at the same time.

It is common in ditch water. A third magnifier shews it very distinctly.

Cyclidium corpore elliptico.
The Cyclidium, with an elliptic body.

This is one of the larger Animalcules. It is of an elliptic, or, as we are apt to express it, an oval figure: it's length is about once and a half it's diameter, and both ends are equally large and obtusely rounded: it's colour is a dusky brown, with a cast of the olive; it's skin is not smooth, but as it were granulated, and the traces of the intestines are very visible in it. It is considerably transparent, and usually appears flattish, sometimes more distended; it is very quick in it's motions, which are always first sideways, and then progressive.

It is frequent in the infusions of vegetable bodies, particularly in a cold infusion of bruised henbane seeds. In the film that appears after three or four days on this infusion, they are so amazingly numerous, that a little drop of it shews to the third magnifier, of a double microscope, many hundreds of them rolling over one another.

Cyclidium corpore ovato.
The Cyclidium, with an oval body.

This is not of the number of the smallest of the Animalcules, though much inferior in size to the former. It is of a tolerably regular ovate figure, representing a hen's egg in miniature; both ends are rounded, but that where the head is situated is much smaller than the other. The surface of the whole Animalcule is perfectly smooth; it's skin colourless and pellucid, and the traces of the intestines are easily distinguished in it, in form of roundish, aggregated spots, which reflect a variety of colours in most lights. It is not very nimble in it's motions; a second magnifier shews it very distinctly.

It is found not unfrequently in ditch water, but much oftener in the cold infusions of vegetables, as in that of white poppy seed.

PARAMECIA.

ANIMALCULES which have no visible limbs or tails, and are of an irregularly oblong figure.

Paramecium corpore sublongo, utrinque obtuso, volubili.
The Paramecium, with an oblong, voluble body, obtuse at each end.

This is of the number of the larger Animalcules. It is of an oblong figure; it's length is equal to about three times it's diameter, and it's colour a very pale brown, with a faint admixture of greenish. The head and tail are both obtuse, but they are smaller than the rest of the body; the skin is thin, smooth, and transparent, and the traces of the intestines are distinctly visible through it. It is tolerably swift in it's motions, and is able to twist and turn about it's body, and to double and fold it up in various manners, so as to appear of a very different shape from it's real one; and it has hence unluckily imposed upon many of the microscope observers, who have figured it as of different species under this diversity of appearances.

It is not uncommon in ditch water; a third magnifier in the double microscope shews it very distinctly.

Paramecium corpore sublongo, caput versus attenuato.
The Paramecium, with an oblong body, smallest toward the head.

This is not of the smallest kind of Animalcules. It's length is about three times and a half it's diameter; it's hinder extremity is large and rounded, from this it grows gradually smaller all the way, but particularly near the head; it is much thicker

than

than in any other part, so as to have the appearance of a neck: the head, at the extremity of this, sometimes is thrust out, and appears roundish, in form of a button; but more usually it is drawn in, and the end is as it were truncated. It's skin is smooth and pellucid; the traces of the intestines are easily seen through it: it's motions are slow and irregular, and the upper part of the body is frequently incurvated.

It is found in many vegetable infusions: in none more plentiful than in that of the dried leaves of yarrow. A fourth magnifier will shew it; a third gives all it's parts distinctly.

Paramecium corpore sublongo, medium versus angustato.
The Paramecium, with an oblong body, narrowest toward the middle.

This is one of the smaller Animalcules, but not of the most minute kind. It's length is equal to about twice and a half it's diameter; it is somewhat flatted, and is considerably larger at the tail than the head, and rounded at both these extremities; but what is singular in it's figure is, that it has a cincture, as it were, toward the middle, but nearer the head than the tail, which gives it an appearance of being divided into two portions. It's surface is smooth, it's colour brownish; it is considerably transparent, and the traces of the intestines are very visible in it. It is tolerably nimble in it's motions, and is apt to throw it's body into various odd positions.

It is frequent in vegetable infusions; that of pepper often affords it, and a cold infusion of the nux vomica scarce ever fails of shewing numbers of it.

Paramecium corpore oblongo graciliore.
The Paramecium, with a slender, oblong body.

This is by much the slenderest of all the Animalcules of this genus, but it in no degree approaches to the figure of the capillary eels. It's length is equal to about five times it's diameter, and it has somewhat of the general figure of a leech; it is flat and thin; it's surface is perfectly smooth, and it's skin so transparent, that the intestines are beautifully seen through it. It is thickest and broadest in the middle, and terminated obtusely at both ends; it is brisk and nimble in it's motions, and is very distinctly seen with a third magnifier.

I have met with it in great abundance in an infusion of zedoary root, and often in ditch water.

CRASPEDARIA.

ANIMALCULES without any visible limbs or tails, but with an apparent mouth, and a series of fimbriæ round it, in manner of a fringe.

Two species of this genus have been figured by some of the microscopical writers, but without names, or any accurate descriptions.

Craspedarium corpore suborbiculari.
The Craspedarium, with a roundish body.

This is an Animalcule of the smaller kind. It's shape is nearly orbicular, but somewhat flatted, and, as it were, truncated slightly at the head, and at the opposite extremity: it's skin is perfectly smooth, and very transparent; the traces of the intestines are beautifully seen through it. The hinder extremity appears closed, but the anterior one or head has an evidently open and large mouth, surrounded by a beautiful fringe or series of slender, flexile filaments, which are in continual motion, and by that means draw lesser bodies, such as the little round animalcules, before described, and others, about the opening of the mouth, for the readily supplying the creature with food. The motions of this Animalcule are languid, and it's general posture is with the head upwards. The fringe about the mouth is in a continual vibratory motion.

It is frequent in shallow, muddy waters; a second magnifier shews it distinctly.

Craspedarium corpore subovato.
The Craspedarium, with an oval body.

This is a somewhat larger Animalcule than the former, but, if many hundred times bigger than it is, would still be undistinguishable by the naked eye. It's figure approaches to that of an egg, but it's head is situate at the large, not the smaller, end; the tail extremity is pointed at it's termination; the other is rounded, but truncated: the skin is smooth, and very transparent, and the traces of the intestines are easily seen. At the larger extremity there is a considerable opening, surrounded by a short but very thick fringe. The filaments composing this are in continual motion, and draw a multitude of lesser Animalcules continually into the creature's mouth.

It is frequent in that coloured water which runs from large dunghils; a third magnifier shews it very distinctly.

Craspedarium corpore cylindrico.
The Craspedarium, with a cylindric body.

This is of the number of the larger Animalcules. It is oblong, thick, and inflated; it's length is equal to about four times it's diameter; it's skin is perfectly smooth and colourless, and so transparent, that the traces of the intestines are easily seen: it's hinder extremity is rounded, and it's anterior truncated; there is a large opening visible at this extremity, which serves as a mouth, and is surrounded by a series of filaments, forming a long and delicate fringe. These are in continual motion, and, as they vibrate about, reflect a variety of colours: the creature is slow in it's motions, nature having provided it with this fringe about it's mouth, to draw in it's food, by the motion it makes in the water, without it's running about in search of it.

It is frequent in rain water, that has stood several days, and in many of the vegetable infusions.

ANIMALCULES.

Class the Second.

CERCARIA.

Those which have tails but no visible limbs.

BRACHURI.

ANIMALCULES with tails shorter than their bodies.

Brachurus corpore orbiculato, cauda brevissima.
The Brachurus, with an orbicular body, and a very short tail.

This is one of the smaller Animalcules. It is of a round shape, inflated and transparent; it's skin is perfectly smooth, thin, and colourless; the lineaments of the intestines are seen very distinctly through it; it has no visible mouth: but at the extremity of the body, opposite to that which goes forward, it has a short tail scarce equal in length to one sixth of the diameter of the body: this little appendage is in almost continual motion, and, when strictly examined, is found to be forked at the extremity. The creature moves very swiftly about; and often, when it has fixed itself by this tail, gives the body a rotatory motion backwards and forwards, with an almost incredible swiftness.

It is frequent in pond water, particularly in such as has been shaded by a covering of duckweed.

Brachurus corpore ovato.
The Brachurus, with an oval body.

This is of the number of the larger Animalcules. It's figure is that of a hen's egg; it's skin perfectly smooth and transparent, so that the lineaments of the intestines are evidently seen through it. It's larger end is the head, and always is directed forward in it's progressive motions; at the other, or smaller extremity, there is a tail equal to three fourths of the body in length, thicker at it's insertion, and gradually growing smaller from thence to the extremity, where it terminates in a single, not a forked, point. The creature often fixes itself by the extremity of this tail; and at other times it's motions are but sluggish.

It is one of the Animalcules, figured by the microscopic writers, as the inhabitants of pepper-water, and is frequent also in many other infusions of vegetable substances: it has been said to resemble the tadpole, but improperly.

Brachurus corpore rotundiore, cauda fimbriata.
The Brachurus, with a roundish body, and a fimbriated tail.

This is one of the smaller Animalcules. It is of a roundish figure, but somewhat approaching to an oblong or elliptic; both extremities are of equal bigness: it is very transparent, yet it is scarce possible to distinguish any traces of the intestines in it. At the extremity of the body, opposite to that which goes forward in the creature's progressive motion, and which consequently is the head, there is a short and tolerably thick tail. It is equal to about half the length of the body, and is all the way of the same thickness: it is naturally hollowed at the extremity, and by that means alone is capable of being fixed very firmly; but, as if this were not sufficient for it's security and other purposes, a series of filaments, in form of a fringe, are occasionally thrust forth from it, which hold the creature with great firmness in it's place. I scarce remember a more pleasing observation in the use of the microscope, than has occurred to me, while writing this description. I had the creature before me under the microscope, and in the same drop of water one of the last species of the former genus. This larger Animalcule was vibrating it's fringe about the mouth, and drawing into it the lesser creatures that came in his way. This smaller Animalcule I first observed attempting, while at some distance, to force itself out of the reach of the current made in the water, by the motion of this fringe; when that was found impracticable, and it was drawn nearer the destroyer, it fixed itself down by it's tail, in it's usual naked state; but when this also proved fruitless, and the motion of the water removed and brought it still nearer the place of destruction, it once more fixed itself by the tail, and, throwing out it's fringe of filaments, became so securely fastened by their means, that no effort of the enemy afterward could move it.

It is frequent in many of the vegetable infusions, particularly in that made of the calamus aromaticus root: it requires a second magnifier to see it's parts distinctly.

Brachurus corpore subrotundo, posterius lunato.
The Brachurus, with a roundish body lunated behind.

This is not one of the very smallest of the Animalcule tribe. It's body, in the whole, somewhat approaches to a roundish figure, but it is truncated, as it were, at the head, and lunated or formed into the shape of a crescent at the opposite extremity, the points or horns of the crescent pointing downwards; between these is placed a short tail, of a conic figure, forked at the end, and in continual motion. The whole Animalcule is colourless, and perfectly transparent, resembling nothing more than a thin skin filled with water: it is very quick in it's motions.

It is frequent in ditch water, and in rain water that has stood exposed for some days: a third magnifier shews it very distinctly.

Brachurus capite rotundo, corpore ovato in caudam desinente.
The Brachurus, with a roundish head, and an oval body terminating in a tail.

This is one of the smaller Animalcules, but it is extremely singular in it's figure. It's head is little and round, and is separated from the body by an evident neck; the body is largest at that part where the neck is inserted, and from thence gradually tapers to the tail, which is not, as in all the other species, affixed to the body; but the body itself, as in the serpent and lizard kind, evidently is continued into, and terminates in it. It is pellucid, but of a somewhat brownish colour; the tail is of about half the length of the body, and the head is of about one third of it's bigness. It's motions are very quick and various; a second magnifier shews it distinctly: it's tail is neither forked nor hollowed at the end, but terminates in a point, and seems rather of use to it in it's motions, than to fix it.

It is one of the Animalcules sometimes found in pepper-water, and very frequently in an infusion of cress-seed.

MACROCERCI.

ANIMALCULES with tails longer than their bodies.

Macrocercus corpore rotundo, cauda tenuissima.
The Macrocercus, with a round body, and a very slender tail.

This is of the number of the smaller Animalcules. It's body is of a very regularly orbicular figure, distended, and perfectly transparent: the skin is very thin and colourless, and the lineaments of the intestines are very visible: there is no mark to distinguish the head, but, at that part of the body opposite to that which goes foremost in the creature's progressive motions, there is a tail of at least four times the length of the body, and very slender; the creature is not very quick in it's motions.

It is found in great plenty in an infusion of galangal root, and often in pepper-water.

Macrocercus corpore utrinque angustato et truncato, cauda longissima.
The Macrocercus, with the body small and truncated at each end, and with a very long tail.

This is an extremely singular animalcule, and is not of the smallest size. It's body is of a figure approaching to elliptic, but runs out into a kind of neck at each extremity much smaller than the other part, and is truncated at the end: in the truncated end of the anterior part is probably situated the mouth of the animal; from the center of the opposite end there grows a slender tail, equal to at least eight times the diameter of the body. This is usually carried straight, but the creature has a power of drawing it up into a convoluted or spiral form at pleasure.

This species is found in infusions of cubebs and ginger; sometimes, but more rarely, in pepper-water.

Macrocercus corpore globoso, cauda crassiore.
The Macrocercus, with a round body, and somewhat thick tail.

This is one of the very smallest of the Animalcule class. It's body is perfectly globular; it's skin smooth, white, and glossy, but not transparent: it's tail is about twice the length of the body, and is considerably thick, and in continual motion: the animal itself is extremely brisk and lively.

It is contained in vast abundance in the semen of the stag and goat kind.

Macrocercus, corpore subovato depresso, cauda extremitatem versus attenuata.
The Macrocercus, with a suboval depressed body, with a tail very small toward the end.

This is also one of the minutest of the Animalcule class. The body is of an oval figure, depressed or flatted, and somewhat larger toward the head, than at the other extremity. It's surface is smooth and glossy; it's skin white and semitransparent, and it's tail about six times the length of the body; considerably thick at it's insertion, but gradually growing smaller towards the extremity, where it terminates in a scarce visible filament.

This is abundant in the human semen, and in that of all the monkey tribe. It requires a first magnifier to shew it distinctly.

Macrocercus, corpore longiore subcylindrico.
The Macrocercus, with an oblong subcylindric body.

This also is an Animalcule of the minutest kind. The body is oblong, and nearly cylindric, but somewhat smaller at the hinder extremity: it's length is about three times it's diameter; it's surface is perfectly smooth; it's skin white, and semitransparent: it's tail is very long, thickest at it's insertion, and gradually smaller all the way to the extremity: it's motions are very swift, and it's tail frequently twisted and convoluted.

It is found in the semen of all the dog kind, as also in that of the cat, the hedgehog, the mole, and the bat.

Macrocercus corpore articulato.
The Macrocercus, with an articulated body.

This is also of the number of the most minute Animalcules. It's body is oblong, and composed, as it were, of three or four parts, or articulations, each of a roundish figure, and joined in a series to one another; the whole is, in length, equal to three or four times it's diameter, and exhibits much the same figure that a soft cylindric body of any kind would, if tied round with a thread, in two or three places, at equal distances: the tail is of six or seven times the length of the body, and is very thick at the insertion, and gradually smaller to the extremity. The skin of the whole Animalcule is white, soft, glossy, and semitransparent.

It is found in the semen of the horse, the boar, and some other animals.

Macrocercus corpore cylindrico longiore, in caudam tenuissimam desinente.
The Macrocercus, with a very long cylindric body, terminating in a slender tail.

This is also of the number of the smallest Animalcules: it's body is long, slender, and of a conic figure; it is of ten times it's diameter in length, and is thickest at the head, where it appears truncated; and whence it is continued smaller and smaller to the tail, which is a continuation of the body, is equal to about twice it's length, and is all the way extreamly slender, but particularly toward the extremity. The skin of the whole is smooth, the colour white, and the body somewhat transparent.

It is found in the semen of the toad, and of the lizard, and serpent kinds: It requires a first magnifier to distinguish it.

Macrocercus corpore orbiculato depresso.
The Macrocercus, with a depressed orbiculated body.

This is a very minute Animalcule: it's body is of a rounded figure, but depressed or flatted, and it's tail is equal to ten times it's diameter in length: the skin of the whole is whitish, with a cast of yellow, smooth and glossy, but not at all transparent: it is very nimble in it's motions; and the tail, which is very slender at the extremity, is frequently twisted about in a very surprising manner.

It is found in the semen of the drone bee, and of many other insects.

ANIMALCULES.

Claſs the Third.

ARTHRONIA.

Thoſe which have viſible limbs.

SCELASIUS.

AN Animalcule which has viſible legs.

Scelaſius corpore ſubovato.
The Scelaſius, with the body of a ſubroval figure.

This is one of the ſmaller Animalcules: it is of the ſhape of an egg, larger at one end, ſmaller at the other, and rounded at both: it's ſkin is perfectly ſmooth, very thin, of a pale olive colour, and ſo tranſparent, that the lineaments of the inteſtines are eaſily ſeen through it. The whole body, excepting about the head, which is at the ſmaller end, is ſurrounded with a multitude of fine ſhort and ſlender bodies, having the appearance of hairs; but they are in reality ſo many legs, by means of which it creeps along the bottom, or crawls up the ſlender ſtalks of water plants: beſides thoſe which ſurround the body, there is alſo a double ſeries which run down the middle of the belly, from the head to the oppoſite extremity; and, by means of theſe, it will climb up any yet ſlenderer body.

It is very common in ditch water, and is leſs quick in it's motions than moſt of the other animalcules of whatever claſs.

Scelaſius corpore ſubcylindrico utrinque attenuato.
The Scelaſius, with a ſubcylindric body, ſmall at each end.

This is one of the larger Animalcules: it is of an oblong figure, largeſt in the middle, and gradually ſmaller toward each end: it's colour is whitiſh, with a tinge of green: it's ſkin ſmooth, and the whole ſo tranſparent, that the lineaments of the inteſtines are eaſily viſible. It appears hairy all round it's edges; but theſe hairs are ſo many legs, and it has alſo four ſeries of the ſame kind of hairs or legs running along it's belly.

It is frequent in ditch water, among the roots of duck-weed; it's motions are not very quick, but it ſeems in continual ſearch after food; a third magnifier will ſhew the Animalcule very diſtinctly; but, to examine it's limbs, there requires a ſecond, or, indeed, to ſee them in perfection, a firſt.

BRACHIONUS. Wheel Animals.

AN Animalcule which has an apparatus of arms for the taking it's prey. The apparatus, which nature has furniſhed theſe creatures with, has been greatly miſunderſtood by the microſcopical writers; they have ſuppoſed it a kind of wheels, and have thence named the creatures that are poſſeſſed of it wheel animals.

Brachionus corpore conico ſubæquali.
The Brachionus, with a conic even body.

This is one of the ſmaller Animalcules: it appears, when in a ſtate of reſt, in form of a plain ſmooth body, of a conic figure, obtuſe at the poſterior extremity, and open at the anterior, of a duſky olive colour and ſemitranſparent. When it puts itſelf in motion, it protrudes, from the open extremity, a part of it's naked body, to the whole of which this outer conic body ſeems to be but a kind of caſe or ſheath, ſerving it as a defence from injuries. From the extremity of this exerted part of the body, the creature ſoon after thruſts out two protuberances, which give it the appearance of a double head; and in each of theſe is diſcovered an apparatus in a continual motion, appearing a rotatory one, but, in reality, it is only a vibratory one, very quick repeated. Each of theſe protruded bodies has ſix arms inſerted into it, and theſe it continually ſhuts and opens over one another, in the manner of the ſhutting and opening of the human hand. Each of theſe arms is furniſhed with a double ſeries of fibres at it's edge, which, being expanded, as the arm is thrown down, and drawn in, as it is moved up again, cauſe it to ſpread to a conſiderable breadth, and make a great motion in the water, as it is thrown down; but to make very little reſiſtance, as it is drawn up again: a repeated motion of this kind has, indeed, ſomething of the appearance of that of a wheel, and has all it's effects; it forms a current in the water, and draws in with it multitudes of the ſmaller Animalcules, which ſerve as food to this creature. When it's hunger is aſſwaged, it draws in the protuberances, to which theſe arms are fixed, and, frequently retracting it's whole body into the caſe again, appears a lifeleſs conic body as at firſt.

It is frequent in the dry mud lodged in water-pipes, and on the tops of houſes: it will remain in it's inactive ſtate in this dirt many months, but, at any time, will put itſelf in motion, on adding a drop of water to a little of the dirt under the microſcope. Moſt of the late writers on the microſcopic diſcoveries have figured this Animalcule; but their deſcriptions, ſuch as they are, are ſo very different from what an accurate examination furniſhes, that I would almoſt ſuſpect they were meant of of ſome other creature, did they not agree in the more obvious particulars, and in the ſingularity of the place where it is uſually found.

Brachionus corpore conico toruloſo, cauda tricuſpidi.
The Brachionus, with a conic torulous body, and a tricuſpidate tail.

This is alſo one of the ſmaller Animalcules: It is of a conic figure, but it's body ſeems compoſed of four or five joints: it's length is about five times it's diameter: it's colour a pale brown, and it's ſkin very thin and tranſparent: it's tail is terminated by three points, by means of which, it fixes itſelf very firmly to any ſolid ſubſtance; and from it's other extremity it protrudes occaſionally two protuberances, each furniſhed with ſix arms, by means of which, it makes a current in the water for the getting it's prey.

It is not uncommon in ſtanding waters, about the roots of duck-weed: a ſecond magnifier ſhews it very diſtinctly.

Brachionus corpore breviore campaniformi, cauda brevi.
The Brachionus, with a ſhorter bell-faſhioned body, and a ſhort tail.

This is a very ſmall and very ſingular ſpecies: it is perfectly tranſparent, and ſeems indeed ſcarce any thing more than a membrane of extream fineneſs, including a quantity of a clear fluid, and with it the inteſtines: it is wholly pellucid and colourleſs: it's body is of a bell-like ſhape, but nearly as broad at the baſe, as at the mouth: the tail is ſhort and forked, and grows from the center of the baſe: the mouth is open, and is furniſhed with two of thoſe apparatus's of arms, each compoſed of ſix, which it uſes as the firſt deſcribed ſpecies, to draw it's prey to it, by making a current in the water.

It is frequent in clear waters, about the ſtalks of the ſmaller water plants: it requires a ſecond magnifier to ſhew it in any degree of perfection.

Brachionus corpore breviore campaniformi, cauda longiore.
The Brachionus, with a ſhort campaniform body, and a longer tail.

This alſo is a very minute Animalcule: it's body is very regularly of the ſhape of a bell, narroweſt at the baſe, and broadeſt at the mouth where it opens; and is furniſhed with two apparatus's, each conſiſting of ſix arms, as in the other ſpecies. It is clear, pellucid, and colourleſs, and the lineaments of the inteſtines are eaſily traced through the ſkin; it's tail is many times longer than it's body; and, as the young are produced, they often adhere, by the extremities of their tails, to the parent's tail, and by this means form a cluſter adhering to one another, and ſeeming to be all produced from one original tail or ſtalk at the bottom.

It is frequent in ponds, about the ſtalks of the water plants.

Brachionus

Brachionus corpore latiore inferne lunulato.
The Brachionus, with a broader body lunulated in it's lower part.

This is also one of the minutest of the Animalcule class: it's body is equal in breadth to it's length; it's form is not equally distended into a roundness, but flattish. It's mouth has a very large opening, and there are two of the apparatus's of arms lodged in it, each consisting of six, as in the others, and used for the same purposes. The lower part of the body is lunated, or formed into a kind of crescent, the horns of which are very long and point downwards; from the midst of the body, between these, there arises a short tail; it is forked at the extremity, and is in continual motion. The whole creature is perfectly pellucid, and of a faint greenish colour: It seems indeed like the last, but composed of a more delicate membrane including the intestines, and a clear fluid about them, seeming the same with the circumambient water. All these species, when hungry, vibrate their arms, as described under the first of them, and by this means form a current of water, which brings with it the defenceless, and yet more minute, Animalcules into their mouths.

This species is frequent in ponds in autumn: the green foam, which is often driven by the winds at that season to the shores, contains multitudes of them.

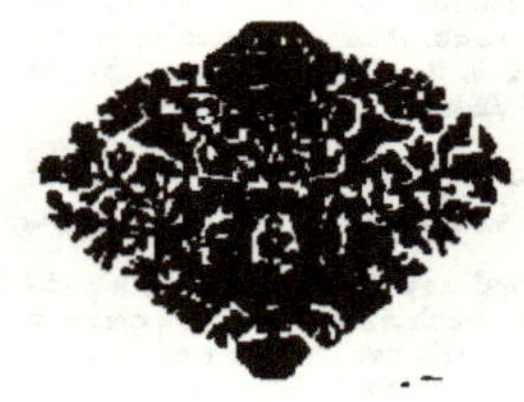

THE

ANIMALCULES

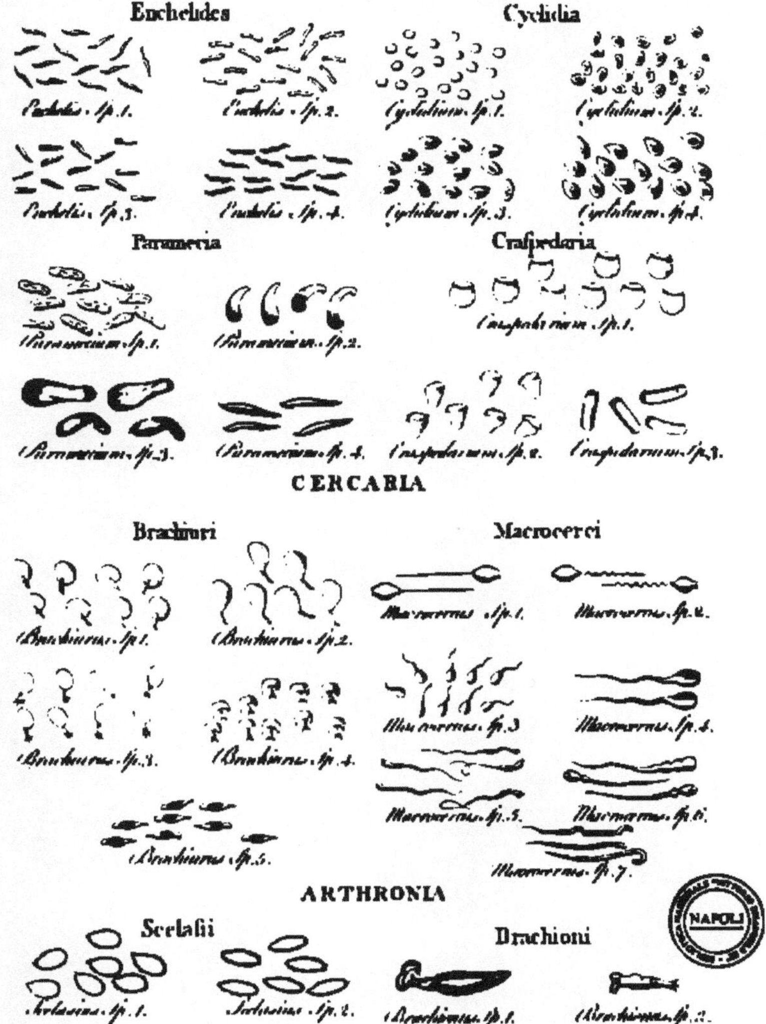

THE HISTORY OF ANIMALS.

PART I.

BOOK II.

Of INSECTS.

INSECTS are animals, whose bodies are neither regularly covered with hair, feathers, or scales, as in the generality of other animals, but either with a hard, and as it were horny, substance, or with a soft and tender skin; and of which the far greater part, that is, all that have a hard covering to their bodies, have on their heads antennæ, or, as they are usually called, horns or feelers.

Of the Insects, some have no wings, others have: they may hence be conveniently arranged into two series, under the names of Apteria and Pteraria.

INSECTS.

Series the First.

APTERIA.

Those which have no wings.

OF the Insects of this series, some have no limbs, others have; they are, according to this distinction, naturally arranged under two classes, under the names of Anarthra and Podaria.

APTERIA, *or Insects having no wings.*

Class the First.

ANARTHRA.

INSECTS which have their bodies covered with a soft skin, and have no limbs.

This class comprehends the Insects commonly called *Worms* and the *Larvæ.*

CHÆTIA

THE Chætia is an Insect resembling a hair, or a piece of fine thread: it's surface is smooth; it's body rounded, and very slender in proportion to it's length; and it's mouth small and placed horizontally; the jaws are both of the same length, and obtuse at their extremities.

Of this genus there is only one known species.

CHÆTIA. — The Hair-worm, called also the Guinea-worm.

This Insect, when full grown, is ten or twelve inches, or more, in length, and of about the thickness of a horse-hair: it's skin is smooth and somewhat glossy, and not at all furrowed: it's colour is a pale yellowish-white all over, except the head and tail, which are black and glossy.

It is not unfrequent in our fresh waters, and in most of the other parts of the world. Gesner calls it Vitulus aquaticus; Merret, Seta aquatica; Linnæus, Gordius; and the medical writers, Dracunculus, and Vena Medinensis: our common people call it the hair-worm, and imagine it to be an animated horse-hair. Dr Lister gives a very particular account of it's origin, and says, that he has found many of them alive, in the body of the common black beetle. This is the worm, that, in Guinea and some other of the hot countries, gets into the flesh of the natives, and occasions great mischief; though authors in general seem not to have found this out: with us, though frequent enough in waters where people bathe, it never attempts this.

ASCARIS.

THE Ascaris is a small Insect of an oblong form, slender, and covered with a very tender and soft skin: it's body is of a rounded figure, and is smaller at the ends than in the middle.

Of this genus there are but two known species.

Ascaris capite minore. *The small-headed Ascaris.* — The Ascarides.

This is a very small worm: it's length is usually about half an inch; it's thickness, that of a small pin; it is of a pale red colour, and smooth surface; and it's head is small and sharp, and it's tail pointed; it is of a tender structure, and easily crushed to pieces and destroyed.

It is frequent in the mud at the bottom of rivers, and is sometimes found in the earth in spring, but it perishes there before summer; wherever it is found, it is always in vast abundance, clusters of millions usually being met with entangled together.

It is one of the worms that infest the intestines of the human species, as well as of other animals: it's general place is in the rectum, but it has been found in the vagina of women who have long laboured under a fluor albus, and of girls in a chlorosis.

Ascaris capite majore depresso. *The Ascaris, with a larger depressed head.*

This grows to about two thirds of an inch in length; it's skin is soft and tender; it's colour, a very pale red. It's head is nearly equal, in diameter, to the thickest part of it's body; but it's tail is pointed, as in the other species.

I met with it last summer in the mud of Mr Apreece's fish-ponds at Washingley in Huntingdonshire, and never found it elsewhere.

LUMBRICUS.

THE Lumbricus is an Insect of the larger kind, of an oblong form, considerably thick, rounded in shape, and covered with a soft and tender skin, marked with annular ridges and furrows.

Lumbricus levis.
The smooth Lumbricus. **The Earth-worm.**

This Insect, when full grown, is often ten inches in length, and more than a third of an inch in diameter. It's colour is a dusky red, and it's skin is formed into rings, but is smooth and soft to the touch.

It is common every-where, at little depths, under the surface of the earth; and, in the different periods of it's growth, varies so extreamly in colour and external appearance, that people, less acquainted with the variations of this kind in animals, have made four or five different species of it, under the names of the lesser worm, the brown worm, the pale red worm, the dunghil worm, &c. It is frequent also in the human intestines, and in those of some other animals; and in this state has been supposed a different species, and called by a new name. Ray has described it three times over, as three different species, under the names of Lumbricus major, Lumbricus minor, and Lumbricus intestinorum teres. Every worm has the separate parts of generation of both sexes, and they rise out of the earth to copulate; each at the same time impregnating the other, and being itself impregnated. The moles feed on them, and they are in continual dread of those destroyers: hence it is, that on stamping upon, or otherwise shaking, the earth where they are, they crawl out on the surface to avoid those creatures, which they suppose occasion the motion.

Lumbricus scaber.
The rough Lumbricus. **The Sea-worm.**

This species grows to a foot, or more, in length, and to the thickness of a man's finger. It is of a pale red colour, and is composed of a number of rings or annular joints, as it were, in the manner of the other; but the skin is scabrous, and all these rings are covered with little prominences, so that it is extreamly rough to the touch.

It is an inhabitant of the mud about the sea-shores, and serves for food to many kinds of fish; I have met with surprizingly large ones about the Bognor rocks in Sussex.

TÆNIA.

THE Tænia is an Insect of an oblong form; the body is composed of evident joints or articulations, in the manner of the links of a chain, or beads of a necklace.

Tænia plana.
The flat Tænia. **The Tape-worm.**

This grows to a surprizing length, frequently to that of several ells. It's body is not rounded, but flat, and is composed of articulations of about a third of an inch long each, and of about two thirds of their length in breadth; the skin is smooth, and the colour is whitish.

It is found in the human intestines, and in those of many other animals, as well fish as quadrupeds. Linnæus found one of them once in the mud of a spring, but very probably it had been voided by some animal; if that were it's native place, it would be met with oftener there. The medical writers call it Tænia, and Lumbricus latus.

Tænia teres.
The round-bodied Tænia.

This grows to two inches long, though it is more usually met with of about half an inch or three quarters. It's body is composed of a number of articulations, and is rounded and pellucid; the skin is extreamly thin and whitish.

It is common in the mud of ponds and ditches: I have met with millions of them in the ponds in the Island in the Park this summer.

SICYANIA.

THE Sicyania is an Insect of an oblong form, flat on the belly, and rounded on the back; it's skin soft, and it's mouth large, horizontal, and emarginated, or dented in the middle.

Of this genus there is but one known species.

SICYANIA. **The Gourd-worm.**

This grows to two thirds of an inch in length, though it is more usually met with of not half that size: It's breadth is nearly equal to two thirds of it's length; it's skin is soft and whitish, with a tinge of brown; it is flattish, but somewhat rounded on the back, and has about eight deep longitudinal furrows in two series. The hinder part of it's body is rounded, and the other extremity shews a large mouth; it somewhat resembles the seed of the common gourd in figure, whence it has been called, by the medical writers, Vermis Cucurbitinus.

It is found sometimes in the intestines, and often in the substance of the other viscera, in quadrupeds.

HIRUDO.

THE body of the Hirudo is flatted, but not jointed; and is broader at either end than elsewhere; the skin is soft and glossy.

Hirudo nigrescens flavo variegata.
The blackish Hirudo, variegated with yellow. **The common Leech.**

This grows to the length of two or three inches; it's skin is smooth, of a blackish colour, and edged with a yellow line on each side: it has also some spots of yellowish on the back; but both these and the lines grow faint, and almost disappear, at some seasons. The head is smaller than the tail, which fixes itself very firmly to any thing the creature pleases. It is viviparous, and produces but one young one at a time, which is in the month of July.

This is the species usually employed to draw blood; authors call it Hirudo vulgaris, and Sanguisuga. It is common in shallow waters.

Hirudo nigra abdomine plumbeo.
The black Hirudo, with a lead-coloured belly. **The Horse-leech.**

This is larger than the former species: It's skin is smooth and glossy; it's colour, on the back, black, but with a number of pale greyish spots: the belly is of the colour of these spots, with a tinge of bluish.

It is common in shallow ponds.

Hirudo lateribus attenuatis.
The thin-sided Hirudo. **The Snail-leech.**

This grows to about an inch in length: it is very flat; it's skin smooth and glossy, and it's colour whitish: it's back is very little elevated, and it's sides thin, and as it were edged.

It

It is common on stones, and about the bottoms of puddles. Reaumur calls it Hirudo-Limax, but it has no relation to the snail-family.

Hirudo dorso elevato, cauda latiore.
The high-backed, broad-tailed Hirudo. **The great-tailed Leech.**

This grows to an inch and a half in length. It's skin is smooth and glossy; it's colour a dusky brown: it's back is elevated into a kind of ridge, it's belly flat; it's tail is remarkably broad, and it holds very firmly by it.

It is common on stones in shallow, running waters. Frisch, calls it, Hirudo ore caudaque ampla.

APTERIA,

Or Insects which have no wings.

Class the Second.

PODARIA.

Insects which have no wings, but have limbs.

Division the First.

Those with oblong bodies and numerous legs, or more than six pairs.

IULUS.

THE Iulus is an Insect of an oblong form, and has the body rounded, not flat. The legs are about a hundred on each side; and the eyes are two, and are simple, or not compounded of other smaller ones.

Iulus fuscus pedibus utrinque centum.
The brown Iulus, with a hundred legs on each side. **The Galley-worm.**

This grows to about two inches in length: the body is of a brown colour, and there runs a double line of a ferrugineous hue all along the back; the legs are of a paler brown. The back is rounded, and the belly flat; the whole animal is smooth, and somewhat glossy; the antennæ are short, and are composed of five joints; the extreme one is of a roundish figure.

It is frequent in the North of England, at little depths in the earth. Aldrovand calls it, Scolopendra terrestris minor; Ray, Iulus quartus glaber.

Iulus cinereus pedibus utrinque centum et viginti.
The grey Iulus, with a hundred and twenty feet on each side.

This grows to three quarters of an inch in length, and to the thickness of a crow-quill. It's back is rounded, it's belly flat; it's colour is a pale grey, but it has two pale ferrugineous lines on the back, and it is somewhat paler at the sides than elsewhere, and every joint of the body is there longitudinally striated; the articulations of the body are about sixty; the feet are whitish; the antennæ consist each of five articulations.

It is found under large stones and old trees, but is not common. Ray calls it, Iulo glabro affinis lividis albisque circulis.

Iulus corpore tenuiore flexuoso ruber.
The red Iulus, with a thin, flexuous body.

This grows to an inch and half in length: the body is very slender, and of a reddish colour; the back is but little prominent; the belly flat, so that the whole animal has

has much of a flat look; it is very nimble in it's motions, and turns and twists it's body about with great facility: it's legs are yellowish, with a cast of red, and it has seventy on each side.

It is frequent at little depths in the earth. Aldrovand calls it, Scolopendra vulgaris ac vera; Ray, Scolopendra valde exilis longa.

There are several other species of the Iulus, but they will easily be distinguished by their names; they are, 1. The long and thick, deep brown Iulus. 2. The little, black Iulus. 3. The little, ferruginous Iulus. 4. The large, hollow-bellied Iulus. 5. The large, dusky Iulus. 6. The short-legged Iulus.

SCOLOPENDRA.

THE body of the Scolopendra is long, slender, and somewhat flatted. The eyes are two, and each of those is composed of three others; the feet are from twenty to a hundred on each side.

Scolopendra exalbida dorso depresso.
The whitish Scolopendra, with a depressed back. — **The White Centipes.**

This grows to three inches and a half in length; the body is composed of a great number of articulations; the skin is tough and whitish; the body is of the thickness of a swan's quill: the back is depressed, but the sides rounded; the legs are very numerous, and considerably long.

It is common about the roots of old trees under hedges, and in the earth in dry places. In the East and West Indies, and in many of the warmer parts of Europe. Aldrovand calls it, Scolopendra major; others, Scolopendra alba vulgaris.

Scolopendra corpore tenuiore fusco.
The brown, thin Scolopendra. — **The brown Centipes.**

This grows to more than an inch in length; it's body is flat and thin, and very flexible; it is of a brownish colour: the legs are short, yellowish, and are seventy on each side.

It is extreamly nimble in it's motions, and is common with us in the earth, and under old logs of wood. Frisch. calls it, Scolopendra longa plana; others, Scolopendra vulgaris.

Scolopendra brevis pedibus variis.
The short Scolopendra, with various legs.

This is a small species: it's length is about half an inch, it's breadth that of a wheat straw; it is flat, and of a red colour; the body is composed of a number of articulations, but they are alternately longer and shorter; the legs are only fifteen on each side: the anterior ones are short and robust; the last pair are very long, and give the creature the appearance of a forked-tail.

It is found at little depths in the earth, but is not very common. Ray calls it, Ad Scolopendram accedens triginta pedibus instructa.

The other species are, 1. The smaller Scolopendra, with twenty legs on a side. 2. The great, thick Scolopendra, with twenty legs on a side. 3. The great, brownish Scolopendra, with fifty-eight legs on a side. 4. The smaller, greyish, white Scolopendra. 5. The largest, thick-bodied, pale Scolopendra. 6. The large-headed Scolopendra. 7. The deep-furrowed Scolopendra. 8. The submarine, or sea Scolopendra. 9. The little, oval, white-tailed Scolopendra, no larger than a louse. Most of these are inhabitants only of the hotter countries.

ONISCUS.

THE body is short and broad, and approaching to an oval figure: the legs are seven or eight on each side; the more usual number is seven.

Oniscus cauda obtusa bifurca.
The Oniscus, with a blunt, forked tail.

The common Wood-louse.

This grows to near half an inch in length, and to about half it's length in diameter. The back is somewhat rounded, the belly flat; the colour a bluish grey: it runs nimbly, and, on being touched, rolls itself up into a kind of ball.

It is common about old trees, and under logs of wood and stones. Ray calls it, Asellus asininus sive vulgaris; others, Millepes.

Oniscus cauda bifida, stylis bifurcis.
The Oniscus, with a bifid tail, and the styles of it forked.

The Water Millepede.

This grows to about half an inch in length; the breadth is nearly a quarter of an inch; the colour a pale brownish grey, and the whole body so thin, that it seems transparent: there are seven joints in it, exclusively of the head and the tail, and they are deeply divided. The articulation at the tail is rounded, flattish, and larger than any of the others, and from this there grow two styles, which are each of them divided also into two parts at the end: the legs are slender, and moderately long; they are of a pale brown, and transparent: there are seven of them on each side, and the hinder ones are longer than the others: the antennæ consist each of three joints.

It is common with us in ponds and ditches, sometimes in running waters. Authors call it, Asellus Aquaticus.

Oniscus cauda obtusa integerrima.
The Oniscus, with an obtuse, undivided tail.

The black Wood-louse.

This grows to half an inch in length, and is of an oval figure: it is black all over, except that the edges of the segments are whitish, and there is on each side a whitish spot near the hinder legs: it's skin is tough and glossy; it runs nimbly, but on the slightest touch it rolls itself up into a kind of ball, and will lie motionless a long time: it's legs are short, and it's tail obtuse, and not at all divided.

It is found about the roots of trees in damp places, but is less common than the forked-tailed kind. Ray calls it, Asellus lividus major.

Oniscus cauda subulata utrinque appendiculata.
The Oniscus, with a subulated tail appendiculated on each side.

The Sea Wood-louse.

This is the largest of the Oniscus kind: it grows to an inch in length, and more than half an inch in breadth, and is of a whitish colour: the back is rounded, the belly flat, and the sides sharp, and as it were edged: the legs are seven on each side; the three anterior pairs are small and smooth; the four hinder pairs are larger, longer, and ciliated, or hairy at the sides: there are two pairs of antennæ on the front of the head: the body consists of seven joints, beside the head and the tail: the tail is near three quarters of an inch long; it is undivided, and of a somewhat triangular figure, and has two convex parallel rays on it's under side.

It is found in most of the northern seas, but no where more plentifully than on some of our own coasts. I have met with it about the Yorkshire coast abundantly. Ray calls it, Asellus marinus Cornubiensis alius; others, Oniscus marinus maximus; Klein, Entomon pyramidale.

PODARIA.

Or Insects which have no wings, but have limbs.

Division the Second.

Those which have shorter bodies, and less numerous legs, or fewer than six pairs.

PULEX.

THE body of the Pulex is short, of a roundish figure, and compressed; the legs are three pairs, and are formed for leaping: the eyes are two, and simple; the mouth is bent downward.

Of this genus there is but one known species.

PULEX. **The Flea.**

The body is roundish and obtuse at the end, the head small, and the eyes large; the antennæ are short, and are composed of four joints; the legs are long, robust, and thick toward the insertion, slender and somewhat hairy toward the extremities; the colour is a deep purple, approaching to black. Hook has given an excellent microscopic figure of it, which has been copied by most of that kind of writers since.

It is not peculiar to man, or to quadrupeds; it is very frequent in swallows nests, and on the bodies of some other young birds. Authors call it, Pulex vulgaris, and Pulex ater.

PODURA.

THE body is short and roundish; the tail is crooked, forked, and of use in the leaping motion of the creature: the legs are three pairs, and serve only for walking: the eyes are two, but each of them is composed of eight others.

This genus comprehends the Pucerons of Reaumur, and other of the French writers.

Podura viridis corpore subovata.
The green Podura, with an oval body. **The common Puceron.**

This is about the bigness of a common flea; it's colour is usually a bright green, sometimes a darker, or bluish green; the body approaches to an oval figure, and is largest and most convex in the hinder part: the thorax is very small; the head is obtuse and green: the eyes are very conspicuous; they are bright, and of a deep black; they stand prominent in the front of the head, and there runs also near them a black line on each side. The legs are very slender, and of a whitish green, and are all of the same length: the antennæ are crooked.

It is frequent on the stalks of the common atriplex, and many other plants and shrubs, covering the whole branches in innumerable multitudes. Authors have called it, from it's resemblance to the common flea, Pulex viridis plantarum; the French, Puceron.

Podura antennis longioribus.
The Podura, with long antennæ. **The Toadstool Puceron.**

This grows to the size of a small grain of wheat: it's body is short and roundish; it's skin glossy and black; it's anus forms a remarkable protuberance in the hinder part; the head is small, the eyes are very minute, and the antennæ are equal to the body in length, and are black, but tipped with white at the end.

It is common in our woods on the larger fungi. Linnæus calls it, Podura atra abdomine subgloboso, antennis longitudine corporis apice albis.

Podura cinerea nigro variegata. **The Currant Puceron.**
The grey and black Podura.

This is somewhat larger than a flea: it's body is of an elliptic form; it's head small, it's eyes little, but very bright, and it's legs very slender, and the antennæ are long: it's colour is grey, but it is variegated with lines and spots of black.

It is frequent on the currant bush.

Podura nigra pedibus albis.
The black Podura, with white legs.

This is a very small species: it's body is roundish, and of a deep black: it's head small, and it's eyes scarce discernible: the legs are short, slender, and white, and the fork of the tail is also white.

It is common in our woods about old beeches. De Geer calls it, Podura arborea nigra pedibus furcaque albis.

Podura nigrescens antennis brevibus. **The Water Puceron.**
The black Podura, with short antennæ.

This is of the size of a small flea, and resembles it in colour, being black and glossy, but with somewhat of a tinge of purplish: it's body is roundish; it's head is small; it's eyes very little; it's legs are longer than in most of the other species, and it's antennæ short.

It is common in ponds, and other standing waters; the surfaces of which are sometimes, in calm days in autumn, almost covered with the multitudes of it. De Geer calls it, Podura aquatica nigra.

The other species of the Podura are very numerous; they are, 1. The Podura with antennæ, consisting of many numerous joints. 2. The short-horned Podura. 3. The round-bodied, bright Podura. 4. The long-bodied, larger Podura. 5. The cottony, or downy Podura. 6. The long-legged Podura. 7. The lead-coloured Podura. 8. The small, black-legged Podura. 9. The very small, white Podura. 10. The short-bodied, blue-green Podura. 11. The dusky, greyish Podura. 12. The short-tailed Podura. 13. The spreading-tailed Podura. 14. The large-headed Podura. 15. The long and slender-legged Podura. 16. The slender-horned Podura. 17. The larger water Podura. 18. The long-bodied, bluish, water Podura. 19. The subterranean Podura. Most of these species are very common with us on the branches of various shrubs and plants, and in our ditches and clear ponds.

PEDICULUS.

THE body of the Pediculus is lobated at the edges or sides: the legs are six, serving only for walking, the creature having no power of leaping: the eyes are two, and are simple.

Pediculus capitis. **The common Louse.**
The Pediculus of the head.

The body of this hateful Insect is oblong and lobated, or deeply indented around the sides: the colour is whitish, often streaked or spotted with black: the legs are short, and each is armed with two claws at the extremity: the antennæ are short and jointed; the head is large, and the eyes are small and black.

It's most natural place of habitation is in the heads of children, but it will infest grown people who are nasty, and will descend from the head to the body. It is a mistake to suppose those of the body of a different species; they are only a variety, and are of the same origin with the others. Authors call it Pediculus, and Pediculus vulgaris.

Pediculus pubis.
The Pediculus of the pubes.

The Crab-louſe.

This is ſmaller than the common louſe: the body is ſhorter, and leſs deeply lobated: the legs are ſhort and ſlender; the head is oblong; the antennæ are ſhort, and the eyes are ſituated behind them: the hinder part of the body is covered with ſilvery hairs.

It's natural reſidence is about the pubes, but it will live any where, if there is hair, as in the eye-brows and on the breaſt, and, when once fixed, is very difficultly diſlodged. Authors call it Pediculus ferus, Pediculus inguinalis, and Pediculus morpio.

Pediculus ligni antiqui.
The Pediculus of old wood.

The Death-watch.

The body is oblong, fleſhiſh, and of a pale browniſh white, but there is a brown annular mark on it, and a brown ſpot behind toward the anus: the head is oblong; the eyes are large and yellow; the antennæ are of the length of the whole body; the ſize is about that of the common louſe, and there are ſpots all down the ſides, of a reddiſh colour, one on every ſegment of the body.

It is common in decayed wood that is kept dry, and in old books. Blancard calls it Pediculus ligni; Ray, Pediculo cognatum et aſſimilis. The female makes that odd noiſe, which reſembles the beating of a watch, as ſhe wooes the male: the male alſo makes it in return, and ſometimes begins it himſelf.

Pediculus adonidum.
The ſtove Pediculus

The Stove louſe.

The body is of an oblong, oval figure, and of a whitiſh colour; the tail is bifid; the ſides are thin and ſharp, and there is a prominent ſpot on the edge of each ſegment: the line carried along the back is elevated, and has an obſolete protuberance at every diviſion, and there are the ſame number of ſpots in the ſpaces between that line and the ſides: the legs are ſlender and brown; the antennæ are alſo ſlender, and of the ſame colour.

It is common on the branches of ſhrubs and plants in our ſtoves.

Beſides theſe, there are a great many others which are peculiar to the bodies of the different ſpecies of birds and quadrupeds. Redi has accurately deſcribed and figured many of theſe; they are named, from the bodies of the animals they are found on, Pediculus bovis, Pediculus corvi; the Louſe of the ox, the Louſe of the raven, &c.

The moſt ſingular of theſe are, 1. The Pediculus of the bear. 2. The Pediculus of the lion. 3. The Pediculus of the opoſſum. 4. That of the wild cat. 5. and 6. The larger and ſmaller of the ox. 7. That of the hawk. 8. That of the thruſh. 9. That of the turkey. 10. That of the crane. 11. That of the hen. 12. That of the hedge-ſparrow; theſe all vary in their forms, as do alſo thoſe which infeſt other ſpecies of animals, and are indeed as diſtinct as thoſe animals; but to figure and deſcribe them all would be to make this volume a hiſtory of Lice, not of Inſects in general.

MONOCULUS.

THE body is ſhort, of a roundiſh figure, and covered with a firm cruſtaceous ſkin; the fore legs are ramoſe, and ſerve for leaping and ſwimming; there is but one eye which is large, and is compoſed of three ſmaller.

Moſt of the ſpecies of this genus are reckoned by authors among the microſcopic animals, but improperly; they are ſufficiently viſible to the naked eye.

Monoculus antennis quaternis, cauda recta bifida.
The Monoculus, with quaternate antennæ, and a straight bifid tail.

The crustaceous water Insect.

The creature is very small, and usually of a brownish, sometimes of a reddish, colour: the body is oblong, approaching toward an oval figure, but smaller toward the tail: it has two pair of antennæ; the tail is long, slender, and forked, and under this, on each side, there is frequently seen a large cluster of eggs; the ovaries, in which these are contained, are of a yellowish colour, and together are often equal to the whole body of the Insect in bigness: the eye is large, black, and seated in the very middle of the head.

It is frequent in our ditches, and other standing waters.

Monoculus antennis capillaceis multiplicibus, testa bivalvi.
The Monoculus, with multiple and capillaceous antennæ, and a bivalve shell.

The testaceous water Insect.

The shell, with which this creature's body is covered, is of an oblong, ovated figure, and of a dusky brown colour: when taken out of the water, it shuts close up, and resembles the seed of some plant; when put into the water again, it opens in the manner of the bivalve shells of the shell-fish, and the antennæ appear from one end of the aperture, the legs from the other: the eye is single, large, and black; it's motion is very swift.

It is common in our ditches, and other standing waters.

The other species of the Monoculus are, 1. The crooked-tailed Monoculus, with dichotomous antennæ, commonly called, after Swammerdam, the Pulex aquaticus arborescens: this is small, of a blood-red, and is so numerous in standing waters at some times, as to give a red colour to the whole surface of them; and this has been called turning water into blood, and has been esteemed a portent. 2. The Apus of Frisch, or the Monoculus with a bifurcous tail. 3. The short dichotomous, horned Monoculus, with the flat back. 4. The long-bodied Monoculus. 5. The little-eyed Monoculus. 6. The short and multifid-tailed Monoculus.

ACARUS.

THE body of the Acarus is short and rounded: the eyes are two; the legs are eight in number, and each consists of eight joints.

Most of the species of this genus have been also arranged among the microscopic animalcules, but with no reason; they are all sufficiently visible to the naked eye. The term Acarus is not to be understood in this sense, as restrained to the Insect commonly understood by it, the Mite; that animal is possessed of characters in common with a great number of other Insects, which have been called by other names, but which are all connected by nature, and are therefore of the same genus; some of them have been called spiders, others lice, and others by other names, referring them to genera, to which they have as little alliance in nature as to these. The genus, on bringing these back to it, appears a very numerous one, and consists of some which are inhabitants of the earth, some of waters; some which live on trees, others among stones, and others on the bodies of other animals, and even under their skin.

Acarus casei.
The Acarus of cheese.

The Cheese-mite.

This is a very small species; the body is roundish, but somewhat ovated; the skin whitish and smooth; the head is small; the legs and the parts about the mouth are somewhat brownish, and of a firm, testaceous substance, not soft like the other parts, and there are some long hairs on the body.

It is common in cheeſe, in flour, and in many other ſubſtances. Authors call it, ſimply, Acarus.

Acarus ovinus.
The Acarus of the ſheep.

The Sheep-tick.

The body is flat, of a roundiſh figure, but ſomewhat approaching to oval, and of a yellowiſh white colour, and has a ſingle large round ſpot on the back: the anus is viſible in the lower part of the body; the thorax is ſcarce conſpicuous; the head is very ſmall and black; the mouth is bifid: the antennæ are of a clavated figure, and of the length of the ſnout; the legs are ſhort and black. The animal, in the whole, much reſembles the common dog-tick; but it's ſnout is longer, and it's body never becomes ſo diſtended as that of the other with ſucking, but always continues flat.

It is common on ſheep, and it's excrements ſtain the wool green: it will live in the wool many months after it is ſhorn from the animal. Charleton, Mouffet, and others call it Reduvius.

Acarus Inſectorum rufus ano albicante.
The brown Acarus of Inſects, with a white anus.

The Louſe of the beetle.

This is an extreamly minute Inſect; it's body is round, reddiſh, and covered with a firm and hard ſkin; the head is very ſmall, the neck ſcarce viſible; the legs are moderately long; the anterior pair longer than the others: it has a whiteneſs about the anus.

It is frequent on the bodies of many Inſects, which it infeſts as the louſe does others; it runs very ſwiftly: the humble-bee, and many other of the larger Inſects, are continually infeſted with it; but none ſo much as the common black beetle, which has thence been called the louſy beetle. Liſter calls it, Pediculus ſubflavus ſcarabæis infeſtus; Blankard, Pediculus Inſectorum volatilium; in the Acta Upſalienſia, it is much more properly named Acarus Inſectorum.

Acarus arboreus ruber diſtentus.
The red, diſtended, tree Acarus.

The ſcarlet-tree Mite.

This is a ſmall ſpecies; the body is roundiſh, and the back not at all flatted, as it is in many others; the ſkin is ſmooth, ſhining, and gloſſy, and the whole animal ſeems diſtended, and ready to burſt; the colour is a bright red, but a little duſkier on the ſides than elſewhere: the head is very ſmall, and the legs ſhort: there is on each ſide a ſmall duſky ſpot near the thorax, and a few hairs grow from different parts of the body.

It is very common on trees, particularly on the currant, on the fruit of which we frequently ſee it running.

Acarus petrarum ruber antennis longioribus.
The red ſtone Acarus, with long antennæ.

The Stone-mite.

This is very ſmall, and of a bright red colour; the body is round and diſtended; the head is very ſmall and pointed; the legs are moderately long, and of a paler red than the body: the antennæ are much longer than in any other ſpecies. In the greater number of the others, indeed, they are ſcarce conſpicuous; theſe alſo, and the mouth, are of the ſame pale red with the legs.

It is frequent about old ſtone walls and on rocks, and runs very nimbly.

Acarus aquaticus ruber abdomine depreſſo.
The red water Acarus, with a depreſſed body.

The ſcarlet Water Mite.

This is a ſmall ſpecies: the body is of a figure approaching to an oval, and the back appears depreſſed; it is of a bright and ſtrong ſcarlet colour. The head is ſmall; the legs are moderately long and firm, and are of a paler red than the body.

It is common in shallow waters, where it runs very swiftly along the bottom; Frisch. calls it, Araneus aquaticus ruber parvus; Charleton, Buprestis araneola.

Acarus scabiei.
The itch Acarus.

The Itch Animal.

This is a very small species; it's body is of a figure approaching to oval and lobated; it's head is small and pointed; it's colour is whitish, but it has two dusky semicircular lines on the back. The legs are short and of a brownish colour, and are harder than the rest of the body, and as it were crustaceous.

It is found in the pustles of the itch: authors in general have supposed it causes that disease; but, if this were the cause, it would be found more universally in those pustules. It is more probable that these only make a proper nidus for it.

Acarus pedibus primi paris cheliformibus.
The Acarus, with the first pair of legs cheliform.

The Scorpion Insect.

This is a small species; it's body is roundish and inflated, of a whitish colour, and covered with a thin and smooth skin; the head is very small, and of a dusky colour toward the mouth. The legs are short, and the fore pair have claws like those of the crab or scorpion.

It is common in old rotten wood that is kept dry in old houses. Frisch. calls it, Scorpio araneus.

Acarus terrestris ruber abdomine depresso.
The red land Acarus, with a depressed body.

The little scarlet Spider, or Tant.

This is a small species: it's body is roundish, but a little approaching to oval; the back somewhat depressed; it is of a fine scarlet colour, and covered with a velvety down. The head is very small; the eyes are two and very small; the legs are short and of a paler red, and there is a small black spot near the insertion of the anterior ones.

It is very common under the surface of the earth, and sometimes on herbs and among hay. It is supposed to be poisonous, if swallowed, but we don't seem to have any certain account of such an effect. Lister and others call it, Araneus coccineus exiguus.

Acarus pedibus longissimis.
The long-legged Acarus.

The Long-legged Spider, called the Crested, or Shepherd-spider.

This is the largest of the Acarus kind: it's body is roundish, of a dusky brown on the back, with a duskier spot of a rhomboidal figure near the middle of it: the belly is whitish; the legs are extremely long and slender. On the back part of the head there stands a little eminence, which has on it a kind of double crest, formed as it were of a number of minute spines: the eyes are small and black, and are two in number.

It is very common in our pastures, towards the end of summer. Ray and Lister call it, Araneus cristatus longipes; Mouffet, Araneus longipes; and, notwithstanding it's having but two eyes, it has been almost universally ranked among the Spiders.

The other species of the Acarus are, 1. The mushroom yellow Acarus. 2. The little black garden Acarus. 3. The Acarus, called the louse of the starling. 4. The pine-tree Acarus. 5. The dog Acarus, or dog tick. 6. The holly-oak Acarus. 7. The grey rough earth Acarus. 8. The variegated stone Acarus. 9. The brown-backed willow Acarus. 10. The brown stone Acarus, with scarlet legs. 11. The gregarious mushroom Acarus. 12. The sea Acarus. 13. The great Acarus, or great fly louse. 14. The very small fly louse. 15. The knotty legged Acarus, found on old trees. 16. The marginated-sided, stone Acarus. 17. The greenish white Acarus of the lime-tree.

ARANEA.

THE body of the Aranea is short and of a roundish, oval, or elliptic figure: the eyes are eight in number, and are placed on the hinder part of the thorax: the legs are eight in number, and the creature has a power of spinning.

The species of Aranea, or Spiders, are very numerous; but authors have made them yet more so, by admitting among them Insects of very different genera. Lister has been very accurate in his descriptions, but he has taken in some of the Acari, &c. into the number: 'tis odd, that so striking a character, as the eight eyes of the Spider, should not exclude Insects that had but two.

Aranea corpore longiusculo argenteo virescente.
The long and silvery-green-bodied Aranea. — **The green Spider.**

This is one of the most singular of the Spider class: it is moderately large; it's body is longer and slenderer than that of any other species, and is of a silvery green colour, with an admixture of yellow, variegated with lines, strokes, and spots of black on the back; and with a black line yellow at edges, running all along the belly. It's legs are long and slender, and are generally protended horizontally.

It is common in damp woods, where it squats down, as it were, on the branches of trees, and throws four of it's legs forward, and four backward, extending them straight along the bough. I have often met with it in Charlton wood near Woolwich. Lister calls it, Araneus ex viridi inauratus alvo longiuscula praetenui.

Aranea livida abdomine ovato aquatica.
The bluish, oval-bodied, water Aranea. — **The great water Spider.**

This is a Spider of the larger kind: it's body is oval, and, as it were, truncated at the tail; it is of a bluish black, and has a transverse line and two spots toward the bottom, all hollowed in it. The legs are long, and the joints large: it is furnished at the head with a very terrible weapon, a hard black forceps, vastly larger than any of the European spiders have it, and wholly resembling that of the tarantula; it will lay hold of any thing very firmly with this; whether there be any thing poisonous in the wound it inflicts is not known.

It is found in standing waters, but is not common; it walks on the mud at the bottom of them: I have met with it in some ponds in Essex near Thorndon, and once in the Serpentine river in Hyde-park.

Aranea abdomine ovato hirsuto, cruribus crassiusculis.
The oval hairy-bodied Aranea, with thick legs. — **The Tarantula.**

This is one of the large Spiders, but is not the very largest known: it's body is three quarters of an inch long, and of the thickness of one's little finger. It is usually of an olive brown, variegated with a duskier colour; but in this it varies greatly: it is covered with a short and soft down, or hairiness: the points of it's forceps are very fine and sharp: as to the effects of the poison they convey into the wound they make, there seems yet room for much explanation about it.

This species is a native of Apulia: the generality of writers call it the Tarantula, and Araneus tarantula dictus.

Aranea corpore oblongo nigro aquatica.
The oblong-bodied black water Spider. — **The lesser water Spider.**

This is a large Spider: the body is of an elliptic figure, and black; a white line runs along each side, from the thorax to the tail, where the two meet: the anus is bifurcated: the legs are long, and of a greyish brown.

It is frequent about waters, but it only skims along upon the surface of them.

Aranea

Aranea abdomine flavo, annulo ovali dorsali rubro.
The yellow-bodied Aranea, with a red and oval mark on the back.

The red and yellow Spider.

This is a very singular and very beautiful species: it's body is of an oval form, and about the size of that of the common blue fly; but it is of a beautiful whitish yellow, with an elegant circle, or rather oval mark of red, upon the back: the legs are long, slender, whitish, and transparent, and the joints of the first pair are black; and there runs along the center of the oval mark a brown line.

We have it in our hedges, but not common: it spins a large, but not very close, web. Lister calls it, Araneus albicans corcum coccinum in alvo ovali.

The other species of spiders are, 1. The great brown Phalangium, with the thick forceps. 2. The largest yellowish Phalangium, with the very sharp forceps. 3. The great tawny Spider, with thick legs. 4. The long and slender-legged great Spider. 5. The vast pale brown Spider. 6. The vast variegated Spider. These are all natives only of hotter countries. 7. The common dark house Spider. 8. The great field Spider, with the brown and white body. 9. The great red-bodied Spider, with the white cross. 10. The yellow and white-bodied great Spider. 11. The grey wood Spider, with a triquetrous body. 12. The green and yellow Spider, with black spots. 13. The grey-bodied Spider, variegated with white. 14. The smooth, black-bodied Spider, with the white cross. 15. The yellow-bodied Spider, with white spots. 16. The middle-sized variegated field Spider. 17. The elegantly variegated wood Spider. 18. The black-bodied Spider, with the tops of the legs variegated. 19. The grey Spider, with undulated lines at the insertions of the legs. 20. The brown and variegated Spider, with foliaceous lines. 21. The little grey Spider, with a black spot on the hips. 22. The little bluish Spider, with a denticulated spot at the hips. 23. The long-legged, yellow, hairy Spider. 24. The great grey Spider, with very prominent appendages at the anus. 25. The dusky Spider, variegated with a bright white. 26. The black Spider, with denticulated white spots. 27. The great blackish Spider, variegated with spots of a deeper black. 28. The grey, softer, woolly Spider. 29. The pluio-yellow, or livid Spider. 30. The square, spotted, yellow Spider. 31. The large, black, hunting Spider. 32. The brown Spider, with an obliquely streaked belly. 33. The long-bodied, yellow Spider. 34. The long-bodied livid Spider, with undulated variegations. 35. The brown, slow Spider. 36. The gold yellow Spider, variegated with brown. 37. The grey Spider, with black and silvery white spots. 38. The reddish brown Spider, with white variegations. 39. The green-eyed yellow Spider, variegated with a deeper yellow. 40. The reddish Spider.

SCORPIO.

THE body of the scorpio is of a figure approaching to oval; the tail is long and slender; the whole covered with a firm and somewhat hard skin: the eyes are eight in number; two of them are placed contiguous, and six sideways. The legs are eight, and there are also a pair of claws at the head, and a pointed weapon at the extremity of the tail.

Scorpio e fusco flavescens, pectinum denticulis octo.
The great yellowish Scorpio, with eight denticulations.

The Barbary Scorpion.

This species, when full grown, measures six or seven inches in length; those of four or five inches are common: it's body is covered with a firm skin, of a brownish colour, but with a great deal of a yellow, or flame colour, in it. The body is of an oval figure, and is terminated anteriorly by the head, which is continuous with it: the eyes are small: the legs are eight, and are not very strong; each leg is composed of six joints, and is terminated by a pair of sharp claws; the legs are paler than the body, and have a few hairs on them: the claws arise, one from each side of the thorax near the head; they are large, black, and much resemble the claws of crabs; each is composed of four joints. The body is also composed of a number of joints, divided by denticulations, as is also the tail: this has it's insertion at the extremity of the body,

3 opposite

opposite to the head, and it's last joint is terminated by a pointed weapon bent downward.

This species is a native of Africa: It's wound is of very bad consequence; authors call it Scorpio Barbarus.

Scorpio fuscus, pectinum denticulis triginta.
The brown Scorpio, with thirty denticulations.

The Italian Scorpion.

This, when full grown, is about an inch and a quarter long: It's colour is a dusky brown, often almost blackish; it's body is oval, and it's tail longer in proportion than in the larger species.

It is a native of Italy, and of many of the warmer parts of Europe: I had one brought to me a few years ago, which the person who had it, assured me, he killed in Cane-wood, near Hampstead-heath; but 'tis the first time England has been supposed to produce them. Authors call this Scorpio minor, and Scorpio Italicus.

The other species are the large brown Indian Scorpion, with thirteen denticulations. 2. The smaller Indian Scorpion, with fifteen denticulations. 3. The large African Scorpion, with eighteen denticulations. 4. The black African Scorpion.

SQUILLA.

THE Squilla has ten legs, the foremost pair cheliform, or made for pinching and holding things: the eyes are two, and the tail is foliated.

These are the characters of the Shrimp; and they are the characters also of the hermit, the cray-fish, the lobster, and the crab: it may appear odd, to those less acquainted with the characteristics of natural genera, to find the crab made a species of shrimp, or either that, or the lobster, introduced among the Insects; but nature has been pleased to connect them, and we have no right to disjoin them.

Squilla macroura, rostro supra serrato, subtus tridentato.
The long-tailed Squilla, with the snout serrated above, and tridentated below.

The common Shrimp.

The body is oblong and rounded above; the tail long, the colour whitish, and the legs long; the beak, or snout, distinguishes it from all the other species: It is long, of a lanceolated figure, sharp-pointed, and has eight denticulations above, and three below.

It is a native of our seas, in vast abundance; authors call it Squilla, Squilla fusca, and Squilla vulgaris.

Squilla macroura rostro integerrimo.
The long-tailed Squilla, with a smooth snout.

The smooth-nosed Shrimp.

This grows to the same size with the common Shrimp, and resembles it so greatly, that it has been generally taken for the same species: the body, however, is thicker, and the snout is very short, and is smooth, wholly without the spines or denticulations of the other.

It is frequent in our seas with the other, and they are brought to market together.

Squilla macroura cauda molli, chela dextra majore.
The long-tailed Squilla, with a soft tail, and the right claw the larger.

The Hermit.

This grows to two inches and a half in length: the body is thick and firm; the tail slender, and covered with a tender skin. The legs are slender and long; the anterior ones have claws on them, like the common crabs, or lobsters claws, but that on the right, is always the larger: the creature seeks some wilk, or other shell, deserted by it's original inhabitant; and, introducing it's tail into it, fixes in the habitation, and draws it after it; the tail is always within the shell, and is so firmly fixed, that it would induce any

…ny body unacquainted with these things, to believe it a part of the animal; when it pleases, it can draw in the whole body, and it then continues to stop up the opening of the shell with the larger claw, and lies retired from mischief.

It is common in our seas; the oystermen at Feversham dredge up great numbers of it: authors call it Cancellus; and Bernhardus, Eremita; we, frequently, Bernard the hermit.

Squilla macroura rostro supra serrato, basi utrinque dente simplici.
The long-tailed Squilla, with the snout serrated above, and with a single denticulation at the base.

The Cray-fish.

This greatly approaches to the figure of the common lobster: it's length, with us, when well grown, is about three inches and a half: it's body, about three quarters of an inch in diameter; but, in some other countries, it grows vastly larger, though the species be evidently the same: it's shell, or covering, is firm and strong: it's legs long and slender; and it's tail thick.

It is a native of fresh waters, particularly of our clear brooks: authors call it Astacus fluviatilis; Matthiolus and Bellonius, Gammarus.

Squilla brachyura corpore verrucoso aculeato.
The short-tailed Squilla, with a warty and prickly body.

The Wart-crab.

This grows to the ordinary size of the common crab: it's body, however, is always smaller, in proportion to it's legs and claws, than in that: the shell is hard; the colour whitish, with some admixture of brown, of red, and of black: the body is covered with large, rough, and irregular wart-like protuberances, and the legs, all over, except the tips of the claws, are covered with larger protuberances, which terminate in points, forming a kind of short and very robust spines.

It is a native of the East and West Indian seas, and is caught also about some of the shores of Europe; authors call it Cancer verrucosus.

Squilla brachyura cruribus longissimis.
The long-legged, short-tailed Squilla.

The Spider-crab.

This is a very small species: the body is of an oval figure, and covered with oblong, but not very numerous, spines: the legs are very slender, and like those of a spider, and are equal to five times the diameter of the body in length: they terminate in long, slender, and sharp points; and the claws of the anterior ones are no thicker than the points of the others, only double.

It is a native of the East Indies, and many other places: authors call it Cancer arachnoides.

Squilla clypeata.
The buckler Squilla.

The Molucca-crab, or King-crab.

This is the most singular animal of all this genus: it's body is covered with a hard and firm shell of a dusky brownish colour, and of a figure approaching to oval, but truncated at the lower end: it is composed, as it were, of two parts; the edge of the upper is smooth and even; that of the under, spinose: the body of the animal is small, in proportion to this immensity of shell, and the legs are short; the tail is inserted in the center of the truncated part of the shell, and is very long, slender, and pointed at the end.

It is a native of the seas, about the Molucca's: authors thence call it Cancer Moluccensis.

The other species of this genus are numerous; they may be conveniently arranged under three subdivisions. 1. The smaller long-tailed Squillæ, which are the shrimp kind. 2. The larger long-tailed Squillæ, which are the lobster and cray-fish kind. 3. The short-tailed, or crab kind.

Of the first division, or shrimp-kind, we have, beside the species already described, only three known ones. 1. The larger, long-snouted sea Squilla. 2. The smaller, narrow-snouted sea Squilla. 3. The fresh-water small Squilla.

Of the second, or lobster kind, we have, 1. The common lobster. 2. The thick-horned, slender-bodied Lobster. 3. The short and broad-bodied Lobster. 4. The great sea Cray-fish. 5. The very long-bodied Lobster. 6. The small bodied Lobster.

Of the crab kind, there are, beside the already described, 1. The common large Crab. 2. The rough-bodied, smooth-clawed Squilla, called Cancer Maea. 3. The smooth and long-clawed Crab, called, by Johnston, the female of the common kind. 4. The little squall Crab. 5. The little woolly Crab. 6. The thick-bodied duck Crab. 7. The round-bodied duck Crab. 8. The common, or oval-bodied, duck Crab. 9. The very long-armed duck Crab. 10. The very small-bodied, rough, long-armed Crab. 11. The lunar Crab. 12. The florid Crab. 13. The frog Crab. 14. The prickly and hairy, long-armed Crab. 15. The great, prickly, long-armed Crab. 16. The short-bodied, reticulated Crab. 17. The elliptic-bodied Crab. 18. The smooth, long-legged Crab.

INSECTS.

Series the Second.

PTERARIA.

Those which have wings.

OF the Insects of this series, some have only two wings, others have four: they are hence naturally arranged into two orders, under the names of Diptera and Tetraptera.

PTERARIA.

Insects having wings.

Order the First.

DIPTERA.

Insects which have only two wings and under each of them a style, or oblong body, terminated by a protuberance or head, and called a Balancer.

OESTRUS.

THE mouth of the OEstrus is a simple fissure, without either tooth or proboscis.

OEstrus thorace flavo cingulo nigro, alis nigra fascia variegatis, pedibus pallidis.
The black and yellow-bodied OEstrus, with black variegations in the wings, and white legs.

The Breeze, or Gad Fly.

This fly is nearly as large as the common blue flesh-fly: it's eyes are black and large; it's antennæ have a single long bristle at the extremity of a lenticular joint: the body is yellow, but has a black girdle, as it were, surrounding it; the belly is of a tawny colour, but the last joint black; the tail is long, and bends under the belly; the wings are whitish, and have a black line, and three black spots in each.

The female of this species lays her eggs in the backs of oxen, under the skin, and the worm lives there all the winter: authors have called this, and several other species, indiscriminately, by the name OEstrum.

OEstrus

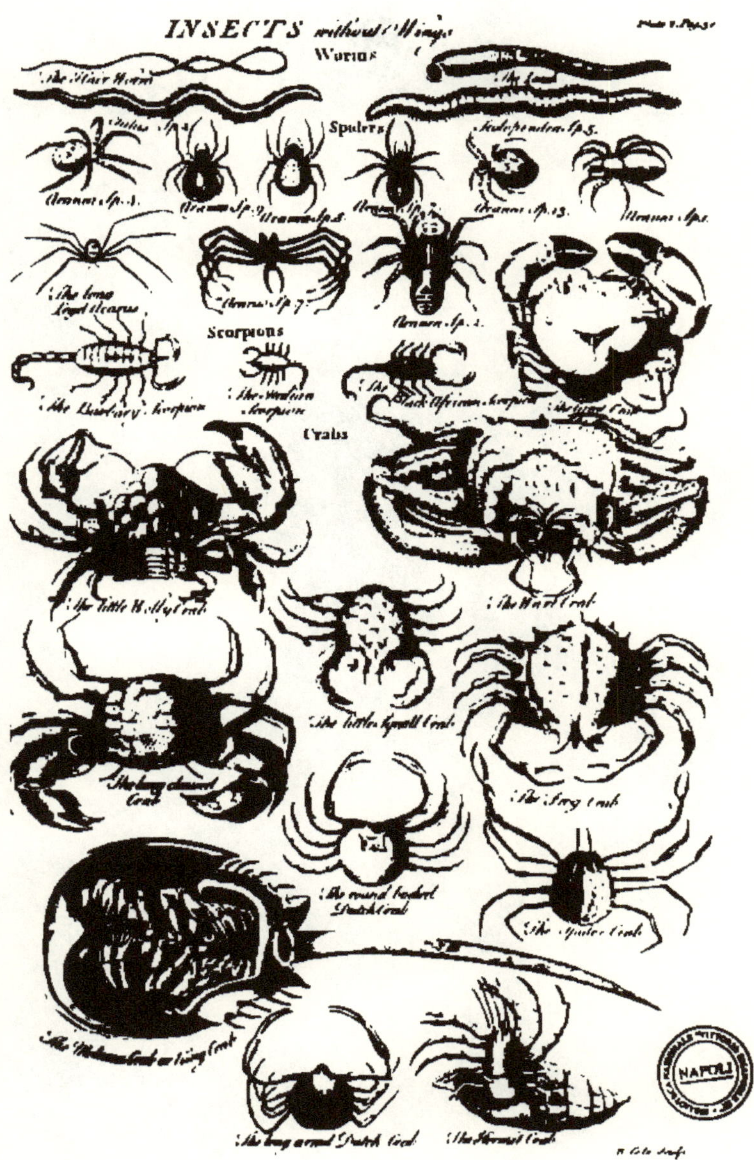
INSECTS without Wings
Worms
Spiders
Scorpions
Crabs

OEstrus niger thorace flavo. **The grey Fly, or**
The blackish OEstrus, with a yellow breast. **trumpet Fly.**

This is a very large fly, considerably exceeding the common blue fly in bigness; the body is of a dusky grey colour, approaching to black, and is smooth, except about the thorax, where there are a great number of yellow, long hairs: the wings are large and pellucid; the body is oblong, and the eyes are large and black: the female deposits her eggs in the nostrils of sheep, deer, and some other animals; and the worms produced of them live either there, or in the frontal sinus of the animal, to the time of their maturity. Authors call it, OEstrum rhangiferinum Laponicum, and OEstrum sinus frontis ruminantium. We call it the grey fly from it's colour, or the trumpet fly from the noise it makes in the heats of summer, attended to by Milton in his Lycidas,

What time the grey fly winds her sultry horn.

It is frequent about our hedges in autumn.

The other species of the OEstrus are, 1. The larger, bee-like OEstrus. 2. The smaller, round-bodied OEstrus. 3. The large, black and yellow-bodied OEstrus. 4. The black-legged, clear-winged, black and yellow OEstrus. 5. The short-winged, long-bodied OEstrus; the worm of this lives in the intestines of horses. 6. The large, rounder-bodied, plain-winged OEstrus; the worm of this lives under water, and is called Tabanus aquaticus, and Chamæleon. 7. The smaller, oblong, and pointed-bodied OEstrus. 8. The variegated-bodied, short-legged, spotted-winged OEstrus. 9. The black and tawny OEstrus.

ASILUS.

THE head is furnished, by way of mouth, with a snout of a subulated figure, which is simple, and very sharp at the extremity.

Asilus ferrugineus abdomine partim flavo, partim nigro. **The Hornet Fly.**
The ferrugineous Asilus, with the body part black, part yellow.

This is one of the largest of the fly kind; it equals the hornet in size, and has so great an external resemblance to it, that they are easily mistaken for one another: the head is large; the snout long and black; the eyes very prominent: the thorax is very large and gibbous, and is of a dusky colour: the wings, the legs, and the belly, are all of a ferrugineous colour; the body, on it's upper part, is black and yellow; it consists of seven articulations; the three upper ones of which are black, and the four others yellow.

It is frequent in our pastures in autumn; it generally flies low. Authors call it Musca crabroniformis, and Musca rapax major.

Asilus niger hirsutus abdominis tribus extremis articulis albis.
The great, black and white-bodied Asilus.

This is larger than the former species, and even exceeds the hornet in bigness. The body is long and slender; the wings are large and greyish: it is all over hairy and black, except that the front of the head is pale, and the three last rings of the body are white; and, in the male, the belly is somewhat yellowish.

This is a scarce fly; I met with it last year in the Fens in the isle of Ely.

Asilus vespaformis antennis capite longioribus. **The Wasp Fly.**
The vespaform Asilus, with the antennæ longer than the head.

This is of the size of the common Wasp, and very much resembles it in shape and colour: the head is smooth and yellowish; the body is obtuse, and all it's joints are edged with a pale yellow.

It is frequent in our orchards. Authors call it, Musca vespiformis.

The other species of the Asilus are, 1. The long, slender-bodied, great Asilus, called the Lupus, or Wolf-fly. 2. The great, smooth, black, and red Indian Asilus. 3. The black-legged, blue-winged, smooth, yellow Asilus. 4. The blue-bodied Asilus, with black streaks. 5. The brown-legged, white-winged, black, hairy Asilus. 6. The black and reddish-bodied Asilus. 7. The black-winged, smooth, black-bodied Asilus. 8. The black Asilus, with roundish, ferrugineous wings. 9. The black Asilus, with white streaks and spots. 10. The willow Asilus, with white wings, with two transverse, black streaks in them. 11. The smooth, black-bodied Asilus, with grey wings. 12. The smooth, oval, grey Asilus. Many of these wound in a very painful manner; others of them are quite harmless.

HIPPOBOSCA.

THE head of the Hippobosca is furnished, by way of mouth, with a snout of a cylindric figure, obtuse, and formed of two valves; the tongue is setaceous.

Hippobosca pedibus tetradactylis. — **The common**
The Hippobosca, with four claws to the feet. — **Horse-fly.**

This is a moderately large fly; it's body is of an oblong form, rounded at the end; it's colour grey, and it's skin smooth: it's eyes are large, and it's wings large and transparent; and each of it's legs is terminated by four short and sharp claws.

It is very frequent about the bodies of horses and oxen; it adheres extremely firmly wherever it lays hold; it seizes on them principally about the perinæum and groin, and is very troublesome. Authors call it, Hippobosca, Ricinus Volans, and Musca equina vexax.

Hippobosca pedibus sex dactylis. — **The Swallow's**
The Hippobosca, with six claws to each foot. — **Nest-fly.**

This is a small fly, but of a very singular form: the head is small, but from that to the tail it grows gradually broader: the thorax is of a conoid form, or resembles a segment of a cone; the body is broadest at the extremity, and is emarginated or marked with an impression or dent in that part: the wings are long, but remarkably narrow: the legs are all terminated by six short claws.

It was first observed by De Geer, in the nest of a swallow; but I have found it on the necks of horses, and other animals. Reaumur has figured it in his eleventh plate, Vol. iv.

The other species of the Hippobosca are, 1. The long-bodied, dusky-brown Hippobosca. 2. The black, oval-bodied Hippobosca. 3. The round-bodied, short-winged Hippobosca.

TABANUS.

THE head of the Tabanus is furnished, by way of mouth, with a proboscis or snout, and with connivent teeth.

Tabanus griseus abdominis segmentis singulis triangulo albo notatis. — **The great**
The grey Tabanus, with every segment of the abdomen marked with a white spot. — **Horse-fly.**

This is a very large fly; the head is greyish, the eyes large and black; the wings are large, broad, and transparent, but they have some dusky, ferrugineous lines in them: the thorax and the body are both grey, but the back part of the body under the wings is a little yellowish, and in the center of each of the rings, all the way down the back, there is a triangular white spot, with it's point upwards: the thighs are black, and the rest of the legs yellow.

It is frequent in our pastures, and is apt to be very troublesome to horses. Authors call it Tabanus major, and Tabanus vulgaris.

Tabanus fuscus alis maculatis.
The brown Tabanus, with spotted wings.

This is a small fly, scarce exceeding the common house-fly in size: the head, thorax, and body are brown, but there is some yellowishness under the roots of the wings: the wings are large, and are beautifully variegated with black and white spots: the eyes are large, and of a bright shining green, but spotted as it were with black: it is, upon the whole, a very beautiful fly.

We have it in our hedges in hilly places. Ray calls it, Musca bipennis pulchra alis maculis albis amplis pictis; others, Tabanus fuscus alis fasciis maculis nigris, and Alis cinereis variegatis.

The other species of the Tabanus are, 1. The black Tabanus variegated with yellow, and with brown legs. 2. The brown Tabanus with ferruginous sides, and three brown streaks over the eyes. 3. The grey Tabanus, with a transverse line over the eyes. 4. The brown Tabanus, with grey wings, variegated with very small white spots, with green eyes with four brown lines over them. 5. The long-bodied Tabanus. 6. The short-bodied Tabanus, with clear wings. 7. The black, clear-winged Tabanus.

MUSCA.

THE head of the Musca is furnished with a mouth, which has a proboscis or snout, but has no teeth. The species of this genus are very numerous.

Musca alis albis apice nigris.
The white-winged Musca, with a black spot on each.

The head of this fly is red, and large; the body is short, obtuse, and black; the legs also are black; the eyes are large; the wings are white, but each has at it's extremity a round black spot: while it is sitting, it is continually vibrating it's wings.

We have it frequent in our orchards, on the leaves of apple-trees. I have met with it more abundantly in Essex than elsewhere.

Musca hirsuta, thorace, abdominis apice, alarumque basibus fulvis.
The hairy Musca, with the thorax, the end of the body, and the bases of the wings yellowish.

This is a very singular fly; it is one of the largest of this genus: the body is of an oval figure and black, the extremity of it covered with a number of yellowish hairs: the thorax is also covered with the same kind of hairs: the head is black, as are also the legs; the wings are pellucid, and whitish toward the base, and have each a large spot of a ferruginous colour toward the outer edge.

It is met with in some of our woods, but not frequently. I saw a great many last summer at Mr Apreece's, at Washingley in Huntingtonshire, in his woods.

Musca nigra lateribus abdominis testaceis.
The black Musca, with the sides of the abdomen testaceous.

This is a moderately large fly: the body is of an oval figure and obtuse; the thorax is oblong; the head and the eyes are large, the legs are black; the sides are each marked with a very large, pale coloured spot.

It is very common about watery places. Ray calls it, Musca bipennis major diversicolor cauda fasciis nigris obsita.

Musca thorace abdomineque viridi-nitente. **The green Fly.**
The Musca, with the thorax and body of a shining green.

This is a large fly, equal to the common blue flesh-fly in size: the head is black, and the eyes large; the breast and body are of one uniform colour, which is a beautiful green, with something of a brassy yellow, in many lights very bright and shining: the legs are black, and the eyes brown, and there is a double transverse line on the abdomen: the body is oval, and has four joints.

It is common about stinking flesh.

Musca cinerea, thorace quinque maculis nigris, abdomine maculis tridentatis.
The grey Musca, with five black spots on the thorax, and tridentated spots on the body.

This has much the appearance of the common house-fly, but it is not more than half it's size: the eyes are reddish; the thorax is grey, with two smaller and three larger spots of black on it: the body is hoary, and composed of four joints. The first of these has no spot upon it; the second has a tridentated mark at it's base; the third and fourth have also spots of the same kind on them, but the middle denticle is almost wanting.

It is common in our pastures and about hedges, in all parts of the kingdom.

The other species are very numerous, and may conveniently be arranged under five or six divisions.

Of those with variegated wings, beside the first described, there are, 1. The Fly, with black wings tipped with white. 2. The Fly, with two black spots on each wing. 3. The Fly, with white wings, and with a single black dot at the extremity of each. 4. The unguiculated winged Fly, with the wings white, and a black spot in the middle. 5. The black Fly, with the wings variegated on the anterior part with black and white. 6. The Fly, with grey wings spotted with black. 7. The grey Fly, with white unguiculated wings, spotted with brown. 8. The Fly with white wings, with black edges, and black spots. 9. The Fly with white wings, with three brown fasciæ, and a brown terminatory spot. 10. The Fly with white wings, and with four grey fasciæ, and as many smaller running alternately with them. 11. The white-winged Fly, with four fasciæ in the wings, and with five pair of spots on the back. 12. The green-eyed Fly, with white wings, and with the letter S marked in a double line of brown on them. 13. The Fly, with white unguiculated wings, with four brown fasciæ, and with the extremity of the thorax yellow. 14. The Fly, with pale wings, with black veins, and with two transverse, undulated, brown lines, and the tips brown. 15. The Fly, with membranaceous wings spotted with black, and three rows of black dots on the body.

Of the hairy Muscæ, beside the second described species, there are, 1. The black Fly, with the edges of the wings thin, sinuated, and whitish. 2. The common hairy, dung Fly, with a spot on each of the wings. 3. The black Fly, with the base of the abdomen white, and it's extremity brown. 4. The Fly, with a grey thorax, and with the apex of the abdomen white, and the wings marked with a ferruginous spot. 5. The Fly, with a grey thorax and black body, and with a dusky, ferruginous spot on each of the wings. 6. The Fly, with the body white in the anterior part, and black behind, and with white wings with a black spot in them. 7. The hairy Fly, with a yellow thorax, and with a brown spot in the wings. 8. The hairy, grey Fly, with ferruginous wings, and a brown spot in each.

Of those with variegated bodies, beside the third described species, there are, 1. The black Fly, with the bases of the wings ferruginous. 2. The Fly with the grey breast, and with the base of the abdomen marked with a yellow spot, and with the edges of the segments whitish. 3. The black Fly, with the segments of the body, all but the first, yellow, with a black mark in the middle. 4. The Fly, with the yellow breast, and four yellow transverse lines on the abdomen, the first larger than the rest, and interrupted. 5. The Fly, with four yellow streaks on the thorax, and three of the segments of the abdomen in part yellow. 6. The black Fly, with a snow-white body, with two black fasciæ on it. 7. The brown and somewhat hairy Fly, with the edge of the abdomen acute, and with three yellow lines, and a triangular spot. 8. The bee Fly, produced from the long-tailed maggot of the necessary houses. 9. The black Fly, with a velvety body marked with three transverse lines. 10. The black Fly, with the body marked with two yellow belts on the back. 11. The black Fly, with ferrugineous wings, with three white interrupted belts on the back. 12. The brown Fly, with ferrugineous wings, and with the edges of the segments of the body grey.

Of those which in the worm state live on trees and plants, and feed on the Insects on them, there are, 1. The Fly, with the black oval body, and with a pair of lunular marks, and three yellow belts on it. 2. The Fly with an oval body, and three pair of whitish lunulæ: this is called by authors the Proboscis Elephantis, it lives in the worm-state on the pear-tree. 3. The oblong, yellow-bodied Fly, with black transverse lines. 4. The oblong-bodied Fly, with three pair of square, yellow spots. 5. The long-bodied Fly, with three pair of trigonal, yellow spots. 6. The cylindric-bodied Fly, with three pair o white lunulæ on the back. 7. The grey Fly, with four black spots on the back of the body. 8. The oblong-bodied Fly, with the hinder legs the larger. 9. The Fly, with the body marked with three yellow, reflex, circular lines.

Of the shining or gilded Flies, beside the fourth described species, there are, 1. The Fly, with a shining blue breast, and a shining green body. 2. The Fly, with a shining green breast, and a shining blue body. 3. The Fly, with a black thorax and green body. 4. The Fly, with a black breast and blue body. 5. The oblong-bodied Fly, with the head green, the thorax yellow, the body of a copper colour, and the wings marked with a brown spot. 6. The oblong Fly, with a brassy thorax, and the body yellow in the fore part, and black behind, with wings without any spots on them.

Of those resembling the common house Fly, there are, 1. The common house, flesh Fly, with a black, tesselated body, and with oblong streaks of black on the thorax. 2. The black Fly, with a white forehead. 3. The black Fly, with the body smooth and glossy. 4. The smooth, black Fly, with ferrugineous eyes, and the base of the thighs whitish; this is produced from the common maggots in cheese. 5. The smooth black Fly, with the edges of the wings black, and thicker than the other part. 6. The somewhat hairy black Fly, with nervous wings. 7. The grey Fly, with fine black marks upon the thorax, and with tridentated marks on the abdomen. 8. The yellow Fly, with the abdomen brown on the upper part, and three black streaks on the thorax. 9. The yellow Fly, with black eyes.

CULEX.

THE head of the culex is furnished with a siphon or sucker very slender, oblong, and filiform.

Culex fuscus rostro bifurco.
The brown Culex, with a forked rostrum. — The great Gnat.

This is of twice the bigness of the common Gnat, but it much resembles it in shape: it's body is long, slender, and grey; it's wings are large, thin, and clear, and have no spots in them; the snout is prominent, as in the other species of this genus; it's vagina is bifurcated at the top, or formed into two patulous leaves, which are hairy, and of a lanceolated figure, and the rostrum or sucker is slender, and placed between them.

This is a very common species about waters; while in the worm state, it lives in the water. Ray has described it under the name of Tipula domesticæ forte fœmina.

Culex lanigerus alis semifuscis.
The lanigerous Culex, with somewhat obscure wings. — The humble-bee Fly.

This is extremely unlike all the rest of the Culices in form, but the structure of the sucker evidently refers it to the same class with the Gnats: it is very like the common bumble-bee in shape, and even in size: it is covered with an extremely thick down

down of short hairs: it's body is short, rounded, and obtuse; it's colour black, but of a reddish brown at the sides: the trunk or sucker is long, and it always carries it exerted; the wings are part brown, and part whitish, and a great deal of the woolly down upon the body is white, which, with the blackness below, has a very singular effect.

It is common with us in gardens, where it flies about flowers, and sucks the honey out of them, without ever settling upon them, always continuing on the wing. Authors calls it Musca apiformis, and Musca bombyliformis.

Culex alis aqueis maculis tribus obscuris.
The Culex, with aqueous wings marked with three dusky spots.

The little Gnat.

This is considerably smaller than the common Gnat; the body is oblong, slender, and of a dusky brown, approaching to black; the head is large; the eyes are large, and very beautifully reticulated: the wings are long, narrow, and of a faint whitish colour; they have three dusky spots near the outer edge.

It is frequent in our woods and about waters; it's wound is attended with great pain, and the mark remains a long time. Derham calls it, Culex minimus nigricans maculatus.

The other species of the Culex are, 1. The common Gnat. 2. The black Gnat, with a white ring, with black legs, and whitish wings. 3. The black Gnat, with a brown body and white forehead. 4. The larger and thicker-bodied Gnat. 5. The long, variegated-bodied Gnat. 6. The long, grey, and yellowish Gnat. These are all the hitherto known species of Gnats, but the genus is by many authors made much more extensive, by the admitting into it several of the tipulæ, from an inattention to their distinguishing characters.

TIPULA.

THE head of the Tipula is furnished, at the mouth, with certain crooked and articulated tentacula.

Tipula alis exalbidis macula nivea, et lineis fuscis.
The Tipula, with whitish wings variegated with a white spot, and brown lines.

The great Tipula.

This is the largest and the most beautiful of the Tipula kind: it's body is long, slender, and of a greyish brown; the tail is bifid; the legs are long and slender; the head is large, and the eyes very beautifully reticulated; the wings are large and very beautiful; they are of a whitish colour, but have a brown line running along the exterior side, which sends off another smaller line toward the other side, and between these there is a large snow-white spot.

It is not unfrequent about Whittlesea Mere in the Isle of Ely; elsewhere I have not found it. Ray calls it, Tipula maxima alis [illegible] ex fusco et albo variegatis.

Tipula corpore nigro fulvoque variegata.
The Tipula, with the body variegated with black and yellowish.

The painted Tipula.

This is a very beautiful species: the body is long, slender, and beautifully variegated with a deep glossy black, and a bright yellow: the thorax is also of the same fine black, with a number of yellow spots on it: the wings are large and brownish, with some dusky veins, and an obscure spot in each toward the edge; the legs toward the top are also yellowish.

We have this among the willows about Battersea, and in some other places; but it is not very common. Ray calls it, Tipula elegans dorso et scapulis nigris, ventre croceo.

Tipula thorace viriſcente alis exalbidis puncto nigro notatis.
The green-breaſted Tipula, with whitiſh wings ſpotted with black.

The Sea Tipula.

This is a very beautiful ſpecies, but it much reſembles the gnat kind, and has been generally called one of them: the body is oblong and brown; the thorax is large, and of a beautiful green; the whole body and thorax are almoſt entirely ſmooth: the male has plumoſe antennæ; the female only villoſe ones. The worm from which this is produced is long, ſlender, of a bright red colour, and compoſed of twelve joints; it lives in the ſea, and makes itſelf clayey or ſandy caſes of a great length.

The fly is frequent in Suſſex toward the ſea, and in many other parts of England.

The other ſpecies of Tipula are conſiderably numerous; they may be conveniently arranged under two diviſions; the one comprehending the larger Tipulæ, which are more generally underſtood to be ſuch, and ſit with their wings patent; the other containing the ſmaller Tipulæ, generally miſtaken for gnats, and confounded with them by authors, and ſitting with their wings incumbent.

Of the firſt kind, or the larger allowed Tipulæ, there are, beſide the two firſt deſcribed ſpecies, 1. The Tipula, with wings beautifully variegated with brown. 2. The Tipula, with whitiſh wings, with a brown marginal line. 3. The whitiſh-winged Tipula, with white ſcattered ſpots. 4. The grey-winged Tipula, with black lines and ſpots. 5. The black-bodied and black-legged Tipula, with blue wings. 6. The grey-winged Tipula, with brown lines and ſpots in the wings, and the back of the abdomen blackiſh. 7. The grey-backed Tipula, with whitiſh wings with brown ſpots. 8. The black and yellow-bodied Tipula, with whitiſh wings marked with faint ſpots. 9. The black-bodied Tipula, with brown wings. 10. The yellow Tipula, with a brown back. 11. The black-bodied Tipula, with wings of a pale browniſh colour, ſpotted with black.

Of the ſecond kind, or the ſmall Tipulæ reſembling gnats, there are, beſide the third deſcribed ſpecies, 1. The brown Tipula, with the anterior part of the body green. 2. The Tipula, with the anterior legs very large, and formed like antennæ, with a white circle on the body. 3. The Tipula, with the anterior legs antenniform, and tipped with white. 4. The white-legged Tipula, with the wings variegated with white and grey. 5. The green Tipula, with white wings without any ſpots. 6. The yellow-bodied Tipula, with the eyes, the thorax, and the back black. 7. The black Tipula, with the thorax gibboſe, and with the hinder legs large. 8. The ſmooth, black Tipula, with black wings, with antennæ ſhorter than the head, and with the anterior legs furrowed on the inner part. 9. The brown Tipula, with the baſe of the wings grey. 10. The oblong, black, hairy Tipula, with black wings. 11. The black velvety Tipula. 12. The red-bodied Tipula, with white wings, with the exterior rim black. 13. The Tipula, with deflex, grey, ciliated, ovato-lanceolated wings. 14. The black Tipula, with white wings, and with a white ſpot on the anterior part of the abdomen. 15. The blue-winged Tipula, with the inner edge of the wings hairy. 16. The reddiſh-bodied Tipula, with a black thorax. 17. The black-headed Tipula, with the antennæ longer than the body, and with the body of a pale red.

PTERARIA.

Inſects which have wings.

Order the Second.

TETRAPTERA.

Thoſe which have four wings.

THE Inſects of this order are very numerous, and have certain evident diſtinctions in the ſtructure of their wings, and may thence conveniently be arranged under five claſſes. Some of them have the external pair of wings very hard, firm, and

 opake,

opake, ſerving as a kind of vagina or caſe for the others; theſe will make the firſt claſs, under the name of Coleoptera.

Others have the external pair of wings harder and firmer than the others, but more approaching to the nature of the internal, and not perfectly rigid and opake, as in the former. Theſe will make the ſecond claſs, under the name of Scleroptera.

Others have all the wings alike, and have them of a membranaceous ſtructure, but furniſhed with ſtrong ribs or nerves: theſe will conſtitute the third claſs, under the name of Neuroptera.

Others have all the four wings covered with a fine duſt, which duſt, when examined by the microſcope, is found to be compoſed of little regular ſcales, which have been improperly called feathers: theſe will conſtitute the fourth claſs, under the name of Lepidoptera.

And, finally, others have the wings merely membranaceous, compoſed of a fine thin ſubſtance, without any remarkable nerves: theſe will make the fifth and laſt claſs, under the name of the Hymenoptera.

TETRAPTERA.

Inſects which have four wings.

Claſs the Firſt.

COLEOPTERA.

Inſects which have four wings, the external ones perfectly unlike the others, being hard, firm, rigid, and opake, and having the appearance of two valves forming a caſe or vagina for the others; and having the mouth formed of two tranſverſe jaws.

SCARABÆUS.

THE antennæ of the Scarabæus are of a clavated figure, and fiſſile longitudinally: the eggs of all theſe hatch into hexapode worms, from which the beetle is afterwards produced.

Scarabæus cornubus duobus mobilibus, æqualibus, apice bifurcis introrſum ramo denticuliſque inſtructis. **The Stag-beetle.**
The Scarabæus, with ramoſe, forked, moveable, equal horns.

This is a very large and very beautiful creature: it's colour on the back is a duſky brown, nearly approaching to black: it's length is more than an inch, and it's breadth about half an inch, or little more; the horns are equal to the body in length, and ſometimes in the full grown ones much more ſo; they are uſually black, ſometimes of a fine red like coral; they are forked at the ends, and have each a large branch, beſide ſome denticulations on the inſide; they are both of the ſame length, and ſerve the creature, as the claws of crabs and lobſters do thoſe animals, to pinch and hold things faſt with: the eyes are hard, prominent, and of a pale colour; the fore legs are longer and ſtronger than the other pairs.

It is frequent in Kent and Suſſex, and in ſome other parts of England. There is hardly a ſtranger ſight than this creature flying; it does not place itſelf horizontally, as might be imagined in this motion, but flies vertically, though it's motion be horizontal. All the writers on Inſects have deſcribed it. Mouffet calls it Cervus volans; Charleton, Cervus volans platyceros; Wagner, Scarabelaphus; Olearius, Taurus volans. Pliny was better acquainted with the ſtructure and uſe of theſe horns, than ſome of the people who wrote in what are called more improved ages.

Scarabeus maxillis lunulatis, prominentibus, dentatis, thorace inermi.
The Scarabeus, with lunated, prominent, dentated jaws, and a ſmooth thorax.

This is a very large ſpecies: it's length is an inch, it's breadth two thirds of an inch: the thorax is ſmooth, convex, black, marginated, but without any prominences: the outer wings are ſmooth, and of a blackiſh purple, and are marginated; the feet are ſerrated; the lunulated jaws are prominent, black, and have each two teeth.

It is found in hedges, particularly ſuch as have aſh-trees in them, in the rotten part of the trunks of which it generally lives, and ſometimes in the earth, under their roots.

Scarabeus corpore viridi-aeneo.
The ſhining, yellowiſh, green Scarabeus. **The braſs Beetle.**

This is one of the larger beetles: the thorax is ſhort and broad, and the whole Inſect is indeed conſiderably broad, in proportion to it's length: the eyes are ſmall; the legs are long and ſlender; the whole body and external wings are of a very ſhining green, with an admixture of yellow.

It is not uncommon with us in gardens: Wormius calls it Scarabeus chlorochryſos; Merrit, Scarabeus ſmaragdinus; and J. Bauhine, Bupreſtis.

Scarabeus ovatus ater glaber.
The ſmooth oval-bodied black Scarabeus. **The little oval Beetle.**

This is one of the ſmaller beetles: it's eyes are ſmall; it's legs long and ſlender; it's body of an oval form, and of a deep duſky black, not gloſſy or ſhining, tho ſmooth; the external wings are oblong, and lightly ſtriated; the ſtriæ ſcarce viſible, unleſs on a cloſe examination: the antennæ are lamellated.

It is not unfrequent with us in woods; Hornſey-wood abounds with it.

The other ſpecies of Scarabeus, properly ſo called, are, 1. The Scarabeus, with lunated, prominent, dentated jaws, and with the thorax on each ſide tridentated. 2. The Unicorn-beetle, or Scarabeoide, the head crooked, with a ſingle horn on the noſe, with the thorax gibboſe, and the abdomen hairy. 3. The Scarabeus, with the ſhield of the head lunated, with the margin elevated, and with an emarginated horn. 4. The cylindric-bodied Scarabeus, with a ſingle horn on the forehead, and with the thorax truncated before, and quinquedentated. 5. The black-bodied Beetle, with white ſpots on the external wings, called the Fullo. 6. The teſtaceous Scarabeus, with the thorax hairy, with the crooked tail, and with the lateral ſegments of the body white, called the common Tree-beetle, or Cockchafer. 7. The teſtaceous Scarabeus, with a hairy thorax, and with the exterior wings of a pale yellow, with three white longitudinal lines on them. 8. The Scarabeus, with a marginated clypeus, and with the external wings black, with two tranſverſe ſtreaks of red. 9. The black Scarabeus, with yellow hair on it, and with the exterior wings marked with two yellow tranſverſe lines. 10. The common black Dung-beetle, or the ſmooth-bodied black Scarabeus, with the exterior wings ſulcated, and with the clypeus of the head armed with a prominent point of a rhomboidal figure in the center. 11. The blue Beetle, with the back and exterior wings ſmooth and gloſſy, and a prominent rhomboidal body in the center of the clypeus of the head. 12. The Scarabeus, with the head and breaſt blue and hairy, and with the exterior wings grey, and the legs black. 13. The Scarabeus, with the head and thorax black and ſmooth, the exterior wings grey, and the legs pale. 14. The Scarabeus, with the head and thorax of an opake black, and with the exterior wings grey, clouded with black. 15. The Scarabeus, with the head and thorax black, and with the exterior wings and the antennæ red. 16. The black Scarabeus, with pale antennæ. 17. The little black Beetle of the waters. 18. The black Scarabeus, with the exterior wings ferrugineous, behind the legs reddiſh. 19. The bluiſh-black Scarabeus.

DERMESTES.

THE antennæ of the Dermestes are of a clavated figure, and are perfoliated transversely: the species of this genus have been confounded with the Scarabæi by Ray, and most other writers; whence that genus has been greatly increased beyond it's natural bounds, and as greatly confounded.

Dermestes niger elytris antice cinereis.
The black Dermestes, with the exterior wings grey on the anterior part.

This is a small species: it's body is of an oblong form; it's head and thorax black; it's exterior wings grey towards the top, and elsewhere of a dusky blackish brown, but a transverse streak of white.

It is common in the carcasses of dead animals: it is particularly fond of birds, and often destroys the preserved specimens of them, in the collections of the naturalists. Godart calls it Dermestes: Ray, Scarabæus antennis clavatis clavis in angulos divisis quartus.

Dermestes niger, alis exterioribus punctis albis binis notatis.
The black Dermestes, with the exterior wings, each marked with a white spot.

This is a small species: it's body is of an oval figure, and black, as are also it's legs and antennæ; there is on each of the exterior wings a remarkable white spot, so that there appear two such spots on the back, when those wings are closed; if the creature be closely examined, there are also usually found on each wing five other smaller spots of white: when touched, or terrified with a noise, it draws in it's head and legs, as it were, under the shell of the body.

It is found about houses; sometimes in old walls, but more frequently in larders, and other places where there is food.

Dermestes cylindraceus collari crasso subhirsuto, elytris testaceis.
The cylindric Dermestes, with a thick hairy neck, and testaceous exterior wings.

This is a small species also; the body is of a cylindric figure: the breast is thick, roundish, and black, and is somewhat hairy: the exterior wings are oblong, obtuse at the point, and testaceous, and have a black edge, or margin running all round them: the wings are of a whitish brown; the body and legs are black; the antennæ are reddish: the size of the creature is about that of a louse; on being touched, or terrified, it draws itself up into a shorter form, and lies perfectly still.

It is common in houses.

The other species of the Dermestes are, 1. The black Dermestes, with a double undulated white line on the exterior wings. 2. The black Dermestes, with two red spots on the exterior wings. 3. The black Dermestes, with four red spots on the external wings. 4. The smooth, grey, and black Dermestes, with a yellow hairy breast. 5. The testaceous hairy Dermestes, with striated, retuse, dentated, exterior wings. 6. The smooth testaceous Dermestes, with the exterior wings obtuse and hairy behind. 7. The black Dermestes, with the exterior wings grey, edged with black. 8. The plain black Dermestes, with the exterior wings grey. 9. The black oblong Dermestes, with an acute belly. 10. The black Dermestes, with red legs, and with the exterior wings of a blackish brown. 11. The black Dermestes, with red legs. 12. The bluish-black Dermestes. 13. The brownish-black Dermestes, with beautiful, punctuated, exterior wings. 14. The smooth, oblong, testaceous Dermestes, with glossy black eyes. 15. The brown Dermestes, with a lobated, acute thorax.

These are all the species of this genus hitherto known; most of them are very small, and their habitations are extremely different; some of them live in horse-dung; others in decayed wood; others in flesh, and others in decayed vegetable substances: some are

 found

found in plants, while flourishing; and some are inhabitants of the waters; but of the latter there are but few.

CASSIDA.

THE antennæ of the Cassida are slender and filiform, but thickest toward the extremity: the thorax is plain and marginated. The species of this genus are also confounded with the Scarabæi by Ray, and others, and called by that name.

Cassida viridis, ovato-levis, clypeo caput tegente integro.
The green, smooth, oval Cassida, with the clypeus covering the head undivided.

The green Tortoise Beetle.

This is a small but very beautiful Insect: it's body is oval, convex on the back, and and plain, or flat, on the belly: the upper surface of the animal consists of the external wings, and the shield, or clypeus, which are both green smooth and uniform, so as to appear one single crust, divided only by a kind of triangular suture: the belly is black; the head is entirely hid under the shield: the antennæ are pale, but deeper at the top than elsewhere; the legs are of a pale brown, and there is a kind of prominent rim running round the wings; they entirely cover the body, so as not to discover the least part of it.

It is frequent in our gardens, on mint and other herbs: Goedart call it, Testudo viridis; and Frisch. Coccinella clypeata viridis; Ray, Scarabæus antennis clavatis in annulos divisis.

Cassida nigra antennis setaceis, corpore teretiusculo.
The black Cassida, with setaceous antennæ, and a rounded body.

This is a small species: it's body is oblong, rounded, and of a dusky black, not at all glossy or shining: the external wings are oblong, lightly striated, and somewhat flexile: they have several very small hollowed spots: the clypeus is rounded, but somewhat broader than long; it is rough on the upper part, and has a cruciated prominent edge, and sometimes two spots on the sides, toward the hinder part, with yellow hairs on them: the belly is black, but, held in some particular directions to the light, it appears white and silvery.

It is common in houses, where it is often very mischievous, eating holes in cloaths and furniture: if touched ever so lightly, it draws up it's head and legs under it's body, and will never change that posture, whatever torture it is put to, but dies in it.

Cassida nebulosa, pallida, ovalis, clypeo caput tegente integro.
The oval, pale, clouded Cassida, with an undivided clypeus covering the head.

This is a very small species: it's body is of an oval figure; it's colour a pale brown, spotted and clouded all over with a more dusky tinge, in such a manner as to have something of the appearance of the coat, or covering, of a tortoise: the clypeus is of a lunated figure; it is not emarginated before, nor spotted, but of a pale colour: the exterior wings are striated and dotted, the striæ running in flexuous lines: the body is black, and the antennæ are black and very slender.

It is not unfrequent in our beds of baum and mint; most of the writers of Insects have figured it under the name of a Scarabæus or Beetle.

The other species of the Cassida are, 1. The black Cassida, with five striæ on the external wings, and those formed of prominent spots. 2. The black Cassida, with three elevated striæ on the wings, and with the antennæ testaceous before. 3. The black Cassida, with ten elevated striæ, and with striated excavated dots. 4. The black Cassida, with five striæ dentated on each side, and with the clypeus emarginated. 5. The black Cassida, with five smooth striæ, and an emarginated clypeus. 6. The black Cassida, with five elevated lines on the exterior wings, with punctuated spaces between, and

and with the Clypeus in it's anterior part entire. 7. The black Caffida, with a ferruginous clypeus, and an elevated line on the exterior wings. 8. The ferruginous Caffida, with alternate, elevated striæ, and excavated dots on the wings.

HEMISPHÆRIA.

THE antennæ of the Hemisphæria are clavated and entire: the thorax, with the exterior wings, which are marginated, constitutes an hemispheric figure.

Hemisphæria elytris rubicundis, punctis septem nigris.
The Hemisphæria, with reddish elytra, with seven black spots on them.

The Lady-cow.

This is an Insect so well known, as scarce to need a description; as there are two or three other species however of the Hemisphæria, which are often confounded with it on a careless inspection, it will be proper to mention it's distinctive characters: the head is black, and on the forehead there are two white spots: the thorax is black, but whitish on each side toward the edges: the elytræ, or exterior wings, are of a reddish colour; there are on each of them three black spots, and, toward their base, there is one spot common to both, and making the whole number seven; this is black, like the others, but whitish in it's anterior part. The antennæ are very small and clavated, and the animal is entirely black underneath.

It is common in fields and hedges every-where in summer: Petiver calls it Cochinella Anglica vulgatissima five rubra, septem maculis nigricantibus adspersa; and Ray, Scarabæus Hemisphæricus rubens major vulgatissimus.

Hemisphæria elytris rubris, punctis duabus nigris.
The Hemisphæria, with red elytra, with two black spots on them.

This is a small species: the exterior wings are red, and have each in the center one black spot: the thorax is black, but has one larger, lateral, white spot, and two other very small ones near the base, as also two others of the same sort at the insertions of the antennæ: the belly and legs are black, so are also the antennæ.

It is frequent with us on the alder, and many other trees.

Hemisphæria elytris nigris, punctis sex rubris notatis.
The Hemisphæria, with black wings, with six red spots on them.

This is also a small species: the thorax is entirely black: the exterior wings are also of a deep shining black, but each of them has three very beautiful blood-red spots on it; of these, that which is nearest the thorax is the larger.

I have met with this in vast abundance on the maples, in some parts of the North of England; we sometimes see it in the hedges about London, but not frequently: Ray calls it Scarabæus Hemisphæricus minor elytris nigris rubris maculis.

The other species of this genus are, 1. The Hemisphæria, with red wings variegated with white spots and lines. 2. The Hemisphæria, with red wings, with five black transverse lines on them. 3. The Hemisphæria, with red wings, with five black spots on them. 4. The Hemisphæria, with yellow wings, with four round and three oblong spots on them. 5. The Hemisphæria, with red wings, with eleven black spots on them. 6. The Hemisphæria, with red wings, with thirteen black spots on them. 7. The Hemisphæria, with yellow wings, with fourteen black spots on them. 8. The Hemisphæria, with red wings, with fourteen white spots on them. 9. The Hemisphæria, with red wings, with fifteen black spots on them. 10. The Hemisphæria, with red wings, with sixteen white spots on them. 11. The Hemisphæria, with red wings, with twenty white spots. 12. The Hemisphæria, with yellow wings, with two and twenty black spots. 13. The Hemisphæria, with red wings, with four and twenty black spots on them, not at all distinct. 14. The Hemisphæria, with red wings, with numerous black spots, and a longitudinal black suture. 15. The Hemisphæria, with yellow wings, with many confused black spots, and with a longitudinal black

black suture. 16. The Hemisphæria, with black wings, with two and twenty white spots on them. 17. The Hemisphæria, with black wings, with fourteen red spots. 18. The Hemisphæria, with black wings, with four red spots. 19. The Hemisphæria, with black wings, with two red spots. 20. The Hemisphæria, with the wings shorter than the body, and bent in. 21. The hairy Hemisphæria, with the edges of the wings inflex, and with two white spots on the thorax. 22. The hairy Hemisphæria, with the edges of the external wings inflex, and the sutures red.

CHRYSOMELA.

THE antennæ of the Chrysomela are made in form of bracelets, or necklaces of beads, and are thickest toward their extremity: the body is of a figure approaching to oval, and the thorax is oblong and rounded; the species of this genus have been usually confounded also with the beetles.

Chrysomela viridis cærulea.
The blue-green Chrysomela.

This is one of the larger species of this genus, though the very largest of them are but small: it's head is small, it's legs slender, and it's belly smooth: it is very convex and rounded on the back, and it's colour of a mixed blue and green, with a very fine tinge of gold diffused through it: the edges of the external wings are somewhat prominent, and they are marked with a few excavated small spots all the way round: the antennæ, legs, and belly are of a simple green.

It is common in our meadows in May and June.

Chrysomela thorace elytrisque rubris.
The Chrysomela, with the thorax and external wings red.

This is a smaller species than the former, and is less convex on the back: it's head is small and black; it's body is also black, and so are the legs, the lower part of the breast, and the antennæ; but the upper surface of the thorax is red, so are also the external wings.

It is frequent with us on the fallow, and some other species of the willow kind, in June and July.

Chrysomela cæruleo-viridis thorace femoribusque rufis.
The blue-green Chrysomela, with the thorax and thighs reddish.

This is one of the smaller species: it's head and exterior wings are of a beautiful shining green, with a cast of blue in it; and there are a few small hollowed spots on the head: the thorax is very convex, but small; it is of a reddish colour, with a cast of the blue-green in it: the upper part of the legs is reddish, the rest black: the antennæ also are black.

It is frequent on our old willows about Chelsea, and almost in all parts of the kingdom.

The other species of this genus are, 1. The black Chrysomela, with hollowed and contiguous spots. 2. The smooth, black Chrysomela, with the bases of the antennæ yellowish. 3. The purplish-black Chrysomela, with striæ on the external wings, formed of excavated spots. 4. The purplish-black alder Chrysomela, with small excavated spots, placed irregularly. 5. The purplish-black Chrysomela, with the thorax yellow on each side, and with a black spot. 6. The bright green Chrysomela, with the thorax smooth, and a number of contiguous, excavated spots on the external wings. 7. The smooth, green Chrysomela, with the thorax excavated in the anterior part. 8. The smooth, green Chrysomela, with the thorax smooth, and the exterior wings contiguous behind. 9. The obscurely testaceous Chrysomela. 10. The pale grey Chrysomela. 11. The black-bodied Chrysomela, with the thorax and exterior wings grey. 12. The red Chrysomela, with the thorax of a cylindric figure, and compressed on each side. 13. The brassy-yellow Chrysomela. 14. The greenish-yellow bodied Chrysomela, with red wings. 15. The bluish-black Chrysomela, with reddish external wings,

wings, black at the extremities. 16. The Chrysomela, with the thorax red, and of a cylindric figure, with two black spots on it, and with the external wings yellow, with a black cross upon them. 17. The Chrysomela, and with a green thorax, and with red wings, with a blue cross on them. 18. The oblong, black Chrysomela, with the external wings red, and with four black spots on them. 19. The oblong, brown Chrysomela, with the letter S, in white, on the external wings. 20. The Chrysomela, with the external wings brown, and livid behind. 21. The bluish-black Chrysomela, with coal-black eyes. 22. The reddish-brown Chrysomela, with two black spots on the thorax, and several on the wings. 23. The blackish-yellow Chrysomela, with the edges of the exterior wings yellow. 24. The blackish-yellow Chrysomela, with a double line of yellow on the exterior wings. 25. The Chrysomela, with a black thorax, and with the exterior wings red, with a black cross on them. 26. The thin-winged, testaceous Chrysomela. 27. The oblong testaceous Chrysomela, with painted exterior wings. 28. The brown Chrysomela, with a black head. 29. The Chrysomela, with brown wings, with yellowish edges.

CURCULIO.

THE antennæ of the Curculio are affixed to an elongated horny snout.

Curculio subfuscus, elytris fasciis nebuloso-testaceis.
The brownish Curculio, with the exterior wings nebuloso-testaceous.

The Saw-wort Beetle.

This is a small species, and is of a greyish-black colour: the body is thickest and very gibbous on the back part, toward the tail: the wings, thorax, body, and indeed every other part of the Insect, are covered with little excavated spots: there are two obscurely testaceous lines on the thorax, and the apices of the antennæ are yellow.

It is frequent in our woods, and is most commonly found on the Serratula, or Saw-wort; Lister call it Scarabæus majusculus fuscus, e leucophæo nigroque varius.

Curculio niger elytris nigris, fascia duplici alba, basi rubris.
The Curculio, with black exterior wings, red at the base, and with two white lines.

This is about the size of the common ant, and, tho' a creature of a wholly different character, yet carries a very great external and general resemblance to that Insect: the head is small, short, black, and immersed in the thorax: the antennæ are short, clavated, and black: the thorax is rough, and of a reddish colour, but the edge next the head is blackish; the exterior wings are oblong, obtuse, flattish, black, and ornamented with a number of little protuberances; they are reddish at the base, and have a double white line running across them, the one broad, and the other flexuous.

It is common about our pastures all the summer; is runs the swiftest of all the Curculio's. All the writers on Insects have figured it under the name of a Beetle: it is the forty-ninth of Johnston; the twenty-ninth of Ray's history.

Curculio nigro-cærulescens.
The bluish-black Curculio.

This is of about the size of the common flea: the thorax, the head, and the legs are of a pale blackish-blue; the exterior wings have a much deeper tint of the same kind, and are marked with longitudinal striæ, formed of excavated spots.

It is common in our hedges.

The other species of the Curculio are, 1. The common brown Curculio, with acuminated antennæ. 2. The grey Curculio, with a ferruginous spot on each of the exterior wings. 3. The black and white Curculio, with a flat hollowed proboscis, of the length of the thorax. 4. The brassy-brown Curculio, with the snout shorter than the thorax. 5. The grey Curculio, with three striæ on the thorax, and with the snout shorter

shorter than the thorax. 5. The grey Curculio, with three striæ on the thorax, and with the snout shorter than the thorax. 6. The black Curculio, with the snout of the length of the thorax. 7. The oblong, grey Curculio, with obtuse wings. 8. The grey, oblong Curculio, with red legs. 9. The oblong, grey, and black Curculio, with the legs and the antennæ brownish. 10. The dusky green Curculio, with the legs and the antennæ brownish. 11. The blue-green, shining Curculio, with black antennæ. 12. The shining purple Curculio. 13. The clouded grey Curculio, with the anterior legs denticulated. 14. The oblong-bodied, green Curculio, with all the legs denticulated. 15. The roundish-bodied Curculio, with two black, longitudinal spots, and a white thorax. 16. The brownish, black, roundish-bodied Curculio, with a white cordated spot on the back. 17. The oblong, testaceous, reddish Curculio, with the thorax nearly of the length of the wings. 18. The oval-body, coal-black Curculio. 19. The black Curculio, with oblong, exterior wings, and clavated thighs. 20. The black Curculio, with oblong, opake, exterior wings. 21. The black Curculio, with shining, exterior wings. 22. The black Curculio, with bluish-black shining wings. 23. The black Curculio, with white thighs. 24. The grey Curculio, with blackish spots on the exterior wings, and pale yellow legs. 25. The grey Curculio, with reddish legs. 26. The grey Curculio, with pale, longitudinal streaks. 27. The bluish-black Curculio, with four blackish spots on the wings. 28. The blood-coloured Curculio. 29. The black Curculio, with the exterior wings reddish in the middle. 30. The black-bodied Curculio, with reddish wings, and the head elongated behind.

CERAMBYX.

THE antennæ of the Cerambyx are long, slender, and setaceous: the thorax is oblong and rounded, and is in most of the species continued to a point at each extremity.

Cerambyx viridi-cærulescens antennis corpus subæquantibus.

The blue-green Cerambyx, with the antennæ nearly equal to the body in length.

The musk Beetle, or capricorn Beetle.

This is a very large and beautiful Insect: it is all over of a beautiful, glossy, blue-green colour, with a cast of shining gold-yellow in it; the body is blue on the upper part, and the inner wings are black: the legs are of the same bluish-green colour with the rest, only somewhat paler: the thorax is mucronated or pointed at each extremity, and between the two points there are three little tubercles near the wings, and three others still smaller toward the head; the wings are oblong, of a somewhat lanceolated figure, flexile, and have three longitudinal and somewhat elevated ribs on them: the antennæ are nearly of the length of the body; they are composed of a number of joints which grow smaller, as they approach the extremity, the contrary of which proportion is the usual case.

It is found in old willows, but not very frequently; it diffuses a most agreeable smell, somewhat like that of musk: it has been described by most of the writers on Insects. Ray calls it Scarabæus Capricornus dictus major viridis odoratus; Lister, Scarabæus magnus suaviter olens, and Goedart, Cerambyx tertius.

Cerambyx nebulosus, antennis corpore longioribus, punctis quaternis luteis in thorace.

The Cerambyx, with the antennæ longer than the body, and with four yellow dots on the thorax.

The body of this species is about three quarters of an inch in length, and is all over of a grey colour; the exterior wings are obtuse, and have a number of minute hairs on them, and, between and among these are a number of small tubercles; there is a dusky blackish shade across the wings, which in it's hinder part is curved toward the middle: the thorax is pointed at each end, and is very grey, but has four beautiful transverſe yellow ſpots toward it's hinder part: the eyes are black, and there is a black ſpot near the antennæ: the antennæ are equal to five times the length of the body; they are grey, and are compoſed of ten articulations, which are ſhorter, as they are nearer the head; the wings are black, ſtreaked with brown. The female in this ſpecies has an elongation at the anus, which continues the body to a third of the length of the antennæ; but in the males they are five or ſix times as long as the body.

It is found in old wood, but is not common with us. I met with a great number of them in ſome old timber in Lincolnſhire laſt ſummer. Poiver calls it Capricornus Ruſſicus cinereus comibus longiſſimis.

Cerambyx nigricans thorace villoſo cinereo, punctis duobus glabris.
The black Cerambyx, with a hairy, grey thorax, and two ſmooth ſpots.

The body of this ſpecies is oblong, and ſomewhat depreſſed; it's colour is a deep black, with an admixture of a faint grey: the body is covered with a number of ſhort hairs, and with prominent tubercles between them: the whole thorax is hairy; it is black, and the hairs are white, ſo that on the whole it appears grey, but there are two ſmooth prominent ſpots on it's hinder part: the antennæ are ſlender, black, and of about half the length of the body; there is an undulated line on the exterior wings, but ſo faint, that it is ſcarce viſible; the exterior wings are ſomewhat flexile.

It is found among timber, but is not very common with us.

The other ſpecies of the Cerambyx are numerous: they may be arranged under two diviſions, thoſe of the firſt having ſharp protuberances ariſing from the thorax, and thoſe of the ſecond not.

Of the firſt diviſion, beſide the two firſt deſcribed ſpecies, there are, 1. The black Cerambyx, with a ſlatted thorax, the edge tridentated on each ſide, and with coal-black antennæ. 2. The reddiſh Cerambyx, with three longitudinal black lines on the wings. 3. The black Cerambyx, with irregular pale ſpots on the wings, the thorax prickly, and the antennæ longer than the body. 4. The grey Cerambyx, with black ſpots, and a black tranſverſe line on the wings, and with the antennæ of once and a half the length of the body. 5. The grey Cerambyx, with black, abrupt, exterior wings, with white ſpots, and a white tranſverſe line on them, and with the antennæ once and a half the length of the body. 6. The grey and black Cerambyx, with the antennæ of about half the length of the body. 7. The grey Cerambyx, with two yellow, tranſverſe lines on the exterior wings, and the antennæ of half the length of the body. 8. The teſtaceous Cerambyx, with a double, white, tranſverſe line on the wings.

Of the ſecond diviſion, or thoſe which have none of theſe ſpinoſe protuberances on the thorax, there are, beſide the laſt deſcribed ſpecies, 1. The black Cerambyx, with two white, undulated, tranſverſe lines on the exterior wings. 2. The teſtaceous Cerambyx, with a grey, hairy thorax, with two ſmooth, ſhort lines on it. 3. The teſtaceous Cerambyx, with the ſmooth thorax. 4. The brown Cerambyx, with impreſſed ſpots on the thorax. 5. The grey Cerambyx, with the exterior wings ſpotted with black. 6. The black Cerambyx, with a longitudinal, denuated, yellow line, and yellow ſpots on the exterior wings. 7. The fine, violet-coloured, ſhining Cerambyx, with black antennæ, and clavated thighs.

LEPTURA.

THE antennæ of the Leptura are oblong, ſlender, and ſetaceous: the exterior wings are truncated at their extremity, and the thorax is of a ſubcylindric figure.

Leptura nigra, elytris rubeſcenti-lividis.
The black Leptura, with reddiſh-livid, exterior wings.

This is a large ſpecies; it's body is of an oblong figure, and grows ſmaller behind: this, the legs, the antennæ, and all the other parts, except only the exterior wings, are black: the belly, and all the under part of the inſect, is black, but, if viewed in a certain light, it looks white and ſilvery: the exterior wings in the females are of a deep and strong red, without any variegation; but in the males they are of a less intense red, and are black, or grey, at the tips or edges, or have other variegations of those colours in them; the whole surface of the exterior wings in both is ornamented, with a vast number of small excavated dots, and with a few short and fine hairs; the head and thorax are sometimes yellowish.

It is not uncommon with us in woods; those about Highgate and Hampstead frequently afford it. Most of the authors who have written on Insects mention it; Ray calls it Cerambyx capite, scapulis, et antennis nigris elytris flavis extremitatibus nigricantibus; Frisch. Scarabeus arboreus major violaceo-ruber.

Leptura nigra elytrorum lineis transversis flavis, pedibus testaceis.
The black Leptura, with yellow, transverse lines on the exterior wings, and testaceous legs.

This is of an oblong, narrow figure, and it's general colour is a blackish-brown: the upper edge of the thorax is yellow, and there is a yellow spot at the juncture of the exterior wings; and there are some oddly undulated yellow lines placed transversely on these: they are truncated at the apex, and are lightly hairy; the antennæ and legs are of a reddish brown.

We have it in our hedges, and not unfrequently in orchards. Rays calls it Scarabæus medius abdomine oblongo angusto niger, lineolis et maculis luteis pulchre variegatus.

Leptura deaurata antennis nigris, femoribus posticis dentatis.
The gold yellow Leptura, with black antennæ, and the hinder legs dentated.

The body is oblong, and the head small; the colour of the whole Insect is nearly that of copper, but with a fine and strong gilded yellow diffused through it: the head, the thorax, the exterior wings, and even the legs are all spotted, with extreamly minute and almost contiguous dots, which are all excavated or hollow; they are dispersed irregularly over the thorax, head, and legs, but on the wings they are tolerably regularly arranged into ten longitudinal series, forming so many lines: the eyes are black, and the antennæ brown.

It is common with us about waters among sedge and the cyperus grasses; sometimes on the leaves of the nymphææ.

The other species of the Leptura are, 1. The great, dusky, greenish, yellow Leptura, with the antennæ variegated with yellow and green. 2. The black Leptura, with the thorax and the exterior wings reddish. 3. The black Leptura, with the exterior wings variegated with black and livid. 4. The black Leptura, with the thorax, the exterior wings, and the legs purple. 5. The black Leptura, with the exterior wings and the thorax yellow. 6. The black Leptura, with the exterior wings livid, and marked with four black spots. 7. The entirely black Leptura. 8. The black Leptura, with the exterior wings testaceous, with six black spots on them. 9. The black Leptura, with four pair of ferrugineous spots on the exterior wings. 10. The black Leptura, with yellow exterior wings, black at the extremity. 11. The black Leptura, with testaceous exterior wings, with a black cross, black lines, and two black spots on them. 12. The violet and gold-coloured Leptura, with the posterior legs dentated.

CARABUS.

THE antennæ of the Carabus are oblong, slender, and setaceous: the thorax is somewhat convex, marginated, of a cordated figure, and truncated in the hinder part.

Carabus violaceo-niger elytris levibus, subrugosis.
The violet-black Carabus, with smooth but wrinkled exterior wings.

The head of this species is small; the eyes are large and prominent, and the antennæ moderately long, and very slender; the general colour is black, but the edges of the thorax and wings are of a beautiful deep and glossy violet colour: the exterior wings have no dots, nor lines on them, but they have some longitudinal, and even some transverse, wrinkles, but they are not very conspicuous.

It is common with us among putrefying vegetables, and on Dunghills. Ray calls it Scarabæus major corpore oblongo è purpura nigricante.

Carabus purpureo-nigrescens elytris cancave punctatis striatisque.
The purplish-black Carabus, with hollow spots and lines on the exterior wings.

This is a large species: the body is of an oblong form, and moderately thick; the general colour is blackish, with a strong and very beautiful tinge of a glossy purple: there are on each of the exterior wings three lines, formed each of a series of longitudinal spots, which are round and hollowed, and placed about twelve in each series; the lines which separate these rows of spots, are also hollowed.

It is not uncommon with us on dunghills, and about decayed and putrid vegetable substances. Ray calls it Cerambyx purpurea punctata.

Carabus ater pedibus rufis.
The black Carabus, with reddish legs.

This is a small species, very little exceeding a common fly in bigness: the whole Insect is black; the thorax is broad and short: the exterior wings are striated each with eight lines; the legs are of a reddish brown, as are also the bases of the antennæ.

It is not unfrequent with us about dunghills: it has been called Buprestis corpore nigro pedibus rufis.

The other species of Carabus are, of the larger ones, 1. The black Carabus, with the exterior wings convexly punctated and striated. 2. The black Carabus, with greenish wings convexly punctated and striated. 3. The black Carabus, with brassy wings convexly punctated and striated. 4. The black Carabus, with sixteen striæ on the exterior wings. 5. The green Carabus, with obtusely striated and not punctated wings, and with the head and legs of a ferrugineous colour. 6. The black Carabus, with green, obtusely sulcated wings, and with the antennæ and legs black.

Of the smaller Carabi there are, beside the last described species, 1. The elegant shining Carabus, with the head and thorax blue, and the wings purple. 2. The brassy, yellow Carabus, with reddish legs, and six hollowed spots on the exterior wings. 3. The black Carabus, with the legs pale. 4. The pale, testaceous Carabus, with smooth, exterior wings. 5. The black Carabus, with the thorax, the antennæ, and the legs ferrugineous. 6. The Carabus, with the head and exterior wings blue, and the thorax red. 7. The Carabus, with the head and wings black, and the thorax red. 8. The yellowish-black Carabus, with the antennæ and legs black. 9. The black Carabus, with a ferrugineous spot, and a ferrugineous, transverse line on the hinder part of the exterior wings. 10. The black Carabus, with the exterior wings greyish on the anterior part. 11. The black Carabus, with the legs and antennæ also black. 12. The pale, testaceous Carabus, with black, glossy eyes. 13. The black Carabus, with a ferrugineous thorax, and four livid spots on the exterior wings. 14. The grey Carabus, with the head, the belly, and the bases of the wings black.

MORDELLA.

THE antennæ of the Mordella are slender, and have the last joint globose; most of the species have also legs which serve them for leaping.

Mordella

Mordella oblonga atra cauda aculeo terminata.
The oblong, black Mordella, with an aculeated tail.

This is a ſmall ſpecies; it's length is about a ſixth of an inch; it's breadth is not equal to half that meaſure: it's colour is black; it is ſmooth and ſomewhat gloſſy on the greater part of it's ſurface; the head is ſmall and inflected; the exterior wings are not at all ſtriated; the thorax is ſmooth, and very convex; the antennæ are very ſlender, truncated, and articulated: the body grows gradually ſmaller toward the tail, where it terminates in a ſharp point or prickle, which is black, and longer than the extremity of the wings, but is too weak to inflict a wound: the legs are ſlender, and conſiderably long, and it hops with them very nimbly.

It is frequent with us in the flowers of the dandelion and hawkweeds.

Mordella ſubrotunda atra opaca.
The roundiſh, opake, black Mordella.

This is a very ſmall ſpecies: it is ſhorter and thicker than the precedent, and approaches to a roundiſh figure: the head is ſmall, the thorax is elevated, and of a duſky deep black, not at all gloſſy; the exterior wings are of the ſame duſky black, and are ſomewhat ſhorter than the body; the legs are ſlender and long, it leaps very nimbly.

It is common in our gardens among the lettuces and endive: it has been called by ſome Gyrinus nigricans minor.

Mordella ovata, cærulea, nitida, tibiis ferrugineis.
The ſhining, blue, oval-bodied Mordella, with ferrugineous legs.

This is a very ſmall but a very elegant ſpecies; it is ſcarce larger than a flea: it's body is ſhort, and approaches to an oval form; it's thorax and back are both very convex and ſmooth, and it's colour a deep, but very beautiful and gloſſy, blue: it's legs are long; the thighs thick, robuſt, and whitiſh: the lower part of the legs ferrugineous, it hops very nimbly.

It is common in our gardens among the cabbages, while young. Ray calls it Capricornus exiguus ſaltatrix.

Mordella ſubrotunda atro-ænea.
The roundiſh, black Mordella, with a braſſy tinge.

This is one of the ſmalleſt of the Mordellæ; it is conſiderably leſs than a flea, and is all over of a deep, but very gloſſy, black, with a fine tinge of braſſy yellow in it: the belly and legs are of the ſame fine black, but without the yellow: the exterior wings are ſtriated, and the ſtriæ compoſed of five ſmall excavated ſpots.

It is not uncommon in our gardens early in ſpring.

The other ſpecies of Mordella are, 1. The yellow Mordella. 2. The roundiſh, mouſe-coloured Mordella. 3. The roundiſh, black, opake, but dotted Mordella. 4. The roundiſh, black, gloſſy Mordella, with the antennæ and the legs yellow. 5. The oblong, black Mordella, with a ferrugineous ſpot in the middle of the exterior wings, and with the ſides of thoſe wings ferrugineous toward the baſe. 6. The oblong, black Mordella, with the exterior wings yellow all the way down the middle. 7. The reddiſh, brown Mordella, with the anterior legs dentated. 8. The black Mordella, with the exterior wings red at their extremities. 9. The brown, opake Mordella, with punctated, exterior wings. 10. The ferrugineous, brown Mordella, with the thorax depreſſed in the anterior part. 11. The Mordella, with the articulations of the antennæ lentiform, except the extream one, which is perfectly globoſe.

CICINDELA.

THE antennæ of the Cicindela are slender, oblong, and setaceous; the jaws are prominent and dentated; the thorax is of a roundish, but somewhat angulated, figure.

Cicindela supra viridis elytris punctis decem albis variegatis.
The green Cicindela, with ten white spots on the exterior wings.

This is one of the most elegant and beautiful Insects of this class: it's whole upper surface glows with a rich gilded tinge among the green, which is itself an elegant colour, but with this addition strikes the eye in a very uncommon manner: the exterior wings are smooth, glossy, and of this high colour, variegated with ten white spots, some of them roundish, some oblong, and one of a lunated figure, and the extremities of these wings are also white: the thorax is narrow, roundish, and of a very strong green; the head is small, depressed, and finely tinged with the gold colour: the eyes are black and prominent; the mouth also is prominent, and the upper lip is obtuse and white: the upper jaws are prominent, and have several strong teeth in them; the lower are only furnished with one tooth, and that at the extremity: there are two pair of antennæ, the one consisting only of two joints each, the other of four: the legs are very long and very slender, and there is a kind of hard, oval body at the base of the thighs: the longer antennæ are composed of ten joints, and are somewhat shorter than the body.

It is common in our pastures in spring; it runs and flies very swiftly. Ray calls it Scarabæus viridis decem maculis albis.

Cicindela ænea elytris punctis latis excavatis.
The brassy Cicindela, with broad excavated spots on the wings.

This is another very beautiful, though small, species: it's colour is a fine, glossy, metalline-yellow, with some little admixture of green: it's eyes are black and prominent; it's thorax is narrow and rounded: the exterior wings are ornamented with a number of broad excavated spots, in the center of each of which there is a prominent point; these excavated spots are connected into series, by a longitudinal elevated line of a deep black: the bottoms of them are of a fine brassy-yellow; and, as the rest of the wings is green, it is a very singular and very beautiful colour, that is produced by the mixture.

It is common with us about the banks of rivers, where it runs along the wet sand very nimbly. Lister calls it Scarabæus parvus inauratus.

Cicindela elytris viridi cærulescentibus abdomine fulvo.
The Cicindela, with the exterior wings of a bluish-green, and the belly yellow.

The body of this species approaches to an oval figure; the head, the thorax, the antennæ, and the legs are all black: the belly is of a tawny colour, and is yellowest toward the hinder part: the exterior wings are of a very singular blue-green colour, and are variegated with hollow spots: the antennæ are slender, and consist of ten joints.

It is frequent in the woods about Hampstead.

The other species of Cicindela are, 1. The common oblong Cicindela. 2. The larger, shorter-bodied Cicindela. 3. The black Cicindela, with six white spots, and a white transverse line on each of the exterior wings. 4. The black Cicindela, with a reddish thorax, and with bluish-black exterior wings. 5. The Cicindela, with a green thorax, and with the exterior wings ferrugineous toward the top, and of a bluish-black below.

BUPRESTIS.

THE antennæ of the Buprestis are slender and setaceous, and the head is half buried in the thorax, which is of a subcylindric figure.

Buprestis

Buprestis viridi-ænea immaculata.
The yellowish-green Buprestis, without spots.

This is a very beautiful Insect: it's length is about half an inch; it's breadth near two thirds of it's length: the thorax is very large and broad, equalling the diameter of the wings; it is of a green colour, with a yellow brassy cast, and has several hollow spots upon it: the head is in a great measure received into, or immersed under, the anterior part of this; the exterior wings are of a fine, glossy, metalline-yellow, with only a slight tinge of green, and they have each about eight striæ on them.

This species is very common in damp places, near waters.

Buprestis fusco-ænea, glabra, nitida, thorace submarginato.
The brownish brassy Buprestis, with a submarginated thorax.

This is a very small species: the whole Insect is of a glossy brass-yellow, with an admixture of brown: the eyes are moderately large, and very prominent; the antennæ are short: the thorax is short, but it is broad, and hollowed, and slightly marginated: the exterior wings are very bright and glossy, and are finely striated, the striæ being formed of very minute and elegant points: the legs are slender and black, as is also the under surface of the body: the snout is prominent.

It is very frequent among the reeds on the little islands formed by the Thames, about Chiswick and Brentford.

Buprestis æneo-nebulosa antennis clavatis.
The brassy and clouded Buprestis, with clavated antennæ.

This is a small species, it scarce equals a flea in size; the belly and lowest part of it's body is all black; on the upper part it is all over of a brassy yellow, with a mixture of brown in it; and it has a brown spot on the back, which touches both the wings: the exterior wings are very finely striated; the thorax is marked with five oblique furrows: the thorax is large, and the head in a great measure is immersed under it, and the body grows narrower toward the tail.

It is common with us about waters, and sometimes under cool shady hedges.

Buprestis fusco-ænea elytris maculis æneis impressis.
The Buprestis, of a dusky brassy colour, with spots on the wings.

This is a moderately large species; it's head is drawn under the breast, as in the others, and the creature has thence a very odd aspect: the thorax is of a brassy colour, with an admixture of a reddish brown: the external wings also are of the same mixt colour, but with less of the reddish; each of them has four or five striæ not very deep, and on each there are two spots so disposed, that they make a square figure together, when the wings are closed; these are of a bright and unmixed yellow.

It is a native of Virginia, and is sometimes found also with us about the sides of rivers. Petiver has figured it under the name of Cantharus Marianus elegans viridis vaginis sulcatis, signaturis flavescentibus.

The other species of the Buprestis are, 1. The great black Buprestis, or great thick-necked, black, tree Beetle. 2. The greyish-yellow, glossy Buprestis, with the exterior wings attenuated toward the end. 3. The black small Buprestis, with depressed spots on the thorax.

DYTISCUS.

THE antennæ of the Dytiscus are usually slender and setaceous, and it has feet formed for swimming, and muticous: it's habitation is generally also in the water.

Dytiscus antennis fuscis perfoliatis.
The Dytiscus, with brown, perfoliated antennæ.

The great Water Beetle.

This is one of the largest of the European beetles: the body is an inch and half long, and all over of a deep and somewhat glossy black: the eyes are moderately large; the antennæ are short, and the exterior wings are smooth on the surface: the body grows smaller, and becomes almost pointed at the extremity.

It is very common in our ponds and ditches: in it's worm-state it is very large, and very destructive among the smaller water Insects. Authors call it Hydrocanthurus maximus.

Dytiscus niger elytrorum marginibus dilatatis, flavis.
The black Dytiscus, with the edges of the exterior wings dilated and yellow.

This is equal to the former species in size: the head is small in proportion to the body; the eyes are large; the legs long and strong; the edge of the exterior wings is very prominent, especially about the middle, and is there of a yellow colour, which, as all the rest of the Insect is black, has a very singular appearance.

It is common with us in brooks and rivers.

Dytiscus elytris striis viginti dimidiatis.
The Dytiscus, with twenty dimidiated striæ on the exterior wings.

The goggle-eyed Water Beetle.

This is a very large beetle, but somewhat inferior to the two preceding ones in size: the head is large; the eyes are very prominent; the exterior wings have each ten striæ, but they do not run the whole entire length of the wing; there is a smooth space near the extremity.

It is frequent in our ponds and ditches: the striæ are very deep, and the colour of the whole a blackish brown.

Dytiscus cinereus margine elytrorum flavo, thoracis meditate flava.
The grey Dytiscus, with half the thorax and the edges of the wings yellow.

This is a small species, scarce exceeding the common blue flesh-fly in bigness: it's thorax is yellow in the middle, but black at the top and bottom, not at the sides: the exterior wings are of a greyish colour, and have on them a vast number of small, lucid specks, of a yellowish colour; their edges are entirely yellow: there is also a yellow spot of a cordated form, with black edges on the summit; and the apex of the sternum is bifurcated but obtuse.

The other species are, 1. The common, black water Beetle. 2. The great, black Dytiscus, with black, perforated antennæ. 3. The great Dytiscus, with a yellow thorax. 4. The small water Beetle, with brown wings and a black belly. 5. The smaller, round-bodied Dytiscus, with ten longitudinal striæ on the wings. 6. The ovato-oblong-bodied Dytiscus, with the thorax and wings black, and the head and legs reddish. 7. The brown, oval Dytiscus, with the legs, and the head, and thorax reddish. 8. The smooth, short-horned Dytiscus, with obtuse points. This is called by some Pulex aquaticus, the water Flea; it plays very nimbly about the surface of the water, and at all times plunges to the bottom.

ELATER.

THE antennæ of the Elater are setaceous; the body is oblong: the creature, when laid on it's back, has a power of leaping with great force and agility.

Elater fusco-viridi-æneus.
The Elater, of a mixed brown, green, and brassy colour.

This is but a small species; the body is oblong, and not very thick; the male has the thorax and exterior wings very bright, and with a great deal of green in them, and the antennæ are elegantly pectinated: the female is more yellow and brassy in colour, and the thorax is broader, and of a glossy colour, with more green in it than in the wings, and the antennæ are not pectinated.

We find it frequently in June, in the meadows about Paddington.

Elater niger thorace rubro.
The black Elater, with the thorax red.

This is the smallest of all the species of this genus: it is all over black, but the exterior wings have somewhat of a bluish cast in them, and are striated: the thorax is almost entirely red, but it's anterior part is black: it has also a very slight verge of black on the hinder edge, toward the wings: upon the whole, the thorax has the appearance of a large, red, lunulated spot, the hollow part of which is turned toward the head: the antennæ are black, and are not pectinated.

This elegant species is very frequent in some of our pastures under hedges; in other parts of the kingdom it is very scarce. I have a seen a great number of them near Pancras.

Elater totus nigro fuscus.
The Elater all over of a brownish black.

This is a moderately large species; the body is oblong, but is less long in proportion to it's breadth, than in the former: it's colour is uniform throughout, and is a greyish-brown, approaching to black: the head is small, and the exterior wings are smooth, and somewhat glossy.

It is not uncommon in pastures about London. Authors have called it Notopeda atra antennis simplicibus.

The other species of the Elater are, 1. The Elater, with the thorax hairy, and with the exterior wings testaceous, and black at the extremity. 2. The black Elater, with the exterior wings red. 3. The greenish and brassy Elater, with yellow legs. 4. The black Elater, with blue, exterior wings. 5. The black Elater, with the exterior wings red in the anterior part. 6. The black Elater, with the exterior wings livid on the outer edge. 7. The red-breasted Elater, with grey wings. 8. The black Elater, with brown wings, and with the antennæ and legs of a reddish brown.

CANTHARIS.

THE antennæ of the Cantharis are setaceous; the exterior wings are flexile; the thorax is somewhat flatted, and the sides of the abdomen are plicated and papillose.

Cantharis cujus fœmina aptera.
The Cantharis, the female of which has no wings.

The Glow-worm.

The male of this species is a small beetle; it's body is oblong and narrow, and it's colour a dusky black: the female, which is the sex we most frequently meet with, and which we call the Glow-worm, is a small, oblong-bodied Insect; and though she has no wings either of the thinner, or thicker kind, yet shews very evidently, by the scutum and form of her thorax, and by the folds and wrinkles of her body, that she is of this class of Insects: her body consists of eleven joints, or has so many inclsures, the first of which is the incumbent Clypeus of the thorax; this is of a semi-oval figure, flatted, marginated, and truncated at the hinder part: the head is placed under this, and is very small; the three last joints of the body are of a yellowish colour on the under surface, and these appear ignited or flaming in the dark.

It is not uncommon under our hedges, and, if taken carefully up, may be kept alive many days on fresh turfs of grass, and will continue to shine in the dark. Ray calls the male of this species Scarabæus lampyris fordide nigricans corpore longo angusto, sive Cicindela mas; and the other, or female, Cicindela impennis sive fœmina.

Cantharis elytris nigricantibus, thorace rubro, nigra macula.
The Cantharis, with blackish, exterior wings, and with a red thorax, with a black spot.

This is a large species: it is more than half an inch long, and about a quarter of an inch in diameter, and is softer to the touch than the generality of the beetle kind: the head is flatted and black, but under the eyes somewhat reddish: the mouth is small and forcipated; there are a pair of very short and small antennæ at the mouth: the others are about half the length of the body, and are composed of eleven joints; they are reddish toward their origin, and brown all the way thence to the extremity: the thorax is depressed on the hinder part, and cordated; the edges are somewhat prominent, and the whole is reddish, except that there is a black spot in the upper part contiguous to the head: the exterior wings are plain, smooth, of an oblong and linear figure, very soft and flexile, of a silky structure, and of a brownish-black colour: the body is brown, except that the last joint is reddish, and there is a tinge of the same colour all along the sides; the sides are all the way compressed, and the incisures are laid in a plicated manner over one another; they are very soft, and their extremities are mammillary.

It is very frequent with us about houses and under hedges. Ray calls it Cantharus sepiarius major elytris nigricantibus, dorso sive thorace supino obscure rufo.

Cantharis elytris rubris, thorace rubro, nigra macula.
The red-winged Cantharus, with a red thorax, with a black spot.

This is but a small species: the body is black, but the exterior wings are of a bright and elegant red, and the breast is of the same colour, but has a spot of black in it: the antennæ are slender; the exterior wings are very soft, silky, and flexile; the under are thin and brown.

It is not very common with us. I found a great number of them, some years since, at Mount-Sorrel in Leicestershire.

Cantharis elytris nigris fasciis duabus rubris.
The Cantharis, with the exterior wings black, with two red, transverse lines.

This is a very small species, it scarce exceeds a louse in size: the legs and the antennæ are black; the head and the thorax are of a greenish colour; the exterior wings are of a deep, and somewhat glossy, black, and have each two transverse streaks of red on them; the one of these lines is near the base, the other very near the apex or extremity, and the sides of the body are also reddish.

It is not uncommon under our hedges, and in pastures; I have frequently met with it on the alcea, or vervain mallow.

The other species of the Cantharis are, 1. The large, yellowish, or orange-coloured, Cantharis, without spots. 2. The black-winged Cantharis, with a red thorax, and a black spot on it. 3. The red-winged Cantharis, with a black thorax, with a red spot on it. 4. The brassy-green Cantharis, with the exterior wings reddish toward their edges. 5. The brassy-green Cantharis, with the exterior wings red at the tips. 6. The black Cantharis, with the exterior wings yellow at the extremities. 7. The brown Cantharis, with the exterior wings yellowish at the extremity, and the breast reddish. 8. The black Cantharis, with livid wings.

TENEBRIO.

THE antennæ of the Tenebrio are slender, oblong, and filiform: the elytra are joined together, and there are no interior wings.

This genus differs extremely from all the others in this singular deficiency, but the form and structure of all it's other parts refer it to this class.

Tenebrio atra elytris acuminatis.
The black Tenebrio, with acuminated wings. **The stinking Beetle.**

This is a moderately large Insect; it is all over of a coal black: the exterior wings are united all the way along their interior edges: the legs are long and slender; the antennæ are moderately long, and are composed of oblong joints, excepting the last, which is round: the thorax is marginated and punctuated, and the exterior wings are rugose.

It is common about dunghils, and sometimes in houses. Mouffet and others call it Blatta fœtida; Petiver, Scarabæus impennis tardipes. It moves indeed very slowly.

Tenebrio atra elytris pone obtusioribus.
The black Tenebrio, with the exterior wings rounded behind.

This is a moderately large species: it's back is somewhat prominent, and it's head is small: the legs are long, and moderately thick, but it's motion is very slow: the whole body and wings are of a fine deep black, but with a glow of purplish in it.

It is frequent about our hot beds in gardens. Lister describes it under the name of Blatta rotundior violacea.

Tenebrio atra pone rotundata maxillis prominentibus.
The black Tenebrio rounded on the hinder part, and with prominent jaws.

This is a small species, scarce exceeding the common large fly in size: the colour is a deep, but not glossy, black; the legs are long, and the thighs clavated: the antennæ are slender, and moderately long.

It is not uncommon with us in woods, among the heaps of decayed and half rotten branches; sometimes we meet with it in hedges, but more rarely.

MELOE.

THE antennæ of the Meloë are slender and filiform: the exterior wings, or elytra, are dimidiated; and, as in the former genus, there are no interior ones.

Of this genus there is but one known species, which has been called by authors Proscarabæus.

MELOE. **The oil Beetle.**

This is a large beetle: the body is oblong and moderately broad, but flattish, not very thick: the legs are long; the exterior wings are dimidiated: they are soft and coriaceous, or of a substance resembling leather, and are rugose on the surface, and are black, but not at all glossy: the whole creature is of the same black colour, and is soft and unctuous to the touch: the body is longer in the female than in the male, throwing itself beyond the extremity of the exterior wings: if touched, it immediately exsudates drops of a very clear and limpid oil from all the joints of the legs.

It is very common in the summer months in our dry hilly pastures. Authors call it Proscarabæus, and Cantharus unctuosus; some, Scarabæus mollis.

NECYDALIS.

THE antennæ of the Necydalis are setaceous: the exterior wings are dimidiated, and there are interior or membranaceous ones.

Necydalis

Necydalis elytrorum apice puncto flavo notato.
The Necydalis, with a yellow ſpot on the extremity of the elytra.

This is a ſmall Inſect, ſcarce exceeding a common louſe in ſize: the head is ſmall and black, but the jaws are yellow: the thorax is yellow, and is marginated: the exterior wings are black, but browniſh in the middle, and terminated by a yellow ſpot: the interior wings are black, and are of twice the length of the exterior, and lie in a cruciform manner over one another: the antennæ are nearly equal to the body in length, and are yellowiſh at the baſe, but black elſewhere: the body is brown, but ſomewhat yellowiſh at the ſides; the lower part of the legs alſo is yellowiſh.

It is not a very common ſpecies, but we meet with it ſometimes on the ranunculus's, both in our fields and gardens. I met with a conſiderable number in June laſt, near Boſton in Lincolnſhire.

Necydalis elytris apice lineola alba notatis.
The Necydalis, with a white line on the extremity of the exterior wings.

The head is black; the thorax is alſo black; it is oblong, and ſomewhat depreſſed, and has two elegant ſpots of white on it: the exterior wings are grey; they are angulated toward the baſe, and are ſcarce half ſo long as the body; they ſeparate at the extremities, and have each a white line in that part: the interior wings are half naked; they are croſſed, and are not drawn in under the others: the antennæ are twice as long as the body; they are of a grey colour; the legs alſo are grey, and they are thick in the upper part.

We have it in our hedges not very unfrequently. Some have called it a Cimex, others a Forficula, or Earwig, but both erroneouſly.

F O R F I C U L A.

THE tail of the Forficula is a kind of forceps, capable of pinching: the exterior wings are very ſhort, or dimidiated: the antennæ are ſetaceous; and the interior wings are wholly covered or drawn under the exterior ſhort ones.

Forficula alis interioribus apice macula alba notatis.
The Forficula, with the interior wings marked with a white ſpot.

The common Earwig.

The antennæ of this well-known Inſect are long and ſlender; they conſiſt of thirteen or fourteen articulations: the covering of the thorax is flat, and is truncated in the anterior part, and rounded behind; it is black in the middle, and paler round the edges: the exterior wings are of a pale reddiſh brown; the interior ones are ſomewhat prominent beyond the extremities of the others, and their extremity has a white oval ſpot: the body is of a reddiſh brown, and is naked, and is armed at the extremity with a pair of crooked, connivent, ſharp claws, as it were, which form a kind of forceps.

It is very common about fruit and other vegetable ſubſtances. Authors call it Forficula, and Auricularia.

Forficula alis elytris concoloribus.
The Forficula, with the interior wings of the ſame colour with the exterior.

The leſſer Earwig.

This ſpecies is but about half the ſize of the former or common kind: it's body is of a bright cheſnut colour; it's head and thorax black; the exterior wings are alſo of a cheſnut colour, and ſo are the interior ones, which are ſomewhat prominent from under them: the forceps at the tail is uſually ſomewhat erect; the legs and the belly are paler than in the former, and the antennæ conſiſt only of ten articulations.

It is not uncommon with us about gardens and in hedges.

STAPHYLINUS.

THE antennæ of the Staphylinus are slender and filiform; there are two vesicles situated above the tail; the exterior wings are dimidiated and short; the interior ones are covered by them.

Staphylinus ater glaber maxillis longitudine capitis.
The black, smooth Staphylinus, with the jaws of the length of the head.

The common Staphylinus.

This is an inch in length: the head, thorax, and exterior wings are of a deep shining black and smooth; but the exterior wings are sometimes variegated with grey; they are short and obtuse: the body is of a deep, but less glossy black than the head; the legs are long and black: there are two hard, firm, long, and very sharp horns in the front of the head.

It is frequent with us under shady hedges. Authors call it Staphylinus, and Staphylinus major.

Staphylinus rufus elytris cæruleis capite abdominisque apice nigris.
The reddish brown Staphylinus, with blue wings, and the head and tail black.

This is a very elegant, though small Insect; it is about the size of a common ant: the whole body is of a pale red, with an admixture of brown in it; but the head, and the three or four last rings of the body, are black: the exterior wings are of a strong blue, without any variegation; the legs are reddish, but their joints black: the antennæ consist of eleven joints, and are of a pale colour, except at the extremity, where they are black.

It is a scarce species; we sometimes meet with it among the water-plants, at the edges of brooks and rivers.

Staphylinus pubescens cinereus nigro-nebulosus.
The hoary Staphylinus, of a greyish colour, clouded with black.

This is a moderately large species: it's body is oblong, and in great part naked; it is of a greyish colour, with a brassy tinge, and with clouded spots of black visible through it, in considerable numbers: the head is large and flatted; the mouth is forcipated; the antennæ are composed of nine joints: the exterior wings are short, and of a greyish-black, with a brassy cast: the tail is furnished with two plumose hairs; and, on pressing the body, two white hooks are protruded from the same part.

It is not unfrequent with us on dunghils.

Staphylinus niger thorace æneo, elytris atro-cæruleis.
The black Staphylinus, with a brassy thorax, and blackish-blue wings.

This is about half an inch in length; the body is slender and flatted: the head is black and glossy; the wings are of a deep blue, with an admixture of black, and are also bright and glossy; they are of a metalline hue, and have a number of very minute dark spots: the antennæ consist of nine articulations; the colour of the whole Insect sometimes varies into a yellowish or brassy one, but the spots on the thorax, when accurately examined, will always distinguish it; they are ten in number, and are hollowed.

We meet with it among the decayed timber in our woods, but not frequently.

The other species of Staphylinus are, 1. The black, hairy Staphylinus, with the thorax and the under part of the belly yellow. 2. The black Staphylinus, with red legs, and red exterior wings. 3. The reddish Staphylinus, with a black head. 4. The ferrugineous Staphylinus, with a black head. 5. The brown Staphylinus, with the lower part of the body, and the legs ferrugineous. 6. The smooth, black Staphylinus,

phylinus, with reddish legs. 7. The smooth, black Staphylinus, with the legs reddish, and the verge of the body yellow. 8. The black Staphylinus, with reddish legs, and the anterior part of the exterior wings grey. 9. The black Staphylinus, with crimson wings. 10. The testaceous Staphylinus. 11. The black Staphylinus, with the thorax, the exterior wings, and the legs subrufescent. 12. The black Staphylinus, with the exterior wings, the legs, and the antennæ ferrugineous. 13. The black Staphylinus, with the exterior wings and the legs brown.

BLATTA.

THE antennæ of the Blatta are setaceous; there are two short horns, as it were, above the tail: the exterior wings are membranaceous; and the thorax is flat, orbiculated, and marginated.

Blatta ferrugineo-fusca, elytris sulco ovato impressis.
The ferrugineous, brown Blatta, with the exterior wings furrowed in an oval form.

The mill Beetle.

This Insect is of the size of the common cricket; it's colour is a deep, ferrugineous brown, approaching to black, resembling that of burnt paper: this colour is uniform, without any variegation: the clypeus or shield that lies over the thorax is plain, oval, and transverse: the exterior wings are of an oval figure; they are somewhat shorter than the body, and are in some degree transparent; each of them has three striæ from the base, the intermediate one elevated, and the interior one hollow and crooked; this, with it's fellow, form an oval space: there are two prickles at the tail, and the legs also are prickly: the male of this species has the exterior and interior wings perfect; the female has only the rudiments of both.

It is common with us in mills, bakers houses, and elsewhere; it only comes out in the dark. Mouffet calls it Blatta molendinaria et pistricaria; Ray, Blatta mollis.

Blatta flavescens elytris nigro maculatis.
The yellow Blatta, with the exterior wings spotted with black.

This is a small species, not exceeding a large fly in size: the clypeus or shield that covers the thorax is membranaceous, patent, of an oval figure, and marginated, and is semi-transparent: the exterior wings are membranaceous and pellucid; they are of a brownish colour, and have elevated striæ, and some black spots on them, but these not very numerous: the legs are of a horny appearance and prickly; the antennæ are long, and there are two articulated horns a little above the anus.

It is not met with, that I know of, in England, nor any where, except in the more northern countries, where it destroys the fish provisions.

GRYLLUS.

THE antennæ of the Gryllus are setaceous: the exterior wings are membranaceous, narrow, and much of the appearance of the wings of some of the fly kind: the thorax is compressed and angulated: and the legs are formed for leaping.

Gryllus pedibus anticis palmatis.
The Gryllus, with the anterior feet palmated.

The Mole-Cricket.

This is the largest of all the European winged Insects: it is two inches and a half in length, and three quarters of an inch in diameter: it's colour is a dusky brown, which is deeper in the male, and more pale in the female: there grow from the extremity of the tail on each side two hairy long bodies, resembling in some degree the tail of a mouse: the body is formed of eight squammated joints, or has so many separate folds, and is of a mixt flesh colour, and brown on the upper part, and of a dusky brown below: the wings are long; they appear narrow and are acuminated, and each has a blackish line running down it: these, when expanded, are found to be much

WINGED INSECTS

FLIES

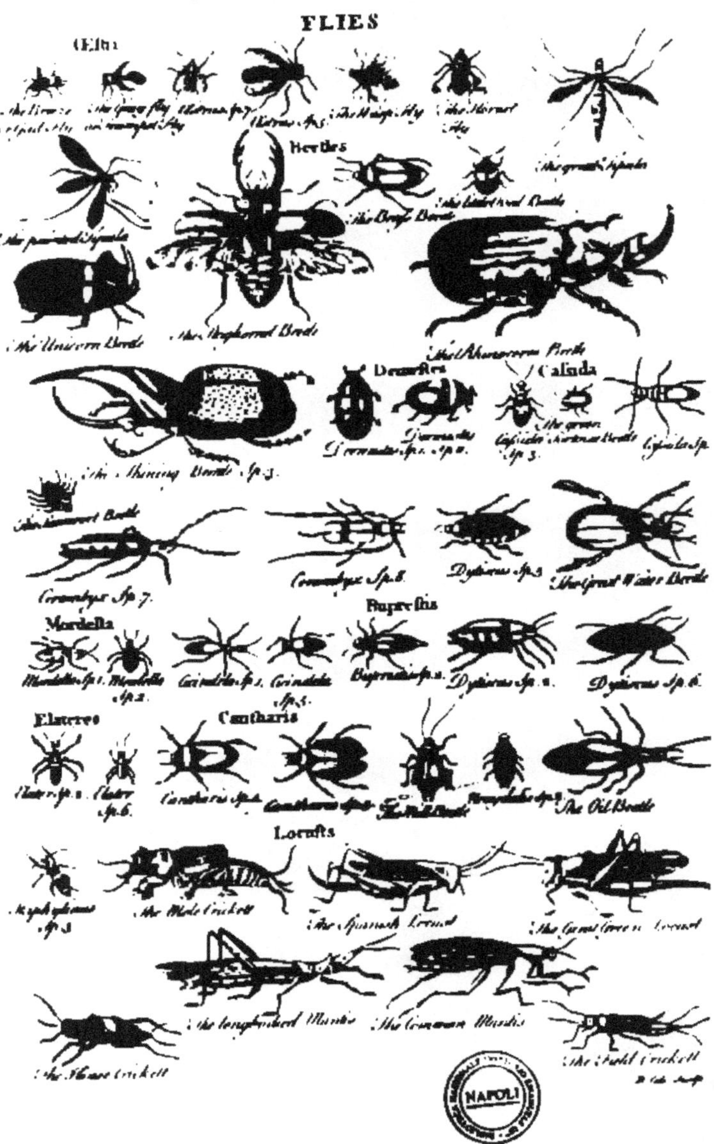

wider than would be naturally imagined; they are in a state of rest, folded in a most accurate and elegant manner: the antennæ lie on the back, and are of about half the length of the wings; the clypeus of the breast is of a firm texture, of a blackish colour, and hairy: the anterior feet are of a very remarkable figure; they are palmated and hairy.

It is common in our fens, and other damp grounds. In Chiswick gardens there always is a great number of them about the walks near the lower part of the canal; authors call it Gryllotalpa, and Talpa Insectum: it is an extremely singular creature, and naturally startles any body at first sight; and what appears the more odd in it is, that it runs backwards as readily and swiftly as forwards; it generally lives under ground, and digs very quick.

Gryllus fusco-cinerascens, alis punctis nigris maculatis.
The Gryllus, of a greyish-brown, with black spots on the wings.

The Spanish Locust.

This is about the length of the former species, but is is not nearly equal to it in thickness: it is of a very singular colour, a deep brown, with an admixture of a whitish grey; and there are a great number of little white dots at the origin of the wings: the exterior wings are beautifully spotted with little dots of black; the interior ones are reticulated: the body is composed of seven or eight joints; the legs are long; the hinder ones are very remarkably so, and very robust, and all of them are reddish on the inner side, and armed with a double series of spines: the eyes are large, and are striated with white.

It is a native of Spain, and other the warmer parts of Europe.

Gryllus incarnatus, femoribus sanguineis elytris virescenti-subfuscis.
The red Gryllus, with crimson thighs, and brownish-green wings.

This is a moderately large species; it considerably exceeds the common grass-hopper in size: the body is of a pale flesh colour on the upper part, and yellowish underneath; the legs are of a deep crimson; the exterior wings are of a close texture, and yellowish colour toward the top; lower down they are reticulated in the manner of the interior ones: the antennæ are of a cylindric figure, and are composed of four and twenty joints: the hinder pair of legs have a blackish hue, and white denticulations, and are terminated by four ungues; besides those of the palm, there is a sharp protuberance in form of a spine on the breast, between the fore legs; this is yellowish; the rest of the body is brown, with a mixture of reddish and bluish.

We have it in our pastures in Yorkshire, and elsewhere in the north of England: it is frequent also in most of the northern countries of Europe.

Gryllus cauda, ensifera recta, corpore subviridi.
The Gryllus, with a straight ensiferous tail, and a greenish body.

The great green Locust.

This is a large species; it is near two inches in length, and about the thickness of a man's little finger; the breast is covered with a firm armature of a triangular figure, which is extended down sideways toward the wings: the colour of the whole animal is a green, not a very bright one, with a considerable admixture of brown: the female has a kind of sword at the hinder extremity of the body; it is formed of two parts, or longitudinal valves.

It is not unfrequent in pastures in many parts of Europe: Ray calls it Locusta viridis major; others only, Locusta major.

The other species of the Gryllus are considerably numerous: of those called Grylli or Crickets, there are, 1. The common house Cricket. 2. The narrow-bodied field Cricket. 3. The large brown Cricket. 4. The little ferrugineous Cricket, with two long hairs at the tail.

Of those called by authors Locusts, there are, 1. The common large Locust. 2. The thick-bodied Locust. 3. The great-legged Locust. 4. The common Grashopper.

hopper. 5. The crooked-tailed Locust, called the Acrigonion, by Lister and others. 6. The praying Locust, or Mantis. 7. The longer-bodied Mantis. 8. The deep, brown, larger Mantis. 9. The flat-bodied marginated Mantis. 10. The doubly marginated Locust. 11. The Locust, with a quadrifid crest. 12. The large African brown Locust. 13. The Locusta talpa, or mole Locust. 14. The Locust, with black rhomboidal spots on the back. 15. The long-bodied Locust, with an oblong and narrow thorax. 16. The long-bodied Italian Locust. 17. The green Locust, with the edges of the wings white. 18. The Locust, with the antennæ of the length of the body. 19. The thick Locust, with elegantly reticulated wings. 20. The Locust with clavated antennæ.

TETRAPTERA.

Or Insects which have four wings.

Class the Second.

SCLEROPTERA.

Insects which have four wings: the exterior flexile, and the interior membranaceous; and which have the aperture of the mouth bent under the breast.

CICADA.

THE antennæ of the Cicada are very short: the snout is bent downward: the wings are cruciated, and the legs are contrived for leaping: the back is convex, and the thorax rounded.

The word Cicada is too generally understood to signify a grass-hopper in the writings of the antients, but this is a great error; they express by it a large four-winged fly of this genus, which is very common at this time in Italy, and many of the other warmer parts of Europe.

Cicada fusca, elytris maculis binis albis lateralibus, fascia duplici interrupta.
The Cicada, with brown wings, with two white spots on them, and a double white line.

The Cuckow Spit Insect.

This is a small species: it's body is oblong and obtuse, it's head large, and it's eyes small: the exterior wings are of a dusky brown, but have two very elegant white spots on them, and a broad, interrupted, transverse double line of the same colour: the rest of the body is of a dusky brown, and the head black; the colour however is not certain and invariable, but sometimes it is all over greyish.

It is common with us on different plants of the lychnis and other kinds, where it is usually covered with a kind of frothy matter resembling spittle, this it does not discharge at the mouth, but at the anus, and from several other parts of the body: all the writers on Insects have described it: Swammerdam calls it Locusta pulex; Petiver, Ranatra bicolor capite nigricante; others, Vermis spumans.

Cicada elytris viridibus, capite flavo punctis nigris asperso.
The Cicada, with green wings, and with a yellow head, with black spots on it.

This is one of the largest of Cicadæ of our colder climates: it is of the length of a large fly, but is very narrow in proportion to that length: the exterior wings are of a deep and very beautiful green, the interior ones are of a bluish-grey: the covering of the thorax is also green, but of a paler green than the wings: the head is yellow, and has two large black spots on the hinder part, and several other smaller at the sides, and there are some transverse striæ on the forehead: the body is of a bluish-grey, and the legs are yellowish.

It is common with us about the water plants in autumn. Ray calls it Locusta pulex tertia; others have made it a species of the Gryllus, but improperly. I met with several of them, this last summer, on the reeds and among the rushes about the edges of Whittlesea-meer in the Isle of Ely.

Cicada elytris flavis linea abrupta, duplici, longitudinali, nigra.
The Cicada, with yellow exterior wings, with a double, longitudinal, abrupt, black line.

This is about the size of the common fly: It is all over of a beautiful yellow, except that, when the wings are close, there appears on each side of the back a longitudinal black line, which is continued or divided towards it's middle, so as to form, as it were, two lines, the one running from the thorax, the other from the tail to the middle, where they are obliquely separated; the two upper parts of these lines unite near the thorax, and there is on each side a black spot on the head and thorax, the two also uniting into one: the antennæ are short; the forehead has several slight transverse furrows on it: the body, when the wings are expanded, is seen to be yellow in the middle, and black at each side.

We have it among the high grass in our pastures in June.

The other species are, 1. The common Italian Cicada, or true Cicada: this is an inch long, and more than half an inch broad, and of a dusky brown. 2. The smaller Italian Cicada. 3. The brown Cicada, with wings spotted with paler brown, and with punctated nerves. 4. The black Cicada, with three equal transverse white lines. 5. The yellow Cicada, with the wings spotted with brown, and with four black spots on them, yellow in the hinder part. 6. The whitish Cicada, with a black longitudinal line on each side. 7. The black Cicada, with white eyes. 8. The black Cicada, with the edges of the exterior wings white. 9. The black Cicada, with a white head. 10. The Cicada, with a bicornute thorax lengthened behind, and with naked wings. 11. The Cicada, with the exterior wings striated with black and white, called, by most authors, the common Locusta pulex. 12. The yellow compressed Cicada. 13. The Cicada, with greenish-yellow wings, with the extremities black and gilded, common on the leaves of the elm, and very nimble in it's motions. 14. The yellow Cicada, with the extremities of the wings white and membranaceous. 15. The Chinese Laternaria. 16. The American Laternaria, or larger yellow compressed Cicada. 17. The Manna Cicada. 18. The least, white Cicada.

CIMEX.

THE rostrum of the Cimex is inflected; the antennæ consist each of four joints: the wings are four, and are cruciated; the legs are formed for running, not leaping; the back is plain, and the thorax is marginated: some of them pass a great part of their life without wings, some even the whole, so far as is yet known of certainty.

Cimex subrotundus viridis, margine undique flavo.
The roundish green Cimex, with a yellow border round it.

The green and yellow Bug.

This is of the size of the common fly, but flatted: the body is of a rounded figure, but somewhat approaching to oval: the antennæ are slender and green, and the lowest joint of them is very small: the head, the thorax, and the exterior wings are of a bright green, and somewhat scabrous: the belly is green, but the back of the body is black; the snout has four joints; the tongue is setaceous, and is immersed longitudinally in the snout; a yellow margin, or edge, surrounds the whole body.

It is frequent with us in gardens, about the esculent plants.

Cimex ovatus cæruleſcenti-æneus puncto albo rubrove ſignatus.
The braſſy-blue Cimex, with a white or red ſpot on the back.

This is of the ſize of the common large blue fly, but it is more flatted: it's whole upper ſurface is of a beautiful blue, but with a braſſy tinge in it: there is a longitudinal line on the thorax, which, at it's lower extremity, terminates in a tranſverſe one: the extremity of the ſcutellum is either red or white, and there is alſo a ſpot of the ſame colour on each ſide of it; the legs and the antennæ are black, and the body underneath is alſo of a bluiſh-black: the ſpots on the male are white, and thoſe on the female are red.

It is found on the Dipſacus, and ſome other large plants, and is common to us, and to moſt parts of America: Ray, and Sir Hans Sloan, have deſcribed it.

Cimex oblongus rubro nigroque variegatus alis fuſcis, immaculatis.
The oblong-bodied, red and black Cimex, with plane brown wings.

The Henbane Cimex.

This is an extremely beautiful Inſect: It is of an oblong ſhape, and it's body is much narrower than in either of the former; the head is black, except in the very middle, where it is of a bright red: the thorax is black at it's anterior extremity, and red every-where elſe, except that it has two angulated black ſpots behind; the exterior wings are red, but they have each a black ſpot in the middle, ſo that, when they are cloſed, there are two black ſpots on the back: the ſcutellum is black, but it's extremity red; the body under the ſcutellum is black, the reſt of the body is red, but juſt near the anus black: the wings are brown and ſtriated, but have no white ſpots on them; the antennæ and the legs are black, and the trunk or ſnout conſiſts of four joints.

It is frequent with us on the ſtalks and leaves of the common Henbane, or Hyoſcyamus; Liſter, Ray, and Petiver have all deſcribed it under the names of Cimex Hyoſcyami, and Cimex Hyoſcyamoides.

The other ſpecies of the Cimex are conſiderably numerous, and as ſome of them have oblong, and others rounder bodies, they may conveniently be enumerated under two ſeries, according to that obvious diſtinction. 1. Of thoſe which have rounder bodies, or are of a figure approaching to round, there are, 1. The common houſe Bug, or Cimex lectularius, ſo troubleſome about our beds; there are authors who pretend that this ſpecies has wings at a certain period. 2. The oval-bodied black Cimex, with an arcuated trunk, and capillaceous antennæ. 3. The black Cimex, with a flat body imbricated, and notched at the ſides. 4. The grey Cimex, with the edge of the body ſpotted with black. 5. The grey oval Cimex, with the thorax obtuſely angulated on each ſide, and with red legs. 6. The grey oval Cimex, with the thorax on each ſide angulated and acuminated, with reddiſh antennæ. 7. The red Cimex, with brown wings. 8. The black oval Cimex, with the exterior wings black and white, and the interior ones white. 9. The oval Cimex, attenuated in the anterior part, with a greyiſh-white body, and red antennæ. 10. The grey oval Cimex, with clouded wings, and the edges of the body variegated with black and white. 11. The oval Cimex, with livid wings, brown at the extremity. 12. The greyiſh clouded Cimex, with three elevated lines on the ſcutum, and the antennæ black at the extremity. 13. The roundiſh, black and red Cimex, with the head and wings black.

2. Of thoſe of an oblong and narrower form, there are, 1. The reddiſh, brown, unſpotted Cimex, with the thorax terminating in a point at each end. 2. The oblong green Cimex, with a cordated yellow ſpot on the ſcutellum, and a double black ſpot on the exterior wings. 3. The red and black Cimex, with brown wings ſpotted with white. 4. The grey Cimex, with a cordated yellow ſpot on the ſcutellum. 5. The green Cimex, with a deep green cordated ſpot on the ſcutellum, and two brown ones on the wings. 6. The grey unſpotted Cimex, with ſetaceous antennæ. 7. The brown and white clouded Cimex of the poplar. 8. The rubiginous Cimex, with blood-red ſtriæ on the wings. 9. The black Cimex, with red legs, and with brown and white wings. 10. The brown and yellow Cimex, with the anterior thighs dentated. 11. The black Cimex, with the exterior wings variegated with white. 12.

The

The black Cimex, with the wings of a greyish brown, and a rhomboidal black spot on each. 13. The black Cimex, with the antennæ and the legs yellow. 14. The black Cimex, with the exterior wings very finely striated, and the anterior ones marked with a yellow spot at the extremity. 15. The black Cimex, with the antennæ terminated each by a hair. 16. The black Cimex, with grey wings black in the anterior part, and with the interior wings white behind. 17. The whitish Cimex, with white sides. 18. The red-legged, black Cimex, with the exterior wings variegated with brown and and yellow. 19. The black-horned, pale-brown Cimex, with a white line on the thorax. 20. The black Cimex, with the exterior wings white, and the antennæ of a bluish colour, but with the lowest joint black. 21. The very narrow Cimex, with very long antennæ, variegated with brown and white. 22. The very narrow Cimex, with the forelegs extreamly short; this is very frequent skimming on the surface of waters. 23. The small, black, water Cimex, with the anterior legs very short. 24 The very slender, whitish Cimex, with all the legs long. 25. The Cimex, with clavated antennæ, and with the edges of the exterior wings and thorax beautifully reticulated and punctuated.

NOTONECTA.

THE rostrum or snout of the Notonecta is inflected: the antennæ are very short; the wings are four, and are cruciated; the legs are formed for swimming.

Notonecta grisea elytris griseis margine fusco punctatis.
The Notonecta, with the exterior wings greyish, but spotted with brown at the edges. **The common Boat-fly.**

This is more than half an inch in length, and about a third of it's length in breadth: the head is obtuse and yellow; the eyes are brown, the thorax is large, and is of a yellow colour, and somewhat pellucid: the scutellum is black, and has a fine velvety gloss in it: the exterior wings are of a yellowish grey, and are spotted with black all round the edges; the interior ones are whitish and pellucid: the fore feet are short, and the middle ones longer: the hinder ones are much longer than any of the others, and are formed for swimming; the trunk or snout is long and sharp-pointed; the antennæ are very short, and consist only of two joints: the belly is black and hairy.

It is common with us in fish-ponds and other waters, and, when taken out, will give a very painful wound with it's trunk. Authors call it, Notonecta vulgaris, and Cimex aquaticus angustior.

Notonecta alis exterioribus pallidis transversè striatis.
The Notonecta, with pale and transversely striated elytra. **The brown Boat-fly.**

This is near half an inch in length, and is somewhat broader and more depressed than the former: it's head is yellow; it's legs also are yellow: the thorax and the exterior wings are brown, and have a multitude of fine, slender, transverse striæ on them, of a pale yellow colour; the under part of the body is yellowish, and the eyes are black.

It is common with us in waters, and has an ill smell. Authors have called it Notonecta compressa fusca.

Notonecta cinerea pedibus posterioribus longissimis.
The grey Notonecta, with the hinder legs very long. **The little Boat-fly.**

This is extreamly small, scarce arriving to half the bigness of the common louse: it's colour is all over a pale whitish-grey; it's back is flatted, and has a longitudinal line running along it: the anterior legs are remarkably short, the hinder ones as remarkably long: the antennæ are very short.

It is not unfrequent with us in shallow, running waters.

The

The other species are, 1. The great, black Notonecta of the East Indies. 2. The black and white, smaller Notonecta. 3. The broader, brown Notonecta. 4. The little yellowish, striated Notonecta.

NEPA.

THE rostrum, or snout, of the Nepa is inflected: the antennæ are formed into a kind of claws; the wings are four, and are cruciated: the legs are only four.

Nepa abdominis margine integro. **The Water**
The Nepa, with the edge of the abdomen undivided. **Scorpion.**

This is a large Insect; it is near an inch in length, and about half it's length in breadth; it's body is of a kind of elliptic form, very flat and thin, and it's tail long and pointed: it's head is small; the antennæ have much the appearance of legs or arms, resembling the claws of a crab, or scorpion, but they have no sharp points: the breast is somewhat acute underneath; the tail is formed of two capillaments, which easily separate: the legs are slender; the body is composed of several joints, and the anus is remarkably large: the trunk is long and sharp; the exterior wings are of a deep, blackish, olive colour; the interior ones white or reddish.

It is common in ponds, and is a very great destroyer of the other water Insects. Most of the writers on these subjects have described it; Mouffet calls it Scorpius aquaticus; others, Scorpio palustris, and Araneus aquaticus.

Nepa abdominis margine serrato.
The Nepa, with the edge of the body serrated.

This is about a third of an inch in length, and is broader in proportion than the former species: it's back is somewhat elevated, and it's colour is a pale brown, with a tinge of the olive: it's head is small, and the eyes black, the whole edge of it's body is finely serrated: it's claws are smaller, in proportion, than in the former, but of the same form.

We have it in great abundance in our trout rivers in Buckinghamshire.

CHERMES.

THE rostrum or trunk of the Chermes is situated under the breast: the abdomen is pointed at the hinder extremity; the wings are four, and lateral; the legs are formed for leaping.

The Insect, called Kermes in the shops, is not of this genus, but of the same with Coccus Polonicus, hereafter to be mentioned under the name of Coccus.

Chermes alni. **The Alder-bug.**
The Chermes of the alder.

This is a small Insect; it's body is of an acuminated figure, it's colour white; it's eyes are lateral and prominent; it's legs are six in number; it's tail is acuminated and bifurcate, and over it there is a prominent style: the antennæ are variegated; the mouth is situated under the belly, between the insertion of the first and second pair of legs, and is blackish; the wings are four, and are white, variegated with brown veins. This Insect, in the worm-state, is about the bigness of the louse; it's body is flatted, and it's colour green and spotted; the belly is obtuse and elevated behind, and is blackish toward the extremity; the antennæ are slender, straight, and part black, part white; there are a number of white, reflex hairs, which have their origin at the tail, and cover almost the whole body.

It is very common on the alder.

Chermes abietis.
The Chermes of the fir. **The Fir-tree Bug.**

This is a small species; it is of an oblong figure, and whitish colour: the head is small, the eyes are moderately large and prominent; their colour brown, with a small speck of black between them; the antennæ are setaceous, and very minute, and there is a quantity of a woolly matter growing at the tail: the wings in the flying state are four; they are thin and whitish.

It is common on the fir and pine-kinds, and sometimes on the yew and other trees: on the trunks, and sometimes at the extremities of the branches, of these, there are found protuberances resembling strawberries in figure, and covered with a kind of squammulæ or scales; under and between these there are found many of these little Insects, and usually near their base the parent animal considerably larger: the whole protuberance resembles a green strawberry, and is surrounded with leaves at it's base.

Chermes graminis.
The Chermes of grass. **The grass Bug.**

This is a moderately large species: It's body is somewhat depressed and broad, and it's head very obtuse: this and the thorax are grey, and are variegated with several white lines; the antennæ are white, except at the tips, where they are blackish; the wings are four; they are not cruciated, as in the cicadæ, a species of which genus this Insect might at first sight be very naturally taken for, but are deflex as in the phalenæ; the body is blackish, but the segments white at their edges; the legs are whitish, and the wings transparent, but ornamented with a number of brown lines, which make various anastomoses with one another.

It is found on the stalks of grass in our meadows in June, and is frequent about Hampstead.

The other species of the Chermes are, 1. The Chermes of the elm, inhabiting it's curled leaves. 2. The Chermes of the cerastium, living in it's monstrous heads. 3. The Chermes of the maple. 4. The Chermes of the birch. 5. The Chermes of the pine, living on it's unaltered branches. 6. The Chermes of the willow. 7. The Chermes of the nettle. 8. The Chermes of the ash. 9. The Chermes of the poplar. 10. The Chermes of the apple-tree. 11. The Chermes of the chamædrys. Whatever others may hereafter be described, may be all named distinctively in this manner, from the trees and plants they are found on.

APHIS.

THE trunk of the Aphis is reflex: the body is formed into two horns behind; the wings are four and erect, or they are wanting; the legs are formed for walking not leaping.

Aphis ribis.
The Aphis of the currant. **The currant Louse.**

This is about the size of the common louse, and is of a brownish green colour, and generally lies flat on the leaves or branches; the legs are green, and the joints of the knees usually stand up above the body, and are brown: the antennæ are longer than the body; they are slender, straight, black, and have a joint at which they are bent near the head: the hinder part of the thorax is blackish, and there are three transverse and nearly contiguous lines on the hips: the sides of the body are variegated with some small black spots: the wings are placed erect; they are compressed and whitish, but have some black veins in them: two of them are very small: the anus is prominent, and has two setaceous appendages shorter than the wings.

It is very common on our currant-bushes, living usually on the under side of the leaves.

Aphis aceris.
The Aphis of the maple.

The maple Louſe.

This is ſomewhat larger than the former ſpecies, and it's body is more ſlender; it has ſcarce any viſible horns or appendages behind; thoſe which there are, are very obtuſe; the back of the body is covered with little rough protuberances, and has on it's hinder part a large black ſpot of a cordated form; the head and the thorax are black in the middle, and the antennæ are ſetaceous at their extremity.

It is frequent on the common ſmaller maple in May and June.

Aphis tiliæ.
The Aphis of the lime-tree.

The lime-tree Louſe.

The body is of an oblong form, and has four ſeries of black ſpots on the back; on the thorax there is alſo on each ſide a lateral line, which is black: the legs and the antennæ are alternately black and white; the wings are aſcendent and compreſſed, and are black on the lower edge, and on the upper they are variegated with ſome black ſpots.

It is common on the branches and leaves of the lime-tree. Authors have called it Pediculus arboreus, the tree Louſe.

The other ſpecies of Aphis are, 1. The Aphis of the elm. 2. The Aphis of the carrot. 3. The Aphis of the elder. 4. The Aphis of the dock. 5. The Aphis of the roſe. 6. The Aphis of the nymphæa, and other water-plants. 7. The Aphis of the ſerratula. 8. The Aphis of the thiſtle. 9. The Aphis of wormwood. 10. The Aphis of the jacea. 11. The Aphis of the birch. 12. The Aphis of the pine. 13. The Aphis of the cucubalus.

COCCUS.

THE trunk of the Coccus ariſes from the breaſt; the body is ſetoſe behind: the wings are only two; they are placed erect, and only the males have them.

This genus contains the gall Inſects and progal Inſects of Reaumur.

Coccus ilicis.
The Coccus of the ilex.

The Kermes.

The female of this ſpecies, which is what we know by the name of Kermes in the ſhops, is, when full grown, of a roundiſh figure, and of a deep purpliſh-blue colour, covered with a fine whitiſh, or greyiſh, duſt, like that on the ſurface of a ripe plum; in this ſtate it is not eaſy to diſtinguiſh it's limbs, or indeed it's natural form; it's being diſtended by young at this period alters, and, in a manner, deſtroys it's figure; it adheres in this ſtate to the leaves and young ſhoots of the ilex, and is collected thence for the ſhops; before this period it runs about on the branches, and has it's form more regular; it's colour is red, and it's body oblong; it's legs and it's antennæ are at that time diſcoverable, and it has uſually a kind of downy or cottony white matter about it. The male is a very ſmall fly, which would ſcarce be ſuppoſed to belong to the ſame ſpecies, if it were not ſeen impregnating the females: it's body is oblong, it's head ſmall, it's eyes little and black, and it's wings whitiſh, and full of brown and ſomewhat rigid nerves.

The female, in the full grown ſtate, is gathered in vaſt quantities for the uſe of the dyers, for their ſcarlet colour, and to be kept in the ſhops as a cordial; at that period it appears rather like an excreſcence than an animal, and has indeed been in general ſuppoſed to be ſuch.

It is frequent on the ilex or holm-oak in Provence, and ſome other places.

Coccus radicum purpureus.
The purple Coccus of the roots of plants.

The Poland Scarlet-grain.

This ſpecies, when full grown, is of the bigneſs of a ſmall pepper-corn, or thereabouts: it's figure is roundiſh; there is little appearance of limbs, or indeed of the form of an animal: It ſticks firmly to the root of the plant on which it feeds, and it's colour is a deep purple, with ſo much of the blue in it, that it approaches to a violet tint, and it has round about it, at the baſe, a rough cup, as it were, in ſome degree, reſembling the cup of an acorn. Such is the ſtate in which it is gathered for uſe, and in this ſtate there are found ſometimes three or four of them on the ſame root, ſometimes only one. From theſe round forms, which appear excreſcences on the root, there afterwards are produced a great number of young; theſe have all the regular appearances of animals, though the parent wants almoſt all of them: they are ſmall, oblong, and ſomewhat flatted; their bodies are formed of ſeveral ſegments, and they have ſix legs, which are ſhort and ſlender, and a pair of fine antennæ; their colour is purple, but a leſs deep tint than in the parent: When theſe have grown to their full ſize, they fix themſelves to the root of the plant, and, forming this cup about them, they never remove afterwards, but become diſtended with a numerous progeny, and die, leaving the young brood under the caſe they at that time form with their own ſkin: this is the condition of the females: the male of the ſame ſpecies is a ſmall two-winged fly, which is ſeen impregnating the females, and could by no other means be diſcovered to be of the ſame ſpecies.

This Inſect is frequent in Poland, and in ſome other of the northern countries. It is moſt frequently found adhering to the roots of the plant, thence called, by C. Bauhine, Polygonum cocciferum; and, by Tournefort, Alchemilla gramineo folio flore majore. I have met with it in England, at the roots of the common Potentilla, or Argentina; and it has been found at the roots of other plants. It makes an excellent ſcarlet dye.

Coccus Tunæ.
The Coccus of the Tuna.

Cochineal.

The Cochineal, though ſo long known in Europe, has been, till of late, miſtook extremely in it's nature; it was long taken for a vegetable production; a ſeed, or an excreſcence of a plant, and, when it afterwards was acknowledged to be an animal, it was generally ſaid to be of the beetle kind, and very like our lady-cow. We owe the true knowledge of what it is, in a great meaſure, to a number of atteſtations given upon oath on the ſpot where it is produced, and by perſons concerned in the management of it, who obſerved it very ſtrictly, that they might do juſtice in the deciding a wager between two mercantile people on the ſubject. By this accidental method, the European naturaliſts became informed of a thing they would otherwiſe have perhaps continued long in the dark about; and learnt, that the Cochineal we uſe, is the female of an Inſect, the male of which is a fly, but this female a reptile without wings, living on the Tuna or Opuntia, the Indian fig, on the juices of which it feeds.

The female, which we know under the name of Coccinella and Cochineal, is, while living, a little Inſect of an oblong form, rounded on the back, and flat on the belly, of a mixt colour of purple and grey, and, when cruſhed, it yields a rich purple juice. It's body is compoſed of ſeveral rings or joints; it's head is ſmall; it's eyes are little and black; it's legs are ſix in number, they are ſhort and ſlender, and are of a pale fleſh-colour; it's antennæ are ſhort and ſlender, and of the ſame pale red: It never leaves the plant it feeds on, but produces it's young ones on it, and dies: the male is a little fly, with ſlender antennæ, long legs, an oblong body, and white large wings, ſomewhat reſembling thoſe of our butter-flies, not tranſparent as in the common fly.

It is frequent wild in Mexico, and ſome other parts of South America; but the people of Mexico, in particular, find it ſo very advantageous an article of commerce, that they make plantations of the Tuna, on purpoſe for the raiſing it, where they regularly breed and manage their crops: vaſt quantities are annually imported into Europe, and are uſed in dying and in medicine.

Coccus hesperidum,
The Coccus of the orange-tree. **The green house-bug.**

This is a very extraordinary Insect, and has so little the appearance of the animal kingdom, that, though it has been at all times very abundant in our stoves and green-houses, it was very long before it was suspected to be an Insect, or acquired the honourable name of the green House-bug: it is a small creature, of an oval figure and flat; it adheres firmly to the bark or leaves of trees; it's back is a little prominent, and it's belly hollowed; and it is obtuse at the anterior part, and emarginated, or, as it were, bifid at the posterior. What we thus see of it, is a kind of shell or covering, within which, the body of the creature is inclosed: this is small and flat; the legs are six, and they are very slender; the eyes are very small and black, and the antennæ are slender: the whole Insect is of a greenish colour; it can thrust out it's legs at pleasure, and there is a period of it's life, in which it moves about, though but slowly; but, for the far greater part, it adheres firmly to the leaf, remaining in it's place, and sucking the juices: this is the female of the species: the male is a small fly, very inconsiderable in proportion to the female in size: it's body is slender and oblong; it's legs long; it's antennæ short, and it's wings white.

The female is frequent on our orange and lemon trees, and on many of the others, preserved in our stoves and green-houses.

Coccus aquaticus.
The water Coccus.

This is a small species: it's body is of an oval figure, and is elevated and rounded on the back, and flat on the belly: it's colour is brown, and, from it's hinder extremity, it thrusts out a kind of white hairy beard, which is bifid and gelatinous; near this extremity of the body, there is placed an obtuse tubercle; and, near the other extremity, an acute one.

It is common on the leaves of the little water lily, and some other water plants: it very rarely shews it's legs; when it does, they are slender and somewhat hairy: the male is a small fly, with silky whitish wings, spotted with brown.

Coccus betulæ.
The Coccus of the birch-tree.

This is a larger species than any of the former: it's body is oblong, somewhat hairy, and composed of several rings or joints: the legs are short, and the antennæ very slender: it is of a deep dusky olive colour.

It is frequent in our woods, on the beech, the birch, the broad-leaved elm, and several other trees: it is usually found single, affixed to their branches at their divarications: the male is a little fly, with dusky brown wings.

Coccus Insectorum.
The Coccus of Insects.

This differs from the others, as much in shape, as in the place where it is found: it is a very small species: it's body is of an oval shape, and has an acute edge; it is somewhat convex, and of a reddish-brown colour; the surface of the whole body is smooth, but not glossy; the legs are very short, and the antennæ scarce visible.

It is frequent on the bodies of some of the larger beetles, where it passes the greater part of it's life, without changing it's place: there are usually a great number of them on the same Insect.

The other species of the Coccus are, 1. The Coccus of the canary-grass; this is fixed to the roots of that plant, in the manner of the first species. 2. The Coccus of the Jacobæa; this is found on many of the compound flowered plants. 3. The Coccus of the alder. 4. The Coccus of the fig and peach-tree.

THRIPS.

THE rostrum or snout of the Thrips is obscure; the body is of a linear figure: the wings are four; they are incumbent on the back, and are straight.

Thrips elytris glaucis corpore atro.
The Thrips, with bluish wings, and a black body.

This is an extremely small Insect, not nearly equal to a flea in size: it's wings are four, and those very small and delicate; it's legs are six, two situated at the neck, and four on the thorax: the antennæ are very slender, black, and composed of six articulations: the exterior wings are of a greyish colour, very narrow, and hairy at the extremities, and at the edges: it runs very swiftly, and frequently twists it's body into a variety of shapes, but it very rarely flies.

It is frequent in the flowers of the dandelion, and others of that kind.

Thrips corpore atro, elytris albis nigrisque fasciis variegatis.
The black-bodied Thrips, with the exterior wings variegated, with black and white fasciæ.

This species is about the size of a louse: it's body is oblong and slender; it's exterior wings are very beautifully variegated, with alternate transverse streaks of black and white, three of each colour.

It is found very often in the flowers of the larger hawk-weeds, and some other plants of that class: it runs very swiftly, but does not fly, unless hunted very much.

Thrips elytris albis, corpore fusco.
The Thrips, with a brown body, and snow-white wings.

This is a very beautiful species: it is of the size of the former, and appears of a mixt black and white colour; but, when nearly examined, it is found to be of a coal-black on the body, and the exterior wings being of a snow-white, and not perfectly covering the body, give this appearance: the legs are short; the antennæ are very slender, and composed of five joints.

It is found on the Bermudas Cedars, as they are called, properly the American Junipers in our gardens, but is not common.

TETRAPTERA.

Insects having four wings.

Class the Third.

NEUROPTERA.

Those which have membranaceous wings, with nerves and veins disposed in a reticulated form in them.

PANORPA.

THE rostrum, or trunk, of the Panorpa is of a cylindric figure, and horny structure; and there is a weapon of the cheliform kind at the tail.

Of this singular genus there is but one known species.

PANORPA. **The Scorpion Fly.**

This is a fly of a moderate size: it's body is oblong and rounded; it's head is small, and is terminated by a hard, horny, oblong snout, which is prominent downwards: the antennæ are setaceous, black, and composed of no less than thirty articulations: the back is brown, the sides are yellow, the wings are white, but have some dusky spots on them, which are disposed in transverse series, forming a kind of lines; the tail is articulated, and is terminated by a weapon, resembling that of the scorpion, which is bifid: some of the same species want this weapon, but this is only the difference of sex, the females alone having it.

The fly is common in our pastures in July: authors call it Musca Scorpiura.

RAPHIDIA.

THE head of the Raphidia is of a horny substance, and is depressed: the tail is armed with a weapon of a slender form, sharp, horny, and simple, not bifid at the extremity. Of this genus there is also but one known species.

RAPHIDIA.

This is of the size of the scorpion fly, and much resembles it in shape: the head is black, smooth, horny, narrow in the hinder part, and depressed; the thorax is narrow, rounded, and black: the antennæ are slender, setaceous, white, and composed of a vast number of articulations: the body is slender and oblong; it is brown, but is variegated with transverse lines of white: the wings are thin membranaceous, and reticulated with nerves, and have usually each an oblong brown spot toward the edge; they much resemble the wings of some of the smaller libellæ: from the hinder part of the body of the female, there grows a long, sharp, slender, and arcuated weapon.

It is common in July in our meadows, especially about waters.

HEMEROBIUS.

THE palate of the Hemerobius is prominent, and has on each side two tentacula: the wings are deflex and erect.

Hemerobius luteo-viridis, alis aquæis, nervis viridibus.
The yellowish-green Hemerobius, with aqueous wings, with green ribs.

The Golden Eye.

This is a very beautiful Insect: it is about three quarters of an inch in length: it's body is very slender, rounded, and of a greenish-yellow colour; it's wings are very large; they are transparent and colourless in the intermediate spaces; but the nerves, or ribs, are of a fine green, and are so large and numerous, that the whole wing seems composed of them: the eyes are very large, and of a fine gold yellow: the Insect, when crushed, stinks intolerably.

The eggs of this fly are affixed by long pedicles to the leaves of our fruit-trees; they stand in clusters erect, and resemble so many pins stuck into the leaf by the points. I sent a leaf with about forty of these eggs on it, last summer, to our Royal Society, where a favourite member declared them the eggs of a Tipula, and will not be convinced of the error, though this beautiful fly has been produced from those very eggs. The worm they immediately hatch into is what Reaumur, from it's voraciously feeding on the aphides, calls Leo aphidis: the fly itself has been called, by Mouffet, Musca Chrysops; by Lister, Tolmerus; by Petiver, Perla minima mordax oleus; and, by Goedart, Audax and Intrepidus.

Hemerobius formicaleonis.
The Hemerobius of the formicaleo.

The Ant-eater Fly.

This is a ſpecies ſo much more univerſally known in the reptile or worm-ſtate, than in it's more perfect form, that it can no way be ſo expreſſively named as from it: when in the perfect or ultimate ſtate, it is a fly with a long ſlender body, of a brown colour, a ſmall head, with large bright eyes, long and ſlender, pale brown legs, and four large reticulated wings; it greatly reſembles the former ſpecies indeed in all things, but that it is more robuſt, and wants it's agreeable colour.

The formicaleo or worm of this fly is about half an inch in length, and more than a third of an inch in breadth; it's body is of an oval figure, and is covered with little rough papillæ, diſpoſed regularly in five longitudinal ſeries: the thorax is ſmall and ſlender; the mouth is furniſhed with a very long pair of forceps, which incline together at the points, and each jaw has alſo three teeth in it; the legs are moderately long and whitiſh; it walks backward, and inhabits a hollow which it forms in looſe ſand, in the manner of a funnel, where it preys on ants, and other Inſects that fall into it, as almoſt every thing muſt do that comes within the verge of it.

It is common in moſt parts of Europe, except England.

Hemerobius alis albis maculis fuſcis ſparſis, antennis variegatis.
The Hemerobius, with white wings ſpotted with brown, and with variegated antennæ.

This is a ſmall ſpecies: It does not exceed an ant in ſize, but it's body is proportionally longer: the wings are large, white, reticulated with brown nerves, and ſpotted with brown: the antennæ are long and ſlender, and are compoſed of alternate, white and brown articulations, ſo that they reſemble a necklace made of beads of two colours: the legs are ſlender and whitiſh; the eyes are blue, with a braſſy tinge, and there is on each ſide of the upper ſurface of the body a ſeries of little brown ſpots.

It is frequent in our hop-grounds in Kent.

The other ſpecies of the Hemerobius are, 1. The black and green Hemerobius, with aqueous wings. 2. The white and ſpotted-winged Hemerobius, with brown antennæ. 3. The black Hemerobius, with a yellow thorax and yellow body. 4. The larger, brown Hemerobius. 5. The larger variegated Hemerobius. 6. The little, brown Hemerobius.

PHRYGANEA.

THE palate of the Phryganea is prominent, and has on each ſide two tentacula: the wings are incumbent; the worm of it lives under water, in a kind of caſe.

Phryganea alis cæruleo-atris, antennis corpore duplo longioribus.
The Phryganea, with bluiſh-black wings, and with the antennæ twice as long as the body.

This is but a ſmall fly; it's body is oblong, and of a duſky brown; it's legs are ſlender, and of a duſky grey, approaching to black; the wings are of a duſky blackiſh colour, with an admixture of bluiſh and greeniſh: the antennæ are twice or three times the length of the body.

It is frequent with us among the reeds and ruſhes, about waters, and flies in vaſt ſwarms in an evening, making a kind of dancing or vibratory motion in the air.

Phryganea alis venoſo-reticulatis, cauda biſeta.
The Phryganea, with venoſo-reticulate wings, and two hairs on the tail.

This is a conſiderably large ſpecies: the body is long, ſlender, and brown, and is terminated at the tail by two long hairs, having the appearance of antennæ: the

wings

wings are long, and toward the top narrow; they are white, but full of brown nerves. It is frequent with us in May and June, in places where there are waters near.

Phryganea nigra, alis albidis, striatis, albo maculatis.
The black Phryganea, with whitish wings striated, and spotted with white.

This is a fly of about half an inch in length; the body is slender, rounded, and black: the wings are of a lanceolated figure, and are longer than the body, and obtuse; they are whitish, and are full of brown nerves or veins, which form very beautiful reticulations; and there runs a long brown line between the two exterior nerves of each wing: the antennæ are black, slender, and of the length of the body: the thorax is plain, but a little convex in the middle.

It is common with us about waters.

The other species of Phryganea are, 1. The Phryganea, with testaceo-nervose wings, and with the antennæ protended forwards. 2. The grey Phryganea, with the exterior wings grey, with a black spot on the edge. 3. The greyish, testaceous-winged Phryganea, with two longitudinal, white lines, and a white spot on each. 4. The Phryganea, with deflexo-compressed, yellowish wings, with a rhomboidal white spot at the side. 5. The Phryganea, with brown wings with two yellow spots on them. 6. The Phryganea, with reticulated wings, and a naked tail, with the edges of the thorax yellow. 7. The Phryganea, with the upper wings clouded, and the antennæ longer than the body. 8. The leaping Phryganea, with the antennæ of the length of the body, and with a green and white spot on the wings. 9. The black Phryganea, with the hairs of the tail truncated. 10. The brown Phryganea, without spots.

EPHEMERA.

THE Ephemera has two gibbous protuberances on the top of the head, resembling eyes; the tail is furnished with hairs, and the antennæ are short.

Ephemera alis albis reticulatis, cauda bisetа.
The Ephemera, with white, reticulated wings, and two hairs at the tail.

This is a moderately large fly: the body is oblong, slender, and whitish, except some few of the extream articulations, which are brown: the head is small, but there are on it two elevated, lenticular, yellow bodies, having the appearance of eyes, and situated above the eyes: the thorax is compressed; the tail is terminated by two straight slender hairs, of twice the length of the body, and between these there are two short and crooked ones: the legs are snow-white, and the anterior ones are larger than the others: the wings are erect, reticulated, of a fine white, but with a glow of flesh colour, as seen in some lights.

It is frequent with us in the summer months about waters.

Ephemera alis albis striatis.
The Ephemera, with white striated wings.

This is a small species: the body is oblong, slender, and brownish; the tail is obtuse, and has not those long hairs so common to the generality of the Ephemeræ: the wings are large, white, and striated; they have a great many nerves in them, but these do not form any reticulations: there are two prominent tubercles on the head, but they are smaller than the eyes: the male has the thorax black, and the body pellucid; the female has the body of a reddish brown.

It is common about waters in June and July.

Ephemera fusca cauda biseta, alis albis.
The brown Ephemera, with white wings, and a bisetous tail.

This is about the size of the common gnat: the body is oblong and slender, and of a dusky brown; the thorax is black; the legs are long and slender; the wings large and

and transparent, but of a whitish hue: the tubercles above the eyes are very large, and of a deep bluish-black; there are two hairs at the tail, which are equal to the body in length.

It is frequent about waters in June.

The other species of the Ephemera are, 1. The clouded and spotted-winged Ephemera, with three hairs at the tail. 2. The thicker-bodied Ephemera, with two hairs at the tail. 3. The white-winged Ephemera, with a black verge, and with two hairs at the tail. 4. The black Ephemera, with the wings white underneath.

LIBELLULA.

THE mouth of the Libellula is furnished with jaws: the antennæ are short, and the tail is terminated by a kind of forceps.

The species of this genus are considerably numerous; and as they differ considerably in size, and in the positions of their wings, while sitting, they may be conveniently divided into three sets: the first consisting of those which are of a middle size, and which carry their wings erect as they sit; the second comprehending the small ones, which sit with their wings erect; and the third the largest, which sit with their wings plane. Of each of these divisions we shall describe one, and after this the rest will be easily distinguished by their names.

Libellula corpore viridi-cæruleo, alis subfuscis, puncto marginali albo. *The bluish-green-bodied Libellula, with brownish wings, with a white spot at the edge.*

This is a very elegant and beautiful Insect: the body is near an inch in length, and of a beautiful green, with a cast of blue in some lights: the head is large; the eyes are remarkably large and prominent; the legs slender and black; the wings have a faint tinge of a yellowish brown, and have each an oblong, white spot near the edge.

It is frequent with us about waters; the male of this species has a blue body, and even the wings bluish.

Libellula corpore incarnato, alis puncto marginali fusco. *The Libellula, with a red body, and a brown spot on the edge of each wing.*

This is one of the smaller flies of this genus; it's body is nearly as long, indeed, as that of the former, but it is slenderer, and is of a bright red colour; the head is large; the eyes are very large and prominent; the legs are slender: there are some black lines near the segments of the belly; the wings are pellucid and brownish, and have each a deep brown spot near the edge.

It is frequent with us about fish-ponds.

Libellula alis macula duplici marginali. *The Libellula, with a double marginal spot on the wings.*

This is one of the largest of the Libellulæ: it's body is near an inch and an half in length, and is considerably thick; it is largest at the two extremities, and somewhat smaller in the middle: the thorax is thick, of a greyish colour, and hairy on the upper surface: the wings are yellowish toward the base, and whitish elsewhere; they have each two spots, one below the other, of a deep dusky brown; the body is of a shining green, and is somewhat hairy at the sides: the tail is furnished with two appendages, each smaller at the end than in the middle.

It is frequent with us about rivers.

The other species of the Libellulæ of the first division are, 1. The smooth, bluish-green-bodied Libellula, with yellowish, brown wings, with no spots at the sides. 2. The blue-bodied Libellula, with bluish-green wings, brown at the extremities, but not spotted. 3. The blue-green, silky-bodied Libellula, with brownish-yellow wings, with a black spot on each. 4. The greyish-bodied Libellula, with dusky brown wings, with a black spot on each.

Of the second division, beside the second described species, there are, 1. The green and white-bodied Libellula, with a silky body, and a brown, marginal spot on each wing. 2. The pale blue-bodied Libellula, with a black spot on each wing. 3. The blue and grey-bodied Libellula, with a black marginal spot on each wing. 4. The larger yellow and blue-bodied Libellula, with a brown spot on each wing.

Of the third division, or largest Libellulæ, which sit with their wings plane, there are, beside the third described species, 1. The broad and short-bodied Libellula, with wings yellowish at the base, and white elsewhere. 2. The brown-bodied Libellula, with white wings, and a simple tail. 3. The Libellula, with the body yellow at it's sides, and with white wings. 4. The black-bodied Libellula, with the thorax of a bright green, with yellow lines, and with pale wings. 5. The brassy-green Libellula, with pale wings, and with black legs. 6. The grey Libellula, with yellow wings, and a diphyllous tail. 7. The black-bodied Libellula, with the thorax of a yellowith green, variegated with black lines.

TETRAPTERA.

Class the Fourth.

LEPIDOPTERA.

Insects which have four wings, and those all opake, and covered with a fine dust, which, when examined by the microscope, is found to be composed of regular scales, commonly called feathers. The mouth or trunk in this class is usually spiral.

THIS class comprehends the Butterflies and Moths, the Papiliones and Phalænæ of authors. The species are very numerous, and, after they are arranged into two genera, according to this distinction, the Papilio and Phalæna, they may be divided into several other sets, according to certain obvious distinctions.

PAPILIO.

THE antennæ of the Papilio are of a clavated form.

Of the species of this genus some have four, others have six legs; they may be arranged therefore, according to this distinction, into two divisions: of each of these divisions we shall describe three species, after which the others will be easily distinguished by their names.

Papilio tetrapus alis angulatis nigris, margine postico albido.
The four-legged Butterfly, with black wings, with the hinder edge white.

This is a very large Butterfly: the head is small, the body considerably thick and oblong, and both these, and the antennæ and legs, are black: the wings are large, and are angulato-dentated; they are black on both sides, but have a white edge, and each two white spots.

It is frequent in our woods: the caterpillar, produced from it's eggs, feeds on the leaves of the birch-tree.

Papilio tetrapus alis angularis fulvis nigro maculatis, secundariis albo notatis.
The four-legged Butterfly, with yellowish-brown wings, spotted with black, and the secondary ones with white.

The body of this species is oblong, and not very thick; the legs are slender, their colour white; the wings are sinuated, and are of a yellowish-brown without, and spotted with black, having each, at the base, four distinct spots of that colour; they are of a greyish or ash colour on the under side, and the lower pair have each a white spot

2 on

on the center: the ſinuations are ſo deep in the wings of this ſpecies, that they appear as it were jagged: the caterpillar feeds on the nettle, the hop-plant, and the gooſeberry-buſh; it is hairy and whitiſh.

Papilio tetrapus alis rotundatis, dentatis, nigro maculatis, fulvis, ſubtus maculis viginti ſeptem argenteis.

The four-legged Papilio, with roundiſh, dentated, yellowiſh wings, with thirty-ſeven ſilvery ſpots underneath.

This is a very elegant ſpecies; it is but a ſmall fly; it's body is oblong and ſlender; it's wings of a duſky yellowiſh colour, ſpotted irregularly on the outſide with black, and having on the inſide thirty ſeven regular and beautiful ſpots, of a ſilvery white, and two ſilvery lines toward the inner edges.

It is very common in *many* parts of Europe, but not in England.

The other ſpecies of the four-legged Butterflies are, 1. The great, elegant, variegated Papilio, with ſinuated wings. 2. The great, variegated Papilio, with more deeply ſinuated wings. 3. The great, duſky, variegated Papilio, with one of the ſinuations of each wing very deep. 4. The yellowiſh-winged Papilio, with four black ſpots on each of the outer wings. 5. The angulated-winged Papilio, with yellowiſh wings, the primary ones having three black ſpots. 6. The angulated, yellowiſh-winged Papilio, ſpotted with black, and with a blue, variegated eye on each wing. 7. The Papilio, with black, denticulated wings, ſpotted with white, and with an arched ſcarlet band on each ſide. 8. The large Papilio, with yellowiſh, dentated wings, variegated with black and white, and with five little eyes on the under ones. 9. The Papilio, with rounded, dentated, yellowiſh wings, ſpotted with black, and with tranſverſe, ſilvery, and white lines underneath. 10. The large, round, and dentated-winged Papilio, of a yellowiſh colour, ſpotted with black on the upper ſide, and marked with twenty-one ſilvery ſpots underneath. 11. The roundiſh, dentated, yellowiſh-winged Papilio, ſpotted with black, and with nine ſilvery ſpots underneath. 12. The Papilio, with roundiſh, dentated, yellowiſh wings ſpotted with black, with three yellow tranſverſe lines underneath them. 13. The Papilio, with roundiſh, dentated wings, variegated with clouds of black and yellowiſh, and with two eyes on the outer ones. 14. The Papilio, with rounded wings clouded with brown, with an eye and half on the outer ones, and five leſſer eyes on each of the inner. 15. The Papilio, with round, ſmooth, brown wings, the primary ones yellowiſh below, and having each one eye. 16. The Papilio, with roundiſh brown wings, with a reddiſh tranſverſe faſcia, and with ſeven eyes on each ſide. 17. The Papilio, with roundiſh brown wings, with three eyes under the primary ones, and five under the others. 18. The Papilio, with rounded, brown wings, with only a ſingle eye under each of the primary ones, and with white tranſverſe lines under the others. 19. The Papilio, with rounded wings, brown on the upper ſide, and greyiſh underneath, with a white tranſverſe line, and five white eyes on each. 20. The little, beautifully-eyed Papilio. 21. The little Papilio, variegated with blotches of white. 22. The little Papilio, with rounded, elegantly painted wings. 23. The little Papilio, with yellow and black wings. 24. The little Papilio, with wings variegated with blue. 25. The larger, beautifully variegated Papilio, with blue eyes. 26. The great, ſinuated-winged Papilio, with variegations of brown, yellow, and red, on duſky wings. 27. The great, rounded-winged Papilio, with the nerves high and whitiſh. 28. The larger Papilio, with ſlightly ſinuated wings, variegated with blue and red, on a duſky brown. 29. The great, black-winged, variegated Papilio. 30. The great, duſky-winged Papilio, with the outer wings clouded with yellow and blue.

PAPILIONES.

Diviſion the Second.

Thoſe which have ſix legs.

Papilio alis flavo nigroque variegatis, ſecundariis angulo ſubulato maculaque fulva.

The Papilio, with the wings variegated with black and yellow, and with a ſubulated angle, and a yellow ſpot on the ſecondary ones.

THIS is a very large and beautiful Inſect: it's wings are very broad, and the primary or outer ones are of an ovated, obtuſe figure, undivided at the edges, and variegated with black and yellow; there are three large lunar ſpots of yellow near the anterior edge of each; near the poſterior there are eight ſmall ones; and between theſe there are ſeven other ſpots in a regular ſeries; the reſt of the wing is black. The ſecondary or interior wings are denticulated at the edge, and the poſterior angle is carried to a much greater length than the reſt: on the edge of this wing there are ſix lunar yellow ſpots, and a ſeventh near the anus, which is ſurrounded with a beautiful eye, edged with blue; the diſk of this wing is yellow.

The caterpillar of this ſpecies feeds on rue, dill, fennel, pimpernel, and the ſeſeli; it is ſmooth, has ſixteen legs, and occaſionally protrudes two yellowiſh horns.

Papilio alis rotundatis integerrimis, ſecundariis viridi nebuloſis, primoribus lunula nigra.

The roundiſh, undivided-winged Papilio, with the ſecondary wings clouded with green, and a black, lunated ſpot on each of the exterior ones.

This is a very ſingular ſpecies: it is of the ſmaller kind, and it's general colour is white: the antennæ are white: the wings alſo are white, but the exterior ones have the verge brown on the outſide, and of a greyiſh yellow underneath, with a brown, lunated ſpot on the middle of each, diſtinguiſhable on each ſide: the interior wings are clouded with green on the under ſurface; on the upper ſide they are white, except toward the baſe, where they are brown: the eyes are of a greeniſh hue, and ſpotted: the male has in the middle of the upper ſurface of the outer wings a yellow ſpot.

It is common in our meadows in Eſſex and Kent.

Papilio alis rotundatis integerrimis ceruleis, ſubtus ocellis numeroſis.

The Papilio, with roundiſh, undivided wings of a blue colour, with numerous eyes underneath.

This is a ſmall ſpecies, but it is extremely beautiful: the body is ſlender, the legs weak, and the eyes large; the wings are of a beautiful ſky-blue on their upper ſurface, and on the under ſide they have a multitude of black eyes, ſurrounded with circles of red, and a number of other ſmaller, ſimple, black ſpots, and ſeveral yellow ones toward the edges.

It is very common with us on heaths, and about the hedges, on dry, hilly paſtures.

The other ſpecies of the ſix-legged butterflies are, 1. The Papilio, with the under wings anguloſo-dentated, and variegated with white and yellow on the under ſide. 2. The Papilio, with angulated, yellow wings, with a ferruginous ſpot on each. 3. The Papilio, with rounded, white wings, with black nerves. 4. The Papilio, with rounded, white wings, with green nerves. 5. The larger Papilio, with rounded, white wings, the primary ones having two ſpots and black tips. 6. The ſmaller Papilio, with rounded, white wings, the primary ones bimaculated, and the tips black. 7. The Papilio, with rounded, white wings, without ſpots. 8. The Papilio, with rounded, undivided, white wings, with four eyes on the upper ſurface, and ſeven

ſeven on the under ſurface of the ſecondary ones. 9. The Papilio, with rounded, undivided, blackiſh-brown wings, with numerous eyes underneath. 10. The Papilio, with rounded wings, brown on the upper ſurface, and with forty-two black ſpots underneath. 11. The Papilio, with rounded, undivided wings, greeniſh, and not ſpotted underneath. 12. The Papilio, with rounded, browniſh-yellow wings, ſpotted on both ſides with black. 13. The Papilio, with rounded, brown wings, ſpotted on the under ſurface with white. 14. The little brown Papilio, with broad and elegantly variegated wings. 15. The little ſhort-bodied elegantly ſpotted Papilio. 16. The larger, plain, duſky Papilio. 17. The great variegated Papilio, with ſinuated wings. 18. The great variegated Papilio, with undivided wings. 19. The oriental velvety Papilio, with ſhort antennæ. 20. The oriental, bright, variegated Papilio, with ſinuated wings. 21. The greater American, black and yellow Papilio. 22. The great American, brown Papilio, variegated with yellow and reddiſh. 23. The American Papilio, with elegant variegations, and the wings deeply ſinuated.

PHALÆNA.

THE antennæ of the Phalænæ are attenuated to the point, not clavated.

The ſpecies of this genus are very numerous; ſome of them have the antennæ of a priſmatic form; ſome have them pectinated, or made in faſhion of a comb; and, of theſe laſt, ſome have no tongue, and others have a ſpiral one; ſome have the antennæ pectinated, and ſit with the wings flat or plane; others ſit with the wings plane and patent, and have ſimple antennæ, and a ſpiral tongue; ſome have the antennæ ſimple, and the tongue ſpiral, but do not ſit with the wings plane; and, of theſe, ſome have the forehead prominent, others not; others, again, have the antennæ ſimple, and have no tongue.

After deſcribing three ſpecies, we ſhall enumerate the others under theſe ſeveral diſtinctions, as the heads of ſo many natural diviſions.

Phalæna priſmicornis, ſpirilinguis, fuſca, alis inferioribus abdomineque faſciis tranſverſis rubris.

The priſmatic horned, ſpiral-tongued Phalæna, with the lower wings and the body ſtreaked with red.

This is a very large and very beautiful ſpecies: the body is elegantly variegated with black and red, and is conſiderably thick; the wings are very long, but they are narrow, caudated, and acute; they are brown on the upper ſide, and are elegantly variegated with tranſverſe ſtreaks of red underneath: the caterpillar of this ſpecies feeds on the privet, the ſyringa, the aſh, and the willow; it is very large and beautiful; it's colour is a bright green, and it has a large horn.

We find the moth in our gardens.

Phalæna pectinicornis, elinguis, alis cinereo flavoque rufis, margine laceris.

The pectinicornate Phalæna, without a tongue, with brown and grey wings, lacerated at the edges.

This is alſo a very beautiful ſpecies: the body is large; the legs are variegated with annular ſpots of white: the wings are lacerated, and ſeem as if they were eroded in the hinder part; they are variegated with brown and reddiſh on the upper ſide, and have a white obliquely tranſverſe line on them, with one ſpot of white at the baſe, and another near the middle: the wings are clouded on their under part: the antennæ are but ſlightly pectinated, and are white toward the baſe, on the exterior ſide.

It is not uncommon in our woods and hedges: the caterpillar feeds on the ground-ivy, and ſeveral other plants.

Phalæna, ſeticornis, alis patentibus albis, poſt faſcia marginali, inferioribus nigris, punctis quatuor albis.

The ſeticornate Phalæna, with white patent wings, with a marginal faſcia behind, and with four white ſpots on the faſcia, which on the inferior ones is black.

This is a very ſingular ſpecies: it is of the ſize of a ſmall fly: the antennæ are ſimple, ſlender, and pointed at the extremities, and are of about half the length of the body, and the ſeveral articulations they are compoſed of have each a white triangular ſpot on one ſide, ſo that they ſeem, when ſlightly viewed, to have annular marks of white; the upper wings are white; they have a brown edge behind, and a ſingle ſpot in the middle, where they are bluiſh, and a flexuous line of bluiſh-green; on the under ſide they are grey, and have a brown ſpot in the middle: the under wings are white, but they have a black tranſverſe line toward the hinder edge, and, over that, a bluiſh-green flexuous line, and four white ſpots, and in the middle they have a black ſpot; they are whiter on the under ſurface than on the upper. It's caterpillar feeds on the common duck-weed, and makes itſelf a caſe of a convoluted leaf of it, in which it paſſes the chryſalis ſtate on the ſurface of the water.

The fly is common with us about ponds.

The other ſpecies of Phalæna, enumerated under the already mentioned diviſions, are, 1. Of thoſe which have the antennæ of a priſmatic figure there are, 1. The Phalæna, with white antennæ, and with greyiſh, plane, eroded wings, and a ſpiral tongue. 2. The ſpiral-tongued Phalæna, with the wings variegated with green, purple, and yellowiſh. 3. The pointed-horned Phalæna, without a tongue, with depreſſed clouded wings, and with the body variegated with white rings. 4. The ſharp-horned Phalæna, without a tongue, with white lanceolated wings, with black veins. 5. The ſharp-horned, ſpiral-tongued Phalæna, with the upper wings bluiſh, with ſix red ſpots on them, and with the under ones red. 6. The ſharp-horned, ſpiral tongued Phalæna, with pale, deflex, eroded wings, with a brown triangular ſpot ſurrounding a red one, and with the thorax gibboſe.

2. Of thoſe which have pectinated wings, and have no tongue, there are, 1. The Phalæna, with the wings green on the upper ſide, with a bluiſh-green, dentated, and ſinuated tranſverſe lines on them. 2. The Phalæna with brown ſublinear wings. 3. The Phalæna, with plain, rounded, clouded, yellowiſh wings, the under ones having a white longitudinal line on them. 4. The Phalæna, with white, ſemitranſparent, deflex wings, with duſky nerves. 5. The Phalæna, with deflex wings, the upper ones brown, with white undulated lines; the lower ones purple, with ſix black ſpots on each. 6. The Phalæna, with black deflex wings, the upper ones variegated with undulated lines of yellow; the under ones red, with black ſpots. 7. The Phalæna, with white deflex wings, and with the legs and the antennæ black. 8. The Phalæna, with deflex white wings, with black ſpots on them, and with five ſeries of ſpots on the body. 9. The Phalæna, with pale deflex wings, with a tranſverſe line of a deep colour on them. 10. The Phalæna, with deflex grey wings, with two white tranſverſe lines on them; the lower one prominent. 11. The Phalæna, with browniſh grey wings, with blue lines, and blue lunar ſpots, and with a black dot underneath: the caterpillar of this fly feeds on graſs, and is very miſchievous and deſtructive. 12. The Phalæna, with plane wings, the upper ones marked each with a white ſpot at the angle of the anus, and the female having ſcarce any viſible wings. 13. The Phalæna, with deflex, grey, and undulated wings, with ſome obſcure tranſverſe lines, and with the head carried between the protended fore-legs. 14. The grey hairy Phalæna, with obſcure blackiſh lines, and a double pale ſpot on the outer wings. 15. The Phalæna, with grey wings, with two obliquely tranſverſe white lines on the exterior ones. 16. The Phalæna, with bluiſh-green wings, with two obliquely tranſverſe yellow lines on the outer ones. 17. The moth produced from the ſilk-worm. 18. The Phalæna, with wings variegated with grey, yellow, and reddiſh, and lacerated at the edges. 19. The Phalæna, with yellowiſh, angulato-dentated wings, with two brown lines on them, and with a yellow thorax. 20. The Phalæna, with plane greyiſh wings, with an eye and a tranſverſe line of brown on each. 21. The Phalæna, with grey deflex wings, with a double yellowiſh-white ſpot on each.

3. Of thoſe which have pectinated antennæ and a ſpiral tongue there are, 1. The Phalæna, with plane yellowiſh wings, with a red ſpot in them. 2. The Phalæna, with very elegant, gloſſy, bluiſh-green wings for the upper pair, and with the lower pair brown. 3. The Phalæna, with plain, rounded wings, brown on the upper part, but with ſome white at the edge. 4. The Phalæna, with plane, angulated, yellow wings, clouded with brown. 5. The Phalæna, with plane whitiſh wings, and with tentacula of the length of the antennæ. 6. The Phalæna, with deep brown plane wings, variegated with a bright red and white.

4. Of thoſe which have pectinated antennæ, and the wings plain as they ſit, there are, 1. The Phalæna, with a ſpiral tongue and patent dentated wings, with two brown tranſverſe lines on them. 2. The Phalæna, with a ſpiral tongue and grey wings, with two brown faſciæ on them. 3. The Phalæna, with a ſpiral tongue and ſub-pectinate horns, with the wings variegated with tranſverſe ſtreaks of brown and white. 4. The ſpiral-tongued Phalæna, with patent, whitiſh wings, with four tranſverſe marginal lines.

5. Of thoſe with ſimple antennæ and a ſpiral tongue, and ſitting with plain patent wings, there are, 1. The Phalæna, with the wings clouded with brown, and with the anus yellow. 2. The Phalæna, with wings brown on the upper ſide, but yellow on the hinder part, and underneath. 3. The Phalæna, with hoary wings, with two or three tranſverſe ſtreaks, and a black ſpot. 4. The Phalæna, with patent white wings, with a number of irregular black ſpots on them. 5. The Phalæna, with ſnow-white, rounded, patent wings, with ſome very ſmall dots of grey on them. 6. The Phalæna, with patent white wings, all of them nebuloſo-reticulated and ſmooth. 7. The Phalæna, with white patent wings, with a marginal faſcia behind; the lower ones black, with four white ſpots. 8. The Phalæna, with patent white wings, the lower ones nebuloſo-reticulate and ſmooth. 9. The Phalæna, with grey patent wings, with brown linear faſciæ and a black ſpot. 10. The Phalæna, with painted, bright, white wings, undulated with a duſky brown. 11. The Phalæna, with patent wings, with three grey faſciæ, and with a ſingle black dot. 12. The Phalæna, with patent green wings. 13. The Phalæna, with patent, yellowiſh-white wings, with black linear faſciæ. 14. The Phalæna, with patent wings of a whitiſh colour, with a broad, nebuloſe, tranſverſe line, with a mark like a figure of eight within it. 15. The Phalæna, with patent white wings, with the edge every-where brown. 16. The Phalæna, with patent wings of a black clouded colour, with two white tranſverſe lines. 17. The Phalæna, with patent wings, with two white, and two black, tranſverſe lines on them. 18. The Phalæna, with patent wings of a grey, ſpotted with a darker colour. 19. The Phalæna, with black wings, white at the extremities. 20. The ſpiral-tongued, totally white Phalæna. 21. The Phalæna, with patent wings, with two white, and two brown, tranſverſe lines placed alternately. 22. The patent-winged Phalæna, which is produced from the croſt caterpillar. 23. The Phalæna, with very patent, linear, ramoſe wings, the exterior ones divided into two parts, the anterior into three. 24. The Phalæna, with deeply divided brown wings tipped with yellow.

6. Of thoſe which have very ſimple wings, a ſpiral tongue, and have no prominence on the forehead, there are, 1. The little Phalæna, with the upper wings brown, with a red line, and two red ſpots, and with the under wings entirely red. 2. The Phalæna, with incumbent wings, the exterior ones grey and clouded; the lower ones yellow, with a black ſpot near the edge. 3. The Phalæna, with incumbent grey wings, clouded with white and yellowiſh-brown. 4. The Phalæna, with deflex, ferrugineous, grey wings, with a white and a yellow ſpot on them, and with a barbated anus. 5. The Phalæna, with deflex wings, the exterior ones brown, with a figure of greek λ on them. 6. The Phalæna, with depreſſed grey wings, with a brown ſpot and with a figure of the letter C under it; the lower ones white. 7. The Phalæna, with deflex nebuloſe wings, with one or two gold-yellow tranſverſe lines on them. 8. The Phalæna, with depreſſed wings, with a ferrugineous ſpot on each ſide, with gibbous ſhoulders, and with a bunch upon the middle of the back. 9. The Phalæna, with depreſſed brown wings; the upper ones with two pale ſpots on them, and with the apex and the baſe of the thorax gibbous. 10. The Phalæna, with deflex, ferrugineous, grey wings, with two grey ſpots on them, and with the apex, and the baſe of the thorax gibbous. 11. The Phalæna, with deflex hoary wings, and with ſome black marks, like the greek ψ, upon them. 12. The Phalæna, with yellow, deflex wings,

wings, with two oblique, ferrugineous lines on them. 13. The Phalæna, with deflex, black wings, with the body of a yellowiſh-brown. 14. The Phalæna, with incumbent, yellow wings, with two grey obliquely tranſverſe lines, the under one interrupted. 15. The Phalæna, with depreſſed, yellowiſh-brown wings, with a triangular ſilvery ſpot at the edge. 16. The Phalæna, with deflex, whitiſh wings, with a double tranſverſe line of a duſky grey on the upper ones. 17. The Phalæna, with deflex, pale grey wings, with a duſkier line on them. 18. The Phalæna, with deflex wings, brown on the upper ſide, with a white line near the edge.

7. Of thoſe with ſimple antennæ, a ſpiral tongue, and a prominence on the forehead, there are, 1. The Phalæna, with plane, incumbent wings, and with two crooked corniculæ criſtated at the baſe. 2. The Phalæna, with ſpotted wings, and with two ſubulated, crooked corniculæ, black on the upper ſide, and white underneath. 3. The Phalæna, with two ſubulated, crooked corniculæ, and with incumbent wings clouded with black and grey. 4. The Phalæna, with the upper wings terminated behind by a crooked claw. 5. The Phalæna, with the upper wings white, ſpotted with black, and the lower wings brown. 6. The noſed, black Phalæna, with a white head, and with the wings white in the hinder part. 7. The grey Phalæna, with a white ſpot on each ſide the thorax. 8. The hoary, noſed Phalæna, with a grey head, and a black ſpot on the middle of the wings. 9. The noſed, green Phalæna. Moſt of the preceding ones are ſmall, and proceed from the worms which eat cloaths; the laſt, indeed, feeds in the caterpillar ſtate on the willow. 10. The Phalæna, with grey incumbent wings, and with the thicker edge of the lower ones whitiſh. 11. The noſed, black Phalæna, with two white, teſſelated, tranſverſe lines on the wings. 12. The noſed, grey Phalæna, with clouds of black on the wings. 13. The noſed, black Phalæna, with four tranſverſe, ſilvery lines. 14. The noſed, black Phalæna, with three tranſverſe, ſilvery lines. 15. The brown, noſed Phalæna, with an obliquely tranſverſe line of gold-yellow. 16. The noſed, black Phalæna, with the exterior wings of a gilded yellow, and with the antennæ longer than the body. 17. The noſed, white Phalæna, with two tranſverſe lines of a ſilvery-brown on the exterior wings, and with ſeveral irregular undulations behind. 18. The noſed, yellow Phalæna, with four tranſverſe lines of a ſilvery-brown on the wings. 19. The noſed, grey Phalæna, with the wings white in the anterior part, and with a white linear faſcia behind. 20. The Phalæna, with black wings, with a white tranſverſe line on them. 21. The noſed, whitiſh Phalæna, with longitudinal white ſtriæ on the wings, each terminated by a black ſpot. 22. The noſed, whitiſh Phalæna, with a longitudinal ſilvery line on the wings, ramoſe in the hinder part. 23. The noſed, grey Phalæna, with a white tranſverſe line on the back. 24. The noſed, brown Phalæna, with a black ſpot on each ſide. 25. The ſomewhat noſed, clouded, brown Phalæna, with a doubly triangular white ſpot on the back. 26. The noſed, grey Phalæna, with a black ſpot on each ſide. 27. The noſed, black Phalæna, with yellow lines on the wings, with undulated white edges. 28. The noſed-brown Phalæna, undulated with white, and with the under pair of wings brown, with white edges. 29. The deep brown, noſed Phalæna, with variegations of a bright red.

8. Of thoſe which have extreamly ſimple antennæ, and no tongue, there are only ten known ſpecies; theſe are, 1. The Phalæna, with a bicornate head, with white wings, with black tranſverſe lines near the edge. 2. The Phalæna, with a bicornate head, with brown wings, with tranſverſe red ſtreaks. 3. The Phalæna, with a bicornate head, with the thorax yellow, and with the wings plano-incumbent, and of a greyiſh-brown. 4. The Phalæna, with very ſhort antennæ of a reddiſh brown, and with a yellow thorax and deflex wings. 5. The Phalæna, with ſilvery wings, with five tranſverſe lines on them. 6. The Phalæna, with incumbent wings; the upper ones with a yellow edge, and the lower ones all over yellow. 7. The Phalæna, with incumbent, linear, hairy wings. 8. The Phalæna, with yellow wings, with extreamly ſmall and numerous black ſpots on them. 9. The Phalæna, with greyiſh-brown wings, the lower ones whitiſh; the female having ſcarce any viſible wings at all. 10. The whitiſh-green Phalæna, with plane wings.

TETRAPTERA.

Claſs the Fifth.

HYMENOPTERA.

Inſects which have four wings, and theſe all entirely membranaceous.

TENTHREDO.

THE female of the Tenthredo has a ſerrated point, or weapon, at the tail: the worm, produced of the egg, has ſeveral feet.

The ſpecies of this genus have been generally confounded with the Ichneumons.

Tenthredo antennis clavatis, atra; ſegmentis abdominalibus partim ferrugineis.

The black Tenthredo, with clavated antennæ, and with ſome of the ſegments of the body ferrugineous.

This is a large fly: it very nearly equals the hornet in ſize: it's body is black, except that the third, fourth, and fifth joints or ſegments of it are ferrugineous, it is hairy: the wings are very thin and tranſparent; the legs black, but the feet yellow within: the antennæ are clavated and yellow, except the loweſt joint, which is black; there are ſome ferrugineous black nerves in the wings, and they have a browniſh tinge toward the outer edge: the worm feeds on the willow, poplar, and birch, and is ſmooth and green, and has a black liſt down the back, edged with yellow; it has twenty-eight legs, and often rolls itſelf up in a ſpiral form.

Tenthredo atra antennis undecim-nodiis, alis nigro albaque maculatis.

The black Tenthredo, with eleven joints in the antennæ, and wings ſpotted with black and white.

This is of the ſize of a common houſe-fly: it's body, thorax, head, and legs are all black: the antennæ are ſlender; the wings have ſeveral veins in them; the upper ones have two black ſpots, of which that neareſt the thorax is lunated, the other round, and near it, toward the top of the wing, there is a white one.

We meet with this ſpecies frequently in the woods about Harrow on the Hill, and in many parts of the North.

Tenthredo antennis duodecim-nodiis, nigris, abdomine ſubtus ferrugineo, pedibus flavis.

The yellow-legged Tenthredo, with the body ferrugineous below, and the antennæ compoſed of twelve joints.

This is a ſmall ſpecies, ſcarce exceeding a flea in ſize: the antennæ are ſlender and black; the head and thorax are alſo black and gibbous: the body is of an oval figure, and is carinated, ferrugineous beneath, and black towards the anus; the legs are ferrugineous, and the wings are larger than the body, and are whitiſh, and have nothing of the ſpots common on many others of this genus.

This ſpecies is produced in the excreſcence of the common Dog-roſe, called the Bedeguar, and ſpunge or gall of the roſe in the ſhops.

The other ſpecies of the Tenthredo are, 1. The black Tenthredo, with clavated antennæ, and an oval body, and ferrugineous wings. 2. The black Tenthredo, with ferrugineous legs, and with a cylindric cornicle at the anus. 3. The black Tenthredo, with ferrugineous legs, and with the apex of the anus depreſſed and acute. 4. The Tenthredo, with the antennæ formed of ſeven joints, with a yellow body, with the head and the middle of the thorax black, and with an oval ſpot on the wings. 5. The

Tenthredo, with ſeven joints in the antennæ, and with the body yellow, but black behind. 6. The Tenthredo, with ſeven joints in the antennæ, with a yellow body, and with a longitudinal black ſtreak in the wings. 7. The yellow Tenthredo, with a ſingle ferrugineous ſpot in each wing. 8. The black Tenthredo, with yellow legs, and with yellow marks on the breaſt. 9. The black Tenthredo, with ſhort antennæ. 10. The Tenthredo of the willow, the worm of which is of a bluiſh-green, and has the breaſt and the tail yellow. 11. The Tenthredo of the poplar. 12. The black Tenthredo, with ſeven joints in the antennæ, and with the edges of moſt of the ſegments of the body yellow. 13. The Tenthredo, with ſeven joints in the antennæ, and with the back black, with pale, tranſverſe, and arched lines on it. 14. The Tenthredo, with antennæ with eighteen joints, with ferrugineous legs, the hinder pair variegated with black and white. 15. The Tenthredo, with black antennæ, conſiſting of twelve joints, with the body ferrugineous below, the legs yellow, and the wings unſpotted. 16. The Tenthredo, with a braſſy-green thorax, and a gold-yellow body. 17. The braſſy-blue Tenthredo, with pale coloured legs, and unſpotted wings. 18. The black Tenthredo, with the back of the thorax greeniſh, and with legs ſerving it for hopping. 19. The Tenthredo, with a ſhining green thorax, and a brown body, with a pale belt at the baſe, and with yellow legs. 20. The Tenthredo, with a black body, and yellow legs. 21. The black Tenthredo, with white legs. 22. The Tenthredo of the willow-leaf gall. 23. The Tenthredo of the gall of the beech-leaf. 24. The oak gall, or common gall Tenthredo. 25. The oak-leaf gall Tenthredo. 26. The Tenthredo of the imbricated gall of the oak. 27. The Tenthredo of the gall of the ground-ivy. 28. The Tenthredo of the gall of the hairy hawkweed.

The Tenthredines of the roſe, the willow, the oak, the ground-ivy, and ſome others, have been diſtinguiſhed from the others, under the name of Cynips, but they are properly of this genus.

ICHNEUMON.

THE weapon at the anus of the Ichneumon is triple.

Ichneumon ater pedibus rufis.
The black Ichneumon, with red legs.

The common Ichneumon Fly.

This is a moderately large ſpecies: it's body is long, ſlender, and black; it's head alſo, and the thorax, the antennæ, and even the weapon at the tail, are all black: the legs are of a reddiſh colour, and are long and ſlender; the wings are pellucid, and have a black ſpot near the edge: the weapon at the tail is longer than the body, and is very ſlender; it conſiſts of three hairs, as it were; the two outer ones black, and the middle one red.

It is common with us about dry banks and in gardens. Ray calls it Veſpa Ichneumon tota nigra præter pedes qui crocei ſunt; and, in another place, Muſca tripilis corpore tenui et prælongo ſeta a cauda longiſſima.

Ichneumon totus luteus.
The all-yellow Ichneumon.

The yellow Ichneumon Fly.

This is a large ſpecies: the head, body, legs, and indeed every part, are of the ſame uniform yellow; the wings are large and membranaceous, but they have a ſpot of yellow near the edge: the body is of a crooked or arcuated figure, and is narrow at the baſe; the antennæ are equal to the body in length, and are compoſed of a great number of articulations; the eyes are large, and are black.

It is frequent with us about dry banks. Ray calls it Veſpa Ichneumon tota fulva alis amplis.

Ichneumon

Ichneumon tibiis poſticis clavatis.
The Ichneumon, with the hinder legs clavated.

This is a moderately large ſpecies: the head, thorax, legs, and antennæ are all black; the weapon in the tail of the female is long, ſlender, black, and compoſed of three hairs; the two outer ones black, the middle one red; the whole is longer than the body, and has a white tip at the extremity: the body is of a clavated figure; it is affixed to the thorax by a ſlender thread; it is black all over, except at the baſe, where it is reddiſh: the hinder legs are ſmall at the top, but the third joint is very robuſt and large.

It is common with us on heaths. [a] Mouffet calls it Muſca tripilis.

The other ſpecies of the Ichneumon are, 1. The Cotton-fly, or the Ichneumon which makes a web of cottony matter, of the ſize of a pigeon's egg, common on graſs in autumn. 2. The Ichneumon, whoſe worm feeds on the fleſh of the caterpiller of the cabbage, living within it's body. 3. The Ichneumon, whoſe worm feeds on the aphides of the ſerratula. 4. The hairy, bluiſh-black Ichneumon. 5. The black Ichneumon, with the thighs teſtaceous, and the antennæ white in the middle. 6. The black Ichneumon, with ferrugineous legs, and antennæ all over of the ſame colour. 7. The black Ichneumon, with the legs teſtaceous, and with the antennæ pale on the under part. 8. The black Ichneumon, with the hinder legs ferrugineous, and the others black, and with the antennæ white in the middle. 9. The black Ichneumon, with reddiſh legs. 10. The black Ichneumon, with the legs and the top of the thorax white. 11. The black Ichneumon, with the legs reddiſh, and the forehead pale. 12. The black Ichneumon, with reddiſh legs, the top of the hinder ones only black, and the antennæ white in the middle. 13. The black Ichneumon, with ferrugineous legs, except the hinder pair, which are variegated with black and white. 14. The black Ichneumon, with the forehead white, and with white ſpots on the body, and with the legs red. 15. The black Ichneumon, with black legs, and with four white ſpots on each ſide the body. 16. The black Ichneumon, with the body and the legs ferrugineous, and the middle of the antennæ white. 17. The black Ichneumon, with the body and the legs ferrugineous, and with a white ring on the antennæ. 18. The yellow Ichneumon, with the extremity of the body black. 19. The black Ichneumon, with the body ferrugineous, and black at the extremity, and with a white circle on the antennæ. 20. The black Ichneumon, with coal-black antennæ, and with the body ferrugineous, but black at the extremity. 21. The black Ichneumon, with the body entirely ferrugineous. 22. The ferrugineous Ichneumon, with the extremity of the body and the lower part of the thorax black. 23. The black Ichneumon, with the body yellow in the anterior part, and with the joints of the legs black. 24. The black Ichneumon, with yellow legs, and the body yellow on the anterior part. 25. The black Ichneumon, with the anterior ſegments of the body ferrugineous, and the top of the thorax yellow. 26. The black Ichneumon, with brown wings, and with the anterior ſegments of the body reddiſh. 27. The Ichneumon, with the body ferrugineous before, and black behind, and with four white ſpots on it. 28. The Ichneumon, with the anterior part of the body ferrugineous, the hinder part black, with five yellow ſpots on it, and with the antennæ whitiſh on one ſide. 29. The Ichneumon, with the anterior part of the body ferrugineous, the poſterior part black, and without ſpots, and with the antennæ marked with a white circle. 30. The Ichneumon, with the anterior part of the body ferrugineous, and the hinder part black, and the inciſures yellowiſh, with a white circle on the antennæ. 31. The black Ichneumon, with the middle of the body and the anterior legs reddiſh, the lower part of the poſterior ones white. 32. The black Ichneumon, with the legs, and the ſecond and third ſegments of the body yellow. 33. The black Ichneumon, with white wings, with a double tranſverſe line of black on each, the hinder one the broader. 34. The long and ſlender-bodied Ichneumon, with the antennæ of the length of the body, with clavated legs, and with ſetaceous tentacula. 35. The braſſy-green, elegant Ichneumon, with black, ſpiral antennæ. 36. The great, black Ichneumon, with the extremity of the body red.

APIS.

THE weapon in the tail of the Apis is simple, and capable of inflicting a wound.

This genus comprehends the bee, the humble-bee, the wasp, and hornet kind.

Apis thorace nigro, antice rufo immaculato, abdominis incisuris puncto nigro duplici contiguo notatis.

The Apis, with a black thorax, and double black spots on the segments of the body.

The Hornet.

This is by much the largest and the most mischievous of this genus; indeed, of all the Insects of this part of the world: the head is reddish, but the upper lip yellow; the thorax is reddish in the anterior part, otherwise all over black: the segments or joints of the body are outwardly yellow, inwardly black: there are two black spots on each of the middle segments, which coalesce at their bases with the inner blackness of the segments themselves: the legs are greyish.

It is too common in the woods and about houses, in some parts of England: it's sting is attended with great pain and inflammation, sometimes with danger. There is a harmless fly, with only two wings, which greatly resembles this Insect; but the characters given in this description convey the essential differences, beside those in the number of the wings, and the want of a sting. All the writers on Insects have described this under the name of Crabro, or Crabro vulgaris.

Apis nitida thorace viridi-caeruleo, abdomine inaurato.

The beautiful gold-yellow Apis, with a blue-green thorax.

The green and yellow Bee.

This species is of the number of those distinctively called Bees, and is vastly the most beautiful of the whole genus: it is of the shape of the common bee, but smaller: the head and thorax are of a very bright and beautiful blue, with a small admixture of green in it: the body is of a bright yellow, so glossy, that it appears gilded and burnished; the wings are brown; the thorax, and the last segment but one of the body, are dentated behind: the antennae are black, and consist of twelve articulations: it stings in the manner of the common bee.

It is found about old stone walls that have crevices in them; sometimes about clayey ditch banks.

Apis nigra abdomine fasciis quatuor undique flavis notato.

The black Apis, with four yellow circles on the body.

The black and yellow Wasp.

This is one of the species properly of the wasp kind; it is somewhat smaller than the common wasp: the head is black, as is also the thorax; the antennae are also black, and are broader in the middle than elsewhere: there are two yellow spots on the hinder part of the thorax: the first joint of the body is black, with a yellow spot on each side; the four succeeding joints of it are also black, but they have each a yellow edge, forming so many yellow fasciae or belts quite round the body: the last segment of the body is entirely black; the legs are black, but variegated with a paler colour.

It is very frequent about the old stone walls in Northamptonshire, and some other parts of England; but about London I have not seen it: it stings like the common wasp, and the wound is equally painful and inflamed.

Apis hirſuta nigra, thoracis cingulo flavo, ano albo.
The black, hairy Apis, with the cingulum of the thorax yellow, and the anus white.

The great Humble-bee.

This ſpecies is of the number of thoſe commonly diſtinguiſhed by the name of the Humble-bee, and is the largeſt of them all: it is near three quarters of an inch in length, and a third of an inch in breadth: it's body is all over black, and very hairy, only that the hinder extremity about the anus is white, and on the anterior part there is a little yellowneſs: the thorax is black, but has a ſlight line of yellow alſo on it's anterior part; and the legs are black.

It is very common about our hedges, and in gardens and orchards; it is not apt to ſting unprovoked, but, when it does, the wound is very painful. All the writers on Inſects have deſcribed it under the name of Bombylius, and Bombylius vulgaris.

The other ſpecies of the Apis are very numerous: having thus deſcribed one of each of the common diſtinctions, the reſt ſhall be enumerated under the ſame diviſions, and will be eaſily diſtinguiſhed by their names, without farther deſcription.

1. Of thoſe called *Waſps* there are, 1. The common Waſp. 2. The black Waſp, with five of the ſegments of the body yellow, and the firſt very remote. 3. The black Waſp, with the baſe of the thorax yellow, and with four of the ſegments of the body yellow, the firſt very remote. 4. The black Waſp, with the baſe and the apex of the thorax yellowiſh, and with four yellow faſciæ on the body, the third of them interrupted. 5. The black Waſp, with four yellow faſciæ on the body, the three firſt interrupted. 6. The ſmooth, black Waſp, with three yellow faſciæ on the body, the firſt of them very remote. 7. The ſmooth, black Waſp, with three yellow faſciæ on the body, the third very remote, and the firſt joint of the body infundibuliform. 8. The ſmooth, black Waſp, with yellow legs, and with three yellow faſciæ on the body, the firſt interrupted. 9. The Waſp, with the black thorax with yellow lines, and with the ſegments of the body black and yellow at the edges; the firſt and ſecond ferrugineous. 10. The Waſp, with the legs and the jaws yellow, black at the extremity, and with the inciſures of the body ſmooth and black at the edges. 11. The black Waſp, with the jaws and the legs yellow. 12. The black Waſp, with the legs and the front of the head yellow. 13. The black Waſp, with the firſt joint of the body infundibuliform, the ſecond very large and campanulated.

2. Of thoſe called *Bees* there are, 1. The common Bee, which is kept in hives for the ſake of the wax and honey. 2. The ſmooth Bee, of a braſſy black colour without ſpots, and with the legs ferrugineous. 3. The greyiſh, ſomewhat hairy Bee, with a conic body, and with the edge of the ſegments of a yellowiſh-white. 4. The greyiſh, hairy Bee, with a yellow head, and with a black body, with the ſegments marked with white. 5. The hairy Bee, with the head and thorax black, the body reddiſh, and the forehead bicornute. 6. The hairy Bee, with a black head and red body, and with a white forehead. 7. The black Bee, with four white lines on the body, and with a browniſh-yellow, woolly matter underneath. 8. The black Bee, with ferrugineous legs, and with four yellow ſpots on each ſide of the body. 9. The little coal-black Bee, with a red tail.

3. Of thoſe called *Humble-bees* there are the following, 1. The black, hairy Humble-bee, without any variegation. 2. The black, hairy Humble-bee, with a brown tail. 3. The black, hairy Humble-bee, with a yellowiſh, tawny tail. 4. The hairy Humble-bee, with the thorax black, and the body yellow. 5. The hairy, tawny Humble-bee, with the body yellow. 6. The hairy, tawny Bee, with the body black in the middle, and white at the extremity. 7. The long-haired, reddiſh Humble-bee, with yellowiſh variegations. 8. The ſmaller, black Humble-bee, with the extremity of the body of an orange colour. 9. The large Humble-bee all over black, except for a long yellow down upon part of the firſt pair of legs. 10. The great, black Humble-bee, with two tranſverſe, yellow lines, one on the thorax, the other on the middle of the body. 11. The very large, black Humble-bee, with a duſky yellow tail, and a line of the ſame colour on the lower part of the body. 12. The large, black Humble-bee, with a reddiſh tail, and with a greeniſh-yellow faſcia on the ſummit of the thorax. 13. The very hairy, great, whitiſh Humble-bee, with a reddiſh tail. 14. The black Humble-bee, with a reddiſh tail, and two

yellow *faſciæ* on the body. 15. The great, black Humble-bee, with a white tail, and with three tranſverſe lines of yellow on the body. 16. The ſhining, black Humble-bee, with a white tail, and the ſides of the body toward the hinder part white. 17. The great, black Humble-bee, with the whole body ſmooth. 18. The great, black Humble-bee, with a white tail, and with two tranſverſe lines of yellow on the body. 19. The great, black Humble-bee, with a tawny faſcia round the body, and another on the thorax. 20. The black Humble-bee, with two faſciæ on the under part of the body; the one broad and yellowiſh, the other narrow and white.

FORMICA.

THERE is in the Formica an erect ſquamma or ſcaly body, placed between the thorax and the body: the males, as they are called, or the ants of no ſex, have no wings.

Formica maxima.
The largeſt Formica.

The great Ant, or Horſe-ant.

This is the largeſt of all the ant kind, and is nearly of twice the ſize of the common ſpecies; the head is black; the thorax is of a duſky, ferrugineous colour in general, but it becomes black toward the hinder part, and it's extremity is white: the legs are ferrugineous; the ſquamma, which is placed between the body and the thorax, is of a rounded oval figure, pointed at the top, and undivided; the body is brown, and conſiſts of five ſegments.

It is common with us in woods, where it uſually lives in the hollow trunks of decayed trees. Authors call it Formica major, and Hippomyrmex.

Formica rubra minima.
The little, red Formica.

The red Ant.

This ſpecies is as much ſmaller than the common ant, as the firſt is larger: it's head is very ſmall; it's thorax is bulky, and the ſquamma which ſeparates that from the body is of a roundiſh figure, and ſlightly dentated: the legs are ſlender; the wings are very thin and browniſh.

It is not uncommon in the hilly and dry paſtures among the graſs, and on the leaves and ſtalks of the ſmaller plants.

Formica minor atra.
The little, black Formica.

The black Ant.

This is of a middle ſize, between the common and the little red Ant: it's head is large, in proportion to the body: the thorax is flatted, and the body joined to it at a diſtance: the ſquamma, which ſeparates them, is of an oval figure, and undivided at the edges: the legs are longer and ſlenderer than in any other ſpecies.

It is not uncommon with us on heaths, and in dry paſtures.

The other ſpecies of the Formica are, 1. The common Ant or Piſmire. 2. The large, brown Ant. 3. The great, yellowiſh Ant of the Eaſt Indies. 4. The great, black Ant, with a ſmall head. 5. The large American red and black Ant.

Among the ants, as among the bees, and many other of the gregarious Inſects, ſome of the ſame ſpecies are males, others females, and others of neither ſex. The two firſt, in all theſe caſes, are few in proportion; the others make up the number. Among the bees, the males are what we call drones, and are three or four hundred in a ſwarm: the female is often only one in the ſwarm, ſometimes two, rarely more. The drones or males are deſtroyed by the working bees, as ſoon as the time of impregnating the female is over; and, among the ants, even the females are denied a place in the ſwarm. The males and females in the common ant have both wings, but the ants which make the ſwarm, and do the whole buſineſs, have none: the females are largeſt of all; the males are of a middle ſize, and the working or neutral ants are much ſmaller than the males: theſe keep poſſeſſion of their habitation, and as ſoon as the buſineſs of impregnation of the females, and laying the eggs, is over, they drive both the males and females from among them.

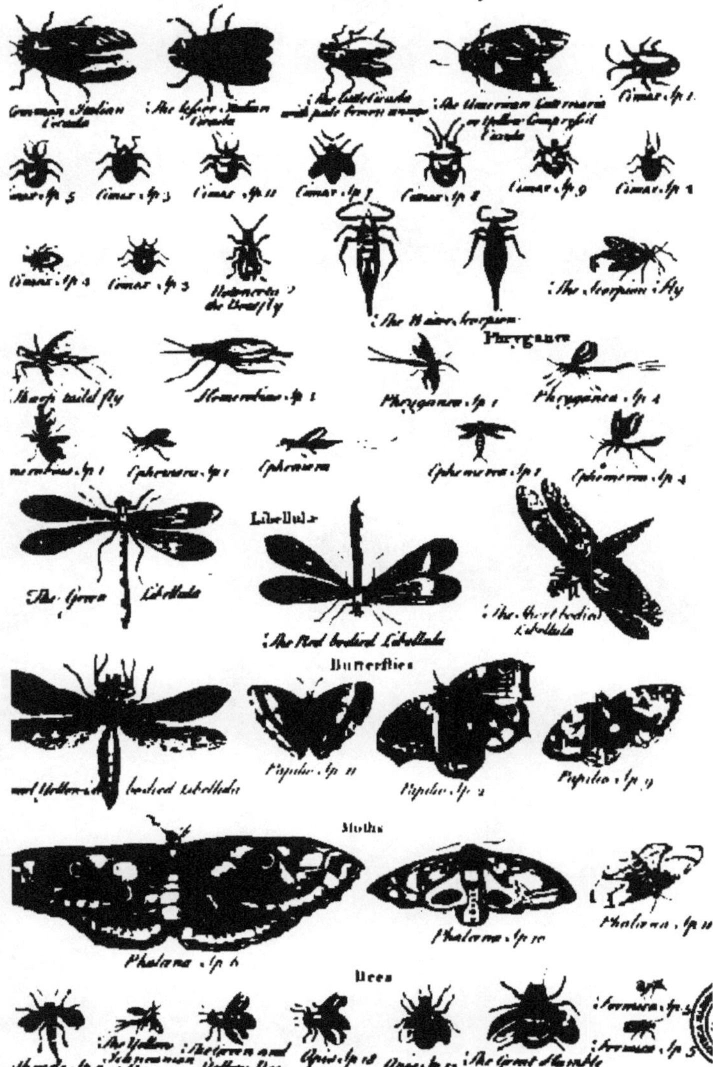
WINGED INSECTS
Cimex Sp. 1
Cimex Sp. 5
Cimex Sp. 3
Cimex Sp. 11
Cimex Sp. 7
Cimex Sp. 8
Cimex Sp. 9
The Scorpion Fly
The Water Scorpion
Phryganea
Sharp tailed fly
Phryganea Sp. 1
Phryganea Sp. 4
Ephemera
Libellulæ
The Green Libellula
The Red bodied Libellula
The Short bodied Libellula
Butterflies
Papilio Sp. 11
Papilio Sp. 2
Papilio Sp. 9
Moths
Phalæna Sp. 6
Phalæna Sp. 10
Phalæna Sp. 11
Bees
The Yellow Ichneumon Fly
The Green and Yellow Bee
Apis Sp. 18
Apis Sp. 53
The Great Humble Bee
Formica Sp. 5

INSECTS.

Series the Third.

GYMNARTHRIA.

Insects which have soft and naked bodies, furnished with limbs.

THE modern naturalists have called these Zoophyta, a term expressing creatures partly of an animal, and partly of a vegetable, nature: it was, indeed, in the earlier ages of natural knowledge, supposed, that there were things of this middle nature between the animal and vegetable world; but as a better acquaintance with the subjects of this study has since sufficiently convinced us, that there are no such creatures, it is extraordinary, that the term should be retained by accurate writers, and that also, while they are perfectly convinced, that there is no participation of the vegetable character in the creatures ranked under it, but that they are as truly and perfectly animals, as any other part of the living creation.

The animals of this series all agree so regularly in their general characters, that there is no dividing them into distinct classes, though the genera are numerous.

LIMAX.

THE body of the Limax is of a figure approaching to cylindric, and is perforated at the side: the tentacula are four in number, and two of them have the appearance of eyes.

Limax ater.
The black Limax. **The black, naked Snail.**

This is a considerably large Insect: it's length is about three inches, and it's diameter half an inch; it's head and tail are smaller than it's middle; the back is raised into a convexity; the belly is flat; the whole body is furrowed and wrinkled very considerably, and is of a deep black colour, except on the belly, where it is paler and somewhat grey. It thrusts out occasionally, at the head, four tentacula, or feelers: the whole body is covered with a glutinous fluid, somewhat like that which naturally covers the eel kind: It is an hermaphrodite, both sexes being in each individual, and both in the coitus impregnate, and are impregnated, at the same time.

It is common in woods, and under hedges, and in our cellars and other cool places: it appears most in damp weather.

Limax subrufus.
The reddish Limax. **The naked, red Snail.**

This is a smaller species than the former: it grows to about two inches in length, and is thinner in proportion than the black snail: it's body is covered with numerous slight furrows or wrinkles, and is of a dusky-reddish colour, except on the belly, where it is whitish or greyish: it is covered with a tough thick mucous matter, and moves very slowly, leaving the mark of the path it has kept in the remains of it's mucilage, which dries into a thin glossy film.

It is common in our woods after rain: Lister calls it Limax subrufus.

Limax flavus maculatus.
The yellow, spotted Limax. **The Amber Snail.**

This, when full grown, is about an inch and a half long: it's back is prominent; it's belly hollowed, and it's head small; the whole surface is slightly wrinkled, and is covered with a mucous juice: the colour is a glossy yellow, with a cast of brown in it; and the whole body is variegated with spots of a greyish colour.

It is very frequent in woods in the North of England: about London, it is rarely met with: Liſter calls it Limax ſylvaticus colore albido maculis inſignitus.

The other ſpecies of Limax, are, 1. The very large, grey Limax, ſpotted with a duſky brown. 2. The little ſhort and thick, grey Limax, without ſpots. 3. The ſmooth-bodied, reddiſh-brown Limax. 4. The ſmall duſky brown, furrowed Limax. 5. The deep chocolate-coloured Limax.

LERNEA.

THE body of the Lernea is of an oblong cylindric figure, and is perforated in the forehead: the tentacula reſemble ears.

Lernea tentaculis quatuor, duobus apice lunulatis.
The Lernea, with four tentacula, two of them lunulated at the top.

This is a ſmall ſpecies: it is about half an inch in length, and of the thickneſs of a ſmall ſtraw: the body is rounded, and of a pale greyiſh-white, gloſſy on the ſurface, and ſomewhat pellucid: it is thruſt out of a kind of coat or ſheath, as it were, at the baſe, which is of a white colour, and reſembles a thick ſkin: toward the other extremity of the body, there are three obtuſe tubercles, one of which is much larger than the reſt: the mouth is ſituated in the anterior part, and near it there are two ſoft and fleſhy proceſſes, and near theſe there is alſo on each ſide another ſoft proceſs, which is lunated at the extremity.

It is found on the ſides of the bream, carp, and roach, in many of our ponds and rivers, in great abundance.

Lernea tentaculis brevibus craſſis ſimpliciſſimis. **The ſea hare.**
The Lernea, with ſhort, thick, and ſimple tentacula.

This is a conſiderably large ſpecies, in compariſon of the former: it grows to two inches and a half in length, and to more than an inch in diameter: it's body approaches to an oval figure, and is ſoft, punctated, of a kind of gelatinous ſubſtance, and of a pale lead colour; from the larger extremity there ariſe four oblong and thick protuberances; theſe are the tentacula; two of them ſtand nearly erect, two are thrown backward.

It is not uncommon about our ſhores: ſome of the writers on theſe ſubjects have called it Lepus marinus; others, Lepori marino congener; and ſome have named it Urticae marinae ſpecies: but, in general, the diſtinctions, conveyed under theſe names, are very little underſtood.

Lernea corpore bifurcato.
The Lernea, with a bifurcated body.

This ſpecies rarely exceeds three quarters of an inch in length: it's body is of a cylindric figure, and moderately thick, and is compoſed of a ſofter matter, contained as it were in a firmer, white, and ſomewhat wrinkled ſkin: the anterior extremity of the body is ornamented with four tentacula of a variable figure; two of them larger at the points than elſewhere; the other two ſmaller: at the oppoſite extremity, the body is forked, or divided into two portions, which are ſhort and pointed.

It is frequent about the Bognor rocks on the Suſſex coaſt, as alſo in Cornwall.

The other ſpecies of the Lernea, are, 1. The larger, whitiſh Lernea, with a double tail. 2. The ſmall, ſoft, and thick, or double-tailed Lernea. 3. The thick Lernea, the Lepus marinus of Clusius. 4. The ſhort, oval, ſharp, and ſingle-tailed Lernea. 5. The oblong-bodied Lernea, with little pointed protuberances on it. 6. The conoidic Lernea. 7. The ſhapeleſs, ſomewhat oval, and flat Lernea. 8. The larger, rounded, depreſſed Lernea. 9. The cingulated Lernea. 10. The oblong, thick, verrucoſe Lernea. 11. The ſhorter and ſmoother Lernea. 12. The Lernea, with a palmated tail. All theſe ſpecies I have met with on our own coaſts; many of them on the Suſſex, ſome on the Yorkſhire ſhores. Authors ſeem to have deſcribed ſome of them, but it is ſo very imperfectly, and uncertainly, that it is hard to aſcertain which particular species is meant, by any of their engravings or descriptions; they will be easily distinguished by these specific names, care being first taken to admit none of a wrong genus among them, but attention first paid to the number of the tentacula.

MEDUSA.

THE body of the Medusa is of an orbiculated figure and convex, and is of a gelatinous substance, and not hairy: the tentacula, or the place, which are in the place of them, are situated in the center of the under part of the animal: authors have described several of the species of this genus, under the names of Urticæ marinæ, and Pulmones marini.

Medusa orbiculi margine sedecies emarginato.
The Medusa, with the rim of the orbiculus emarginated in sixteen places.

The Sea Lungs.

This is a very singular and odd animal: it seems a mere lump, of a whitish semipellucid jelly, and is as easily broken and destroyed by a touch, as the common jellies brought to our tables: it's shape is rounded, rising into a convexity in the middle, where it is therefore thickest, and whence it becomes gradually thinner to the sides; on the under side it is plain, and on this there is visible a rough, or as it were echinated, circle, within which there run eight pairs of rays from the center toward the circumference; and from the center there arise also a number of curled appendages, which are sometimes reddish, but more usually whitish, and a vast multitude of slender filaments: the edge, or the circumference of the body, is regularly divided into eight portions, and each of these is emarginated, so that on the whole verge there are sixteen sinus's.

I have met with this species, in vast abundance, floating on the surface of the water about Sheppey island in Kent, and elsewhere on that coast; great quantities of it are destroyed, by being thrown on shore with the waves, whence it has no power of getting off again; and, in the open seas, many fish skim near the surface, and prey on them: all the writers on this subject have described an animal of this class, under the name of Pulmo Marinus, but so inaccurately that it is hard to say, if they meant this species; their figures also are as bad as their descriptions: in short, this class of animals, in general, are so very confusedly treated of by authors, that the only way to come at a knowledge of them is to disregard what has hitherto been said on the subject. Linnæus's generical distinctions are, indeed, accurate, and, as such, the characters are retained here.

Medusa orbiculo subtus quatuor cavitatibus notato.
The Medusa, with four cavities on the under surface.

The Sea Nettle.

This appears, as floating on the water, to be a mere lifeless lump of jelly: it is of a whitish colour, with a cast of bluish-grey, and is of an orbiculated figure, elevated into a convexity in the middle, on the upper side, flat on the under, and furnished with a fringe of fine and somewhat rigid filaments round the edge, resembling white hairs: on the under surface there are four cavities near the center, each of an arcuated figure, and surrounded with an opake line, formed of about twenty-four parallel points or dots: from the very center of the under side there arise four crooked appendages, which have each a row of hairy filaments on the exterior edge; and on the upper surface there is an appearance of fine vessels of a pale colour.

This species is frequent floating on the surface of the sea, or adhering to rocks about our own coasts; and, when the sun shines on them, they have a very beautiful lucid appearance. Authors have described it, but very inaccurately, under the name of Urtica Marina, or the Sea Nettle, a term used to express it's causing a disagreeable tingling in the hands, when taken up.

Medusa orbiculo cruce alba picto.
The Medusa, with the orbiculus marked with a white cross.

This is a very beautiful little species; it rarely exceeds an inch in diameter, or a third of an inch in thickness: it is a lump of a fine transparent colourless jelly; but, under the full sunshine, sometimes appears as it were on fire, and sometimes shews all the colours of the rainbow: it is quite even and undivided at the edges, and, on the center of the under surface, is beautifully variegated with a white cross, which reaches quite from one side to the other.

It is less frequent than the two former species, but is sometimes found about our coasts. I have met with it in Sussex.

The other species of the Medusa are, 1. The very large, flat Medusa, with eight double punctated lines. 2. The great, convex, and very thick Medusa. 3. The great Medusa, with only four sinuations at the edge, but those very deep. 4. The lesser smooth-edged Medusa. 5. The Medusa, with six emarginated lobes at the edge. 6. The mulтifid-edged Medusa. All these species I have met with about the Sussex and Kentish coasts: authors seem to have meant some of them by their figures and descriptions, but they have expressed themselves so inaccurately, that nothing can be determined with certainty from their accounts.

APHRODITA.

THE body is of an oval figure, and aculeated; and there is a perforation in the middle of the back.

Aphrodita elliptica versicolor.
The elliptic, changeable-coloured Aphrodita.

This is an extremely singular animal; at first sight, it is not easy to make a guess at what it is; it has greatly more the appearance of a fragment of some other body, than of a compleat animal. It is about two inches in length, and more than an inch in diameter in the middle, from whence it grows gradually a little smaller to each end; but is at both extremities obtuse, and the difference in breadth is so little between those and the middle, that it has, upon the whole, the appearance of a tolerably regular elliptic figure: it's back is convex and rounded; it's under part flat: it is of a soft and tender structure, and is covered all over with little tufts of short and slender filaments in manner of hairs; these are shortest on the middle, and longer gradually toward the edges; they are of a mixt purplish and yellowish colour, and in several lights appear of different degrees, between a gilded brown and those two colours, in the manner of the changeable silks. In the middle of the back there is an aperture or hole, surrounded with a tuft of these hairs; and, when the creature is pressed, a liquor, like sea water, is forced out at this aperture, and the whole body becomes flaccid and less convex.

I met with it in great abundance this last summer on the Kentish coast, among rocks: the accounts authors give of these bodies are so imperfect that it is not certain this species has been described by any of them.

Aphrodita subcylindrica variegata. The Sea
The subcylindric variegated Aphrodita. Mouse.

This is a slenderer species than the former, but is more than equal to it in length: the back is rounded; the belly is also somewhat prominent in the middle, not flat as in the other species: it's substance is somewhat firm, and it's two ends are nearly equal to the middle in thickness: it is covered with hairy filaments, which are short on the middle of the back, but longer at the sides; they are all somewhat rigid and firm; those on the back stand erect, like the quills of a porcupine; those on the sides lie flat, and are of a great variety of colours; a beautiful blue and lively green are very distinct in them, and a gold-yellow seems the most predominant colour; on the back they are of a duskier colour, and in many places of a greyish-brown.

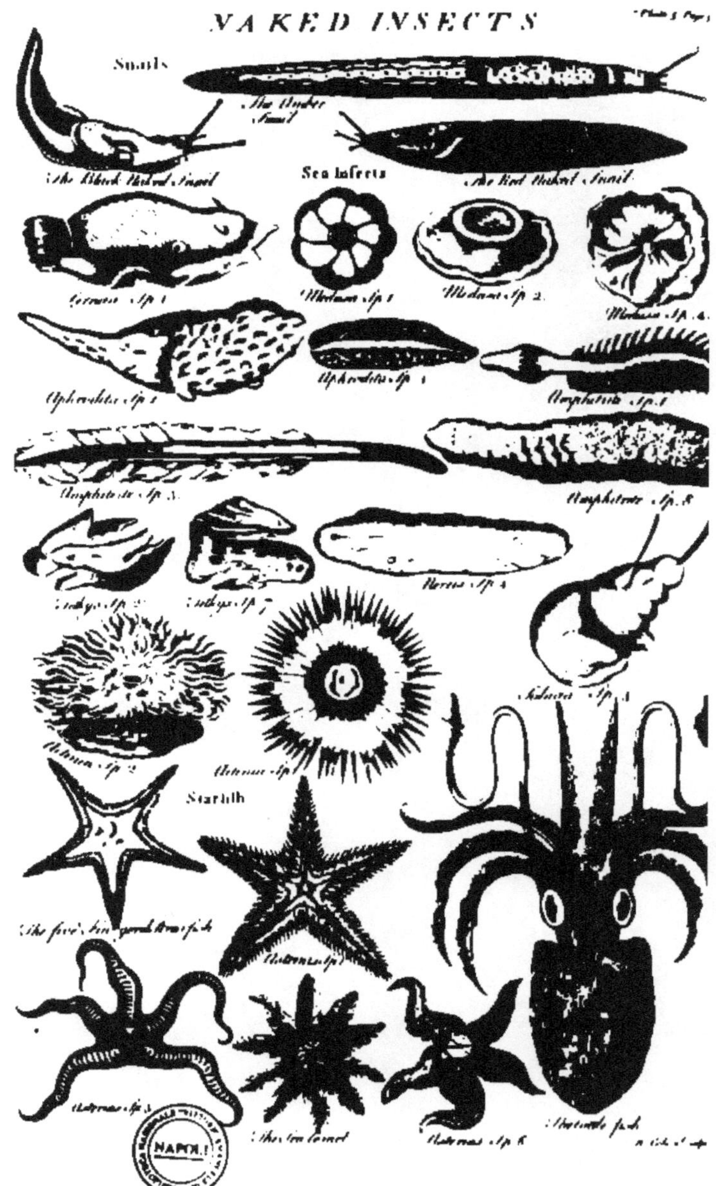
NAKED INSECTS
Snails
Sea Insects
Starfish

We have it about the coaſts of Suſſex, Kent, and Eſſex: moſt of the writers on theſe ſubjects have deſcribed it. Columna calls it Pudendum regale; Bartholine, Vermis aureus; others, Vermis aureus, and Eruca marina griſeo-fuſca; ſome Muſ Marinus, or the Sea Mouſe.

Aphrodita ſubrotunda.
The roundiſh Aphrodita.

This is a moderately large ſpecies: the ſhape is nearly round, the back very convex, and the belly flatted: the ſubſtance is tender and gelatinous: it is covered with a quantity of fine ſhort hairs, of a greyiſh, or bluiſh, colour, with an admixture of yellow; theſe are longeſt at the ſides, and have there more beautiful colours, than in any other part: in the center of the animal on the back, there is a ſmall aperture of an oval figure, out of which water is diſcharged in a ſmall quantity, on preſſing the creature.

It is found on the Kentiſh coaſt, but not frequently. I have not met with it elſewhere.

The other ſpecies are, 1. The large brown Aphrodita. 2. The large, depreſſed, greyiſh Aphrodita. 3. The oblong and ſubcylindric Aphrodita. 4. The ſmaller, bluiſh Aphrodita.

AMPHITRITE.

THE body of the Amphitrite is of an oblong figure, and has a great number of deep ſtriæ, giving the appearance of numbers of lamellæ: it has but one tentaculum, which is of a ſlender and oblong form, reſembling a piece of thread.

Amphitrite ſtriis numeroſis corpore latiore.
The broader bodied Amphitrite, with numerous ſtriæ.

This is a large ſpecies; it grows to an inch and a half in length, and to three quarters of an inch in diameter in the largeſt part: it's ſhape approaches to an obtuſely conic one; it's ſubſtance is tolerably firm; it's colour a bright pale blue, and in the ſun it is very bright and gloſſy: the whole ſurface is covered with a vaſt number of ſlight furrows; the ridges between, which form a kind of lamellæ, like thoſe under the head of a ſmall muſhroom: the tentaculum is near half an inch long, when exerted to it's full length, and is white and ſoft.

It is found in the ſalt water ponds, on the American iſlands, in great abundance: ſome of the old writers have called it a ſpecies of Urtica Marina.

Amphitrite ſubcylindrica ſtriis profundioribus undulatis.
The oblong Amphitrite, with deep and undulated ſtriæ.

This is ſometimes met with of the length and thickneſs of a man's finger, but often ſmaller; it is of a ſubcylindric figure, thickeſt toward the hinder part, but obtuſe at both ends: it's colour is a bluiſh, with a caſt of brown; it's ſubſtance tolerably firm: it has only a ſingle tentaculum, which is ſomewhat thick, and of an inch in length: the whole body is covered with longitudinal furrows, which do not run ſtraight from end to end, but wind about in many places, and, in moſt, are a little undulated.

I have met with it on ſome of the extream rocks of the Bognors, on the Suſſex coaſt, which are almoſt always under water: authors ſeem to have deſcribed it under the name of Epipetrum.

Amphitrite corpore marginato.
The Amphitrite, with a marginated body.

This is a very ſingular ſpecies; it is about two inches and a half in length, and in the broadeſt part, where it is marginated: it's colour is a pale greyiſh-white; it's ſubſtance ſoft and tender; for about a third part of it's length, it is rounded and naked, but the other two thirds are extended in breadth by a marginated edge each way, form-

ed

ed of a kind of skin, or thick membrane: the whole surface is covered with slight undulated furrows.

I have met with it on the Yorkshire coast, but not frequently. Some of the old writers have called it Penna marina.

The other species are, 1. The shorter, marginated Amphitrite. 2. The long and almost smooth Amphitrite. 3. The long, slender, and digitated Amphitrite. 4. The thick, conic Amphitrite, small at one extremity. 5. The pediculated, rough Amphitrite. 6. The small and rounded Amphitrite obtuse at both ends. 7. The large, obtuse, and very deeply furrowed Amphitrite. 8. The slightly furrowed Amphitrite, called Adamas marinus.

TETHYS.

THE body of the Tethys is formed, as it were, of two lips, with an oblong cartilaginous body between them: there are four tentacula, which have the form of ears, and two perforations in most species near the tentacula.

Tethys rugosa corpore longiore.
The oblong, rugose Tethys.

This grows to two inches and a half in length; it's substance is tough, but not hard; it's colour a dusky grey, with an admixture of brown: the divisions of the body are large and thick; the intermediate part of a cylindric figure, but buttoned, as it were, at the end: the tentacula are broad and short, and two of them are much shorter than the others.

We have it on the coasts of Cornwal, and in some other places. Authors have very confusedly described it under the name of Holothurium and Tethys, among several other bodies; some of them species of this genus, others of others, under the name of Tethyos variæ species.

Tethys verrucosa corpore breviori.
The verrucose Tethys, with a shorter body.

This is a small species: the divisions of the body are thick, somewhat flat, and of a firm texture; the intermediate part is of an oval figure, more firm and verrucose, or covered with small, obtuse protuberances; the colour of the whole is a bluish-grey: the tentacula are all four of equal length, and are slender and thin; the apertures are large.

This is frequent about the coasts of Italy, but is not met with on our own. It has been imperfectly figured and described by authors, among their various species of Tethys.

Tethys corpore ovato-glabro.
The smooth, oval-bodied Tethys.

This grows to about an inch and half in length, and more than an inch in diameter in the largest part: the divisions of the body are broad, short, and thick: the intermediate part is also thick, short, and rounded at the end: the tentacula are thick and obtuse, and the perforations very small: the surface of the whole is smooth, and the texture soft and tender.

This is found on our own coasts, but not frequently. I have met with it in Sussex.

The other species are, 1. The large, oblong, smooth Tethys. 2. The large, verrucose, thick Tethys. 3. The irregular-bodied Tethys. 4. The roundish-bodied, furrowed Tethys. 5. The short, hairy Tethys. 6. The pyriform Tethys. 7. The broad, flat Tethys. 8. The small, grey, hairy Tethys.

NEREIS.

THE body of the Nereis is of a cylindric figure; and the tentacula are four in number, but two of them are usually very short, often scarce perceptible.

Nereis

Nereis tenuior lævis.
The slender and smooth Nereis.

This is one of the smaller species: the length is about two inches, the thickness not more than that of a large earthworm; the substance is gelatinous, and the colour a bluish-white, when the sun shines on it very bright and glossy: the whole surface is smooth, and the belly a little flatted, otherwise the whole body is rounded; there are four tentacula, two of which are half an inch long, the other two not more than a third of an inch, and the latter are thicker than the others.

It is found in the American islands in salt-water ponds. Authors have mentioned what they call a Scolopendra marina lævis, which seems to be this species; though, what should induce them to give it the name of Scolopendra, is not easy to say.

Nereis subhirsuta.
The hairy Nereis.

This is about two inches and a half long, and about three quarters of an inch in diameter: it's substance is firm and tough, and it's colour a pale brown: it has many little clusters of hairs growing on the back, and yet more numerous ones at the sides; they have a kind of fleshy protuberances to which they grow, and, while the creature is in the water, they stand erect; at other times they are flaccid; the colour of the hairs is a mixt brown and yellow, and is very glossy.

It is found on our own coasts, and by some authors has been called also a Scolopendra marina, and that with somewhat more appearance of reason than the former, though in reality an animal of a quite distinct class.

Nereis tuberculosa.
The tuberculose Nereis.

This species grows to three or four inches in length, and, in the largest part, to about an inch in diameter: the substance is soft and slender, and the colour a mixt grey and brown; there are no hairs on it, but there are at several distances a kind of papillæ or fleshy tubercles, obtuse at their tops, and paler than the rest of the surface: the tentacula are four, but two of them are so short, as to be scarce distinguishable.

It is frequent on the Sussex coast, and has been figured by some of the writers on these subjects.

The other species are, 1. The small, smooth, thick Nereis. 2. The tuberculose Nereis, with the tubercles pointed at the ends. 3. The wrinkled Nereis. 4. The large, soft Nereis.

SALACIA.

THE body of the Salacia is of an ovato-oblong form, and the tentacula are numerous, and disposed in little clusters.

Salacia superficie undulata.
The Salacia, with an undulated surface.

This is a tolerably large species; it's length is an inch and three quarters, and it's thickness about an inch: it is largest at the naked extremity, where it terminates in an obtuse and rounded form; from this it grows gradually, but not regularly, smaller to the other extremity, where it terminates in a rounded, but not very thick, end: the substance is tolerably firm; the whole surface is undulated and uneven; the colour is a brownish-grey; there are three clusters of tentacula toward the smaller extremity. On pressing this, and several of the other species of this and many of the preceding genera, there is a thin fluid discharged from their natural aperture, after which, being less full, they appear flaccid, till replenished again: this has occasioned those of the oblong form, in general, to be confounded by authors, under the whimsical names of Priapus marinus, and Mentula marina.

Salacia oblonga glabra.
The smooth, oblong Salacia.

This is about two inches and a half long, and an inch and a half in diameter: the surface is perfectly smooth; the colour a very pale grey, and the substance soft and tender: it is larger at one extremity than at the other, but obtuse and rounded at both; it's whole surface is smooth and glossy, and, when the sun shines on it, is very bright: toward the smaller extremity there are several clusters of tentacula.

It is common on the coast of Portugal.

The other species of the Salacia are, 1. The small, furrowed Salacia. 2. The small, punctated Salacia, with short tentacula. 3. The larger, thick Salacia, with numerous tentacula. 4. The great, gelatinous Salacia.

ACTINIA.

THE body of the Actinia is of a naturally cylindric, but variable, figure: the tentacula are very numerous, and are arranged in several series about the mouth, which is placed at one of the extremities of the body.

Actinia tentaculis versicoloribus, cauda tricuspidi.
The Actinia, with changeable-coloured rays, and a tricuspidate tail.

This is an extremely elegant species: it lodges itself in little cavities of rocks, and of the larger sea-plants of the stony kind, and only appears on their surface, when all is quiet about it: it's body is naturally of an oblong figure, and equally thick in all parts, and seems, in a moderate state of extension, to be about half an inch long; but the creature, at pleasure, contracts it almost to nothing, and extends it to a great length: it's tail is divided into three parts, or, as it were, terminated by three points; these it fixes into some cracks of the body it inhabits, and by that means fixes itself very firmly in it's place. When all is quiet about it, it advances it's other extremity, at which is the mouth, to the surface of the stone, and displays all round it a great number of tentacula, formed like so many rays, and disposed in two or three series successively shorter than each other; these are in a continual, vibratory motion, and by that means draw small animals that inhabit the water into it's mouth for food: the whole animal is of a pale flesh colour, except these tentacula, which are of a conic figure, and have a beautiful variety of colours, red, yellow, and blue, and many others: when they are fully expanded, they give much of the appearance of some of the flower of the compound, flowered plants.

It is not uncommon on the coasts of the American islands, and in the salt-water ponds on several of them.

Actinia corpore ventricoso, cauda simplici, tentaculis planis.
The Actinia, with a ventricose body, a simple tail, and plane tentacula.

This is a larger, and, if possible, a more elegant species than the former: it's body, in what seems it's state of greatest rest, is oblong, rounded, and somewhat bellied in the middle: it has six ridges running longitudinally at equal distances, from end to end of it, and the tail is simple; between these ridges there are several rounded, depressed spots, of a paler red than the rest; the ridges also are pale, the rest of the body of a strong flesh colour. It buries itself, in the manner of the former, in a cavity of some rock, and at the surface displays it's tentacula in three or four distinct series round the mouth, which is a large opening, of an oval figure, at the extremity of the body: the tentacula are successively shorter from the outer circle to the inner one; they are all of a pyramidal figure, thin and smooth, and tinged with a beautiful variety of colours.

It is found on the coasts of America and of the East Indies, in the cavities of rocks.

This,

This, or some other species of this genus, has been very pompously described by Mr Hughes, in the Philosophical Transactions, and in his History of Barbadoes, under the name of the *Sea Anemone*; but, from his unacquaintance with the characters and distinctions of the animals of this class, it is not easy to say, with certainty, which it is that he means.

Actinia tentaculis capillaceis.
The Actinia, with capillaceous tentacula.

This is a very small species: it's body is oblong, slender, and whitish, seldom exceeding a third of an inch in length: it's tail is simple; it's tentacula are numerous, but all of the same length, and disposed in a double series; they are very narrow, and are in a continual, vibratory motion.

It is frequent in the regular, stellar cavities of the madrepora, and in those of several of the other stony sea-plants, as also in the holes of rocks and small stones. There are not wanting authors of great credit, who assert that the madrepora, the corals, and the generality of what we call the sea-plants, are not vegetables, but the work of Insects of this kind, and of other of these genera, the Nereides, Polypes, and Medusæ. It is very evident, that these Insects are found frequently in the cavities of those plants, but they are also as frequently found in those of rocks and pebbles; and it would be as just to suppose those greater and smaller irregular masses of stony matter fabricated by them, as these regular and beautiful plants.

The other species of this genus are, 1. The larger Actinia, with numerous, variegated, conic rays, and a smooth body. 2. The great, white, simple, coloured, furrowed-bodied Actinia. 3. The little, white Actinia of the stony plants, with few tentacula. 4. The little, flesh-coloured Actinia, with very numerous tentacula.

BIOTA.

THE body of the Biota is of a cylindric figure, but variable: the tentacula are arranged in a single series, round the aperture of the mouth, at the extremity of the body.

The generality of authors have called the species of this genus Polypes, a very absurd and improper name for an animal, which has no legs, the sense of it being many legs, or many feet. Linnæus has called the genus Hydra, I suppose from the repullulation or reproduction of the parts, when cut off, which put him in mind of the heads of the imaginary serpent so called; but there seems an affectation and quaintness in this, which the pure simplicity of natural history does not allow. The principle of life, in the several parts of this creature, any fragment of which will become an entire animal in a few days, is, indeed, very remarkably strong; and I have rather chosen to express it by the generical name, by the simple word, Biota, from the Greek Bios, Life, than by so affected a term as Hydra.

Biota corpore subcylindrico, tentaculis longissimis.
The Biota, with a subcylindric body, and very long tentacula.

The Polype.

This is the species of this genus: the surprizing properties of which were discovered by Monsieur Trembley, and which, having been first known and kept alive in glasses among us, obtained, by way of eminence, the name of *the Polype*, the other species being distinguished by their different characters.

There is scarce an animal in the world, which it must be more difficult to describe, than this, as it has scarce any thing constant in it's form: it varies it's whole figure at pleasure, and is frequently found beset with young in such a manner, as to appear ramose and divaricated; these young ones adhering to it in such a manner, as to appear parts of it's body.

When simple and in a moderate state, as to contraction, or dilatation, it is an oblong animal, slender, pellucid, and of a pale reddish colour: it's body is somewhat smaller toward the tail, by which it affixes itself to some solid body, and larger toward the other

other extremity: at the larger extremity there is a large opening, which is the mouth, and round about this are placed the tentacula; they are eight in number, and are uſually extended to about half the length of the body; the creature can at pleaſure contract it's body into a much ſmaller compaſs, or diſtend it into a larger; it can reduce it to a ſhort button, as it were, or extend it a much more conſiderable length; but the tentacula are yet more diſtenſible. I have had ſome of them, that have extended them to ſeven inches in length, in which ſtate they become of an almoſt unconceivable fineneſs; and the dimenſions of the glaſs I kept the creature in, ſeemed at that time to ſet limits to the extenſion, which nature had not: by means of eight of theſe tentacula, or arms, as they are commonly called, thus expanded into a circle of more than a foot diameter, the creature feels every thing that can ſerve it as food, that ſtirs in that portion of the water in which it lives, and, ſeizing the prey with one of them, calls in the aſſiſtance of the reſt, if neceſſary, to conduct it to it's mouth.

There does not appear to be any thing analogous to copulation in this animal, in order to the producing it's ſpecies. A ſingle one, put by itſelf into a glaſs of water, will, after a little time, produce from it's ſides, in different places, ſeveral young ones; and theſe, while they yet adhere to the parent animal, will alſo have others produced from their ſides; ſo that it is not uncommon to ſee a parent animal loaded with ten, or even twenty young ones, and their progeny, before any drop off.

What is however the moſt ſingular property, in the increaſe of this Inſect, is the reproduction of the mutilated parts, and indeed the growth of the whole from any ſegment; not only the tentacula, or arms, are reproduced after being cut off, but the tail, or the head, with the whole apparatus of the tentacula, will be reproduced on the body after cutting off; and the part cut off will alſo reproduce the body; in fine, there is no part of the Inſect, which, on being ſeparated from the reſt, will not reproduce all that is wanting to make it a perfect animal; ſo that the ſhort way to have a number from one Polype is to cut it into ſeveral pieces, each of which will ſoon be a perfect animal, and will produce it's young ones in the method already mentioned.

This ſpecies is frequent in Holland; we have it alſo in our ditches, but leſs frequently than ſome of the other kinds. It's food is the common aſcaris, or ſmall red worm, abundant in the mud of the Thames, and moſt of our ponds and rivers.

Biota corpore craſſiore tentaculis brevioribus.
The thicker-bodied Biota, with ſhorter tentacula.

This is a ſmaller ſpecies than the former, and, in it's ſtate of moderate contraction, is ſhorter and thicker in proportion than that: it's uſual length is half an inch; it's head is larger than in the former; and it's colour, a duſky greeniſh, or olive: it's tentacula are uſually not more than a third of the length of it's body; it can indeed extend them to ſeveral inches, but not fully to the dimenſions of the other.

This ſpecies is very common with us in the little drains, cut through our meadows for carrying off the water: it adheres by the tail to ſticks, and to the ſtalks of the water plants: it has all the properties of the former.

Biota tenuior albeſcens tentaculis capillaceis.
The ſlender, whitiſh Biota, with capillaceous tentacula.

This, in a ſtate of moderate contraction, is equal to the firſt ſpecies in length, but it is very ſlender, and all the way of the ſame thickneſs: the colour is white: the tentacula are extremely minute; they are alſo white, and may be diſtended to a great length.

It is not unfrequent in our trout rivers, adhering to the ſtalks of the water plants. It feeds on ſmall worms, and other Inſects, and has all the qualities and properties of the others.

Biota corpore brevi, tentaculis brevibus.
The Biota, with a ſhort body, and ſhort tentacula.

This is a very ſmall ſpecies: it's body is ſhort, and ſomewhat thick in proportion to it's length, and is white: the tentacula are very fine and very ſhort; they are diſtenſible to a ſmall degree, but not at all in the manner of the others.

It is frequent in the natural cavities of that species of the alcyonium, commonly called Fucus tubam lineam sericeamve texturam æmulans, and in some other sea plants: Linnæus, and other of the modern systematists, have gone so far as to suppose those plants not to be truly vegetables, but cases formed by these Insects for their own lodging.

The other species of the Biota are, 1. The little Biota of the Corallines. 2. The large Biota of the red Coral. 3. The pale, slender Biota, of the several Corallines. 4. The small, deep, red Biota, of the Tubularia. 5. The great oval-bodied sea Biota. 6. The great, round-bodied, sea Biota.

SEPIA.

THE body of the Sepia is of an oblong figure, and depressed: there are ten tentacula; two of which are longer than the others, and are pedunculated:

Sepia corpore ovato oblongo.
The Sepia, with an oblong, oval body.

The Ink-fish, or Cuttle-fish.

This is a large species: it's body is often six inches in length, and three and a half in diameter, though the greater number are found smaller: the body is of a somewhat oval, but oblong, form: it is broadest toward the head, and grows smaller to the extremity, where it is obtusely pointed: it is supported by an oblong, light, and spungy substance, of a friable texture, and lined with a light fungous pith; this is what our silversmiths use under the name of Cuttle-bone; and it is also received in the shops as an ingredient in tooth-powders: the head of the animal is large, and somewhat compressed; and from it rise the ten tentacula: eight of them are of a pyramidal figure, and somewhat more than equal to the body in length: the other two are much longer; they consist of a thick pedicle, terminated at the extremity by an oval body, considerably larger than any part of the other: the inner surface of these oval bodies, and that of the other tentacula along their whole course, are furnished with a number of hollow tubercles, or protuberances, formed for seizing and holding things fast.

It is frequent in the European seas, but is not common about our coasts: when in danger of being taken, it emits a black liquor like ink out at it's mouth in considerable quantity, which obscures the water about it, and gives it an opportunity of escaping: all the writers on these subjects have described it under the names of Sepia and Sepia vulgaris.

Sepia Tentaculis pedunculatis longioribus, corpore angulato.
The angular-bodied Sepia, with long pedunculated tentacula.

This is a larger species than the former: it's body is often eight inches in length, and four and a half in diameter in the broadest part; and the long, or pedunculated, tentacula are, in that case, ten inches in length: the body is oblong and angulated: it is narrow toward the upper extremity; from thence it becomes broad near the bottom; and, from this broadest part, continues decreasing to a point at the base: there runs a high and sharp ridge all down it's middle, and the sides are somewhat sharp: the head is small in proportion to the body; the eight common tentacula are of a pyramidal form, but very narrow in proportion to their length; and the other two are terminated by oval bodies, as in the other species: the insides of these, and of the tentacula also, as in the other, are furnished with a number of tubercles, hollowed at their tops.

It is frequent in the warmer parts of Europe, but rare with us: authors call it Loligo.

The other species are, 1. The long and slender-bodied Sepia, with eight very short and two very long tentacula. 2. The short and roundish-bodied Sepia, with two of the tentacula very long.

TRITON.

THE body of the Triton is oblong: the rostrum at the mouth is of a spiral form: the tentacula are fourteen in number, and twelve of them are cheliferous.

Triton corpore subovato, tentaculis membrana ad basin connexis.
The Triton, with a subovate body, and with the tentacula connected by a membrane at the base.

This is a very singular animal: it is of about half the size of the common Sepia, it's body seldom exceeding three inches in length: it is of an oval figure, and somewhat compressed: it's base is the larger part of the oval, from whence it gradually becomes smaller to the head: the head is oblong and rounded, not compressed; and there rises from it a thick and tough membrane, which connects all the tentacula at their bases, in the manner of the web between the toes of the feet of our water fowl; two of the tentacula are simple, oblong, and of a conic figure: the other twelve are cheliform at their extremities.

I have been favoured with two specimens of it from Italy; elsewhere I have not heard of it.

Triton corpore graciliore.
The slender-bodied Triton.

This does not exceed four inches in the length of the body, and is not more than two in diameter in the broadest part: it is largest toward the head, and obtusely pointed at the other extremity; the head is small; the tentacula all short and slender; the two simple ones are not more than an inch in length: the other twelve do not much exceed two inches, but they are beautifully cheliform at their extremities.

This also is found on the coast of Italy.

Triton tentaculis longissimis.
The Triton, with very long tentacula.

This is a smaller species than the last; the body is oval, and obtuse at the extremity: the head large and inflated, and the rostrum beautifully spiral: the two simple tentacula are about an inch and a half in length; the twelve others are at least four inches, and are cheliform at their extremities.

This also is found on the shores of Italy.

ASTERIAS.

THE form of the Asterias is that of a radiated star: the mouth is situated in the center on the under part; and the anus in the center on the upper part: the tentacula are extremely numerous, and in a manner cover, either the whole upper surface of the body, or the extremities of the ramifications.

Asterias radiis quinis latiusculis asperis.
The Asterias, with five, broad, rough rays.

The five-fingered Star-fish.

This is one of the most frequent of the Asterias kind: it is usually met with of about five inches in measure, from the tip of one ray to that of the opposite one: it's body, as the central part is usually called to distinguish it from the rays, is small and somewhat elevated on the upper part; the rays are of the thickness of a man's finger at the base, and grow gradually smaller to the extremity; the whole is of a pale whitish-colour, with an admixture of brown, and a faint cast of red; the whole upper surface, both of the central part, and of the rays, is covered with short tentacula.

It is common about our own coasts, and has been described by all that have written on these subjects, under the names of Stella marina vulgaris, and Stella marina Plinii.

Aſterias radiis tredecim obtuſis.
The Aſterias, with thirteen obtuſe rays.

This is a very ſingular ſpecies: it has ſcarce any diſtinguiſhable central part or body; the rays ſtand ſo cloſe, that they cohere for more than half their length, only ſeparating toward their points; they are thick, ſhort, and obtuſe; the diameter of the whole animal is about two inches and a half, and it's colour a reddiſh-brown: it is very tender and friable, and the whole upper ſurface is covered with ſhort tentacula.

It is not uncommon on the coaſts of the Eaſt Indies, and ſome other places.

Aſterias radiis quinque tenuioribus.
The Aſterias, with five ſlender rays.

This is a ſmall ſpecies, the rays ſeldom exceeding ſingly an inch and a quarter in length: the body is ſmall, the rays are of a conic figure; they are pointed at the extremity, and are all over covered on the upper ſurface, with fine ſhort and ſlender tentacula: it is of a browniſh-white colour, and very tender and friable.

It is frequent about the coaſts of Italy, and in moſt of the ſeas that waſh the warmer parts of Europe.

The other ſpecies of Aſterias are, 1. The broad-bodied, ſhort-rayed, five-rayed Aſterias. 2. The large, obtuſe, and ſhort-rayed Aſterias. 3. The broad and undulated-rayed Aſterias, with thick and pointed tentacula. 4. The broad, ſhort, and flat-rayed, five-rayed Aſterias, with ſhort, hair-like tentacula. 5. The Aſterias, with five ſhort, obtuſe, elliptic rays. 6. The Aſterias, with five, long, ſlender, and undulated, hairy rays. 7. The Aſterias, with five, broad, hairy, undulated rays. 8. The Aſterias, with five, narrow, thick rays, with the tentacula as it were prickly. 9. The Aſterias, with five, thick, narrow, and very ſharp-pointed rays. 10. The Aſterias, with five, broad, hollowed, and pointed rays. 11. The Aſterias, with five deeply furrowed rays. 12. The Aſterias, with twelve ſhort-pointed rays. 13. The Aſterias, with ten hairy and undulated rays. 14. The Aſterias, with ſeven narrow and long rays. 15. The Aſterias, with ſeven broader and tuberculoſe rays. 16. The Aſterias, with three long, ſlender rays. 17. The Aſterias, with numerous, hollowed rays. 18. The Aſterias, called the Sea Comet by authors. 19. The lumbrical Aſterias. 20. The narrow and hairy-armed Aſterias. 21. The great Aſterias, with ramoſe rays, called the Magellanic ſtar-fiſh, and the basket-fiſh: the extremities of the rays of this ſpecies are ſubdivided to an almoſt inconceivable fineneſs; and the creature, when it extends them fully, forms a circle of near three feet in diameter: the fragments of the rays of this fiſh furniſh the foſſile entrochi.

THE

It is frequent in New Spain, and has been met with about the settlements of the Dutch at Surinam.

Cæcilia rugis profundioribus nonaginta et octo.
The Cæcilia, with ninety-eight deeper rugæ.

This is a small species, but is very singular in it's figure: it's body is somewhat flatted; the back appearing depressed, the belly but little prominent, and the sides rounded; the head is also depressed, but is short, and almost rounded; the upper lip hangs over the jaws all the way round the opening of the mouth, and has two short, somewhat thick, and truncated tentacula growing to it: the opposite extremity of the animal does not go off into a tail, but is, as it were, truncated and obtuse: the colour of the whole animal is a pale brown, with some slight admixture of tawny in it: the belly is a little paler than the back: there are no scales, nor any other armature on it, only a rough skin, which is tolerably smooth on the back and belly, but has a number of deep, oblique furrows, forming so many wrinkles at the sides; it moves but languidly; it's food is worms, and other small insects, and young frogs.

It is frequent in many parts of New Spain.

Cæcilia rugis numerosissimis, capite angustiore.
The Cæcilia, with very numerous rugæ, and a narrow head.

This grows to about fifteen inches in length: the head is small, narrow, and somewhat depressed; the upper lip is prominent beyond the rest of the mouth, and forms at the extremity a kind of point: the tentacula are very short, but thick and truncated: the body is rounded; the back very convex, and the belly flatted but little: the hinder extremity is truncated, and gives no appearance at all of a tail: the colour of the whole animal is a dusky livid-blue, with an admixture of grey and of black; it is palest on the underside: the back and belly are almost smooth, but the sides have a vast number of slight wrinkles, which run almost straight up toward the back; they are deepest and plainest toward the middle, and grow so faint toward the extremities, particularly toward the head, that they are not easily counted. I counted more than an hundred and forty of them, on a specimen preserved in spirits, the only one I have seen. I purchased this among the others of the late Duke of Richmond's Museum, and remember it's having been sent, about six years ago, to his Grace from Georgia.

There is but one other known species, the black, short Cæcilia, with eighty-two rugæ.

AMPHISBÆNA.

THE body of the Amphisbæna has a number of circular annuli, surrounding it from the head to the extremity of the tail; so that it seems composed of a number of narrow and somewhat rounded rings, applied close to one another, and having deep furrows between them.

Amphisbæna annulis corporis ducentis et viginti sex, caudæ sedecim.
The Amphisbæna, with two hundred and twenty-six annules on the body, and sixteen on the tail.

This is a very singular and extraordinary Serpent, and has at first sight much of the appearance of a monstrous earth-worm: it's length, when full grown, is two feet and a half; it's thickness is very considerable, the diameter being more than an inch: the body is almost round, only that the belly is a little flatted, as is also the back, but very slightly, as it approaches the tail; the sides are as it were inflated: the head is obscurely triangular, broad, flatted at the base, and continued thence to a somewhat obtuse point at the extremity: the mouth is large, and the upper jaw somewhat hangs over the other; the whole head is covered with a firm, whitish skin, with a slight tinge of a yellowish-brown, and has several irregular, transverse, and oblique furrows on it both above and below: immediately behind the head begin the annules or rings of the body; but the two or three first of them are obliterated on the upper surface, so that they are most distinctly counted on the belly: the colour is a very pale brown, with a faint admixture of a reddish tawny on the back: the sides are paler, and the belly is very pale, and almost whitish. The whole body is formed, as it were, of rings connected sideways to one another; they are rounded, and about equal to the diameter of a straw; they have nothing of a scaly appearance, but are covered with a firm and tough skin, and their surface is smooth; but they have oblique furrows all round them, at about a quarter of an inch distance, and cruciform ones at the sides: the aperture of the anus is very large, and is placed within less than two inches of the extremity of the body; from this to the extremity of the tail there are sixteen rings: the tail does not terminate in a point, but has an obtuse extremity, much resembling a head, but smaller and less pointed than the real head: it is of the bigness of the end of a large finger, and has none of the rugæ or wrinkles on it that cover the rest of the body, but is covered with the same kind of firm skin that invests the head, but on this part it has no wrinkles or furrows of any kind.

The specimen, from which I have formed this description, I met with among the other Serpents which I purchased at the sale of the late Duke of Richmond's Museum. It is a native of Surinam.

Amphisbæna annulis corporis centum et nonaginta, caudæ viginti duo.
The Amphisbæna, with a hundred and ninety rings on the body, and twenty-two on the tail.

This is a Serpent of an extreamly different form from the former; it's length is more than two feet, and it's thickness not more than that of a man's finger: the back and belly are both somewhat flatted; the belly very much so, but the sides are round and inflated, as in the other: the colour is a deep, disagreeable brown, with an admixture of orange colour toward the sides: the belly is of a pale brown; the head is large, flatted, and obtuse; it is covered with a very firm membrane, on which there are a few irregular furrows: the tail is thick and obtuse, and it's extremity is of the bigness and form of the head, and is covered with the same kind of skin, wrinkled in the same manner: they are both of a colour, much paler than the rest of the body; and, as the creature lies at it's length, it is not easy to say which is the head, which the tail: the whole body is composed of annules or rings, as it were, joined side to side; they are nearly of the breadth of a straw, and are somewhat flatted.

It is a native of Peru and Mexico, and of some parts of the East Indies; in which last part of the world it has obtained, among the English, the name of the two-headed worm.

Amphisbæna annulis corporis centum et octoginta duo, caudæ triginta.
The Amphisbæna, with the annules of the body a hundred and eighty-two, and those of the tail thirty.

This grows to more than two feet in length, and to the thickness of a man's finger: it's body is rounded, the back very convex, and the belly scarce at all flated; the sides inflated and prominent: the colour is a deep, livid, greyish-blue, with some cloudings and variegations of a deep black; the belly is bluer and paler than the rest: the head is of a figure approaching to the half of an elliptic; it is oblong, but obtuse and rounded at the extremity: the upper jaw all the way over hangs the lower, and the opening of the mouth is very large; the whole head is covered with a smooth and glossy membrane, having the appearance of a thick parchment, with a few oblique and not very deep furrows on it: the rings of which the body is composed are rounded, and of the breadth of a small straw; they have oblique lines, at small distances on them, and cruciform ones all down the sides: the extremity of the tail is somewhat smaller, and more rounded than the head, but it much resembles it in shape; it is obtuse, and is covered with the same smooth skin that invests the head, but it has no wrinkles on it.

This species is a native of Surinam, and of some other of the warmer parts of America. I met with it among the Duke of Richmond's collection; his Grace had it from Carolina.

The other known species of the Amphisbæna are, 1. The great, brown Amphisbæna, with a flat head. 2. The lesser, brown Amphisbæna, with the head and tail both rounded, and extreamly alike in form. 3. The slender, bluish Amphisbæna.

4. The

4. The slender, variegated Amphisbæna, with a small head. 5. The thicker, variegated Amphisbæna. 6. The great, livid, and black Amphisbæna. 7. The larger, smooth, and flat-headed Amphisbæna. Most of these are the produce of the warmer parts of America, and have not yet been described or mentioned by any of the authors, who have written on these subjects.

ANGUIS.

THE under parts of the body and of the tail of the anguis are both covered all over with squammæ or scales, without any scuta.

Anguis squammis abdominalibus centum & triginta quinque, caudalibus æquinumeris.
The Anguis, with the squammæ of the abdomen a hundred and thirty-five, and those of the tail the same number.

The Slow-Worm.

This is a small species, and, of all the European Serpents, has least the appearance of one: it grows to about a foot in length, and to the thickness of a man's little finger, or hardly so much. It's colour is a deep, dusky, greyish-brown on the back, and livid on the belly: the scales are small and compact; the head is small, and of a form approaching to triangular, but obtusely pointed, somewhat flatted, and of a paler colour than the body: the tail is obtuse; the opening of the mouth is but small; the squammæ of the abdomen are small but rigid, and the anus is at a very considerable distance from the extremity of the tail.

It is frequent with us in gardens and pastures, where it lives principally under ground feeding on worms: all the writers on animals have described it. Ray calls it Cæcilia, Typhlinus Græcis; and others, Typhlus and Cæcilia.

Anguis squammis abdominalibus centum & viginti, caudalibus octodecim.
The Anguis, with the squammæ of the abdomen a hundred and twenty, those of the tail eighteen.

This is a small species: it grows to about fourteen inches long, and to the thickness of a man's little finger: the colour is a dusky bluish-grey on the back, and a deep, shining, iron-grey on the belly, the head is large and flatted; the opening of the mouth wide, and the whole head covered with a grey skin, of the firmness of parchment, with a few oblique and irregular furrows on it: the tail is small, and terminates in a point: the whole surface of the back, is covered with very small and compact scales: the anus is situated greatly nearer the tail than in the former species, and is very large.

This is a native of the warmer parts of America.

Anguis squammis abdominalibus ducentis quinque, caudalibus triginta octo.
The Anguis, with the squammæ of the abdomen two-hundred and five, and those of the tail thirty-eight.

This grows to two feet in length, but is slender, seldom much exceeding the thickness of a man's little finger: it's colour is a dusky-grey, with some variegations of a paler colour on the back; and the belly is of a deep, or almost black, hue, very bright and glossy: the head is large, and of a kind of oval form, obtusely pointed at the extremity, and carried in form of a segment of a circle to the beginning of the back: it is elevated and convex, especially on the hinder part, and is covered all over with a tough and firm skin, of a bluish-grey colour, somewhat glossy, and marked with a few furrows, dividing it into spaces of an irregular figure: the tail is smaller than other part of the body, but it terminates obtusely, not in a point: the scales are small, of an oval figure, and fit very close and firm together.

It is frequent on the island of Borneo, and in some other parts of the East Indies.

The other species are, 1. The little brown Anguis, with a pointed tail. 2. The little grey Anguis, with a very small mouth, and an obtuse tail. 3. The larger, variegated

riegated Anguis, with obtuse, elliptic scales. 4. The grey and yellow Anguis, with larger scales. 5. The small scalled Anguis, called Scytale. 6. The long and slender, variegated Anguis, with a broad, flat head. 7. The large, variegated Anguis, with the head very convex, and pointed at the extremity. 8. The larger, brown Anguis, with a black belly. 9. The larger grey and white Anguis, with an obtuse tail. 10. The little grey Anguis, with the back and belly of the same colour. 11. The grey, American Anguis, with very small scales.

COLUBER.

THE abdomen, or under part of the body, of the Coluber is covered with scuta, and the under part of the tail with squammæ or scales.

Coluber scutis abdominalibus centum & septuaginta septem, squammis caudæ octaginta quinque.

The Coluber, with the scuta of the abdomen a hundred and seventy-seven, and the squammæ of the tail eighty-five.

The Necklace-snake.

This is a very singular and very beautiful species: it's back is of a plane, simple, blackish colour: it's belly of the same hue, but with an admixture of bluish and more glossy: the sides of the belly are very elegantly variegated with regular series of white spots, which arises from the scales being white at their edges in that part: the head is large and somewhat flatted: the opening of the mouth is large, and the teeth are arranged in a double series, but they are all of a length: the throat is of a bright white; the sides of the upper part of the head also are white, and have several beautiful black lines on them: behind the head, there is a fine chain, as it were, of a bright and beautiful yellow; this gives it the appearance of an ornament round it's neck, and occasioned the English name of the Necklace-snake, and the Latin one of Torquata: the scales of the back have each of them an elevated stria, or ridge, along the middle: the nostrils are not prominent: it grows to three feet in length, and to a moderate thickness in proportion, but it varies sometimes very considerably in the colouring: the back, in some, is entirely black; in others, the scales which cover it are grey, and some of them only have black extremities, or tips: in some also, in the place of the circle behind the head, that part is convex and black, and has a large white spot in it.

It is frequent with us, especially about waters; it swims extreamly well, and, though it frequently does this with it's head above water, yet it will, at pleasure, go altogether under the surface, and twist itself at leisure about the stalks of the water plants, where it will remain a long time. Most of the writers on these subjects, have described it: Ray calls it Natrix torquata; others simply Serpens, Anguis, or Coluber.

Coluber scutis abdominalibus centum & quadraginta quinque, squammis caudæ triginta sex.

The Coluber, with the scuta of the abdomen a hundred and forty-five, and the squammæ of the tail thirty-six.

The Viper.

This is the most poisonous and mischievous in it's bite, of all the European Serpents: it grows to near three feet in length, and to a considerable thickness in proportion: the principal, or ground, colour of the body is a dusky grey; all along the back, there runs a broad brown line, which is dentated on each side; and, on each side of this, there is a kind of continued bluish line, formed of a series of spots of that colour, one of which is situated in the space formed by every denticulation in the back line: the belly is of a bluish-black, very bright and glossy, resembling the colour of high-polished, sanguined steel; and, when closely examined, there is found a small dot of a deep black at the apex of every scale: the head is large and flattish; the throat is of a pale colour, and the mouth is large, and the edge of the upper lip is whitish: the iris of the eye is of a flame colour; the pupil black; and there is a blue space, forming an acute angle, which separates the head from the longitudinal line on the back; these are the general characters of the viper, but it's colours vary so extreamly, that there is no determining any thing with certainty from them; there often is a cast of greenish in the grey, that is the

the general colour, and sometimes the whole body is of the colour of the belly, a fine, glossy, purplish-blue, very deep, and approaching to black.

It is frequent under warm hedges, and on heaths: common sallad-oil, applied externally, and at the same time swallowed internally, has been found to be a remedy for the mischiefs occasioned by it's bite; but, before this was publickly known, we have seldom met with any fatal consequences from it's bite, tho' country labourers have often suffered it.

The other species of the Coluber are very numerous: they are, 1. The Coluber, with the scuta of the abdomen two-hundred and fifty, and the squammæ of the tail thirty-five, called the leeanifcata. 2. The Coluber, with the scuta of the abdomen two-hundred and nineteen, and the squammæ of the tail one-hundred and ten. 3. The *Cracoatl*, or Coluber, with the scuta of the abdomen two-hundred and twenty, and the squammæ of the tail one-hundred and twenty-four. 4. The *Aparbyraatl*, or Coluber, with the scuta of the abdomen two-hundred and seventeen, and the squammæ of the tail one-hundred and eight. 5. The *Prtela*, or Coluber, with the scuta of the abdomen two-hundred and nine, and the squammæ of the tail ninety. 6. The Coluber, with the scuta of the abdomen one-hundred and ninety-six, and the squammæ of the tail sixty-seven. 7. The *Nela*, or Coluber, with the scuta of the abdomen one-hundred and ninety-tree, and the squammæ of the tail sixty. 8. The *Annulata*, or Coluber, with the scuta of the abdomen a hundred and ninety, and the squammæ of the tail ninety-eight. 9. The Anguis Æsculapii, with the scuta of the abdomen one-hundred and ninety, and the squammæ of the tail forty-two. 10. The Coluber, with the scuta of the abdomen one-hundred and eighty-four, and the squammæ of the tail fifty. 11. The *Sibon*, or Coluber, with the scuta of the abdomen one-hundred and eighty, and the squammæ of the tail eighty-five. 12. The Coluber, with the scuta of the abdomen one-hundred and sixty-five, and the squammæ of the tail twenty-four. 13. The *Albertalla*, or Coluber, with the scuta of the abdomen one-hundred and sixty-four, and the squammæ of the tail one-hundred and fifty. 14. The *Hippo*, or Coluber, with the scuta of the abdomen one-hundred and sixty, and the squammæ of the tail one-hundred. 15. The Coluber, with the scuta of the abdomen one-hundred and fifty-five, and the squammæ of the tail ninety-four. 16. The Coluber, with the scuta of the abdomen one-hundred and fifty-two, and the squammæ of the tail one-hundred and thirty-nine. 17. The *Calella*, or Coluber, with the scuta of the abdomen one-hundred and fifty, and the squammæ of the tail fifty-four. 18. The *Æsping*, or Coluber, with the scuta of the abdomen one-hundred and fifty, and the squammæ of the tail thirty-four. 19. The Coluber, with the scuta of the abdomen one-hundred and forty-two, and the squammæ of the tail seventy-four. 20. The *Ammodytes*, or Coluber, with the scuta of the abdomen one-hundred and forty-two, and the squammæ of the tail thirty-two. 21. The Coluber, with the scuta of the abdomen one-hundred and twenty-eight, and the squammæ of the tail forty-six. 22. The Coluber, with the scuta of the abdomen one-hundred and twenty-four, and the squammæ of the tail fifty. 23. The *Cerastes*, or Coluber, with the scuta of the abdomen one-hundred and nineteen, and the squammæ of the tail one-hundred and ten. 24. The Coluber, with the scuta of the abdomen one-hundred and eighteen, and the squammæ of the tail sixty-one.

To these we are to add a number of others, whose scuta and squammæ have not yet been counted, but which may be still distinguished by expressive names, 25. The brown Coluber, with the black and yellow belly, called the Water-viper. 26. The thick and short, flat and broad-headed, black Coluber. 27. The deep brown, large Coluber, with large scales. 28. The brown Coluber, with the reddish belly, called the Copper-bellied Snake. 29. The extreamly long and slender Coluber, with the back of a bright green, and the belly blue. 30. The large greyish-black Coluber, with the scales oblong and pointed. 31. The little, very slender, brown Coluber, with transverse broad streaks of blackish on the back. 32. The very long and slender, brown Coluber, with two longitudinal lines of yellow on the back, and with a bluish-grey belly. 33. The long and slender Coluber, with the back of a pale olive, spotted with black, and the belly of a greyish-blue, spotted in the same manner. 34. The large bluish Coluber, with annular marks, and some irregular spots of yellow, called the Chain-snake. 35. The greenish Coluber, with a bright ridge on the back, and with several black spots along the sides. 36. The very long and slender Coluber, all over of a brown colour, called the Coachwhip-snake. 37. The red Coluber, variegated with flesh colour and brown, called the Corn-snake. 38. The thick and short Coluber,

of

of a pale brown, with large black ſpots, called the Hog-noſed-ſnake. 39. The very long and ſlender Coluber, all over of a fine graſs-green. 40. The entirely blue Coluber, with undulated, deep blue ſcuta on the belly, called the Wampum-ſnake. 41. The brown Coluber, variegated with large and elegant ſpots of a coral red, and a pale yellow, called the Bead-ſnake. 42. The great Coluber, with the neck broad, and ſwelled out behind. 43. The very large Coluber, with a thick neck and ſmall head. 44. The long Coluber, with a very fine pointed tail.

CENCHRIS.

IN the Cenchris there are ſcuta which cover the whole under part of the body, and of the tail: the head alſo is covered with ſmall ſquamæ, and the tail has no appendage.

Cenchris ſcutis abdominalibus ducentis quadraginta, ſcutis caudæ ſexaginta quatuor.

The Cenchris, with the ſcuta of the abdomen two-hundred and forty, thoſe of the tail ſixty.

The Boiguacu.

This is the largeſt of all the Serpent kind; we have ſkins of it, in ſome of our muſeums, of five and twenty feet in length, and are aſſured from the people who have ſeen it, that it grows to between thirty and forty: it's thickneſs is proportioned to it's length: it's head is large, and the opening of it's mouth extremely wide: the whole head, from the extremity of the noſe to the joining with the back, is covered with a peculiar kind of ſcales, much ſmaller than thoſe of the reſt of the body.

It is a native of the Eaſt Indies, and has been met with in the ſouthern parts of America: it will ſeize on very large animals as it's prey, and firſt break their bones, by twiſting it's body round them, and afterwards ſuck them down whole. It frequently lies in ambuſh on trees, from whence it throws itſelf down on any thing that comes within reach.

Cenchris ſcutis abdominalibus centum & nonaginta, caudæ quinquaginta duo.

The Cenchris, with the ſcuta of the abdomen a hundred and ninety, thoſe of the tail fifty-two.

This is a very large Serpent, though inferior to the former in ſize: it grows to fifteen feet or more in length, and is very beautifully variegated in colour: the principal, or ground, colour is a pale olive, with an admixture of green: this is ſpotted with large and irregular blotches of yellow; and there runs an undulated line of a duſkier colour, approaching to black, down the back: the belly is of a bright and ſhining black: the tail is ſhort and obtuſe; the head is covered with ſmall pointed ſcales, of a pale olive colour, which have their extremities a little erect.

It is met with in the woods of the ſouthern parts of America, and ſometimes comes into the huts of the natives, but is not miſchievous: it feeds on the ſmall animals.

Cenchris ſcutis abdominalibus ducentis et viginti quatuor, caudæ ſeptuaginta octo.

The Cenchris, with the ſcuta of the abdomen two hundred and twenty-four, thoſe of the tail twenty-eight.

This alſo is a very large, but not a miſchievous, Serpent; it grows to twelve or fourteen feet long, and to the thickneſs of a man's leg: it's head is ſmall, in proportion to the body, but the opening of the mouth is very large: the ground colour is a duſky yellow, with an admixture of a greyiſh-green in it; this is variegated with a deep line, of a denticulated form down the back, and with ſome fainter lines by it's ſides, between which there are ſeveral large ſpots of a bright yellow, and ſome of a greeniſh-olive: the tail is obtuſe, and the head of a bright green, variegated or ſprinkled as it were with grey; the little ſcales with which it is covered being of an oval form, but pointed ſo ſharply, as to be almoſt prickly; and the body of them being green, but their tip or point grey.

It is common to the Eaſt Indies, and to the warmer parts of America: there were lately ſeveral of theſe diſcovered among ſome American timber, brought over for the uſe of our cabinet-makers.

CROTALOPHORUS.

THE Crotalophorus has ſcuta covering the whole under ſurface of the body and tail; and the extremity of the body is terminated by a kind of rattle, formed of a ſeries of articulated articulations, which are moveable, and make a noiſe, when ſhaken.

Crotalophorus ſcutis abdominalibus centum et ſeptuaginta duo, caudæ viginti uno.
The Crotalophorus, with the ſcuta of the abdomen a hundred and twenty-two, of the tail twenty-one.

The Rattle-ſnake.

This is a very terrible, and, at it's full growth, a very large Serpent: it grows to eight feet in length, with a proportionable thickneſs, and, when of this ſize, weighs about nine pounds: the head is large, broad, depreſſed, and of a pale brown: the iris of the eye is red; the back is of a brown colour, with an admixture of a reddy yellow, and is variegated with a great many irregular tranſverſe liſts of a deep black: the belly is of a pale greyiſh-blue; the rattle is of a firm, and, as it were, horny ſubſtance, and brown colour, and is compoſed of a number of cells, which are articulated one within another: the point of the firſt goes to a conſiderable depth into the ſecond, and ſo on of all the reſt; and, the articulation being very looſe, theſe included points ſtrike againſt the inner ſurface of the rings they are admitted into, and make that rattling noiſe, when the creature vibrates or ſhakes it's tail.

It is too frequent in the woods in the ſouthern parts of America, and ſometimes, though rarely, comes into the huts of the natives, and the houſes of the Europeans ſettled there; the bite is fatal, but it is eaſy to avoid it: the ſnake is ſluggiſh, and moves ſlowly, and never attacks a man, unleſs provoked; and, when diſturbed or injured, they generally give notice, before they bite, by ſhaking their rattle; when a large Serpent of this kind bites with it's full force in a fleſhy part, death is the conſequence in two minutes; ſlighter bites, and from ſmaller rattle-ſnakes, are ſometimes ſurvived. The colour of this ſpecies is not certain or determinate; it is ſometimes deeper, ſometimes paler, and often has a caſt of greeniſh, ſo as to be a kind of olive: the creature frequently alſo caſts it's ſkin, and always is of a different colour from the uſual one, when it firſt appears in it's new one, till the air changes that to the ſame colour with the former.

Crotalophorus ſcutis abdominalibus centum et ſexaginta quinque, caudæ viginti octo.
The Crotalophorus, with the ſcuta of the abdomen a hundred and ſixty-five, of the tail twenty-eight.

The leſſer Rattle-ſnake.

This is an extremely beautiful, though very miſchievous, Serpent, it's bite being no leſs fatal than that of the common rattle-ſnake, when of the ſame ſize: it grows to about four feet in length, and to an inch and a half in diameter: it's head is very large, broad, and depreſſed, and it's neck ſmall and ſlender; the opening of the mouth is very wide; the body is thickeſt about the middle, and from thence grows gradually taper again toward the tail, at the extremity of which is placed a rattle exactly like that of the former ſpecies, only ſmaller: the general colour of the back of this ſpecies is a pale dirty brown, with ſome faint admixture of grey; it is variegated in a very beautiful manner, with clouds of a duſky reddiſh, and with large, irregular ſpots, of a deep black: the belly is of a deep and gloſſy blue.

It is less frequent in America than the former, or ordinary rattle-snake; but it is sometimes found in the very houses of the inhabitants.

The European remedy of oil of olives, rubbed on the wounded part over burning charcoal, has been tried to the bite of this serpent in America with success, particularly, last year, to a wound in the foot of a woman; but as we have instances of people there, who have survived it's bite, and even that of the other under favourable circumstances, without any assistance, it is hard to say exactly how much may have been owing to the remedy. The French were at the pains, soon after the efficacy of the oil was published in England, to make a great number of experiments on animals bitten by vipers, some of which were dressed with the oil, others not; but the result did not argue much in favour of the remedy.

Amphibious Animals.

Series the Second.

Those which have legs.

The number of legs in these is invariably four, but they are very differently formed and situated, so as in the several genera to serve to very different purposes.

LACERTA.

THE body of the Lacerta is oblong and rounded; the legs are four, and the hinder part is terminated by a tail.

Lacerta pedibus inermibus fissis, manibus tetradactylis, plantis pentadactylis, cauda ancipiti. **The Water Newt.**
The two-edged-tailed Lacerta, with four toes on the anterior, and five on the hinder feet.

This grows to about four inches in length, and to the thickness of a man's finger: the back is of a deep shining brown; the belly of a bright and glossy yellow, and sometimes whitish, always spotted irregularly with brown: the head is small, and the extremity of the nose obtuse; the tail is compressed and flatted, and has two edges which stand perpendicularly, not horizontally: in the male, the back and the tail are both dentated; in the female, they are smooth: the throat in the male is variegated with dusky spots; the parts of generation in both are prominent: the legs are short and lateral; the fore ones are divided into four, the hinder into five toes; and, in the male of this species, the toes are edged with a membrane.

It is very common with us in fish-ponds, and other standing waters, and is voraciously eaten by the ducks and other water-fowl. All the writers on animals have described it under the names of Lacerta aquatica and Salamandra aquatica.

Lacerta cauda ancipiti, pedibus triangulatis, palmis tetradactylis, plantis pentadactylis palmatis. **The Crocodile.**
The edged-tailed Lacerta, with triangular feet, the anterior with four toes, the posterior with five.

This is the largest of all the Lizard kind, to which, notwithstanding this strange disproportion in size, it evidently belongs. It grows to twenty-five feet in length, and it's thickness is that of a man's body: it's colour is a dusky and a disagreeable brown; it's head is large, it's eyes are small, and the opening of the mouth is vastly wide: the back is elevated into a ridge, which is continued to the extremity of the tail, and is deeply dentated; the lower part of the tail has also it's ridge opposite to the other, which gives it a two-edged appearance: the legs are short and thick, and the fore feet have only four toes each, the hinder ones five.

It

It is produced in greateſt abundance in the torrid Zone, and fourteen degrees more north; they are frequent in America, to the latitude of 33, but are rare beyond this; and this nearly anſwers to the north of Africa, where they are always found in great numbers. They frequent ſalt-water rivers principally, but they are alſo found about the large freſh ones, and even about large lakes: they generally lie hid among the reeds and ruſhes, waiting quietly for the approach of men or animals, which they ſeize and drag into the water, always taking this method of drowning them firſt, that they may ſwallow them without trouble or reſiſtance: it's general food, however, is fiſh; but even theſe, as well as other animals, it is obliged to take by ſurpriſe, as it can, in purſuit, only run ſtraight forward; the joints of the back-bone being ſo ſtiff, that it turns but very ſlowly and difficultly: they lay a number of eggs, which are not larger than thoſe of a turkey; theſe they depoſit in the ſand, on the edges of rivers, and never are at any pains to take care of them afterward: the ſun's heat hatches the young, and they are no ſooner out of the ſhell, than they inſtantly get into the water, where they are a prey to all kinds of devourers, fiſh, fowl, &c. and are even ſwallowed in great numbers by their own ſpecies. In the colder of the countries they inhabit, they lie torpid all the winter in caverns, near the ſides of rivers. The hinder part of their belly and tail are eaten by the Indians and Africans; the fleſh is white, and of a kind of perfumed favour. It has been deſcribed by all that have written on animals, under the names of Crocodilus and Lacertus maximus. We call it the Crocodile; and in thoſe countries where it does not grow to it's full dimenſions, as in the colder ones, the Alligator.

Lacerta cauda tereti longa pedibus pentadactylis, criſta gulæ dentata, ſutura dorſali denticulata.

The long and rounded-tailed Lacerta, with five toes to the feet, and the creſt of the throat and dorſal ſuture dentated.

The Guana.

This ſpecies of Lizard very much reſembles the crocodile in it's general form; it grows to five feet in length, though thoſe of two or three are much more frequently met with: the body is very large and rounded; the head is ſhort, the eyes are large, the creſt of the gula is dentated, and all down the back, and along one third of the tail, there runs a dentated and prominent line, formed, as it were, of a number of triangular, pointed ſpines: the tail is long and not compreſſed, but rounded; the legs are moderately long, and the toes are very long and ſlender; the mouth is armed with extremely ſmall teeth in great numbers, and has a kind of bony beak, with which it bites very ſeverely; the colour is a duſky olive, with more or leſs of the green in it, and with ſome prominent veins of a paler hue ſeen through the ſkin.

It is a native of the hot countries; the woods about the tropics abound with it; it gets into hollow trees, or the crevices and leſſer caverns of rocks. They lay a great number of eggs, which are covered with a ſoft and flexible tough ſkin inſtead of a ſhell, and are left in the ſand to be hatched by the ſun's heat, without any farther care of the parent animal: they feed wholly on vegetables, and are themſelves an excellently well-taſted food; their eggs alſo are delicate: the natives of many parts of South America, in a great meaſure, live upon them; they have dogs that hunt them, and often kill them; they go out with theſe dogs on the expedition frequently: what are killed they eat as their proviſions for the time; what they take alive they ſew up the mouths of, to prevent their biting, and carry home: they are a ſlow animal, and make no uſe of their legs in ſwimming, only puſhing themſelves along by the tails: they are ſo impatient of cold, that they rarely appear out of their holes, unleſs in bright and ſun-ſhiny warm weather. Moſt of the authors who have written on theſe ſubjects have mentioned it; they call it Senembi, Guana, and Iguana.

Lacerta cauda tereti longa, pedibus pentadactylis, dorſo ſtriato.

The blue tailed Lizard.

The long and rounded-tailed Lizard, with five toes to the feet, and a ſtriated back.

This is a ſmall but an extremely beautiful ſpecies; it rarely grows to more than ſix inches in length, and is ſlender in proportion: the head is ſhort, broad, and tinged faintly with reddiſh; the body is of an olive brown, with four or five longitudinal lines of yellowiſh running from the head to the beginning of the tail: the tail itſelf is long, rounded, not flatted, and of a beautiful blue: the legs are ſhort, and the claws or toes long and ſlender.

It is a native of America, and is very frequent in many places: It is ſuppoſed to be poiſonous, but there does not ſeem to be any great foundation for this opinion.

The other ſpecies of the Lacerta are, 1. The common, brown, land Newt, or Lacerta. 2. The long-legged and thick-bodied, ſmall, green Lacerta. 3. The long and ſlender, beautiful, deep green Lacerta. 4. The very long-tailed, brown Lizard, called, from the fierceneſs of it's aſpect, the Lion Lizard. 5. The Lacerta, with the two-edged tail with the anterior feet with four divided toes, and the poſterior with five palmated ones, and the belly freckled: this is called the Rana piſcis, and is ſuppoſed to be produced by a change of a frog into a fiſh, though it neither is the one, nor ever was the other. 6. The Lacerta, with the tail verticillated with denticulated ſcales, and with five toes to the feet; this is called the Cordyle. 7. The Lacerta, with a verticillated tail, with the feet divided into five unguiculated claws, and the ſcales of the body ſquare. 8. The Lacerta, with a rounded, verticillate tail, with five unguiculated toes to each foot, called the common Lizard. 9. The Lacerta, with the rounded tail, and with only three toes to each foot, called the Chalcides. 10. The Lacerta, with a long, rounded tail, with muricous feet, with four toes on the anterior, and five on the poſterior. 11. The Lacerta, with a long and rounded tail, with five toes on the feet, and with the creſt of the gula and the ſummit of the back ſmooth, called the Principalis, and the ſmooth-backed Guana. 12. The Lacerta, with a long, rounded tail, with five toes on each foot, with the gula ſubcriſtated, and dentated in the anterior part, and with the back ſmooth, called the Marmorata. 13. The Lacerta, with a long, rounded tail, with five toes on each foot, the back denticulated on the anterior part, and with the hinder part of the head and the neck aculeated. 14. The Calotes, or Lacerta, with a long and rounded tail, and with five toes on each foot, with the back dentated on the anterior part, and the head denticulated behind. 15. The Ameiva, or Lacerta, with a long, rounded tail, and five toes on each foot, with the abdomen covered with ſcuta, and with no creſt. 16. The Gecko, or Lacerta, with a rounded and moderately long tail, with five criſtated toes imbricated on their under part on each foot, and with the body verrucoſe, or covered with ſmall, rough protuberances. 17. The Barbara, or Lacerta, with a rounded and moderately long tail, with five toes on each foot, with ſmooth, round, greyiſh ſcales on the back, and brown ones at the ſides. 18. The Lacerta, with a rounded and ſhort tail, with five toes on each foot, two or three of which grow together; this is the *Chameleon*. 19. The Lacerta, with a ſhort and rounded tail, with unarmed feet, the anterior ones having four toes each, and the poſterior five. 20. The Teguixin, or Lacerta, with a long and rounded tail, with five toes on the feet, with no creſt, but with plicated ſides. 21. The Lacerta, with a long, rounded tail, with five toes on the foot, and with a longitudinally ſtriated back. 22. The Lacerta, called by many the Tarantula. 23. The flying Lizard, called by many the Dragon.

The dragons figured and deſcribed by authors would come in this place, or follow the Lizard, if they had, indeed, any place in nature, but that is not the caſe; the love of talking of ſtrange things, in ſome writers, and the credulity of others, have given birth to theſe monſters of the vitiated fancy. Aldrovand has figured two dragons under the name of Hydræ; the one from Geſner, the other from his own fancy, each with two legs; and another ſtranger monſter, which he calls a Baſiliſk, with eight legs, with a body like a hog, the tail of a lizard, and the head of a bird of prey, with a crown on the ſummit of it: and Parey and others have added dragons with

with four legs, and dragons without any legs at all; ſome with wings, ſome without; ſome with two, and one of Petty's, indeed, with horns. What we ſee under the names of dragons ſhewn at fairs, and even kept in the cabinets of ſome collectors, are in general formed by art out of the ray-fiſh; and the others, figured and deſcribed ſo pompouſly by theſe authors of general credit, have no better a foundation in nature.

RANA.

THE body of the Rana is broad and ſhort, and has no tail; the legs are four.

The Bull-frog.

Rana manibus tetradactylis fiſſis, plantis pentadactylis palmatis, maxima.

The great Rana, with the anterior feet with four divided toes, the poſterior with five webbed ones.

This is the largeſt of all the Frog kind; when the limbs are extended, it meaſures near two feet: the body is often eight inches long, and four or five in breadth; the fore legs are ſmall, but the thighs of the hinder ones are as thick as a child's arm: the head is large; the opening of the mouth very wide; the eyes large and prominent; the pupil has a yellow circle round it; the iris is red, and there is another yellow circle at the verge of it: the ears are large, and of a circular form; they are ſituated ſideways of the eyes, and a little behind them, and are covered with a fine thin membrane: the colour is a duſky brown, with an admixture of green, ſpotted all over very thick with oblong, black ſpots: the belly is of a duſky whitiſh, with a caſt of yellow, and is alſo faintly ſpotted.

It is a native of the northern parts of America: in Virginia the ſides of the hills almoſt all produce ſprings; and near the mouths of theſe there are uſually found a pair of theſe frogs, which, when frighted, leap into the little baſon, uſually made by the ſpring near the opening: they leap to a ſurpriſing diſtance, and their croaking is ſo loud, that it reſembles the roaring of a bull, heard at ſome diſtance; 'tis hence they have obtained the name of the Bull-frog: they are very voracious, and frequently ſwallow young ducks, and other water-fowl, before they have ſtrength enough to ſhift for themſelves.

The Tree-frog.

Rana palmis tetradactylis plantis pentadactylis, digitis ſummitate excavatis.

The Rana, with four toes on the anterior, and five on the poſterior, and all hollow at the end.

This is a very extraordinary ſpecies of frog; the name of a Tree-frog has an odd ſound, yet this kind is never found hardly off of trees, and is formed by nature for living very eaſily on them: it's body is about an inch and a half long, and three quarters of an inch in diameter, ſomewhat rounded, and the back in particular very convex: it is of a beautiful bright and gloſſy green, but has on each ſide a longitudinal ſtreak of yellow; all the other parts are of the ſame bright green with the body, without any variegation; the head is obtuſely pointed; the eyes are large, black, and gloſſy; they ſtand prominent, and have a yellow circle round them: the legs are ſlender, the hinder ones conſiderably long; the toes are of a very peculiar form and ſtructure; they are rounded and hollow, opening at the extremities, in the manner of the mouth of a leech: 'tis by means of this peculiar mechaniſm, in this part, that they are able to hold their place on the leaves, and they are ſo well conſtructed for this, that they will inſtantly faſten the animal on a plate of a glaſs, or of the moſt highly poliſhed metal: the creature leaps very nimbly, and to a vaſt diſtance.

They are frequent in North America, and in ſome of the northern parts of Europe, particularly in Sweden: they live on trees and plants of various kinds, but particularly on thoſe which have large leaves; and their poſt uſually is on the under ſide of the leaf, where they are hid from the birds that otherwiſe would devour them: they are uſually

quiet

quiet by day, but, as night comes on, they are ſeen inceſſantly hopping from branch to branch, and catching inſects. Moſt of the writers on animals have deſcribed this ſpecies. Linnæus calls it Rana viridis; Charleton, Johnſton, and others, Ranunculus viridis; and Geſner, Dryopetes.

Rana dorſo pullifero.
The Rana, with the young coming from the back.

The Surinam Toad.

This is a large and moſt extremely ſingular ſpecies; it's body is about four inches and a half long, and near three and a half broad: the head is large, and of a pyramidal form: the opening of the mouth very wide; the eyes are not very large, and ſtand forward on the head: the fore legs are ſlender, and not very long: the feet of them have each four toes, which are connected together; the hinder legs are very large, robuſt, and ſtrong, and much longer than the fore ones; they have five toes to each foot, connected together by a membrane or web: the colour of the whole animal is a deep duſky brown, variegated with irregular lines and ſtreaks of black, and with ſmall ſpots. The firſt appearance of the young is in this creature a very amazing circumſtance; the back of the female is found elevated in many places into little tubercles of an obtuſe form, and from theſe the young ones make their way very ſmall, but in their perfect form. This is worthy a very ſtrict attention, and it is much to be wiſhed, that ſome perſon of abilities may have an opportunity of examining thoroughly into it on the ſpot.

The ſlaves on the Surinam ſettlements eat this ſpecies, and eſteem it a very delicate food. Moſt of the late writers have named it under the title of the Surinam Toad, or Frog, and Pipa.

The other ſpecies of the Rana are, 1. The common Frog. 2. The green and yellow American Frog. 3. The large, brown American Frog. 4. The common Toad, or the great African grey Toad. 5. The little, brown Frog, with the extremities of the toes rounded. 6. The little Frog, with the joints of the toes protuberant. 7. The Rana, with the hinder feet ſubpalmated, and the pollex ſhort and broad. 8. The round-toed Rana, with the body narrow behind. 9. The Rana, with an obtuſe anus ſpotted on the under part. 10. The large, green, and yellow Eaſt Indian Frog.

TESTUDO.

THE Teſtudo has four legs, and it's body is covered with a firm ſhell.

Teſtudo unguibus acuminatis palmarum plantarumque quaternis.
The Teſtudo with, acuminated ungues, four on the hinder, as well as fore feet.

The Caret Tortoiſe, or Hawk's-bill Turtle.

This grows to two feet, or more, in length, and to nearly as much in diameter in the ſhell; the head and hinder legs protended, it meaſures conſiderably more, but it never arrives at the enormous ſize of ſome of the other ſpecies: the ſhell is of a figure approaching ſomewhat to oval, but hollowed in the form of a ſegment of a circle at the upper extremity, and terminated in a ſharp and even prickly point, when perfect, at the other; it's edges towards the top are ſmooth, but round; in the lower half, they are indented, and the ſerratures are deep, and the points ſharp and prickly: the whole ſhell is divided into a number of irregular ſpaces by ſeveral oblique and angulated lines, and is of a brown, with a tinge of reddiſh in it, variegated with ſpots and blotches of a duſky colour: the neck is oblong and thick, the head large, and of a ſomewhat oval form; the eyes are large, and the mouth formed, in ſome meaſure, like the beak of a bird of prey, the upper jaw falling over the other: the fore legs are longer than in any other known ſpecies, and the hinder ones are robuſt and broad; the whole is of the ſame kind of brown colour with the ſhell.

This

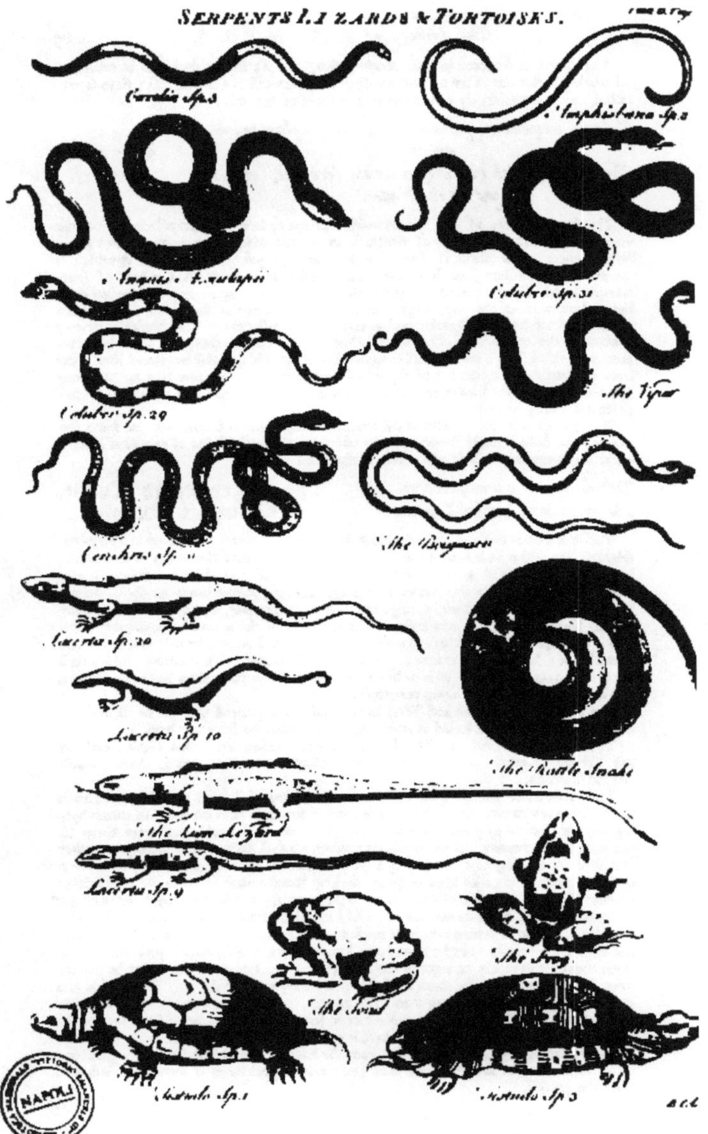
SERPENTS LIZARDS & TORTOISES.
Coluber Sp. 31
Coluber Sp. 29
The Viper
Lacerta Sp. 20
Lacerta Sp. 10
The Rattle Snake
Lacerta Sp. 9
The Frog
The Toad

This species is frequent in the American seas, and it's shell is that used in ornamental works, under the name of Tortoise-shell; authors call it Caretta. It's flesh is well tasted, but is considerably inferior to that of some of the others.

Testudo unguibus palmarum duobus, plantarum unicis.
The Testudo, with two ungues on the fore feet, and one on the hinder.

The green Turtle.

The English name of this species might be apt to lead the unexperienced into the error of supposing the colour of the shell, or of the whole animal, green; but it is a West Indian name given it, from the fat having a greenish appearance at table: it grows to a vast size; we had one lately dressed at a tavern in London, of four-hundred and eighty pounds weight: the shell is of a figure approaching to oval, lightly hollowed at the top, to give room for the motions of the neck; and dentated all round the sides, but slightly, and at considerable distances. The hinder extremity is obtuse; the colour is a dusky brown, shaded with a still darker, and the whole surface is divided into a number of irregular spaces by oblique and angulated lines; the head is roundish; the opening of the mouth not very large; the fore legs are of a considerable length; the hinder ones short and broad, and the colour of the whole a disagreeable dusky brown.

This species is frequent about the American islands, but it breeds only in some few places: the flesh is finely flavoured, and esteemed the best of that of any kind; this is what we commonly in England mean by the word Turtle.

Testudo capite magno subovato.
The great oval-headed Testudo.

The logger headed Turtle, or Caouanne Tortoise.

This is a species easily distinguishable at sight from the others: it grows to a very considerable size; the shell is of a more regularly oval figure than that of any of the others; it is often two feet and a half long, and two feet broad near the top, whence it gradually grows smaller to the hinder extremity, where it terminates in an obtuse point; it is of a dusky brown colour, variegated with a darker hue, and is divided into more regular spaces than that of any other kind: the head of the animal is remarkably large, and it has a great fierceness in it's aspect: the legs are short and broad; and the whole animal is of a dusky brown colour: this is the worst tasted of all the kinds, and it's shell is of little value; whence it is suffered to live more at peace than any of the others, and consequently is greatly more numerous.

It is a native of the East and West Indies, and is the greatest traveller of all the Tortoise kind: it has been found at open sea, at vast distances from any land.

The other species are, 1. The Testudo arcuata, called the Trunk-turtle; and, by the French, Caffre. 2. The little land Testudo. 3. The great, thick, small-headed Testudo.

The Americans find so good account in the taking of Turtle, that they have made themselves very expert at it; they watch them from their nests on shore in moon-light nights, and, before they reach the sea, turn them on their backs, and leave them till morning; the creature can by no efforts recover it's first posture, and consequently they are sure of finding as many as they have thus turned: at other times they hunt them in boats, with a peculiar kind of spear, striking them with it through the shell, either as they lie on the surface of the water, or underneath it. When they have once got sight of a Turtle, patience and attention seldom fail to make them sure of it: the creature dives, as soon as it perceives itself pursued; but, as it must every now and then rise to the surface for breath, they by this means know where it is: when tired with flying from them, it generally plunges to the bottom at once, where, if the water be not too deep for the length of the handle of the spear, they easily strike it. The spear is of a very simple form, it is no more than a peg of iron, two or three inches long, fastened to the end of a handle of ten or twelve feet in length, in the manner of the iron of the harpoon used in the whale-fishery: there is a cord fastened to the iron, and, when the stroke is given, the head separates easily from the handle; but, the cord remaining fastened to it, the creature is managed by it with great ease, especially if it have been first tired with the pursuit.

THE HISTORY OF ANIMALS.

PART III.

Of SHELL-FISH.

SHELL-FISH *are animals with a soft body, covered by, or included in, a firm, hard, and, as it were, stony covering, composed of one or more parts, and more or less moveable at the animal's pleasure.*

The genera of the Shell-fish are extremely numerous, and the species, under many of them, are also very much so. The animals, included in these hard, external cases, have most of them the characters of one or other of the genera of the Gymnarthria before described, and might be reduced under the same genera with the naked ones; or the whole series of Shell-fish might be arranged according to their characters; but as these characters are few, and the bodies themselves very numerous; and as the external coverings or shells are extremely different in their form and structure, and these differences are sufficiently obvious, determinate, and numerous; it will tend more to the making natural history an easy and an universal study, to arrange this part of it according to the differences of the shells themselves, than those of the included animals. There is this farther consideration also in favour of this arrangement, that the bodies of the animals themselves are very rarely seen, and never preserved in collections, whereas the shells make an eminent figure in them; and that many of them have been only met with, empty of the included animal.

I shall not so far disregard, indeed, the form and structure of the included animal, as to leave the reader in the dark as to what it is; but, as all of the same genus are the same in this respect, it will be sufficient barely to mention which of the former genera of Gymnarthria *they belong to; after the generical characters of the shell. Thus I shall not omit to say, that the animal inhabiting the patella is a limax; that the animals inhabiting the several shells of the cochlea, the nerita, the buccinum, the turbo, the trochus, the voluta, the murex, the purpura, and the lyra, are all limaces also; and that the inhabitants of the concha Veneris and auris marina are all of the same genus: that the inhabitant of the dentalium is a* Nereis, *that of the nautilus a* Sepia, *that of the ostrea, the pecten, and the like, a* Tethys, *and that of the concha anatifera and balanus a* Triton.

The mention of this, at the head of the genus, will convey a sufficient idea of what the animal is that inhabits the outer case, which is the object of our more immediate attention; and I flatter my self, that it will appear much better to pay this secondary regard to the form of the body of the creature, than, for the sake of arranging those of the same structure in the less obvious part together, to have brought into one genus the auris marina, the concha Veneris, the buccinum, the voluta, and the patella.

The

The Shell-fish, on this plan of arranging them according to their shells or coverings, not to the different structure of the included animal, may be divided into three series; the first containing those formed of only one piece; these I shall call the simple ones; others have called them univalve ones, but the structure of the shell is in many, by no means, referable to a valve. The second series containing those which are formed of two parts or valves, under the name of bivalves; and the third, under the name of multivalves, comprehending those whose shells are composed of several valves or parts.

SHELL-FISH.

Series the First.

Simple Shells.

Those which consist only of one single shell, and have no hinge.

PATELLA.

THE Patella is a simple shell of a conic, or other gibbose, figure, with a very wide opening at the bottom; always applying itself firmly to some solid body, which serves it in the place of another shell: the animal inhabiting it is a limax.

The summit of the patella is in some acute, in others obtuse; in some depressed, and in some it is perforated; and in others turned down, or otherwise crooked.

Patella vertice acuto, decem costis instructa.
The sharp-topped Lepas, with ten striæ. — **The striated Limpet.**

This is a very beautiful species; the full grown ones are an inch broad at the mouth, and three quarters of an inch high; the base is not exactly rounded, but somewhat oblong, and the top sharp or pointed; the colour of the outside is a dusky brown, with a cast of olive; it is elevated in ten places by so many small ridges, placed at nearly equal distances, and is rough all over: the striæ are most conspicuous toward the mouth, and become fainter, as they approach the top; where they are obliterated, and the sharp point, and a little of the shell below it, is smooth: the inside of the shell is beautifully variegated with brown, yellow, and whitish, disposed in the form of irregular circles, and in the whole shews a great general resemblance of the auricula flowers.

It adheres very firmly to the rocks in many parts of the East Indies, and elsewhere.

Patella radiata costis septenis, septem aculeos ad oram efformantibus.
The radiated Patella, with seven ribs, forming seven prickles at the rim. — **The starry Limpet.**

This is a very elegant and very singular species: it is but little elevated; it's mouth is of a figure approaching to elliptic, and it's base is about an inch long, and two thirds of an inch broad; it's summit is pointed, but the height is hardly half an inch, and the point does not rise exactly in the center of the shell: on the outer surface there are seven very elevated ribs, placed at considerable and nearly equal distances; they run strong and distinct to the very point of the summit, and are continued to the verge, where they run out each into a point beyond the rest of the verge, and form so many prickly rays, with hollows in form of segments of circles between them: the colour on the outside is whitish, variegated with clouds and spots of black, especially about the ribs; within it is beautifully variegated.

It is found adhering to rocks about the islands of Sumatra, Borneo, and elsewhere, in the East Indies: it has been sent also from some of the American shores, and some other places.

Patella glabra subovata vertice depresso.
The smooth, oval-mouthed Patella, with a depressed top.

The oval, smooth Limpet.

This is a large and beautiful species; it is at the base two inches and a half long, and near two inches broad at one extremity, whence it becomes gradually smaller to the other, where it is also obtuse and rounded; it's height from the center of the base to the summit is about an inch and a quarter; the top is somewhat depressed, and stands nearer the largest end of the oval than the other; the rim is even, and the whole surface is smooth; the colour is a dusky whitish, elegantly marbled, and variegated with a deep and blackish clouding.

It is found adhering to the rocks on the shores of some of the warmer countries in Europe, but not on the English coast, so far as has been yet observed.

Patella concamerata rostro in summitate curvato.
The concamerated Patella, with a crooked rostrum on the summit.

The beaked Limpet.

This is a tall species; it's mouth is rounded, and the verge smooth and even; it's diameter is about an inch, and it's height nearly as much: the outer surface of the shell has no ridges, but has some roughnesses disposed irregularly all over it; it is bellied half way up, where it begins to form a rostrum or beak; this is somewhat pointed, and bends down: the inside of the shell is concamerated, and the division terminated by a kind of point; it is not very elegantly coloured, but this structure is singular.

We have it from the shores of some of the American islands.

Patella concamerata rostro ad extremitatem parvula.
The concamerated Patella, with a short rostrum toward one end.

The chambered Limpet.

This is of an oblong or irregularly elliptic figure at the base; it's length at the mouth is about an inch and a quarter, it's breadth hardly half an inch; both ends are nearly equal in breadth, and the verge is somewhat undulated: about half the aperture from one end, which may be esteemed the head, is covered by a thin crust or shell, connected to the end and along the sides, but irregularly marked at the verge: the height of the shell is about half an inch; it's rostrum or beak is short, and seems only a kind of button; it does not stand in the center of the shell, but just at that end which we call the head, and near the verge of the shell. The whole outer surface is rough and warty, and the inner smooth, but neither very beautifully coloured.

We have it on some of the European shores.

Patella elliptica depresso vertice perforato.
The elliptic and depressed Patella, with the top perforated.

The great, oblong Limpet.

This is a large species; it's length at the base is near three inches; it's diameter an inch an half at the largest end, the other is a little smaller, but 'tis so little, as not to give it a title to the oval: the edge is smooth and even; the height of the shell is about an inch; it's surface is almost smooth, but there are a number of longitudinal rays, and some transverse or circular ones also visible on it: it's colour is a dusky brown, and it is thicker and more robust than many of the others: it's top is not much elevated, and is as it were cut off; an aperture or hole appearing there, which is of an oblong figure, and seems to be formed of two round ones joined together: the colour on the outside is a dusky grey; within it is whitish, but with a few variegations.

We have it from the East Indies, and it is also common to some of the warmer parts of Europe.

After having deſcribed one of each of the more ſingular kinds of the Patellæ, the others, when arranged under the diviſions, mentioned in the generical character, will be eaſily diſtinguiſhed by their names. 1. Of the conic kind, with the pointed ſummit, beſide the firſt deſcribed ſpecies, there are, 1. The ſmooth, pyramidal, and mucronated Patella. 2. The pyramidal, deeply-furrowed Patella. 3. The ridged and circled, grey Patella. 4. The common brown Patella. 5. The ſmooth, tall, ſharp-pointed Patella. 6. The ſmooth Patella, with circular lines very fine, and at ſmall diſtances. 7. The low, but ſharp-pointed, Patella. 8. The oval-mouthed, ſharp-topped Patella. 9. The ſlightly, ſtriated Patella. 10. The highly ridged and dotted Patella, with an aculeated edge. 11. The high-ridged, but not dotted, round-mouthed Patella, with an aculeated edge. 12. The conic, circular-ridged Patella.

2. Of thoſe with an obtuſe or rounded top there are, 1. The large, rounded-mouthed Patella, with high and very numerous ridges, and lower between them with a very broad-rounded top, and a denticulated edge. 2. The ſmaller, obtuſe-topped Patella, with a few high ridges. 3. The obtuſe-topped, ſtriated Patella. 4. The obtuſe-topped, ſmooth Patella, with circular lines. 5. The obtuſe-topped Patella, with longitudinal and circular lines.

3. Of thoſe with depreſſed ſummits there are, beſide the deſcribed ſpecies, 1. The furrowed and marbled Patella. 2. The elliptic, fine-ſtriated Patella. 3. The round-mouthed Patella, with the top reſembling a nipple. 4. The depreſſed-topped Patella, with elevated ſtriæ, and with a dentated edge. 5. The depreſſed topped, finely ſtriated Patella, with a ſmooth and even edge. 6. The ſtriated and prickly Patella, with a depreſſed top. 7. The broad Patella, with a depreſſed top, called the tortoiſe-ſhell, ſhield Limpet. 8. The depreſſed-topped Patella, elegantly ſpotted with ſcarlet. 9. The depreſſed-topped Patella, radiated with white and red. 10. The circular-lined, elegantly variegated Patella, called the goat's eye Limpet. 11. The depreſſed-topped Patella, with the inſide of a fine bright red. 12. The depreſſed-topped and deeply ſtriated Patella, with an undulated edge.

4. Of thoſe which have the top formed into a kind of roſtrum and crooked, and are not concamerated within, there are, 1. The little, white, long-beaked Patella. 2. The round-mouthed, ſmooth, and long convoluted-beaked Patella. 3. The round-mouthed, ſhort-beaked, deeply-ſtriated Patella. 4. The long and obtuſe-pointed, beaked Patella, with an undulated edge. 5. The ſhort and obtuſe-beaked, faſciated Patella.

5. Of the concamerated kind, beſide the deſcribed ſpecies, there are, 1. The tall, ſmall, chambered Patella, of a whitiſh colour, and ſmooth ſurface. 2. The broad and low concamerated Patella, with a ſhort beak, and very deep ſtriæ. 3. The oblong or elliptic-mouthed Patella, with a large chamber. 4. The round, volutated-chambered Patella. 5. The elegant, Chineſe, concamerated Patella. 6. The tall Patella, with a ſtyle ariſing from the ſummit on the inſide. 7. The concamerated Patella, with the vertex irregularly elongated. 8. The depreſſed, ſemi-concamerated Patella. 9. The low, concamerated Patella, with a deeply ſinuated edge. 10. The ſtarry, edged, concamerated Patella.

6. Of thoſe which have the top perforated, beſide the deſcribed ſpecies, there are, 1. The elliptic Patella, with an elliptic large hole on the ſummit. 2. The deeply-furrowed, elliptic Patella, with an oblong and irregular perforation. 3. The low, conic Patella, with light, longitudinal, and circular lines, and with a ſmall, round perforation at the ſummit. 4. The very bright and rounded-ridged Patella, with an undulated mouth, and an oblong perforation, ſeeming to be formed of two round ones, conjoined. 5. The tall and deeply-ridged, oval Patella, with an oval, large perforation at the top. 6. The low and deeply-ridged Patella, with a ſmaller perforation. 7. The low, oval-mouthed, denticulated, ſharp-ridged Patella, with a ſmall perforation. 8. The low, circular-lined, and ſmooth-edged Patella, with a large, irregular, but roundiſh aperture. 9. The convoluted-beaked Patella, with a round perforation at the ſummit. 10. The elliptic, high-ridged Patella, with a double top, and a ſmall perforation in each. 11. The great, oval, ſmooth, whitiſh Patella, with a very ſmall, oblong perforation. 12. The great, oval, deep-lined Patella, with a large perforation. 13. The great elliptic Patella, with longitudinal and circular lines, and with a large perforation, ſeeming made up of two round ones.

HALIOTIS.

THE Haliotis is a simple shell without any hinge, and formed all of one piece, of a depressed figure, very patent at the mouth, having an approach to the spiral form at the summit, and having several perforations on the lateral part of the disk. The animal inhabiting this shell is a limax.

It has obtained the name *Auris marina*, from it's resemblance to the figure of an ear; some of the old writers, as Aldrovand and Rondeletius, have arranged it with the limpets, and called it Patella fera; and some others have called it Ormus. Lister has arranged it among the turbinated shells, because of the spiral turn at the summit; but this is by no means sufficient to give it a place among those shells.

Haliotis margine æquali, foraminibus septem.
The Haliotis, with an even edge, and with seven holes. — **The great Ear-shell.**

This is a very large species; it's length is often five inches, it's breadth nearly three, but it's height, at the utmost, not more than three quarters of an inch: it's figure approaches irregularly to an oval; the end, where the spiral turn is placed, is the largest; the other a little smaller; the back or hinder edge is thick, and is turned so as to form a kind of lip; the other or anterior edge is thin, simple, and even: the upper surface of the shell is of a brown colour, rough and uneven, with a kind of undulated lines; the part toward the head is thickest, and the spiral turn short and depressed: along the back part of the shell, near the thicker edge, there are placed a series of holes or perforations, of a roundish figure; the number of those which are open is seven; there are the marks or vestiges of more toward the top of the shell, but they are closed up, and partly obliterated: as the shell grows in length, there are from time to time new holes formed in the extended part, and for every one of these there is one of the old ones filled up, or obliterated, the number nature allots to be open being only seven: the inside of the shell is of a beautiful pearly hue, variegated with a great many bright and elegant colours, as viewed in different directions to the light; and, the outer coarse crust of the upper surface being eaten off with acids, that exhibits the same pearly hue, with the same variety of colours; there are very frequently found a kind of roundish protuberances, in the form of warts, adhering to the inside of the shell, which have much the appearance of pearls, and are sometimes cut out by our workmen, and used in ornaments: real pearls are also sometimes formed in this species.

It is frequent in most parts of the world; Europe and America both produce it, but it is no where so abundant as in the East Indies.

Haliotis angustior margine undulato, foraminibus octo.
The narrower Haliotis, with an undulated edge, and eight holes. — **The long Ear-shell.**

This is a very beautiful species: it grows to three inches in length, and not more than an inch and a quarter in breadth in the largest part; it's height is no where half an inch: the head is large and thick, and the spiral turn is very beautiful and fair; the back of the shell forms a regular and even lip, and the anterior edge is thin and undulated: the inside of the whole shell is of a beautiful pearly hue, with fine variegations of colours; the outside is smooth, but somewhat undulated from the spiral turn, and is of a greenish colour, variegated with a brownish red: there is a long series of holes on the back edge of the shell, and eight of them are always open; the others toward the head are closed.

It is found on the rocks on the Malabar coast, and is washed ashore in some other parts of the East Indies; we have also had some very fine ones from the island of Ormus, in the Persian Gulph.

Haliotis rugosa foraminibus sex. **The striated**
The rugose Haliotis, with six holes. **Ear-shell.**

This grows to three inches and a half in length, and to nearly two inches in diameter toward the head; the height is no where more than three quarters of an inch: the outside of the shell is of a dusky brown, and is elevated into a number of slight and irregularly undulated ridges, which seem all to take their origin from the turn of the spiral end, though the eye can by no means trace them so far: the back is thick, and forms a kind of lip; the anterior edge is thin, and somewhat undulated; the hinder extremity is often more so: there is a long series of holes near the back or thicker edge of the shell, but there are only six of them open: the inside of this species is of a beautiful pearly blue, like the rest, and there have been found real and fine loose pearls in it.

It is found on the rocks about our own coasts, and in most other parts of Europe, and is washed on the shores, in some places, in vast abundance.

The other species of Haliotis are, 1. The smooth, brown, large Haliotis, with seven holes. 2. The smooth, oblong, and narrow, reddish Haliotis, with seven holes. 3. The smooth, narrow, green Haliotis, with only six holes. 4. The smooth but undulated, green and brown Haliotis. 5. The scabrous, brown, oblong and narrow Haliotis. 6. The smaller, brown, oval Haliotis. 7. The greenish, oval Haliotis.

STOMATIA.

THE Stomatia is a simple shell without any hinge, and formed of one piece: It's figure is depressed and flat; it's mouth the most patent of all the shells, the limpet only excepted: it has a short spiral turn running into the mouth at the head, and has no perforations in any part of the surface.

Authors have confounded this shell with the Auris marina, or Haliotis; but with how much impropriety is evident, from the want of the perforations, which are the distinguishing character of that genus, and from the difference in the spiral turn of the shell.

The animal inhabiting this shell is a nereis; and in this also it differs from the haliotis, that and the patella being inhabited by a limax.

Stomatia ovata fasciis obscurioribus.
The oval Stomatia, with almost obliterated fasciæ.

This is about an inch and a half in length, and near an inch in breadth in the largest part: it's height is no where half an inch; the whole shell is very thin, and the edge all round is thin and even; the head is the smaller extremity, and has a short spiral turn running into the cavity of the mouth; the opposite end is somewhat large, and is hollowed in the manner of a spoon; the outer surface of the whole shell is of a pale tawny-brown, paler toward the edges than in the middle; the inside also is of a pale brown, and has nothing of the pearly hue of the haliotis; there is no mark of any perforation in any part.

It adheres very firmly to the rocks, on the shores of many of the American islands, and in several other places.

Stomatia gibbosa striis numerosissimis tenuibus.
The gibbose Stomatia, with very numerous fine striæ.

This is about two inches in length, and an inch and a half in diameter, and is nearly of the same breadth all the way; the upper surface is full of extremely numerous, fine striæ, and is of a dusky chocolate colour: it is more elevated or gibbous than the former kind, and the edge is somewhat thick all round, and is whitish; the inside of the shell is of a pale brown, with no particular gloss or beauty.

It is sent us from the American islands.

Stomatia margine sinuato.
The sinuated-edged Stomatia.

This is the largest species of this genus; it is four inches in length, and near three in breadth at the broader end, and is near an inch in height in the most elevated part: the upper surface is of a very deep chocolate colour, variegated with white; is is full of obtuse, undulated ridges, which are of a pale and almost whitish colour on their summit, and of a deep purplish brown on the sides, as is also the intermediate part of the shell; the inside is of a pale brown, with a tinge of a violet red; the spiral turn at the head is small, and the edge is all the way round-sinuated, the depressions running between the extremities of the elevations.

It is found on the coast of Malabar, and no where else, so far as is yet known: few specimens of it have yet got into Europe, and most of those are in Holland.

The other species of the Stomatia are, 1. The small, purple, smooth Stomatia. 2. The small, brown, undulated, and ridged Stomatia. 3. The larger, smooth, purple Stomatia, with fine striæ.

DENTALIUM.

THE Dentalium is a simple shell, having no hinge, and formed only of one piece: it is of a figure approaching to cylindric or conic, and is sometimes crooked, sometimes straight; sometimes closed at one end, sometimes open at both: the animal inhabiting it is a nereis.

The name Dentalium has been given to this genus, from the great resemblance some of the species have to the dentes canini of animals: they have also been called Antales and Entalia by some, and by many, from their tubular form, Tubuli marini.

Dentalium subcylindricum, striatum, et annulatum.
The subcylindric, striated, and annulated Dentalium.

This is a singular shell; when most perfect, it has much the appearance of a fragment only of some of the other species: it is about two inches in length, and of the thickness of a swan's quill; it's colour is a greyish-white, often, but not always, variegated with green, sometimes with a tinge of reddish: it has about ten deep, longitudinal furrows running all the way down it's outer surface, and the ridges between them are rounded; there are usually also three annules or circles passing round at unequal distances, and interrupting, but not wholly discontinuing, the longitudinal furrows: the shell is thin, and easily crushed or broken.

It is common on the shores of Italy, and in some other parts of Europe.

Dentalium conicum, inflexum, læve. **The Dog, Tooth-shell.**
The smooth, crooked, conic Dentalium.

This grows to about an inch in length, and to the thickness of a small goose-quill: the figure is conic, largest at the mouth or opening, and thence becoming gradually smaller to the point, where it is sharp, and has no opening: the whole external surface is smooth, and naturally polished; and the colour is a fine bright white, like that of the finest porcelain; sometimes, but more rarely, it is tinged with purple, sometimes with a bright red or brown.

It is frequent on many of the coasts of Europe, and is also common to the East and West Indies; vast quantities are washed on shore on the coast of Portugal.

Dentalium conicum, inflexum, striatum, et fasciatum.
The conic, crooked, striated, and fasciated Dentalium.

This is a considerably large species; it grows to four inches in length, and to the thickness of a child's finger: the largest part is at the opening, or mouth, and it thence becomes gradually smaller to the other extremity, where it is terminated by a sharp point, without any perforation; it is not straight, but lightly curvated: it's sur-

face

face is furrowed, with eight deep longitudinal lines arranged in pairs, with a vacant space between them: the ridges are rounded, and there are two broad, annular marks surrounding the shell toward the base, and one other much narrower toward the point: the colour of the whole shell is a dusky grey, and it's surface not at all glossy.

It is found on the shores of Italy and elsewhere; and is very frequent, also, fossile, in the mountains of Sardinia.

Dentalium conicum utrinque apertum.
The conic Dentalium, open at each end. — **The great sea-pipe.**

This is six or seven inches in length, and three quarters of an inch in diameter near the mouth: the opening of which is rendered still larger, by a kind of irregular lip which extends it to an inch and half in breadth; from this extremity to the other it diminishes all the way in thickness; and the opposite end is hardly more than equal to a goose-quill, and is open: the whole surface is of a dusky brown, and there are about twelve circular or annular ridges on it; they are not much elevated above the rest of the surface, and those toward the ends are less so than the others.

It is frequent in the German ocean, and is washed on shore in many places in sufficient abundance, but is seldom found entire, being a thin and brittle shell, easily broken by striking against others: when perfect, the larger extremity is not found wide open as in the others, but has an obtusely conic body placed in it, pierced with a great number of small apertures. By long lying on the shores it becomes white, and is very rarely found with this part remaining on it.

The other species of the Dentalium are, 1. The subcylindric, smooth, oblique, and thin Dentalium, called the Dentalium of the shops by many. 2. The irregularly cylindric, flexuous, and contorted Dentalium, found on other shells and corals. 3. The cylindric and variously contorted Dentalium, found in wood immersed in sea-water. 4. The large, striated, conic Dentalium. 5. The large, smooth Dentalium. 6. The straight, conic Dentalium. 7. The long and slender, smooth Dentalium. 8. The short, smooth, open Dentalium.

NAUTILUS.

THE Nautilus is a simple shell, having no hinge, formed of one continued piece, rolled, as it were, into a spiral form, and having it's cavity divided into a great number of cells, by transverse partitions, each of which has a perforation, and is continuous to the others, by means of a siphunculus carried the whole length of the shell.

The animal inhabiting this curiously formed shell is a Sepia.

Of the Nautili, some are thick and strong, others thin, light, and brittle; some are aurited, others are not so; and some are smooth on the external surface, and others furrowed.

Nautilus crassus apertura oblongo-ovata.
The thick Nautilus, with the aperture of an oblong, oval figure. — **The common, thick Nautilus.**

This is a beautiful shell, and grows to a very considerable size, often measuring ten or twelve inches the long way, when we meet with it entire; but the lip is thin, and very liable to be chipped and broken, and most of those brought to us have met with more or less of this kind of mischief: when entire, the lip stands very high, and the opening of the mouth is large, and of an oval, oblong figure; from this part there runs a kind of tail, which turns within the hinder part of the opening of the mouth, and is continued in a spiral form for several circumvolutions, but within the body of the shell, and therefore not visible, but by cutting it vertically open: the shell is considerably thick, heavy, and firm; it's outside is of a brownish hue, variegated with a paler colour; the inside is of the most beautiful mother of pearl colour imaginable: the opening of the mouth does not go deep into the shell, but in it's hinder part there is a hole, which is the first opening of the siphuncle of the inner part; from this place the

the whole cavity of the shell is divided into a multitude of cells, not more than a third of an inch deep in general; the divisions are made by transverse plates of the same pearly shell that lines the mouth, and the pipe or siphuncle, the mouth of which is visible in the base of the first aperture of the shell, is continued regularly, through all the cells, to the very extremity of the shell. The cells are forty in number, and they grow gradually smaller from the first to the innermost of all.

It is found in the Archipelago, and in the Eastern seas; and is very frequent in the fossile state in the clay pits and stone quarries of most parts of England, though not a native of our seas. This and other the like observations prove the vast distance to which creatures were carried from the places of their original residence, at the time of the universal deluge; the only period at which, and the only catastrophe under which, the shells of the East, and the ferns of America, which we find so frequent in this island, could have been brought hither.

Nautilus crassus apertura subrotunda.
The thick Nautilus, with a roundish aperture.

The little, thick Nautilus.

This species is, in the whole, of a rounded form, when entire, and rarely exceeds an inch and a half in diameter: the surface of the whole shell is smooth, and the opening of the mouth large, and approaching to round, but filled up in part behind, by the turn of the spiral part of the shell into it: the colour on the outside is a pale tawny-brown, with lines of a darker brown on it: the inside is of the same fine, pearly hue with that of the former: the mouth is open to a much greater depth in proportion, than that of the other kind; and at it's bottom has an aperture, which runs through a multitude of cells, into which the inner cavity of the shell is divided as in the other.

This is a native of the Persian gulf, and of some other parts of the East.

Nautilus tenuis, auritus, dorso denticulato.
The thin, aurited Nautilus, with a denticulated back.

This is a most extremely elegant species: it grows to eight or ten inches in diameter, and it's shell is scarce thicker than a strong paper: the opening of the mouth is extreamly large, and is of a figure approaching to elliptic, but truncated at the hinder part, where the spiral turn of the shell enters it: the whole rim or verge of the mouth is undulated, and the hinder extremity of it is furnished with two appendages of shell, called ears, one on each side the turn of the shell, which does not nearly fill up the space between them: the whole shell is of a beautiful snow-white, and it's outer surface is beautifully marked with undulated ridges, with circular lines dividing them into spaces, and with a multitude of tubercles: the back is formed all the way into a hollow, from the center of which there rises a denticulated ridge.

This species is frequent in the seas about the East Indies, and in some parts of America: our sailors have sometimes the curiosity to pick them up, as they are in their boats, and the creatures sail by them; these are usually brought over entire to us; but what others we had, used to be greatly damaged, till of late, that shells are become a regular part of commerce, and people are employed to collect them as carefully as any other article of trade.

Nautilus tenuis, non auritus, dorso sulcato.
The furrowed-backed Nautilus, without ears.

The paper Nautilus.

This is a very large and a very elegant species; it is all over of a perfect snow-white, and is not thicker than a piece of strong paper; it measures often near a foot the largest way, but it is thin, and, as it were, compressed: the opening of the mouth is very large, but it is narrow, and there are no ears, but the corners of the hinder part of the mouth, between which the spiral turn of the shell enters the cavity, are high and sharp: the whole outer surface is elegantly ridged with rounded and undulated lines, with hollow spaces between them: the back has a furrow hollowed all the way along it, and the ridges, which make it, are sharp and serrated.

It is found in the Eastern seas, and in some of the warmer parts of Europe; in fine weather there are vast numbers of them seen together, sailing about on the surface of the water, near the shores.

Nautilus tenuis, non auritus, dorso sulcato et undulato.
The thin Nautilus without ears, and with a sulcated and undulated back.

This is a smaller species than the former, but it is a very beautiful one; it measures about five inches the longest way, and is between three and four in depth, but it is very thin, and, as it were, compressed: the opening of the mouth is very long, but it is narrow, and it has no ears, nor even the sharp corners of the former species; the colour is a yellowish-white, and the substance very thin and delicate; the whole surface is ridged in a transverse, undulated manner: the back is furrowed and undulated, and the two ridges of it are raised into broad and obtuse denticulations.

This species is sometimes found about the shores of the American islands, but rarely.

The other species of Nautilus known recent at this time are, 1. The little papyraceous Nautilus, with a ridged back. 2. The little, thick Nautilus, with an oval mouth. 3. The little, umbilicated Nautilus, with furrowed sides. 4. The aurited, thin Nautilus, with a broad mouth. 5. The prickly, aurited, thin Nautilus. 6. The thin Nautilus, with serrated prickles.

These are all that have been found recent in our seas, but we have many other species preserved in the fossile world, which probably are, at this time, the inhabitants of the deep seas, whence no accident ever brings them on shore: of these kinds are the Nautilus armatus, or mailed Nautilus; and the great, thick Nautili of our stone quarries, which differ from the common thick Nautilus: all the species of the cornua Ammonis also belong to this genus; as do also the orthocerAtitæ, and the other concamerated shells found fossile, and figured and described in the former volume of this work: some of these are quite straight, and of a conic form; others have their ends more or less spirally twisted; but all of them have their cavities divided in this manner into many cells, with a siphuncle.

COCHLEA.

THE Cochlea is a simple shell, having no hinge, formed of one single piece, of a spiral form, and containing only one cell, or having it's whole inner cavity undivided: it is of an umbilicated form.

The species of this genus are very numerous, and have certain distinctions, according to which they may be conveniently arranged under several divisions: the clavicle in some is depressed, in others it is elevated; in some the surface is smooth, in others it is striated; in some it is rough, and in others tuberculose, furrowed, or jagged: these are, however, less regular marks of distinction, than are discoverable in the opening or mouth; this, 1. in some is quite round, or nearly so; 2. in others it is semicircular, or semioval, being truncated, as it were, or otherwise terminated by a straight ridge behind; and, 3. in others it is depressed, or formed of a semicircle, crushed together, as it were, or in part filled up by one of the turns of the shell entering into it.

1. Cochleæ, with round, or nearly round openings or mouths.

Cochlea elata tuberculosa ore rotundo aureo.
The tall, tuberculose Cochlea, with a round, yellow mouth.

The gold-mouthed snail.

This is a very elegant shell; it grows to about two inches in height, and an inch and a half in diameter: the thickest part is toward the middle; from thence it grows a little smaller one way to the mouth, and gradually tapers the other way to a point at the extremity: it's surface is deeply furrowed with circular or spiral lines, which seem plicated, and are somewhat irregular, and there are a great number of tubercles on it, disposed regularly in five series; they are moderately large, and are obtuse at the extremity: the surface is variegated with a deep brown, and a paler colour; the extremities of the tubercles are palest; there is much variation in the external colour, but the

the mouth is always of a fine, bright, gold-like yellow within; it is round, and is edged all the way about with a narrow lip.

It is thrown on shore in great abundance in many parts of America, and is found also in some parts of Europe.

Cochlea subulata echinata et laciniata. **The echinated**
The subulated, echinated, and laciniated Cochlea. **Snail.**

This is an extremely beautiful shell; it grows to about an inch in height, and to nearly two inches in diameter in the largest part, which is near the mouth: it does not run up to a point at the upper extremity, but the clavicle is, as it were, flatted for two or three turns; after these the shell lengthens, and extends in diameter till near the mouth, where it again grows somewhat smaller, and the mouth itself is nearly round, and is edged with a narrow lip: the whole external surface of this species is, as it were, echinated; there arise prominences too large for the name of tubercles all over it, following the spiral turn; some of these are simple, a third of an inch long, and pointed at the extremity; others are, as it were, complex, or formed of three or four of these together: these last are flatted and undulated, and are dentated at their extremities: the general colour of the external surface of the shell is yellowish, but these prominences are of an elegant rose colour, sometimes with a tinge of violet: the inside is entirely white, and of a fine, pearly appearance.

This is brought from the shores of many parts of the East, and has also been met with in the American seas. There is indeed no class of animals that is so general, or so much at large, in the place of their habitation, as the shell-fish.

Cochlea elata profunde fasciata. **The belted Snail.**
The deeply fasciated, elate Cochlea.

This is a large species: it grows to two inches and a half in height, and to very nearly as much in thickness; it consists of four very distinct, spiral turns, of which that next the mouth is vastly the largest; the mouth is very large, and is nearly round, but somewhat approaching to oval; the upper extremity is obtuse, though very small; the whole surface is very deeply furrowed in lines, following exactly the spiral turn of the shell; and the prominences between these are so many broad belts of a rounded, not ridged, figure at their tops: the colour of the external surface of the whole shell is a mixt brown and grey; the inside is a bluish-white, with very much of the pearly appearance.

We frequently meet with this shell stripped of it's outer crust in our cabinets, when it has undergone this artificial change; the whole is of a beautiful pearly white, and is very beautiful. It is frequent on the coasts of the American islands, and in some other places.

Cochlea elata lævis fasciata. **The smooth, rib-**
The elate, smooth, fasciated Cochlea. **band Snail.**

This is one of the tallest, or longest, in proportion to it's breadth, of all the Cochleæ: it is two inches long and little more than an inch in diameter; it consists of five spiral turns, and is terminated at the top by a very sharp point: the whole external surface is smooth; there is not the least protuberance, nor so much as the least line or furrow on it: it's colour is a dusky, but not disagreeable, brown, paler in some places than in others, and it is variegated with a number of broad and very beautiful streaks in the form of ribbands, following the volution of the shell; these are of a very elegant reddish-brown, and are variegated in some places with whitish; between these there are often other narrow and fainter lines of the same kind, and the whole is very beautiful.

We have it on the shores of many parts of Italy.

Cochlea depressa lævis.
The smooth, depressed Cochlea.

The White Cornu Ammonis Snail.

This is a very beautiful shell, and in it's general shape differs extremely from the two preceding; it consists of five spiral turns, but it is perfectly flat, the clavicle not at all elated, but the point is sunk lower than any other part of the shell, in consequence of it's being more slender; the diameter of this shell is about two inches and a half, and it's thickness in the largest part not more than half an inch: it's mouth is round, and the shell is there largest, and goes from thence gradually tapering all the way to the central point, the surface is smooth, or at the utmost has only the appearance of a few obliterated fasciæ: the colour is entirely white, except that the eye of the volute is brown.

It has something of the external figure of the Cornua Ammonis, which we find fossile, but their cavity is divided by cells in the manner of the nautilus; this is entire, so that it is quite of another genus.

Cochlea elata cancellata et tuberculosa.
The cancellated and tuberculose, elate Cochlea.

The Warty Snail.

This is a small species; it scarce ever exceeds three quarters of an inch in length, or about half an inch in diameter: it consists of four spiral turns, and it's top is obtusely pointed; the whole outer surface of the shell is marked with spiral lines, and with others crossing them: these are both disposed at such small distances, that they form a kind of chequer of small squares, in the centers of many of which there stand little tubercles, in the manner of warts: the shell is of two colours; the upper part of it is of a pale brown, and the tubercles on it are of a whitish colour, and make a very pretty appearance; the lower part has a tinge of greyish, and the protuberances are large, obtuse, and of the same colour: the mouth is nearly round, and both that and the whole inner surface of the shell are of a fine pearly white.

It is frequent on the coasts of America, and is found also in several parts of Europe.

The other species of the round or fully open-mouthed Cochleæ are, 1. The great, fasciated, and tuberculose Cochlea. 2. The large, smooth, grey, and white Cochlea. 3. The large, green Cochlea. 4. The variegated, smooth Cochlea, called by many the serpents skin-snail. 5. The oval, umbilicated Cochlea. 6. The flat, belted Cochlea. 7. The villated Cochlea. 8. The deeply-furrowed Cochlea, with sharp ridges. 9. The green and yellow Cochlea. 10. The fine, tall, chesnut-coloured, unvariegated Cochlea. 11. The reddish, brown, variegated, smooth Cochlea. 12. The greenish, spotted Cochlea, with the depressed clavicle. 13. The little, entirely green Cochlea, called the Emerald shell. 14. The greenish, Chinese Cochlea, with two white lines. 15. The brown, Chinese Cochlea, with a white cross. 16. The violet colour Cochlea. 17. The yellow and white-streaked Cochlea, with an extended lip. 18. The rough and furrowed Cochlea, with a silvery mouth. 19. The umbilicated and granulated Cochlea. 20. The depressed Cochlea, every-where armed with pointed tubercles. 21. The small, elegant, red Cochlea. 22. The simply, fasciated, brown and white Cochlea. 23. The smooth, marbled, and clouded Cochlea. 24. The more deeply, fasciated Cochlea, with undulated plates. 25. The lesser, smooth-clouded, obtuse-pointed Cochlea. 26. The high-ridged, clouded Cochlea. 27. The roundish and somewhat depressed Cochlea, with clouded variegations. 28. The tall, slender, fasciated Cochlea. 29. The tall, slender, and sharp-pointed, smooth Cochlea. 30. The tuberculose Cochlea, with obtuse prominences. 31. The depressed Cochlea, with pointed, simple spines at the middle of every turn.

2. *Cochleæ, with the opening or mouth of a half round figure, or truncated behind. This division of the Cochleæ comprehends the neritæ of authors, which are genuine Cochleæ.*

Cochlea levis fusca ore magno.
The smooth, brown Cochlea, with a great mouth.

This is a beautiful species; it grows to about an inch in diameter, and as much in height: it's surface is smooth and even, and it's colour a bright agreeable brown, gradually growing paler toward the edges of the shell, till it becomes almost whitish: the clavicle is but little exerted, and the mouth is very large; it is of a very regularly semicircular figure, the anterior part of it extending to a true segment of a circle, and the hinder being formed into a straight line, cutting it off, as it were; this part has a small lip, and an umbilicated hole behind it; the whole inner surface of the shell is whitish.

It is a native of the East Indies.

Cochlea fasciis tribus elevatis.
The Cochlea, with three elevated fasciæ.

The three-ribbed Snail.

This is a singular and beautiful species: it grows to an inch and half in diameter: the body is large; the clavicle small, spiral, and depressed. The whole is smooth, except that there run three lines or fasciæ, which rise above the surface, which are narrow, but not very sharp, along the whole body, dividing it, as it were, into so many spaces: the ground colour of the shell is a reddish-brown, and the variegations, which run in the form of irregular clouds, are of a paler colour: the mouth is large and semicircular, and both that and the whole inner surface of the shell are of a pearly hue.

It is frequent on the coasts of South America, and is also brought sometimes from the East Indies.

Cochlea subrotunda levis ore lato.
The roundish, smooth Cochlea, with a wide mouth.

This is a small species, and is the roundest in it's figure of any of this genus: the body is large, and is about equal to it's depth in it's diameter: the clavicle is small, and forms a depressed spiral turn at the end: the mouth is very large, and is somewhat broader than it is deep; it is regularly semicircular at the anterior part, but the part of the shell that cuts it off behind, is continued forward toward the middle, so as to fill up a part of the opening; it is in this place formed into a kind of lip, and has an umbilicated hole behind it: the whole surface is smooth, and of a greyish-white, variegated with two irregular zones, formed of a number of clouded spots, rough and lacerated, as it were, at the edges: the inside of the shell is of a pearly white.

This species is frequent in the East Indies, where it is not uncommon to find the Cancelli, or Hermit Crabs, while young in them.

Cochlea labio posteriore bidentato.
The Cochlea, with the posterior lip bidentated.

The toothed Nerite.

This is one of those Cochleæ which have been injudiciously separated from the others, under the name of Neritæ: it grows to about an inch in length, and three quarters of an inch in diameter: the body is large, and of an oblong inflated figure; the clavicle is small, and forms a depressed spiral turn: the colour of the outer surface of the shell is grey, variegated with a deeper iron grey, which is almost black; the variegations lie in irregular clouds, not very large: the inside is whitish or pearly; the mouth is large and semicircular; the hinder lip is dentated, or formed about the center with two oblong and obtuse teeth, which are sometimes reddish towards their extremities, sometimes all over: in this case, our collectors call it the Bloody-tooth Nerite, but they give the same name also to the next species, and to several others under the same accidents.

Cochlea profunde fasciata, ore utrinque bidentato.
The deeply fasciated Cochlea, with the mouth bidentated each way.

This is a very elegant species; it grows to about an inch in diameter, and is nearly round; the clavicle is very small and depressed: the whole outer surface of the shell is formed into elevated broad fasciæ, which are only separated by narrow furrows, and run all the way parallel to one another: the colour is a pale whitish-grey, variegated in a beautiful manner with black: the inside of the shell is white and pearly; the mouth is of an irregularly semicircular figure: the anterior lip is broad, thick, and prominent, and the posterior, or transverse, one is very broad; each of them has two denticulations in the form of teeth, which are sometimes white, like the rest of the inner surface of the shell, but often red, and as it were bloody.

It is very frequent in the East Indies, and has been met with on some of the European coasts.

Cochlea elevata spinosa.
The elevated, prickly Cochlea. **The prickly Snail.**

This is a small, but a very singular and elegant, species: it's length, or height, is about three quarters of an inch; it's diameter hardly half an inch; it consists of four turns, and it's surface is formed into perpendicular, obtuse ridges, with small hollowed spaces between them; round the larger turns, it has two series of long and slender spines, very sharp-pointed; the outside of the shell is beautifully variegated with a deep reddish-brown, and a paler colour approaching to white.

It is a Chinese shell; we have it in some of our cabinets.

Cochlea elata fusco-nebulata.
The elated Cochlea, clouded with brown. **The clouded Snail.**

This is about an inch and a half in height, or length, and an inch in diameter: it consists of five spiral turns, and the extremity of the clavicle is small and obtuse: the whole external surface is smooth; no rising, nor any fasciæ or protuberances, appear on it, nor even so much as a streak of any kind: the ground colour is a pale brown, and the variegations, which are disposed irregularly in small clouded spots, are of a very dark brown; the mouth is large, and of a tolerably regular figure of the semicircular kind; that and the whole inside of the shell is silvery.

The other species of this genus are, 1. Of those generally called Cochleæ by authors. 1. The large, broad-mouthed Cochlea, with an oblong umbilicus. 2. The roundish, smaller-mouthed Cochlea, with the apex but little exerted. 3. The roundish, smaller-mouthed Cochlea, with the apex depressed. 4. The fine, roundish, umbonated Cochlea. 5. The thin, roundish, ivory Cochlea. 6. The thick, ivory, white Cochlea. 7. The small, orange-coloured, large-mouthed Cochlea. 8. The great, brown Cochlea.

2. Of those called by authors Nerites, or Neritæ, there are, 1. The common, smooth, yellow Nerite: this varies in colour, being sometimes white, sometimes brown. 2. The lesser, white Nerite. 3. The larger, fasciated, thick-lipped Nerite. 4. The larger, slightly striated, broad-lipped Nerite. 5. The striated and pustulated Nerite. 6. The fasciated Nerite, with black and yellow spots. 7. The deeply furrowed Nerite. 8. The brown and yellowish spotted Nerite. 9. The brown and greyish spotted Nerite. 10. The rostrated jasper Nerite. 11. The small, greenish, oblong Nerite. 12. The small, yellow, oblong Nerite. 13. The reticulated Nerite. 14. The larger, prickly Nerite. 15. The greyish Nerite, spotted with red. 16. The greyish-brown Nerite, spotted with black. 17. The black, red-streaked, and spotted Nerite. 18. The lightly striated, green Nerite. 19. The larger, black and yellow Ne-ite.

3. *Cochleæ, which have the mouth flatted or depressed, and in part filled up with the turn of the shell: these are generally of a conic figure: they are generally broad at the base, and, in most, the clavicle is elated; some have it depressed, and are umbilicated;*

licated; and most of them are remarkable for the pearly splendor of their inner surface. Authors have called the greater part of the species of this division Trochi.

Cochlea conica tuberculosa.
The conic, tuberculose Cochlea.

The rough Trochus.

This is a large and very elegant shell: it is often three inches in height, and two in breadth at the base: it's form is pretty regularly conic; it's mouth is moderately large: it consists of about six spiral turns, and is terminated at the top by an obtuse end. It is elegantly beset with tubercles: they are large, broad at the base, and obtuse, and are disposed in five or six series, and the rest of the surface is undulated: the colour is a beautiful grey, variegated with blackish.

It is a native of the East Indies; we have it in great abundance sometimes from China: the inner surface is of a fine pearly hue, and it is a common custom with our collectors of pretty shells to have the outer coat of this taken off with acids; after which, a little polishing gives it the entire pearly appearance throughout.

Cochlea conica laevis,
The smooth, conic Cochlea.

The smooth Trochus.

This is a large species; we meet with it of three inches in height, and more than two in breadth: it consists of about seven distinct spiral turns, and gradually becomes smaller all the way from the base to the top, where it is terminated by an obtuse end: the mouth is moderately large, and is of a silvery white within, as is also the whole inner surface of the shell: the outer surface is tolerably smooth, and is of a whitish colour, variegated, with irregular rays of reddish and brown.

It is an East Indian shell, but we have it also from some other places.

Cochlea conica transversim radiata.
The transversely radiated conic Cochlea.

The wavy Trochus.

This is about two inches in height, and an inch and a half in diameter at the base; it consists of about seven spiral turns; the two lowest disproportionately large and flattish, and the others very small; the extremity is an oblong, obtuse point: the whole shell is of an elegant white, the outside approaching to the porcelain white, and the inner surface being perfectly pearly: there are a great number of transverse ridges on all the parts of the shell; they are considerably elevated above the rest of the surface, but they are not ridged, but rounded irregularly at the summit.

It is a native of China, and some other parts of the East.

Cochlea conica spinosa.
The prickly conic Cochlea.

The prickly Trochus.

This is one of the most elegant of this genus. It is about two inches in height, and an inch and a quarter in diameter at the base; it consists of about ten spiral turns, but the four or five upper ones are very indistinct: the lower ones are large, and, round the edges of the three or four larger, there stand regular series of sharp and tolerably strong prickles: the surface of the lower part of the shell is somewhat irregularly raised into a few transverse ridges, and has many fine lines running principally in pairs in the course of the volutions: the outside is of a beautiful chestnut brown; the spines paler, and the lines darker than the rest.

It is a native of the East Indies: we have it principally from China.

Cochlea depressa laevis.
The smooth, depressed Cochlea.

The French-horn Snail.

This is an elegant species, but of an extremely different form from all the preceding; they are all elevated into a conic figure; on the contrary, this is depressed and flat, and the extremity of the clavicle, is sunk within the rest of the surface. It is of an orbiculated figure, and about two inches in breadth, and half an inch thick: it's it's mouth is a segment of a circle depressed, and filled up in a great measure by the succeeding turn of the shell; it consists of four or five volutions: It's colour is usually yellowish, sometimes white.

It is frequent on the shores in many parts of the East Indies, and in some other places.

Cochlea parum elevata margine volutionum serrata.
The little elevated Cochlea, with the edge of the volutions serrated.

The serrated Snail, or Spur-shell.

This is a very beautiful, as well as very singular, shell: it is about two inches in breadth at the base, and not much more than an inch in height. It is composed of about six spiral turns or volutions, and each of these has a sharp edge which is beautifully serrated, the denticulations rising into a kind of spines, which are broad at the base, and pointed at the extremity: in many positions, the shell gives a fine resemblance of a spur with large rowels; the clavicle is depressed at the top; and the mouth of the shell is considerably large; the colour varies; it is sometimes of a beautiful pale yellow, sometimes perfectly white; the inner surface is always of a beautiful pearly hue.

It is common in many parts of the East Indies, and has not yet been met with elsewhere.

Cochlea conica elata undulato-tuberculosa.
The somewhat conic, elate Cochlea, with tuberculous undulations.

This is about an inch and a half in diameter, and nearly as much in height: it consists of about six spiral turns, and is terminated at top by a little roundish button; it's surface all over is elevated into a kind of undulated, transverse lines; the central points of which are raised into blunt tubercles: the colour on the surface is a greenish-grey: the more prominent parts, and the summit of the clavicle, are paler than the rest, and sometimes whitish.

It is common to the East and West Indies; but, in Europe, we have not yet found it.

The other species of the flat-mouthed Cochlea are, 1. Of those which have the apex of the clavicle exerted. 1. The elegantly spotted, brown and grey Cochlea. 2. The conic Cochlea, with red and white spots. 3. The elegant, punctulated, conic Cochlea. 4. The large, green, conic Cochlea: this is often stripped of it's outer green coat, and is then perfectly pearly throughout. 5. The smaller, knotty Cochlea. 6. The smooth, greenish Cochlea. 7. The smooth, reddish Cochlea. 8. The undulated, yellowish Cochlea. 9. The slightly, undulated, greyish Cochlea. 10. The dusky-grey, lower Cochlea. 11. The spotted, black and white, lower Cochlea. 12. The streaked, yellow and black, lower Cochlea. 13. The Cochlea, with elevated ribs, and a pointed top. 14. The tuberculose and prickly Cochlea. 15. The beautifully variegated Cochlea, called the Stair-case Shell. 16. The white Cochlea, with depressed ribs. 17. The very rough, grey, and black Cochlea. 18. The dentated, Cochlea. 19. The Spur-shell, with short spines. 20. The gold-yellow Spur-shell, with a silvery white umbo. 21. The totally-yellow, long-spined Spur-shell.

2. Of those which have the apex of the clavicle more plain, or even depressed; there are, beside the described species, 1. The brown and grey Cochlea, with an expanded mouth. 2. The flat Cochlea, with white and brown variegations, in broad irregular lines. 3. The small depressed Cochlea, with a dentated mouth. 4. The larger, dentated, mouthed Cochlea. 5. The yellow Cochlea, hollowed in the center. 6. The yellowish, broad Cochlea, with the apex a little exerted. 7. The whitish Cochlea, with brown lines. 8. The brown, flat Cochlea, with yellow and white lines. 9. The white, flat Cochlea, with yellow spots and lines.

BUCCINUM.

THE Buccinum is a single shell having no hinge, formed of one piece, and that shaped in some degree like a horn, or other wind instrument: the belly of the shell is distended; the aperture of the mouth is large, wide, and elongated; the tail is more or less long, and the clavicle more or less covered.

Lister has made so large a genus of this, that it comprehends almost all the other shells of the univalve class; but the species, under these limited characters, are sufficiently numerous, and a multitude of others, that would have bred confusion, are by these restrictions thrown out, and made the species of the several succeeding genera, all of which are in some degree related to one another.

Buccinum tenuius tuberculosum rostro elongato.
The slender tuberculous Buccinum, with an elongated rostrum. **The Spindle-shell.**

This is a very singular species: it's length is four or five inches, and it's diameter, in the thickest part, not more than an inch: this thickest part is nearly in the middle of the shell, for the mouth has a rostrum or snout for it's lip, which is continued almost to an equal length one way, with the clavicle the other: the shell consists of about ten spiral volutions, each of the five larger, or lower ones are elevated into a kind of ridge in the middle, which ridge is formed of a row of short and obtuse tubercles: the rostrum which goes from the mouth is slender, and is spirally radiated: the general colour of the outer surface is white, but the tubercles are yellowish, as are also the radiations and lines between the several series of them.

It is found on the shores of many of the American islands.

Buccinum tenuius læve rostro fisso.
The smooth and slender Buccinum, with a split rostrum. **The Mitre-shell.**

The general length of this species is about three inches: it's diameter is three quarters of an inch: it consists of about five volutions, the lower one of which is at least equal in length to all the rest together: the ground colour of this shell is white, and it is beautifully variegated with irregular clouds and spots of a dusky reddish: it is perfectly smooth and even: the mouth is long and narrow, and the upper extremity of it is, as it were, split.

This was first brought from the East Indies, but it has since been found to be a native also of the American seas, and of many other places.

Buccinum tenuius læve, rostro oblongo, labio exteriore denticulato.
The smooth, slender Buccinum, with an oblong rostrum, and the outer lip denticulated.

This is also a very singular and beautiful shell: it's length is five or six inches; it's diameter, in the largest part, scarce more than an inch, and it's rostrum equal to about half the length of it's tail. It consists of twelve spiral volutions: the lowest of these is largest, and somewhat bellied; it opens, at the front, into an oblong mouth, the anterior lip of which has four or five large denticulations on it: and the joining of the two lips is continued into a rostrum or snout, of a long and slender figure, and pointed at the end: the surface of the whole shell is smooth, and the colour white.

It is very frequent on the shores of our American islands.

Buccinum ſublongum tranſverſim radiatum.
The ſublong and tranſverſely radiated Buccinum.

This is a very ſingular and beautiful, and a very ſcarce, ſhell; it's length is about four inches, and it's diameter, in the broadeſt part, nearly two. It is compoſed of ſeven ſpiral turns, and theſe are ſeparated, as it were, one from another, and are flatted on their ſides, not bellied as in moſt others: the colour of the outer ſurface is yellow, and it is all over tranſverſely radiated with thick and ſomewhat prominent, rounded ſtreaks: the opening of the mouth is oblong and large, and the extremity of the roſtrum obtuſe.

It is a native of the Eaſt, and is ſeldom met with: it ſeems to be an inhabitant of the deep ſeas, and only to be thrown up in great ſtorms.

Buccinum brevius tuberculoſum ore magno.
The ſhort tuberculous Buccinum, with a large mouth.

This is a beautiful ſpecies: it's length is about two inches, and it's breadth, in the largeſt part, nearly as much: the mouth is oblong and very large, and the outer lip broad; the extremity of the mouth is a little ſplit. The tail is ſhort and thick; it conſiſts of about ſix volutions: the outſide of the ſhell is of a browniſh colour, with ſome variegations of red and whitiſh, and is beſet with a great number of ſhort and obtuſe tubercles, diſpoſed in ſeveral ſeries.

It is a native of the Weſt Indies, and is frequent in the cabinets of our collectors.

Buccinum brevius læve ventricoſum.
The ſmooth, ſhort, ventricoſe Buccinum.

This is a very ſingular ſhell, in that the volutions go the contrary way to thoſe of the generality of theſe, and the mouth in conſequence, is turned a contrary way. It is about an inch and three quarters long, and near an inch in diameter in the thickeſt part. It conſiſts of ſix volutions, the loweſt of which is much the largeſt, and is inflated, as it were; the mouth is ſmall, and of a figure approaching to an oblong oval: the whole ſurface of the ſhell is ſmooth, and it's colour is yellow.

It is a native of the American ſeas; we have it from Barbadoes.

Buccinum brevius læve labio protenſo.
The ſhort, ſmooth Buccinum, with an expanded lip.

This is an inch and a half in length, and more than an inch in diameter in the broadeſt part: it conſiſts of about four ſpiral turns, and the extreme one is thick, and very obtuſe at the top: the mouth is oblong and large, and the lip is expanded to ſome breadth: the outer ſurface of the whole ſhell is ſmooth, and is of a pale browniſh colour, elegantly variegated with a great number of yellow zones.

It is found on the ſhores of many parts of the Eaſt Indies, and is met with foſſile in France very perfect.

Buccinum tenuius cancellatum.
The ſlender, cancellated Buccinum.

This is about two inches and a half in length, and is no where more than an inch in diameter: it is compoſed of ſeven volutions, and the four extreme ones are very ſmall, the laſt terminating in a fine point: the mouth is oblong, moderately large and reddiſh; the lip is narrow, and is covered at the bottom, and ſplit a little at the top: the whole outer ſurface of the ſhell is ornamented with longitudinal ridges, elevated pretty high and rounded; theſe croſs the volutions, and ſtand ſomewhat cloſe to one another; and there are a number of broad lines alſo following the volutions of the ſhell, which, croſſing the ridges, divide the whole ſurface into ſmall ſquares, in a beautiful manner.

It is frequent on the ſhores of the American iſlands.

Buccinum breve ventricosum ore angusto dentato.
The short, smooth, ventricose Buccinum, with a narrow dentated mouth.

This is one of the most singular species of this genus: it is an inch and a half in length, and near an inch and a quarter in diameter: it consists of about nine volutions, but the lower one makes, alone, almost the whole shell, the others being all comprised in the space of the third of an inch; the lower, or great volution, is of a figure approaching to oval, but too small at the larger end, to answer exactly to that figure: it's mouth is small, oblong, and narrow; the lip is carried at each end a little beyond the aperture, and forms a ridge; and the mouth is dentated on both sides: the colour is a pale brown, variegated with a deeper colour of the same kind, and, in some places, with a faint whitish.

It is a native of the East Indies, and is common to some other places.

Buccinum crassum ore magno simplici.
The thick Buccinum, with a great simple mouth.

This grows to four or five inches in length, and is two inches and a half in diameter in the thickest part: it consists of six spiral turns or volutions, and is terminated by a small but obtuse end: the lower volution, which forms, as it were, the body of the shell, is ventricose and large; the mouth is large and oblong, and is quite simple, the lips being thin, and not at all edged: the colour is a beautiful whitish, variegated with red, yellow, and brown, in an elegant manner, and the whole surface is smooth, and has a fine natural polish.

It is a native of the West Indies.

Buccinum ovato-oblongum læve ore longo angustissimo.
The smooth, ovato-oblong Buccinum, with an oblong and very narrow mouth.

The Midas Ear-shell.

This is one of the most singular of this varying genus; it is three or four inches long, and between two and three in diameter; the figure is oblong, and, with the length of the aperture of the mouth, has some resemblance to the ear of an ass, whence it has obtained it's common name of Midas's Ear. It consists of six volutions, but the lower one alone makes up almost the whole shell; this is of a figure approaching to oval: the mouth reaches almost the whole length of it, and is very narrow, appearing scarce more than a slit, somewhat broader at the bottom than elsewhere: the lip is rounded, and, toward the bottom, has a kind of reduplication: the surface is perfectly smooth, and the colour a pale brown.

It is a native of the American seas.

Buccinum brevissimum rostro sublongo cavo.
The very short Buccinum, with a long, hollowed snout.

This is about two inches and a half in length, and near two inches in diameter, but, excepting for the snout or rostrum, it is much thicker than it is long: it is composed of about six spiral volutions, but they are very short and flatted, and the extremity is obtuse; the body of the shell is large; the mouth wide, and of a somewhat oval figure; and the extremity of the shell continued into a kind of snout, considerably long, slender, and hollowed: the lip is thick and wrinkled, and the whole surface of the shell is very deeply wrinkled, and is full of low, oblong, protuberances, which seem, indeed, little more than the higher elevations of the ridges between the wrinkles.

It is a native of the American seas.

The other species of the Buccinum are considerably numerous; they are, 1. The great, long-tailed, and oblong-mouthed Buccinum. 2. The great, brown and white, contabulated and tuberculous Buccinum. 3. The small, brown, furrowed, long, Buccinum. 4. The long, brown Buccinum, with a duplicated and dentated lip. 5. The long Buccinum, with a narrow body, called the Babylonian Tower-shell. 6. The oblong Buccinum, variegated with red spots 7. The long-pointed Buccinum, called

called the Perſian veſt. 8. The long, radiated Buccinum, with broad, blackiſh ſpots. 9. The faſciated and undulated Buccinum, with a dentated, hinder lip. 10. The thick Buccinum, with oblong and ſpotted tuberoſities. 11. The long Buccinum, with brown and blue ſpots. 12. The ventricoſe, ſtriated, and ſpotted Buccinum. 13. The Buccinum, with the convoluted or pulvinated clavicle. 14. The long, ſtriated Buccinum, with three tuberous eminences. 15. The long, coſtated, and ſtriated Buccinum. 16. The little Buccinum, with a ſhort tail, and an expanded mouth. 17. The ſhort, hairy Buccinum. 18. The grey, wide-mouthed Buccinum. 19. The ſhort, ſpotted, and lineated Buccinum. 20. The undulated and ſtriated, ſhort Buccinum. 21. The ſhort Buccinum, with a great number of acute tubercles, regularly diſpoſed all over it. 22. The depreſſed, umbilicated Buccinum, with the lip and the columella both dentated. 23. The ſhort, yellow, umbilicated Buccinum. 24. The ſhort, alated, and punctated Buccinum. 25. The ſhort, tuberculoſe Buccinum, with two very high ribs. 26. The rough, furrowed Buccinum, with an oval mouth, and thick lip, and with ſome prominent ribs. 27. The long Buccinum, with a dentated clavicle, called the Chineſe Tower-ſhell. 28. The long Buccinum, with very elegant red and white faſciæ. 29. The long Buccinum, with the clavicle and the lip both dentated. 30. The long Buccinum, with ſhort, undulated variegations. 31. The long, brown Buccinum, with the columella dentated. 32. The long Buccinum, variegated with many brown and yellow faſciæ. 33. The long Buccinum, with red lines and red ſpots. 34. The great triton-ſhell, or inflated Buccinum. 35. The long Buccinum, with an expanded lip, and with many annular, broad lines. 36. The ſtriated Buccinum, with rows of prickly ribs. 37. The reticulated Buccinum, with oblong tubercles. 38. The ſpotted Buccinum, with the clavicle irregularly ſtriated, and the extremity red. 39. The ſhort Buccinum, with the crooked beak. 40. The ſhort, rough Buccinum. 41. The diſtorted Buccinum, with the columella and the lip both dentated. 42. The ſhort Buccinum, with the rim of the mouth expanded. 43. The ſhort Buccinum, with the rim of the mouth prominent, but leſs expanded. 44. The Buccinum, with low and obtuſe tubercles. 45. The rough Buccinum, with elated tubules. 46. The tuberculous Buccinum, with a broad, dentated mouth, and a long, crooked, and furrowed beak. 47. The yellow, tuberous, and ſtriated Buccinum, with the columella and lip ſtriated, and the clavicle depreſſed. 48. The red-mouthed Buccinum, with the clavicle erect. 49. The ſtriated Buccinum, with oblong tubercles. 50. The ſtriated, gold-yellow Buccinum. 51. The elegantly variegated Buccinum, with white, yellow, and brown ſtriæ.

TURBO.

THE Turbo is a ſimple ſhell, having no hinge, and formed of one continued piece; it is long and ſlender, of a figure more or leſs regularly conic, always terminating in a very long and fine point; the mouth is narroweſt toward the baſe, and is sorted; the ſhape is irregular: in ſome it is oblong in others, rounded, broad, or depreſſed; in ſome dentated, in others ſmooth.

Some authors have called theſe Strombi, and many have confounded them with the Buccina. In Engliſh we call them Screw-ſhells; the French, Vis.

Turbo craſſior et ventricoſior ore ſubovato.
The thick ventricoſe Turbo, with an oval mouth.

This is a large ſpecies; we meet with it ſeven inches long, and near two inches in diameter in the largeſt part: it conſiſts of about fourteen volutions, the ten larger are ſufficiently diſtinct; the four extream ones rather confuſed; the loweſt volution is the largeſt, all the others gradually diminiſh from this to the extremity, where it is pointed; they are all rounded on the ſurface or ventricoſe: the ſurface of the ſhell is ſmooth, and the ground colour white; it is beautifully variegated with yellow rays, and with elegant and broad faſciæ, formed of a mixture of blue and brown lines, interrupted by irregular ſpirals of the ſame colours: the mouth is large and wide; the lip thin, and not dentated.

It is abundant in America, and is found alſo in ſome of the European ſhores.

Turbo gracilior spiris ventricosis ore parvo rotundo.
The slender Turbo, with ventricose spires, and a small round mouth.

The Needle-shell.

This is a very elegantly formed shell; it's length is five or six inches; it's diameter, in the thickest part, not much more than half an inch; it is largest at the extremity, where the mouth is, and hence becomes gradually smaller to the other, where it terminates in a point; it consists of about fifteen spiral volutions, and they are all bellied or inflated, as it were, their middle rising very high, and the divisions between them deep and narrow: the mouth is small, and is nearly round; it is quite simple, having only a thin duplicature, with a kind of notch in the place of the ear in the other; the surface of the whole shell is perfectly smooth, there is not so much as a line marked any where on it; it's ground colour is white, and it is variegated with yellow.

Turbo gracilior spiris ventricosis ore ovato.
The slender Turbo, with ventricose spires, and an oval mouth.

This is a small species; it's length is about two inches; it's diameter, in the thickest part, hardly so much as a third of an inch; it consists of about fourteen volutions, of which the lowest is much the largest, and the rest decrease gradually to the extreme one, which is very small and pointed; they are all rounded or prominent on the surface, and the furrows are deep that separate them: the mouth is moderately large and oval, and is edged but with a thin lip.

It is very common on the shores of the American islands.

Turbo spiris ventricosis, costis plurimis elevatis.
The ventricose-spired Turbo, with numerous, elevated ribs.

This grows to an inch and a half in length, and it's largest volution is more than a third of an inch in diameter; this, which is the lowest, is much bigger than the succeeding one; the others are gradually smaller to the top, where it terminates in a point: there are about seven of these volutions; they are all rounded or bellied, and are beautifully ribbed, each of them having a number of straight, perpendicular, and very elevated ribs running down it at small distances: the whole shell is white, as are also those ribs; the mouth is round, moderately large, and furnished with a thick lip.

It is very frequent on the shores of Barbadoes.

Turbo conico spiris planis numerosissimis striatis.
The conic Turbo, with plane, striated, and very numerous spires.

The Telescope-shell.

This is the most regularly conic, in it's figure, of all the shells of this genus; it's length is about four inches, it's diameter at the base is an inch and a quarter, and from thence it gradually becomes smaller to the summit, where it is terminated by a fine point; the volutions are broad toward the base, but they become afterwards very narrow, and are very numerous, but indistinct; the volutions are all flat, and are spirally striated; the colour is a plain brown, deepest in the thickest part, and almost white at the point: the mouth is very oddly formed, it is flat and not large, and the lip runs each way to some distance beyond it, in the form of a ridge.

It is found in the East Indies, and in America: some parts of Europe also have afforded fair specimens of it.

Turbo gracilior spiris spiraliter striatis.
The slender Turbo, with spiral lines on the volutions.

This is four or five inches long, and scarce half an inch thick in the first volution, from this it gradually becomes smaller to the extremity, where it terminates in a point: the volutions are all rounded, and the lines between them somewhat deep; they are very beautifully

beautifully striated all the way, with deep, continuous furrows following the turn of the shell: the mouth is small, and irregularly oval; the lip thin; the colour is yellow, but it is sometimes very pale, and in shells that have lain long on the shore, after the death of the animal, it is often white.

It is found in great abundance on the shores of Barbadoes, and other of the American islands; very beautiful ones have also come from the East Indies.

Turbo spiris angustis prominentibus distantibus.
The Turbo, with distant and prominent spires.

This is a considerably large and very singular species; it's length is about five inches, and it's thickness, in the largest part, or lowest spire or volution, about three quarters of an inch; from this it gradually diminishes to the other extremity, where it terminates in a point: the volutions are about seventeen, and are of a very singular kind; they are very high, extreamly narrow at the surface, and have very deep furrows between them: It's mouth is irregularly rounded, and the lip not large, but a little auriated; it's colour is whitish, but with some tinge of yellow and of red.

It is a native of the East and West Indies; the finest are from China.

Turbo spiris angustis planis bullatis.
The Turbo, with narrow flatted, and bullated spires.

This is an extreamly elegant species; it's length is about four inches; it's diameter, in the largest part, about half an inch; from this, it gradually tapers to a fine point: it is composed of about fourteen spires, but the extream ones are very indistinct; they are all flatted and undulated in a very elegant manner; but what is most singular of all in it is, that there runs a series of bullated dots, or protuberances, obtuse and low, and of a paler colour than the rest of the shell, along the edge of each spire: the mouth is large, of a very irregular form, and projects from the rest of the shell; the colour of the whole is a dusky brown.

It is an oriental shell, and has not yet been met with elsewhere.

Turbo verrucosa ore lato depresso. The caterpillar shell.
The verrucose Turbo, with a broad, depressed mouth.

This is about two inches and a half long, and near three quarters of an inch in diameter in the largest volution; it consists of about twelve of these, and terminates in a point: the several volutions, particularly the larger, are irregularly beset with obtuse, and not very tall, protuberances: the mouth forms a detached part from the rest of the shell, connected to it by a kind of neck: it is broad and depressed, and the lower lip is a little turned back, and both extremities are auriated: it is elegantly coloured; the ground is pale, and there are a great many variegations of a darker colour in it, disposed very beautifully; the protuberances are bluish, and give a very happy variation to the rest.

It is a Chinese shell.

Turbo ore longo hiante.
The Turbo, with a long, gaping mouth.

This is three inches long, and of the thickness of one's little finger; it is extreamly singular in it's structure, scarce at all resembling any of the other species of it's genus; it is, indeed, largest at the end where the mouth is, and thence becomes gradually smaller to the other extremity; but then it consists only of three volutions, of which the first constitutes almost the whole shell; the mouth is near two inches long, and is narrow at the lower part, but very gaping toward the extremity of the shell, and the lip is alated; it is of a pale brownish colour, very beautifully variegated with a deeper brown, and with a reddish tawny; and it's surface is perfectly smooth, except that at the rim of the second volution there are a few tubercles.

It is an East Indian shell; some have been said to be brought from the American shores, but, I am afraid, on no very good foundation.

Turbo

Turbo fasciis spiralibus latis, ore parvo.
The Turbo, with broad, spiral fasciæ, and a small mouth.

The ribband, Screw-shell.

This is about two inches and a half long, and a little more than a third of an inch in diameter at the larger end, whence it gradually decreases till it terminates in a point: the volutions are about ten in number, and are somewhat rounded, but not greatly so; the lines which divide them are very small; the mouth is of a figure approaching to oval, but somewhat irregular, and is small, in proportion to the bigness of the shell; there run all the way round the volutions a number of broad fasciæ or belts, from five or six to two in number; these follow the whole course of the shell, and stand but at small distances from one another; they are of a darker colour than the rest of the shell, and are very even and determinate at their edges.

It is frequent on the shores of the American islands.

Turbo muricata ore anguste oblique.
The muricated Turbo, with a narrow, oblique mouth.

This is two inches long, and a third of an inch in diameter at the base; it is formed into about fourteen volutions, they are flattish, or somewhat hollowed inward, toward the base; their outer edge is more prominent, and is all the way armed with a series of very robust and large spines, with somewhat obtuse extremities: the mouth seems a detached part from the body of the shell, and is connected by a kind of neck; it is of a very singular figure, oblong, oblique, and narrow.

It is not uncommon on the coasts of Italy.

Turbo ore oblongo labiato, spiris ad marginem crenulatis.
The oblong-mouthed Turbo, with the spires crenulated at the edges.

This is about two inches long, and a third of an inch thick at the base: the mouth is small, of an oblong figure, widest in the middle, and furnished with a broad lip: the shell consists of about fifteen volutions, the first of them is somewhat ventricose, the rest are flat, and they are all crenulated at the edge: the colour of the shell is a faint brown, with some variegations of tawny and reddish, and with some regular series of little black dots: the crenulated edges also are paler than any other part, and make a variegation.

It is not uncommon on the American shores, and has been found in some parts of Europe.

Turbo crassior spiris ad marginem profunde crenatis aurita.
The thick, aurited Turbo, with the volutions deeply crenated at the edge.

This is three inches in length, and in the largest part is not less than an inch in diameter: it consists of about eleven volutions, which are all, except the first, flat, and have their edges very deeply and very elegantly crenated: the whole surface of the shell is smooth; the colour is a pale whitish, variegated with irregular spots of a darker hue, and the extremities of the spires, where the denticulations are paler than any other part, and sometimes are quite white: the mouth is oblong and narrow, and is remarkably aurited.

It is a native of the East Indies; we have also had it from the shores of Carolina.

The other species of the Turbo are, 1. The very slender Turbo, with a long, narrow mouth, and a rugose columella. 2. The elegant, slender Turbo, with a large lip, and with blue variegations. 3. The slender Turbo, with longitudinal streaks of yellow, called the Owl-shell. 4. The thicker Turbo, with variegations in the form of punctated circles. 5. The very slender Turbo, with a great variety of dark spots, and dark lines. 6. The slender, whitish Turbo, with a great variety of lines and spots of a tawny yellow. 7. The white, larger, reticulated, and granulated Turbo. 8. The variegated, slender Turbo. 9. The slender Turbo, with a dentated mouth, and rugose columella. 10. The contabulated and elegantly fasciated Turbo. 11. The

belted

belted Turbo, called the Child in swaddling cloaths. 12. The pyramidal Turbo, with a depressed mouth. 13. The thicker, white Turbo, elegantly lineated with yellow. 14. The beautifully, variegated Turbo, called the Chinese obelisk. 15. The elegantly variegated Turbo, with circles of close tubercles. 16. The oblong, thick, Chinese Turbo, with fine lines, and little granulations between them. 17. The whitish, long-mouthed, simple Turbo. 18. The plane, yellowish Turbo. 19. The virgated and lineated, brown and yellow Turbo. 20. The long Turbo, with a large, depressed, auried mouth. 21. The white, rostrated Turbo, with many spires and tubercles on the edges of them. 22. The large, oval-mouthed, yellow and white Turbo. 23. The narrow-spired Turbo, with elegant variegations of black, yellow, and red, in lines. 24. The greyish Turbo, with an elegantly variegated clavicle. 25. The ribband Turbo, with purple fasciæ on a bright yellow ground. 26. The deeply-sulcated, round-mouthed Turbo. 27. The whitish, sinuated, many-spired Turbo. 28. The elegantly, variegated, white and yellowish Turbo. 29. The slender Turbo, with furrowed lines. 30. The brown Turbo, with striated lines. 31. The slender, sorised Turbo. 32. The very slender, black, and yellow Turbo.

VOLUTA.

THE Voluta is a simple shell, having no hinge, formed of one piece, and of a figure approaching to conic, but short; the clavicle being usually depressed, in all very short: the mouth is long, perpendicular, and narrow: the animal inhabiting this shell is a limax.

Authors, who have written on these subjects, have in general called these shells Rhombi; but as Rhombus is the name of a mathematical figure, which has nothing to do with that of this shell, as not at all resembling it in shape, it is better cancelled than preserved.

Some of the most beautiful, as well as of the scarcest and dearest shells we know, are of this genus.

Voluta fascia lata flava, linea punctata insignita.
The Voluta, with a broad, yellow fascia, with a punctated line in it.

The Admiral Shell.

This is one of the scarcest and one of the most beautiful shells we are acquainted with; the extream rarity of the species occasions people picking up a shell of it, under whatever disadvantages of decay or wear; and hence it is that we see some, the colours of which are so faded, that we are apt to wonder at the price they have been sold at, and the admiration with which people speak of them; but we ought to see a perfect and fine specimen of it, before we pass our judgment.

It is about two inches long, and near an inch in diameter toward the head; from this part to the extremity of the mouth it decreases gradually in size, so as to form a kind of cone with an obtuse point; the clavicle is carried up from the same part much diminishing in diameter, and terminates in an obtuse point; it is nearly equal to one third of the length of the body of the shell, which is longer than the generality of the Volutæ have it: the ground colour of the shell is a very bright and elegant yellow, very much resembling the ground colour of the Syenna marble; the variegations, however, take up so much room, that this ground colour is not seen in the proportion of more than a third of the surface: there is a circle of it visible at the head, of about the breadth of a large straw; below this there are three broad fasciæ, or circles of elegantly varied colouring; the lowest of the three is somewhat broader than the others, and they are all separated by fine lines of the ground colour. Under these three circular lines, the fine yellow of the ground colour shews itself again, in the form of a broad fascia; and in the center of this there runs a narrow, punctated line, of the same colours with the other variegations; this line is what distinguishes, and is the characteristic of, the Admiral shell, the Vice-admiral having most of it's other characters, but wanting this: under this broad yellow fascia there runs another equally broad, of the same variegations; and, from the verge of this to the extremity of the shell, the yellow or ground colour prevails again: the clavicle is elegantly variegated with the same colours, in a beautiful irregularity, and with a peculiar brightness.

It is an East Indian shell, and is highly esteemed, whenever it falls into the way of collectors.

Voluta fascia lata flava simplici.
The Voluta, with a broad, simple, yellow fascia. **The Vice-admiral Shell.**

This is an extreamly elegant shell, and, as it much resembles, so it very nearly comes up to the beauty of the former species: it is somewhat more than two inches in length, when full grown, and is about an inch in diameter at the head; the clavicle is rather taller than in the admiral, and has about ten volutions: the ground colour is a bright and strong gold yellow, and the variegations are of the same beautiful admixture of colours with those of the other, but they have more white in them: there is a line of the ground colour or yellow at the head, of the breadth of a straw; below this there is a circular line of the variegations, about equal to it in breadth; under this is a somewhat narrower line of the yellow, but not pure; and under that a very broad line of the variegations: below this there stands a broad line or fascia of yellow, as in the admiral, but without the punctated line, which in that shell runs through the middle of it; after this there is another broad fascia of the variegations, and then the point of the shell, which is also yellow: the clavicle is very beautifully clouded with the variegations.

This is also a native of the East Indies, and bears a very great price in auctions, and a very high rank in the cabinets of collectors.

Voluta albida variegata, fasciis duabus luteis.
The whitish, variegated Voluta, with two yellow fasciæ. **The false Admiral-shell.**

This is a beautiful shell, and is shewn in some collections as the admiral-shell, but it is vastly inferior to that elegant species: it is about an inch and a half in length, and half an inch in diameter at the head; the body is conic, though but very little tapering; and the clavicle is not quite so tall in proportion, as in the two former: the ground colour of this shell is a dusky whitish, variegated with several faint colours; and there are two broad fasciæ of a beautiful yellow surrounding it, the one near the point, the other a little higher than the middle: the whole surface is smooth, and the mouth narrow.

It is a native of the East Indies, and is principally brought to us from China; when it is in perfection, it is a very elegant shell, but the colours are very apt to be decayed and dead.

Voluta rubra maculis magnis albis variegata.
The red Voluta, with large, white spots. **The Tiger-shell.**

This is a very elegant shell: it is about two inches and a half in length, and an inch and a quarter in diameter; it's mouth is very long, but narrow; and it's clavicle has only about four turnings, and is very little elevated: the ground colour of the shell is a dusky red, and it is spotted all over with large, irregular blotches, as it were, of white; some of them are oblong, others irregularly angulated and indented.

It is a scarce shell, we have it only from the East Indies; the next species is by many called by it's name, but is greatly inferior to it.

Voluta flava maculis minoribus albis variegata.
The yellow Voluta, with smaller, white spots. **The yellow Tyger-shell.**

This is also a very elegant shell, but it is greatly inferior to the former: it's length is about two inches and a half; it's thickness somewhat more than an inch and a half at the head, and it thence gradually diminishes, in a conic form, to the aperture of the mouth: the clavicle consists of about six volutions, and is carried to a fine point at the top, but it is not very high: the ground colour of the shell is yellow, and it is very beautifully variegated with white spots; these are of an irregular figure, and not very large, few of them exceeding the bigness of a pea.

It

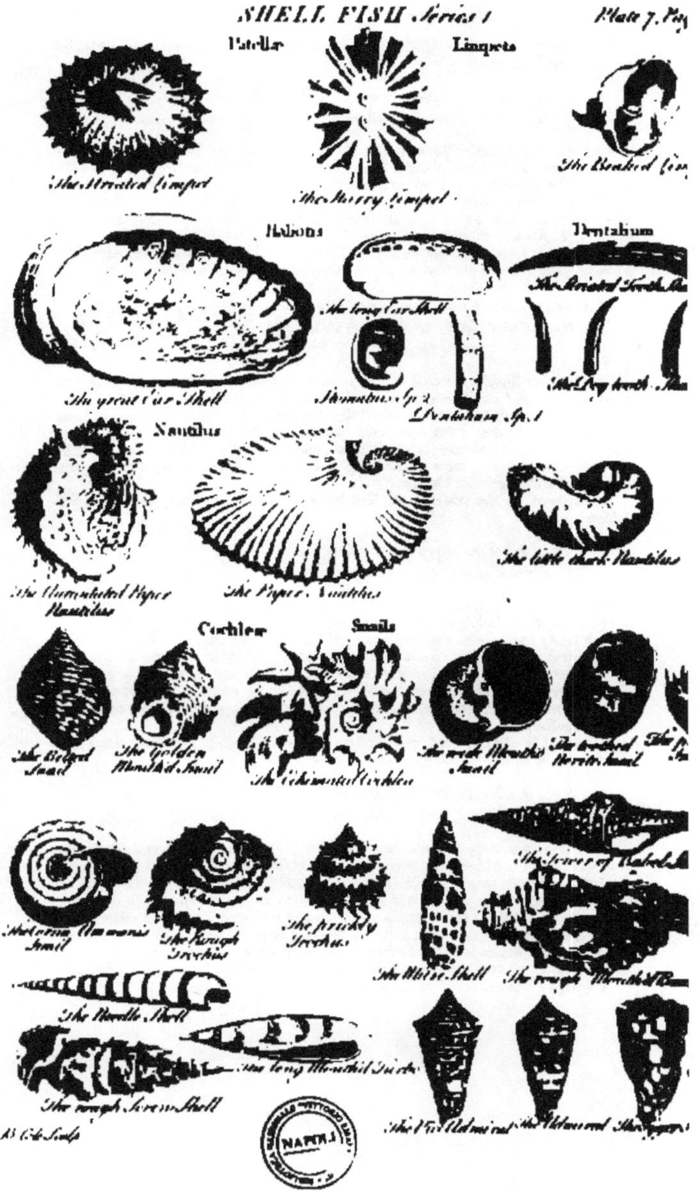
SHELL FISH Series 1
Plate 7. Pag
Patellæ
Limpets
The Striated Limpet
The Starry Limpet
The Beaked Lim
Haliotis
Dentalium
The great Ear Shell
The long Ear Shell
Stomatius Sp 2
Dentalium Sp. 1
The Striated Tooth She
The Dog tooth Sh
Nautilus
The Undulated Paper Nautilus
The Paper Nautilus
The little thick Nautilus
Cochleæ
Snails
The Belted Snail
The Golden Mouthed Snail
The Echinated Cochlea
The wide Mouth'd Snail
The toothed Nerite Snail
The Cornu Ammonis Snail
The Rough Trochus
The prickly Trochus
The Mitre Shell
The Tower of Babel Sh
The rough Mouth'd Buc
The Needle Shell
The long Mouth'd Turbo
The rough Screw Shell
The Vice Admiral
The Admiral
B Cole Sculp

It is a native of the East Indies, and is frequently kept in collections, under the name of the Tyger-shell.

Voluta albescens maculis nebulosis subrubentibus.
The white Voluta, with clouded, reddish variegations.

This is a very beautiful species; it is about two inches, or little more, in length, and more than an inch wide at the head: the clavicle is moderately exerted, but it is obtuse at the extremity, and has only about four or five volutions, and those not distinct: the ground colour is a faint white, and the variegations are of a dusky reddish; they are large, and are of an extremely irregular figure, very indistinct at the edges, and disposed without any order on the shell.

We have it both from the East and West Indies, and it is a shell of value; the French, from the odd distribution of the variegations, call it the Spectre-shell.

Voluta coronata zonis latis flavis variegatis.
The coronated Voluta, with two broad, variegated zones.

The Crown Imperial-shell.

This is a very singular, as well as a very beautiful, shell: it is three inches long, and barely an inch and a half in diameter at the top: the clavicle is so depressed, that in a front view of the shell it is not at all visible: the head is surrounded with a very beautiful series of tubercles, pointed at their extremities: the ground colour is pale, and there are two broad and very beautiful zones running round it; the one near the head, the other toward the other extremity; these are of a fine yellow, but variegated in an elegant manner with black and white.

It is frequent on the coasts of the East Indies, but few arrive here that are in perfection.

Voluta albescens nigro variegata.
The white Voluta, variegated with black.

The Hebrew Letter-shell.

This is a smaller shell than most of the others of this genus: it rarely exceeds an inch and a quarter in length, or three quarters of an inch in diameter at the top; it's body is pretty regularly of a conic figure, and it's clavicle exerted, and has many volutions, five at least, and is obtuse at the extremity: the ground colour of the shell is a pearly white, and it is all over variegated with large and very irregular marks of black; they are disposed in about four series on the body, and in a single row on each of the volutions of the clavicle, and are supposed to have some resemblance of the Hebrew characters.

It is found in the East and West Indies; and when perfect, and finely coloured in the ground, is an elegant shell.

Voluta albida maculis fuscis et purpureo-caeruleis variegata.
The whitish Voluta, variegated with brown and purplish blue spots.

This is a very elegant and a very large shell, for one of this genus: it is often near four inches in length, and more than two in diameter at the head; from hence the body tapers but very gradually, and is large and obtuse at the extremity; and the clavicle the other way, though it is formed of seven or eight volutions, is not much elated, and is obtuse at the extremity: the ground colour of the shell is white, but it is very elegantly variegated with spots of different sizes, disposed principally in circles round it; there are no less than twenty or thirty of these circles on the shell, and the spots which form them are some of them brown, and others of a deep purplish blue.

We have it from the coast of Guinea.

Voluta semi-coronata superficie undulata.
The semi-coronated Voluta, with an undulated surface.

This is the most singular of all the shells of this genus, and is not without it's beauty: it more approaches to the crown imperial than to any of the other species, but differs abundantly from that in many respects: it is about two inches and a half in length, and is near an inch and three quarters in diameter at the head: the verge of this is deeply indented, so as to form a kind of corona or crown, with the denticulations, but they are short, low, and obtuse at the points: the clavicle consists but of a few volutions, and is obtuse at the point; the whole surface of the shell is undulated, or is marked with a number of longitudinal, but not straight, furrows, placed at the distance of a straw's breadth from each other, and the ridges between these form a kind of undulations: the colour of the shell is white, and the variegations are irregular, and of a faint brown.

We have it from the African coast.

Voluta tenuior claviculo longa.
The slender Voluta, with a long clavicle.

This is a very singular and very beautiful shell: it is about two inches in length, and hardly three quarters of an inch in diameter at the head: from this part the body forms a kind of cone, tapering very regularly to a point; and the other way the clavicle is long, very slender, and formed of ten or twelve volutions, terminating in a sharp point; this singular structure of the shell is attended with a singular colouring: the ground colour is white, and shews itself on the body, in the form of three zones, considerably broad and variegated, with purple spots and lines; between these are two other zones, which are also broad, and are of a beautiful orange colour.

We have it from the East Indies but rarely.

Voluta punctata fasciis tribus angustis oculatis.
The punctated Voluta, with three narrow, oculated fasciæ.

The Butterfly-shell.

This is one of the most elegant of this beautiful genus: it is three inches long, and about an inch and three quarters broad in the largest part, or at the head; the body from this forms a tolerably regular cone, tapering very gradually, and but obtusely pointed; the clavicle has five or six volutions, and is pointed at the extremity, but it is not very prominent: the ground colour of the shell is yellow; it is all over beautifully variegated with small, brown spots, formed into regular, round series: there are three very beautiful fasciæ or bands round the body, and one narrower near the head; these are formed of continued large spots, formed of a deeper brown, a paler brown, and white, and they resemble nothing so much as the spots in the form of eyes, upon the wings of some of the butterflies.

This is an East Indian shell, and is a very rare one; we have it in some of our cabinets.

The other species of the Voluta are, 1. The orange-coloured Voluta, variegated with purple. 2. The large-headed, short Voluta, called the Turnep-shell. 3. The brown, elegantly-lineated Voluta. 4. The pale Voluta, with flame-coloured spots. 5. The elegantly-spotted and variegated Voluta, called the Guinea-shell, or the Speculation-shell. 6. The reddish, fasciated, and striated Voluta. 7. The yellow Voluta, variegated only with small spots. 8. The brown Voluta, with two broad zones of white. 9. The dusky brown Voluta, called the Bat-shell. 10. The large, whitish Voluta, with small yellow spots, and innumerable punctules of the same colour. 11. The black-spotted Voluta, called the black Leopard-shell. 12. The yellowish Leopard-shell. 13. The reddish Leopard-shell. 14. The fasciated Voluta, with yellow and whitish spots. 15. The agate-coloured, virgated Voluta. 16. The yellow Voluta, with a white fascia. 17. The elegantly-variegated Voluta, called the Onyx-shell. 18. The greenish Voluta, variegated with numerous spots, and with two variegated fasciæ. 19. The coronated Voluta, variegated with brown. 20. The coronated

nated Voluta, with obscure fasciæ. 21. The coronated Voluta, with marbled, black variegations. 22. The striated and coronated Voluta. 23. The cancellated and elegantly variegated Voluta.

CYLINDRUS.

THE Cylindrus is a simple shell, without a hinge, formed of one continued piece, and of a figure approaching to a cylinder: when it departs any thing from that, it is not largest at the head as the voluta is, but, toward the middle, the mouth is always long and narrow.

The animal inhabiting the Cylindrus is a limax.

The clavicle in this shell is, in some species, continuous with the body, in others it is divided from it by a kind of circle, and in some it is coronated: the columella is smooth in most, but in some it is rough.

Cylindrus crassior coronatus albescens, fusco variegatus. The thick, coronated Cylindrus, of a silvery white colour, variegated with brown. — **The Brocade-shell.**

This is a large and elegant shell; it is three inches and a half in length, and near two and a quarter in diameter, and is nearly of the same thickness in the whole body, only diminishing a little at the two extremities; the head is denticulated or coronated, but the denticulations are not sharp; the clavicle has four or five turns, and is terminated by a point: the ground colour of the shell is a silvery white, and it is beautifully variegated with a bright brown, in fine irregular lines, clouds, and blotches; the whole surface of the shell has a naturally fine polish, and the ground-work and variegations have much the appearance of a fine brocade.

We have it from the coasts of the warmer parts of America, and sometimes from those of Africa.

Cylindrus crassior albescens, caeruleo et fusco variegatus. The thick, white Cylindrus, variegated with blue and brown. — **The tulip, Cylinder-shell.**

This is a very beautiful and a very scarce shell: it is about three inches long, and two in diameter, but it is not so regularly of a cylindric figure as some of the others; it is largest a little below the head, and thence grows gradually smaller to the other extremity: the clavicle has ten or twelve volutions, but it is not very much elated, and terminates in an obtuse point: the ground colour of the shell is whitish, but the variegations, which are partly blue, and partly brown, are very beautifully thrown into irregular clouds, and with larger and smaller spots.

It is a native of the East Indies, and is but seldom brought over to us, especially in any tolerable state of perfection, and brightness of colour.

Cylindrus tenuior purpureo et albo variegatus. The slender Cylindrus, variegated with white and purple. — **The Porphyry-shell.**

This is about two inches and a half in length, and an inch and a quarter in thickness; it is pretty regularly of a cylindric figure, and the clavicle is short and obtusely pointed: the colour of the shell is a pale whitish, with a cast of red, and it is all over clouded with variegations of a deeper red, approaching to purple: this purple, though not the ground colour, takes up the far greater part of the surface, and the white, which is the true ground colour, is scarce any where seen without a greater or lesser tinge of the red in it: the purple is, in most parts of the shell, formed into irregularly longitudinal and dentated lines.

It is brought to us from the coasts of South America.

Cylindrus tenuior albidus fusco variegatus.
The slender, whitish Cylindrus, variegated with brown.

This is three inches and a half in length, and scarce an inch and a quarter in diameter; it is of a tolerably regular cylindric figure, but is somewhat smaller toward the point than elsewhere, and a little, though very little less at the head, than somewhat lower down: the clavicle has five or six volutions, and is pointed; the ground colour is whitish, but is very elegantly variegated with brown, disposed in narrow and irregular lines, and in great irregular spots, covering a very considerable part of the surface. We have it from the coasts of South America: the Dutch call it the Brunette.

Cylindrus albescens zonis literatis.
The white Cylindrus, with lettered zones. **The Letter-shell.**

This is one of the smaller Cylinders: it rarely exceeds an inch and a quarter in length, and is less than half an inch in diameter: the clavicle has four or five volutions, and the body of the shell is split at the other extremity, by the continuation of the mouth: the ground colour is white, which is preserved tolerably pure over a great part of the body, but near each extremity there is a broad zone; these are not of one continued colour, but are lineated and variegated in such a manner, that fanciful people have supposed themselves able to make out evident letters, in particular, the capitals B, D, and A frequently repeated in them.

We have it both from the East Indies, and South America; but few of those brought over are in high perfection.

Cylindrus tenuior fusco et albo variegatus.
The slender Cylindrus, variegated with brown and white.

This is an extremely elegant shell: it grows to three inches long, and about an inch and a quarter in diameter: it's figure is nearly cylindric, the two ends being only a very little smaller than the middle; the clavicle has four or five volutions, but is not much elated, and terminates obtusely: the colours are only two, a bright white, and a pale tawny brown; they each take up so great a share of the surface, that it is not easy to say which is the ground colour; and they are disposed in a beautiful manner in denticulated lines: the whole surface has a fine natural polish, and the colours have a singular brightness.

It is brought to us from South America.

The other species of the Cylindrus are, 1. Of those which have the clavicle continuous with the body, or not separated by a depressed circle; 1. The elegant, yellow Cylindrus, called the Gold-brocade shell. 2. The elegant, white Cylindrus, called the Silver-brocade shell. 3. The pale, lemon-coloured, variegated Cylindrus. 4. The fasciated, large Cylindrus. 5. The Cylindrus, blotched with yellow and white. 6. The cancellated or reticulated Cylindrus. 7. The Cylindrus, with the variegations representing feathers. 8. The elegant Cylindrus, variegated with blue. 9. The granulated Cylindrus, with a great multitude of spots, and little punctules. 10. The elegant, pale, yellow, granulated Cylindrus. 11. The dusky brown, and silvery-white, variegated Cylindrus.

2. Of those which have the clavicle separated, as it were, from the body of the shell by a circle, there are, 1. The elegantly variegated, brown and yellow Cylindrus. 2. The pale Cylindrus, with clouded and lineated brown spots. 3. The white Cylindrus, with dusky blackish, nebulous spots. 4. The pale brown and purple Cylindrus.

3. Of those which have the clavicle coronated, there are, 1. The fine, brown and white, tawny Cylindrus. 2. The fine, silvery, variegated Cylindrus. 3. The pale brown, variegated, glossy Cylindrus. 4. The silvery and brown Cylindrus.

4. Of those which have the columella not smooth, as it is in all the others, but wrinkled or rugose, there are, 1. The greenish, olive Cylindrus. 2. The greyish, agaty Cylindrus, with variegations on the lower part. 3. The white Cylindrus, variegated with purple and brown. 4. The purple Cylindrus, spotted with white. 5. The

The dark, blackish, olive Cylindrus. 6. The yellow-olive Cylindrus. 7. The elegantly variegated, olive Cylindrus. 8. The olive Cylindrus, with the lower part variegated with brown. 9. The elegant, violet, Panama Cylinder. 10. The white Cylinder, with yellow lines.

MUREX.

THE Murex is a simple shell, without any hinge, formed of a single piece, and beset with tubercles or spines: the mouth is large and oblong, and has an expanded lip, and the clavicle is rough.

The animal, which is the inhabitant of this shell, is a Limax.

The clavicle is in this genus sometimes exerted, sometimes depressed; and the mouth in some of the species is dentated, and in others smooth; the lip also in some is digitated, in others elated, in some laciniated; and the Columella in some is smooth and even, and in others rugose.

Murex per totum echinatus albescens nigro variegatus.
The whitish murex, echinated all over, and variegated with black.

The hedge hog Murex.

This is a very singular and very elegant shell: it is three inches and a half in length, and about two and a half in diameter in the largest part: the clavicle is exerted and pointed, and the body of the shell approaches to a conic form. It's colour is whitish, and it's whole surface is wrinkled with circular furrows; it is surrounded also with a number of series, of long, erect, robust, and sharp spines; and the clavicle has also protuberances of the same kind on it's several volutions, but they are not so long, and are more obtuse at the end: these spines are all black, and there are also some other black variegations on the intermediate parts of the shell, which give it a great deal of beauty, as well as singularity, in it's appearance.

It is brought to us from the coast of Africa, and from the East Indies.

Murex variegatus corpore echinato, clavicula laevi.
The variegated Murex, with the body of the shell echinated, and the clavicle smooth.

This is two inches and a half in length, and near two inches in diameter in the broadest part: it's ground colour is a brownish-white, but it is variegated with a very deep colour, seemingly formed of a mixture of brown, olive, and purple. On the body of the shell, there stand three rows of spines, placed at considerable distances from one another, but connected by a ridge; the clavicle has about five volutions, and is smooth; the dark colour is disposed in such abundance on the shell, that it seems at first view the ground colour: the spines are white, and add greatly to the beauty of the shell.

It was first brought to us from the African coast, but we have it now from some of the shores of the Mediterranean.

Murex spinis excavatis, corpore rugoso inermi.
The Murex, with hollowed spines, and with a naked rugose body.

This is a very singular and very elegant species: it is about two inches and a half in length, and, in the largest part, is nearly an inch and a half in diameter: the clavicle consists of about nine elegantly distinct volutions, the two lower of which have each a series of imperfect spines on them; the body of the shell is bellied, or distended, and has a number of deep longitudinal furrows on it: toward the top, or end, next the clavicle there are two series of spines placed near one another; they are short, of a conic figure, and hollowed: the body of the shell is then naked, till within half an inch of the other extremity, where there is another single row of them.

It is brought to us from the East Indies, but it is a very rare shell.

Murex

Murex heterostrophus clavicula depressa.
The heterostrophus Murex, with a depressed clavicle.

The left-handed Murex.

This is also a very singular species. It is near three inches in length, and about an inch and a half in diameter at the head: from thence it becomes gradually smaller to the extremity, where it is terminated in a sharp point: the mouth is very large, it reaches the whole length of the shell, and is very wide at the head; the clavicle has about six volutions, but it is depressed: the colour is a pale whitish-brown, with a cast of reddish; but the great singularity of the shell is, that it is all the way turned in a contrary direction to the others, and consequently it's mouth opens the wrong way.

It is a rare shell; we have it from the South Seas.

Murex labio expanso rugosus.
The rugose Murex, with an expanded lip.

The Spider-shell.

This is about three inches in length, and, measuring from the most extended part of the lip, is nearly as much in diameter: the colour is a pale yellowish-brown, with some variegations: the clavicle is long, and has about six volutions, which are broad and elegant: the whole body is elevated, at different distances, into a kind of rounded ridges; the lip is very far expanded, and these ridges are all continued through it, and even project beyond the rest of it's edge, forming what have been called a number of feet; and the species, from that, the many-footed Spider-shell: The mouth is very large, and the extremity of the shell small and turned up.

We have it in great plenty from the West Indies; and it has been found, though smaller and less beautiful, on the coasts of the Mediterranean.

Murex tuberculosus labio quinque dentato, extremitatibus attenuatis.
The tuberculose Murex, with five teeth on the lip, and attenuated ends.

The Scorpion-shell, commonly called the Spider-shell.

This is an extremely singular species. It is five inches in length, and more than three in diameter: it's general colour is yellowish, but it is slightly variegated with a duskier colour; the mouth is very long and wide, and the inside of the lip and the columella are sometimes only reddish; but, when the shell is in it's greatest perfection, they are variegated with a fine violet colour, and white: the clavicle is short, and in a front view of the shell is not at all seen: the whole surface of the shell is covered with tubercles, and, at each extremity of the mouth, is carried into a long and slender point: the whole verge of the lip is crenated in an irregular manner, and there are five oblong protuberances, resembling those of the two ends of the mouth.

It is brought to us from the coasts of the East Indies, and some parts of America.

Murex labio protenso corniculis sex longioribus armata.
The protended-lipped Murex, with six long cornicles.

The six-legged Spider-shell.

This grows to three inches in length, and two and a half in diameter: the clavicle has about six volutions, but they are not very distinct: the general colour is a tawny brown, variegated with some darker clouds, and with a little flesh colour and white: there are a few obtuse and low tubercles on different parts of the body; and from the back of the protended part of it there arise six long and very remarkable protuberances, resembling so many horns; of these, those next the head of the shell are longest, but those toward the opposite extremity are usually sharper-pointed, and somewhat crooked; the mouth is very large, and of an elegant flesh colour within.

We have this species out of the Mediterranean; the French call it the Araigna, and the Lambis; It is not common in our cabinets.

Murex costatus clavicula tuberculosa.
The costated Murex, with the clavicle tuberculose.

This is a large and a very singular species: It is four inches or more in length, and more than two in diameter at the head; from thence the body gradually grows smaller to the extremity, where it is split a little way, by the continuation of the mouth: the clavicle is long, and has about eight volutions, of which, the lower ones have each a series of tubercles, which on the first and second make a kind of corona: the whole surface of the body shell is elevated at small distances into longitudinal ribs, and is of a yellowish-brown; very beautifully variegated with lines of a darker colour, disposed so, as to give it much the appearance of the veins of wood; whence the French have called it Le Bois vené.

It is a rare shell; we have it from the shores of South America.

Murex obscure costatus, zonis striatis.
The obscurely costated Murex, with striated zones. **The ribbed Music-shell.**

This is about two inches long, and is near an inch and three quarters in diameter: the body of the shell is short, and is elevated at considerable distances into several broad and low ribs: the clavicle is long, elegant, and formed of six or eight volutions. The ground colour of the shell is a whitish-brown, and is elegantly surrounded with three or four zones, each formed of four or five slender and even, black lines, with spots of blackish and reddish between them, resembling very much the lines in which music is written, with the marks of crotchets, &c. A French writer has gone so far, as to prick down a tune, which he pretends to have found on a shell in one of the cabinets in that kingdom; and there is thus much in his favour, that, though it make a very pretty figure on the shell, it is but bad music.

We have this species from the East Indies.

Murex costa simplici majuscula dorsali.
The Murex, with a single, large, dorsal rib.

This is one of the smaller Murices: it does not exceed an inch and a half in length, and is about an inch in diameter: the body of the shell is inflated, and approaches to an oval form: the opening of the mouth is long and wide, and the lip is bordered with a thick edge: the clavicle is short, and has about five volutions; the one bellied or rounded: along the middle of the back, there runs a longitudinal rib, which is large and thick, rounded at the summit, and very like the thick edge of the lip: the rest of the surface is smooth, and the colour is a fine horn colour, variegated with brown.

We have this from several parts of the East Indies, but it is not common in our cabinets.

Murex tuberculosus labio protenso bidentato.
The tuberculous Murex, with a protended, bidentated lip.

This is a small species; it rarely exceeds an inch and a half in length, and, measuring from the most protended part of the lip, it is at least as much in diameter: the clavicle is long, and composed of five volutions; the lower one so continuous with the body of the shell, that it is not easy to determine where it ends: the colour is a dusky brown, with a few paler variegations; the inside of the mouth is whitish, or reddish: the whole surface of the body of the shell is surrounded with round tubercles, small, and disposed in series; there are about four or five of these series, and the tubercles are paler than the rest of the shell: the mouth is very wide, and the lip protended a great way, and armed with two points; the one at the extremity, the other toward the middle.

Murex tuberculis maximis craſſis obtuſis.
The Murex covered with large, thick tubercles.

This is a ſmall ſpecies, but the tubercles on it are remarkably large; it is rarely more than an inch and a quarter in length, and is equal to that at leaſt in diameter, including the ſpines in the meaſure; the clavicle is ſhort, and conſiſts of a conſiderable number of volutions: the tubercles are arranged in four or five ſeries, and are large and obtuſely pointed: the whole ſurface is variegated with a violet blue, a deep purple, and brown.

We have it from the Eaſt Indies; but moſt of thoſe we receive, having been long on the ſhore without the animal, are deficient in the colours.

The other ſpecies of the Murex are very numerous, and may be conveniently arranged under certain ſubdiviſions, according to their ſeveral particular variations.

1. Of thoſe which have the body of the ſhell beſet with tubercles and ſpines, there are, beſide the before deſcribed, 1. The Murex, with a depreſſed clavicle, the body covered with large, black ſpines. 2. The long, clavicled, greyiſh Murex, with black ſpines, and a plicated ſurface between them. 3. The bluiſh, echinated Murex, with a depreſſed clavicle. 4. The brown Murex, with four ſeries of obtuſe ſpines. 5. The whitiſh, plicated Murex, with two rows of ſpines. 6. The brown and blue Murex, with three ſeries of ſpines. 7. The yellow Murex, with the whole ſurface covered with ſhort ſpines. 8. The whitiſh Murex, with yellow, obtuſe, and low protuberances, with a purple mouth and the lip dentated, and the columella plicated. 9. The white, dentated-mouthed, echinated Murex, with black ſpines. 10. The Muſic-ſhell, with a wrinkled columella. 11. The pale Muſic-ſhell, with faint lines and ſpots. 12. The pale, brown, elegantly variegated, ſlaty Murex, with the columella wrinkled. 13. The variegated Murex, with the clavicle covered, and ſurrounded with ſharp tubercles. 15. The white, ſtriated Murex, with the clavicle furniſhed with very long ſpines. 16. The yellow ribbed and furrowed Murex, covered on all parts with pointed tubercles. 17. The tuberculoſe, ſtriated, and umbilicated Murex, with a reddiſh clavicle. 18. The large, reddiſh, and yellow Murex, covered in all parts with long, hollow, and pointed tubercles.

2. Of thoſe which have ſmooth bodies, and the clavicles a little rough, and have a crooked roſtrum, there are, 1. The triangular Murex, or the helmet-ſhell of Rondeletius, with a dentated mouth, and the lip plicated. 2. The red, turban Murex, with the lips expanded both ways. 3. The narrower-mouthed, helmet Murex of Bonani. 4. The horn colour Murex, with the mouth ſcarce at all dentated. 5. The large, brown Murex, variegated with a tawny yellow. 6. The grey, ſtriated, helmet Murex. 7. The whitiſh, helmet Murex, variegated with yellow, undulated lines. 8. The elegant, agate-coloured, helmet Murex, variegated with regular, brown marks and ſpots. 9. The bluiſh, helmet Murex, with reddiſh, brown, flexuous ſtriæ. 10. The little, elegant, deeply-ſtriated, helmet Murex, with duſky lines.

3. Of thoſe which have remarkably digitated lips, and are called Spider-ſhells by writers, beſide the ſpecies deſcribed at large, there are, 1. The large Murex, with crooked fingers, called the male Spider-ſhell. 2. The large Murex, with more ſtraight and ſlender fingers, called the female Spider-ſhell. 3. The horned or many-fingered Spider-ſhell. 4. The ſhorter ſpider Murex, with ten fingers. 5. The oblong and wide-mouthed ſpider Murex, the heptadactylus or ſeven-fingered kind of Pliny. 6. The little, ſlender, five-fingered Spider-ſhell. 7. The larger, broad, four-fingered, ſpider Murex. 8. The ſpider Murex, with ſix hollowed fingers. 9. The ſpider Murex, with the aperture of the mouth ſurrounded with very numerous, minute inciſures. 10. The purple ſpider Murex, with ſtraight fingers. 11. The large ſpider Murex, with the fingers hooked in the manner of the beak of a bird of prey. 12. The ſpider Murex, with the lip folded ſo, as to form five appendages, and variegated with blue, white, and brown.

4. Of thoſe which have the lip alated and laciniated, there are, 1. The large, alated Murex, with a crooked roſtrum, and the lip red within, called the aſs's Ear-ſhell. 2. The triangular Murex, with deep furrows and tubercles on the ſurface, called the hog's Ear-ſhell. 3. The large Murex, with a red mouth, and with the columella black. 4. The Murex, with the mouth black on all parts, and ſtriated. 5. The brown and white-mouthed Murex. 6. The broad, deep, brown Murex, called the

the Turtle-ſhell. 7. The great, wide-lipped, ear Murex, of Rumphius. 8. The Murex, with a very wide, red, laciniated lip, and with an aculeated clavicle. 9. The Murex, with a narrower, deeply-jagged lip, and with an aculeated clavicle. 10. The variegated, verrucoſe Murex, with a laciniated and very thick lip. 11. The variegated, tuberculoſe Murex, with a thin and laciniated lip. 12. The yellow Murex, with a thin, laciniated lip, and a gibboſe clavicle. 13. The lead-coloured, belted Murex, with the lip folded. 14. The ſmooth Murex, with a thick and folded lip, and with a dentated columella. 15. The yellow, tuberculoſe Murex, with a folded lip dentated on one ſide, and ſpotted on the other. 16. The yellow Murex, with an irregular ſpotted rib running tranſverſely from the upper edge of the lip to the oppoſite ſide of the ſhell. 17. The grey, coſtated Murex, with the lip crenated on the ſide of the columella. 18. The white, coſtated, and contabulated Murex.

PURPURA.

THE *Purpura* is a ſimple ſhell, having no hinge; formed of one continuous piece, and covered from the top to the bottom with ſpines, with tubercles and umbo's; with the mouth ſmall, and approaching to a round figure, and with a ſhort clavicle; but uſually, with the other extremity, protended to a conſiderable length.

The *Purpuræ* have, in general, been confounded with the murices, but very improperly, as the genera are ſufficiently diſtinguiſhed, were it only by the ſhape of the mouth, and as the ſpecies under each are conſiderably numerous. The murices are, in general, ſhells of conſiderable beauty, but the *Purpuræ* are yet more ſingular and elegant.

Purpura flaveſcens roſtro longiſſimo, ſpinis longis arcuatis armata.
The yellow, long-beaked Purpura, with long and ſomewhat crooked ſpines.

The thorny Woodcock-ſhell.

This is an extremely elegant ſhell; it is about four inches in length, ſometimes more; it's body is ſhort, of a figure approaching to oval, and ventricoſe, and the clavicle is alſo ſhort, but diſtinct in it's volutions: the roſtrum or ſnout, protended from the extremity of the mouth, is of nearly twice the length of the reſt of the ſhell, and both this and the body are armed with four ſeries of very long ſpines, ſome of them more than an inch in length; theſe are diſpoſed in regular, longitudinal ſeries, and are ſlender, pointed at the ends, and the greater part of them ſomewhat crooked: the mouth is nearly round, but it's opening is continued in the form of a ſlit up the roſtrum of the ſhell: there are ſometimes regular ſeries of ſhort ſpines between thoſe of the longer; ſometimes there are only a few of them, and thoſe diſpoſed with leſs order: the natural colour of the ſhell is a tawny yellow, with an admixture of brown, but, with lying expoſed to the ſun and air upon the coaſt, it bleaches into a whitiſh hue: many of thoſe we receive are thus altered in colour, and moſt of them have the ſpines more or leſs injured.

It is a native of the American ſeas, and is not unfrequent in our cabinets.

Purpura flaveſcens variegata tuberculoſa roſtro longo.
The variegated, yellowiſh Purpura, with tubercles and a long beak.

The common Woodcock-ſhell.

This, when entire, is, in the whole, four or five inches long: the body of the ſhell is ſhort and inflated, and of a figure approaching to oval; the clavicle is thick and oblong, and the roſtrum or ſnout is ſlender, and equal to about once and a quarter the length of the whole ſhell beſide: the mouth is round, not very large, and is ſurrounded with a narrow lip of a reddiſh colour; the roſtrum or ſnout is continued from the extremity of this, and the opening runs all along it, in the form of a fiſſure: the whole external ſurface of the ſhell is covered with irregular, large, tubercles: it's ground colour is a duſky yellow, but it is variegated with brown and grey.

It is brought to us from many different places, but the most beautiful are from the East Indies.

Purpura rostro brevi, spinis sex seriarum laciniatis.
The short-beaked Purpura, with six series of laciniated spines.

The Endive-shell.

This is a species extremely different from the two former; it's length is about two inches, and it's diameter, including the spines, is an inch and a half: the body is of a figure approaching to oval; the clavicle is moderately long, and the rostrum short, or without that protended part, which gives so remarkable an appearance to the others: there are, on the surface of this shell, six longitudinal series of spines or oblong protuberances, some of them more than a third of an inch in height, and all broad and jagged at the top: the body of the shell is white, and these are black, either all over, or at least at the ends.

The shell is found in great abundance on the shores of the American islands; but it frequently is blanched to a white all over, before we receive it.

Purpura rostro brevi, spinis expansis tribus seriebus dispositis.
The short-beaked Purpura, with expanded spines ranged in three series.

The Caltrop-shell.

This is three inches long, and, including the spines, is nearly as much in diameter: the body is large at the head, and gradually diminishes thence in diameter, so as to approach toward a conic figure; the clavicle is moderately long, and has about six volutions; the whole surface of the shell is deeply furrowed in a transverse direction, and is ornamented with three series or rows of spinose protuberances: these are from half an inch to a third of an inch in length, and are half as broad as they are long; they are longitudinally furrowed on their surface, and are broad at their extremities and hollow: the general colour of this shell is a dirty whitish.

We have it from the shores of the American islands, and from the Mediterranean.

Purpura rostro sublongo, spinis brevibus pungentibus armata.
The somewhat-long-beaked Purpura, with short, pungent spines.

This is about an inch and a half in length, and an inch in diameter: the body is ventricose, and of a rounded figure; the clavicle is moderately long, and has about five volutions; and at the other extremity is placed the beak, which is slender, hollowed, and of about two thirds of the length of the body; the whole surface of the shell is armed with a number of short and not very thick spines; they are of a conic figure, and are some of them hooked; they stand in eight or ten rows, and are continued to the rostrum, and, though with less regularity in the disposition, to the clavicle; the shell is of a whitish colour, and, when quite in perfection, the tips of the spines, and some of them entirely are black; this colour soon blanches away, as the shell lies on the shores; and, consequently, the most of what we have are altogether white.

Purpura rostro sublongo arcuato, spinis capillaceis.
The somewhat-long-beaked Purpura, with capillaceous spines.

The hairy Purpura.

This is about two inches in length, and an inch and a quarter in diameter in the largest part: the mouth is of a roundish figure, and small; the clavicle is moderately long and thick, it has about four or five volutions; the rostrum or beak at the opposite extremity of the shell is about equal to a third of the body of it in length, and is moderately thick, of a conic figure, obtuse at the end, and somewhat crooked; the whole surface of the shell is covered with high-raised ribs; they stand in a longitudinal direction at small distances, and there are furrows running round the shell, which transverse them; on the surface of the body of the shell there stand almost innumerable spines; they are very small, and resemble segments of bristles of some animal: the colour of the whole shell is a dusky, greyish, white, and the lining of the mouth is white.

The other species of the Purpura are, 1. The spinose Purpura, with a flesh-coloured mouth, and three rows of fine spines. 2. The yellow Purpura, with three rows of large spines. 3. The whitish Purpura, with three series of very broad and deeply jagged spines. 4. The furrowed Purpura, with yellow variegations, and with three series of short, broad, and ragged spines. 5. The large Purpura, with five rows of very broad spines laciniated and expanded at their extremities, like the feet of a frog. 6. The Purpura, with six series of short and somewhat obtuse spines. 7. The Purpura, with six series of foliaceous spines deeply laciniated, and with the edges of the segments curled. 8. The lesser, prickly Purpura, with three rows of robust, but not very sharp, spines. 9. The lesser, rough Purpura, with five rows of low spines. 10. The Purpura, with a long clavicle and long snout, and with several series of obtuse tubercles. 11. The long-snouted, smooth Purpura, covered only with low, rounded prominences. 12. The little-spotted, tuberculose, moderately long-snouted Purpura. 13. The little, moderately long-snouted Purpura, with low tubercles. 14. The thick, costated Purpura, with yellow variegations, and a crooked beak. 15. The marbled, costated, and tuberculose Purpura, with very elegant amethystine lines on the body. 16. The elegantly variegated, costated, and marbled Purpura, with prickly protuberances. 17. The straight-beaked, grey Purpura, with three series of capillaceous spines. 18. The crooked-beaked Purpura, with setose fimbriæ. 19. The short-beaked Purpura, with a depressed clavicle, and the body armed with sharp spines. 20. The bristly Purpura, called the sea Porcupine by some writers. 21. The elegant purple and white Purpura, called by many the Porphyrites, or Porphyry-shell.

DOLIUM.

THE Dolium is a simple shell without any hinge, formed of one continuous piece, which makes a body of a figure approaching to round, distended, and, as it were, inflated: from the resemblance of the body of this shell to a vessel for the containing fluids, the genus has been named Dolium; others have called them Conchæ globosæ; but this is less determinate, as they are many of them of an oblong, not a rounded, figure. The animal inhabiting this shell is a limax.

Some of these have the mouth dentated, others smooth; and in some the clavicle is moderately long, though in most it is depressed.

Dolium subovatum spiraliter costatum.
The subovate Dolium, with spiral ribs.

This is a very elegant shell: it is about two inches and a half long, and nearly as much in diameter in the largest part: the clavicle is oblong, and pointed at the extremity, and is in such a manner continued from the body of the shell, that it is not easy to say where it begins; the other extremity of the shell is formed into a short rostrum, which turns a little up: the whole shell is tumid, and inflated, as it were, and largest near the head, and it's whole surface is covered with a number of elevated ribs, of the breadth of a straw, separated only by spaces of about the same diameter: the ribs are yellowish, and the spaces between them whiter, and both are spotted irregularly with a deeper yellow.

It is found on the shores of many parts of the East Indies, and has been sometimes also brought from America.

Dolium subovatum læve, eleganter variegatum. The Partridge-shell.
The smooth, subovate Dolium, with elegant variegations.

This is about two inches and a half in length, and near two inches in diameter, and is a thin shell, seemingly inflated; the clavicle is moderately long, and has four volutions; these are thick, inflated, or rounded, and the lowest of them is separated from the body of the shell by a hollow line; the other extremity of the shell termi-

 nates

tures without a beak: the mouth is large and reddish within; the whole external surface of the shell is perfectly smooth, and is of a brown colour, variegated in an elegant manner, with a deeper brown and a grey, and in the whole has something of the appearance of the plumage of the partridge.

It is brought to us from the East Indies, and sometimes from the shores of the American islands.

Dolium ellipticum longitudinaliter costatum.
The elliptic, longitudinally costated Dolium.

The Harp-shell.

This is one of the most beautiful shells of this genus: it is about two inches and a half long, and a little more than an inch and a half in diameter: the body is, however, all the way distended in such a manner, as to give it the perfect figure of the others of this genus: the clavicle has five volutions; the lowest of them, or that next the body of the shell, is large, the others very small, and the top pointed; the whole surface of the body of the shell is ornamented with large and elevated ribs; they stand at such distances, that the spaces between them are equal to twice or three times their own diameters; the colour of the shell is a deep brown, and it is variegated with a paler brown and with white, in a very beautiful manner.

We have it from both the East and West Indies, but the most beautiful are from China.

Dolium ellipticum læve coronatum.
The smooth, coronated, elliptic Dolium.

The Ethiopian-crown.

This is about three inches in length, and two in diameter; it's figure is oblong, and approaches to an elliptic, but is somewhat smaller at each end than in the middle: the mouth is long and wide, the extremity of the shell a little split by it: the clavicle is short, and obtuse at the extremity; it has about four volutions, and the lower one, as well as the upper edge of the body, are elegantly coronated, or deeply dentated, the denticles arising into regular, even, and conic points: the surface of the shell is tolerably smooth, but has the marks of many longitudinal lines on it; it's colour is a pale brownish-yellow.

We have it from the coast of Africa, and from the East Indies.

Dolium ovatum læve, clavicula depressa, ore magno.
The great-mouthed, smooth, oval Dolium, with a depressed clavicle.

This is a shell considerably differing from any of the former; it is about an inch and a half long, and an inch in diameter at the larger end, whence it gradually grows smaller to the other, where it is still very obtuse, and somewhat split by the extremity of the mouth: the mouth itself is of the whole length of the shell, and is extremely wide and gaping; the lip is thin, and the columella has two or three foldings or indentings near it's lower extremity: the clavicle is depressed in such a manner, that it is scarce visible in most directions of the shell: the surface of the whole body is white, and is elegantly variegated with clouded spots of yellow.

We have it from the American shores.

Dolium subrotundum læve, ore latissimo.
The roundish, smooth Dolium, with an extreamly wide mouth.

The thin Gondola-shell.

This is also a very singular species; it is about an inch and a half in diameter, measured either way; it's figure approaches to round, but the two extremities of the mouth are extended beyond the line of the other parts of the shell, and spoil the figure, with an appearance of two ears. This is one of the most rounded of this whole genus, but in front it appears almost all mouth: it's colour is a pale dirty grey on the outside, and a dead white within.

We have it on some of the shores of the Mediterranean.

Dolium

Dolium suberosum tuberculosum.
The suberal, tuberculose Dolium.

The Mulberry-shell.

This is about an inch in length, and three quarters of an inch in diameter in the largest part, which is not at the end, but nearer the middle: the clavicle is moderately long, and has three or four volutions; the opposite extremity is a little split by the mouth; the surface of the body is elegantly ornamented with oblong and moderately large tubercles, disposed in several spiral series; there are usually six or seven series of them; these are usually black, which on the paler ground of the shell have a very beautiful effect.

It is brought to us from the coast of the Mediterranean.

Dolium bullatum rostro brevi adunco.
The bullated Dolium, with a short, crooked beak.

The Pearled-snail.

This is about two inches in length, and an inch and a half in diameter in the broadest part; the body of the shell is of a rounded figure; the clavicle is moderately long, and has about four volutions; the mouth is long, and is terminated, at the lower end of the shell, by a short, thick, and somewhat crooked beak: the body of the shell is elegantly ornamented with a number of bullated protuberances, round, moderately large, and resembling pearls; they are disposed in regular series in a spiral form, and there are usually five of those series on the body of the shell, and one of a smaller, surrounding the lowest volution of the clavicle; the mouth is large and dentated; the general colour of the shell is yellow, but the tubercles are somewhat paler than the rest.

It is brought to us from the East and West Indies, and is a very elegant shell.

The other species of the Dolium are, 1. The round, thin, umbilicated Dolium. 2. The white-furrowed Dolium, with yellow streaks. 3. The yellow-furrowed Dolium, with spotted streaks. 4. The white, thick, furrowed Dolium, with dentated lips. 5. The striated and spotted Dolium, with a rugose columella. 6. The oblong, yellow Dolium, with a depressed clavicle. 7. The oblong, white Dolium, with a depressed clavicle. 8. The smooth, oblong, brown Dolium, with a depressed clavicle. 9. The oblong, variegated Dolium, with a depressed clavicle. 10. The variegated, contabulated Dolium, with a longer clavicle. 11. The oblong, costated, and umbonated, brown Dolium. 12. The elegant Dolium, with thirteen bright, red ribs. 13. The reddish, oblong Dolium, with fourteen narrow ribs. 14. The broad-ribbed, yellow and brown Dolium; the four last are called by our collectors Harp-shells. 15. The elegant, umbonated Dolium, called the Persian-shell, or the Panama purpura. 16. The smooth Persian-shell, with white lines. 17. The striated Dolium, with brown and white spots. 18. The Dolium, with an oblong and somewhat crooked rostrum, called the Fig-shell, with the depressed clavicle. 19. The rostrated Dolium, with a beautiful amethystine tinge, called the Turnip-shell. 20. The long-beaked, striated, yellow Dolium. 21. The thick, yellow Dolium, with lines and spots irregularly disposed. 22. The white Dolium, with lines and spots of brown, more regularly distributed over it. 23. The grey, thick Dolium, called the Boat-shell and Seamew. 24. The oblong, greenish Dolium. 25. The oblong, reddish Dolium. 26. The thin, white, papyraceous Dolium. 27. The yellowish Dolium, with four brown fasciæ. 28. The capillaceously, striated Dolium. 29. The white boat Dolium, umbilicated on each side. 30. The blue, fasciated, round Dolium, yellow within. 31. The greenish, or olive-coloured, round Dolium.

PORCELLANA.

THE Porcellana is a simple shell, without any hinge, formed of one piece, and of a gibbous figure on the back; the mouth is long, narrow, and dentated on each side.

The animal inhabiting this shell is a limax.

Porcellana

Porcellana albida, utrinque rostrata, intus flavescens.
The white Porcellana, yellow within, and beaked at each extremity.

This is of an oblong figure, and very gibbose: it's length, including the two rostra, is about three inches, and it's diameter, in the middle, is nearly two inches: it's colour is white on the outside, without any variegation, and yellow within; the mouth is large, and is continued in the form of a rostrum or snout at each end.
We have it from the African coast, and also from some parts of the East Indies.

Porcellana oblonga, utrinque rostrata, oculata.
The oblong, oculated Porcellana, rostrated at each extremity.

The Argus-shell.

This is an extremely elegant shell; it is about three inches long, two inches in diameter, and somewhat less than that in height, being less gibbose than most of the other kinds: the mouth is wide, and it's lips are continued at each extremity, beyond the verge of the shell, so as to form each way a kind of broad and short beak: the ground colour of the shell is yellowish, but there are three brown fasciæ running over it of considerable breadth, but not very distinct or deep; and the whole surface is ornamented, as well on these fasciæ as elsewhere, with multitudes of round spots, composed of an out-line not filled up, and resembling the eyes, which we see on some of our butterflies wings.
We have it from the coast of Africa and elsewhere; very fine ones have lately been brought from the East Indies.

Porcellana claviculata fusca, albo variegata.
The claviculated Porcellana, with white variegations.

The Map-shell.

This is a very elegant species: it is about two inches and a half in length, and nearly as much in diameter, and it's back is very gibbose: at the head there is a short clavicle placed a little above the extremity of the mouth, and composed of about four imperfect volutions; the other end of the shell is obtuse: the general colour is brown, but there are some elegant and irregularly undulated lines of white on it, which, with the other spots and clouded marks of the same colour, give the whole surface very much of the appearance of a map: the mouth is dentated or spinated, and about it the shell is paler than elsewhere, often whitish.
We have this from the East, and from the coast of Africa.

Porcellana subcærulea fasciata, rostrata et claviculata.
The bluish, fasciated Porcellana, with a clavicle at one end, and a beak at the other.

This is about two inches in length, and nearly an inch and a half in diameter, and is very gibbose and rounded on the back: at the head there stands a small conic clavicle, formed of about four volutions; the two lower ones imperfect, and terminating in a sharp point: at the other extremity there is a very short but broad rostrum: the general colour of the shell is bluish, but there are on it two, or sometimes three, fasciæ, of a greyish brown.
We have it from the shores of many of the American Islands, and sometimes from the coast of Africa.

Porcellana subovata, utrinque longissime rostrata.
The suboval Porcellana, with a long rostrum at each extremity.

This is a small species; it's body is of an oval, or nearly oval, figure, about three quarters of an inch in length, and half an inch in diameter, and considerably elevated or gibbose on the back: at each extremity there is a rostrum or beak of two thirds of the length of the body, and of the thickness of a large straw; the colour of the whole

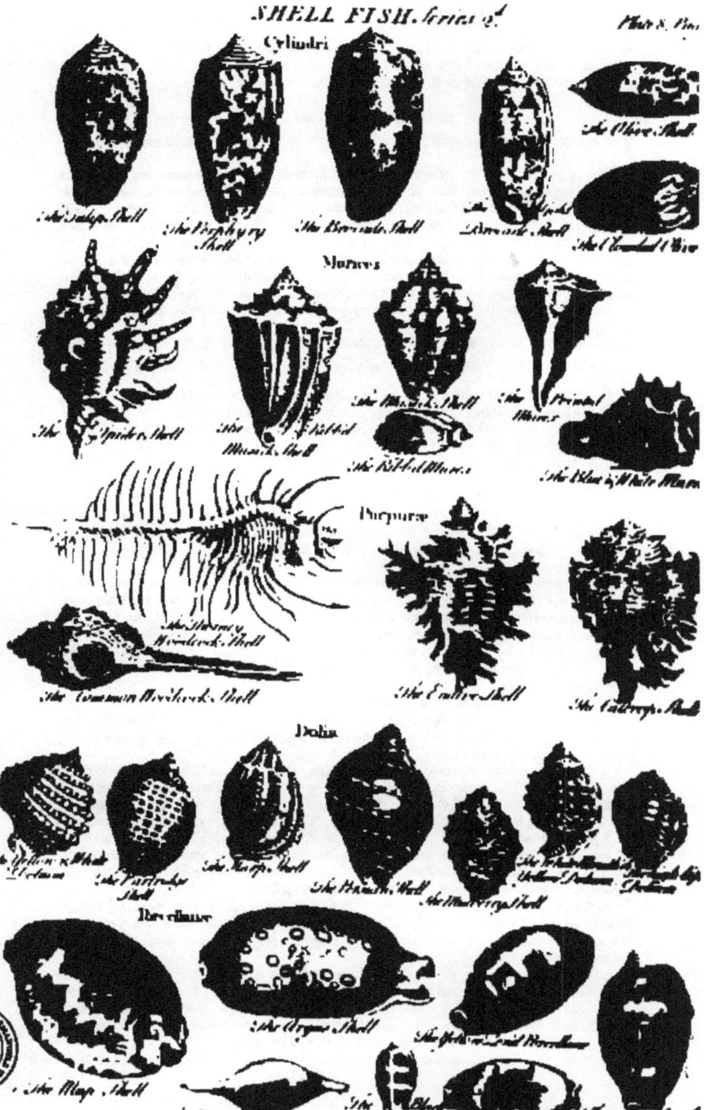
SHELL FISH. Series 2d
Cylindri
The Olive Shell
The Porphyry Shell
Murices
The Spider Shell
The Ribbed Murex
The Thorny Woodcock Shell
The Common Woodcock Shell
Purpuræ
Dolia
The Partridge Shell
The Argus Shell
The Map Shell

whole shell is white, and it's surface is smooth, and naturally polished, as is also that of most of the species of this genus.

We have this from the East Indies, and from the coast of Africa.

Porcellana utrinque rostrata, alba, fusco variegata.
The white Porcellana, variegated with brown, and rostrated at each end.

This is a small, but an extremely elegant, species; it is barely three quarters of an inch in length, and about half an inch in diameter, and is rostrated or beaked at each end, but in a very different manner; the rostrum at the head of the shell is short and broad, and seems no more than a continuation of the mouth, surrounded with a lip: that at the opposite extremity is longer, slender, and truncated; the body of the shell has a fine white for it's ground colour; and the variegations, which are of a bright brown, are disposed in crooked and irregularly angulated lines, so that they have something of the appearance of the marbled figures on coloured paper.

We have it from the shores of the American islands in considerable plenty.

Porcellana utrinque rostrata bullata.
The bullated Porcellana, rostrated at each end. — The Small-pox Porcelain.

This is also a small species; it does not exceed three quarters of an inch in length, and it's diameter is about half an inch: the body of the shell is of an oval figure, and very gibbose, or elevated on the back; it is rostrated at each extremity: the beak at the head is the longest, and both are split at the extremity: the whole surface of the shell is covered with elegant, round protuberances, of a fine white colour, imitating pearls.

We have it from the coast of Africa.

Porcellana grisea macula singulari magna fusca.
The grey Porcellana, with a single, large brown spot. — The beetle Porcelain.

This is about an inch in length, and two thirds of an inch in diameter: the back is gibbose, and considerably elevated; the mouth is wide for that of one of this genus, and is continued in a rostrated form at each extremity of the shell; at the head the rostrum is oblong, but truncated; and at the other extremity it is very short, and turned up: the colour of the shell is white, but on the summit of the back there is a large, irregular spot of brown, in a rude manner, resembling a beetle, or some other animal of that form.

We have it from the coast of Guinea.

The other species of the Porcellana may be arranged under four divisions, as they are, 1. Conglobated and thick: 2. Pyriform, or oval and thin: 3. Oblong and thick; and, 4. Irregularly gibbose.

1. Of those which are conglobated and thick, there are, 1. The elegantly-lettered Porcellana. 2. The tyger-spotted Porcellana. 3. The snake Porcellana. 4. The small, flatted Porcellana. 5. The small, flatted, and, as it were, fimbriated Porcellana. 6. The elegantly punctuated Porcellana. 7. The simple, greyish-white Porcellana. 8. The short and rostrated, elegant, Chinese Porcellana. 9. The simply, violet-coloured Porcellana. 10. The reddish Porcellana. 11. The elegantly variegated Porcellana. 12. The guttated Porcellana. 13. The broad, depressed Porcellana, elegantly variegated, and called the Tortoise-shell Porcellana.

2. Of the pyriform and thinner Porcellanæ there are, 1. The wide-mouthed Porcellana, elegantly spotted with yellow. 2. The brown Porcellana, with two zones, and with a wide mouth. 3. The elegantly variegated brown, white, and yellow arcuated-mouthed Porcelain.

3. Of those which are of an oblong form and thick, there are, 1. The short and very gibbose, small, oculated Porcellana, or Argus-shell. 2. The irregularly ocellated Porcellana, called the false Argus. 3. The greenish Porcellana, with numerous, pearly, round, and small tubercles. 4. The brown Porcellana, variegated with four red zones. 5. The brown Porcellana, with four yellow zones. 6. The elegant Porcellana, with amethystine zones. 7. The green-spotted Porcellana. 8. The greyish-brown,

brown, virgated Porcellana. 9. The blue, oblong Porcellana. 10. The blue, oblong, ſpotted Porcellana.

4. Of thoſe which have irregular gibboſities, there are, 1. The milk-white Porcellana, with a pale red lip, and red protuberances. 2. The white, gibboſe, ſmooth-mouthed Porcellana. 3. The yellow, gibboſe, ſmooth-mouthed Porcellana. 4. The dentated-mouthed Porcellana, with ſeven gibboſities on the ſurface, commonly known by the name of the Guinea-money. 5. The large, gibbated-backed Porcellana.

SHELL-FISH.

Claſs the Second.

BIVALVES.

Thoſe which have the external covering or ſhell compoſed of two parts, or valves.

OF theſe there are ſix genera: 1. The Oſtreæ or Oyſters. 2. The Chamæ. 3. The Mytuli or Muſcles. 4. The Cardiæ or Heart-ſhells, among which are the Cockles. 5. The Pectines or Eſcallops, uſually called Scollop-ſhells; and, 6. The Solenes or Razor-ſhells.

OSTREA.

THE Oſtrea is a ſhell compoſed of two valves; the lower one hollowed on the inſide, and gibboſe without; the upper one more flat: both compoſed of a multitude of laminæ or cruſts, and uſually ſcabrous, or rough, on the outer ſurface.

The ſhells of ſome of the Oſtreæ are ſmooth, thoſe of others furniſhed with tubercles, or with ſpines, or deeply furrowed and plicated: the figure of moſt is roundiſh, but in ſome it is quite irregular.

Oſtrea coſtata, foliata, roſea et alba.
The red and white, coſtated and foliated Oſtrea.

The foliated Oyſter.

This approaches to the ſhape of the common oyſter; it is about three inches in diameter, and the lower valve is deep, the upper one nearly flat: it is all the way, on both valves, furrowed deeply in a longitudinal direction, with rounded but irregular ridges between the furrows; theſe are numerous, but the greater part of them are naked; about every fifth only is furniſhed with a ſeries of very elegant foliaceous prominences, which are flat, curled, and jagged at the edges; they are from half an inch to three quarters in height, and are on both valves, but moſt fair and entire on the upper: the ground colour of the ſhell is a roſe colour, with ſome admixture of purple, variegated with white; the prominences are principally white.

It is a native of the Eaſt Indies.

Oſtrea tenuis lævis ſubrotunda.
The thin, ſmooth, roundiſh Oſtrea.

The onion-peel Oyſter.

This is a ſpecies very ſingular in it's thinneſs and delicacy; it is from an inch to two in diameter, and is extremely different, in the ſeveral ſpecimens, in the depth or gibboſity of the valves; ſometimes both are nearly flat, ſometimes one is much hollowed, and ſometimes both are ſo: the ſurface of both valves is ſmooth and gloſſy, and the colour whitiſh, with a tinge of red or green on the outſide, and a fine pearly white, with more or leſs of the ſame tinge within: it is frequently undulated in an irregular manner on the ſurface, and has a large roundiſh opening near the hinge.

It is a native of the Eaſt Indies, and of ſome parts of Europe.

Oſtrea ſubovata tenuior ſinuata.
The thin, oval, ſinuated Oſtrea.

The Tree-oyſter.

This has been by many ſuppoſed to be the ſame with the common oyſter, only differing in it's manner of growing; but later obſervations have ſhewn, that in places where it has the ſame accidents attending it's formation, as the common oyſter, it ſtill preſerves it's diſtinct form: it is about two inches long, and an inch and a half in diameter at the larger end: the hinge is placed at the ſmaller extremity, which terminates obtuſely, and from thence it gradually becomes larger to the oppoſite end, where it is rounded and ſinuated at the edge: the whole ſurface of the ſhell is deeply furrowed in a longitudinal direction, and is ſomewhat rough, though not nearly ſo much ſo as that of the common oyſter: the colour is brown, with a faint admixture of reddiſh on the outſide, and a pearly white within. When this ſpecies is produced in the neighbourhood of robuſt ſea-plants, or other ſmall and rounded bodies under water, it adheres to them, and there will grow, from the ſeveral parts of the ſhell where it touches them, a kind of clavicles or claſpers, which ſeize upon, or meet round the thing, holding the ſhell faſt to it.

It is very frequent in the Weſt Indies, where it abounds on their mangrove-trees covered by the tides, and has thence obtained the ſtrange name of the Tree-oyſter. To this has been owing the report of oyſters growing on trees in that part of the world.

Oſtrea anguſtior capite tranſverſo.
The narrow Oyſter, with a tranſverſe head.

The Hammer-oyſter.

This is by far the moſt ſingular of all the Oyſter-kind, and is indeed one of the moſt extraordinary ſhells in the world: it's ſhape is that of a hammer, or rather a pickax, with a very ſhort handle, and a long head: the body of the ſhell, which ſtands in the place of the handle of the inſtrument from which it is named, is about four inches long, and three quarters of an inch in diameter: what is called the head, and anſwers to the head of the pickax, is five or ſix inches long, and, except where it joins to the body, is ſcarce any where more than half an inch in diameter; it ſtands tranſverſely or croſſwiſe to the body, and is irregularly formed, uneven at the edges, and terminates in a narrow but obtuſe point at each end: the hinge is at the lower end of the body; the valves open all the way from each end to this part, and, though of this ſtrange figure, they ſhut in a remarkably cloſe and elegant manner: the edges of the body and head have often great irregularities and protuberances on their ſurface, and at their edges; and are deeply furrowed and plated all over, the lines running in different and very irregular directions: the colour is a deep brown, with a tinge of a violet purple on the outſide, and a pearly white, with ſome faint tinge of purpliſh within.

It is a native of the Eaſt Indies, and is a very rare ſhell; when entire, which it very rarely is, it ſells at a great price: ſix guineas were, laſt winter, given at an auction for an imperfect one.

Oſtrea horrida ſpinis erectis et aduncis.
The Oſtrea, with numerous, erect, and crooked ſpines.

The great, prickly Oyſter.

This is a large and very elegant ſpecies: it is of a figure approaching to oval, four inches long, and at the larger extremity near three in diameter; at the ſmaller, where the hinge is, it is about an inch and a half: both ſhells are gibboſe and hollowed, but the under one moſt ſo: the whole ſurface of each is furrowed longitudinally, but ſomewhat irregularly, and is covered with a vaſt number of ſpines; they are robuſt, ſharp-pointed, and of various lengths, from half an inch to a tenth of an inch; they ſtand cloſe and crowding upon one another, and ſome of them are ſtraight, others crooked: the colour of the outſide of the ſhell is a dirty red, on the inſide it is white and pearly.

It is a native of the coaſts of Africa, and we have it alſo in ſome other places.

Ostrea conica spinis ad oram undulatis.
The conic Ostrea, with undulated spines at the rim.

This is about two inches in length, and an inch and a quarter in breadth at the larger extremity, from whence it gradually becomes smaller to the other, at which the hinge is placed: the body of the shell is rough and undulated, and has a few short spines toward the edges, but rarely any about the middle; toward the rim, at the larger extremity, there stand three or four rows of long, robust, and sharp-pointed spines; they are not straight but undulated, or bent backwards and forwards, and make a kind of thick fringe, those of the upper and lower valve meeting, and hiding the joining of the two parts of the shell: the body of the shell is of a dirty white, but the spines are of an elegant purplish-red, and make a very beautiful appearance.

It is frequent on the shores of South America.

Ostrea foliata imbricata albescens.
The whitish, foliated, and imbricated Ostrea.

This is one of the most extraordinary shells of this genus; it's figure is oval; it's length about three inches and a half; it's hinge is at the smallest end, and it's diameter at the other is two inches and a half: it is deeply sinuated at the edge, and the surface is all over covered with transverse, foliated protuberances, of an undulated form: they are flat, uneven, and notched at the edges, and there are usually six or seven series of them placed at tolerably equal distances one over another: the general colour of the shell is white, but there are variegations of a bright rose colour on it.

It is a native both of the East and West Indies, but is not frequent in the latter place.

Ostrea hians globosa echinata.
The echinated, globose, gaping Ostrea. **The hedgehog-oyster.**

This is a very singular and pretty species; it is about an inch in diameter, as much in length, and nearly as much in depth: the shells are both gibbose, or nearly hemispherical, and are so shaped, the upper one being smaller than the under, that they never can close or shut, but always leave a considerable opening: the whole shell is furrowed longitudinally, and is very thick-set with short, crooked spines, the points of them all turning toward the hinge: the colour is a tolerably clear white, without any variegation.

It is frequent in the East Indies, but rarely comes over perfect.

Ostrea dentata crassa.
The thick, dentated Ostrea. **The Cocks-comb Oyster.**

This is a thick and coarse species, but it has it's singularity, that may very well stand in the place of beauty: it's diameter in breadth is about four inches; it's depth, from the hinge to the verge of the mouth, hardly three; it is elevated into three, four, or more very prominent and sharp ridges on the surface, with very deep, angular furrows between them; these ridges are extended also at the verge, beyond the rest of the shell, and form a deeply dentated edge; the colour on the outside is a deep brown, with a tinge of violet; within it is of a pearly white: the surface is somewhat rough, with transverse furrows; and the opening of the shell, though so deeply sinuated, is formed so nicely, that it closes with great exactness.

It is a native of some of the coasts of the Mediterranean: we have both this and the hammer-oyster fossile in the sand-pits on Blackheath; and both, among many other shells, in the stone of which St Paul's Church is built.

The other species of Oysters are to be arranged under two or three divisions, and will then be easily understood by their specific names, without farther descriptions.

1. Of the more flat and smooth kind, there are, 1. The common Ostrea. 2. The little, greenish, oriental Ostrea. 3. The great, broad Ostrea, which produces the genuine pearl, and is thence called the Concha Margaritifera, and Pearl-shell. 4. The great,

great, oblong, and very flat Pearl-shell. 5. The small, round, variously coloured, smooth Ostrea, with an opening near the top; this is very uncertain in it's colour, varying into red, yellow, brown, and green, and has been described under these by some, as so many different species. 6. The small, oval, variegated Ostrea. 7. The large, thin, and smooth Ostrea, with an undulated surface, called the Saddle-oyster. 8. The thin, elegantly spotted Ostrea. 9. The elegant, red, striated, and lineated Japonese Ostrea. 10. The caudated, oval-bodied Ostrea, called the Swallow-shell.

2. Of those which have the surface convoluted or plicated, but have no spines, there are, 1. The large, oval, sinuated, and plicated Ostrea. 2. The thin, round, deeply plicated Ostrea. 3. The irregularly tortuous Ostrea, called the Leg-oyster. 4. The large, thick, tabulated Ostrea, with an elegantly sinuated edge.

3. Of those which are of a more globose figure, or have both shells gibbose and armed with spines, there are, 1. The large, very gibbose Ostrea, with rounded and not very numerous spines. 2. The gibbose and rounded Ostrea, with numerous flatted spines. 3. The less gibbose, flame-coloured Ostrea, with sharp spines. 4. The reddish, gibbose Ostrea, with numerous short white spines. 5. The deeply furrowed, purplish-red Ostrea, with numerous whitish, flat spines. 6. The gibbose, rounded Ostrea, with red and livid spines. 7. The great, plicated Ostrea, with a few flat spines. 8. The elegantly striated Ostrea, with a few spines.

4. Of those which are gibbose and foliated, or have flat and jagged eminences in the place of spines, there are, 1. The white, rounded, scaly Ostrea, with small but elegantly sinuated foliations. 2. The smaller gibbose Ostrea, with foliations forming tubules at the ends. 3. The yellow, gibbose, and elegantly foliated Ostrea. 4. The less, gibbose, white and purple foliated Ostrea. 5. The little, globose Ostrea, with very long foliations.

5. Of those which are of an oblong figure, and have either spines or obtuse protuberances, there are, 1. The narrow, sinuated Ostrea, with short and few spines. 2. The narrow, sinuated Ostrea, with numerous, obtuse tubercles. 3. The elegant Ostrea, with white foliations, edged with delicate purple spines. 4. The variegated, red and white Ostrea, with large spines. 5. The conic, deeply furrowed, spinose Ostrea.

CHAMA.

THE Chama is a shell formed of two valves, which are both convex, or gibbose, and equal, and which, when shut, leave always an opening in one part.

Some of the Chamæ are perfectly smooth, some are striated, and some rugose, and even spinose; some of them are oblong, others roundish; some equilateral, and others inequilateral: the several species have very considerable differences in shape, but the generical characters are obvious and immutable.

Chama depressa levis, fusca, lineis irregularibus nigris.
The smooth, depressed Chama, variegated with irregular black lines.

The Arabian Shell.

This is a large species; it is three inches in diameter from one extremity to the other, and two and a half from the hinge to the opposite verge: the head, where the hinge is placed, is not in the center of the top, but near one side: the whole surface of the shell is perfectly smooth and even, and has a fine natural polish: it's ground colour is a pale brown, and it is variegated with lines and streaks of black; they are narrow, angulated, and disposed across one another in a beautiful, irregular manner: their accidental meetings sometimes form resemblances of the Arabic or Arabian characters, and thence the shell has obtained the name of the Arabic or Arabian Chama.

It is frequent on the coasts of many parts of Europe.

Chama gibbosa, striis transversis numerosissimis exarata.
The gibbose Chama, with very numerous, transverse striæ.

This is about two inches and a half in diameter, either way, and the shells are both so much elevated, that it's thickness is not less than an inch: It is of a beautiful yellow colour without any variegation, and the whole surface is deeply furrowed: the furrows run transversely in a semicircular direction, and there are small and sharp ridges between them.

It is frequent on the coasts of the American islands, and is found also in some parts of Europe.

Chama reticulata ovata rotunda.
The roundish, oval Chama, with a reticulated surface.

This is two inches and a half in length from the hinge to the opposite edge, and two and a quarter in diameter in the broadest part, which is near the edge; the whole surface is covered with somewhat deep, longitudinal furrows, with rounded ridges standing very close between them; and there are also a number of transverse lines, though much fainter than the others, running in a circular direction, and forming, with the others, a kind of net-work: the whole shell is of a dusky whitish colour.

We have it on some of our own shores, and those of the Mediterranean abound with it.

Chama truncata margine spinosa.
The truncated Chama, with a spinose edge.

The Venus Shell, or Concha Veneris.

This is a shell of a very singular figure; it is about an inch and a half in length, and as much in diameter toward the larger end: the valves are both convex, and are deeply striated in a longitudinal direction: the hinge is placed at the extremity, where the shell is rounded and prominent, and the end that should have gone the other way seems truncated or cut off: the opening is covered with a very elegant lip, propagated from each side; these are wrinkled, and of a beautiful reddish colour, with some white among it; they do not unite perfectly in the middle, but leave an oblong aperture, and there stand at the farther edge of each, or round about the truncated end of the shell, a series of long, slender, and beautiful spines.

It is found on the shores of the American islands, and, when perfect, is valued at a high rate by our collectors, but the spines are usually broken in the carriage.

Chama truncata profunde striata margine serrato.
The truncated, deeply striated Chama, with a serrated edge.

This is about an inch in length, and as much in diameter; the head is rounded and small, and one extremity is rounded, the other truncated, as in the former species: the whole surface of the shell is furrowed with deep lines, with a few broad ridges between them; the truncated end is closed by two whitish, wrinkled lips, leaving only a small, oblong aperture between them, and the edge is surrounded by a dentated or serrated margin, not by a series of little spines, as in the former: the ground colour of the shell is a dead white, but it is elegantly variegated with a glossy brown.

We have it both from the East Indies, and the American islands.

Chama truncata margine levi.
The truncated Chama, with a smooth edge.

This is an elegant shell; it is about an inch and a half long, and as much in diameter: the head, where the hinge is placed, is small and obtuse; the whole surface of the shell is smooth, and has an elegant, natural polish; it's colour is white; the truncated end is large, and the lips leave but a small aperture between them; they are very much wrinkled, and are of a brownish colour, with a strong tinge of violet.

It is a native of the East Indies; we have not yet received it from any other part of the world.

Chama

Chama truncata obscure fasciata.
The obscurely fasciated, truncated Chama.

This was one of the first known species of the truncated Chamæ, and obtained very early, in these studies, the name of Concha Veneris: it is now distinguished by the name of the oriental Concha Veneris; but, in contradiction to the generality of oriental species, it is vastly inferior in beauty to the occidental Concha Veneris, or the spinose-lipped kind described before: it is two inches and a half in length, and as much in diameter; the head, where the hinge is placed, is small and rounded, and the truncated part is furnished with two lips, which are so large, as to meet near the top, leaving only a small opening just under the hinge, but they gape again lower down: the whole surface of the shell is furnished with obscure fasciæ, eight or ten in number, and the lips are smooth at the edge, or have no spines, nor so much as a dentated margin.

It is frequent on the shores of the East Indies, but it is not greatly valued by our collectors.

Chama truncata levis variegata.
The smooth, variegated, truncated Chama.

This is a large and very beautiful species; it is two inches and a half in diameter, and more than two inches from the hinge to the opposite edge; both the valves are gibbose in a considerable degree, and the truncated end is furnished with a pair of lips, which leave only a small, round aperture under the hinge: the surface of the whole shell is perfectly smooth, and has an excellent natural polish: it's ground colour is a snowy white, and it is variegated with very elegant streaks and blotches of a violet purple.

It is frequent in the American seas.

Chama suborbiculata variegata levis.
The smooth, variegated, roundish Chama.

This is near two inches in diameter, and is not of the truncated kind; the head is large, obtuse, and rounded, and both shells are convex; they shut tolerably close, all the way round, except that they leave a small roundish aperture near the head on one side: the colour of the shell is yellow, and it is elegantly variegated with red spots; there is not the least ridge or furrow on any part of it's surface, and it has a fine natural polish.

It is brought to us principally from the East Indies, though we have it in some parts of Europe.

Chama rugosa subrotunda capite oblongo.
The roundish, rugose Chama, with an oblong head.

This is an inch and a half in length, and about an inch and a quarter in diameter; it's colour is a simple brownish-yellow, without any variegations: the valves are both considerably gibbose, and have a number of deep, transverse furrows, with some roughnesses and obtuse protuberances on the ridges between them: the head is elevated and elongated, and terminates in an obtuse point; this has as little beauty as any of the Chamæ.

It is a native of our own seas, and is also sometimes brought from America.

Chama suborbiculata profunde striata.
The deeply striated, suborbiculate Chama.

This is a small species; it measures about three quarters of an inch in diameter, and somewhat less than that from the hinge to the opposite verge: the colour is a tolerably pure white, and it is variegated with elegant, blood-red spots: the whole surface is ornamented with a kind of longitudinal ribs, running between deep and broad furrows; there are also some transverse lines, but they are much fainter.

It is a native of the Mediterranean, and has been found also in great perfection on the Irish shores.

Chama labiata subsphærica striata.
The roundish, striated, labiated Chama.

This is one of the most singular of the Chamæ: it has, in general, the structure of the truncated ones, but the truncated part is not flat or straight as in them, but swells into a roundness, and the opposite part of the shell is shorter than in any of the others; by this means it departs greatly from their figure, and approaches toward roundness: the valves are very gibbose also, and the shell is almost as deep as it is long or broad; it is all over of a dusky white; the whole surface, except the lips, is striated transversely; the lines are small, close, and elegant; the lips are almost smooth; what inequalities they have are owing to some oblong wrinkles, but these are faint, and are often quite wanting.

It is found on the shores of some of the American islands, but is very rare in our collections.

The other species of the Chama are, 1. Of the oblong and transversely striated kind; 1. The elegant, very lightly striated, brown and black Chama. 2. The elegant, flat, purple and white Chama. 3. The elegant, brown and white, flat Chama. 4. The pale brown Chama, with narrow, black streaks. 5. The deeply furrowed Chama.

2. Of those which are oblong, and are longitudinally striated or costated, there are, 1. The grey, lightly striated Chama. 2. The elegant amethystine Chama. 3. The lightly striated, yellow Chama. 4. The almost smooth, red Chama. 5. The beautifully variegated, finely striated Chama.

3. Of those which are of a figure approaching to round, and are equilateral, there are, 1. The lightly striated, spherical, whitish Chama. 2. The reticulated Chama, or basket-shell. 3. The reticulated and granulated Chama. 4. The common, transversely ridged, deep Chama. 5. The rugose or deeply furrowed, brown Chama. 6. The smooth Chama, with irregularly angulated lines of black. 7. The elegantly punctated Chama. 8. The clouded brown and white Chama. 9. The red and white, fasciated Chama. 10. The thin, plain, yellow Chama. 11. The elegantly fasciated, violet Chama. 12. The white, thick, plicated Chama. 13. The snow-white, rounded Chama, called Chamapelodes by authors. 14. The smooth, somewhat flatted Chama, called by Bellonius the Chama Pelorides. 15. The blackish, depressed Chama, called the Chama Glycymeris of Ælian. 16. The very deeply furrowed, rounded Chama, called the Chama Trachea. 17. The oblong, smooth, brown and white Chama, called by some the Chama Peprium. 18. The smooth, variegated Chama, called the Chamelaia. 19. The rough Chama, called the Tyger's-tongue Chama. 20. The fine, thin, white, reticulated Chama.

4. Of those which are truncated and inequilateral, beside the species described above, there are, 1. The elegant violet-coloured Chama, with an oval aperture. 2. The grey-lipped, smooth Chama. 3. The brown-lipped, smooth Chama. 4. The truncated Chama, with smooth lips, and longitudinal striæ on the body. 5. The smooth-lipped, truncated Chama, with transverse ridges. 6. The large, smooth, truncated Chama, of an elegant purple within. 7. The large, grey, truncated Chama, with a mixture of purple and white within. 8. The small, grey Chama, with the inside of a pearly white. 9. The variously, striated, truncated Chama.

MYTULUS.

THE Mytulus is a shell composed of two valves, of an oblong figure, and shutting close all the way: the valves are both convex, and of a similar shape. The animal inhabiting it is a Tethys.

Of the Mytuli some are of a conic figure, others oblong, and equal at both extremities; some are smooth on the surface, some rough; and some are much deeper than others.

Mytulus major utrinque angulatus.
The large, angulated Mytulus. The Carolina Muscle.

This is four or five inches in length, an inch and a half in diameter, and nearly as much in thickness; the head or place of the hinge is situated very near one of the extremities of the shell; this end is the more obtuse, the opposite one is more pointed; the two valves are exactly of the same shape, and they shut every-where regularly close; they are deep, and have their middle elevated into an irregular, angular gibbosity; the colour of the outside is yellow, with a faint tinge of purple in some places; within it is of a pearly hue, with much of the purple or violet toward the edges.

It is brought from South America; some very fine ones brought from Carolina have given it the general name of the Carolina Muscle: we have it fossile in England.

Mytulus depressior tegmine coriaceo.
The depressed Mytulus, with a coriaceous covering. The coated Muscle.

This is one of those species which have been called Conci and Telline by authors, fond of distinctions where there were very little differences: it is two inches in length, the transverse way, and not an inch from the hinge or head to the opposite verge; the hinge is situated much nearer to one end than to the other, and the shell is elevated into a rounded protuberance in that part; the surface is rough, and of a dusky brown; it is covered with a thick, coriaceous coat, furnished with a multitude of short hairs.

It is frequent in the Mediterranean.

Mytulus levis variegatus gibbosior.
The smooth, variegated, gibbose Mytulus.

This also is one of the conci of authors; it is two inches from end to end, and near an inch and a half from the hinge to the opposite verge of the shell; it's thickness is, at least, three quarters of an inch, both the valves being very convex: the hinge is placed near one end, which is large and rounded; the other is flatter, and terminates in a kind of edge: the ground colour of the shell is whitish, but it is elegantly variegated with yellow and brown.

It is a native of the American Seas; we have it very frequently fossile in England.

Mytulus angustior fasciatus et variegatus.
The narrow, fasciated, and variegated Mytulus.

This is a very elegant and very singular shell; it is three inches long, and about one inch in diameter: the valves are both considerably convex, and consequently the shell is as thick as broad. It's largest part is near the middle, a little toward the lower end; from thence it gradually becomes smaller to each end, at the head or upper extremity, at which the hinge is placed; it is terminated by a small, obtuse button; at the opposite extremity it is more obtuse: the whole lower part is marked with transverse lines or fasciæ, somewhat broad, and sunk lower than the rest of the shell; from the middle to the top the surface is more smooth: the ground colour is a yellowish-red, and there are a number of beautiful variegations of a deep purple, on several parts of it, but mostly near the sides.

It is a native of the South Seas, and is very rare in our collections; the French cabinets have some fine ones: the colours are faint on most of these, but that is owing to their having been long washed on the shores, before taken up; they are naturally very bright.

Mytulus ovato-oblongus variegatus levis.
The smooth, ovato-oblong, variegated Mytulus. The purple, Magellanic Muscle.

This is four inches long, and toward the extremity is two and a half in diameter; from this part it gradually grows smaller to the head where the hinge is situated, which is is narrow and somewhat pointed; both the valves are gibbose, and the shell considerably deep: the ground is a deep and fine violet purple; the variegations are brown and whitish, or grey.

It is found on the coasts of the southern parts of America; it is met with in some of our collections, but is not common.

Mytulus ovato-conicus, maximus, rugosus, et striatus.
The ovato-conic, great, striated, and rugose Mytulus.

The Pinna Marina.

This is one of the largest of the bivalve shells; it is frequently two feet long, and near one foot in breadth: the valves are neither of them very deep, so that the cavity of the shell is shallow, in proportion to it's great extent; it is small and narrow at the head; from thence it descends growing gradually larger for two thirds of it's length, and forms so far a part of a cone: but from thence to the extremity, though it continues still growing larger, it assumes a rounded figure, and at that end forms the larger extremity of an oval: the colour of the outside is an olive brown; within it is partly of a pearly hue, partly reddish: the external surface is rough and elevated in many places into a kind of squammae, and is furrowed longitudinally from near the top to the bottom.

It is a native of the East Indies, and of some other parts of the world.

Mytulus conicus squammosus et echinatus.
The conic, squammose, and echinated Mytulus.

The prickly Pinna.

This is six or seven inches in length, and three in diameter at the base, from whence it gradually becomes smaller to the opposite extremity; it does not swell into a roundness, as the former species toward the base, but the extremity is, however, a little bellied out, not plainly truncated: the colour of the shell is a pale olive on the outside, and a fine pearly white within; the external surface is deeply furrowed in a longitudinal direction; the furrows run at some distances, and the elevated parts between them are consequently broad: they have a kind of squammae or scales rising on them, which terminate, many of them, in prickly points.

It is a native of South America; we have it also sometimes from the Asiatic and African shores.

The other species of the Mytulus are numerous, and are called by authors by three names: the Pinnae Marinae, Musculi, and Tellinae.

1. Of those called Pinnae Marinae, that is, such as are flat, oblong, and terminated at one extremity by a point, there are, 1. The large, blackish-brown, Magellanic Pinna. 2. The lesser, very deeply striated Pinna. 3. The striated and elegantly variegated Pinna. 4. The large, smooth, olive-coloured Pinna. 5. The narrow, paler brown Pinna. 6. The narrow and somewhat deeper Pinna. 7. The elegant blue Pinna, striated deeply in the lower parts; this is a very scarce shell. 8. The small, rose-coloured, variegated Pinna. 9. The grey, deeply striated Pinna. 10. The large, pale brown, lightly striated Pinna, of Newfoundland. 11. The smaller and smoother Pinna, of the Canada lakes. 12. The reddish-grey, large, rough Pinna. 13. The dusky, red, aculeated, and deeply furrowed Pinna. 14. The lesser, deep brown Pinna, called Pinna Tridacna by Rondelet. 15. The rounder-ended Pinna, called the Duck-billed Pinna. 16. The very long, narrow, truncated, and elegantly ribbed Pinna.

2. Of those which have the valves deeper, or more gibbose and equilateral, and are called Musculi and Muscles, there are, 1. The greyish-white, very thin-shelled Muscle. 2. The bright, silvery-white Muscle. 3. The oblong Muscle, which buries itself in stones, and is called Pholas. 4. The shorter and more clate, black, saxatile Muscle or Pholas. 5. The brown Muscle, which never closes it's shell, but has a large fleshy body, and a trunk. 6. The lesser, open Muscle, with a deep brown, obtuse shell. 7. The dusky-brown, obtuse-ended, American Muscle. 8. The small and elegantly variegated Muscle. 9. The deep red Muscle.

3. Of those which have equal extremities, and are of an oblong figure and plane, and are called by many writers on these subjects Tellinae, there are, 1. The smooth, large, violet-coloured Tellina. 2. The smaller, purple and white, elegantly variegated

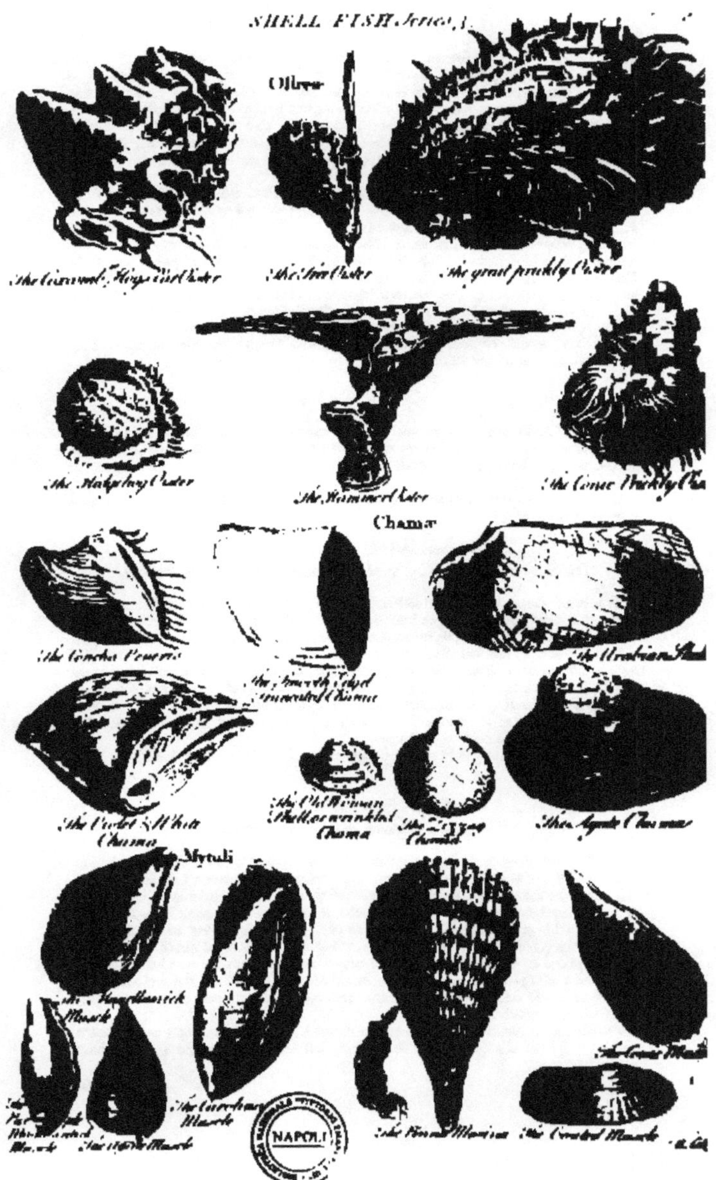
SHELL FISH
Ostrea
the great prickly Oyster
The Hammer Oyster
Chamæ
The Old Woman Shell, or wrinkled Chama
The Agate Chama
Mytuli
The Crested Muscle

gated Tellina. 3. The smooth, flat Tellina, with elegant white fasciæ. 4. The purple coloured, deep brown, hairy Tellina. 5. The shorter, coriaceous Tellina, of the New England lakes. 6. The broader and flatter Tellina, of the Azores islands. 7. The narrow, oblong, coriaceous Tellina, from the banks of Newfoundland. 8. The deep brown, round-ended, coriaceous Tellina. 9. The pointed-ended, coriaceous Tellina: all these coriaceous kinds, when cleaned of their outer rough coat, shew elegant variegations, red and white, blue and white, and the like, and are usually met with in that state in our cabinets.

4. Of the oblong Mytuli, with the extremities unlike, which are also called Muscles by authors, we have the following, 1. The reddish-brown, rostrated Mytulus. 2. The yellow, oblong, and rostrated Mytulus. 3. The narrow, rounded, and deep Mytulus, called the Sheath-muscle. 4. The long-beaked, less hollowed Mytulus. 5. The ordinary, deep blue Mytulus. 6. The oblong, flatted, and rough Mytulus, called the Cat's-tongue Shell. 7. The flatted, red and white, variegated Mytulus. 8. The yellow, plicated Mytulus. 9. The flat, oval Mytulus, called the Leaf-muscle by Rumphius. 10. The oblong, white Mytulus, with a granulated surface. 11. The reddish, transversely striated Mytulus. 12. The violet-coloured Mytulus, with the top striated. 13. The oblong, reddish or yellowish, flat Mytulus, with longitudinal striæ continued from the top to the base of the shell.

CARDIA.

THE Cardia is a shell formed of two valves, and resembling the figure of a heart at cards: the valves are equal and gibbose, or otherwise elated; they have no ears or appendages near the head, and are always either furrowed, imbricated, or spinose.

Of the Cardiæ, some approach to a globose, others to a triangular figure; others are irregularly oblong.

Cardia tenuis albida costis decem excavatis.
The thin, white Cardia, with ten hollowed ribs on the surface.

This is an extremely singular and elegant shell; the valves are rounded and very deep; it is three inches in length from the hinge to the opposite verge, about as much in diameter, and but little less in depth: the head, where the hinge is placed, is prominent and obtuse, and has much of the figure of the upper part of a heart at cards; there run along each valve, from the head to the verge, ten ridges or costæ; they are very high, of a triangular figure, edged at the top, and continued in the form of denticulations beyond the verge of the shell: they are hollow, but their cavity does not communicate with that of the shell; the whole is of a beautiful white, and so thin, that it is transparent, when held up to the light.

We have it from the coasts of many parts of the East Indies, and sometimes from South America.

Cardia profunde sulcata spinosa.
The deeply sulcated and spinose Cardia.

This is a very large and elegant species; it is three inches and a half long, and at least as much in diameter, and both valves are hollowed and elevated so much on the back, that the thickness of the shell, when closed, is not much inferior to it's length: the head is rounded and large, and the beaks of the two valves meet in such a manner over the hinge, as to give the appearance of a heart: the colour of the shell is a greyish-white; it's surface is very deeply furrowed in a longitudinal direction; the furrows are placed at some distance, and the ridges between them are rounded: the prominent back of the shell is often smooth or naked, but toward the end and the edges there stand a vast number of long, sharp, and robust spines, some of them straight, others a little crooked.

It is not uncommon on the shores of the American islands: I brought more than a hundred weight of this species in a fossile state, last summer, from the stone pits near Yaxley.

Cardia profunde ſulcata apicibus plano ſeparatis.
The deeply furrowed Cardia, with the tops ſeparated by a plane.

This is a very ſingular ſpecies: it is two inches in diameter, and about an inch and three quarters from the head to the oppoſite verge: the head is formed of two beaks, one to each valve; and theſe are rounded and turned inward, as in the others, but they do not meet or touch; a little below them there is a plane or flat part of the ſhell, in which is the hinge, and which keeps them from meeting: the hinge itſelf is of a very beautiful ſtructure; the notches of it are as fine as thoſe of a file: the valves are equal, and are extremely thick and heavy; they are white, and of the conſiſtence and appearance of marble: they are deeply furrowed in a longitudinal direction, or from the head ſtraight downwards, the furrows widening as they recede from the head, ſo as to extend over the whole breadth of the ſhell; the ridges between theſe are low and rounded, and have frequently tubercles on them: there is ſomething in this ſhell that approaches, in ſome degree, to that odd ſpecies the Noah's Ark-ſhell.

This is a native of the American Seas; I have met with it foſſile in ſand-pits about Canterbury.

Cardia flava coſtata et elegantiſſime imbricata.
The yellow, elegantly ribbed, and imbricated Cardia.

Our cabinets ſhew very few more beautiful ſhells than this Cardia: it is a large ſhell; it meaſures four inches in diameter, and more than three and a quarter from the hinge to the oppoſite verge: the valves are equal, and are conſiderably gibboſe, ſo that the cavity within the ſhell is large; the ſurface of each valve is raiſed into five very large ribs, which are rounded at their tops, and ſtand at conſiderable diſtances; theſe run from the hinge to the oppoſite verge of the ſhell: the whole ſurface is covered with elegant tranſverſe plates or flakes, which run from edge to edge, the croſs-way holding their courſe uninterruptedly over theſe ribs, and over the hollows between them: theſe plates are thin, ſharp, and elevated at their rims above the ſurface of the ſhell, and are placed in an imbricated manner over one another: the verge of the ſhell is undulated and indented, or rather ſinuated, and the beaks joining at top form a heart: the colour of the external ſurface is a ſtrong and elegant yellow; the inſide is of a pearly white.

We have it from the ſhore of the Eaſt Indies; ſometimes from Africa.

Cardia truncata, coſtata, et aliquantulum ſpinoſa.
The truncated, coſtated, and ſomewhat ſpinoſe Cardia.

There are no ſhells that have, in the ſame genus, ſo great a variation of ſhapes as the Cardiæ: this is extremely different in figure from the others hitherto deſcribed, in having one of it's ends truncated in the manner of the chamæ; it is two inches and a half in diameter, and about as much from the hinge to the oppoſite verge: the verge is rounded, and the valves are equal and very gibboſe: one ſide from the head runs out into a kind of obtuſe or rounded extremity, the other is truncated, and the mutilated part filled up by two perpendicular lips: the ſurface of the ſhell is ornamented with about ten large and rounded ribs; the cavities between them are narrow, and longitudinally furrowed, and, on ſome of the ribs, eſpecially toward the verge, there ſtand a few ſhort and obtuſe ſpinoſe protuberances: the ground colour of the ſhell is a greyiſh-white, but it is elegantly variegated with a fine coral red.

We have it from the Eaſt Indies, but it is not common in our collections: I have beautiful ſpecimens of it foſſile from the ſhores of ſome of the rivers in the Eaſt Indies, the ſubſtance agate.

Cardia globoſa echinata margine dentato.
The globoſe, echinated Cardia, with a dentated edge.

This approaches conſiderably to the ſhape of the common cockle, or pectunculus vulgaris of authors: it is about two inches in length, as much in diameter, and at leaſt

as

as much in thickness, when closed: the beaks at the top turn in over the hinge, and meet so, as to form a kind of heart-like figure, which is favoured also by the make of the whole shell, when viewed sideways: the whole surface is deeply furrowed, and the ridges are continued at the edge, so as to form a dentated verge; they are all the way up furnished also with a kind of hollow, pyramidal squamæ, pointed at the ends: the ground colour of the shell is whitish, but it is variegated with spots of red and of yellowish.

This is not unfrequent on the shores of the Mediterranean, and in many other parts of Europe.

Cardia sulcata apicibus distantibus.
The sulcated Cardia, with the beaks distant.

This is a very singular species; it is about an inch and a quarter in length, and as much in diameter: the valves are equal, and they are large toward the base, and considerably elated and hollowed, but at the top they form each a beak, longer, slenderer, and more contorted than ordinary; these are placed at a distance, when the shell is closed, and make a very singular appearance: the whole surface of the shell is deeply furrowed in a longitudinal direction; the ridges are not much elevated, and the colour of the whole shell is a dusky white; sometimes, though rarely, it is variegated with brown, and with a yellowish-red.

It is a native of the American Seas.

Cardia compressa elegantissima marginata. **The Heart-cockle.**
The elegant, compressed, and marginated Heart-shell.

This is a most extremely elegant species, and is so tender and delicate, that, though we often had seen fragments of it, it was not till shells bore a price as merchandize that we met with it entire: it is about an inch and a quarter in length, and an inch in it's largest diameter, which is not as usually from the opposite edges, but from the ridged and marginated backs of the shells: the valves are extremely elevated in the back, and that not in the rounded manner of the rest, but they form a sharp ridge, which is surrounded with a narrow margin different from the rest of the shell; they are pointed at the bottom, and have each it's twisted beak at the top, and, when closed, make a much more perfect figure of a heart than any of the former: the whole surface is furrowed, and the ribs between the lines are not much elevated; it is extremely thin, and it's colour is a plain and tolerably bright white on the outside, and a brighter and purer white within.

We have it from the shores of the American islands, and sometimes from the East Indies; those from the latter place are more rare, but they do not at all excel the occidental ones in beauty.

Cardia gibbosa dorso acuto marginato, variegata. **The Venus Heart-shell.**
The sharp-backed Cardia, with elegant variegations.

This is also an extremely elegant species, and, if it wants any thing of the delicacy of structure of the former, it excels it in the variety of it's colouring: it is an inch and a half long, and an inch and a quarter in diameter, measuring from the ridge of the back of one of the valves to that of the other: the figure is very regularly that of a heart at cards; the bottom is pointed; the top formed of the joining of two beaks, which indeed go a little over one another; and the edges of the two valves, where they join, are denticulated: the shell is thin and light, and is elegantly furrowed; the ridges rising between the lines are rounded, and there runs each way a margin round the back of the valves, from the extream turn of the beaks to the point: the ground colour of the shell is white, but it is elegantly variegated with lines and spots of a deep and bright purple, and sometimes with a few fainter clouds of yellow.

We have it on the coasts of some of our American islands, and it has also been brought from the East.

The other species of the Cardiæ are these: 1. Of those of a somewhat globose figure, there are, 1. The large, yellow, furrowed Cardia. 2. The greyish-brown, spinose

ſpinoſe Cardia. 3. The ſnow-white, deeply-furrowed Cardia. 4. The large, ſquammated, globoſe, white Cardia. 5. The thick, white Cardia. 6. The elongated, irregularly depreſſed Cardia. 7. The giant, white, echinated Cardia, with ſhort ſpines.

2. Of thoſe which approach near to a triangular form, there are, 1. The elegantly reticulated and ſpotted Cardia. 2. The ſtrawberry Cardia, or tuberculoſe, reddiſh cockle. 3. The white, elegantly ſtriated Cardia, with a finely denticulated edge. 4. The deeply furrowed, and elegantly variegated Cardia. 5. The more obſcurely variegated, ſtriated Cardia. 6. The ſtriated and bullated Cardia.

3. Of thoſe which are compreſſed and pointed at the baſe, and more regularly than any of the others reſemble a heart at cards, there are, 1. The ſnow-white Cardia, furrowed deeply within. 2. The little, purple Cardia. 3. The little, white, and protuberant Cardia. 4. The tall and very compreſſed Cardia.

4. Of thoſe which have the ſurface ſquammoſe or tubulated, and imbricated, there are, 1. The great, protuberant Cardia, with erect, broad, and roſe-coloured laminæ. 2. The compreſſed and deeply imbricated, brown Cardia. 3. The bright, red Cardia, with diſtinct and elegant white imbrications. 4. The white, large, imbricated Cardia, variegated with red.

5. Of the irregularly oblong Cardiæ, there are, 1. The furrowed, reddiſh Cardia, called the Noah's Ark-ſhell. 2. The yellow and white, ſlightly coſtated Ark-ſhell. 3. The elegantly variegated, more ſlightly ſtriated Ark-ſhell. 4. The thin, white, canaliculated Ark-ſhell.

6. Of the more regularly figured and leſs cordated ſubgloboſe kinds, there are, 1. The common browniſh, white, furrowed, and denticulated edged Cardia. 2. The larger and leſs denticulated-edged Cardia. 3. The little, red, deeply ſtriated Cardia. 4. The little, variegated, furrowed Cardia. 5. The large, ſomewhat oblong, deeply furrowed Cardia. 6. The larger variegated, deeply furrowed Cardia. 7. The flatter, oval, variegated Cardia. 8. The deeply furrowed, umbonated, flatter Cardia. 9. The roundiſh, ſinuated-edged, red and white Cardia. 10. The long and narrow, deep, coſtated Cardia. 11. The yellow and brown, coſtated and ſtriated Cardia. 12. The ſimply, ſtriated, elegantly variegated Cardia. The ſix former of this laſt diviſion are commonly called by authors Cockles or Pectuncles; and the reſt are called Pectines inauriti or ſcallops, without ears to the ſhell.

PECTEN.

THE Pecten is a bivalve ſhell, ſhutting cloſe all round, uſually of a depreſſed or flatted form, and always aurited, or having one or two proceſſes, called ears, iſſuing from the head of the ſhell near the hinge: theſe ears are in ſome large, in others ſmaller, and in ſome ſo minute, that it requires a nice examination to diſcover them; they are, however, a part of the generical character of the ſhell, and nothing, that is abſolutely without them, is to be received as a genuine ſpecies.

The greater part of the Pectens are ſtriated or coſtated; the ribs or ridges, running in ſtraight and even lines like the teeth of a comb, whence the genus has been named, but ſome of the genuine ſpecies depart from this character.

Pecten coſtatus, variegatus, auriculis magnis æqualibus.
The coſtated and variegated Pecten, with large and equal ears.

This is an elegant ſpecies; it's length is about two inches and a half, it's breadth two inches and a quarter, and the valves are very flat, or but little hollowed; it is rounded and ſinuated at the verge, and thence becomes ſmaller to the head, where it terminates in an oblong point; but from each ſide in this part there is propagated an auricle, which is continuous to the edge of the head, and runs down a third of an inch farther on the ſhell; theſe are what are called the ears of the ſhell; they unite at the top, and in that part riſe a little above the level of the head: the whole ſurface of the valves is ornamented with ribs; there are about twelve on each, and they are broad and rounded at the top, and have ſpaces equal only to about half their own diameter between them; the ears alſo are ſtriated and coſtated: the ground colour of the ſhell is white, but it is elegantly variegated with brown, in beautiful large ſpots.

We have it in great abundance on ſome of our own ſhores.

Pecten coſtatus, ſtriatus, et tuberculoſus ruber auriculis inæqualibus.
The red, coſtated, ſtriated, and tuberculoſe Pecten, with unequal ears.

This is a very large and fine ſpecies; it is four inches long, about as much in diameter, and of a figure approaching to circular, but deeply and regularly ſinuated round the edges, and but little elevated on the back: the colour of the ſhell is a deep red, with an admixture of a purpliſh-brown; it is very rare there are any variegations on it; when there are, they are of a paler red or whitiſh: there are nine large and regular ribs on the ſurface; they are broad, depreſſed, and ſtriated, and have alſo ſeveral undulated protuberances on them, and ſome more regular ones, which are rounded, and ſome of them open at the top: the ears are large, but one of them is conſiderably more ſo than the other.

It is frequent on the ſhores of ſome parts of the Mediterranean.

Pecten echinatus auriculis inæqualibus.
The echinated Pecten, with unequal ears.

This is an elegant ſpecies, and is one of thoſe called by authors Semi-aurite, there being only one of the ears large, the other ſeeming but a rudiment of one: it is of an oblong and ſomewhat ovated form, but flat; it's length is two inches and a quarter; it's breadth, in the largeſt part, which is about the middle, is about an inch and three quarters; from this part it continues to form a rounded and ſlightly ſinuated verge one way, and the other way it grows ſmaller, till it forms an obtuſe point; on one ſide of this there ſtands a very large and fair ear, on the other there appear, as it were, only the remains of a truncated or broken one; but, when the ſhell is ever ſo entire, this ear is never of more than a fourth part of the length of the other: the whole ſurface of the ſhell is formed into fine and ſlender ribs; there are about twenty-ſeven of them on it, and the ſpaces they leave between them are very inconſiderable: each of theſe ribs is lightly ſtriated, and is beſet with a number of ſharp prominences, or ſhort ſpines from the verge, to two thirds of the length of the ſhell: the ground colour is a beautiful red, and the ſpines and part of the ribs are white.

It is not unfrequent on the coaſts of many parts of the Mediterranean.

Pecten rubeſcens flavo et albo variegatus, coſtis paucioribus.
The variegated Pecten, with fewer ribs. — The Ducal-mantle Shell.

This is the moſt elegant of all the family of the Pectens: it is three inches long, and nearly as much in breadth; the verge is regularly and beautifully ſinuated, and the head is furniſhed with two large and beautiful ears, nearly of the ſame bigneſs: there are about thirteen ribs on the ſurface; they are broad, and ſomewhat elevated in the middle, and leave fair ſpaces between them: the ground colour of the ſhell is a ſtrong and beautiful red; the edges are orange-coloured, and the ſurface is every-where beautifully variegated with yellow and white: the head of the ſhell is ſomewhat paler than the reſt, and the ears are very beautifully variegated.

It is a native of the European Seas, but is not common.

Pecten rubeſcens, coſtis depreſſis, albo eleganter variegatus.
The low-ribbed, red Pecten, variegated with white. — The Iriſh Scallop.

This alſo is an extremely elegant ſpecies: it is about two inches long, and nearly as much in breadth, and has on the ſurface about fifteen ribs; they are broad, depreſſed, and placed at nearly equal diſtances from one another: the valves are very little elevated, and the ears are moderately large, and one a little bigger than the other: the ground colour of the ſhell is red, but this, in all the degrees, between the deepeſt purple and the paleſt fleſh colour, in the ſeveral ſpecimens: it is ſometimes uniform and uninterrupted, but more frequently it is variegated in an elegant manner, with white diſpoſed in irregular blotches.

It

It is found in great abundance on the Irish coasts, and is a very great ornament to the grottoes and other shell-works of our ladies.

Pecten flavescens parvus elegantissime striatus.
The little, yellow, elegantly striated Pecten.

This is a very singular species: it's length is about an inch; it's breadth, in the largest part, full as much, and it's head or summit pointed, but ornamented with a pair of large and fair ears, very nearly equal in size: the verge, at the opposite extremity, is rounded, and is even not sinuated as in most of the other species: there are on the surface about six ribs; they are very broad, depressed, and stand at small distances, and both these and the intermediate spaces are striated in a very elegant manner: the colour of the whole shell is a brownish-yellow, sometimes a little variegated with a fainter yellow or with white, but this very rarely.

It is found on the shores of several parts of the Mediterranean, and makes a very singular figure in the cabinets of our collectors.

Pecten tenuissimus fuscus, transversim striatus, auriculis parvis æqualibus.
The small-eared, very thin Pecten, with transverse striæ.

This is the most singular of all the Pectens: it is about two inches and a half in diameter, and is nearly round; the edges are every-where even, not at all indented or sinuated, and the head is furnished with two small and very regular equal ears: the valves are but little elevated, and there is not, on either of them, the least appearance of those ribs, which are almost universal to the other species of this genus: the shell is extremely thin, light, and brittle, and is of a pale brown colour; it's whole surface is elegantly striated, but the striæ do not run longitudinally as in the other Pectens, but transversely, or in somewhat rounded lines, beginning near the head where they are least distinct, and continued in larger circles to the bottom of the shell.

It is a native of the American Seas.

The other species of the Pecten are considerably numerous, and most of them beautiful.

1. Of those which have two ears, both of considerable size, whether equal or not, there are, 1. The large, rounded, depressed Pecten, called the red Ducal-mantle. 2. The large, rounded, costated and striated, yellow Pecten, called the yellow Ducal-mantle. 3. The narrower-ribbed and very elegantly variegated Pecten. 4. The gold-yellow, broader-ribbed Pecten, of the Caspian Sea. 5. The very large, reddish Pecten, or common escallop. 6. The great, variegated, blush-red Pecten. 7. The elegant, red, furrowed Pecten. 8. The yellow and white Pecten, called the Umbrella-shell. 9. The white and spotted Pecten, with narrow and numerous ribs. 10. The yellow, high-ribbed Pecten. 11. The reddish Pecten, with both the valves hollowed. 12. The pyriform or depressed, ovated Pecten. 13. The elegant, pale red and white, narrow-ribbed Pecten. 14. The elegantly marbled and polished Pecten. 15. The rough Pecten, with a multitude of yellow spots. 16. The yellow and purple broad-ribbed Pecten.

2. Of those which have one of the ears very much smaller than the other, or such as are called the semi-aurited Pectens by authors, there are, 1. The black, aculeated Pecten. 2. The red, spinose Pecten, with short spines. 3. The grey, tuberculose Pecten. 4. The orange-coloured, echinated Pecten. 5. The variegated, echinated Pecten, with broad ribs. 6. The white, smooth Pecten.

3. Of those which have both auricles so small, that they appear wanting, and are therefore called earless, or inaurite Pectens, there are the following, 1. The rough, red and brown Pecten. 2. The oblong, white, subechinated, or rough Pecten. 3. The yellow, costated Pecten, with the laciniated verge. 4. The elegantly variegated Pecten, with the variegated verge. 5. The thick, blue and yellow, variegated Pecten, with rounded ribs resembling cords. 6. The smooth, elegantly variegated Pecten. 7. The white, rounded Pecten.

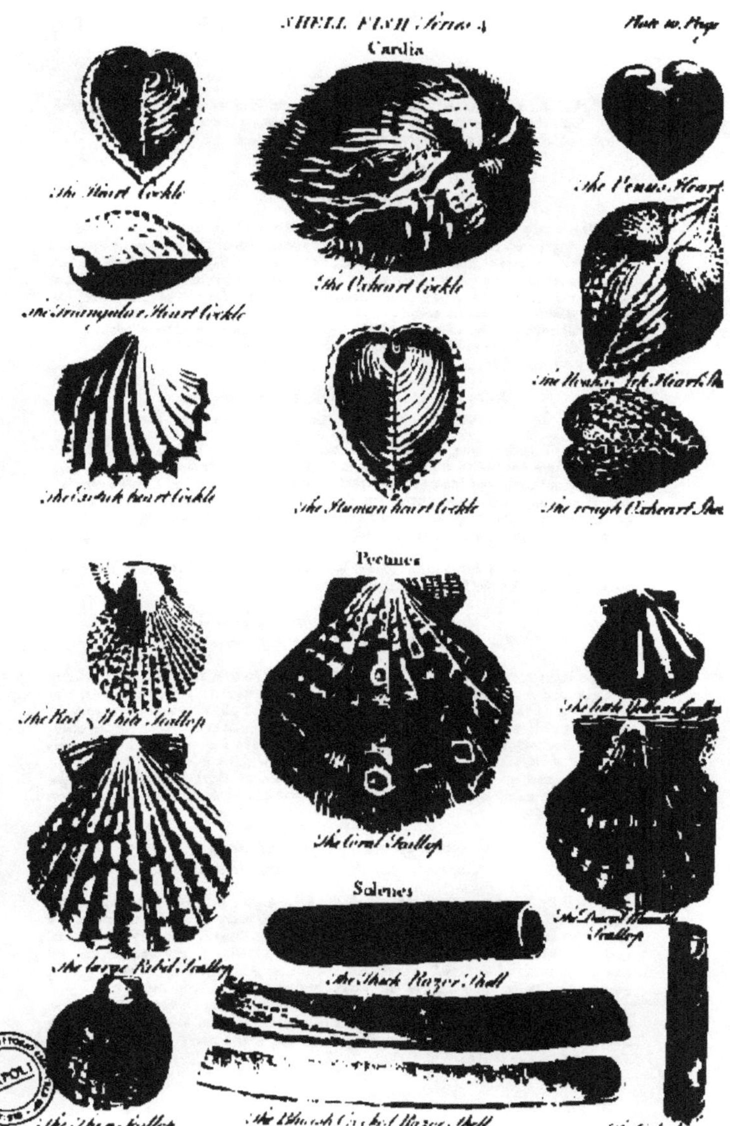

SHELL FISH
Cardia
The Heart Cockle
The Oxheart Cockle
The Triangular Heart Cockle
The Human heart Cockle
Pectines
The Red & White Scallop
The Coral Scallop
Solenes

SOLEN.

THE Solen is a bivalve ſhell, of an oblong and ſomewhat rounded figure, with both the extremities open. The animal inhabiting it is a Tethys. Some of the ſpecies are ſtraight, and others are crooked.

Solen rubeſcens craſſior.
The thick, red Solen.

This is a very beautiful ſpecies: it is about three inches long, and of the thickneſs of a man's finger; it's ſurface is ſmooth, only that toward one of the extremities there are a few arcuated lines of a ſomewhat furrowed ſtructure, and ſeeming to expreſs the terminations of ſo many plates of the ſhell, begun at the thicker end, and not continued quite to the other: the whole ſhell is conſiderably thick for one of this genus, but the end where the plates are entire is thicker than the other; both extremities are open, to the full extent of the diameter of the ſhell: the colour is an elegant pale red, varied, in degree, from the ſtrongeſt damaſk roſe colour to the fainteſt fleſhy tinge.

It is a native of the Eaſt Indies.

Solen arcuatus cæruleo et fuſco variegatus.
The arcuated Solen, variegated with brown and blue.

This is ſix inches long, and three quarters of an inch in diameter; it is of equal thickneſs from one end to the other, but is not ſtraight as the former ſpecies, and moſt of the other Solens, but ſomewhat arcuated or bent in the manner of a bow: the ſurface is naturally ſmooth and gloſſy, but there are the extremities of ſo many broken and otherwiſe imperfect plates always ſeen on it, that it never is tolerably uniform; it is variegated with brown, and a beautiful violet-blue on the outſide, and is of a pearly white within.

We have it on our own coaſts; and it is common alſo on moſt others of the European, as well as on the North American, ſhores.

Solen rectus tenuior fuſco et albo pictus.
The ſlender, ſtraight, brown and white Solen.

This is three inches long, and ſcarce a third of an inch in diameter; it is perfectly ſtraight, and is open to it's full extent at both extremities: it's ſurface is rendered irregular, as in the others, by the extremities of imperfect plates, otherwiſe it would be ſmooth and of a natural fine poliſh throughout: the ground colour is an elegant olive-brown, but where the imperfect plates terminate, and in ſome other places, it is varied with white: the inner ſurface is throughout of a pearly white.

It is frequent on the ſhores of the Eaſt Indies, and of ſome parts of Europe; the antients have called this the female Solen, and have named the common larger brown Solen the male.

Solen papyraceus purpuro-violaceus.
The violet-purple, papyraceous Solen.

This is by far the moſt elegant of all the Solen kind; it is four inches long, near half an inch in diameter, and is equally open at both ends, to the full extent of the ſhell: it's whole ſubſtance is not thicker than that of a ſheet of tolerably thick paper, and is ſo brittle, that it is eaſily cruſhed to pieces: it's ſurface is rendered irregular, by the terminations of imperfect plates, and it's colour is a moſt elegant bluiſh purple, approaching to the deep tinge of the violet, but with more of the red in it.

It is found on the ſhores of the iſland of Ormus, in the Perſian Gulph, and in ſome parts of the Eaſt Indies; the perfect ſpecimens are ſuch, as have been carefully dug out of the ſands, under two or three feet water at low tides: thoſe found on the ſhores are uſually broken.

The other species of the Solen are, 1. The large, brown, common Solen, called the Razor-shell and Sheath-shell, and by some authors the male Solen. 2. The very slender and much arcuated, grey Solen. 3. The brown and reddish, straight, thick Solen, red on the inside. 4. The snow-white, straight Solen. 5. The elegant, American, variegated Solen. 6. The blue and brown Solen, called the Onyx-shell. 7. The long, thick Solen, called the Pipe-shell. 8. The slender, brown and red, somewhat crooked Solen, called the Finger-shell. 9. The thicker, variegated Solen, called the Reed-shell. 10. The very long and slender Solen, of a deep brown, variegated with a paler. 11. The foliated Solen, called the Sabre-shell. 12. The short, thick, olive-coloured Solen.

SHELL-FISH.

Class the Third.

MULTIVALVES.

Such as have the outer covering or shell composed of more than two pieces, or valves.

OF these there are only four genera: 1. The Balani. 2. The Pollicipedes. 3. The Conchæ Anatiferæ, as they have been usually called, by us the Pentelasmes; and, 4. The Pholades.

BALANUS.

THE shell of the Balanus is of an oblong figure, approaching to that of an acorn, open at the mouth or top, and composed of several portions or valves, from six to twelve in number, not moveable or loose as in the other bivalve or multivalve shells, but fixed to one another by an intermediate substance.

The animal which inhabits this shell is a Triton, a genus described before in it's place.

Balanus major striatus ore ampliore.
The great, striated Balanus, with a large mouth.

This is the largest of all the Balani, and that in which the figure, structure, and parts of the shell and it's inclosed animal may be the most distinctly seen: it is about an inch and half in height, and more than an inch in diameter in the largest part, which is a little lower than the middle: the base is broad, and is firmly affixed to some solid substance; the top is nearly of equal diameter with the base, and is wide open; the mouth is not regularly round, nor is it's verge even, some part of it standing up irregularly above the rest: the shell is composed of twelve valves or laminæ, of a slender and somewhat pyramidal figure: they touch one another or join at the base, but, as they grow narrower toward the top, they recede from each other, and at the very summit are considerably distant: they are not loose or moveable at the animal's pleasure, but are united firmly to one another, by an intermediate, shelly matter, somewhat thinner than themselves, and of a paler colour: the valves or ribs of this shell are striated longitudinally, and are of a brownish-red: the intermediate substance is of a paler hue, with less of the brown in it, and is irregularly undulated in a transverse direction: the inner surface is of a whitish colour, with an admixture of flesh colour and blue.

The included animal has an armature composed of four pieces, of a triangular form at it's head, and from the center of these there issue the arms of the creature: these four pieces of armature serve it at other times for the closing the aperture, at which foulnesses of various kinds would get in, to the great annoyance of the animal.

It is frequent on the shores of the East and West Indies, adhering to rocks, and to the shells of the larger shell-fish; sometimes to the bottoms of vessels that have lain long in harbour: great numbers of them are usually found affixed side by side to one another.

Balanus

Balanus fuſcus ore ampliſſimo campanulatus.
The bell-faſhioned, very large-mouthed Balanus.

This is a very ſingular ſpecies; it is uſually found in vaſt cluſters or congeries, fixed to one another at the tops, and forming a kind of reticulated ſurface in the whole, with large cavities underneath, between the bodies of the ſeveral individuals, which ſerve as a happy receſs for the habitation of ſea-worms of many kinds: the ſhell is narroweſt at the baſe, where it is affixed to ſome ſolid body, and thence becomes gradually wider to the very top, where it is broad, open, and as it were bent a little down at the verge, ſo as to exhibit, in the whole, ſome reſemblance of the figure of a bell: it is compoſed of eight firm and hard portions, of a conic form, very diſtant at their ſummits, but connected by a firm, ſhelly ſubſtance, ſomewhat thinner than their own: the colour of the whole is a browniſh-grey; the ribs more grey, and the intermediate matter more brown.

We have large cluſters of this ſpecies ſometimes from the American iſlands, affixed to wood that has lain long under water; but they uſually come to us imperfect and injured by the carriage.

Balanus humilis ore amplo.
The low Balanus, with a large mouth.

This is about three quarters of an inch high, and more than half an inch in diameter; it uſually is found in large cluſters, and the ſeveral individuals that form theſe are ſo cloſely connected, that their inner ſurface is much more conſpicuous than the outer: this is of a pale whitiſh-grey, with an admixture of brown, and a ſlight tinge of red; the ſeveral variegating colours are ſeen much more diſtinctly in ſome parts, than in others. When the cluſter is broken, and a ſingle ſhell examined, it's outer ſurface ſhews that it is compoſed of ſix parts or valves, connected together by an intermediate, firm, ſhelly ſubſtance: the valves or ribs are thick, ſtriated, and of a pale red, with an admixture of grey and of brown: the intermediate matter of the ſhell is of a yet paler red, with nothing of the brown in it: the largeſt part of the ſhell is near the baſe; the verge is a little contracted from this greateſt diameter, but is ſtill very large: the whole ſhell has ſomewhat of the appearance of a drinking cup, and has been thence called the Chalice or Cup-ſhell.

It is not uncommon on the rocks and harder ſea plants about the American ſhores.

Balanus ellipticus griſeo rubeſcens, ore anguſtiore.
The narrow-mouthed elliptic, greyiſh-red Balanus.

This is an extremely elegant ſpecies; it uſually grows in cluſters of twenty, or more, together, and has a very ſingular and pretty appearance: they do not all ſtand upright, as in moſt of the other ſpecies, but are very cloſe at the baſe, and more diſtant toward the top, ſo that they uſually form a cluſter of a hemiſpheric figure, of which the extream ones are conſequently in an inclined, and ſometimes almoſt in an horizontal, direction: the ſingle ſhell, ſeparated from the cluſter, is about half an inch long, and nearly a third of an inch in diameter: it's thickneſs is nearly equal all the way, but at the baſe it is ſomewhat narrower than elſewhere, and the top is contracted, and the edges are drawn inward, ſo as to leave but a ſmall opening by way of mouth: the ſhell is compoſed of twelve valves or portions, which are connected indeed by an intermediate ſubſtance, as in the others; but they ſtand ſo cloſe all the way, that the intermediate ſubſtance is very little, and, the valves being more elevated, it ſcarce appears at all: the valves are all deeply ſtriated, and the colour is a duſky grey, with a conſiderable admixture of reddiſh in it.

We have this ſpecies from the Eaſt Indies, and from ſome parts of America.

Balanus

Balanus tenuior valvis elevatis, ore angustiore.
The narrow-mouthed, slender Balanus, with elevated valves.

This is a very singular species; it is three quarters of an inch high, and not more than a third of an inch in diameter in the largest part, which is a little lower than the middle: the base is somewhat smaller than this part, and the summit is much more so; the verge of the shell is contracted, and the aperture or mouth small: it is formed of eight valves, connected by an intermediated thinner substance: the valves form so many ribs, as it were, approaching very near one another at their sides; but, their form rendering the intermediate substance visible, it is distinctly seen, though not broader than in the former species, in which it is scarce visible at all: the ribs are of a rounded form, considerably elevated in the middle, and are of a brownish-grey colour; the intermediate matter is of a pale grey, without any admixture of the brown.

This species is extreamly common in the East Indies, and on the coast of Africa: it has also been brought from some of the European shores.

Balanus tenuior valvis striatis purpurascens, ore angusto.
The purple, narrow-mouthed, slender Balanus, with striated valves.

This is three quarters of an inch high, and little more than a third of an inch in diameter: it is largest at the very base, and thence grows gradually smaller all the way to the top; so that, where there are large clusters of it, there are found tolerably large spaces between the extremities, which become gradually smaller all the way toward the bases of the shells: each shell is composed of only six valves or ribs, which are broadest at the bases, and there touch one another; but, growing narrower all the way, they are connected by a thin but sufficiently firm shelly matter, which toward the top makes by much the greater part of the substance: the valves or ribs, as they are called by many, are flatted, and deeply, but not very regularly, striated, in a longitudinal direction: the intermediate substance is undulated transversely; the verge is somewhat contracted toward the center, so as to form a narrow mouth: the whole shell is of an elegant purple colour, but the ribs are deeper, and have more of the blue, the intermediate matter is paler, and has more of the red.

We have it from the American shores, but it is not common.

Balanus tenuior striatus griseo-albescens, ore angusto.
The greyish-white, striated, slender Balanus, with a narrow mouth.

This is a small species; it rarely exceeds a third of an inch in length, and is hardly equal to a third of it's length in diameter: it is broadest at the base, and thence becomes gradually smaller to the top, where it is contracted inward a little, so as to form but a narrow mouth: the colour is a greyish-white, and the whole shell so thin, that it is easily crushed and broken: it is composed of only six valves, which are largest at the base, and thence become gradually smaller to the top; they are striated in a longitudinal direction, and the striæ are deepest and most distinct toward the top; in the lower part they are scarce visible: the intermediate matter is also irregularly undulated, but it is in a direction contrary to that of the valves.

This species is frequent about our own shores, on stones, and on the shells of other shell-fish.

Balanus tenuior ore angusto, valvis angustissimis rotundatis flavis.
The slender Balanus, with a narrow-mouth, and with the valves rounded and yellow.

This is about half an inch in height, and is barely a quarter of an inch in diameter in the largest part, which is toward the middle; from this part to the base it becomes gradually a little smaller, and toward the summit more so: it's colour is a dusky brownish-yellow on the ribs or valves, and a paler yellow, with less of the brown in it between them: the valves are twelve in number; they are very narrow, of a rounded figure, and stand at very small distances.

It

It is a native of the East Indies, and frequently comes to us in great abundance, on the bottoms of our India ships.

The other species of the Balanus are, 1. The large, open, brown Balanus. 2. The large, open, grey, and reddish Balanus. 3. The low, broad, and undulated Balanus, called the Turban-shell. 4. The elegantly striated and variegated Balanus, called the Tulip-shell. 5. The broad-mouthed, purple and brown Balanus. 6. The broad-mouthed, grey Balanus. 7. The tall, broad-mouthed, elegantly variegated Balanus. 8. The very low, brown, open Balanus.

Of the narrower-mouthed kind there are, 1. The little, low, striated, narrow-mouthed Balanus. 2. The little, elegant, red, narrow-ribbed, small-mouthed Balanus. 3. The slender, tall, broad-ribbed, deep purple Balanus. 4. The narrow, high-ribbed, small-mouthed, orange-coloured Balanus. 5. The thicker, grey Balanus, with a contracted and corrugated mouth. 6. The thick, corrugated-mouthed, purple and grey Balanus.

POLLICIPES.

THE Pollicipes is a Shell-fish of the multivalve kind; it has a long, thick, and fleshy pedicle, smaller at the base, and largest at the top, on the summit of which stands the shelly covering of the body of the animal; this is composed of a considerable number of shelly laminæ, of different shapes and sizes, but all together forming a triangular body, from the opening of the two sides of which the creature thrusts out it's arms.

The animal inhabiting this singularly constructed shell is a Triton, described in it's place.

Pollicipes subcæruleus pediculo corpore longiore.
The bluish-grey Pollicipes, with the pedicle longer than the body.

This is a very singular and remarkable animal: the pedicle is an inch long, and at the top it is a third of an inch in diameter, and near a quarter of an inch at the base; it is of a tolerably firm fleshy structure, and is covered with a tough and thick skin, granulated on the surface, and furrowed irregularly in a transverse direction: at the top of this stands the shelly covering of the body of the animal; this is about half an inch long, and of a pyramidal figure; it is at the base equal in diameter to the top of the pedicle, and it terminates in a tolerably sharp point: it is composed of an uncertain number of valves or pieces, the middle ones the largest, the rest smaller, and surrounding them at their bases; they are all of them of a pyramidal figure, and shut very closely and exactly, unless when the creature chuses to throw out it's arms: the pedicle is of a brownish colour, the shelly part a mixt bluish, grey, and white.

This species usually is found in great clusters together, and is not uncommon on the shores of the North of Ireland, and in many other parts of Europe: the pedicle, when cut, is found to be composed of a fine white and firm fleshy matter, like the flesh of a lobster; this becomes red on boiling, and is a pleasant and wholesome food. It is eaten in most places, where the creature is found in any plenty, and has much of the taste of the cray-fish.

Pollicipes subruber pediculo longo graciliore.
The reddish Pollicipes, with a long and slender pedicle.

This species is, in the whole, about two inches in length; the pedicle is more than an inch and a half, the shelly part about a third of an inch: the pedicle is about a quarter of an inch in diameter, and is equally thick all the way up: it's surface is deeply wrinkled in an irregular manner, but most of the wrinkles run transversely; it's colour is a pale whitish-red, and it's substance fleshy, but less firm than in the former species: at the top of this stands the shelly part, which is of a pyramidal figure, but short in proportion to it's breadth, and is composed of an irregular and uncertain number of parts or valves, some large, others very small, but all of a pyramidal figure: they shut very close, and are of a whitish-red colour.

This species is found on the coasts of the North of England, but it is not common; it usually adheres in clusters to wood that lies under water.

Pollicipes albescens pediculo corpore breviori.
The white Pollicipes, with the pedicle shorter than the body.

This is about three quarters of an inch in length; and, in the largest part, which is at the head of the pedicle, it is about half an inch in diameter: the body is of a pyramidal form, and is somewhat more than equal to the pedicle in length; it is composed of a great number of oblong, pyramidal pieces of different diameters, and of a whitish colour: the pedicle also is white, and is thickest at the top, whence it goes all the way smaller to the base.

This species is usually formed in considerable numbers together, and is not uncommon on some of our northern coasts.

Pollicipes pediculo longissimo distento.
The Pollicipes, with a very long and distended pedicle.

This is a very singular species, and makes a very different figure from all the rest; it's pedicle is the most conspicuous part of it: this is four inches long, and distended to three quarters of an inch in diameter; it resembles a portion of a gut of some animal, filled with a reddish fluid: this is of equal, or nearly equal, diameter all the way, and is affixed at one end to some solid body, and at the other is terminated by a shelly part, like the bodies of the former species; this is about three quarters of an inch long, and is nearly equal in diameter to the pedicle at it's base, but grows gradually smaller to a point at the top: it appears small, in proportion to the pedicle, and is of a reddish colour, and formed of a great number of valves.

We meet with this on the Lancashire coast in great abundance; the fluid, filling and distending the pedicle, is of a gelatinous substance and insipid taste.

Pollicipes purpureus pediculo breviore distento.
The purple Pollicipes, with a shorter, distended pedicle.

This is somewhat like the preceding species, but it is larger in the body, and shorter and slenderer in the pedicle: the body or shelly part is an inch long, and half an inch in diameter at the base: it is composed of an irregular number of thin, pyramidal pieces of various sizes, and of a purplish colour, variegated with white: the pedicle is two inches and a half long, and half an inch in diameter: it's surface is smooth, and it resembles, like the former, a thick membranous bag, filled with a reddish fluid: this is a thick, gelatinous matter, of an insipid taste: the membrane, which incloses it, is tough and firm, but it is not very thick.

We have this also on the Lancashire coast, and on some other of our shores; it usually is found singly, not in clusters, but often many near one another.

The other species of the Pollicipes are, 1. The oblong, yellow Pollicipes, with a short, fleshy pedicle. 2. The oblong, variegated Pollicipes, with a long conic, fleshy pedicle. 3. The white, long, and slender, pedicled Pollicipes. 4. The bluish, short, and thick, distended-pedicled Pollicipes. 5. The bluish, longer and thicker distended Pollicipes. 6. The slender pedicled, distended, red and white Pollicipes.

The fish or animal, contained within the shelly part of the Pollicipes, is very like that in the following genus, called by authors Conchæ Anatiferæ; but the arms are longer and more deeply fringed, so that they resemble feathers still more than those of that genus. The error of supposing birds to be produced out of these has also extended itself to these, and they are called Goose-shells and Barnacle-shells in the North of England, as well as the others.

PENTELASMIS.

THE Pentelasmis is a genus of animals, composed of a shelly body, affixed to a fleshy and soft pedicle: the body is composed of five valves, and the pedicle is sometimes short, and in other species considerably long.

The

The animal inhabiting the fleshy body of this genus is a Triton: the arms of this creature, being long, slender, and fimbriated, have, while they are expanded in the water, somewhat of the appearance of feathers; and, as the bodies themselves, usually adhere, by means of their fleshy pedicles, to old wood, and the trunks of trees fallen into the sea. It became first an opinion, that a bird of the goose kind was produced from them; and afterwards it was affirmed, that the shells themselves originally grew on the trees, in the manner of their fruit: from this arose the opinion of the barnacle or brent goose being the produce of a tree; and there have not been wanting authors, who, though they had seen the fallacy, and known in what manner only it was that these kind of shells were found on trees, have propagated it, by delivering their accounts in such words, as gave room to believe they had seen them growing on the trees in the manner of their fruit.

Pentelasmis subrubens pediculo annulato.
The reddish Pentelasmis, with an annulated pedicle. **The Goose-shell, or Barnacle-shell.**

This is the species described, by the generality of authors, under the name of Concha Anatifera, and is that which was originally supposed to produce the goose; the others, both of this and the preceding genus, were afterwards described in it's place, and under it's name. It is a tender and brittle Shell-fish; it's length is about an inch, and it's diameter three quarters of an inch: it is composed of five broad and angulated valves or parts, which together form an oblong body, shutting up tolerably close in all parts, except when the fish has a mind to thrust out it's arms: this shelly part is of a pale red or flesh colour, variegated with white; and it adheres to a neck of an inch long, and of about a fifth of an inch in diameter, by means of which it is affixed to old wood, or to stones and sea-plants, or any other solid substance that lies under water: this pedicle or neck, as some call it, is of a brownish colour, and annulated structure, and is firm, tough, and fleshy.

We usually meet with vast numbers of this species together, on the old piles, and other pieces of wood that have been long fixed under water, and not unfrequently on the bottoms of ships that have been long in harbour: sometimes the larger stones, at small depths, are covered with them.

Pentelasmis albescens pediculo granulato crasso.
The whitish Pentelasmis, with a thick, granulated pedicle.

This species is, in the whole, about an inch and a half in length, including the pedicle: the body is formed of five broad and irregularly angulated shells, and is of an oblong figure, somewhat flatted, and obliquely truncated at the extremity: they are of a whitish colour, with a faint admixture of blue; the pedicle is very thick and short; it is not more than half an inch in length, and is continuous to the whole lower extremity of the body, and equal to it in diameter: it is of a brownish colour and fleshy texture; it's surface is slightly annulated, the annules, being narrow, scarce distinguishable, and nicely connected to one another; but the whole surface is granulated or chagrined.

We have this in several of the Northern Seas; it usually grows to the branches of the larger sea-plants, and, as it is frequently in considerable numbers together on the same plant, some have figured the whole together, and made their readers understand the plant as the ramose pedicle of the shell.

Pentelasmis violaceus pediculo longiore et tenuiore.
The violet-coloured Pentelasmis, with a long, slender pedicle.

This is about three quarters of an inch in length in the body, and half an inch in diameter; it's figure is oblong, and obliquely truncated at the upper extremity, where it is smallest; and it is composed of five irregularly angulated and flat pieces: it's colour is a fine, deep, violet-blue, variegated with a paler greyish-blue, and with white; the pedicle is fleshy; it is two inches long, and is not much thicker than a whipcord, so that it is scarce able to support the shell erect.

We have it in the North about our coasts of Yorkshire and elsewhere.

The

The other species of the Penselasmis are, 1. The very short pedicle, obtuse, bluish Penselasmis. 2. The short and thick pedicled, very small-bodied Penselasmis. 3. The long and distended-pedicled, very thin-shelled Penselasmis, called by the people of Lancashire the Sea-Sausage. 4. The long and distended-pedicled Penselasmis, with an oblong, obliquely, truncated, blue and white, very elegantly variegated body. 5. The small but distended-pedicled Penselasmis, with a flesh-coloured, oblong body.

All these species are found on old piles and trees fallen into the sea, on the northern coasts of our own kingdom: they all throw out their fimbriated arms, at times, from the body of the shell, and the common people, supposing these to resemble feathers, keep up the tradition of the goose-shell, by declaring, that all these produce, at a certain period of the year, young birds of that genus.

PHOLAS.

THE Pholas is a Shell-fish of the multivalve kind, and, like the former species, has it's outer or shelly covering made up of five pieces: these, however, are not, as in that genus, thin flaky scales, as it were, but are of the general form and structure of the other shells: they are also very singular in their proportions and arrangement; two of them are large, and resemble those of the ordinary bivalve shells in all respects; and these seem, on a slight view, to compose the whole shell, which therefore appears of the bivalve kind; but a further examination discovers the other three portions or valves, if they may be so called; these are small, and serve to close up an opening left by the irregular meeting of the two principal valves.

The animal inhabiting the Pholas is a Tethys.

The name Pholas has been very greatly misapplied by the writers on natural history, and great errors and confusion have arisen from this. The Pholades, properly so called, inhabit holes wonderfully made in stones, and other solid bodies: the corals of the larger kinds sometimes serve as nidus's for numbers of them, and the bottoms of our ships frequently afford specimens of them; they lie at different depths in these recesses, and thrust out a kind of proboscis to the orifice: this singular quality in the Pholades has been understood by many as peculiar to them; and, in consequence of this, every shell-fish found living in the bodies of stones, &c. has been called a Pholas; but there are bivalve shells of the common muscle and chama kinds, which live in the same manner: these we are to separate out of the genus of the Pholas, where too many have placed them, as the structure of a shell, not it's habitation or manner of living, is to determine it's genus.

Pholas albescens oblongus cancellatus et asper.
The oblong, white, cancellated, and rough Pholas.

This is two inches in length, and about an inch and a quarter in diameter in the broadest part, which is toward the middle, but nearer the head than the other extremity: the whole shell is composed of five pieces or valves; two of these are large and lateral; these have all the appearance of the valves of the ordinary bivalve Shell-fish; behind these there are two other dorsal valves; these are very small, somewhat broad and short, and beside these there is a fifth a single one, which is very long and narrow, and is extended under the hinge: the colour is a whitish, with a faint admixture of a yellowish or brownish among it; and the surface of the two large valves is deeply furrowed, both in a longitudinal and transverse direction, so that the whole appears cancellated or divided into little squares: the whole surface is rough, like a file to the touch. This species usually lives in holes in stones, and in none so plentifully as in the large masses of the Septaria, or Ludus Helmondi, that lie about many of our shores: there is no way for the creature to get out of it's habitation, for the aperture at the surface of the stone is usually not larger than the diameter of a goose-quill; the creature thrusts out a kind of proboscis thus far, but no other part of it's body can ever come to the surface of the stone.

This species is frequent about our own coasts, and in most parts of Europe; the manner in which it makes it's way into the stones, which by the smallness of the external aperture must be done, while it is very little, and the method of it's enlarging the cavity, as it grows, are unknown. The easy solution has been the supposing the creater only clay or mud when the creature got into it, but to have petrified and become stone afterwards: the Ludus Helmontii, which is the residence of these fish on our own coasts, oftener than any other species of stone, seemed to favour this opinion, as many have erroneously supposed that stone to be petrified clay, before they were informed of this additional appearance in favour of the conjecture: but the hardest stones are found pierced by them in other places; and, about the coasts of Italy, pieces of wrought marble are frequently taken up after wrecks, which, if they have lain only a few years in the sea, are usually spoiled by the holes made in them by these creatures. We had lately a specimen of a very hard stone presented to the Royal Society here, in which were living shell-fish; these were determined to be Pholades, but they were, in reality, one of the mytulus kind, described already in it's place: the general opinion was, that the stone had grown about these creatures; nor was it without difficulty that I persuaded that learned body, that the creatures had a power of making their way into hard substances, while hard, by producing some pieces of oak, in which there were several of them lodged.

Pholas albescens valvis majoribus brevibus, subovatis.
The white Pholas, with the larger valves short and suboval.

This species is singular, on account of the smallness of the shell in proportion to the body of the animal: in most of the Pholades, the shell is not large enough to surround and compleatly inclose the creature, the stony sides of the cavity it is lodged in answering that purpose in it's place; but, in this particular kind, the larger valves bear no proportion to the bulk of the body: they are of a whitish colour, of a suboval figure, somewhat hollowed, and longitudinally striated; their surface is rough, and their texture tender and delicate; the two smaller valves are placed at the back of the others, in a contrary direction; and the fifth or last valve is long, slender, and convex, and covers the whole hinge: these three have a radiated surface, but they are smoother than the others.

This species is frequent on our Kentish and Sussex coasts; I have met with it in great abundance about Shepey island, and among the Bognor rocks in Sussex.

Pholas subrubens profunde sulcatus et asper.
The reddish, deeply furrowed and rough Pholas.

This is four inches long, and about two in diameter: the two principal valves are oblong, largest near the middle, and deeply hollowed; they are very thin and delicate; their colour is a pale but elegant red, often variegated with white, and their surface is deeply furrowed with parallel, longitudinal, and not very distant lines; the spaces between rise up into ridges, and their surface is rough; the other three valves are small, and are placed at the back of these; two are oval, and one is oblong and very slender.

This species is found in stones about the American coasts, and not unfrequently makes it way into the larger corals, and other stony plants, and into several other solid substances.

Pholas angustior tenuissimus et asper.
The narrow, thin, and rough Pholas.

The West Indian File-shell.

This is the largest known species of Pholas: it is six or seven inches long, and, with that length, not more than two inches in diameter: the two large or principal valves, which are all that we usually meet with, are of a very tender and delicate structure, and are hollowed, and of a beautiful white: they are smooth on the inner surface, but on the outer they are deeply furrowed, both in a longitudinal and transverse direction, and are rough to the touch in the manner of a file: the three other valves are, as in the former species, two short and rounded at the extremities, and the other very long, very narrow, and hollowed; this runs along the back, covering the whole hinge.

This species is brought to us from the American coasts; it lives in hard stones, but is seldom met with, except loose on the shores, when it has been dislodged by some accident, as the breaking of the body that inclosed it: for this reason a single valve is all that is usually found, and often that not entire.

The other species of the Pholas, properly so called, are, 1. The short and very rough Pholas, of a greyish colour. 2. The short, truncated Pholas, with deep furrows. 3. The long and smoother-shelled Pholas. Authors have mentioned many other species under this name, but they are shells of other genera, only called so, from their being found in stones.

SHELL-FISH.

Series the Second.

Those which inhabit the fresh waters.

AMONG the Shell-fish already described as inhabitants of the seas, there are many species that occasionally make their way up the larger rivers, and some that will live in water which is only salt while the tides flow: but there are a whole series of others, which are truly and properly natives of the fresh waters only, and which die on being put into sea-water. Many of these are inhabitants of standing waters, as ponds and lakes; others of our brooks and rivers; they are all referable to the genera of sea Shell-fish, and no formal distinction of them therefore will be necessary here. The system of this work would have pleaded for the joining them to the sea species, as the place of living makes no regular distinction, but the custom of the present age, and the intent of rendering this part of natural history as little confused as possible, have pleaded for the separating these from the sea-shells, and the land-shells from these, but this only in place; the generical names are continued from the former part of the work, and the scientific reader has only to add these to the others, as so many more species.

PATELLÆ, LIMPETS.

Patella subovata rostro elevato, incurvato.
The suboval Limpet, with an elevated and crooked rostrum.

THIS is a shell of no great beauty: the generality of the fresh-water ones are, indeed, in this respect, greatly inferior to these of the sea, but none more so than those of this genus: this is a very thin shell; the figure is oval, and depressed all over, except in the middle, where it rises into a rostrum or beak; in other parts it is but very little elevated above the surface of the body to which it adheres: the rostrum is carried to a height equal to two thirds of the diameter of the shell, and is somewhat crooked and obtuse at the extremity: the whole shell is little more than of the extent of a man's finger-nail, and it's colour is a dusky brown, with a tinge of olive; it is very thin, and easily crushed.

We have it on stones, in many of our brooks in Northamptonshire.

Patella elevata subrotunda umbonata.
The roundish, elevated, and umbonated Patella.

This is a smaller species than the former; it rarely exceeds a third of an inch in diameter; it's figure is roundish, and it is considerably elevated in a gibbose manner, and, though not rostrated or beaked as the former, yet is terminated at the top by a round umbo or button, which gives that part a singular appearance: it is a very thin and delicate shell; it's surface is tolerably smooth, and it's colour an olive brown, deeper on the outside than within.

It is found in some of the rivers in Leicestershire: I remember to have met with it sticking to small pebbles, in a little brook that runs behind the house of Mr Calderot, at Calthorp in that county; as also about Loughborough, and in some other places.

Patella elliptica vertice perforato.
The elliptic Patella, with a perforated top.

This is about a quarter of an inch in diameter, and about half it's diameter in height: it is of an oblong figure, rounded at both ends, and of a dusky brown colour: the shell is very thin, and is easily crushed to pieces; it's verge is even, and it's summit is naturally perforated; the aperture is small, but it is of an oblong figure, and seems as if formed of two round holes, which had broke their way into one another.

We have this species about London; the ditches in Tuttle-fields afford it in considerable plenty.

The other species of fresh-water Limpets are, 1. The conic, radiated, yellowish-brown Patella. 2. The depressed, deep brown Patella. 3. The low Patella, with the umbo near one edge. 4. The little, gibbose, pale brown Patella.

COCHLEÆ, SNAILS.

Cochlea spiralis clavicula parum elevata, ore rotundo.
The spiral Snail, with the clavicle a little elevated, and with a round mouth.

THIS is a very pretty shell; it's figure is roundish, it's diameter three quarters of an inch: it's shell is considerably firm and solid, and is smooth on the surface; it is formed of about four volutions, and the clavicle in the center is elevated above the rest of the surface: the colour is a greyish-white, and the outer or larger volution has a streak of black running all along it in the course of the spiral, but it loses itself, as it gets into the next volution: the mouth is round, and the lip edged with a rim considerably thicker than the rest of the shell, and of a brighter white; this is a rare shell.

I have met with it in the lakes upon the hills, in the North of England.

Cochlea spiralis clavicula depressa, ore etiam depresso.
The spiral Cochlea, with a depressed clavicle, and depressed mouth.

St Cuthbert's horn-shell.

This is also a pretty shell, it is of the spiral, flatted kind; it's contour is round, and it's diameter about half an inch: the shell is firm and solid, and the colour a fine glossy brown, with a tinge of the olive: it consists of two or three volutions; the clavicle is depressed in the center, and the mouth is, in part, filled up by the next volution of the shell: the lip is only a narrow rim or fillet, thicker, and of a paler colour than the rest of the shell.

We have this species in great plenty in our brooks and rivers; the ditches about London abound with it.

Cochlea elatior clavicula longiore, ore semirotundo.
The taller Cochlea, with a long clavicle, and a half-round mouth.

This is one of the most elegant and beautiful of the fresh-water Cochleæ: it is about an inch high, and three quarters of an inch in diameter in the body; the mouth is of the half-round kind, being in part filled up by the succeeding volution of the shell entering it: it is surrounded with a thin rim, of a pale yellowish colour; the body of the shell is rounded and smooth; the clavicle has four volutions, and terminates in a point: the ground colour of the whole shell is a dusky yellow, and this is elegantly variegated with an olive brown, disposed in irregular spots and clouds.

We have this species in the river Nen, near Peterborough, and in some other of our deep and not very rapid waters, but it is not common; the colour is sometimes simply a pale yellow, but this is an accidental variety.

Cochlea

Cochlea corpore subovato, clavicula elata obtusa variegata.
The variegated, oval-bodied Cochlea, with an elate, obtuse clavicle.

This is also an elegant species: it is about three quarters of an inch in height, and near half an inch in diameter: the body of the shell is large, inflated, as it were, and of a figure approaching to oval, but pointed at the extremity of the mouth; the clavicle has three volutions; they are large and rounded, and the summit is obtuse; the ground colour of the shell is a greyish-white, and it is variegated with regular fasciæ or zones of a deep brown.

We have this in most of our large rivers, often in ditches.

Cochlea fusca ore ampliore, clavicula lata acuta.
The large-mouthed, brown Cochlea, with an elate, acute clavicle.

This is about half an inch high, and very little less in diameter: the body is of an oval figure, and the clavicle is long and pointed at the summit; the colour is a dusky brown; the shell is not so thick as many of the others, and the mouth is remarkably large, and edged by a thin rim, by way of lip.

We have this very frequent in our brooks and ditches.

Cochlea clavicula depressa ore angusto labiato.
The narrow and labiated-mouthed Cochlea, with a depressed clavicle.

The grey, fresh-water Nerite.

This is a very pretty shell: it is of the number of those Cochleæ distinguished by the name of Nerites, and is not the least elegant of that kind; it is about a third of an inch in length, and near as much in diameter: the body is of a form approaching to oval; the clavicle is depressed, and is situated near the smaller extremity on one side; the mouth is narrow, and has a little lip on each side; the general colour is a pale grey, but it is variegated with a deeper iron grey, approaching to black, in little irregular spots; and the whole surface is lightly furrowed.

We have it in some of our large rivers, particularly in the North of England, but it is not common.

Cochlea ore angusto labiato cancellata.
The cancellated Cochlea, with an oblong, narrow mouth.

The cancellated, fresh-water Nerite.

This is also a very beautiful species; it is about half an inch in length, and nearly as much in diameter at the larger extremity: it's figure approaches to oval, and it's clavicle is short, depressed, and has but about two volutions: the mouth is narrow, and has a lip on each side, but is not dentated either way: the whole surface of the shell is furrowed with longitudinal lines, and with others crossing them, so as to form a cancellated or kind of lattice-worked surface: the ground colour is a pale grey, almost white, but it is irregularly variegated with blackish, sometimes disposed in large clouds, sometimes in the form of small spots.

We meet with this species in some of our swift and clear brooks, and smaller rivers; I have frequently seen it in the trout rivers about Uxbridge.

Cochlea subrubens ore angusto labiato.
The reddish Cochlea, with a narrow, labiated mouth.

The red, fresh-water Nerite.

This is about half an inch in length, and is more than a third of an inch in diameter; the colour is throughout a pale, but not bright, red; there is a duskiness in it, which brings it nearer to the colour of a very pale red brick, than to any of the diluted purple or blossom colours; the substance of the shell is more firm, and is considerably thicker than in most of the others, and the surface is nearly smooth; there run two or three lines, according to the course of the volution of the shell; but they are so faint, that

that they are feldom diftinguifhed; the mouth is narrow, and the lips are of a paler red than the reft of the fhell, and are fmooth: the clavicle is fhort, and turns only about twice near one edge, at the fmaller end of the fhell.

We have it in the rivers in Northamptonfhire in abundance.

Cochlea conica, ore depreffo, grifea.
The grey, conic Cochlea, with a depreffed mouth.

The grey, frefh-water Trochus.

This is a fmall fpecies, but is exactly like the large conic Cochleæ, or, as they are ufually called, Trochi, in figure; it is of a tolerably regular conic fhape, and has four volutions; it's height is about a third of an inch, it's diameter nearly as much at the bafe, and is terminates in an obtufe, little umbo: it's colour is a pale grey, often approaching to whitifh; it's fubftance very thin and tender: it's mouth is moderately large, and of a depreffed form, or in great part filled up with the fucceeding volution of the fhell.

We have it in feveral of our large rivers; I have received very fair fpecimens of it from the Avon, and have met with it in the Thames about Richmond.

Cochlea conica ore depreffo coftata et variegata.
The coftated and variegated conic Cochlea, with a depreffed mouth.

This is a fmall, but an elegant, fhell; it's height is, at the utmoft, not more than half an inch; it's diameter at the bafe is more than a third of an inch; it's mouth not fo large as in the former fpecies, and it's fummit pointed: it is of a very regularly conic figure, and confifts of five volutions; it is of a fomewhat firmer fubftance than the former fpecies, but is ftill eafily crufhed: It has a rifing in form of a rib, of a rounded figure in the middle of the firft or loweft volution, which follows the whole turn of the fhell, but grows fainter, as it approaches nearer the top: the ground colour is a faint grey; and the variegations, which are difpofed in the form of clouds and lines of an irregular form, are black.

We have this fpecies in plenty in the North, in our deep rivers; elfewhere I have not met with it: it often is found on the ftalks of the water-lilies.

Cochlea conica fufca.
The brown, conic Cochlea.

The brown, frefh-water Trochus.

This is a very fmall fhell, and has no great beauty: it's height is hardly a quarter of an inch; it's diameter at the bafe is not more than a fixth of an inch, and it's colour is a pale brown, with fome faint caft of yellowifh, and without any variegations: the fhell confifts of four volutions, and thefe are rounded and fmooth on the furface; the top is obtufe, or terminated by a kind of umbo: the fubftance is extreamly thin and tender, fo that the leaft touch deftroys it.

It is frequent in our ditches, where no other fpecies of the conic kind is found: we have it in abundance in the ponds and ditches in Tothil-fields, and about Batterfea and Wandfor.

The other of the frefh-water Cochleæ are numerous.

1. Of the depreffed kind we have, 1. The brown, ftriated Cochlea, with the clavicle a little raifed. 2. The elegant, brown, and yellow Cochlea. 3. The larger, pale, olive-coloured, depreffed Cochlea, with a larger mouth. 4. The larger, grey and black, depreffed Cochlea, with a white, thin lip. 5. The leffer, firm, greenifh, depreffed Cochlea. 6. The leffer, thick, yellowifh, depreffed Cochlea, with few volutions. 7. The leffer, deep brown, depreffed Cochlea, with a great mouth.

2. Of the ventricofe or bellied kinds, with a half-round mouth and exerted clavicle, there are, 1. The large, brown and yellow, elegantly variegated Cochlea. 2. The little, thick, deep brown Cochlea, with a large mouth. 3. The thin, fmall, olive-coloured Cochlea. 4. The greenifh, oval-bodied Cochlea. 5. The elegant, pearly white Cochlea. 6. The variegated, brown and grey Cochlea. 7. The very fmall, grey Cochlea, lineated with black.

3. Of the conic Cochleæ, called Trochi, befide the already defcribed fpecies, there are, 1. The large, plane, brown, fharp-pointed Cochlea. 2. The large grey, and

 black,

black, conic Cochlea. 3. The lesser, olive-coloured Cochlea, with transverse or oblique lines of brown. 4. The little, green and brown, obtuse-pointed Cochlea.

TURBINES, SCREW-SHELLS.

Turbo gracilior lævis albescens.
The smooth, slender, whitish Turbo.

THIS is a very elegant little shell; it's length is more than an inch, and it's thickness at the bottom not more than that of a goose-quill: it grows regularly taper from this part to the summit, where it terminates in a sharp point: it consists of about fifteen volutions, which are flatted on the surface, and have no variegations, only the line that marks their separation from one another is of a pale brown; the substance of the shell is very tender, and it's colour a pale whitish-grey; it's mouth is small, and of the depressed kind.

I met with many thousands of this species together, about a cluster of flags in one of the lakes, on one of the hills in Lancashire, which of them I do not distinctly remember. I have since received specimens of it from France, where it abounds in the Seine, and some other rivers; the French specimens are whiter than ours, but no way else different.

Turbo ore oblongo, margine volutionum denticulato.
The long-mouthed Turbo, with the edge of the volutions denticulated.

This is a very elegant species; it's length is about three quarters of an inch; it's diameter at the base is about a third of an inch, and it's whole structure formed of about thirteen volutions: the mouth is oblong, narrow, and edged with a furrowed and irregular lip; the volutions are flat, and their surface is smooth, but the upper edge of each, or the continued edge of the whole, is deeply and elegantly denticulated: the colour of the whole shell is a very pale grey, with somewhat of a pearly tinge in it, but it is rarely quite pure on the surface.

We have it in the river Nen near Peterborough, and in some other of our large rivers, and sometimes in ponds.

Turbo margine volutionum bullato.
The Turbo, with the edge of the volutions bullated.

This is a very elegant species: it's height is an inch; it's diameter at the base at least a third of an inch, and it's mouth of an irregularly oblong figure: it's substance is very thin and tender, and it's colour a dead whitish; it is composed of about seven volutions; they are broad, flat, and perfectly smooth on the surface; but at the verge, where the other species has it's denticulations, this has every-where a series of little rounded protuberances or bullæ, which give it an extreamly elegant figure.

It is a very rare species; I have met with it in some of the rivers in Leicestershire and Yorkshire, but not elsewhere: it is frequent in the ditches in Holland, from whence I have received very beautiful specimens.

Turbo albescens volutionibus tumidis glabris.
The whitish Turbo, with smooth, tumid volutions.

This is an inch and a half long, and of the thickness of a swan's quill at the base; the mouth is of a roundish figure, and the summit obtuse: the shell consists of about fourteen volutions; they are all, except the four last, very distinct, and are rounded or bellied, and consequently have the line that separates them much sunk in below their surface: the whole is of a tender structure, and a whitish colour.

I met with it last summer in Whittlesea-meer in Cambridgeshire, and have not seen it elsewhere.

Turbo

Turbo tenuissimus volutionibus vix elevatis.
The very slender Turbo, with the volutions scarce elevated.

This is an inch and a half in length, and is the slenderest, in proportion to it's length, of all the shells of this genus: it's mouth is oblong and narrow, and it's summit finely pointed: the shell consists of ten or twelve volutions, but they are so little elevated, that it rather appears surrounded with a spiral line, than really separated into so many portions: the whole is very thin, and of a tender structure; the colour is a very pale brown, often whitish, and the line between the volutions is of a deeper brown.

We have this in some of our large ponds, but it is far from common.

The other species of fresh-water Turbines are, 1. The tall, slender, variegated Turbo. 2. The short, thick, whitish Turbo, with indented volutions. 3. The little, yellowish, round-mouthed Turbo. 4. The little, brown Turbo, with a small, detached, round mouth. 5. The larger, round-mouthed, brown Turbo.

BUCCINA, TRUMPET-SHELLS.

Buccinum tenuissimum fuscum ore magno.
The very thin, brown Buccinum, with a large mouth.

THE extream thinness of this shell, in proportion to it's size, readily distinguishes it, without any further characteristic: it is an inch and a quarter long, half an inch in diameter in the body, and not thicker in it's substance than a piece of the finest paper: the mouth is very large and oval; the body of the shell is oblong and tumid, and the clavicle has about four volutions: the colour is a brown, with a tinge of greenish or olive.

It is very frequent in our ponds and ditches.

Buccinum subrubens angustius ore magno.
The slenderer, large-mouthed, reddish Buccinum.

This is an inch long, and about a third of an inch in diameter: the body of the shell is of an elliptic figure, and the clavicle has four volutions, and is terminated by a little button: the mouth is very large and oblong; the shell is of a thin and delicate structure, easily broken, and it's surface is every-where smooth; it's colour is reddish, with an admixture of white.

We have it in some of our large rivers, where the water is deep, and the current but slow.

Buccinum striatum ore magno ovato.
The striated Buccinum, with a great, oval mouth.

This is a very elegant species; it's length is about three quarters of an inch; it's breadth is about a third of an inch, and the substance of the shell extremely thin: the body is large, in proportion to the rest; it makes up, in a manner, the whole shell: the mouth is very large and oval; the clavicle is small, and has only three volutions; it terminates in a sharp point: the colour of the shell is a very pale brown, and it's surface is not smooth, as in the preceding species, but is very beautifully furrowed in a longitudinal direction, both on the body and clavicle, but much more strongly on the body than elsewhere.

We have it in some of our large ponds, but not frequent; it is more common in the West of England than elsewhere.

Buccinum angustius clavicula longiore, ore ovato minore.
The narrow Buccinum, with a longer clavicle, and a small oval mouth.

This is about half an inch in length, and is not so much as a third of it's length in diameter: the body is small and rounded; the clavicle is long, and very elegantly formed

formed of six volutions; the mouth is of an oval figure, but it is little, and is edged with a pale, narrow lip: the whole surface of the shell is smooth, and the colour a pale brown, with a tinge of olive; the volutions of the clavicle are tumid or rounded, and are all very distinct and beautiful.

We have this in great plenty in Hackney river, and in several other of the shallow and swift rivers in most parts of England.

Buccinum breve ventricosum ore parvo rotundo.
The short, tumid Buccinum, with a very small, round mouth.

This is about half an inch in length, and is as much in diameter; it's figure is very singular, and wholly unlike that of all the other species: the body of the shell is less distinct from the clavicle, than in the generality of this genus; and the whole shell appears, as if it had been depressed or crushed together from the summit to the mouth; the body is of once and a half it's length in diameter, and the mouth, which is small, and perfectly and regularly round, is connected to it, as it were, by a neck: the clavicle has only three volutions, and those short, but very distinct, and it terminates obtusely; the substance of the shell is thin and tender, and the colour a pale brown.

I first met with it in the little river that runs under the town of Mount Sorrel in Leicestershire; it is not common with us.

Buccinum angustius margine volutionum dentato.
The narrow Buccinum, with the edges of the volutions dentated.

This is an elegant species; the length is an inch and a half, and the diameter is barely half an inch; the substance of the shell is moderately thick, and it's colour a reddish-brown: the body is not large in proportion to the rest of the shell, and it's figure is oval; the mouth is large and oblong; the clavicle is very large; it consists of five volutions, all broad, flat, and distinct, and the edge all the way elegantly denticulated; the summit is obtuse: it is a rare species.

I found it in several of the rivers of Warwickshire, but not abundantly in any of them, and in very few others.

Buccinum heterostrophum ore ovato.
The oval-mouthed, heterostrophous Buccinum.

This is a very singular species; the volutions running in a contrary direction to those of the others, and the mouth consequently standing the wrong way: the shell is half an inch long, and nearly as much in diameter, and is so extreamly thin and tender, that it is not easy to touch it without breaking it: it's colour is a pale brown, with a tinge of yellow; the body is very large, and of an oval form; the mouth also is large, but it is not nearly of the size that some of the first species have it, it's opening not reaching the whole length of the body; the clavicle has only three volutions, and those small but distinct, and it's summit pointed.

We have it in some of the shallow ponds and ditches about Tothill-fields, and in our fishponds in many places.

The other species of fresh-water Buccins are, 1. The little, thicker, brown Buccinum. 2. The larger, thicker, deep brown Buccinum, with a small mouth. 3. The large, brown Buccinum, with a short clavicle. 4. The smaller, greyish Buccinum, with a pointed clavicle. 5. The larger and thicker, small-mouthed Buccinum.

DOLIA, GLOBOSE SHELLS.

Dolium fuscum ore maximo expanso.
The brown Dolium, with a very large, expanded mouth.

THIS is about three quarters of an inch long, and half an inch in diameter: the body of the shell is tumid, and, as it were, inflated, and the clavicle is very short and inconsiderable; it consists of four very minute and indistinct volutions, and is pointed at the extremity: the mouth is extreamly large, and the lip expanded

a great

a great way: the whole surface of the shell is smooth, and it's substance extremely thin; it's colour is a dusky brown on the outside, and a paler whitish brown within.

We meet with it in the sea-ditches in Lincolnshire in vast abundance, and sometimes in the muddy waters about London.

Dolium subovatum ore longiore et angustiore.
The oval Dolium, with a longer and narrower mouth.

This is near an inch in length, and two thirds of an inch in diameter; it's substance is extremely thin, and it's colour a dusky brown: the body approaches to an oval figure, but it is somewhat bent in toward the extremity: the clavicle is very short and inconsiderable; the mouth is large, but less so than in the former species; it is indeed as long as in that, but the lip is not expanded in that manner, so that it is much narrower, and approaches to an oval figure.

We have this in some of the ditches about London in very great abundance.

Dolium subrotundum flavescens ore ovato.
The yellow, round Dolium, with an oval mouth.

This is about half an inch in length, and as much in diameter; the substance of the shell is extreamly thin and delicate, and it's colour a pale yellow: the body is of a roundish figure, and the mouth is large and oval; the clavicle is very small.

We have this in the ponds in Northamptonshire very common; I have also, some years, seen many of them in the ditches in Tothil-fields.

Dolium subovatum albidum ore amplo.
The white, oval Dolium, with a large mouth.

This is about three quarters of an inch long, and half an inch diameter: the body is of a form approaching to oval, but indented toward the verge of the mouth; the clavicle is very inconsiderable, but it is sharply pointed at the extremity: the mouth is large and roundish, but with some approach to the oval: the colour of the whole shell is a beautiful white.

It is frequent in some of our shallow rivers, where there are many weeds.

The other species of fresh-water Dolia are, 1. The short, round-mouthed, olive-coloured Dolium. 2. The oval, yellowish, thicker Dolium, with a very small clavicle. 3. The oblong, wide-mouthed, reddish Dolium, with a somewhat longer clavicle. 4. The short and depressed, oval-mouthed Dolium. 5. The very small, blackish Dolium.

AMMONIÆ, CORNU AMMONIS SHELLS.

THOUGH we have not, among the almost infinite variety of the Sea-shells, any one of the figure and structure of the fossile Cornu Ammonis, we have several small shells of the fresh-water kind, which in all respects agree with them, especially with the more common kinds.

Ammonia variegata ore majore.
The larger-mouthed, variegated Ammonia.

This is one of the most elegant among the whole series of fresh-water shells: it is of a rounded, depressed figure; it's diameter is about an inch, and it's thickness toward the mouth a third of an inch: it consists of about four volutions, the part nearest the mouth being largest, and the body growing gradually smaller, all the way of the spiral, afterwards to the center, where it forms a little umbo or button: the mouth is somewhat larger than a plain segment of the shell would form, there being a little elevated lip all round it: the surface is obscurely undulated in a transverse direction; the colour is a pale grey, variegated with brown, in very irregular spots and lines: when the shell is broke, it's cavity is found to be not simple and uniform, but divided into a number of cells, in the manner of that of the Nautili, and of the fossile Cornus Ammonis, by transverse plates or partitions, with a siphuncle or little pipe running through each; it is a rare shell.

I remember to have seen it, some years since, in the river near Wisbech, and also in some parts of Leicestershire; but, in the latter place, it was smaller and less variegated.

Ammonia fusca glabra ore depresso.
The brown, smooth Ammonia, with a depressed mouth.

This is about three quarters of an inch in diameter, and in the thickest part is of the bigness of a small goose-quill; it consists of about four volutions: the shell is thin and easily crushed; the mouth is of a somewhat depressed form, the second volution of the spiral entering a little way into it: the colour is a beautiful glossy brown, without any variegations, except that there run a kind of obscure lines, somewhat darker than the rest, marking the places of the divisions on the inside of the shell: these are formed of a thin, whitish, shelly matter, and stand at about a line distance from one another, dividing the cavity of the shell in the manner of that of the nautilus.

I have met with this in some large ponds in Derbyshire, and not elsewhere.

Ammonia subflavescens minor glabra.
The smooth, little, yellowish Ammonia.

This is about a third of an inch in diameter, and in the largest part is not thicker than a crow-quill: it consists of about three volutions, terminating with a little umbo in the center: the whole surface of the shell is perfectly smooth and even, and it's colour is a bright brown, with a considerable tinge of yellow in it.

We have this in some of our fish-ponds about London.

Ammonia pallida costata.
The pale, ribbed Ammonia.

This is about half an inch in diameter, and it's largest end is not much thicker than a crow-quill: the shell is of a tender substance, and it's colour very pale; it consists of three or four volutions, and is all the way transversely ribbed; the ribs are low and rounded, and mark the places where the plates are situated within, which form the divisions between the several cells.

We have this, in great abundance, in the fen-ditches in Lincolnshire.

The other species of the Ammonia are, 1. The common, little brown Ammonia. 2. The common, larger brown Ammonia, with a rim round the mouth. 3. The larger, white Ammonia. 4. The large, yellow Ammonia.

FRESH-WATER SHELLS.

Division the Second.

BIVALVES.

These are much fewer in number than the simple, or, as they are usually called, Univalve shells; and more of the genera of the sea kinds are wholly unknown among them.

CHAMÆ.

Chama depressa levis albescens.
The white, smooth, depressed Chama.

THIS is about three quarters of an inch in breadth, and nearly of the same measure from the hinge to the opposite verge: the valves are both equally hollowed, but it is not much; they are very thin, when joined, about the edge, and the gibbosity

gibbosity toward the cardo does not give them any very great depth: their surface is perfectly smooth, and the colour is a dusky whitish, without any variegations.

We have this species in some of our large rivers, but it is not common: it is abundant in the ditches of Holland.

Chama gibbosior albida variegata.
The gibbose, whitish, variegated Chama.

This is about half an inch in breadth, and is nearly as much from the hinge to the opposite verge; the shells are equally and considerably deep; they are slightly fasciated or marked with obscure, circular lines, at small distances; the ground colour of the shell is a dusky whitish, but it is variegated with olive-coloured spots.

We have it in our rivers of Northamptonshire and Warwickshire in great abundance; it is frequent also in the rivers of Germany, but there it is more beautiful, being frequently variegated with red and green; but the species is still the same.

Chama rotundior fasciata subrubens.
The reddish, round fasciated Chama.

This is about an inch and a half in diameter, and as much from the hinge to the opposite verge: the shells are equally hollowed, but not greatly so; they are lightly fasciated, and are of a pale brownish-red on the outside, and of a very elegant, pearly hue within.

We have this in some of our large fish-ponds, but it is not common with us: I have had a great many specimens of it from France, some of them of this reddish colour, but most of a pale brown, with little or no red in it.

Chama subovata subcærulea.
The bluish, oval Chama.

This is about half an inch in length from the hinge to the opposite verge, and is somewhat less than that from side to side, which is not the usual case with the Chamæ: it is broad and rounded at the verge, and at the top it runs up smaller, but terminates obtusely; it is elegantly and finely striated in a longitudinal direction; the striæ are very fine and close; the colour is a pale bluish-grey.

We have this in some of our rivers, but it is not frequent with us.

The other species of fresh-water Chamæ are, 1. The larger, flat, and smooth Chama. 2. The smaller, flat, and obscurely fasciated, yellowish Chama. 3. The smaller, striated, reddish, hollowed Chama. 4. The small, white, fasciated, gibbose Chama. 5. The great, brown, rough Chama, of a pearly white within. 6. The great, olive-coloured, depressed Chama, with a more even surface. 7. The larger, olive-coloured, flat Chama.

MYTULI, MUSCLES.

Mytulus tenuissimus virescens. — The green river Muscle.
The extreamly thin, greenish Mytulus.

THIS is an extreamly elegant shell: it is near an inch from corner to corner, and is about three quarters of an inch from the hinge to the opposite verge of the shell; it is considerably gibbose and deep, and approaches to the figure of the tellina of authors: the whole surface is lightly fasciated; the colour is a pale, elegant green, and the shell is as thin as the finest paper.

I met with several of this species in the lakes on the mountains in Yorkshire, and other parts of the North; we have it from the rivers of France and Germany also, but it is so tender and delicate, that it is usually broke or injured in the carriage; it is, indeed, hardly possible it should escape accidents.

Mytulus fasciatus albescens variegatus.
The whitish, variegated, fasciated Mytulus.

This species also approaches to the figure of the tellina of authors: it is about an inch in measure from the extremities of the sides, but is not much more than half an inch, from the hinge to the opposite verge of the shell; it's colour is whitish, variegated with brown; the valves are both considerably gibbose, and the surface is pretty deeply fasciated.

We have this in most of our large rivers.

Mytulus angulatus angustior fuscus.
The narrow, brown, angulated Mytulus.

This is near an inch long, and not half an inch in breadth; the valves are both very deep, so that it's thickness is nearly equal to it's breadth: they are not even on the surface, but are irregularly angulated, and raised into two or three ridges; the substance of the shell is tolerably thick and strong; the colour a pale brown, with a tinge of the olive.

We have it in some of our ponds, but it is not common.

The other species of the fresh-water Mytulus are, 1. The little, purplish, thin Mytulus. 2. The white, oblong, smooth Mytulus. 3. The brown, oblong, smooth Mytulus. 4. The broad and flat, slightly fasciated Mytulus. 5. The pale red, thin, striated Mytulus. 6. The larger, dusky greyish Mytulus. 7. The large, brown, thick Mytulus. 8. The great, broad, smooth, brown Mytulus, pearly within, called the Horse-muscle, or the Pond-muscle. 9. The greater, oblong, brown Mytulus, pearly within. 10. The large, yellowish, smooth Mytulus, pearly within. These three last are common in our ponds, and are by many indiscriminately called the Horse-muscle.

CARDIÆ, HEART-SHELLS.

Cardia profunde striata triangularis.
The triangular, deeply striated Cardia.

THIS species has some resemblance to the common cockle eaten at our tables, but more to some of the fossile Conchæ Anomiæ: it is about half an inch broad, and nearly as much in length, from the hinge to the extream verge; it is broad at the bottom, whence it grows smaller to the top where the hinge is, and, though rounded at the verge, it by this means approaches to a triangular figure: the whole surface is deeply striated or furrowed; the ridges between the furrows are narrow and rounded, and the colour is a pale brownish, almost white, with a faint tinge of reddish in it.

We have this in considerable abundance in some of our large rivers.

Cardia subglobosa livida.
The bluish, rounded Cardia.

This is an elegant shell, and, when viewed sideways, has as much the appearance of the figure of a heart at cards, as any of the marine kinds: it is about an inch in length, and is as much in diameter from side to side, and but little less in thickness: the whole surface is elegantly, but not very deeply, striated, and the colour is a livid or dusky bluish.

I met with a few specimens of this in the river Nen, near Water Newton in Northamptonshire; elsewhere I have not seen it.

Cardia fasciata subrubens.
The reddish, fasciated Cardia.

This is also an elegant shell for one of the fresh-water kind; it is about half an inch in length, from the hinge to the opposite verge, and is nearly as much in diameter:

meter: both the valves are deeply hollowed, and, when closed and viewed sideways, have much the appearance of the figure of a heart at cards: the substance of the shell is tolerably firm and solid, it is lightly fasciated, and the intermediate surface is smooth, and has a fine polish; the colour is white, with a tinge of red.

We have this species in some of our large rivers, but it is not common.

The other species of the fresh-water Cardiæ are, 1. The larger, smooth, brown Cardia. 2. The larger, pale red, striated Cardia. 3. The little furrowed, whitish, round Cardia. 4. The little, fasciated, olive-coloured Cardia. 5. The lesser, flat, furrowed Cardia, called the fresh-water Scallop. 6. The larger, flat, furrowed Cardia.

These are all the known species of the bivalve fresh-water shells: as to the several genera of the multivalves, we have not yet met with a single species of any one of them, that is an inhabitant of any water but such as is salt.

ANIMALS living in SHELLS.

Series the Third.

These which live at land.

COCHLEÆ TERRESTRES, LAND-SNAILS.

Cochlea subrotunda, fusca, et nebulata, clavicula elatiore.
The round-bodied, brown, clouded Snail, with an elated clavicle. **The Pomatia.**

THIS is the largest of the snail kind known among us: it is more than an inch in height, and as much in diameter; the shell is considerably firm and strong, and the colour a deep dusky brown, variegated with clouded spots and oblique streaks of a paler brown; and sometimes, but not always, there runs a spiral line of the same brown, following the volutions of the shell: the body of the shell is rounded, and the mouth nearly round, but a part of it filled up by the succeeding volution of the shell: the clavicle is elated, and has four volutions.

We have this species in some parts of Surry, but it is not frequent: it is not properly a native of England, but of the warmer parts of Europe; there are people who remember the time of a large quantity of them being brought over from France, for the use of a person of quality in that part of the kingdom, to whom they were prescribed as a medicine: they came over alive, and were turned loose in the garden, and from that parcel all the country about has been furnished with them; and many other places from thence.

Cochlea subrotunda, undulata, et fasciata, clavicula obtusiore.
The roundish, undulated, and fasciated Snail, with a more obtuse clavicle. **The great Garden-snail.**

This is the largest of our ordinary snails, but it is considerably smaller than the former: it is about three quarters of an inch in height, and as much in diameter; the body of the shell is rounded; the mouth is large, and approaches to a round figure, but is in part filled up by the succeeding volution of the shell; the clavicle has four turns, and is obtuse at the extremity: the colour of the shell is a dusky brown; there runs a broad simple fascia or belt along the body, following the spiral turn of the shell, and sometimes there are two other faint ones: above and below this middle fascia, there run several broad and short oblique lines or clouds of a different brown.

It is extremely common in our gardens, and under hedges.

Cochlea subrotunda flavescens fusco variegata.
The yellow, round-bodied Snail, variegated with brown.

This also is a large species, though inferior in that respect to the Pomatia: it is about three quarters of an inch in height, and is fully as much in diameter: the body is rounded, as in the two former species, but the mouth is more depressed; the clavicle has four volutions, and is terminated by a little round button: the shell is considerably thick and firm, and it's ground colour is a dusky yellowish; it has a single broad fascia, of a deep brown, following the turn of the shell, and placed just in the middle of the body; it has also some other faint variegations of brown, and the mouth is surrounded with a thick rim of white.

We have this under some of our hedges in the West of England, but it is not common.

Besides these three species, we have, of the common large Snail kind, 1. The great, brown and white Snail, with a depressed clavicle. 2. The great, brown Snail, with scarce any variegations, with an elated clavicle; and, 3. The large, bluish Snail.

2. Of the common smaller Snails we have a very considerable variety, and most of them very elegant, even the plain-coloured ones not being without considerable beauty.

Cochlea lutea labio pallido.
The plain, yellow Snail, with a yellow lip.

This is barely half an inch in height, and is about as much in diameter: the body is not so distinct from the clavicle as in the larger species, but the whole shell may be said to have five volutions or turns: the form is less elevated than that of the larger kind, but the clavicle rises from the rest of the shell, and terminates obtusely; the mouth is large, and approaches to round, but it is depressed, and in part filled up by the succeeding volutions of the shell: the colour of the whole is a plain, but bright, yellow, with no variegation, only the lip or verge of the mouth is of a paler hue than the rest, and sometimes whitish.

We have this every-where in our gardens and hedges.

Cochlea flava fasciis singularibus.
The yellow Cochlea, with single fasciæ.

This is a very beautiful shell; it is nearly half an inch in diameter, and almost as much in height: it consists of five volutions, and it's mouth is of a depressed figure, and is surrounded by a whitish rim or lip: the colour of the shell is yellow, but there runs a single broad fascia or belt of a deep, purplish-brown colour, in the center of every volution following the whole spiral turn of the shell.

We have this species as frequent as the former, in our gardens and hedges.

Cochlea flava fasciis triplicibus.
The yellow Cochlea, with triple fasciæ.

This is also a beautiful species; it is near half an inch in diameter, and as much in height: the mouth is roundish but depressed, and the whole shell is composed of five volutions, the clavicle terminating in an obtuse umbo or button: the ground colour is yellow, but there runs a triple fascia, of a purplish-brown, round the whole volution of the shell, so that there appear in the center of each turn three lines, the middle one broad, and the other two narrower.

This is frequent with us in gardens and hedges.

Cochlea flava fasciis quaternatis.
The yellow Cochlea, with quaternate zones.

This is about half an inch in diameter, and is nearly as much in height: the whole shell consists of five volutions, and the top or extremity of the clavicle is obtuse, and formed into a little umbo: the mouth is large, rounded, but depressed, and edged with a rounded, thin rim, of a whitish colour: the general colour of the shell is yellow, but there run on each volution four narrow lines of a purplish colour; they follow the whole turn of the shell, and stand in the midst of all the spiral rounds.

This is frequent with the former species in our fields and gardens.

Cochlea

Cochlea flavescens fasciis quinatis.
The yellowish Cochlea, with quinate fasciæ.

This is half an inch in diameter, and nearly as much in height: the shell is thinner than in most of the other species, and is easily burst or crushed to pieces: the clavicle terminates in a little, whitish umbo; and the mouth, which is large and depressed, is surrounded with a thin rim or lip of a whitish hue: the colour of the shell is a pale whitish-yellow, but in the center of the body there run five narrow fasciæ or zones: they are continued, though in a less determinate manner, to the top of the shell, running along the center of each turn.

This species is as frequent as the others, in our fields and gardens.

The other species of the smaller Snails are, 1. The plain, white Cochlea. 2. The white Cochlea, with a single zone of a purplish-brown. 3. The white Cochlea, with ternate fasciæ or three zones on each volution. 4. The white Cochlea, with quaternate zones. 5. The white Cochlea, with quinate zones, or five streaks on each volution. 6. The plane, reddish Cochlea. 7. The reddish Cochlea, with a single fascia. 8. The reddish Cochlea, with three fasciæ. 9. The reddish Cochlea, with four fasciæ. 10. The reddish Cochlea, with five fasciæ. 11. The pale brown Cochlea, with a thicker shell. 12. The fine, thin, very pale brown, and transparent Cochlea. 13. The brown Cochlea, with single zones. 14. The brown Cochlea, with ternated zones. 15. The brown Cochlea, with four zones. 16. The brown Cochlea, with five zones. 17. The thin-shelled, plane, yellow Cochlea, with a brown lip. 18. The brown Cochlea, with a purple lip. 19. The very thin, reddish Cochlea, with a purple lip.

These are the species of the smaller Cochleæ I have hitherto observed; they are all to be met with about Battersea, near the gardens, in great abundance; and, on throwing them into boiling water, the animal is readily taken out, and the shells are fit for laying up.

3. Of the depressed, or broad and very little elevated snails, these are the following species.

Cochlea depressa carnea, fusco radiata, ore marginato.
The depressed, flesh-coloured Snail, radiated with the brown, and with a marginated mouth.

This is an extreamly beautiful species; it is near an inch in diameter, and little more than a third of an inch in height: it is formed of four volutions, and the clavicle, which is scarce at all elated, terminates in a rounded extremity; the mouth is large, rounded, and edged, with a round and thin rim; it is expanded to considerably more than the diameter of the shell: the colour is a faint reddish or flesh colour, and it is elegantly radiated with a deep purplish-brown.

I have met with it on the bushes in Lancashire, but very rarely; in Germany it is frequent.

Cochlea depressa, albida, purpureo-fusco radiata.
The whitish, depressed Snail, variegated with rays of purplish-brown.

This is a very depressed or flatted kind; it measures about three quarters of an inch in diameter, and it's height or thickness is not more than a third of an inch: it consists of about three volutions, and has a rounded but depressed mouth, filled up in great part by the succeeding volution, and edged with a thin rim: the colour of the whole is a pale greyish-white, but it is faintly radiated with lines and clouds of an obscure purplish-brown.

We have it on the bushes and in gardens in Yorkshire, but it is less frequent in other parts of the kingdom; in Germany and France it is common.

Cochlea albida, depressa, ore subrotundo labiato et dentato.
The whitish, depressed Cochlea, with a roundish, labiated, and dentated mouth.

This is a very singular species; it's breadth is about half an inch, and it's height or thickness not more than a third of an inch: it consists of three or four volutions, and the clavicle terminates obtusely, and is not at all elated: the mouth is situated on the under part of the shell, and is more than a quarter of an inch in length, and almost as much in breadth, being of a figure nearly rounded; it has a somewhat broad lip, of a pearly white colour, and is slightly dentated on each side: the colour of the whole shell is white, without any variegation.

We have it in Charlton-forest in Sussex, elsewhere I have not seen it living; but I have specimens of it from Italy, where it is frequent.

Cochlea depressa cornea fasciata.
The horn-coloured, depressed, fasciated Cochlea. **The Jamaica, Ribband-snail.**

This is an extreamly elegant species; it is an inch and a half in diameter, but so very much depressed or flatted, that it's utmost height or thickness is not half an inch: it consists of three or four volutions; the outer one considerably large, the others smaller in proportion, as they approach the center: the clavicle is scarce at all elevated, and terminates obtusely: the mouth is rounded, but in part filled up with the succeeding volution of the shell: the ground colour of the shell is a pale brown, and the other white, having much the appearance of a ribband, whence the name of the shell.

We have not this species a native with us, but it is sent frequently from Jamaica, where it is common.

Cochlea albida, depressa, ore expanso labiato et profunde dentato.
The white, depressed Cochlea, with a wide, labiated, and deeply dentated mouth.

This is a very singular and a very elegant species; it is about an inch and a quarter in diameter, and a third of an inch, or little more, in thickness: it consists of four elegant volutions, and the clavicle is not at all elevated, but terminates obtusely: the colour is white, but there runs an orange-coloured fascia along all the volutions in most of the specimens, but in some this is wanting. The most extraordinary circumstance in this species is the mouth, which is not a simple opening, terminating the last volution of the shell, as in the others, but is placed transversely, and has a very large and broad lip, covering on the lower part two or three volutions of the shell, and elegantly and deeply dentated on both sides.

This is an American species; we have it principally from Jamaica.

BUCCINA, TRUMPET SHELLS.

Buccinum flavescens ore rotundo minore.
The yellow, land Buccinum, with a round, small mouth.

THIS is a species of no great beauty, but very singular in it's form; it is about three quarters of an inch in height, and less than half an inch in diameter in the largest part: the body of the shell is of an oval figure; and the mouth, which is continued from it's extremity by a kind of neck, is small and perfectly round: the clavicle has four volutions, and terminates obtusely; the colour of the whole shell is a dusky yellow.

It is very frequent in the woods in Germany, and in some other parts of Europe.

Buccinum albescens fasciis violaceis.
The whitish Buccinum, with violet-coloured zones.

This is an inch in height or length, and three quarters of an inch in diameter in the largest part of the body: the mouth is large, and nearly round; the body of the shell is inflated, as it were, and the figure approaches to oval: the clavicle has four volutions, and terminates in a little round button; the ground colour of the shell is a pale grey, but it is variegated with elegant zones of a deep violet blue, variegated with a bright white.

It is a native of many parts of Europe, but not of England; we have specimens of it from Italy, and from the South of France.

Buccinum tenue ventricosum variegatum.
The thin, bellied, variegated Buccinum.

This is a very large and a very elegant species; it is three inches in length or height, and near two in diameter in the largest part of the body; the mouth is oblong and very large; the body of the shell is tumid, and approaches to an oval figure; and the clavicle has about four volutions, and terminates obtusely: the shell is extreamly thin and light, and is variegated in an elegant manner, with a deep glossy brown, and a bright white, and frequently there is some purplish among it.

We have this from the American Islands, where it is frequent in the woods, and makes a very odd appearance, as it crawls up the trees, and along the hedges.

Beside these three species there are, of the larger land Buccina, 1. The elegant, brown Buccinum, variegated with purple and white. 2. The large, tall, thin Buccinum, with elegant variegations of white and flesh colour. 3. The pale, brown Buccinum, variegated with a deeper brown and white; and, 4. The simple, brown, or olive land Buccinum.

2. Of the minute land Buccina, there are the following species.

Buccinum fuscum ventricosum ore rotundo.
The brown, ventricose Buccinum, with a round mouth.

This is a very neat and elegant little species, though wholly without variegations; it is about a third of an inch in height, and a quarter of an inch in diameter in the largest part: the body is inflated or ventricose, and of a rounded figure; the clavicle has four volutions, and terminates obtusely; the mouth is small and round: the colour of the shell is a fine chesnut brown; the whole surface bright and glossy, but without any the least admixture of any other colour.

I have met with this about the decayed stumps of trees in Charlton-forest, and in the Arbor Vitæ grove at Goodwood.

Buccinum conicum ore ovato.
The conic Buccinum, with an oval mouth.

This is a very minute species; it's length is about a tenth of an inch, and it's diameter, in the largest part, is not more than equal to a third of it's length: it's figure is pretty regularly conic; the mouth is in a manner detached from the body of the shell, and affixed to a kind of neck, and is of an oval figure, and surrounded with a narrow lip or rim: the body grows gradually smaller from the base to the top, and is formed of about six volutions; the summit is pointed: the colour of the whole shell is a pale yellowish.

I have met with this in some of the woods in Yorkshire, among the damp moss at the bottoms of old trees; it is sometimes of a deeper, sometimes of a paler, yellow.

Buccinum conicum albescens, ore majore ovato.
The whitish, conic Buccinum, with a larger oval mouth.

This is about an eighth of an inch in length, but it is a very elegant and regularly figured shell; it's diameter, in the thickest part, is equal to about a third of it's length: it's mouth is large, and of a tolerably regular oval figure, and projects a little from the verge of the shell; the whole consists of five volutions, and the extremity of the clavicle is formed into a fine point: the colour of the whole shell is a greyish-white, with an elegant natural polish.

I have met with it in Rockingham-forest in Northamptonshire, among the moss about trees.

Buccinum heterostrophum conicum ore ovato minore.
The small, oval-mouthed, heterostrophous, conic Buccinum.

This is also a very elegant and singular little shell; it approaches greatly to the figure of the two former, but it's volutions are turned the contrary way, and consequently it's mouth opens in a contrary direction to them: it is about a sixth of an inch in length, and it's diameter, in the largest part, is equal to about a third of it's length; the mouth is small, and of an oval form, and is surrounded by a narrow, rounded lip: the whole shell consists of five volutions, and the clavicle is carried up to a fine point; the colour of the whole is a glossy brown, without any variegation.

I have met with this several times in the West of England: they have a way of building walls of rough stone, by way of inclosures, in this part of the kingdom. I first observed this shell, on taking off a specimen of the Umbilicus Veneris from one of these walls near Shepton Mallet in Somersetshire; several of them were fixed to the lower part of the stalks of that plant.

DOLIA, GLOBOSE SHELLS.

Dolium subrotundum ore maximo labiato.
The roundish Dolium, with a very large, labiated mouth.

THIS is a very delicate and tender shell; it is indeed so thin and brittle, that it is not easy to touch it without crushing it to pieces: it is about half an inch in length, and nearly as much in diameter; the body is rounded and inflated, and the mouth extremely large, and of a somewhat oval figure, surrounded with a narrow lip: the clavicle is short; it has not more than three volutions, and terminates in a sharp point: the colour of the whole shell is white, without any variegation.

I have seen this species sometimes in the West of England, but it is not common: in the South of France it is frequent.

Dolium elatius ore ovato nudo.
The taller Dolium, with an oval, naked mouth.

This is three quarters of an inch high, and but little more than a quarter of an inch in diameter in the largest part: the body is of an oval form; the mouth is pretty regularly oval, and is not nearly so large as in the former species, and has no lip: the clavicle has three volutions, and terminates in fine a sharp point: the colour is an elegant pale yellow, much resembling that of the peel of a lemon; and the shell is so extremely thin, that it is not easy to handle it without destroying it.

It is frequent in the woods of France, but we have it not native in England.

The other species of Dolia are, 1. The larger, thicker, shelled, brown Dolium. 2. The larger, thicker, shelled, white and reddish Dolium. 3. The large, very thin, shelled, bluish Dolium. 4. The lesser variegated yellow and brown Dolium, with the shell extremely thin. 5. The lesser, brown and purple Dolium, with a somewhat thicker shell. 6. The lesser, thick, white Dolium, with a blue rim to the mouth.

TURBINES,

TURBINES, SCREW-SHELLS.

ALL the known land Turbines are very minute.

Turbo angusta fusca ore elongato.
The slender, brown Turbo, with a narrow, oblong mouth.

This is about a quarter of an inch long, and is hardly a fourth of it's length in diameter at the bottom: from this part it grows gradually smaller to the summit, where it is terminated by a very sharp point: the mouth is narrow, and is carried up a great way into the first volution of the shell; the whole shell consists of about eleven of these, and is somewhat deeply furrowed by the spiral line, which separates them; the colour of the whole is an elegant pale yellow, but the lining of the mouth is white.

We have it in some of our woods about London, among the damp moss, about the stumps of decayed trees.

Turbo brevior volutionibus striatis.
The shorter Turbo, with the volutions all striated.

This is not longer than a full-grown barley-corn, but it is thicker than the generality of the Turbines, in proportion to it's length: the mouth is of an oval figure; the largest part upwards, which is the contrary of the mouth of the preceding species; the top is obtusely terminated; the whole shell is formed of about twelve volutions, not very distinctly separated, but all of them finely striated in a spiral direction: the colour is an elegant chesnut brown.

We have it in some of our old hedges, among the damp moss that grows about the stumps of the trees, and in the ditches.

Turbo angustior margine volutionum pectinata.
The slender Turbo, with the edge of the volutions pectinated.

This is the most elegant in it's structure of all the land Turbines; it is about a sixth of an inch in length, and is hardly a third of it's length in diameter: the mouth is narrow, of a form approaching to oval and labiated; the whole shell consists of about twelve volutions, and the rim or edge all the way is deeply and elegantly denticulated, or rather pectinated, the denticulations being deeper than what are usually expressed by that name: the colour is a dusky grey, but toward the top it is whitish.

I met with this species in considerable abundance some years since, among the great tree lungwort, on the old beeches, in Charlton-forest in Sussex; elsewhere I have not seen it.

Turbo volutionibus tumidis ore subrotundo.
The round-mouthed Turbo, with tumid volutions.

This is about an eighth of an inch in length, and nearly a third of it's length in diameter: the mouth is round, and has no lip surrounding it; the shell is formed of about nine volutions; they are elevated and tumid, and the line between them is consequently sunk very deep: the colour of the whole is a yellowish-brown, only toward the top, or smaller extremity, the brown wears off, and the colour more approaches to a whitish-yellow.

This species also I met with in Charlton-forest, and have not seen it elsewhere.

The other species of land Turbo's are, 1. The very slender, yellow Turbo. 2. The very slender, whitish Turbo, with punctuated volutions, and a narrow mouth. 3. The slender, brown Turbo, with a round mouth, and a very shallow line between the volutions. 4. The short, thick Turbo, with a very sharp point, and a long slit in the place of a mouth. I am apt to believe more species will be discovered; these have been principally of my own finding in damp woods, while I was investigating the mosses.

Animals living under the defence of shelly coverings.

Series the Fourth.

CENTRONIÆ. Sea hedge-hogs.

THE Centroniæ are animals living under the defence of a shelly covering, formed of one piece, and furnished with a vast number of spines, moveable at the animals pleasure.

The animals of this series have been called by authors Echini Marini, Sea Hedgehogs, from some resemblance, in the prickly surface of the shell, to the coat of the land animal of that name; but this is a very poor characteristic from which to name a class of animals, whose whole structure is full of essential marks of distinction.

It has been warmly disputed, between authors of note in these studies, whether the Centroniæ, or Echini Marini, as they call them, belonged properly to the testaceous or the crustaceous animals, that is, whether they were to be arranged among the ordinary shell-fish, or among the crab and lobster kind; It is easy to determine the controversy, by observing that they belong to neither: their characteristics, the structure of their bodies, and even the form, use, and intent of their several exterior parts, usually comprehended under the term, shell, are wholly different from those of all other creatures; and though all the species are referable to the same class, and, indeed, in strictness to the same genus, yet they are a distinct series of animals by themselves, and demand to be placed accordingly.

Centronia subglobosa papillis minoribus. The round
The roundish Centronia, with small papillæ. Sea-egg.

This is a considerably large species; we usually meet with it of about an inch and a half in height, and as much in diameter, but it often grows to three times that size: it's figure is nearly round, but somewhat depressed at the bottom; it's covering, which, as in the shell-fish, is the thing principally demanding our attention, is formed of one single piece, and is of a shelly structure and hardness: in the center of the base there is a large, roundish aperture, at which is placed the mouth of the animal; and on the summit there is another aperture, smaller than the former, but of a like roundish figure, out of which the creature voids it's excrements: the colour of the whole shell is a dusky reddish, with an admixture of whitish, and often of greenish: it is covered with an innumerable multitude of papillæ, which are small, and disposed in regular lines from the upper aperture to the lower: ten of these lines, disposed into five pairs, are formed of papillæ larger than the rest, and make a pretty figure on the shell; they are all elevated, more or less, above the rest of the surface of the shell, and give origin to a multitude of spines, which arm the whole surface, and give the creature, while living, and under water, that appearance of the hedge-hog, from which it has been named Echinus: the spines are about a third of an inch long, and very slender; they are of a reddish colour, and are pointed at the extremities; they are all moveable, by means of muscles affixed to their bases, but they are apt to fall off, when the creature is dead, on the slightest touch: beside these spines, which are more than two thousand in number on this species, there are about twelve hundred apertures in the shell, out of which the creature thrusts so many fleshy tentacula or filaments at pleasure, by means of which it fixes itself to the bottom: the spines serve it, on the contrary, by way of legs, to move forward with, and for it's defence, and several other purposes; the natural posture of the creature is, with the broad or depressed part, called the base, toward the bottom: at the aperture, in the center of this, is situated the mouth, which is furnished with five sharp teeth, fixed at the extremities of five little bones, in the center of which is placed the tongue; the base of this is a caruncle of a rounded form, and the whole is fleshy: from the hinder part of the mouth there begins an intestinal canal, which is carried in several volutions round about the inside of the shell, to which it is fixed by a multitude of fine fibres; it finally terminates at the aperture, in the top of the shell.

This

Thus much of the general structure of the creature may serve also as part of the descriptions of the succeeding species, for they all agree in general in these particulars.

This species is frequent in our seas, and in most of the other parts of Europe; it becomes white, by long lying on the shore.

Centronia subglobosa spinis maximis quadratis.
The roundish Centronia, with very large, square spines.

This is an extremely singular species, and, when perfect, makes a very odd figure; but the spines are so easily displaced, that we very rarely meet with it so; the body of shell, exclusively of the spines, is about two inches in diameter, and as much in height, but, with these appendages on it, measures nearly three times as much; it has a large aperture at the base, where the mouth is placed, and a smaller at the summit, both of a roundish but somewhat dentated figure: the whole surface of the shell is beset with large tubercles or papillæ, though they are much less numerous than in those which have them smaller: on these are affixed the spines; they are an inch and a half in length, and nearly a third of an inch in diameter in the largest part: they are largest near the middle, and have four angles, which give them a square figure at the base, where they are affixed to the papillæ of the shell; they have a kind of head, formed so as to fit the papillæ; at the other end they are obtuse: the colour of the body of the shell is a deep brown; the spines are of a paler brown, somewhat approaching to whitish.

We have it from the coasts of the American islands, but it is very rare even there, and is seldom taken up without the loss of many of it's spines; and the greater part of the rest are usually lost, before it comes to us; so that we usually meet with only the body of the shell, with a loose spine or two that have fallen off from it.

Centronia subglobosa spinis maximis cylindraceis.
The roundish Centronia, with very large, rounded spines.

This is an extremely elegant species; the body of the shell, which is all we usually meet with, is of a figure approaching to roundish, but much depressed at the base: it has a large opening at the center of the base, and a smaller on the summit, both approaching to a rounded figure, but with dentated edges: the papillæ on this species are but few in number, but they are extremely beautiful; they are of the breadth of a silver two-pence, and are of a figure approaching to that of a segment of a sphere; at their top there stands a little obtuse umbo, and round about the verge of each there is a circle of little round protuberances: these papillæ are placed in double series, each series having about ten of them; and these are separated by broad, undulated fasciæ or bands, running from the upper orifice of the shell to the lower: the colour of the shell is a faint red, when recent; but we usually meet with it white, from it's having lain long on the shores. This is all we usually see of this species in our cabinets; but, when the creature is perfect and living, there is affixed to each of these papillæ a spine of two inches in length, and of the thickness of a child's finger; these have a kind of head at one end, where they have been fixed to the papillæ, and at the other they are obtuse; they are of a rounded, not angulated, figure, and are nearly of the same thickness all the way, only a little largest toward the middle.

This is a native of the Red-sea and of the Persian Gulph; the finest we have seen have been taken up about the shores of the island of Ormus.

Centronia subglobosa spinis variis subcæruleis.
The roundish, bluish Centronia, with differently shaped spines.

This is also a very elegant species; the body is of a roundish figure, but flatted at the base, and the aperture at the top is very large, nearly equal to that at the base in diameter: the whole surface is very elegantly diversified with papillæ of various forms and sizes; these are divided into certain series by five longitudinal fasciæ, which are elegantly formed, and are full of little perforations; between these there stand several series of larger and of smaller papillæ, all following the direction of the bands, and all very regularly arranged: this is all that we usually see of this species in our cabinets,

 but,

but, when living, it makes a much more singular appearance; all the larger papillæ are furnished with spines of a rounded figure, obtuse, and an inch long; on the smaller papillæ stand spines of half the length of the others, very slender, and pointed at the extremity, and from the holes in the bands the creature throws out a kind of arms, or fleshy filaments, to hold itself in it's place.

It is a native of the Eastern seas, but we have the naked shell frequent in our cabinets; it is bluish, when recent, but becomes white in keeping.

Centronia depressa cordata spinis capillaceis.
The depressed, cordated Centronia, with capillaceous spines.

This is a large species; it's figure is oblong and cordated; it's length is three inches; it's breadth at the larger end, where it is sinuated so as to give it a cordated form, is two inches and a quarter, and it's depth not more than an inch and half: the aperture where the mouth is placed is large, and in the center of the base; that for the anus is at the extremity of the body, opposite to the cordated end, not on the summit, as in the rounded species: there run from the center of the upper part of the shell four rays, each formed of a double series of lines, and each narrow at the base, broader in the middle, and narrow again at the point: the rest of the surface is covered with small, roundish papillæ; the colour is whitish, and, when the creature is alive, there stand a multitude of short, slender, and capillaceous spines on the papillæ, and fleshy filaments are thrust out from the apertures or perforations at the edges of the lines or rays.

We have it from the American shores.

Centronia depressa plana margine digitato.
The depressed, flat Centronia, with a digitated edge.

This is about two inches in diameter, and is not more than a sixth of an inch in thickness or height: the shell is very thin, and of a whitish colour; it is rounded, and in some part the verge is entire and undivided, but in other parts it is deeply sinuated or cut in, so that the parts between the indentings resemble fingers: it has a beautiful configuration of four rays on the surface, and has a number of minute papillæ, not covering the whole surface, but only disposed in series, on which, when the creature is living, there stand very fine, slender spines.

We have this on some of the American shores: the number of the digitations in the shell is uncertain and irregular; it is usually from six, or seven, to twelve.

Centronia depressa plana disco quinquies perforata.
The depressed, plain Centronia, with five perforations on the disk.

This is also a very singular species; it is one of those which are flatted into the form of a thin cake: it's circumference is round; it's diameter not less than two inches, but it's thickness is not more than that of a crown-piece: it's colour is whitish, and it's surface furnished with a number of papillæ in regular series, and very minute: there are four elegant rays formed of perforated double series of lines running from the center toward the verge, in this species; but what is most singular is, that there are between these, toward their extremities, five oblong, oval perforations, a third of an inch in length, and half their length in diameter, pierced quite through both surfaces of the shell.

We have this from the shores of the American islands; when perfect, there stand on the papillæ a great number of fine, slender, and pointed spines.

Centronia depressa ad marginem biperforata.
The depressed Centronia, with two perforations near the edge.

This is another of the flat kinds; it is of a rounded figure, and about an inch and a half in diameter: it's whole thickness is not more than that of a crown-piece; it's colour is white, and the edges are undivided; from the center of the upper surface there run five double series of lines, making so many short, broad, obtuse, and somewhat oval rays; and toward the edge, on the anterior part of the shell, there are two oblong

and

and narrow perforations quite through the body of the shell. While living, the creature is furnished with a great number of very fine spines, like segments of bristles; but these fall off very easily, so that we rarely see any of them.

It is a native of the American Seas; we have it frequent in our cabinets.

Centronia depressa biperforata margine digitata.
The flat, biperforate Centronia, with a digitated edge.

This might, at first sight, be esteemed only an accidental variety of the preceding species, but the disposition and form of the rays on the summit shew that it is truly a distinct species: it's figure approaches to round, but it is somewhat irregular; it's diameter is two inches and a half, and it's thickness, including both the shells and the hollow for the habitation of the animal, is not more than a fifth of an inch; part of the verge is even and undivided, but part is deeply and irregularly sinuated, and forms about ten digitations: on the opposite or undivided part of the shell there are two perforations; they are about a third of an inch in length, and half their length in diameter: the digitations in this, and the other species which have them, seem formed of these perforations, continued to the edge of the shell: in the center of this species there are five narrow, long, and sharp-pointed rays, formed of double series of lines, punctuated as in the others; there are also some annular lines at small distances, running all over the surface; and, when the creature is living, there are many series of fine, slender spines.

We have it from the shores of the American islands.

Centronia ovata depressa margine undulato sexies perforata.
The oval, depressed Centronia, with an undulated edge, and six perforations.

This is another of the flat kinds; it is about two inches in length, and an inch and half in breadth at the larger end; it is considerably smaller at the other, but is obtuse at both, and is not more than a fifth of an inch in thickness: the whole verge is undulated, but is not digitated in any part, and toward the verge there stand, at pretty regular distances, six oblong and narrow perforations: from the center there run five rays; they are oblong and narrow, and are formed, as in the others, of double rows, of punctated perforations: when the shell is recent, there stand several series of fine, slender, and pointed spines on it; but these are usually lost, from the shells preserved in our cabinets.

We have it from the American islands, principally from Barbadoes.

Centronia subglobosa spinis aduncis fasciculatis.
The roundish Centronia, with crooked and fasciculated spines.

The Sea-apple.

This is one of the roundest of this whole genus: the depressed part at the base is small, and the rest of the figure is very regularly spherical; the aperture for the mouth is at the base, and that for the anus at the summit of the shell. When we see the shell naked, as it usually is, in our collections, we only can observe a number of small papillæ on it's surface, thrown together in little clusters; but, when the creature is living, it is all over covered with little clusters of spines; these are about a third of an inch in length, and are slender, crooked, and pointed at the extremities: this is a very rare species.

It is sometimes brought from the Mediterranean, but few of the spines are preserved.

Centronia globosa spinis acicularis.
The globose Centronia, with needle-like spines.

The Needle-shell.

This is one of the most singular of the whole genus; the body is almost perfectly globose; the apertures are one at the base, and the other at the summit, and the whole shell, as we usually meet with it, which is without the spines, is covered with very minute,

minute, reddish papillæ: when it is perfect, there grow from all these papillæ straight, long, slender, whitish spines, an inch in length, and not thicker than a small needle; so that the whole has much the appearance of a round pincushion, in which somebody had fixed as many fine needles as it was possible, with their points outward.

It is a very rare shell, and is a native only of the Eastern Seas.

The other species of Centronia, or Echini Marini, are numerous, and may be arranged under different divisions.

1. Of the roundish or subglobose ones, called by Klein and some others Cidares, beside the species already described, there are, 1. The great, round Centronia, with minute papillæ. 2. The great, round Centronia, with very prominent papillæ. 3. The little, round Centronia, with elevated papillæ. 4. The lesser, globose Centronia, with high papillæ. 5. The globose Centronia, with variolated papillæ. 6. The little, globose Centronia, with larger, variolated papillæ.

2. Of the cordated kind, called by Klein Spatangi, and Spatagoides, there are, beside the already described, 1. The high-backed Centronia, with five narrow rays. 2. The little, ovated Centronia, with the sinus very deep. 3. The little, high-backed Centronia, with a shallow sinus. 4. The large and somewhat depressed Centronia. 5. The large, very broad, and thin-edged Centronia. 6. The large, depressed Centronia, with five elegant broad rays.

3. Of the flat kind, called Placentæ by Klein, beside the already described species, there are, 1. The oval, flat Centronia, with two perforations. 2. The round, flat Centronia, with three perforations, and the edge somewhat digitated. 3. The flat, undulated-edged Centronia, with four perforations. 4. The rounded, flat, somewhat digitated-edged Centronia, with seven perforations.

Klein, who has been at great pains to arrange these bodies, has divided them into a number of other genera, but they may be all conveniently arranged under one or other of these divisions. Beside the known recent species, we meet with several others in the fossile state, and with the spines of other equally unknown kinds, many of which are of a very singular figure.

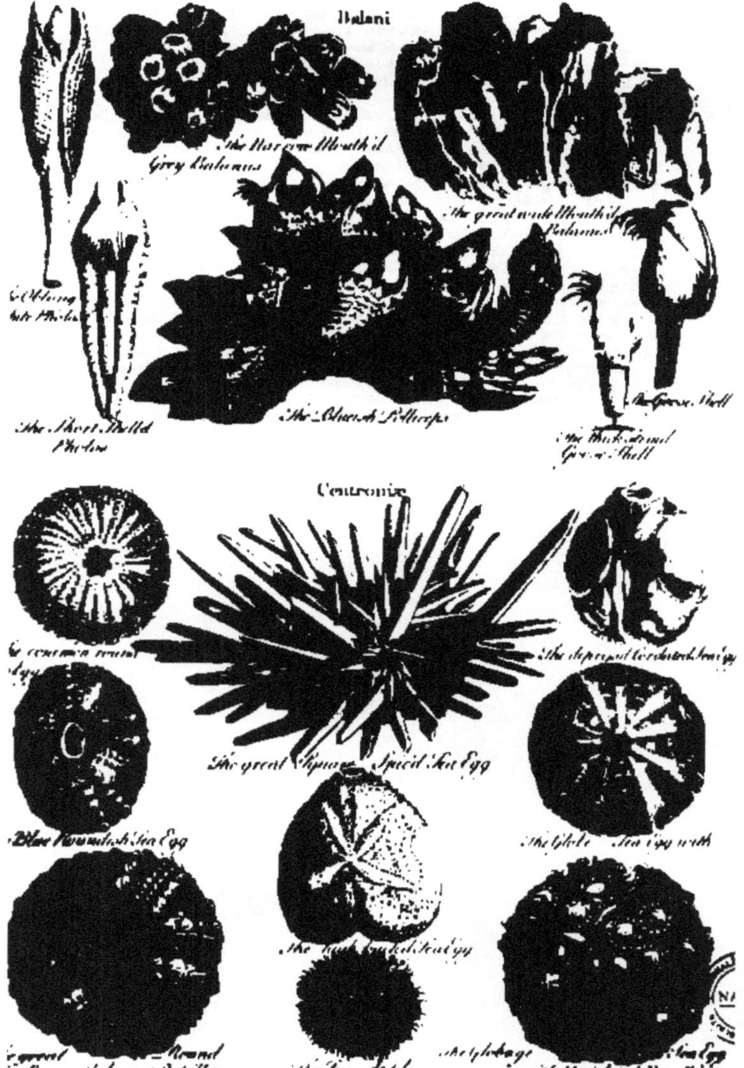
Balani
Grey Balanus
The Goose Shell
The Sea Apple

THE HISTORY OF ANIMALS.

PART IV.

FISHES.

FISHES *are animals which have no feet, but have always fins.*

Fish respire either by means of their branchiæ or gills, or by lungs, like those of land animals; they inhabit the waters, and perform their motions, either by means of their fins, or by the flexuous agitation and turning of their bodies. There are some species which will occasionally quit the water, and come on land; and some which fly for a time, by means of their pectoral fins, which are large, and formed for that purpose.

The several kinds of fish have certain obvious and invariable characters, according to which they may be arranged into certain series, and under these into separate classes, composed of the different genera which agree in their general characters, though they differ in such as are sufficient to distinguish them, under these, into the ultimate arrangement.

The first division into series is established on the difference in the situation of the tail, and in the structure of the fins, and of the parts about the gills.

The generality of fish have their tails placed in a vertical direction, but some have them horizontal, as all the cetaceous kind; these last form only a single series. Of the others, or those which have the tail placed perpendicularly, some have the rays or parts which support the fins of a bony structure; others have them only cartilaginous; the last of these make only one series, comprehending the ray-fish, the sturgeon, and the like. Of those which have the rays of the fins bony, some have the branchiæ ossiculated, and others have none of these ossicles about them. The fish of the latter kind form also only one series; but those which have the branchiæ ossiculated, have the rays of the fins also inoffensive, and others have them pungent or prickly.

On these distinctions, the whole body of fishes are divided into five series.

The first, under the name of Malacopterygii, *contains those which have perpendicular tails, the rays of the fins bony and not pungent, and the branchiæ ossiculated.*

The second, under the name Acanthopterygii, *comprehends those which have the tail perpendicular, and have the rays of the fins bony, and sharp or prickly, and the branchiæ ossiculated.*

The third, under the name of Branchiostegi, *comprehends those which have perpendicular tails, and have the rays of the fins bony, and the branchiæ not ossiculated.*

The fourth, under the name of Chondropterygii, *comprehends those which have perpendicular tails, and have the rays of the fins not bony but cartilaginous.*

The fifth, under the name of Plagiuri, comprehends those which have the tail not perpendicular, but placed in an horizontal direction. This is so remarkable a character, that there needs nothing farther to distinguish the series.

The Fishes, comprehended under some of these series, admit no farther subdivision into classes, all of them agreeing in their more general characters; but those of others, and particularly the very numerous ones arranged under the Acanthopterygious and Malacopterygious series, have natural and obvious classical distinctions under those of the series, and above the generical.

Malacopterygious Fishes, or those whose fins are not prickly.

Class the First.

MESOPTERYGII.

Those which have only one back fin, and that situated in the middle of the back.

SYNGNATHUS.

THE body of the Syngnathus is long and very slender; sometimes it is rounded, but in most of the species it is angulated: the fins are in some species four, but in others there is only one: the head is of an oblong figure, and compressed; the jaws are closed together at the sides, and the mouth has only a small opening, which is quite at the extremity: the coverings of the gills are composed, each, only of one simple and very thin bony lamina.

Syngnathus teres pinnis pectoralibus caudaque carens.
The rounded-bodied Syngnathus, with no pectoral nor tail fins.

The Sea-Adder.

This, of all the known fishes, has the least appearance of one, and serves properly to connect the insect and fish tribe: our people in Cornwal, who have given it the name of the Sea-adder, express their sense of it's resemblance to the serpent-kind, but it much more, in reality, resembles the worm-kind: it's length is about nine inches, and it's thickness is hardly that of a common earth-worm: the head is very long, slender, and compressed, and the opening of the mouth is very small; the lower jaw is somewhat longer than the upper, and is moveable, and naturally turns up so, that the opening of the mouth seems to have a kind of operculum: the body is of a rounded but somewhat compressed figure, and it's colour is a dusky grey, with a tinge of greenish: the eyes are small, and are placed very backward in the head; their iris is yellow: the nostrils are two on each side; they are small apertures near the eyes, but are not easily seen, unless in the full-grown fish; there are no teeth: the coverings of the gills are not open at the lower edge, as in other fishes, but there are two apertures higher up in the neck, at the summit of these: they are very small openings, and scarce exceed the holes made by thrusting a large needle through a piece of paper: there are no scales, but the body is naked, and is divided into a kind of rings, like the body of an earth-worm: the anus is nearly in the middle of the body, but a little nearer the tail than the head; there is no fin at the tail, but it terminates in a fine, slender point, scarce thicker than a large thread: there is only one fin, and that is placed on the middle of the back; it is formed of a thin, whitish membrane, supported by thirty-four fine, slender rays: the branchiæ appear to be four on each side, or rather there runs a pulmonary viscus of a reddish colour, down each side of the throat.

This singular fish is an inhabitant of the Northern Seas in considerable abundance; we have it in great plenty on the coasts of Yorkshire, and other places. Willughby calls it Acus Lumbriciformis; and Ray, Ophidion Lumbriciforme. The people in Yorkshire call it the Sea-worm; but it is more generally known in England by it's Cornish name of the Sea-adder, though the former is the more expressive.

Syngnathus

Syngnathus corpore medio heptagono, cauda pinnata.
The Syngnathus, with the middle of the body heptangular, and a fin at the tail.

This is a very ſmall, as well as a very ſingular, ſpecies: It rarely exceeds four inches and a half in length, and it is of the thickneſs of a large gooſe-quill: the head is long and ſlender; the opening of the mouth is very ſmall, and is placed at the extremity of a kind of roſtrum or ſnout, formed by the elongation of the head: the eyes are very ſmall, and the coverings of the gills are ſmall, and are not round, but oblong, and placed high up toward the back of the head: the body is ſlender, but it is not rounded, as in the former ſpecies, but angulated; from the head to the anus it is heptangular, or has ſeven ridges; from the anus to the extremity of the back fin it is hexagonal, and from this part to the extremity of the tail it is quadrangular or ſquare, and grows gradually ſmaller, all the way to the end: the whole body is covered with a kind of laminæ or plates, of a figure approaching to quadrangular, and ſtriated; it's colour is a reddiſh-brown: the noſtrils have two apertures on each ſide, placed juſt before the eyes; the pectoral fins are two; they are roundiſh and very ſmall; they have fourteen or fifteen rays in each, and ſtand ſomewhat nearer the belly than the back of the fiſh; there are no belly-fins; on the back there is one fin; it ſtands lower on the body than the anus; it has thirty-five rays which are not divided at their tops, and are all of the ſame length: the tail terminates in a radiated fin, which is ſupported by ten rays, not divided at the extremities; there is juſt below the anus a very ſmall fin, ſupported by three rays.

We have this ſpecies about our own coaſts, particularly on the weſtern part of the Iſland, but we have no Engliſh name for it. Seſner calls it Typhle; Ray and Willoughby have named it, Acus Ariſtotelis ſpecies altera major.

Syngnathus corpore medio hexagono cauda pinnata.
The Syngnathus, with the middle of the body hexangular, and the tail pinnated.

The Needle-fiſh.

This is about five inches in length, and is more than a third of an inch in diameter; the head is long, narrow, and of a compreſſed form; the opening of the mouth is ſmall; the eyes are very ſmall, and the apertures of the noſtrils are two on each ſide, and are placed very near them: the body is hexangular from the head to the anus, and from thence to the extremity of the tail it is quadrangular or ſquare: there are two ſhort and roundiſh pectoral fins; there is one fin on the lower part of the back, and the tail is terminated alſo by a fin, which, when expanded, is ſhort and truncated; and, finally, there is an extremely ſmall fin near the anus: there are no viſible lines on the ſides; the anterior part of the roſtrum or head is very ſlender, and compreſſed.

We have this alſo frequent about our ſea-coaſts; I have met with it in great abundance about the ſhores of Yorkſhire; all the writers on fiſhes have deſcribed it. Aldrovand and others call it Acus Ariſtotelis; others, Typhle marina, and Nerophidion; Gillembourg, Serpens marinus; and Pliny, Sphondylus. We call it the Needle-fiſh.

The males and females in this, and the former ſpecies, are eaſily diſtinguiſhed; the females carrying a kind of long bag, reaching from the anus half-way to the tail; there are numerous eggs in this at the proper ſeaſon; they are full of the ſize of rape-ſeed, and of a whitiſh colour: they neither are ſo numerous, nor at all reſemble, in their arrangement or diſpoſition, the eggs of the generality of the other fiſhes of this claſs: this may indeed be, in ſome degree, ſaid to be viviparous, for the rudiment of the fœtus is, in ſome meaſure, animated, before it is diſcharged from the veſica or bag.

Syngnathus

Syngnathus corpore quadrangulo, pinna caudæ carens.
The ſquare-bodied Syngnathus, with no fin at the tail.

The Sea-horſe, or Hippocampus.

This is an extremely ſingular ſpecies; we ſee it dried frequently in the ſhops of druggiſts and country apothecaries, who keep it as a curioſity; and people, not acquainted with this genus of fiſhes, would take it for any thing rather than a fiſh: it's tail uſually curls up in the drying, and the head, bent down, has ſome rude reſemblance of that of a horſe; 'tis hence that the name ſea-horſe has been given it: as to the Greek name Hippocampus, thoſe who gave it did not mean to expreſs any thing of the figure of a horſe by it; but, as they called the great ant Hippomurmex, and other large ſpecies both of animals and vegetables were expreſſed by the word Hippos prefixed to their name, they expreſſed this, which they took, as they met with it dried, to be of the caterpillar kind, by the name of Hippocampus, the great caterpillar.

It is five inches long, and in the largeſt part is about half an inch in diameter; the head is long, ſlender, and compreſſed, and forms a kind of ſnout; the mouth is ſmall; the eyes are very ſmall, and there are two little apertures near them on each ſide, which are the openings of the noſtrils: juſt behind the orbits of the eyes there ſtands on each ſide a ſhort protuberance, pointed at the extremity; and there are three other protuberances on the hinder part of the head, near the opening of the gills; the beginning of the back is acute, and the anus is ſituated nearer the extremity of the tail than the head, which is not the caſe in the others: the body, from the head to the anus, is of a quadrangular figure, but the diviſions are not equal, the loweſt being broadeſt, and the upper one narroweſt of all; the lower one is, indeed, not flat but prominent in the middle, and, if cloſely examined, is found to have three ſmall angles or ridges, ſo that this part of the body, accurately ſpeaking, is heptangular: from the place of the anus to the end of the back fin the body is hexangular, but it is rather depreſſed than compreſſed ſideways in this part: from the end of the back fin to the middle of the tail the figure is quadrangular or ſquare, and, for a little way thence, it is pentangular; and, finally, from this part to the extremity it is ſquare again, but the angles or ridges are placed contrariwiſe, in this part, to their direction on the body; the hinder part of the body is very ſmall and ſlender, in proportion to the reſt; the whole body is covered with bony laminæ: there are ſeventeen of theſe from the head to the anus; from the anus to the extremity of the tail there are about forty-five: the extremity of the tail is pointed, and has no appearance of a fin; there is a long but low fin on the back, ſupported by thirty-ſix rays, and there are two ſhort and rounded pectoral fins: the colour of the whole fiſh, while living, is whitiſh, with a caſt of green, but it becomes brown in drying.

It is a native of the Mediterranean, but is not very frequent. All the naturaliſts have mentioned it, though many of them have been much at a loſs to gueſs to what claſs of animals it belonged; which is indeed the more excuſable, as very few have met with it otherwiſe than dried. It is generally called Hippocampus: Ray has called it Hippocampus Aculeatus; and ſome authors have called it Zideach. Theſe are all the known ſpecies of the Syngnathus, or, as authors in general expreſs it, of the acus kind.

COBITIS.

THE head and the body of the Cobitis are both of a compreſſed figure, and there are certain cirri or beards at the mouth: the body is ſpotted, and the back fin, and thoſe of the belly, are at the ſame diſtance from the extremity of the head.

Cobitis tota glabra, maculoſa, corpore ſubtereti.
The ſmooth, ſpotted Cobitis, with the body ſomewhat rounded.

The Loache.

This is about three inches in length, and of the thickneſs of a man's little finger; it is very ſmooth and ſoft to the touch, and is of a duſky olive brown, ſpotted with a deeper

deeper colour on the back and fides, and pale on the belly; the head is compressed; the body is somewhat compressed also, but it is so little, that it has a rounded look: the eyes are large, and their iris is yellow; the beards are short, white, and fleshy; the back fin has eight rays; the pectoral fins have each of them twelve: the belly fins have seven rays each, and the pinna ani has six.

We have this species in some of our shallow brooks; all the writers on fishes have described it. Rondeles and Gesner call it Cobitis barbatula; Ray and Willughby, Cobitis fluviatilis barbatula; and some authors, Fundulus and Grundulus. It always keeps at the bottom, among the loose pebbles, and often gets under large stones.

Cobitis aculeo bifurco infra utrumque oculum.
The Cobitis, with a bifurcated spine under each eye.

The bearded Loach.

This is about three inches and a quarter in length, and about half an inch, or somewhat less than that, in diameter, in the largest part: the head is of a depressed figure and oblong, and the rostrum is bent somewhat downwards; the body is also of a depressed form and thin, and is somewhat broader toward the head than elsewhere, but it is nearly equal in breadth all the way: the upper jaw is longer than the under, and closes over it, receiving it in the manner that the lid of a box does the body: the opening of the mouth is very narrow; there are six cirri or beards about it; two of them are placed on the upper part of the mouth, two on the lower, and two on the sides; they are all slender, oblong, and white; the nostrils are in the middle, between the eyes and the mouth: the eyes are small, and are situated on the sides of the upper part of the head; their iris is white in the lower part, and of a dusky colour above: on each side of the head near the nostrils, and below the anterior part of the eye, there stands a little bone of a whitish colour, bifid at the top, and prickly; the exterior point is shorter than the interior: the branchiæ are four on each side: their coverings are oblong, and are open sideways only near the pectoral fins, not on the lower part: there are no perceptible teeth in the mouth, nor is there any distinctly visible linea lateralis: the scales are roundish and very small; they are so thin, and laid so close on one another, that they are not visible in the recent fish, but in the dried specimens of it they are distinct enough: the back is of a dusky colour; the sides are variegated with many whitish lines, and a number of very minute, black dots; the belly is white.

The pectoral fins are short, and have each eight rays; the first of these is undivided; the second, which is the largest of all, is bifid or ramose at the top, as are also all the others: the colour of the fin is white; there is one fin on the back, which is variegated with dusky spots; this has nine rays in it; the first is the largest and undivided, the rest are bifid or ramose at their top; the belly fins are white, and have seven bones or rays each: the pinna ani stands a little behind that aperture; it has eight rays, the two last very small and indistinct: the tail is short, rounded at the extremity, and spotted with black; it has eighteen bones or rays: there is a little prominence in this species between the anus and the origin of the pinna ani.

The heart is small, not exceeding a grain of hempseed in size: the liver is long and red; it lies upon that part of the intestine, which performs the office of a stomach, and is joined to it by a membrane.

This lies in slow rivers, and sometimes in lakes and ponds. Rondelet, Aldrovand, and others call it Cobitis aculeata; Willughby and Ray, Cobitis barbatula aculeata; Schoonveldt, Tænia cornuta.

Cobitis cærulescens, lineis utrinque quinque nigris longitudinalibus.
The blue Cobitis, with five longitudinal black lines on each side.

The fossile Mustela.

This is a species very singular in it's manner of living, as well as in it's figure; it is five inches long, and somewhat more than half an inch in diameter: the head is short, broad, and obtuse; the belly is smooth, and of a bluish colour, with ten black longitudinal lines running down it, five on each side of the back: about the mouth there are placed a number of cirri or beards; they are not fewer than ten, and are whitish

and

and slender: the pectoral fins are short, and have each eleven rays; the ventral fins have five rays, the dorsal seven, and the pinna ani seven.

It is frequent about the rivers and lakes in many parts of Germany; and, where the shores are sandy, it will work it's way under the sand to a great distance from the pond, and is there dug up by the people, who know how to distinguish the places by certain signs, and thus take numbers of them. It has been mistaken, by most authors, for a species of the Mustela; and, from this singular circumstance of it's being dug out of the shores, has been called, by Gesner and others, Mustela fossilis. Jonston calls it Piscis fossilis; and Schoneveldt, Fossilis; the Germans call it Misgurn or Fisgurn.

CYPRINUS.

THE mouth of the Cyprinus is totally smooth; but within the fauces there stand, on the lower part, two hard and serrated bones, which serve in the place of teeth; and in the upper part of the fauces, overagainst these, there is placed another bone, which is smooth. The air-bladder is constricted in the middle, as if it were tied with a thread; the branchiostege membrane has three bones on each side.

The two jaws are in most species pretty exactly of the same length, but in some the upper one is a little longer than the lower.

Cyprinus iride, pinnis que ventris et ani plerumque rubentibus. **The Roach.**
The Cyprinus, with the iris and the belly fins usually red.

The general length of this fish, when full grown, is about nine inches, but there are waters in which it becomes considerably larger; it's more frequent size is from two to four inches: the head is small, flatted on the upper part, and of a mixed olive and dusky colour: the back is somewhat ridged in the anterior part, and convex lower down; the belly is narrow and flat, except that the space between the ventral fins and the anus is a little contracted: the opening of the mouth is small; the jaws are both of the same length, but in the larger fish the upper, when the mouth is shut, appears a little longest: the iris of the eye is usually red, but sometimes it is yellowish, or of a silvery hue; but this principally in the young fish: there are five teeth situated within the fauces; the lateral line bends downward toward the belly: the scales are large, glossy, and placed in a regular, imbricated manner; they are of a silvery white on the sides, but on the back they are of a duskier colour.

The branchiae are four on each side, and have each a double series of very minute tubercles: there is only one fin on the back, and this has thirteen rays; the two first of these are undivided, the rest ramose: the pectoral fins are whitish, and have each fifteen rays; the belly fins are red, but in the small fish sometimes yellowish; they have each nine rays: the pinna ani is also naturally red, but sometimes yellowish; it has twelve or thirteen rays: the tail is of a pale colour, and is forked; it has nineteen rays: the vertebrae are forty-four.

It is frequent in all our fresh waters; our rivers, brooks, lakes, and ponds, all abound with it. Willughby and others call it Rutilus, or Rubellus fluviatilis; Schoneveldt, Rubellio and Rubiculus; Rondeletius, Phoxinus; the Germans, Rotang, Rotage, and Rotel; the Swedes, Mort.

Cyprinus iride, pinnis omnibus, caudaque rubris. **The Rarfe.**
The Cyprinus, with the iris of the eye and all the fins and tail red.

This is a large species; it somewhat resembles the roach, but it is narrower, in proportion to it's length: it is, when full grown, ten inches in length, and not more than three in breadth in the largest part: the head is small and convex on the upper side; the opening of the mouth is small; the teeth stand deep in the fauces, at the beginning of the stomach: the opercula of the gills and the lateral line are crooked, and run parallel with the belly: when the mouth is open, the lower jaw appears somewhat longer than the upper; but, when it is shut, they are found to be both of a length: the iris of the eye is of an orange scarlet; the branchiae are four on each side, and the cranium is of a dusky brown colour: the anterior part of the back is ridged, but not sharp, and behind the fin it is also rounded or convex: the anterior part of the belly is flat, but between the ventral fins and the anus it is contracted into an edge; the scales are large, and of a beautiful silvery colour on the sides, but they are of a dusky tinge on the back, and somewhat yellowish toward the belly: the tail and all the fins, except the pectoral ones, are red; the pectoral fins are brown, and have sixteen rays; the ventral fins are as red as blood, and have ten, and sometimes eleven, rays in each: there is only one fin on the back, and that is red on the upper part, but of a mixed, duskish, and hoary colour below: the pinna ani is of a paler red than the others, but extremely bright and beautiful; it has fourteen rays, sometimes fifteen: the tail is forked, and of a bright red, and has, beside the little extreme rays, nineteen long and large ones: the vertebræ are thirty-seven in number; the long ribs fourteen or fifteen.

This is frequent in the rivers and lakes of Sweden and Denmark, and many other parts of Europe. Ray and Willughby call it Bramis affinis; others, Erythrophthalmus, or the Red Eye; the Swedes, Sarfe.

Cyprinus cirris quatuor ossiculo tertio pinnarum dorsi et ani uncinalis armato.
The Cyprinus, with four beards, and with hooks on the third ray of the back fin.

The Carp.

The general length of a well grown fish of this species is about eleven inches; the head is of a dusky blue colour and glossy; the back rises into a kind of ridge in it's anterior part; lower down it is more convex: the belly is all the way flat, and is considerably broad, and the body is rounder or thicker than in most others of this genus: when the mouth is shut, the upper jaw appears a little longer than the under, and both are tinged a little yellowish at the edges; the mouth is moderately large: there are, on the upper jaw, four cirri or beards, two on each side; the lower of these is the larger, and is situated at the end of the opening of the mouth; it grows to three quarters of an inch in length, and is of a yellowish colour; the upper cirrus is shorter and of a blackish hue: the pupil of the eye is round, and looks blue; and the iris is of a mixed silvery and gold colour: the coverings of the gills are striated, and are of a yellowish colour.

The scales of this fish are large, and of an irregularly angular figure, pentangular, or quadrangular, or otherwise; those which are situated in the middle of the sides, and form the linea lateralis of the fish, have each of them a small hole bored through them; the others have nothing of this: the scales all vary in colour, according to the age of the fish and other accidents; they are sometimes of a silvery white alone; sometimes they are yellowish, or greenish, or shaded with olive colour, or with black; they are placed in a regularly imbricated manner, and are very bright and glossy: the lateral line is straight, and is nearer the back than the belly; the back fin is blackish; the ventral fins are of a pale colour, with a tinge of red, as are also the pectoral ones: the pinna ani and the tail are red, or of a mixed colour of reddish and blackish: the pectoral fins have each sixteen rays, the ventral ones each nine; the back fin is long, and has twenty-four rays, and the pinna ani has nine: the tail is forked, and has nineteen long rays. The size that Carp will grow to in some places is surprising; they will live a long time out of water, which is owing to the coverings of the gills shutting very close: the heart is of a roundish, but somewhat angulated, figure; the spleen is flatted and angulated; the ovaries are two, and are very large, as are also the vesiculæ seminales: the liver is divided into three lobes; the gall-bladder is large and blue, and has a duct running to the orifice of the stomach: the kidnies are two; they are laid close to the back-bone, and have two manifest ureters running to the urinary bladder, which is situated close to the extremity of the intestine.

The vertebræ of the Carp are thirty-seven, the ribs thirteen or fourteen.

The Greeks call the Carp Cyprinos and Cyprianos; and the Latin writers, in general, from them, Cyprinus. Some have called it Carpa, Carpo, and Carpera, and some Carpenus; Paulus Jovius calls it Rayna and Bulbarus.

Cyprinus

Cyprinus pinnis omnibus nigrescentibus, pinna ani ossiculorum viginti septem.

The Cyprinus, with all the fins black, and with twenty-seven rays in that of the tail.

The Bream.

This is a considerably large species, and is much broader, in proportion to it's length, than any of the others: a well-grown Bream is twelve or fourteen inches long, and the body is flat and thin, though very broad: the back is raised into a sharp ridge in the anterior part, but lower down it is more rounded; the belly, in it's anterior part, is flat, but, from the ventral fins to the anus, it is contracted into a sharp ridge: the head is short, thick, and obtuse: the mouth is to be opened by pressing the lower jaw, and the upper is then seen coming, as it were, out of a case: when the mouth is shut, the upper jaw appears somewhat the longer; the mouth is roundish and small; the eyes are round, and stand at the two sides of the head: the pupil is round, the iris is usually in general silvery, but it is sometimes tinged in part with yellowish or red: the coverings of the gills are also of a silvery colour, mixed with some yellow, and some reddish: there are certain ducts or foramina, which form lines on each side above the nostrils, as also on the lower jaw under the eyes: the branchiæ are four on each side, and each has a double series of unguiform tubercles: the nostrils stand nearer to the eyes than the extremity of the rostrum; the teeth are five, and are situated in the fauces: the branchiostege membrane and the lateral line are crooked, and nearer the belly than the back

The scales are large, and are placed in a regularly imbricated manner; they are of a silvery white on the belly, with some faint admixture of red, and on the sides they are silvery, but with a tinge of yellowish or greenish; on the back they are of a dusky colour, but they have still somewhat silvery even there; the opening of the mouth is large; in the full grown fish it will easily take in a man's thumb.

The fins are all of a blackish colour, with an admixture of greyish: the pectoral fins are greyish toward their origin, and black at the extremity; these have seventeen rays each: the ventral fins are also whitish just at the base, but black in every other part; they have each nine rays: the back fin is throughout of a greyish-black, and has fourteen rays; and the pinna ani is black throughout, and has twenty-seven rays: the tail is very forked; it is of a greyish-black, and has nineteen long rays; the vertebræ are twenty-four, and the long ribs are fifteen on each side.

It is very common in our fresh waters, particularly in large rivers. Aldrovand and Jonston call it Cyprinus latus; Bellonius and Charleton, Abramus; others, Brasmus, Prasmus, and Bresma; the Germans are fond of it at their tables; we esteem it a very poor and coarse fish.

Cyprinus oblongus macrolepidotus, pinna ani ossiculorum undecim.

The oblong Cyprinus, with large scales, and eleven rays in the pinna ani.

The Chub.

This is somewhat of the general figure of the carp, but narrower and thinner, in proportion to it's length: a well-grown fish of this species is a foot long: the head is large; the opening of the mouth wide; the upper jaw a little longer than the under: the scales are very large and beautiful, and are of an angular figure: the tail is forked; the back fin has nine rays; the pinna ani eleven: all the fins are bluish, and the iris of the eye is of a mixed silvery and yellow colour: the lateral lines run parallel with the belly.

It is common in our rivers, particularly in deep holes worn by the currents, and under the stumps of trees growing on the banks. Varro and Columella call it Squalus; Pliny, Squalius; the more modern writers, Cephalus and Capito, and Cephalus fluviatilis; some, Dobula; the Germans call it Alet.

Cyprinus mucoſus totus purpureo-nigreſcens cauda æquali.
The purpliſh-black, mucous Cyprinus, with an even tail. **The Tench.**

This ſometimes grows to an immoderately large ſize, but a moderately grown fiſh of it is ten or eleven inches in length; it's breadth, with this length, is about three inches, and it's thickneſs more than two: the head is ſmall, and the opening of the mouth, in the full grown ones, will admit the end of a man's finger: there are no teeth in the jaws, but there are five on each ſide deep within the fauces; the coverings of the gills are compoſed of four laminæ and three curvated bones; the lateral line is crooked, and runs near the belly; there are many little foramina, forming a kind of irregular lines, on the ſurface of the head: the eyes are ſmall, and their iris is red; the apertures of the gills are leſs open, than in the generality of theſe fiſh: the gills are four on each ſide, and each has a double ſeries of pectiniform oſſicula: the back riſes into a rounded form, and is not ridged, as in ſome of this genus; the belly is flat and broad: the ſcales are very ſmall, and are of an oblong figure; they adhere very cloſely to the body of the fiſh, and are on the back black, and on the ſides of a purpliſh, or ſometimes a yellowiſh, colour, mixed with black; the belly has a whitiſh tinge, but there is ſo much of the black diffuſed on every part of it, that the general colour is that: the ſcales of this fiſh are externally covered with a thick, mucous matter, ſo that the fiſh is ſlippery in the manner of an eel; the fins are all blackiſh, as is alſo the tail; the pectoral fins are roundiſh, and have ſeventeen rays; the belly fins are alſo rounded at the end, and have eleven rays; there is only one fin on the back, which has twelve rays, and the pinna ani has eleven: the tail is not forked but even, and has nineteen rays.

We have this in ſtanding foul waters, in large ditches and fenny ponds. The antients called it Phalon, Trulla, and Gnaphæus; the moderns, Tinca and Tenca; and ſome of them, Phycis fluviatilis and Merula; the Germans, Schley.

Cyprinus oblongus maxilla ſuperiore longiore, cirris quatuor, pinna ani oſſiculorum ſeptem.
The oblong Cyprinus, with the upper jaw longeſt, and with four beards. **The Barbel.**

This is a large ſpecies, a moderately grown fiſh of it being near a foot in length: it is not ſo broad, in proportion to it's length, as the tench or carp, but rounder-bodied: the head is large, and the back part of it high, whence it ſlants off to the extremity, where it is ſharp; the upper jaw is longer than the under, and there are four white, fleſhy, ſlender beards: the eyes are moderately large; the back is elevated and ridged; the lateral line runs down the middle of the ſide; the belly is flat, and ſomewhat broad, and, when the fiſh is laid on it, the mouth touches the ground: the tail is forked.

It is frequent in our large rivers; we frequently catch it in the Thames of three or four pounds weight, but ſome are ſeven or eight pounds. The antient as well as modern writers have called it Barbus and Barbulus; ſome Barbo, Barbatulus, Mugil barbatus, Mullus fluviatilis, and Muſtus fluviatilis; the Germans call it a Barbe and Barble.

Cyprinus quincuncialis pinna ani oſſiculorum viginti.
The five-inch Cyprinus, with twenty rays in the pinna ani. **The Bream.**

This is one of the ſmaller ſpecies, but it is a very pretty fiſh; it rarely grows to a larger ſize than is expreſſed in it's name: the greater number caught are but three or four inches, and it is moderately broad, in proportion to it's length, but has not much thickneſs: the head is ſmall and ſomewhat acute; the mouth is moderately large, and, when it is open, the under jaw is ſomewhat longer than the upper: the eyes are large, and their iris is of a ſilvery white in the lower part, but in the upper there is ſomewhat of a yellowiſh tinge in it: the branchiæ are four on each, and the teeth are not placed in the jaws, but deep in the fauces: the ſcales are large, and placed in an imbricated manner, but they adhere so closely, that they easily fall off on the sides and belly; they are of the most beautiful silvery white, but on the back they have somewhat of a bluish tinge: the fins are all of a whitish colour; the pectoral ones are perfectly white, and have fourteen or fifteen rays; the belly fins are whitish, and have nine rays; the back fin is whitish, and has nine or ten rays; the pinna ani is also whitish, and has one and twenty rays; the tail is forked and whitish; the back is convex all the way; but the belly, between the ventral fins and the anus, is contracted into a ridge.

It is a very common fish in our rivers, and usually is found in vast numbers together. Ray, Willughby, and the rest call it Alburnus; Jonston, Albula; and Schonevelde, Albula minor; we call it the Bleak, and the Germans, Bleig and Weissfisch.

Cyprinus quincuncialis maculosus maxilla superiore longiore, cirris duobus.

The small, spotted Cyprinus, with the upper jaw longest, and with two beards.

The Gudgeon.

This is also a small species; it rarely grows to more than four or five inches in length, and has less breadth with this length than any other species of this genus, the body being of a rounded form, and somewhat depressed: the head is large and depressed; the upper jaw is longer than the under; there is on each side a single, oblong, white beard at the angle of the mouth: the eye is large, and the iris of a silvery colour, with some admixture of yellow: the back is all the way rounded and thick; the belly is flat; the anus is placed in the middle, between the belly fins and the pinna ani: the lateral line is black, almost straight, and runs along the middle of the sides: the scales are large; they are of a dusky colour on the back, and whitish on the belly and sides; there are seven or eight large black spots running longitudinally along the middle of the back, from head to tail; on the sides also, near the lateral lines, there run about nine large spots, of a blackish colour, but these, in some of the fish, are scarce visible; and, beside these, there are a great number of smaller black spots on the head, back, and sides, and on the fins and tail: the back fin is small, and has ten rays; the pectoral fins have each fifteen rays; the pinna ani is white, and has nine or ten rays; the tail is broad, and is a little forked; it has nineteen rays; the vertebræ are forty.

We have this in most of our shallow rivers, where the water is clear, and the bottom gravelly or sandy. The antients call it Gobio and Gobius; the later writers, Gobius fluviatilis and Fundulus; some, Gobius non capitatus; the Germans call it Gresling and Grundling; the French, Goujon and Vairon.

Cyprinus admodum latus et tenuis pinna ani ossiculorum quadraginta.

The broad and thin Cyprinus, with forty bones in the pinna ani.

The Lake Bleak.

This is a very singular species; it grows to six or seven inches in length, and is of a depressed form, very broad and thin; the head is very small and flatted; the skull is of a yellowish brown colour, and is bright, and in some degree pellucid: the eyes are large, and the iris of a silvery colour, except that in the upper part, over the pupil, it is yellowish: the scales are very small, and all of them of a silvery colour, and all the fins are whitish; the pectoral fins have sixteen rays each; the belly fins have each nine or ten; the back fin has ten or eleven: the pinna ani is large and broad, and has forty, and sometimes forty-one, rays; the tail is very forked, and has nineteen rays.

This species is frequent in the lakes of Germany, and some others parts of Europe, and though as unlike the bleak, as two fish of the same genus can easily be to one another, it has been confounded by many with it. Aristotle and the other Greek writers call it Balerus, and from them Rondeletius, Jonston, and others, Ballerus; some Bullerus, and others Blicca and Plestya; the Germans call it Blick and Honerk; and the French, Bordeliere. We have it not in England, though it is, in a manner, universal in other parts of Europe.

Cyprinus

Cyprinus pinna dorsi ossiculorum viginti, linea laterali recta.
The Cyprinus, with twenty rays in the back fin, and the lateral line straight.

The Charax.

This is one of the short and broad Cyprini; it frequently grows to eight or nine inches in length, sometimes to considerably more: the head is broad and short; the opening of the mouth is small; the end of a man's little finger can hardly be introduced into it in the full-grown fish: the body is depressed, but it is considerably thick, not thin as in the bliccа; the upper and under jaw are of the same length; there are four or more teeth on each side deep on the fauces, but none in the jaws: the eyes are small; their iris is of a dusky yellow, and there are some black spots in them: the back is elevated, and from the head to the back fin it is ridged: the belly in general is flat, but the space between the belly fins and the anus is somewhat edged: the branchiæ are four on each side; the lateral line is nearly straight, and stands in the middle of the sides: the scales are large and beautiful; they are placed in an imbricated order, and are of a fine gold yellow on the sides, but of a somewhat duskier tinge on the back; the fins are all red.

It is very frequent in the large lakes of Germany, and some other parts of Europe, but we have it not in England. The Greeks call it Charax, and some of the Latins have borrowed the same name; some have called it Carassius, and others Carasus and Karass; Willughby and Gesner call it Carassius simpliciter dictus, sive Carassii tertium genus; the Germans call it Karayfche; and the Swedes and Danes, Ruda.

Cyprinus rostro nasiformi prominente, pinna ani ossiculorum quatuordecim.
The Cyprinus, with a nasiform snout, and fourteen rays in the pinna ani.

The Nasus.

This species in figure very much resembles the common dace or dare, and grows usually to about the same size, though sometimes considerably larger: the head is moderately large, and of a singular figure, broad at the top, but running out into a kind of nose at the end: the body is rounded, and the back convex; the belly is flat and broad: the scales are large and beautiful; the lateral line runs near the belly; the fins of the under part of the body have all a tinge of reddish.

It is frequent in the lakes of Germany, and many other parts of Europe. Gesner, Aldrovand, and others call it simply Nasus; others, Nasus fluviatilis, and Nasus Alberti; the Italians call it Savetta; and the Germans, Nase and Nasen.

Cyprinus cubitalis pinna ani ossiculorum quatuordecim.
The large Cyprinus, with fourteen rays in the pinna ani.

The blue Chub.

This is a very handsome and bright-looking fish; the head is large, smooth, glossy, and of a deep olive colour, with a tinge of purplish: the mouth is small, and the eyes are large and bright; the body is of a rounded form; the scales are very large, and stand in an imbricated manner; the colour is a silvery white, with a considerable tinge of a purplish blue, more conspicuous on the back than in any other part: the fins are of a pale colour, and the tail is forked.

It is very frequent in the rivers of most parts of Europe. Gesner and Willughby call it Capito cæruleus; others, Cephalus fluviatilis cæruleus; others, Lesus and Jesus; The Germans, Scheert and Koppen.

Cyprinus magnus crassus argenteus.
The great, silvery, thick Cyprinus.

The Corvus piscis.

This is one of the largest of the Cyprini; it grows to two feet in length, and is a remarkably strong and lively fish: the head is broad and short, and the nose obtuse; the eyes are large; the back is elevated and convex, and the body is rounded, not flat; it's diameter is about a fifth part of it's length: the scales are large; they are placed in an an imbricated manner, and are of a fine silvery brightness, but with an admixture of yellowish: the fins are of a pale colour, and the tail is forked.

It is frequent in the rivers of many parts of Europe; it loves deep water. Gesner and Aldrovand call it the Capito rapax, and Capito fluviatilis rapax; Schonevelde calls it simply Rapax; and many, Corvus Piscis.

Cyprinus novem digitorum corpore, longiore pinna ani radiorum decem.

The small, long-bodied Cyprinus, with ten rays in the pinna ani.

The Dace.

This is the slenderest and most rounded in the body of all the Cyprini, and is a very clean and pretty fish: it's usual length is about six or seven inches, though it will grow much larger: the head is moderately large, and of a dusky olive colour; the eyes are large, and the iris silvery; the mouth is small: there are no teeth in the jaws, but a few deep in the fauces: the scales are large, and elegantly arranged in an imbricated manner; the fins are of a pale colour.

It is frequent in our swift and clear rivers and brooks. Ray, Willughby, &c. call it Leuciscus; Rondeletius and Gesner, Leucisci secunda species; others, Albicilla and Albicula.

Cyprinus rostro acuto, oculis magnis.

The Cyprinus, with large eyes, and an acute rostrum.

The Albus Piscis.

This is also one of the rounder-bodied species; the head is large and broad at the base, but the extremity terminates in a sharp snout: the eyes are remarkably large, and the opening of the mouth is small; the back is convex; the fins are all of them of a dusky blackish colour; the scales are large and perfectly white: it is frequently caught of a foot or fourteen inches in length, and sometimes, though rarely, grows to more than two feet: it's flesh is very white, but is the worst tasted of that of all the Cyprini.

It is caught in some of the lakes of Italy, and in standing waters in many other parts of Europe. Salvian calls it Alburnus and Albus Piscis.

These are the more remarkable species of the large genus of the Cyprinus: the others are, 1. The oblong Cyprinus, resembling the roach in figure, with ten rays in the pinna ani; this is called the Gresling and Grislagine. 2. The large, small-mouthed, and small-eyed Cyprinus, called the Vrow-fish. 3. The yellowish Cyprinus, with red fins, called the Rud and the Finscale by us, and by authors, Orfus fluviatilis: this is common in our rivers, and resembles the roach, but that it is broader; some of the Latin writers have called it Rutilus and Rubellio; and others, Capito ruber and Capito subruber. 4. The anadromous Cyprinus, called by the Italians the Zerte and the Blicke; this, at certain times of the year, goes into the sea, but it breeds in rivers. 5. The Cyprinus, called by authors the Sargus, and by some the the Leuciscus prior Rondeletii: this is common in the markets at Rome, and is there called Lascha; the French call it Gardon. 6. The Cyprinus, called the lesser Squalus by authors, and by the Germans the Hasle, or Hasseler. 7. The little, brook Gudgeon, called by the Germans the Wapper. 8. The common, little Cyprinus, known by the name of the Minim, or Minow. 9. The somewhat larger, scaly Minow. 10. The large Cyprinus, called by Salvian Picus, or Pigho. 11. The small Cyprinus, with five and twenty rays in the pinna ani, called, by the Swedes, Biork and Biorkfisch. 12. The two-inch Cyprinus, with red eyes, and with nine rays in the pinna ani. 13. The yellow-eyed Cyprinus, with the tail and the belly fine red, called by the Swedes Id. 14. The Cyprinus, with the lower jaw longest, and elevated at the summit, and with fifteen rays in the back fin; this is called by the Swedes Asp. 15. The acute-nosed Cyprinus, with a ridged back, and with twenty-four rays in the pinna ani, called by the Swedes the Asp. 16. The three-inch Cyprinus, with a broad, orbicular body and very large scales, called by Gesner and Bellonius the Bubalca.

CLUPEA.

THE belly of the Clupea is remarkably acute, or is formed of ſuch a peculiar arrangement of the ſcales, that it appears ſerrated: the back fin is ſomewhat nearer the head than the ventral or belly fins are, and the branchioſtege membrane contains eight bones.

This genus comprehends the herring, ſprat, anchovy, &c.

Clupea maxilla inferiore longiore maculis nigris carens.
The Clupea, with the lower jaw longeſt, and without any black ſpots.

The Herring.

This is the common herring, eaten under various forms at our tables: it may ſeem needleſs to ſome to enter into the particulars of the deſcription of a fiſh ſo well known; but the moſt vulgarly known things are often the leaſt underſtood, and I am apt to believe there are very few who diſtinguiſh this fiſh properly from the others of the ſame genus. It is very often confounded with the ſhad and pilchard; and moſt people ſuppoſe ſo diſtinct a ſpecies as the ſprat to be only a herring not yet grown to it's ſize.

The length of the herring is from five to eight inches, it's breadth little more than an inch; it's head is compreſſed, and ſomewhat acute in the anterior part, and the upper part of it is hollowed: the mouth is very large; in the opening of it the roſtrum is a little elevated, and the maxillary bone of the upper jaw, which covers the lower on each ſide, is protruded a great way forward: the noſtrils are very conſpicuous; they have a double aperture, but the anterior one is very ſmall; they ſtand nearer to the extremity of the roſtrum than to the eyes: the eyes are large, and the iris is ſilvery.

In the extremity of the lower jaw there are ſome very ſmall teeth, and in the upper one others ſo very minute, that they may be eaſily overlooked, unleſs on a careful inſpection; the lateral bone of the upper jaw, which covers the other on each ſide, is ſlightly ſerrated at the edge: in the middle of the anterior part of the palate there is a large area, beſet with teeth in two ſeries, running longitudinally; and, finally, the tongue, which is acute and looſe on the lower part, and is of a blackiſh colour, is armed with a great number of teeth directed inwards: there is uſually an elegant red, or violet-coloured ſpot, on each ſide, at the extremity of the coverings of the gills; theſe coverings are, excepting for this ſpot, throughout of a ſilvery colour, and are compoſed of three or four bony laminæ, and eight oſſicles, ſomewhat arcuated in figure, and connected by a membrane.

The lateral line runs ſtraight, and near the back, but it is not very conſpicuous: the ſcales are very large, and of a ſilvery white; they are placed in an imbricated manner, but they are very looſely fixed, and eaſily fall off: the back is of a duſky bluiſh colour, and is more blue in ſpring than at other times; the ſides and the belly are of a ſilvery white: the whole belly, from the gills to the anus, is rough, and is contracted into an edge; the back is rounded or convex; the fin on the back ſtands near the middle, and is of a whitiſh colour; it has nineteen rays; the pectoral fins are whitiſh, and ſtand low; they have each eighteen rays; the ventral fins are very ſmall and white; the pinna ani is near the tail, and has eighteen rays; the tail is of a greyiſh colour, forked, and furniſhed with eighteen rays.

The heart is of a ſquare figure; the liver is ſmall and angular; the ſtomach is, as it were, double; the air-bladder is long, ſlender, and protended all along the abdomen, and is ſimple: the ribs are thirty-five on a ſide.

It is extremely common in our ſeas, whence great part of Europe has been long ſupplied. It was known from the earlieſt times; Ariſtotle and the other Greeks call it χαλκὶς, Chalcis; Pliny borrows the ſame name; Bellonius calls it Chalcidis ſpecies, Gera, Erica, and Orica; others, Alec and Halec; Paulus Jovius, Aringa; and many, Harengus; the Swedes call it Sill, and the Germans, like us, Herring.

It varies in ſize, being ſmaller in ſome ſeas than in others: the ſmaller kind has been diſtinguiſhed by the Greeks under the name of Ara; and, by Bellonius, under that of Celerinus; others call it Membras; and ſome Strommingus, or Stromlingus, from the Swediſh, Stromming.

Clupea minor maxilla inferiore longiore, ventre acutissimo.
The little Clupea, with the lower jaw longest, and the belly very acute.

The Sprat.

The sprat has been very generally, but erroneously, supposed a herring, not grown to it's full size: the length is about four or five inches, the breadth somewhat more, in proportion, than in the herring, and the thickness of the back proportionably also less: the head is compressed; the eyes are large, and the iris a silvery white, but sometimes tinged with red; there is a spot on each side, near the extremity of the coverings of the gills, and no other on any other part of the body: the back fin has seventeen rays; the pectoral fins have also each seventeen rays; the ventral ones have six or seven each, and the pinna ani has nineteen.

The belly, from the gills to the pinna ani, is of an extremely acute figure, much more so than in the herring; and the points of the scales of the same fish, in the sprat and herring, are very different in figure, those of the sprat being very acute, whereas those of the herring are obtuse: the spurious ribs in the sprat are nine on each side, the vertebræ are forty-eight; we find that number in the smallest, and no more in the largest.

This is very frequent also in our seas, and is brought in surprising quantities to London.

Clupea apice maxillæ superioris bifido, maculis nigris utrinque.
The Clupea, with the upper jaw bifid at the extremity, and spotted with black.

The Shad, or mother of the Herrings.

This very considerably resembles the common herring in many particulars, but it is sufficiently distinct from it in others: the head is compressed, and somewhat hollowed at the top; the eyes are large, and their iris of a silvery white; the opening of the mouth is large, and is furnished with a great number of small but sharp teeth: the lower jaw is a little longer than the upper, and the upper jaw is bifid or forked at the extremity, but the division is not deep: on each side, below the coverings of the gills, and near the back, there is a large roundish black spot; and from this there run in a straight line toward the tail, and a little above the lateral line, four, five, or six other smaller spots: the lateral line is straight, but it is not very conspicuous; the back is of a mixed, bluish, greenish, and silvery colour; the upper part of the head has a cast of yellowish, and the rest of the body is of a fine silvery white: the scales are very large; they are roundish, soft, and loosely fastened, so that a little touch takes them off: the scales of the belly, from the gills to the anus, are robust, and, as it were, prickly; they are from thirty-five to thirty-seven in number, and are of a very singular figure; they are continued from the sides into a long point, and on the lower part, both on the fore and hinder sides, they are terminated by a short spine or prickle; the posterior spine is the shorter of the two, and forms the serrated and prickly carina of the belly: all the fins, except that of the back, are whitish; the pectoral ones have fourteen or fifteen bones or rays; the ventral ones nine; the dorsal one eighteen or nineteen; the pinna ani has twenty-two or twenty-three: the tail is very forked, and has nineteen rays; and, at the base of each of the pectoral and ventral fins, there is a singular, oblong, large scale.

The body is broader and thinner, in proportion to it's length, than in the common herring; the back is rounded, and the length of the fish is more than equal to four times it's breadth.

We have this in our seas; and in May, and the beginning of June, it makes it's way up rivers in vast abundance. It was well known to the antients: The Greeks called it Θρίσσα, Thrissa; and the Latins, Trissa and Alausa; Bellonius calls it Trichis and Pulchella; Salvian and others, Alosa; Charleton, Alosa major; and Bellonius, Alosa minor; others call it Alosa fluviatilis and Clupea; Albertus, Aristotus, and Paulus Jovius, Laccia. When of different sizes and growth, it has been described by authors also under two or three other names; Salvian calls it Agone and Acone; Charleton, Sarachus and Sarachinus; and the Harengus minor and the Sardina of authors are only

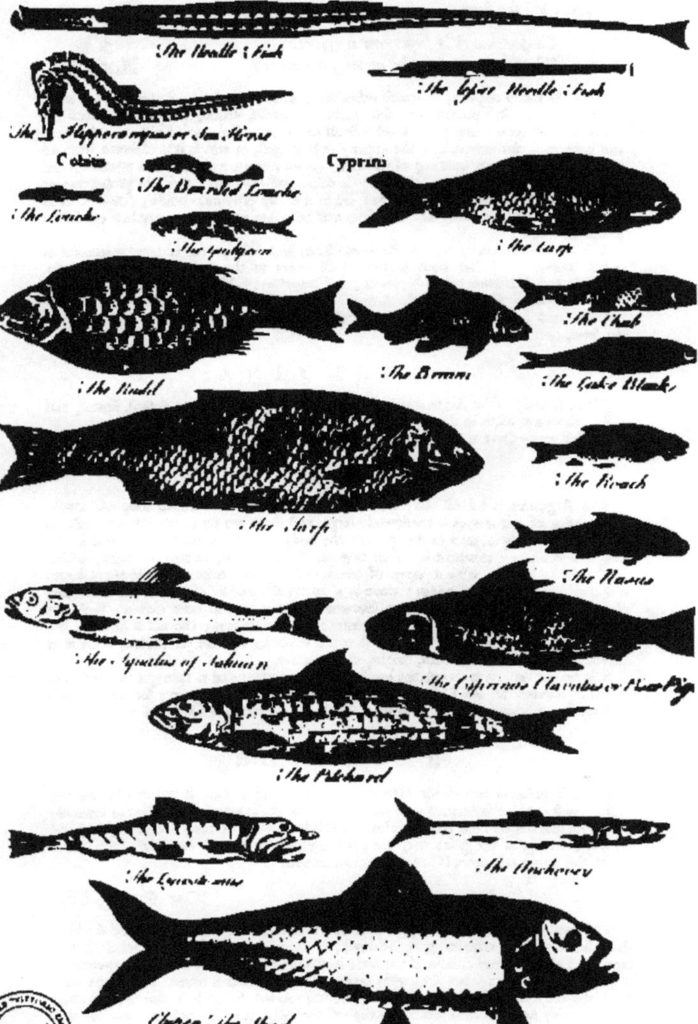
Syngnathi
Cobitis
Cyprini
The Roach
The Pilchard
The Anchovy

only names for it, under the different circumstances of growth and other particulars, the *species* still the same.

Clupea maxilla superiore longiore.
The Clupea, with the upper jaw longest.

The Anchovy.

This, in many respects, so much resembles the common sprat, that it is not a wonder that fish is often pickled and sold under it's name, among people not much acquainted with such things: the head is small and compressed, broad at the hinder part, and pointed at the extremity; the upper jaw is longest, in which it is different from all the other Clupeæ: the opening of the mouth is very large, and the teeth numerous and minute: the back is convex; the body is thin, and tolerably broad, in proportion to it's length; the tail is forked; the fins are of a dusky brownish-white; the back is of a mixed bluish and brown, and the sides and belly are white; the vertebræ are forty-five or forty-six.

It is frequent in many of the European Seas, and is imported in great quantities in pickle among us. It has been known at all times to the writers on these subjects. Aristotle and Athenæus call it [illegible], and [illegible]; Ælian, [illegible], and Ælian, [illegible]. From these the Latins have called it Engraulus, Engraulis, Encrasicholus, and Lycostomus; some have called it Halecula, a name given by others to the sprat.

ARGENTINA.

THE body of the Argentina is oblong, and of a rounded, not flatted, figure, and there are teeth in the palate, and on the tongue.

Of this genus there is but one known species.

ARGENTINA.

The Argentina is a small fish; it's head is oblong, and the rostrum long and small; the opening of the mouth is moderately large, and there are six or eight sharp teeth on the tip of the tongue, and in the palate: the body is of a rounded figure, and is oblong and slender; the whole is of an elegant, silvery colour, except the back, which is of a dusky brown, with a tinge of bluish and of olive colour: the eyes are large, and their iris of a silvery colour: there is a fin on the middle of the back, consisting of ten rays: the pectoral fins have fourteen rays, the ventral ones eleven; these are placed over-against the back fin: the pinna ani has nine rays; the tail is forked; the anus is placed very near the tail; the stomach is black, and the air-bladder is conic at each end, and is of a beautiful, bright, silvery white.

It is caught in great abundance in the Mediterranean, and is brought to market at Rome. Rondeletius and Gesner call it Sphyræna parva, and Sphyrænæ secunda species; Willughby, Pisciculus Romæ Argentina dictus; the French call it Hautin.

EXOCÆTUS.

THE pectoral fins of the Exocætus are very long; there is but one fin on the back, which is situated very backward: the scales are very large; the branchiostege membrane on each side contains ten ossicles; three of these are broad, and hid by the coverings of the gills; they are not easily numbered with exactness.

Of this genus also there is but one known species.

EXOCÆTUS.

The Flying-fish.

The head is very flat on the upper part, but is somewhat compressed toward the hinder extremity; the anterior part, however, is more depressed, and the whole is not large, not equalling the middle of the body in diameter: the body is oblong, approaching to a rounded figure, but somewhat compressed: the back is broad, and all the way flat; the belly also is flat in the anterior part, but toward the tail it is convex: the anus is placed very near the tail; the extremity of the rostrum is somewhat acute, and the

opening

opening of the mouth is not large: the lower jaw, when the mouth is open, appears somewhat longer than the upper; the nostrils are large, and stand nearer to the eyes than to the extremity of the rostrum; there is a kind of appendage in the middle, which makes each seem as if formed of two apertures: the eyes are very large and round; their iris is of a silvery colour, and the pupil is round: the apertures of the gills are not large; there is in each jaw a single row of very minute teeth; those on the under jaw in particular are so very small, that they are scarce visible: the scales are large, hard, smooth, and whitish; they are, in general, of a semicircular, some of a somewhat angulated, figure, and toward the base or anterior part they have three, four, or more, large striæ; they stand thick together, and, as it were, in straight rows: the colour on the back is a dusky brown; the belly and sides are of a fine silvery white; there is no lateral line in the accustomed place on the sides, but in the lower part of the body, at the sides of the belly, there run two genuine lines formed of scales, perforated in their center, and, in all respects but the place, answering to the characters of the lateral lines in other fishes.

The pectoral fins, in this fish, are extreamly singular; they are situated just below the extremity of the covering of the gills on the sides, but elevated toward the back; they are not affixed horizontally, but in an inclined plane at the base, and are so long, that they equal the whole body of the fish, reaching to the beginning of the tail; they are not very broad, and each of them has seventeen nodose rays, ramose at the ends: the membrane which connects these, and forms the fin, is smooth on the upper side, but underneath there are hollows between the several rays: the ventral fins are situated on the lower part of the belly, not far from the anus; they are of an oblong figure, white, and stand at a great distance from one another; each of these has six rays, all of them ramose at the extremity: the pinna ani is white and small; it has eleven short and soft rays: the dorsal fin is not large, and is situated toward the hinder part of the body; it is whitish in colour, and has eleven short and soft rays: the tail is very forked, and has fifteen long rays: the branchiæ are four on each side, and are formed as those of the pearch, each having a double row of tuberculous and somewhat rough apophyses.

It is caught in the Mediterranean, and some other seas. The antient Greeks called it Ἐξώκοιτος and Ἄδωνις, Exocœtus and Adonis; and the Latins borrowed both these names from them: some have called it Exochinus, and other Mugil alatus and Hirundo piscis; some have thought the Hirundo piscis and the Exocœtus different, but without reason; the Italians call it *Pesce Rondine*; and Salvian take great pains to prove, that it was the Χελιδών, Chelidon, of the antient Greeks.

The use it makes of it's pectoral fins is very singular; they serve it for flying, but this only in a limited manner; the fish has a power of throwing itself into the air from the surface of the water, and, when it is there, it suspends itself, and moves forward very nimbly by the motion of these fins, which serve it as wings; but when they become dry, as they soon do in the air, they are unfit for any farther service of this kind, and the creature drops into the water again. It finds a means to escape the pursuit of some larger fish, by means of these wings; but this often exposes it to new danger, for, when in the air, it becomes the prey of the sea-birds.

CORE GONUS.

THE dorsal fin of the Coregonus is placed nearer the head than the ventral ones are: the teeth are very small; in many species they are so remarkably so, that they are hardly visible: the branchiostege membrane on each side has seven, eight, nine, or ten bones in it.

Coregonus edentulus maxilla inferiore longiore.
The Coregonus, with no visible teeth, and with the lower jaw longest.

The head is oblong, somewhat acute and compressed, and in the upper part is somewhat pellucid; the eyes are large, and their iris is of a silvery colour: the lateral line is straight, and is nearer the back than the belly; the scales are moderately large, and of a silvery colour; they are oblong, smooth, and but slightly affixed to the body: the back is all the way convex; the belly is flat from the place of the pectoral to that of the ventral

tral fins, but from the ventral fins it is convex; both the jaws are wholly without teeth, nor are there any on the palate, but the tongue has a number of very minute ones on it: these are to be felt at all times, but they are scarce distinguishable to the sight in the recent fish, though, when dried, they appear sufficiently plain: the lower jaw is subacute, and is protended a little beyond the other; the mouth is considerably large; the top of a finger may be introduced into that of a full-grown fish, without difficulty. The branchiostege membrane has seven, or sometimes only six, bones, and they are of a somewhat broad figure: the scales are of a roundish figure, but somewhat oblong, and are placed in an imbricated manner; they are but loosely fixed, and fall off on the slightest touch: the back of the fish is somewhat bluish; the whole body beside is of a fine silvery white: the back fin is of a greyish colour, and has thirteen or fourteen rays; the pectoral fins are whitish, but somewhat black at the extremities; they have each sixteen rays: the belly fins are white, and have each twelve rays; the pinna ani is whitish, and has fifteen; the tail is forked, and has nineteen long rays.

This species is frequent in the lakes and other large waters in Germany, Sweden, and many other parts of Europe, but we have it not in England. Ray, Willughby, Gesner, and the other writers on fishes call it Albula minima; the Germans call it Stint; and the Swedes, Sikloja, Blicta, and Mockin.

Coregonus maxilla superiore longiore, plana, pinna dorsi ossiculorum quatuordecim.

The Coregonus, with the upper jaw longest and plane, and with fourteen rays in the dorsal fin.

This is a fish that varies very considerably in it's size, and some other particulars, under the different circumstances of growth and place, and has been thence described four or five times over, by most of the Ichthyologists, under as many distinct names. It's general length, when moderately grown, is about ten inches; the body is oblong and narrow, in proportion to this length; the head is subacute and compressed, and is somewhat pellucid in the upper part; the rostrum is somewhat acute, and is a little prominent beyond the lower jaw: the back, from the head to the fin, is somewhat plane, but from that part to the tail it is much flatter, and somewhat broad: the belly is flat and broad all the way, from the gills to the tail; the lower jaw is smaller than the upper, and, when the mouth is shut, is covered by it: the opening of the mouth is small; the rim of the lower jaw, and the whole palate, are smooth, or without teeth, but there are some teeth on the verge of the upper jaw; there is but a single series of these, and they are all slender and but soft: the tongue also is furnished with a vast number of very minute teeth, and the fauces are also armed with small ones, both above and below, to the extremity of the smaller gill: the nostrils stand in the mid way between the eyes and the extremity of the rostrum; they have each two apertures, the anterior smaller and round, the posterior larger and oblong: the eyes are large; the pupil is bluish, and of an ovato-acute figure; the iris is silvery: the coverings of the gills are of a silvery colour, with an admixture of yellow; they consist of four laminæ, and of eight or nine broad ossicles, connected by a membrane; the lateral line is straight, and runs nearer the back than the belly of the fish: the scales are moderately large; they have somewhat of a bluish-grey tinge on the back; in all the rest of the body they are of a silvery white; there are none on the head or coverings of the gills: there are two fins on the back; the first is of a greyish-black, and has fourteen rays; the hinder one is only a membrane, and has no rays to support it: the pectoral fins are of a greyish-white, but black at the tips; they have each sixteen rays: the belly fins are of a whitish colour, but bluish at the extremities, and have each twelve rays: the pinna ani is whitish at the base, and black in the other part, and has seventeen rays: the tail is forked, and has twenty long rays.

This species is frequent in the Baltic, and other of the Northern Seas, and about the mouth of large rivers; it's flesh is very white and well-tasted. Authors have called it by the several names of Lavaretus, Albula nobilis, Bezola, and Albula cærulea, Albula parva, and Albula lacustris and farra; under these names they have, in general, described it five times over, as of so many distinct species.

The Grayling, or Umber.

Coregonus maxilla ſuperiore longiore, pinna dorſi oſſiculorum viginti trium.
The Coregonus, with the upper jaw the longeſt, and twenty-three rays in the back fin.

This is a large and handſome fiſh; it's general length, when moderately grown, is ſomewhat more than a foot: it's body is but narrow, in proportion to this length, but it is moderately thick: the head is depreſſed, and is plane at the top, and acute at the end: the back is convex, and in the anterior part is ſomewhat acute; and the belly, from the gills to the pectoral fins, is plane; the teeth are ſhort, but they are ſharp; thoſe in the upper jaw are longer than any of the other; they are of a ſomewhat conic figure, and there is only one ſeries of them; in the lower jaw alſo there are only one ſeries, and theſe very ſmall; in the anterior part of the palate, at the ſides, there are a number of very minute teeth placed in a right line: in the very apex of the palate there is a little area, of a roundiſh figure, which is covered with minute teeth; and, finally, there are two other ſmall areolæ in the fauces, covered in the ſame manner with teeth: the tongue is ſmooth; the upper jaw is ſomewhat longer than the under, and the opening of the mouth is large: the noſtrils have each two foramina, and there are ſome ducts viſible on the head above the eyes: the iris of the eye is of various colours, a ſilvery white, a blackiſh and greeniſh; the pupil is of a bluiſh-black, and of an oval figure, but pointed on the anterior part: the branchioſtege membrane on each ſide has ten flat and ſomewhat broad oſſicles: the lateral line is ſtraight, and is ſituated ſomewhat nearer to the belly than to the back: the ſcales are large and beautiful; they are of a ſemicircular figure, and adhere very firmly; they are hard and ſtrong: on the belly they are of a whitiſh colour on the ſides; they are of a ſilvery white, with a tinge of green, and on the back they are of a duſky grey; they are not placed in the uſual, ſimple, imbricated manner, but they run in ſtraight lines, and form about twenty ſuch; the very tail has ſmall ſcales on it, of the ſame figure with the others, but ſofter and not adhering ſo firmly: the back fin is large; it ſtands nearer to the head than to the tail, and has twenty-three *rays*; the membrane which connects theſe is blackiſh, but it is ſpotted and ſtreaked with red, and toward the hinder part is variegated often with greeniſh; there is, beſide this, on the extremity of the back, a kind of pinniform, membranaceous appendage, but without rays: the pectoral fins are brown, and have each ſixteen rays; the belly fins are reddiſh, and have each twelve rays; the pinna ani is alſo reddiſh, and has fourteen rays; the tail is forked, and ſomewhat reddiſh, and has nineteen long rays: the vertebræ in this fiſh are forty-nine, and the ribs on each ſide thirty-four.

We have it in our large rivers, particularly in the North of England, in great abundance; and it is frequent alſo in the rivers and lakes of moſt other parts of Europe. The Greeks call it [illegible] and [illegible], and from them the Latins, Thymus and Thymallus; ſome Tithymallus, and ſome Tunallus; others call it alſo Umbra and Umbra fluviatilis, and often deſcribe it as another ſpecies under this name; the Italians call it Temelo, and the Germans Aſch; the Swedes call it Harr.

The Hautin, or Outin.

Coregonus maxilla ſuperiore longiore conica.
The Coregonus, with the upper jaw longeſt and conic.

This is a ſmall ſpecies; it rarely exceeds eight or nine inches in length, and is but narrow, in proportion to that length: the head is oblong, and the roſtrum of a conic figure: the eyes are large, and the noſtrils have each two foramina, the anterior the ſmaller: the ſcales are moderately large, and of a figure approaching to round; they ſtand in an imbricated manner: the back fin is of a duſky colour, and has thirteen rays; the pectoral fins are of a greyiſh colour, and have each ſeventeen rays; the ventral fins are alſo greyiſh, with a tinge of black, and have twelve rays; the pinna ani has fourteen rays; the tail is forked, and has nineteen: there are no teeth either in the jaws or fauces, but the tongue is rough.

This ſpecies is frequent in the rivers and lakes in Germany, and moſt other parts of Europe; the rivers in Flanders produce it in great abundance. Ray, Willughby, and Gesner call it Oxyrynchus, and some have called it Sphyræna fluviatilis; the people of Flanders call it Hautin and Outin, and we have borrowed the same names.

Coregonus depressus capite brevi et obtuso.
The depressed Coregonus, with a short, obtuse, head.

The Amboina Coregonus.

This species is of a very singular figure, extremely different from all the other Coregoni, though the generical characters refer it to them: the head is short, and the rostrum obtuse and rounded; the body is depressed, broad, and thin; the back between the head and the fin is somewhat ridged, and the belly is also ridged: the opening of the mouth is moderately large; the lower jaw is longer than the upper, when the mouth is open, but, when it is shut, they are nearly equal; the nostrils are large, and have each two foramina; they are situated about the middle way, between the eyes and the extremity of the rostrum: the eyes are roundish and moderately large, and their iris reddish: the branchiostege membrane contains five or six slender bones; the apertures of the gills are small; there is a single row of short but robust teeth in the lower jaw; there are two series of the same kind of teeth in the upper jaw, and the tongue, palate, and fauces are smooth: the scales are large, white, of a semicircular form, and placed in an imbricated manner: the whole fish is of an elegant and bright silvery white, except the back, which is somewhat dusky.

There is only one fin on the back; this stands near the middle, and has twelve rays, but they are soft and weak: the pectoral fins have each thirteen weak rays; they stand near the belly; the ventral fins are narrow and oblong, and have each nine or ten rays: the tail is very large, and deeply forked; the gills are four on each side, and are of a very elegant structure.

We had the first accounts of this fish from Amboina, but it has since been met with in many parts of the East Indies, and, as some have affirmed, in South America; but this seems uncertain.

OSMERUS.

THE back fin and the belly fins are placed at the same distance from the head in the Osmerus, though not in the Coregonus, or Salmo: the teeth are large and strong, and are placed in both jaws, and also on the tongue and palate: the branchiostege membrane on each side has seven or eight rays.

Osmerus radiis pinnæ ani septemdecim.
The Osmerus, with seventeen rays in the pinna ani.

The Smelt.

This is a beautiful little fish; it's length is five or six inches, and it's breadth not great in proportion, but the thickness is considerable: the head is of an oblong figure, and somewhat acute; the opening of the mouth is large: the back is convex all the way; the belly is somewhat flat; the lower jaw is a little longer than the upper: the nostrils stand in the middle, between the eyes and the extremity of the rostrum; they have each two apertures: the eyes are large and round; the pupil is round and black; the iris is of a silvery white, but tinged a little with bluish toward the upper part: the upper part of the head is of a greyish colour, with a tinge of green, and is spotted all over with minute dots of black; the scull is so pellucid toward the upper part, that the lobes of the brain are easily seen through it: the coverings of the gills are of an elegant, shining, silvery colour, and seem composed only of two laminæ, but, when strictly examined, there are found many more: the lateral line is straight, and it runs very near the back: the colour on the back is various; on the top it is of a greyish or hoary appearance; a little below this it is of a beautiful green, and under this colour there runs a line of blue; from this there appears a bright silvery hue, which is continued all over the rest of the body, except that on the lower part of the sides there is somewhat of a faint violet tinge: the scales are moderately large, soft, of an oval figure, placed in an imbricated manner, and they easily fall off: all the fins, excepting only that forming the tail, are white: there are two on the back; the first or principal one has eleven rays; the hinder one is rather a membranaceous appendage than a fin, having no ray at all: the pectoral fins have each of them eleven, and

sometimes

sometimes twelve, rays; the ventral fins have eight rays each; the pinna ani has seventeen rays; and the tail, which is of a dusky greyish colour, and very forked, has nineteen long rays: the whole body, especially toward the back, is very bright, and in some degree pellucid; and the fish, when fresh taken, has an extremely agreeable smell: the vertebræ are fifty-nine; the ribs are thirty-five on a side; the air-bladder is simple, and largest in the middle.

We have this valuable fish in the Thames, and in other of our rivers, which communicate with the sea. Ray, Willughby, and the rest of the Ichthyologists call it Eperlanus; some, as Jonston and Schonveldt, Spirinchus and Stinchus; it varies in size in different places, and has been hence, by some, described under two names, as if two distinct species.

The Tarantola fish.

Osmerus radiis pinna ani undecim.
The Osmerus, with eleven rays in the pinna ani.

This is of the size of the smelt, or larger: the body is rounded, and the belly flat; it's thickness is about that of a man's thumb, when it's length is seven inches: the belly is white; the head is flatted on the upper part, and there is a little furrow between the eyes: the opening of the mouth is very large, and the rostrum is acute: there is in each jaw a single series of long teeth on each side of the palate; there is also a single series disposed in a right line; there are two series on the tongue, and the apertures of the branchiæ are very large: the back fin has twelve rays, and there is, toward the extremity of the back, a membranous appendage of the appearance of a fin, but without any rays: the pectoral fins have each thirteen rays; the ventral fins are longer by much than the pectoral ones; they stand a little below those, and have eight rays: the pinna ani has ten rays: the tail is very forked: the scales are large, and are placed in an imbricated manner; they are of a roundish figure, and adhere tolerably firmly.

This species is frequent in the Mediterranean; it is brought to market in Rome and elsewhere under the name of the Tarantola. It was well known to the antients; Aristotle and Ælian have mentioned it under the name of Σαῦρος; Athenæus calls it Sauris Speusippi; Salvian and others call it Saurus; others, as Rondelet and Gesner, Lacertus peregrinus, and Lacertus maris Rubri.

SALMO.

THE Salmo has large, sharp, and strong teeth in both jaws, and on the palate, tongue, and fauces: the back fin is placed nearer the head of the fish than the ventral ones; the body is, in most of the species, variegated with spots, and the branchiostege membrane contains ten, eleven, or twelve bones.

The Salmon.

Salmo rostro ultra maxillam inferiorem sæpe prominente.
The Salmo, with the rostrum extending beyond the lower jaw.

The Salmon is an inhabitant both of the sea and rivers: the head is small, in proportion to the body, and, when the mouth is shut, is of a conic figure; the opening of the mouth, however, is large: the rostrum is evidently prominent, when the mouth is shut, beyond the lower jaw: the nostrils have each a double aperture, and are situated a little nearer to the eyes than to the rostrum; the eyes are round, and their Iris of a silvery colour, with a faint admixture of green: the pupil is black, and is rounded on the hinder part, and protruded on the anterior part into a subacute angle: the coverings of the gills are of a silvery colour, and are composed of two, or rather of four, bony laminæ, and of twelve broad and somewhat crooked bones, connected by a membrane, and there are some irregular black spots on them: the lateral line is very straight, and runs somewhat nearer the back than the belly; and there are some black spots about the line, both above and below it, but they do not stand close: the scales are moderately large, and are placed in an imbricated manner; they are largest on the back, and are there black; on the sides and belly they are smaller, and of a silvery colour: the back is convex, and the belly somewhat broad and flatted.

There

There is a single series of teeth in the upper and under jaw, placed at the verge, and very sharp, and among these there are some smaller than the rest, and moveable; and there are more teeth in the upper, than in the lower, jaw: at the sides of the palate there are two series of teeth in longitudinal lines, and in the anterior part of the interstice, between these, there are a few smaller teeth: the palate is entirely smooth, but deep in the fauces; there are some very sharp teeth turning inwards: the tongue is thick, and has on it a few sharp teeth, disposed, as it were, in two series, and pointing inward.

There are two fins on the back: the one has fifteen rays; the other, which is the hinder, has no rays at all, but is a thick, blackish, fatty, and merely membranaceous appendage, in form of a fin: the pectoral fins are blackish at the extremities, and have each fourteen rays; the belly fins are whitish, with a little black on the outside; they have each nine rays, and sometimes ten, and there always stands a large fleshy and squamose apophysis at the top of each of these: the pinna ani is white and bluish, and has twelve or thirteen fins; the tail is forked, or rather is hollowed in form of a small segment of a circle; it is of a blackish colour, and has nineteen rays.

We have this in many of our large rivers, particularly in the North of England, in vast abundance. Pliny and all the Latin writers call it simply Salmo. Cassiodorus Ancharago; and Albertus and Cuba, Esox and Esox. Aldrovand and some few others call it Salmo nobilis; some, Salmo vulgaris and Salmo.

Salmo maxilla inferiore paulo longiore, maculis rubris.
The Salmo, with the lower jaw somewhat largest, spotted with red. **The Trout.**

This is not an inhabitant of the sea or rivers indifferently, as the salmon, but is entirely a river fish: the head is small and sharp; but the opening of the mouth is very large: the teeth are large, numerous, and sharp; there are three series of them on the palate, and one series in each of the jaws; the three series in the palate run nearly parallel in longitudinal lines, and the middle one is largest: the lower jaw is visibly longer than the upper, though the difference is but little: the body is thick, and not broad, in proportion to it's length; the back is of a dusky blackish; the sides of a fine silvery hue; and there are, both above and below the lateral line, a number of elegant red spots; these, in the months of May and June, are of the finest bright crimson, but at other times the red is less glowing: the tail is not forked, but is hollowed, as it were, in form of a segment of a circle: it is a very beautiful fish, and is, with justice, greatly esteemed at our tables.

We have it in most of our shallow rivers, where the current is swift. Authors in general call it Trutta and Truta; some Trutta fluviatilis. Albertus calls it Truta, and Isidore, Varius. Some have called it Thecla; and others Fario, Salar, and Forella.

Salmo vix pedalis, pinnis ventralibus rubris, maxilla inferiore paulo longiore.
The little Salmo, with the belly fins red, and the lower jaw somewhat longest. **The red Charr.**

This is a very elegant fish: it's general length is nine or ten inches, and it's body is not broad, but slender and somewhat rounded: the back is convex; the belly is flat: the head is small, and of a somewhat conic figure; the opening of the mouth is large; the teeth are numerous and sharp, but there is no middle series of them on the palate: the colour of the back is a dusky brown, approaching to black; on the sides it is a fine silvery white, but there are several broad spots of a dusky colour, both above and below the lateral line; the belly is somewhat flatted, and is of a reddish colour; there are some whitish spots between the dusky ones: the lateral line is straight; the scales are moderately large, and stand in an imbricated manner, and the tail is hollowed rather than forked.

We have this species in the cold lakes, on the tops of our high hills in Westmoreland, and other parts of the North of England. Linnæus found it in the cold lakes in Lapland, in which no other fish, nor hardly so much as an insect, could live, where not so much as a single water-plant was to be found: it is there in such abundance, that the Laplanders feed in a great measure on it. The generality of authors call it Umbla minor, and Umbla lacustris minor; Aldrovand, Umbla minor alba.

The other species of Salmo are, 1. The Salmo spotted with grey, and with the extremity of the tail even; this is called the Grey Salmon, and simply the Grey. 2. The smaller Salmo, with five series of teeth in the palate; this is called the Charr, and the gilt Charr in Westmoreland, where it is very common. Gesner calls it Salmo vel Trutta Benaci lacus; the generality of other authors, Carpio lacus Benaci. 3. The broader-bodied Salmo, with black and red spots, and with an even tail; this is frequent in the North of England, where it is called Scurf and the Bull Trout; Charleton calls it Trutta taurina; and Johnston and Willughby, Trutta Salmonata. 4. The yellow-backed Salmo, with yellow spots and a forked tail: this is called by authors Salmarinus and Salamandrino; it's head is shorter and more rounded than that of any other species, and it seldom exceeds a foot in length. 5. The forked-tailed Salmo, with the lateral lines bent upwards; this is called by authors the Umbla major and Umbla prior, and the Salmo lacus Lemani. 6. The oblong Salmo, with two series of teeth on the palate, and with the spots only black; this is called by the Germans, who have it in great abundance, Huch; authors call it Trutta piscinaria and Trutta fluviatilis altera. 7. The forked-tailed Salmo, with only black spots, and with a longitudinal furrow in the belly: this is the species which authors in general call the Trutta Salmonata and Parvus Salmo; others call it Trutta lacustris; Gesner, Trutta magna lacustris; and Aldrovand, Trutta lacustris, sive Trutta lacus Benaci. 8. The smaller Salmo, with only bright red spots.

E S O X.

THE body of the Esox is of an oblong figure: there is a fin, not very large, at the extremity of the back toward the tail: the branchiostege membrane has fourteen bones.

Esox rostro plagio-plateo.
The Esox, with a depressed rostrum. — The Pike.

This grows to a very considerable size in waters, where there is plenty of food; we meet with it of three feet in length in Whittlesea Meer, and in some of the rivers in Northamptonshire; but from fourteen inches to two feet is it's more usual size: the head is of a very odd figure; it is oblong and obtuse at the extremity, from the eyes to the extremity of the rostrum it is depressed, and the rest of it reaching on the hinder part, from the eyes to the back, is compressed: the body is oblong, and of a figure approaching to quadrangular: the back, from the head to the fin, near the extremity, is flatted; the sides are also flatted; and the belly, from the gills to the anus, is also flatted: the lower jaw projects a little beyond the upper; the opening of the mouth is large, and the nostrils are very conspicuous; they have a double aperture, and are situated just before the eyes: the eyes are large, and their iris is yellow, though often clouded with blackish, greenish, and other colours: the pupil is oval and bluish; there are twelve ducts or foramina in the head; six of these are situated behind the eyes, two between the apertures of the nostrils, and two before them, and two between the eyes; beside these, there are also several other smaller ones on the verge of the lower jaw, and about the hinder part of the head.

The teeth are very numerous and very regularly arranged; there is a single series of them in the verge of the lower jaw; the anterior ones of these are smaller; the hinder ones large, and all of them alternately fixed and moveable; the upper jaw has no teeth at the sides, but about the extremity there is a row of small ones; there are three series of teeth on the palate; these run parallel to one another in a longitudinal direction; the middle one of these consists of very small teeth; the two outer ones are composed of larger, all of them pointing inward and moveable: the tongue is a little bifid at the extremity, and is furnished with a number of small teeth, and the fauces behind the tongue are also furnished in the same manner: there are two oblong bones in the hinder part of the fauces, furnished with teeth; and above these there are four others, which are also armed in the same manner.

The

The lateral line is straight; the scales are of an oblong figure, but rounded at the ends; and the colour of the fish on the back is a blackish olive, but on the sides a pale silvery white, with a cast of yellowish or greenish: the fins on the breast and belly are of a whitish-yellow; the others are of a more dusky brownish-yellow, spotted with black: there is only one fin on the back, which is situated near the extremity of the body; this has twenty-one rays; the pectoral fins are oblong, and have each fifteen rays; the ventral fins have each eleven rays, and the pinna ani has eighteen; the tail is forked, and has nineteen long rays.

This is very frequent in our ponds, ditches, and rivers; it is an extreamly voracious fish: it's common food is the smaller fish of the roach and similar kinds, but it eats also frogs, snakes, rats, and even young water-fowl. All the writers on fishes have described it; Ælian, Oppian, and Athenæus call it Oxyrinchus and Oxyrinchus Nili; the Latin writers, in general, Lucius.

Esox rostro cuspidato, gracili, subtereti, et spithamali.
The Esox, with an extreamly long, slender, rounded, and pointed rostrum. **The Gar-fish.**

This is a fish of an extreamly singular figure; and, though evidently of the Esox or pike kind, yet has been, on account of the slenderness of it's body, ranked by many authors among the acus, a fish as different as well can be from this genus: it is one of the slenderest fish we are acquainted with, in proportion to it's length; it grows to a foot and a half long, and is not thicker than a man's finger: the body is compressed, but the back is somewhat convex, and the belly flat: the head is extreamly long, and the opening of the mouth is continued all the way up to the eyes, and is beset with a vast number of teeth, many of which are moveable: the end of the rostrum is somewhat pointed, and the lower jaw is longer than the upper: the back is blackish, or of an extreamly deep olive, and the sides are somewhat yellowish.

We have it frequent about our own coasts; it is a very nimble swimmer, and is extreamly voracious. Aristotle and Ælian call it Βελόνη and Βελόνη θαλάττια; Oppian and Athenæus call it Ἀκὶς; Pliny describes it under the name of Belone; Albertus and Cuba call it Abaniger; and Wotton, Acus sive Belone; most of the Latin writers call it Acus piscis, Acus vulgaris, and Acus Oppiani; we call it the Gar-fish, and, in some places, the Horn-fish.

Esox maxilla superiore longiore, cauda quadrata.
The square-tailed Esox, with the upper jaw longest. **The great Gar-fish.**

This greatly approaches to the former in shape, but it is larger and somewhat thicker-bodied, in proportion to it's length: it grows to more than two feet long, and to the thickness of a man's thumb: the head is very long and slender; it is of a rounded figure, except toward the extremity, where it is depressed: the opening of the mouth is surprisingly large, being continued the whole length of the head: the eyes are small, situated very high, and their iris yellowish: the back is broad and flat; the sides but little prominent; and the belly also flat, from the gills to the anus: the scales are small, oblong, and rounded at their ends; they are placed in a very elegant, imbricated manner, and stand extreamly close.

We have this species also about our own coasts, but it is not so frequent as the former. I met with several of them, a few years since, among the Bogner rocks in Sussex. Lister, and after him Willughby and Ray, call it Acus maxima squamosa; confounding it with the other acus of the Syngnathus kind.

ECHENEIS.

THE head of the Echeneis is of a depressed figure, broad and flat on the upper part, and marked with a number of rough, transverse ridges, by means of which it fixes itself to any solid body: the body is oblong and rounded, but somewhat depressed; there is an oblong fin on the hinder part of the back: the branchiostege membrane has about nine ossicles in it.

Of this singular genus there is only one known species.

ECHENEIS. The Remora, or Suck fish.

This singular fish grows to about nine inches in length, and to more than two inches in diameter in the largest part of the body, which is that near the head; it thence becomes gradually smaller to the tail: the head is broad and obtuse; the lower jaw is somewhat longer than the upper: the back is convex; the belly is flat; the sides are rounded, and the colour of the whole fish is a dusky brownish-grey: the mouth is moderately large, and is very full of teeth; the fins are six, beside the tail: there are two pectoral and two ventral ones, the pinna ani, and the dorsal one, which is single.

The great singularity of this fish is in the structure of the upper part of it's head; there are, on the surface of this, twenty-two elevated, rough lines, running in a transverse direction, and divided down the middle by one longitudinal one: by means of this structure, the fish applies itself firmly to any solid body that it pleases; it is frequently found sticking to the bottoms of ships, and often to large fish, particularly to the shark kind. There have been most idle and romantic stories of one of these fish stopping a vessel under full sail, by applying itself to it's side, or stopping a shark in the same manner, at it's pleasure: but to name such follies is to contradict them.

It is frequent in some of the European Seas, as well as in the American. Aristotle, Ælian, and Oppian call it Ἐχενηΐς; and the Roman writers have called it, from them, Echeneis; some of the later naturalists have called it Remora and Remeligo; some Iperuquiba, it's Brasilian name; and others, Achandus: the bottoms of ships stationed in the West Indies are frequently, in a manner, covered with them.

CORYPHÆNA.

THE head of the Coryphæna is very obtuse before, and runs with a declivity almost perpendicular from the vertex to the mouth: the body is of a depressed form: the fins are seven; one of them, which is situated on the back, is extended from the head to the tail: the branchiostege membrane contains on each side five bones, beside two others, which lie under the bony operculs, and therefore are not to be easily distinguished.

Coryphena cauda bifurca.
The Coryphæna, with the forked tail. **The Dolphin.**

This is a very beautiful fish: it grows to six or seven feet in length, and the body is considerably thick, though not very broad in proportion: the head is large, broad, and short; the rostrum is rounded; the eyes are large; the nostrils have each two apertures, and are situated a little below them; the body is of a somewhat depressed form; the back is broad, and has an elegant fin running along it all the way; the colour is a dusky olive on this part, but on the sides and belly it is of a silvery white: the pectoral fins have each twenty rays; the ventral ones have only six rays each: the tail is very forked, and has eighteen long rays, besides a number of short ones at the sides, not easily counted.

This is a native of the American Seas, and some other parts of the world. It was well known to the Ichthyologists of all times. Aristotle and Oppian call it Ἵππουρος; Athenæus, Ἵππουρος; Ray, Willughby, and most of the modern, as well as the antient Latin, writers, call it Hippurus; others, Lampugo, Equiselis, and Equitelis; some call it Dorado and Auratus piscis; the Brasilians, Guaracapema; we call it a Dolphin, but there is great confusion in the English names; we call one of the cetaceous fishes also by the same name. This is the Dolphin our painters figure, but they do it vilely.

Coryphena palmaris, pulchre varia, dorso acuto.
The small, beautiful, variegated Coryphæna, with a sharp back. **The Razor fish.**

This is a very beautiful little fish, and is, in it's general shape, as well as in size, extremely different from the former species: the head is short, and the rostrum obtuse: the body is compressed and moderately broad, in proportion to it's length; the back is ridged, and the whole body is elegantly painted with a variety of colours: the

lateral

FISHES Series the 2d.

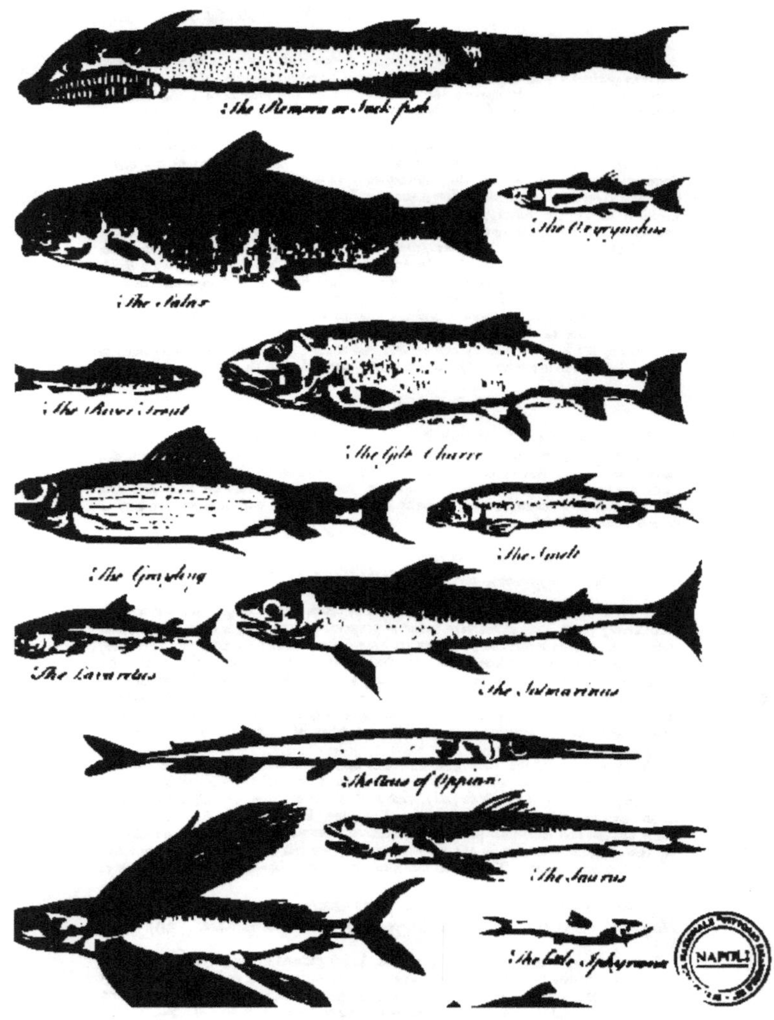

lateral line is tolerably ſtraight, and runs very near the back; the anus is ſituated nearer to the head than to the tail: the eyes are ſmall, and ſtand high in the head; the teeth are large, and the front of the jaws is well furniſhed with them; the tail is broad and equal.

This ſpecies is frequent about the coaſts of Minorca, and among the iſlands of the Archipelago. Authors in general call it Novacula piſcis; Salvian, Peſce pettine; and ſome Peſten, and Peſten Romanorum.

There is but one other known ſpecies of this genus, which is the Coryphæna with an undivided tail, and with the lateral line crooked; this is deſcribed by authors under the name of Pompilus.

AMMODYTES.

THE body of the Ammodytes is oblong and ſlender, and is of a rounded, but ſomewhat depreſſed, figure; there are no belly fins: the head is of a depreſſed form; and the branchioſtege membrane on each ſide contains ſeven bones, but they are in great part covered by the opercula of the gills.

Of this genus there is only one known ſpecies.

AMMODYTES. The Sand-Eel, or Grig.

The head of the Ammodytes is ſmall, much narrower than the body, and of a compreſſed figure, and acute at the forepart: the body is long and ſlender, and is a little compreſſed: the lower jaw is narrow and acute, and projects a great way beyond the upper, and the opening of the mouth is large: the noſtrils have each a double aperture, and ſtand in the middle, between the eyes and the extremity of the roſtrum: the eyes are moderately large, and their iris of a ſilvery colour: the laminæ covering the gills are large and bony; the apertures of the gills are large: a lateral line is ſtraight, and runs along the middle of the ſide, at about an equal diſtance between the back and the belly: this, however, when ſtrictly examined, is not exactly and properly what we mean by a lateral line in fiſh, but marks only the interſtices of the muſcles; on a nice examination there will be found another line, beſide this, running from the head to the tail, near the back, and owing it's appearance to the peculiar ſituation of the ſquamulæ; this is properly the lateral line. There run along the belly alſo three lines in a longitudinal direction; they have their origin at the pectoral fins, and the middle one terminates at the anus; the other two run to the tail.

The ſcales are extremely ſmall on the back; they are of a greyiſh colour, with an admixture of brown, on the ſides. The belly is of a ſilvery white. Moſt of the authors, who have written on theſe ſubjects, have ſuppoſed this fiſh to have no ſcales at all, their ſmallneſs having cauſed them to be overlooked. The pectoral fins are oblong, narrow, and very ſmall; they are ſituated immediately under the gills, and have each twelve rays; there are no ventral fins: the back fin is very long; it begins juſt behind the head, and reaches very nearly to the tail; it has fifty-three or fifty-four rays: the pinna ani is alſo long; it is extended from the anus nearly to the tail, and has twenty-ſeven or twenty-eight rays: the tail is ſhort, and is a little forked at the extremity; it has fifteen long rays: the anus is placed much nearer the tail than the head.

There are no teeth in the jaws, nor on the palate, but on the upper part of the fauces there are two oblong bones, which have very ſmall teeth on them: the tongue is oblong and narrow, and is looſe on the lower part, and ſmooth.

We have this very frequent with us, and call the larger or fuller grown fiſh Sand-eels, and the ſmaller ones Grigs; but we have a way of confounding this with the eel, by calling the young of that fiſh by the ſame name grig. Many of the modern authors have got our Engliſh name, but they ſpell it Sandils: Schonveldt calls it Tobianus; and Geſner, Ammocætus, Enchelyopus, and Ammodytes; Charleton calls it, Anguilla de arena; and Boccone, Cicirellus Meſſanenſis.

PLEURONECTES.

THE eyes of the Pleuronectes are both placed on one ſide of the head, and this is ſometimes the right, ſometimes the left ſide: one ſide of the fiſh is always white, the other variouſly coloured or obſcure: the branchioſtege membrane contains on each ſide ſix bones; they are of a cylindric figure, and in the lower part, in the middle between theſe, there is another pair; theſe are contracted together, and are ſcarce conſpicuous: the eyes are uſually ſituated in the right ſide of the head, and there is, in ſome ſpicies, a ſhort ſpine at the anus.

Pleuronectes oculis, et tuberculis ſex a dextra capitis, lateribus glabris, ſpina ad anum.
The Pleuronectes, with ſix tubercles on the right ſide of the head, and a ſpine at the anus.

The Plaiſe.

The head and the body are both flatted and depreſſed, and the ſides are very compreſſed and thin: the back and the belly, immediately under the head, are aſſurgent and acute: the mouth is ſmall, and obliquely ſituated, with reſpect to the back: the lower jaw is a little longer than the upper; there is a ſingle row of obtuſe teeth in each jaw; in the upper part of the fauces there are alſo two bones, covered with the ſame kind of ſhort, obtuſe, and granulous teeth, and each conſiſting of three parts; there are three other ſmall bones alſo on the lower part, furniſhed with teeth in the ſame manner; but the whole palate and tongue are ſmooth: the noſtrils are ſituated juſt before the eyes; they have each two apertures, and the anterior one has it's valve: the eyes ſtand both on the right ſide of the head; they are very near one another, and ſcarce protuberant at all: there are ſix bony tubercles between the eyes and the lateral line on the right ſide; they are arranged in a ſtraight line, and the fifth, from the eye, is the largeſt: the lateral line is nearly ſtraight, and divides the body into two; the whole body is ſmooth; there are, indeed, ſcales on the ſides and on the head, but they are very ſmall, and adhere extreamly firmly: the right ſide of the fiſh is brown, and is elegantly variegated with round, bright, red ſpots; the back fin and the pinna ani are alſo, in the ſame manner, ſpotted with red: the left ſide is entirely white; this is what we uſually call the belly of the fiſh.

There is only one fin on the back, but this reaches the whole length of it, from the eye to the tail: it has ſeventy-ſix or ſeventy-ſeven rays: the pectoral fins are placed on the ſides below the lateral line; they have twelve rays each: the ventral fins are placed immediately under the pectoral ones, or a little more anteriorly; they have each ſix rays: at the beginning of the pinna ani there is a robuſt, thick, and acute ſpine bent forwards; the pinna ani itſelf is very long, and is extended nearly to the tail; it has fifty-five rays: the tail is oblong, and is not exactly even at the extremity, but the middle is ſomewhat longer than the reſt; it has twenty rays: the branchiæ are four on each ſide; the vertebræ are forty-three, the ribs thirteen.

We have this in our large rivers toward the ſea, and about the ſea-coaſts in many other places. Athenæus calls it Ψῆττα, Cuba; and ſome others, Plais. Bellonius, Ray, and Willughby, Plateſſa; Aldrovand and Johnſon, Paſſer levis; Rondeletius, Quadratulus and Paſſeris alia ſpecies. We call it a Plaiſe, and the Germans, a Pladiſe and a Scholle.

Pleuronectes oculis a dextra, linea laterali aſpera, ſpinulis ſupinæ ad radices pinnarum.
The Pleuronectes, with the eyes on the right ſide, the lateral line rough, and ſpinules at the fins.

The Flounder.

This is in general a ſmaller fiſh than the plaiſe, though in ſome places it grows to a tolerable ſize; with us it's general ſtandard is about ſeven inches: the head is ſmall; the mouth is narrow, and the teeth obtuſe: the lower jaw is a little longer than the upper, and each has a ſingle row of teeth: the eyes protuberate very much; they are ſituated on the right ſide of the head: the right ſide of the body is of a duſky olive brown,

brown, often marbled, and often variegated with large, regular, round, black ſpots; the left ſide, or, as we vulgarly call it, the belly, is white, but it is ſometimes ſpotted with yellow: the lateral line is in the lower part ſtraight, but in the upper it is ſomewhat arcuated: the ſcales are very ſmall, and adhere cloſely and firmly to the ſkin, which is thick and ſmooth; and the whole, indeed, appears ſo ſmooth, that on a ſlight view there would not be ſuppoſed to be any ſcales at all on the fiſh: the ſcales at the lateral line, on the right ſide, or, as we call it, on the back, are rendered rough by certain ſmall tubercles, and particularly about the curvature of that line at the pectoral fins: on the right ſide alſo there are, at the roots of the dorſal fin, and the pinna ani, certain bony tubercles ſomewhat ſtarred, and furniſhed with ſeveral little ſpines, which turn their points backward: the back fin has from forty to forty-two rays; the pinna ani has forty-two or forty-three; the pectoral fins have each twelve, and the ventral ſix: the tail is oblong, of a ſquare figure, and equal at the extremity, not forked nor pointed; the vertebræ are twenty-five; the ribs on each ſide are eleven.

We have this ſpecies in vaſt abundance in the Thames, and many other of our large rivers toward the ſea. Ray, Willughby, and Bellonius call it Paſſer fluviatilis, vulgo fleſus; Charleton, Paſſer niger; and Geſner, Fleſus and Flesſelus. We uſually call it the Flounder, ſometimes, the Fluke or Bul.

Pleuronectes oculis a dextra, totus glaber.
The Pleuronectes, with the eyes on the right ſide, and the body ſmooth. **The Turbot.**

This grows to a very conſiderable ſize, and is one of the moſt eſteemed fiſh at our tables: the head is ſmall, in proportion to the ſize of the body; the eyes are large and protuberant; the noſtrils have each two apertures, and are ſituated under the eyes; the right and left ſides, or, as we uſually expreſs ourſelves, the back and belly of this fiſh, are both ſmooth. It is conſiderably thick and fleſhy, and the right ſide, which is uppermoſt, as it ſwims, is of a pale reddiſh-brown, and the other white: the dorſal fin reaches from the head to the tail, and has a hundred and five rays: the pectoral fins have each fifteen rays, and the ventral ones ſix; the pinna ani has ſeventy-nine rays.

We have it in our ſeas in great abundance. Ray and Rondelet calls it Hippogloſſus; Geſner, Hippogloſſus, id eſt, Bugloſſus maximus in oceano; Charleton calls it Paſſer Britannicus; and Schonevelde, Paſſerum genus maximum. We call it in ſome places a Hollibut, but more uſually a Turbot.

Pleuronectes oculis a ſiniſtra, corpore glabro.
The ſmooth-bodied Pleuronectes, with the eyes on the left ſide. **The Pearl.**

This is a moderately large ſpecies, but it is not ſo thick and fleſhy, in proportion to it's ſize, as the turbot, nor is it's fleſh ſo well taſted or ſo firm: the head is ſmall, and the opening of the mouth larger than in moſt of this genus: the eyes are protuberant, and ſtand on the left ſide of the head, not on the right, as in all the preceding ſpecies: the teeth are ſmall and obtuſe; the eyes are large and prominent: the right ſide or back is of a duſky brown, and the other ſide of a fine, bright, pearly white, with a tinge of bluiſh in it.

We have this on our coaſts in ſome plenty; it is ſometimes brought to market in London; it uſed to be called the Pearl, but of late our fiſh-mongers have found it more to their intereſt to ſell it under the the name of the Hollibut, on which foundation they are in the right, who diſtinguiſh the hollibut from the turbot. Ray and Willughby call this ſpecies Rhombus non aculeatus ſquamoſus; Ælian calls it ſimply [illegible]; and moſt of the old Latin writers, after his example, Rhombus; Geſner and ſome others call it Rhombus levis; and Bellonius, Rhombus alter Gallicus; our people in Cornwall, Lug-a-leaf.

Pleuronectes

Pleuronectes oculis a ſiniſtra, linea laterali utrinque aculeata.
The Pleuronectes, with the eyes on the left ſide, and the lateral line prickly on both ſides.

This is a moderately large ſpecies: the head is ſmall; the opening of the mouth ſmall; the eyes are protuberant, and the noſtrils are ſmall, but each has two apertures; the teeth are obtuſe, and form a kind of bony granules; and the eyes are on the left ſide of the head, not as uſual on the right: the general length of the fiſh is about ten inches; the upper or right ſide is of a duſky brown, the under or left ſide is white: the lateral line is prickly on both ſides; on the left ſide there are ſeventy-one prickles, and on the right fifty-eight: the back fin has ſixty-ſix rays; the pinna ani has fifty; the pectoral ones have each nine, and the ventral ones ſix.

We have this in abundance in our ſeas. Ariſtotle calls it ψῆττα; Ælian and Oppian uſe the ſame name. Ovid and all the old Latin writers call it Paſſer; Paulus Jovius, Rhombus; Geſner calls it Rhombus Bellonii; and moſt other writers, Rhombus aculeatus.

Pleuronectes corpore oblongo, ſquammis utrinque aſperis. **The**
The long-bodied Pleuronectes, with rough ſcales on both ſides. **Soal.**

This is of an uncommon figure for a fiſh of this genus; it is of an oblong ſhape, tending to elliptic, but ſmaller at each end, and is thin, in proportion to it's extent, though conſiderably broad: it's uſual length is from five or ſix to fourteen inches, and it's breadth not nearly half it's length: the head is obtuſe, and the mouth ſmall; the upper jaw is longer than the under, and the eyes are on the left ſide; the mouth, though ſmall, is formed in a ſingular and odd manner, and is very deep cut; the eyes are not protuberant, and are placed at a greater diſtance than in moſt other ſpecies.

The teeth are very numerous; they are ſmall, ſlender, and moveable, and are placed in ſeveral ſeries in the fauces; there are two roundiſh and large bones in the upper part, and two other ſmaller on the lower, which are more oblong; theſe are all beſet with teeth; and there are a great number of ſhort white hairs, as it were, above and below the mouth, on the white ſide of the body, but there are none of theſe on the other: one ſide of the body, which is the uppermoſt in ſwimming, is of a browny, browniſh colour, the other is of a pure and elegant white: the lateral line is very ſtraight, and is nearer the back than the belly; but it is continued to the head itſelf, and there, above the gills, it riſes perpendicularly: the ſcales are ſquare, but ſomewhat oblong; they are ſmall, whitiſh in colour, and have ſeveral little ſpinules at the extremity; theſe cover the white as well as the brown ſide of the body.

The anus of this fiſh is very oddly ſituated; it is between the ventral fins, juſt at the opercula of the gills: the lateral fins are very ſmall; one of them is white, the other is often half-way black; they have each eight or nine bones or rays: the ventral fins ſtand very forward; they have only five rays each, and are very ſmall: the back fin reaches from the head to the tail, and has ninety-one rays; theſe are inclined toward the tail, and there are often ſome black ſpots on the membranes of theſe fins: the pinna ani is very remarkably long, it runs from the anus to the tail; it has ſeventy-four rays; and both this and the back fin are, in part, inveſted with ſcales, which all turn toward the tail; there are no tubercles at the baſes of the fins; the tail is ſomewhat oblong, but it is rounded at the end, and has fourteen rays.

This is frequent in moſt of the European Seas, and is deſervedly in great eſteem at our tables; the antients were as well acquainted with it as we are, and valued it as much. Athenæus calls it βούγλωσσον; Oppian, βούγλωσσος; Ovid, Pliny, and moſt of the other Latin writers, Solea; Rondelet calls it Bugloſſus; Varro, Lingulaca; and moſt of the other writers, Bugloſſus ſive Solea.

Pleuronectes oculis a dextra, squammis asperis, spina ad anum.
The Pleuronectes, with the eyes on the right side, with rough scales, and a spine at the anus.

The Dab.

The body is very broad, in proportion to it's length, and is compressed and thin; the head also is compressed: the back rises into an acute ridge; the belly also is ridged: the mouth is somewhat obliquely situated, and it's opening is very small, and the lower jaw is somewhat longer than the upper; there are a great number of teeth in the upper jaw; in the lower there are a much smaller number, and of these many are moveable: the eyes are protuberant, and on the right side; and their iris is usually yellow, sometimes of a milky white.

The lateral line is elevated into an arch, the convex part toward the back, near the pectoral fins; in the other part it is straight, and runs along the middle of the body, dividing it into two halves: the scales on this species are large for those of one of this genus, and are armed with a number of little spines at the edges, especially on the right side; and the rays of the dorsal fin, and the pinna ani, are on each side invested with scales. The right side is of a brownish-grey colour, and is variegated with a great number of small, pale, yellow spots; the left side is wholly white; the back fin has seventy-eight or seventy-nine rays; the pectoral or lateral fins have each twelve rays; the ventral ones have only six in each, and the pinna ani has forty or forty-one, and there is a spine at the anus: the tail is oblong, and is nearly even at the extremity; it has eighteen rays: the two extreme ones of these are small, the others are long; the middle ones are forked half the way down; the vertebræ are thirty-nine.

This is frequent in our seas, and is a well-tasted fish, but not at all equal to the former. Ray, Willughby, and the other Latin writers call it Passer asper sive squammosus; Bellonius, and after him Gesner, Limanda; Charleton, Citharus; the French call it Limande, and we the Dab.

The other species of the Pleuronectes are, 1. The Pleuronectes, with the eyes on the right side, and the anus placed on the left side, and with very sharp teeth in the mouth; this has been called Pola by Bellonius, by some Linguatula, and by the greater number of writers Cynoglossus. 2. The Pleuronectes, with the eyes on the left side, and with the body rough; this is the Rhombus squamosus of Charleton, and the Rhombus maximus asper of Ray. 3. The Pleuronectes, with the eyes on the right side, and with a rough, heavy body, and the lateral fins scarce conspicuous. This is a native of the Seas about Amboina, and has not been described by any of the authors who have hitherto written on fishes.

STROMATEUS.

THE body of the Stromateus is extremely broad, thin, and compressed; the dorsal fin is single, and runs all along the back, and there are no belly fins. Of this genus there is but one known species.

Stromateus.

The Fiatola.

The body is of an oblong form, considerably broad, and very thin; the head is small, and the opening of the mouth not large: the back is ridged, and so is also the belly; the tail is very forked: there are teeth in both jaws, and also on the palate, but the tongue is smooth and broad: the back fin is very long, and has forty-six rays: the pinna ani has thirty-four rays; the pectoral fins have each twenty-five rays, and the fish has no belly fins at all: the colour of the body is a deep, dusky brown, but it is variegated in an elegant manner, with lines and streaks of a variety of colours.

It is frequent in the Mediterranean, and is brought to market at Rome. Athenæus calls it Στρωματεὺς, and from him most of the Latin writers have called it Stromateus and Stromatheus; Bellonius calls it Callichys and Fiatola; the Venetians call it Lucerne; and the Italians, Lampuga.

GADUS.

THE head of the Gadus is uſually compreſſed, but in ſome few of the ſpecies it varies from this, and is depreſſed: the back is furniſhed uſually with three fins, but in ſome ſpecies only with two: the branchioſtege membrane on each ſide contains ſeven bones, of a ſomewhat cylindric figure.

Beſide the difference in the number of the back fins, ſome of the Gadi have cirri at their mouths, and others have not: they may, from theſe obvious differences, be arranged under two or three diſtinct diviſions.

GADI. *Diviſion the Firſt.*

Thoſe which have three fins on the back, and have no cirri.

The Whiting.

Gadus dorſo tripterygio, ore imberbi, maxilla ſuperiore longiore, albus.

The white Gadus with no beards, and with three fins on the back, and the upper jaw longeſt.

The head of this ſpecies is compreſſed, in general; but, when the mouth is ſhut, it is, toward the extremity, a little depreſſed; the body alſo is compreſſed: the back is convex; the anus is at a great diſtance from the tail, and is indeed very near the head: the colour of the whole fiſh is a ſilvery white, except that on the back there is an admixture of a blackiſh tinge, which gives it a hoary appearance: the ſcales are very ſmall, and are roundiſh and white; the upper jaw ſtands prominent beyond the under, inſomuch that, when the mouth is ſhut, the teeth of it fall over the lower: the noſtrils have each a double aperture, and are placed high, being much nearer to the eyes than to the extremity of the roſtrum: the eyes are very large, and the iris is ſilvery; the pupil is very large and bluiſh: the teeth are very numerous; there are ſeveral ſeries of them in the upper jaw, but thoſe of the outer ſeries are longer than any of the others, but they are not all alike in magnitude; in the lower jaw there is only one ſeries, but the teeth are in this alſo very unequal in ſize; in the anterior part of the palate there is a bone furniſhed with a number of ſmall teeth, forming two ſides of a triangle; and in the upper part of the fauces there are two little bones, of a roundiſh figure, and in the lower part two others that are more oblong; theſe four are all furniſhed with a number of teeth: the middle of the palate is ſmooth, as is alſo the tongue; there are nine or more puncta on each ſide of the upper jaw, and there are no beards at the mouth; the lateral line is of a duſky or blackiſh colour, and is crooked, and runs much nearer the back than the belly; and there is on each ſide of the body a large black ſpot, at the origin of the pectoral fins.

The pectoral fins are of a greyiſh colour, and have each twenty-one bones or rays; the ventral fins are ſituated more forward than the pectoral; they are of a whitiſh colour, and have each ſix rays: there are on the back three fins of theſe; the firſt is triangular, and has twenty-one rays; the pinnæ ani are two, and have, the firſt thirty-three, the ſecond twenty-three, rays; they are both of a whitiſh colour, and the firſt is of an oblong figure and larger, the ſecond ſhort and ſmall: the tail is even at the extremity, and is of a blackiſh colour; it has thirty-one rays; the vertebræ are fifty-four.

This ſpecies is frequent in our ſeas, and is much eſteemed at our tables. The modern writers of fiſhes have all deſcribed it, but they have been ſtrangely confuſed in their naming it; they all make it an Aſellus; but Ray and Willughby call it Aſellus mollis major; Charleton, on the other hand, Aſellus mollis minor; Rondeletius calls it Aſelli ſecunda ſpecies; Schoneveldt, Aſellus candidus primus; Aldrovand, Aſellus minor alter; Geſner and ſome others have alſo called it Merlangus.

Gadus dorso tripterygio, maxilla inferiore longiore, linea laterali recta.
The Gadus, with three back fins, with the lower jaw longest, and the lateral line straight.

The Cole-fish, or raw Pollack.

This species greatly approaches to the common whiting in shape and size: the head is compressed, and the body is also somewhat compressed, but it is thick, in proportion to it's length: the eyes are large; the mouth also is very large, and it is furnished with a great number of sharp teeth: the back is convex; the lateral line is white; the general colour of the fish is blackish, but it is elegantly variegated with a dusky yellow: the tail is somewhat forked, but it is very little; the lateral line is very broad: the first of the three back fins has fourteen rays, the second has twenty, and the third twenty-two; the pectoral fins have each eighteen rays, and the ventral ones have only six: there are two pinnæ ani; the first of them has twenty-two rays, and the second nineteen.

We have this in our seas, especially about the northern coasts. Our common people in Northumberland call it the raw Pollack and the black Pollack; Ray, Willughby, and Aldrovand, all call it Asellus niger; Bellonius and Gesner borrow one of our common people's names, and call it Piscis Colefish Anglorum; Schonevelde turns this into Latin, and calls it Asellus niger sive carbonarius; and Charleton, Asellus niger sive mollis nigricans.

Gadus dorso tripterygio, maxilla inferiore longiore, linea laterali curva.
The Gadus, with three back fins, with the lower jaw longest, and the lateral line crooked.

The Whiting Pollack.

The usual length of this fish is from eight to thirteen inches, and it is considerably thick, in proportion: the head is compressed; the opening of the mouth is large, and the teeth are numerous and sharp: the eyes are large, and their iris is silvery, and the pupil is large and grey: the back is convex; the sides are somewhat rounded; the lateral line is of a dusky colour, broad and conspicuous, and is crooked: there are three fins on the back; the first of these has only eleven rays, the second has nineteen, and the third has sixteen: the pectoral fins have each sixteen rays, and the belly fins have only six each; the pinnæ ani are two; the first of these has sixteen rays, and the second has eighteen.

This is frequent in our seas, and is taken in great abundance on the northern coasts. Ray and Willughby call it Asellus Huitingo-Pollachius, a strange barbarous name, formed of the English one; Schonevelde calls it Asellus virescens; and the before-mentioned authors, not observing that this was the same fish, have described it again from Schonevelde, and, under his name, as another species. This is an error, by which the species of fish have been strangely multiplied.

GADI. *Division the Second.*

Those which have three back fins, and have cirri or beards at the mouth.

Gadus colore vario, maxilla superiore longiore, cauda æquali.
The various-coloured Gadus, with the upper jaw longest, and the tail even.

The Cod-fish.

THIS species grows to a very considerable size, three or four feet in length being not uncommon: the head is compressed behind, but in the anterior part, when the mouth is shut, it appears rather depressed: the body is very thick, the back convex, and the sides rounded: the eyes are large, and the nostrils have a distinct, double aperture; the belly is somewhat prominent, and the anus is situated high: the upper

upper jaw is longer than the under, and at the extremity of the under jaw there is a cirrus or beard white, moderately thick, of a fleſhy colour, and about half an inch in length: the teeth are very numerous; there are ſeveral ſeries of them in the upper jaw, but the exterior ſeries is longeſt: in the anterior part of the palate there is a bone of a lunated figure, which is furniſhed with teeth; and in the fauces there are two roundiſh bones above, and two oblong ones below, all four beſet with very ſharp, though not very tall, teeth; there is a furrow between the head and the firſt fin of the back, which is very obſervable in the larger fiſh: the lateral line is broad, and of a whitiſh colour; it is much nearer the back than the belly, and it is ſomewhat crooked: the ſcales are extremely ſmall, and are of a roundiſh, or elſe of a ſomewhat oval, figure; they are of a whitiſh colour: the back and part of the ſides of the fiſh are of a duſky colour, but there are a number of ſmall ſpots, of a blackiſh, or ſometimes of a browniſh or yellowiſh, colour on the back, the ſides, and the head.

The belly is white, but it has ſometimes a number of black ſpots on it; there are three fins on the back, and they are all of a blackiſh colour: the firſt of theſe has fourteen or fifteen rays; the ſecond has eighteen, nineteen, or twenty rays; and the third has ſeventeen or eighteen: there are two pinnæ ani; the one has eighteen, nineteen, or twenty rays; and the other or hinder has ſixteen, ſeventeen, or eighteen: the pectoral fins have each twenty rays, and the belly fins have only ſix each: the tail is ſmall, in proportion to the bulk of the fiſh, and is even at the end: the vertebræ are fifty-three, and the ribs eighteen on a ſide.

This ſpecies is frequent in our ſeas, and in thoſe of moſt other parts of the world, and is of vaſt uſe, as proviſion, both freſh and preſerved in ſalt, in which laſt ſtate it becomes a very conſiderable article of commerce. Ray, Willoghby, and Schoneveldt call it Aſellus varius ſive ſtriatus; Jonſton and others, Aſellus varius; we call it the Cod and Cod-fiſh; the Swedes Sum Torſk.

Gadus corpore albicante, maxilla ſuperiore longiore, cauda parum biſurca.
The white Gadus, with the upper jaw ſomewhat longeſt, and a forked tail.

The Haddock.

This grows to a conſiderable ſize: the head is large, and of the ſame figure with that of the former ſpecies: the noſtrils are conſpicuous, and have a double aperture; the eyes are large; the opening of the mouth is alſo wide, and is furniſhed with a great number of teeth; the upper jaw ſtands out to ſome diſtance beyond the under one, and at the extremity of the under one there is a ſmall and ſingle cirrus or beard: the back, from the head to the firſt fin, is contracted at it's top into a kind of ridge; the belly is flat, and of a fine, elegant, milk white: the lateral line is nearly ſtraight; it runs near the back, and is of a black colour; the reſt of the body is of a greyiſh-white, but the back is duſkier than the reſt, and has no variegations.

We have this ſpecies alſo in our own ſeas in great abundance, as well as in thoſe of moſt other parts of the world; the ancients were very well acquainted with it. Pliny calls it Callarias; and Charleton recounts the ancient names of Callaris Galerides and Galaxias; others call it Piſcis Capitoſus; Bellonius, Æglefinus and Ægrefinus; Geſner, Rondelet, and others, Aſellorum tertia ſpecies Eglefinus; Schoneveldt, Aſellus minor; and Turner and ſome others, Onos ſive Aſinus antiquorum; we call it the Haddock; the Danes, Kole.

Gadus longitudine ad latitudinem tripla pinna ani oſſiculorum triginta.
The Gadus, with it's length equal to three times it's breadth, and with thirty rays in the pinna ani.

The Whiting Pout.

The head of this ſpecies is large, and it's upper jaw is a little longer than the under: on the extremity of the under jaw there is a ſingle cirrus or beard of ſix or ſeven lines in length: the back is convex; the body is, in it's general ſhape, like that of the

the rest of the Gadi, but it is broader, in proportion to it's length, than any of the others, the breadth equalling one third of that measure: at the roots of the pectoral fins there is, on each side, a single, blackish spot: the tail is black at the extremity, as are also the back fins; and the first of the pinnæ ani is frequently bluish: the body of the fish is of a whitish colour, obscured with a faint tinge of blackish: the back is brown; the lateral line is brown, or sometimes blackish; it runs very crooked, and is much nearer the back than the belly: the scales are large and white; they are of a roundish but somewhat oblong figure, and some of them angular: the mouth does not open so very wide, as in some of the other species.

The pectoral fins are of a pale reddish-brown colour, and have each nineteen rays: the ventral fins are white, and have each only six rays. There are three fins on the back; the first of them has thirteen rays; it is taller than the other two, and is of a somewhat triangular figure: the second is longer than the others, and has twenty-four rays; the third or hindermost has twenty-one rays: there are two pinnæ ani; the first is of an oblong figure, and has thirty-one rays; and the other is, in a manner, continuous to this, and is placed exactly over-against the third back fin; this has twenty-one rays: the tail is even at the extremity, and it's rays are all of a length.

The anus in this species is very near the head; it is, in a well-grown fish, not three inches from the extremity of the rostrum, which is barely a third of the length of the fish: nine inches in length, and three in breadth, are the usual standards of this species, but it sometimes grows much beyond these: there are about nine puncta on each side, on the lower jaw.

We have this species in our own seas; it is caught in great abundance on the coast of Cornwal, and in many other places. Willughby and others call it Asellus mollis latus; Charleton, Asellus barbatus; and we, the Pouting, the Whiting-pout, and sometimes simply the Pout. Authors have described, as another species, what they call the Asellus luscus, and what our country people in Cornwal call the Bib, or the Blind; but, on the strictest examination, this appears not to differ from the species here described.

Gadus corpore sescunciali, ano in medio corporis. **The**
The inch and half Gadus, with the anus in the middle of the body. **Poor.**

This is the smallest of all the Gadi: it's general appearance, at first sight, would lead an inaccurate observer to suppose it only the young of some of the other kinds; but, on examination, it appears perfectly distinct from them all; and even the common people of Cornwal, who frequently meet with it, distinguish it, and give it a peculiar name: it's usual size is an inch and a half in length, and it very rarely indeed exceeds two inches: the head is large, in proportion to the body; the eyes are prominent; the opening of the mouth is very large, and is well furnished with teeth; and the nostrils have each a double aperture, and are situated at some little distance below the eyes: there are nine puncta on each side on the jaws; the back is convex, and of a dusky brown; the belly is somewhat flat and white, and the anus stands at about an equal distance from the extremity of the head and the tail.

There are three fins on the back: the first has twelve rays; the second has nineteen rays, and the third seventeen: the pectoral fins have each thirteen rays, and the ventral ones have six; there are two pinnæ ani; the first has twenty-seven rays, and the second has seventeen: the tail is moderately large, and is even, not divided or hollowed at the end; the lateral line is crooked.

This species is very frequent in our seas, and in those of most other parts of Europe and elsewhere. Ray and Willughby call it Asellus mollis minor, and Asellus omnium minimus; Rondeletius calls it Anthiæ species secunda: we call it the Poor and the Power; the Venetians, Mollo; and the Massilians, Capelan.

GADI. *Division the Third.*

Those which have only two fins on the back.

Gadus dorso dipterygio, maxilla inferiore longiore.
The Gadus, with two back fins, and with the lower jaw longest.

THIS is a considerably large species; it frequently grows to two feet in length, and sometimes to considerably more; but it is the slenderest of all the species, in proportion to it's length: the head is large and broad; the body is rounded; the back is convex; the belly is somewhat flatted, but less so than in many of the other species; the sides are full and prominent; the colour on the back is a dusky greyish, with an admixture of brown; the sides are of a silvery grey, with some admixture of the same brownish colour, but in small proportion; the belly is of a milky white; the tail is large, and is even at the extremity, not at all forked or hollowed.

There are two fins on the back; the first has only nine rays; the other, or hinder one, is much longer, and has forty; the pectoral fins have twelve or thirteen each, and the ventral ones seven; the pinna ani has thirty-nine rays; all the fins are of a greyish colour, with more or less admixture of black.

We have this species frequent in our seas. The antients were acquainted with it; they call it Ὄνος and Ὄνος θαλάττιος; and some of them Γάδος; Pliny calls it Bacchus; Bellonius and Gesner, Merlucius; Ovid, Varro, and many other of the Latin writers call it simply Asellus; Salvian, Asellus minor; Charleton, Asellus fuscus; and others, Asellus primus sive Merlucius. We call it the Hake; the Italians, Merluzzo; and the French, Merlus. Salvian's figure has a strange inaccuracy in it; it gives two pinnæ ani, and but one back fin, the contrary of which is in nature, the pinna ani being single, and the dorsal ones two.

Gadus dorso dipterygio, ore cirrato, maxilla superiore longiore.
The cirrated Gadus, with two back fins, and with the upper jaw longest.

The Ling.

This species in some degree approaches to the hake in shape, but it is sufficiently distinct in many particulars; the head is large, broad, and depressed, and the opening of the mouth very wide, and furnished with a great number of sharp teeth: the eyes are large, but not very prominent; the nostrils have each a double aperture, but they are not very conspicuous: the body is very long, in proportion to it's thickness, and is not flat but rounded: the back is convex, and of a deep, blackish, brown colour; the sides are prominent, and of a paler brownish-grey, and the belly is flatted, and of a fine milky white: the first of the two back fins is short, and has only fifteen rays; the other is greatly longer, and has no less than sixty-five rays: the pectoral fins have each fifteen rays, and the ventral ones only six each: the pinna ani is tough and long; it has sixty-two rays.

This is frequent in many of the European, as well as the American, Seas. Charleton calls it Molva major and Morhua major; Ray, Willughby, and Schonevelde, Asellus longus; the Swedes, Langa; and we, Ling.

Gadus sulco ad pinnam dorsi primam ore cirrato.
The cirrated Gadus, with a furrow at the first back fin.

The Whistle-fish.

This is a small species; it's usual length is about eight inches, and it's thickness not great, in proportion: the head is large, and somewhat depressed; the opening of the mouth is large, and the eyes are prominent; the back is convex, and of a dusky brown, with an admixture of an iron grey: the head is of the same colour, but more glossy, and with somewhat of an admixture of olive colour: the sides are of a paler and somewhat silvery grey, with no brown in it; and the belly is somewhat flatted and white; there is a remarkable furrow on the anterior part of the back, out of which the first of the two back fins arises: this has about twenty rays; the second is much

much longer, and has fifty-ſix rays; the pinna ani has forty-ſeven rays, and the ventral fins have only ſeven rays each: the tail is large, and is rounded at the extremity.

We have this very frequent in our own ſeas, and it is abundant in the Mediterranean, and in many others. Bellonius calls it Gaſus Venetorum ſive Aſellorum alterа ſpecies; Rondeletius, Geſner, Aldrovand, Willughby, and Ray call it Muſtela vulgaris; Jonſton ſimply, Muſtela; Schonevelde, Muſtela altera: the Venetians call it Donzelino and Sorge; and we, the Whiſtle-fiſh.

Gadus dorſo dipterygio, ore cirrato, maxillis æqualibus.
The cirrated Gadus, with two fins on the back, and the jaws equal.

The Eel-pout.

This is a long and ſlender ſpecies: the head is large, broad, and depreſſed; the eyes are not very large, nor ſo prominent as in many of the other ſpecies: the body is long, ſlender, and not flat, but of a rounded figure: the back is convex; the ſides are rounded, and the belly is alſo ſomewhat prominent and rounded: the back is of a duſky browniſh-grey, with a tinge of olive colour in it; the ſides are of the ſame colour, only fainter, and both are elegantly clouded and variegated; the belly is white and ſimple: the whole fiſh is ſoft, lubricous, and ſlippery, in the manner of an eel: there are two fins on the back; the firſt of them is ſhort, and has thirteen rays; the ſecond is long, and has twenty-ſix rays: the pectoral fins have each twenty-one rays; the ventral ones have ſeven rays each, and the pinna ani has fifty-five rays.

We have it in plenty in ſome of our larger rivers. Aldrovand, Willughby, and Ray call it Lota; Geſner, Lota Gallis dicta; and Jonſton, Lota Gallorum; Bellonius and Geſner call it Strinſia ſive Botatriſſa; Iſidore and Cuba, Barbotha; Olaus Magnus Barbocha; Salvian calls it Botatria and Triſcus; and Hildegard, Abropa: moſt of the other writers have called it Muſtela fluviatilis and lacuſtris.

The other ſpecies of the Gadus are, 1. The Gadus, with three fins on the back, with a cirrated mouth, and with the tail nearly even at the extremity, with the firſt ray of it ſpinoſe or prickly. This is the ſpecies which the Danes call Cabbing, and the Swedes Cabila, and which many of the writers on fiſhes have called Morva, or Morhua, or Molva vulgaris, a name ſome have underſtood, as ſignifying the common cod-fiſh, but this erroneouſly. 2. The long, ſmaller-mouthed, Eaſt Indian Gadus.

ANARHICAS.

THE body of the Anarhicas is compreſſed, and the head, on the contrary, is ſomewhat depreſſed; there are no ventral fins: on the back there is only one fin, which is very long, and extends almoſt to the tail: the pinna ani is alſo very long, and the tail is diſtinct and ſquare: the branchioſtege membrane contains ſix, and ſometimes more, oſſicles.

Of this genus there is but one known ſpecies.

ANARHICAS.

The Wolf-fiſh.

This is a very ſingular fiſh; it grows to four or five feet in length, and is conſiderably thick, in proportion: the head is very large, and of a depreſſed figure; the opening of the mouth is enormouſly wide, and there are a vaſt multitude of large and ſtrong teeth in the jaws, on the palate, and deep in the fauces: the eyes are very large and prominent; they ſtand very high on the head, and have a look of a peculiar ſierceneſs: the noſtrils ſtand at a moderate diſtance below them, and have each a double aperture; the back is convex, and the ſides are prominent: the whole body of the fiſh is ſoft and lubricous, in the manner of that of an eel, and is elegantly variegated with ſeveral colours: the pectoral fins are very large, and of a rounded figure: the back fin is very long, but not remarkably tall; the pinna ani alſo is long and low.

It is ſometimes, but rarely, caught in the Mediterranean, and has been accidentally taken upon our own coaſt. I remember to have ſeen a large one at Goodwood, brought as a preſent to the late Duke of Richmond from the Suſſex coaſt, where it had been caught by ſome fiſhermen. It is an extremely voracious and bold fiſh, ſeizing upon almost any thing, and destroying whatever is once fixes it's jaws upon. Schonevelde calls it Lupus marinus nostras; Jonston, Lupus marinus; Ray and Willoghby, Lupus marinus nostras et Schonevelde; Gesner calls it Anarrhicas, Scansor, and Rhein-fisch; we the Wolf-fish.

MURÆNA.

THE body of the Murena is long, slender, and rounded, or subcylindric; in some species the fins are three, in others they are four, and in some again there is only one: at the very extremity of the rostrum there are two short tubes or foramina, one on each side; these are the anterior apertures of the nostrils; the branchiostege membrane on each side contains ten slender and crooked bones, but the skin of the fish is thick and firm, so that, till it is taken off, they are not easily discovered.

Murena unicolor maxilla inferiore longiore. — The
The simple-coloured Murena, with the lower jaw longest. — Eel.

The head of the eel is depressed in the anterior part, but in the hinder it is rounded, or, if any thing, a little compressed; it is small, and acute at the extremity: the body is long and rounded, toward the extremity it is somewhat compressed; it's whole surface is smooth and slippery, being covered with a thick, viscous matter; it seems to have no scales, but, when the skin is dried, there are found some: the lower jaw is somewhat prominent beyond the upper; there are, in the extremity of the rostrum, two little tubes, one on each side; these are open at the extremity, and pervious all the way; they are no other than the anterior foramina of the nostrils assuming that form: the two posterior foramina are situated at a great distance from these; they are situated, indeed, just before the eyes, one under each: the eyes are small, round, and covered with a thick and but little pellucid membrane; the iris is of a dusky red; the pupil is very small and black; in both jaws there are a great number of little foramina or ducts; there are about sixteen of these in the lower, and many more than that in the upper: there are several series of teeth, both in the upper and under jaws; in the anterior part of the palate there is a bone situated longitudinally, and contiguous to the teeth in the jaw; this is covered with teeth, but the rest of the palate is smooth: in the upper part of the fauces there are two bones covered with teeth; and in their lower part, toward the gills, there are also two more, which are of a longer figure than these, and are not quite so rough: the tongue is smooth, and is immoveable, being supported along it's middle by a rigid bone.

The coverings of the gills are not open, either above or below, but they have a little aperture on each side, near the pectoral fins: the lateral line is straight; in the anterior part it rises toward the back, and in the lower, or from the middle to the tail, it divides the body regularly in half, and there is a single row of puncta toward the lower part of the line: the colour of the back and sides is naturally dusky and blackish, but sometimes coppery, and sometimes greenish; the belly is either of a silvery white, or yellowish; the anus stands nearer the head than the tail.

The eel has three fins; the pectoral ones are two; they stand one on each side, near the aperture of the gills, and are small, and of a blackish colour; each has eighteen or nineteen rays: the back fin is single; it begins at a considerable distance from the head, and surrounds the body, going round the tail, and coming up again continuous as far as the anus; this is but low, and is supported by a vast number of bifid rays: the extremity of this fin, which forms what may be called the tail, is neither round nor square, but subacute: the branchiæ are four on each side; they are full of elegantly divided blood-vessels on the convex part, and on the concave they have no apophyses: the vertebræ are a hundred and sixteen; the ribs are short, and adhere but slightly to them.

We have this species in all our fresh waters, in ponds, ditches, and rivers. The antients called it Ἔγχελυν; the moderns, Anguilla: the Swedes call it Al; the Germans, Ahl; and we, the Eel. About two feet is it's general standard with us, but there have been occasionally caught some vastly larger.

The Conger, or Sea-eel.

Muræna ſupremo margine pinnæ dorſalis nigro.
The Muræna, with the upper edge of the dorſal fin black.

This has greatly the external appearance of the eel, but exceeds it extremely in ſize: we frequently ſee it of five or ſix feet long, and of the thickneſs of a man's thigh, and ſome have been caught vaſtly larger: the head is acute; the opening of the mouth is large, and the teeth are extremely numerous and ſharp: the eyes are ſmall, and covered with a thick membrane; their iris is of a ſilvery white, and the pupil ſmall, round, and black: the pectoral fins are ſmall; and the lateral line runs ſtraight, and is broad and white: the colour of the back and ſides is a deep olive; the belly is of a ſilvery white, ſometimes tinged yellowiſh or browniſh: the back fin ſurrounds the whole hinder part of the body; it is ſupported by a vaſt number of rays, and is of a greyiſh olive colour, except at the edge, where it is blackiſh.

This ſpecies is caught at ſea, and in the mouths of large rivers. Athenæus and Oppian call it Γόγγρον; the Latin writers in general Congrus and Conger; ſome of them Gonger, and ſome Gryllus.

The flat-tailed, Sea-ſerpent.

Muræna roſtro acuto, maculis albis vario, margine pinnæ dorſalis nigro.
The Muræna, with the ſnout ſharp and ſpotted with white, and with the edge of the back fin black.

This has much of the general reſemblance of the eel, but the body is leſs rounded; it grows to between three and four feet in length, and, when of this ſize, is as thick as a man's wriſt: the head is ſmall, and the roſtrum acute; the colour is a duſky greeniſh olive, but it is variegated with a number of white blotches: the opening of the mouth is very wide, and the teeth are numerous, ſharp, and larger than in the common eel: the eyes are ſmall and round; the iris ſilvery, and the pupil is black and round: the iris has ſometimes a red tinge, but this is leſs common: the pectoral fins are ſmall, and are ſituated juſt by the apertures of the gills: the body is ſomewhat compreſſed all the way, but in particular from the anus to the tail it is very remarkably ſo: the back is of a yellowiſh colour, but there are ten elegant, ſilvery ſpots running along the middle of it: the belly is of a ſilvery white; the back fin ſurrounds the tail, and comes up to the anus; it is ſupported by a great number of rays, but they are ſlender, and the membrane that covers them is ſo thick, that they are not eaſily counted.

This ſpecies is frequent in the Mediterranean, and has ſometimes been caught in our ſeas; I remember to have ſeen ſmall ones often on the coaſt of Yorkſhire. Ariſtotle, Ælian, and Athenæus call it Μύρον; Pliny, Smyrus and Myrus; moſt of the Latin writers, after him, have called it Myrus; Gaza calls it Myrus; Ray, Willughby, and many others call it Serpens marinus alter caudâ compreſſâ.

The Sea-ſerpent.

Muræna exacte teres, cauda acuta, apterygia.
The cylindric Muræna, with the tail naked and acute.

This has vaſtly more the appearance of the ſerpent-kind than the former ſpecies, but both of them are properly and punctually Murænæ: as the ſpecies hitherto deſcribed have only three fins, this, from a diſcontinuity of the ſurrounding fin, has four; it grows to five feet in length, and to the thickneſs of a man's wriſt: the head is ſmall, and the roſtrum acute, but the opening of the mouth is very large, and is furniſhed with a vaſt number of very ſharp and ſtrong teeth, of unequal ſizes: the eyes are ſmall; their iris is of a gold yellow, the pupil round and black: the colour of the head is a duſky olive, with an admixture of grey: the body is not at all compreſſed, but perfectly rounded; the back and ſides are of a duſky olive, and the belly of a ſilvery white: the lateral line is broad and pale, and is not ſtraight as in the others, but very ſinuous and crooked: the pectoral fins ſtand juſt at the apertures of the gills; they are ſmall, and they have each ſixteen rays: the back fin does not ſurround the body, or indeed reach to the extremity of it, but is terminated at ſome distance above it: the pinna ani is placed over-against this, and terminates in the same manner, short of the extremity of the body, so that the tail is naked; it is rounded, and terminates in a sharp point: the pectoral fins are of a pale brownish-grey; the back fin and pinna ani are of a deeper grey, with a tinge of bluish, and are edged with black: the whole fish has much the appearance of the others of this genus, but for the singularity of the naked tail; which, together with the roundness of the body, obtained it the name of the sea-serpent.

It is frequent in the Mediterranean, and is sometimes caught about the coasts of France. Aristotle and others of the old Greek writers call it Ὄφις θαλάττιος; and from them the Latins in general Serpens marinus. Most of the writers on fishes have figured it, but few of them correctly.

Muraena teres, gracilis, maculosa, cauda tereti cuspidata apterygia.
The slender, spotted Muraena, with a pointed, naked tail.

The spotted Sea-serpent.

This is one of the slenderest, in proportion to it's length, of all this genus: it grows to four feet in length, and is not thicker than a well-fed eel of two and a half: the head is small; the mouth is large, and the palate, as well as jaws and fauces, are all furnished with strong and sharp teeth, which point inwards: the eyes are small, the nostrils very conspicuous; the whole body is rounded, not compressed: the back and sides are of a dusky livid colour, with some tinge of yellowish in it, and are spotted in an irregular manner, with a pale whitish: the belly is also whitish; the pectoral fins are small, and placed just at the apertures of the gills; the back fin is low and long, as is also the pinna ani; they are both of them whitish, as are also the pectoral ones, and none of them are at all spotted: the tail is naked, and is not compressed, but rounded, and terminates in a sharp point.

This is frequent in the Mediterranean, and in some other seas; I have seen it caught on the Sussex coast. Ray after Lister calls it Serpens marinus maculosus.

Muraena pinnis pectoralibus carens.
The Muraena, with no pectoral fins.

The Murena.

This is a very singular species; it has properly only one fin, which is the pinna dorsi surrounding the tail, and running up to the anus: it grows to about two feet, and to the thickness of a well-fed eel of the same length: the head is small; it's extremity acute, and the opening of the mouth very large; there are much fewer teeth than in the other Muraenae, there being only one row in each jaw: the eyes are small; their iris is yellow, and the pupil round, small, and of a bluish black: the back and sides of the fish are variegated with a deep brown, a blackish, and a yellowish tinge, and the belly is whitish: there are no pectoral fins, but the dorsal one is long, and surrounds the tail.

This species is frequent in the Mediterranean, and in many other seas. Aristotle, Aelian, Oppian, and Athenaeus call it Μύραινα; and from them the Laun writers in general Muraena; some, Murena, but this improperly; Columella calls it Flora.

OPHIDION.

THE body of the Ophidion is long, subcylindric, and has three fins: the branchiostege membrane contains seven bones; they are oblong, slender, and somewhat crooked, and are very difficultly distinguished, unless the fish have the skin first taken off.

Ophidion cirris quatuor in maxilla inferiore.
The Ophidion, with four beards on the lower jaw.

This grows to the size of a moderately large eel, but much smaller ones are greatly more frequent: the head is small, and the nose pointed: the opening of the mouth is not very large, but the teeth are numerous and strong; the upper jaw has a few

few spots of a pale colour on it, and from the under jaw there hang four beards or cirri; they are short, of a cylindric figure, and whitish: the eyes are small, and the apertures of the nostrils are not so distinct as in the eel kind; the back is convex, and the sides prominent, so that the whole body appears rounded; the upper part of the back is of a blackish hue; the lower part, toward the sides, of a livid tinge, with an admixture of olive, and the sides themselves, especially toward the belly, are principally of an olive colour; the belly is white: the pectoral fins stand at the opening of the gills, and are small and obtuse: the back fin surrounds the tail, and returns up to the anus, which is situated at a considerable distance above it; it is low, but has a great number of rays, all of them bifid at the summit.

This is frequent in the Mediterranean, and is sometimes caught on the coasts of Sweden and Denmark. Pliny calls it Ophidium; Rondeletius, Ophidion; Gesner, Aldrovand, and Jonston, Ophidion Plinii; Bellonius calls it Gryllus vulgaris asselli species; and Pliny, in another part of his work, describes it a second time under the name of Pisciculus Congro similis.

Ophidion cirris carens.
The Ophidion, without beards.

This grows to about a foot and a half long; but the greater number that are met with are much smaller: the head is small, depressed a little toward the crown, and acute at the extremity; it's colour is a bright olive, with a strong admixture of yellow: the eyes are small; their iris is yellow, and the pupil is small, round, and blue: the back is rounded, as indeed is the whole body; there are no beards on the under jaw, and the opening of the mouth is not nearly so large as in the former: the pectoral fins are short, and stand near the apertures of the gills; the back fin is long, and goes round the tail up toward the anus: the back is of a dark colour, approaching to chesnut; the sides are yellow, and the belly is whitish.

It is frequent in the Baltic, and in some other seas. Rondelet calls it Ophidion flavum sive Ophidion imberbe; Willughby and Ray, Ophidium alterum sive flavescens.

ANABLEPS.

THE Anableps has only one small fin at the extremity of the back: the branchiostege membrane contains six bones.

Of this genus there is but one known species, which needs therefore no further description.

GYMNOTUS.

THE Gymnotus has no back fin: the branchiostege membrane contains only five bones.

Of this genus also there is only one known species; this is the Carapo of Marcgrave, and is sufficiently distinguished by the generical characters, without any further description.

F I S H E S.

Class the Second.

ACANTHOPTERYGII.

Fishes which have the tail perpendicular, and have the rays of the fins bony, and some of them prickly, and have the branchiae ossiculated.

BLENNIUS.

THE head of the Blennius is compressed, usually obtuse in the anterior part, and declivous* from the eyes to the extremity of the mouth; the body also is of a compressed figure, and more or less variegated in colour: the jaws are covered with large

large lips; there is only one fin on the back, and that is extended from the head to the tail, or nearly ſo: this ſometimes contains ſimple, and often aculeated, rays, and ſometimes there are one or two appendages on the front of the head, toward the eyes: the ventral fins are placed very forward, and contain only two rays; the eyes are covered with a ſkin: the branchioſtege membrane on each ſide contains ſix diſtinct bones.

Blennius ſulco inter oculos, macula magna in pinna dorſi.
The Blennius, with a depreſſion between the eyes, and a large ſpot on the back fin.

The Butterfly-fiſh.

This is an extremely pretty fiſh; it grows to ſix or ſeven inches in length, and is moderately thick, in proportion: the head is large, and there are two ſmall pinnules on it, one over each eye, but theſe are ſometimes wanting: the eyes are large, and their iris is red: the opening of the mouth is moderately large; the body is ſomewhat compreſſed, but not much ſo; the colour is an elegant mixture of blue, green, and grey, but the laſt is the predominant one: the lateral line is tolerably broad and crooked; the back fin is very long, but not very high toward the hinder extremity; ſome of the rays of it are very ſharp and prickly, others leſs ſo; and there is a large and beautiful ſpot on it, like thoſe ſpots which we call eyes, in the wings of our butterflies, whence the fiſh has it's name: the tail is of a rounded figure, and has twenty-ſix rays: the pectoral fins have each twelve rays; and the ventral ones, which are ſituated anteriorly, have each only two rays: the pinna ani has ſeventeen rays.

This ſpecies is taken in great abundance in the Mediterranean, and, when freſh caught, is extremely beautiful in it's colours; but they go off in a great meaſure, as it dies; the ancients were very well acquainted with it. Athenæus calls it [illegible]; and Oppian, [illegible]; Pliny calls it Blennius; Salvian and others, Blennus; others call it Blennus Salviani and Blennos Bellonii; the Italians, Meſoro.

Blennius pinnulis duabus ad oculos, pinna ani oſſiculorum viginti trium.
The Blennius, with two pinnules at the eyes, and twenty-three rays in the pinna ani.

The Gattorugine.

This is alſo a beautiful fiſh; the length it uſually grows to is ſeven or eight inches, and it's thickneſs, in proportion, is moderate: the body is covered with an unctuous, lubricous matter, and is of a beautiful variety of colours: the head is compreſſed and obtuſe; the eyes are large; the opening of the mouth is moderately wide, and the anus is ſituated much nearer the head than the tail: the back fin is very long, and has thirty-one rays, ten or twelve of which are rigid and ſomewhat prickly; in ſome of them there is a black ſpot toward the origin, but not in all: the pectoral fins have each fourteen rays; the ventral ones ſtand very forward, and have only two: the pinna ani has twenty-three rays; the tail is even at the extremity, and has twelve rays; there are two pinnules on the head, one placed over each eye; theſe have uſually four rays each.

This ſpecies is frequent in the Mediterranean, and is ſometimes caught about the ſhores of France, though there it is rarely more than four inches in length. Ray and Willughby call it by it's Venetian name, Gattorugine; Willughby, indeed, adds Alaudæ Rondeletio dictæ affinis, Exocæto primo Bellonii ſimilis, ſi non idem, or Piſcis Guttoroſus Geſneri.

Blennius maxilla ſuperiore longiore, capite ſummo acuminato.
The Blennius, with the upper jaw longeſt, and the top of the head acuminated.

The Mulgranoc, or Bulcard.

This is like the other Blennii, a very beautiful fiſh; it's length is about ſix or ſeven inches, and it is moderately thick, in proportion: the head is in general compreſſed, but it is alſo depreſſed on the hinder part, and acuminated on the anterior part of the front, ſo that the whole figure of it is very irregular: the eyes are large, and their iris is red; the opening of the mouth is large, and there are no pinnules over the eyes; the upper jaw falls over the lower, and both are covered with thick lips: the whole body is covered with a lubricous fluid, and the back fin reaches almost from the head to the tail: the ground colour of the fish is a greyish olive, but it is elegantly variegated with tinges of bluish, greenish, and several other bright colours, most of which go off, in a great measure, when it dies: the tail is large, and is circular at the extremity; the back fin has thirty-six rays; the pinna ani twenty-eight; the pectoral fins have eleven rays each, and the ventral ones only two.

This species is caught in the Mediterranean and the Ocean; we have it frequently about the shores of Cornwal, and some other of the western parts of England. Rondelet, Gesner, Jonston, and many others call it Alauda non cristata; Aldrovand, Alauda marina; the Italians call it Galesto; and we, the Bullard.

Blennius crista capitis transversa cutacea.
The Blennius, with a transverse, cutaceous crest on the head. **The crested Alauda.**

This is a very pretty fish, and it's head is of a very singular figure; it is compressed, and somewhat acuminated on the top, though less so than that of the former species; but what gives it the great singularity of appearance, is what authors call the crest; this is a lobe of a triangular figure and cutaceous substance, red at the edges, and paler elsewhere, and is situated between the eyes: the body of the fish is variegated with several bright colours, upon a dusky livid olive: the back fin reaches nearly from the head to the tail; the pectoral fins are larger than in the former species, and obtuse; and the belly fins are placed very forward, and have only two bones each.

This species is frequent in the Mediterranean, and is caught also in many other places. Rondeletius calls it Galerita; Jonston and Aldrovand, Alauda cristata; Gesner and Willughby, Alauda cristata sive Galerita; and Gesner, Alauda cristata sive Galerita prima.

Blennius maculis circiter decem nigris, limbo albo, utrinque ad pinnam dorsalem.
The Blennius, with about ten black spots, with a white edge, on each side, at the root of the back fin. **The Butter-fish.**

This grows to about six inches in length, and an inch in breadth; the body is somewhat compressed, so that it is less than that in diameter: it's whole body is elegantly variegated with a reddish-brown, a deep olive colour, a bright green, and a deeper and more dusky green and white; and on each side, at the root of the back fin, there are ten or twelve elegant, round, black spots, each surrounded with a white limb or verge: the head is small and compressed; the mouth is large and bent upwards; the eyes are small, and are covered with a skin; the iris is of a gold yellow, or sometimes of an orange colour; there is only a single row of teeth in the jaws, but there are some bones also in the fauces: the scales are very minute, and, indeed, on the living fish are scarce perceptible: the pectoral fins are rounded, and have each eleven rays; their colour is an elegant yellow: the ventral fins stand almost in the middle of the breast, and nearly perpendicular under the others; the back fin runs the whole length of the back, and seven or eight of the rays of it are prickly: the pinna ani is long, and of a yellow colour, spotted with brown; it has forty rays, and the two first of them are aculeated; the tail is roundish.

We have this species in our own seas; it is frequently caught upon the coast of Cornwal, where our people call it the Butter-fish. Authors call it the Gunellus Cornubiensium and Lipara.

The other species of the Blennius are, 1. The Blennius, with pinniform and bifid cirri under the throat, and with transverse areolæ on the back; this is called by authors the Lumpen Antverpiæ, the Mustela vulgaris, and the Galea piscis. 2. The Blennius, with the head and back of a brownish-yellow, spotted with black, and with the pinna ani yellow; this is called by authors Mustela vivipara, Mustela marina vivipara, and tertia Mustelæ species.

GOBIUS.

THE ventral fins in the Gobius grow together in such a manner, as to form only one simple one, which has somewhat of the shape of a funnel; and these are situated at the same distance from the extremity of the head with the pectoral ones; there are two fins on the back, and the anterior one is composed of subrigid rays: the scales are rough; the body is oblong, and the head is subcylindric, but somewhat depressed; the eyes are covered with the common skin of the head: the branchiostege membrane on each side contains five very distinct bones; these are unequal in size, for the first or uppermost, and the fourth, are much broader than the others.

Gobius varius pinna dorsi secunda ossiculorum quatuordecim.
The variegated Gobius, with fourteen rays in the hinder back fin.

The Sea-gudgeon.

This grows to eight inches in length, and it is tolerably thick, in proportion; it is very beautifully variegated in colour, the back and sides being elegantly tinged with brown, white, yellow, green, blue, and black: the tail, the back fin, and pinna ani are of a pale blue, and the belly and the coverings of the gills are yellow: the eyes stand very close to each other, and look upwards; they are covered with one common membrane: their iris is yellow, but not purely so, but spotted; the mouth is very large, and there are sharp and strong teeth both in the fauces and the palate, as well as in the jaws: there is a depression or furrow on the back, between the head and the first back fin; this fin has only six rays; the hinder one has fourteen; the pinna ani also has fourteen rays; the pectoral ones each seventeen: the whole body of the fish is covered with rough scales, which, with the beautiful variety of the colouring, make an elegant figure.

This fish is frequent in the Mediterranean, and in many other seas; it is caught, in great abundance, in the Baltic, and is not very unfrequent about the coasts of the West of England: the antients were acquainted with this species. Aristotle calls it Κωβιὸς; and Ælian, Oppian, and Athenæus, Κωβιὸς; Pliny, Columella, and Juvenal call it Gobio; Salvian, Gobio marinus; Bellonius and Gesner, Gobio marinus niger; and Charleton, Gobius, or Cobius, marinus; the Venetians call it Go and Gogus.

Gobius linea lutea transversa in summa pinna dorsalis prima.
The Gobius, with a yellow, transverse line on the top of the first back fin.

The Paganellus.

This grows to about six inches in length, and is moderately thick, in proportion; the body is of a rounded figure, but somewhat compressed: the eyes are small, and stand near one another on the top of the head: the head is not large, in proportion to the body, but the mouth is wide, and is very well furnished with teeth; there is a depression from the back of the head to the beginning of the back fin, but it is less in this species than in the former: the colour of the body is in general paler than in that species, but it has the same variegations; the ventral fins are connected in a remarkable manner together, and have more of the funnel-like appearance in this, than in any other species.

We have this very common about the coasts of the northern parts of England. Rondelet calls it Gobio; and Bellonius and Gesner, Gobius albus; others call it Gobius marinus and Paganellus veterum; Jonston and Aldrovand call it Paganellus sive Gobius major ex Gesnero; and the Venetians, Paganello.

Gobius

Gobius pinna ventrali cærulea, officulis pinnæ dorsalis primæ assurgentibus.
The Gobius, with the ventral fin blue, and the rays of the back fin assurgent.

The Jozo.

This grows to six or eight inches in length, and to about an inch in diameter: the head is thick, but it is somewhat compressed; the body is rounded, but is also a little compressed; the opening of the mouth is large: the eyes are large, and stand near one another on the upper part of the head; their iris is of a silvery white; the pupil is large and blaish; the colour of the body is a pale bluish-grey; the fins are also of a greyish colour, and the tips of them blue; the belly fins in particular, which are beautifully connected together into the form of a funnel, are entirely blue: the scales are small and rough: the lateral line is broad, conspicuous, and black; the rays of the former, or more anterior, of the back fins, stand up beyond the edge of the membrane.

This species is frequent in the Mediterranean; it was very well known to the antients. Aristotle calls it Κωβιὸς λευκὸς; and Athenæus, Κωβιὸς λευκότερος; Rondelet, Gesner, and others, Gobio and Gobius albus, confounding it with the former species; Willoughby calls it Gobius tertius Jozo Romæ Salviani.

Gobius uncialis pinna dorsi secunda officulorum septendecim.
The small Gobius, with seventeen rays in the second dorsal fin.

The Aphya Cobites.

This is a very pretty, though a very small, fish: the head is short and somewhat compressed; the body is rounded, and it also somewhat compressed: the length of the whole fish is but about an inch and a half; the mouth is small, and the eyes also are small: there is a small sulcus or depression behind the head, running to the back fin: the scales are small and rough; the colour of the whole body is a pale yellowish-grey. The first of the two back fins has six rays; the second has seventeen; the pectoral fins have each seventeen rays; the ventral ones have only six rays each, and are connected together as in the other species: the pinna ani has eleven rays; the first rays of the first back fin are rigid and prickly: the fins are all of a pale greyish colour, but those of the back and the tail are variegated with transverse lines of brown.

This species is very frequent in the Mediterranean; it swims in vast shoals about the shores there, as our minnows do in the fresh waters: it was very well known to the antients. Aristotle calls it ἡ Κωβῖτις; and Athenæus, Ἀφύη κωβῖτις; from this name most of the Latin writers have called it Aphya Cobitis; some, Aphya Gobitis and Gobitaria; Gesner and Aldrovand call it Marsio and Marsio Venetorum; the Venetians call it Marsione and Pignoletti. It has been esteemed by many writers to be only the young of some of the larger species; but this is an error, for it never exceeds this size, and it's specific characters are different from those of all the others.

XIPHIAS.

THE rostrum or extremity of the head of the Xiphias is continued forward, with an extremely long point, of a depressed or somewhat flatted figure, resembling the blade of a sword, and of a bony structure: the body is oblong, and of a rounded figure; there are no belly fins, and on the back there is only one fin, which is very long, and lowest in the middle: the branchiostege membrane, on each side, contains only eight bones.

Of this singular genus there is but one known species.

XIPHIAS.

The Sword fish.

This is a very large fish; about fifteen feet in length is the size of a moderately large one, but it is not unfrequently is met with much bigger: the body is of a rounded rather than a flatted figure, and is considerably thick, in proportion to it's length: the back

back is convex, and the sides are rounded; the rostrum is continued into a very long prominence, of the figure of a sword, of a bony structure, and sharp at the point: the lower jaw is acute, and of a somewhat triangular figure; the opening of the mouth is not large; there are no teeth in it: there is but one back fin, but it is of a great length, reaching nearly from the head to the tail; it is lowest in the middle, and has forty-one rays, of which the twenty-five middle ones are very short: the pectoral fins stand low, and have seventeen rays each; there are no belly fins: the pinna ani is lowest in the middle, and has fifteen rays: the tail is extreamly forked, or rather it is very deeply lunated: on each side, toward the tail, there is a great longitudinal eminence, and the anus is situated very near the tail: there are five or six cartilages, with rough surfaces on each side of the jaws, which serve in the place of teeth.

This is frequently met with in the American Seas, and in those of some other parts of the world; it was well known to the antients. Aristotle calls it Xiphias; and Athenæus, Oppian, and Ælian use the same name; the Latins hence call it also Xiphias; and some of them, Ziphius and Zifius; Gesner calls it Xiphias, id est, Gladius piscis; Jonston, Xiphias sive Gladius; Ray and Willughby, Xiphias piscis, Latinis Gladius; and many other writers, simply, Gladius; the Italians call it pesce Spada; and the Genoese, Emperador.

SCOMBER.

THE tail of the Scomber is extreamly forked, so as to represent the figure of a crescent: there are, on each side, one or more eminences in a longitudinal direction, near the tail: there are on the back, in some species, only two fins, in others there are, beside these, a great number of extreamly small ones toward the tail, as well on the upper as the under part: the branchiostege membrane contains seven slender bones; the uppermost of these is in great part covered by the operculum of the gills.

Scomber pinnulis quinque in extremo dorso, spina brevi ad anum.

The Scomber, with five pinnules at the extremity of the back, and a spine at the anus. **The Mackarel.**

The head of the Mackarel is large, and of a compressed figure: the body is also somewhat compressed, but the sides are fleshy and prominent, or rounded; the back is throughout convex, and the belly is also convex, but less so than the back: from the ventral fins to the tail the body grows gradually smaller and narrower, but just above the tail it is round: the mouth is wide; the jaws are, when the mouth is shut, exactly of the same length; but, when it is open, the lower appears somewhat the longest: the rostrum or anterior part of the head is subacute; the upper part of the head is plane, smooth, and blackish: the nostrils have a double aperture, and are placed nearer to the eyes than the extremity of the head; the anterior foramen of each is round, open, and conspicuous; the posterior is placed at a considerable distance from this, and is a kind of transverse fissure, and usually is shut: both have a large cavity within the head, with which they communicate.

The eyes are large, and placed at a distance on the sides of the head; the pupil is round, and the iris is of a mixt silvery, greenish, and blackish colour; they are naked in the middle, but, both before and behind, they are covered with a skin or film.

The coverings of the gills are large, and of a silvery colour, and in the hinder part they do not terminate either in a round or acute extremity, but in a kind of straight line: the apertures of the gills are very large: the teeth are very numerous and sharp; there is a single series of them on the verge of each jaw, and there are also two longitudinal lines, one on each side, at the edges of the palate, in each of which there are two rows of teeth of the same kind: there are also, in the upper part of the fauces, four bones of an oblong figure, all of which are extreamly thick set with teeth; the hinder pair of these are much the largest; there are, on the lower part of the fauces also, two bones of an oblong figure, which are covered with teeth, but they are less rigid than the foregoing: the palate in the middle, and on it's whole anterior part,

part, is smooth; the tongue also is smooth, and it is whitish, and in part loose on the under side.

The lateral line is nearer the back than the belly; it is not perfectly straight, but flexuous, and is placed a great deal above the line of the muscles: the scales are very small; they are whitish, and are placed in an imbricated manner, and adhere extreamly firmly to the skin: the belly and sides below the line are of a fine silvery white; from the line to the back there are a number of elegant, broad, oblique, flexuous lines, of a bluish, a greenish, and a black colour; and the spaces between these are of a silvery hue, with a strong cast of an elegant bluish-green: the outer skin of this fish is thick, and the inner one extreamly thin, and scarce separable from the muscular flesh; yet the whole colouring of the fish, strong and elegant as it is, is lodged, not in the outer, but in this inner skin.

The pectoral fins are blackish, and are placed on the sides, nearer to the back than to the breast; they have each twenty rays: the ventral fins are white, and are placed very near one another; they stand at a little more distance from the head than the pectoral ones, and have each six rays: the back fins are two larger, the one not far from the head; this has also twelve rays; the other is situated at the very extremity of the back, and has eleven or twelve rays: there is a furrow in the back, behind the first fin; behind the latter there stand several small pinnules; the first of these has only one ray, but that is very ramose; the others have two rays each: these pinnules are connected to one another by a low membrane, and they are also connected, by a continuation of the same membrane, to the hinder fin of the back, so that they may, in some degree, be called only a continuation of that fin: the pinna ani is whitish, and has thirteen rays; and the anus is much nearer to the head than to the tail: the tail is of the form of a crescent, and is supported by seventeen rays, and there are on each side two eminences just at the tail. The usual size of the mackarel is about a foot, but is sometimes considerably larger: mackarel of twenty inches are sometimes met with, but they are rare: the vertebræ are large, oblong, and are only twenty-one in number: the ribs are long, and are but slightly affixed to the vertebræ; they are eleven or twelve.

This species is very frequent about our own shores; the antients were well acquainted with it. Aristotle, Ælian, Athenæus, and Oppian call it Σκόμβρος; and from them the Latins, in general, Scomber and Scombrus; the Neapolitans call it Lacerto; the Spaniards, Caballo; we, the Mackarel.

Scomber pinnulis octo vel novem in extremo dorso, sulco ad pinnas ventrales.

The Scomber, with eight or nine pinnules on the hinder part of the back, and a furrow at the belly fins.

The Tunny.

This has very much the shape and general figure of our common mackarel, but it grows to seven feet long: the head is large and somewhat compressed; the eyes are large; the upper and lower jaw are both of the same length: the teeth are numerous; the jaws, the palate, and the bones in the fauces, are all covered with them: there are two large fins on the back, each of which has fourteen rays: the pectoral fins are moderately large, and have each thirty-four rays; the ventral fins have only six rays each; the pinna ani has thirteen; behind the hinder back fin there are eight or nine, and sometimes ten, small pinnules, like those of the common mackarel, on the back; and there are eight smaller ones over-against them on the belly.

The back of this fish is rounded; the sides are fleshy and prominent, and the belly also is somewhat rounded: the lateral line is not perfectly straight: it runs nearer the belly than the back; the tail is very forked; the belly and sides are of a fine silvery white up to the lateral line; above that they have an elegant variety of colouring; and the back is dusky or blackish, with a fine strong tinge of green in some lights.

This species is frequent in the Mediterranean; it is eaten fresh in great abundance, and a much larger quantity is pickled in the manner of herrings, and sent to the several parts of the Mediterranean, where it is a considerable article of commerce: the antients were well acquainted with it. Aristotle, Ælian, Athenæus, and Oppian call it Θύννος; and from them the Latin writers, Thunnus and Thynnus, and some Thinnus: they have, however, from accidental varieties of the fish, described it under

 other

other names, as if of two different species, beside the ordinary one. Under the one of these appearances, Ælian, Athenæus, and Oppian call it Ὄρκυνος; and from them Pliny, Gesner, and Aldrovand, Orcynus; Salvian has figured it in this state, under the name of Limosa. Under another appearance, Aristotle, Athenæus, and Oppian have called it Πηλαμύς; and from hence Pliny, Bellonius, and others have called it Pelamys, and Pelamis vera, Thynnus Aristotelis; Gaza calls it Limaria: and, finally, the old Greeks have also described it over-again, under the name of Κορδύλος; and hence others have called it also the Cordylos and Cordyla. All these names have been given it, only from the error of supposing the species different, when fish have been caught of different sizes, under the full growth.

Scomber linea laterali aculeata, pinna ani ossiculorum triginta.
The Scomber, with the lateral line prickly, and with thirty rays in the pinna ani.

The Horse-Mackarel.

This species is, by the generality of the world, supposed to be the same with the common mackarel; we eat it under the name of the same fish, but distinguish it not to be equal in flavour to what we take to be the same in a higher season; but it is an error; the Horse-mackarel, as we call it, is a very distinct species, nor has the characters, any more than the taste, of the common kind.

It's usual length is about eleven inches, but it will grow to fourteen or fifteen: the head is large and compressed; the eyes are large, and their pupil round; the iris is of a silvery hue, with a mixture of changeable colours in it: the back is convex, as is also the belly; the sides are somewhat prominent, but the body is of a much more compressed form than that of the common mackarel: the lateral line is crooked and serrated, and there is a furrow or depression in the middle of the back: the mouth is very large, and the teeth are numerous and sharp; the jaws, the palate, the tongue, and the fauces are all furnished with them: there are two fins on the back; the first has only eight rays, the second has thirty-four: the pectoral fins have twenty rays each, the ventral ones six, and the pinna ani has thirty; the two first of these are aculeated: the tail is very deeply forked: the back of the fish is of a deep blackish colour, with an admixture of bluish and greenish; the sides are elegantly variegated with blue, green, and black; the belly is of a silvery white.

We have this in great abundance in our own seas; it is caught, in great quantities, a little before the mackarel-season, and is brought to London, where it is sold under the name of the Horse-mackarel; we also, in some places, call it the Scad: the antients were very well acquainted with it. Aristotle calls it Σαῦρος; Ælian, Τραχοῦρος; Athenæus, Oppian, and Galen also call it by the latter name; Paulus Jovius and Salvian call it also Saurus; Bellonius, Lacertus sive Trachurus; Gesner, Lacertorum genus quod Trachurum Græci appellant; others call it simply Trachurus and Lacertus; the Italians, Suaro.

Scomber ossiculo ultimo pinnæ dorsalis secundæ prælonga.
The Scomber, with the last ray of the hinder dorsal fin very long.

The Amia.

This is one of those species which have only the two principal fins on the back, without any of the little pinnules, which in the others run from the hinder fin to the tail: the fish is nearly of the shape of the common mackarel, but it is greatly larger; it's usual length is about three feet; it's head is large and compressed, and the body is also somewhat compressed, but the sides are thick and fleshy: the eyes are large; the anterior apertures of the nostrils are very conspicuous: the back is convex, and of a deep bluish-black: the sides are, on the upper part, elegantly variegated with green, blue, and black; and the lower part of them, as also the belly, are whitish and silvery, tho' often there is a slight tinge of flesh colour: the body grows small toward the tail, and is of a somewhat squared figure in that part: the mouth is large, and well-furnished with teeth; they stand on the tongue and palate in great numbers, as well as in the jaws and fauces.

The pectoral fins have each twenty rays; the ventral fins have only six rays each: the anterior back fin has only five rays, but the posterior one has thirty-four; this is lowest in the middle: the pinna ani has twenty-four rays; this is very much of the figure of the hinder back fin, the tail is very forked.

This species is frequent in the Mediterranean. Aristotle, Ælian, Athenæus, and Oppian call it [illegible]; and three of them also call it, in other places, [illegible], as if it were of a different species; Ælian describes it again also under the name of [illegible]. The Latins in general call it Amia; Isidore and some others, Hamia; Willughby, Amia Salviano; Paulus Jovius, Lechia; and many others, Glaucus. Charleton calls it Glaucus major; Jonston, Glaucus veterum; Gesner, Glaucus major hexacentrus and Glaucus major seu prima species. Ælian does not give this species the name of Glaucus, as the other Greeks do, for his Glaucus is a species of Squalus.

Scomber ossiculo secundo pinnæ dorsalis secundæ altissimo. **The**
The Scomber, with the second ray of the hinder back fin tallest. **Glaucus.**

This is also one of the Scombers which have only the two large back fins, and none of the pinnules toward the tail; this is extremely different from all the other species in shape: the body is broad and thin, and approaches to a rhomboidal form; the head is large and compressed; the back is elevated, and there are three or four remarkable brown spots on the sides; the lateral line is nearly straight, for the most part of it's course; the tail is very broad, and is extremely forked; scarce any other fish has it so much so: the pectoral fins have each eighteen rays, and the belly fins have only five each: the first of the two back fins has only seven rays, and those very short; the first of these is bent forwards, the others all turn backward: the hinder back fin has twenty-six rays; the pinna ani has twenty-seven rays, and two of them are prickly.

It is frequent in the Mediterranean: the generality of authors call it Glaucus; Willughby, Glaucus primus.

M U G I L.

THE scales of the Mugil are large, and cover not only the body but the head itself to it's very extremity, and also the coverings of the gills: the head is of a depressed form in the anterior part, and the body is oblong and compressed: on each side of the head, below the nostrils, there stands a little bone, which is serrated on it's lower part: the eyes are not covered with a skin, and there are teeth on the tongue and palate, but none in the jaws or fauces: the branchiostege membrane on each side contains six officles; these are of a crooked figure, and the upper one, which is the broadest, is covered by the coverings of the gills in such a manner, that only five appear.

Of this genus there is but one known species.

MUGIL. **The Mullet.**

The Mullet grows to about a foot in length, sometimes to considerably more, and is moderately broad, and thick, in proportion; the body is somewhat compressed, especially toward the tail: the head is of a roundish figure, but angulated, and is compressed, in some degree, in manner of the body behind, but in the anterior part it is depressed; it's hinder part toward the back is broad and flat, not assurgent: the anterior part of the back is somewhat angulated, especially toward the head, but from the back fin to the tail it is convex and rounded; the belly, from the head to the ventral fins, is flatted, but from those fins to the tail it is ridged or convex: the mouth is extremely small, and, when shut, the upper jaw is a little longer than the under, and on pressing bends downward; the verge of the under jaw is thin, and of a bright red, and at it's extremity there is a bony prominence, which is received into a cavity in the upper; this rises from the bone of the lower jaw: the apertures of the nostrils are two; they are situated somewhat nearer the eyes than the extremity of the head: the anterior one on each side is small and round; the posterior large and oblong, and placed transversely: the eyes are roundish, and stand at a distance from one another; they are situated near the extremity of the head, and have no covering: the iris is silvery; the pupil is round and black: the coverings of the gills are hard and bony, and the

the apertures are large: there is on each side under the nostrils, in the upper jaw, a little bone, which is furnished with small teeth on it's under part.

The scales of this fish are large, and cover not only the body but the head, and some part of the fins; they are placed in an elegantly imbricated manner, and are smooth, and somewhat hard, but not rigid or rough: on the belly and sides they are of a silvery white, and on the back they are greyish, with a tinge of brown; they are rounded at their extremities, and striated longitudinally: there is properly no lateral line on this fish, but there are ten or eleven longitudinal lines, of a blackish colour, which run parallel through the middle of the scales, from the head to the tail.

There are two fins on the back: the first is small, and stands nearly in the middle of the back; this has five prickles, and sometimes only four: the two anterior ones are largest and robust; the others are softer and flexile: the hinder back fin is at a distance from the other, and stands about the mid-way, between that and the tail; this also is small, and has ten or eleven rays: the first of these is shortest, and is aculeated; the third and fourth are longest; all of them have a direction toward the tail, and cannot be set quite upright: the pectoral fins are placed almost transversely under the gills; they have each eighteen rays; the four first largest, the others very small: the ventral fins are white; they stand nearer the pectoral fins than to the anus, and are firmly joined together by a membrane at their base; these have each six rays, of which the first is shortest and prickly, the rest are ramose and longer: there is at the base of each of these fins a fleshy appendage, covered with scales: the pinna ani is also white, and has twelve rays, or sometimes thirteen; the two or three first of these are prickly, the others ramose, and cannot be raised perpendicularly: the anus is much nearer the head than the tail; the tail is forked or semicircular, and has fourteen long rays.

The mouth is furnished with a number of small teeth: the tongue is cartilaginous, and has a series of them on each side, and on each side of the palate also there is a little bone covered with teeth: on the rim of the upper jaw also, behind the snoutlet, there are a number of fine teeth resembling hairs; but the middle of the palate, the fauces, and the under jaw, are all entirely smooth: the flesh is white and firm, and is very well tasted; it is composed of large muscles, and has no spines among it: the vertebræ are oblong and large, and are only twenty-four in number: the ribs are long and large, and are somewhat compressed; the air-bladder is large and simple; it is extended the whole length of the abdomen, and fixed to the back-bone: the heart is large and quadrangular; the liver is small and undivided, but the gall-bladder is large; the spleen is long, angular, and black; the intestine is three times the length of the fish, and has about ten volutions.

The Mullet was well known to the ancients. Aristotle, Oppian, and Ælian call it Κέφαλος, Κεστρεὺς; and the latter sometimes, [illegible] and [illegible]; Gaza calls it Capito; Paulus Jovius, Rondeletus, and others, Cephalus; Gesner, Cephalus Mugil and Cestreus Mugil; most of the other Latin writers, simply, Mugil; Charleton, Mugil imberbis; and Salvian, Mugilis; the Italians call it Cefalo.

It is frequent in the Mediterranean, and in our seas, and is much esteemed at our tables.

LABRUS.

THE Labrus has large and thick lips, covering the teeth: the teeth are large, and, beside those in the jaws, there are in the fauces three hard and thick denticulated bones, one of them placed below, and two above, which are furnished with teeth: the palate and tongue, however, are smooth; the membrane of the back fin has, in it's anterior part, double rays, growing from the same root or base; of these one is always soft, and the other rigid and prickly: the colour of the body is elegantly variegated; the tail, in most of the species, is undivided, and the scales are large, smooth, and soft; the branchiostege membrane contains six small and somewhat broad bones, when the fish is full grown; but, in the young of most species, there can only be counted five.

Labrus

Labrus roſtro ſurſum reflexo, cauda in extremo circulari.
The Labrus, with the roſtrum turning upward, and the tail circular at the end.

The Wraſſe, or Old-Wife.

This is a very beautiful fiſh; it's uſual ſize is about ten inches in length, and it is conſiderably thick, in proportion: the body is ſomewhat compreſſed, and the back elevated; the ſcales are large and very beautiful, and the whole body is elegantly variegated, and, as it were, painted with red, yellow, and brown in ſeveral ſhades: the head is moderately large; the eyes are large, and their pupil round and black: the palate and tongue are ſmooth, but there are numerous teeth in the jaws and fauces; the back fin has twenty-ſix rays, fifteen of which are prickly; the pectoral fins have fourteen rays each, and the ventral ones only ſix: the pinna ani has thirteen, and three of theſe are prickly; the tail is large, and is ſemicircular at the extremity.

This ſpecies is frequent in the Mediterranean, and we have it alſo in our own ſeas in conſiderable plenty. Rondeletius and Geſner call it Turdus duodecimus Vielle dictus; Ray and Willughby, Turdus vulgatiſſimus, Tinca marina Venetis; Athenæus call it [illegible] and [illegible]. On the coaſt of Cornwall, where it is very common, our people call it the Wraſſe, and ſome the Old-wife.

Labrus viridis linea utrinque cærulea.
The green Labrus, with a blue line on each ſide.

The green Turdus.

This is a very beautiful fiſh; it's length is from eight inches to fourteen, and it is conſiderably thick, in proportion: the head is moderately large; the eyes are large, and the noſtrils have each a very conſpicuous, double aperture: the mouth is not very large, and the lips are thick, and the teeth numerous in the jaws; but the palate, tongue, and fauces are ſmooth: the back is elevated, and the anterior part of the belly is alſo elevated a little into a convexity: the colour of the whole fiſh is an extremely beautiful bright green, variegated with a pale blue line on each ſide; the ſcales are large, and the whole fiſh is ſmooth and ſoft to the touch, having a lubricous fluid all over it: the back fin has thirty rays, and eighteen of them are prickly; the pectoral fins are large, and the tail is undivided: the colour ſometimes varies in this ſpecies.

It is not unfrequently caught almoſt entirely black, and ſometimes ſpotted or tuberculoſe. In theſe two accidentally varied ſtates it has been deſcribed over-again by authors, as if two other ſpecies: in the firſt ſtate Willughby calls it Turdus niger and Merula Salviani; and in the other, Lepus ſive Pforus Bellonii. In it's more uſual ſtate, the antient Greeks have deſcribed it under the name of [illegible] and [illegible], and Salvian calls it Verdone: when of it's black colour, a great many have called it Merula; ſome, Merulus and Turdus niger.

Labrus pulchre varius pinnis pectoralibus in extremo rotundis.
The elegantly variegated Labrus, with the pectoral fins round at the extremity.

The Pea-cock-fiſh.

This is a very beautiful ſpecies: the head is large; the mouth is furniſhed with a great number of teeth in the jaws and fauces, and is cloſed by two thick and fleſhy lips: the eyes are large; the noſtrils have each a double aperture: the anterior round, the hinder oblong and tranſverſe; the ſcales are large, ſoft, and unctuous to the touch: the ground colour of the body is an elegant ſky blue, but it is variegated in a beautiful manner, with red, yellow, and brown, and is one of the brighteſt and gaudieſt-coloured fiſh we are acquainted with: the back fin has twenty-eight rays, and more than half of them are prickly; the pectoral fins are ſhort and broad; the belly fins are narrower, and the tail is not forked, but ſtraight at the extremity.

This beautiful ſpecies is frequent in the Mediterranean, and ſome other ſeas. The generality of writers on fiſhes have, from it's beautiful variety of colours, called it Pavo; Geſner calls it Turdus ſecundus ſive pavo colore ex cæruleo viridi; Bellonius,

 Turdus

Turdus simpliciter dictus, Ray and Willughby, Turdus perbelle pictus, or Pavo Salviani; the Italians call it Papagallo. It is apt to vary a little in colour, and under this variation has been described again by many as another species. Willughby calls it, in this state, Turdus major varius præcedenti similis; and Artedus, in his Synonyma, Labrus ex flavo et cæruleo varius, dentibus anterioribus majoribus.

Labrus oblongus viridis iride aurea.
The golden-eyed, oblong-bodied, green Labrus. **The great green Turdus.**

This is also a very beautiful species: it's head is large, and the lips very thick, firm, and fleshy; the eyes are large; they stand at a considerable distance: the iris is of a bright gold yellow, and the pupil of a bluish black: the body is longer, in proportion to it's thickness, than in any other species: the colour is a deep green, very pure and elegant; it is dusky indeed on the back, and somewhat pale toward the belly; it is on the sides that it shews the true tint: the back fin has thirty-two rays, of which nineteen are prickly: the pectoral fins are broad, short, and obtuse; the tail is large, broad, and not at all forked.

This species is very frequent in the Mediterranean, and in other places. Willughby calls it Turdus viridis major; Ray, Turdus viridis major corpore oblongo. It varies in colour, and is sometimes of a dusky brown, spotted: in this state it has been described by authors, as if another species, under the name of Turdus fuscus maculosus.

Labrus tetraodon virescens cauda bifurca.
The four-toothed green Turdus, with a forked tail. **The Indian Labrus.**

This is a very elegant species, and is considerably different in it's form from the others: the head is large and broad; the mouth is furnished with four large teeth in the front of the jaw, which are covered by a pair of thick wrinkled and furrowed lips: the eyes are large; their iris is of a silvery hue, and the pupil of an oval figure and blue: the nostrils have each a double aperture; the anterior one round, and the other oblong and transverse: the back is elevated; the belly flatted, and the sides are prominent and fleshy: the colour is a deep shining green, blackish on the top of the back, and thence growing gradually paler to the belly: the back fin has thirty rays, fourteen of which are prickly; the pectoral fins are short and obtuse, and the tail is large and forked.

We have this species sometimes sent as a rarity from the East Indies, but the Pavo of the Mediterranean is greatly superior to it in beauty; it is also caught in the Archipelago, and in some other seas. Lister and Willughby call it Turdus viridis Indicus; and Aldrovand, Scarus Creticus.

Labrus palmaris varius dentibus duobus majoribus maxillæ superioris.
The variegated, small Labrus, with two large teeth in the upper jaw.

This is a very elegant little fish; it grows to three or four inches in length, and is considerably thick in proportion: the head is small; the mouth is furnished with two very large teeth, which stand forward, and have their origin both from the upper jaw: the lips close over these, and are thick, fleshy, and firm; the eyes are large, and stand at a considerable distance from one another: the iris is of a flame colour; the pupil is bluish and round; the nostrils have each a double aperture; the lowest is round, the upper one oblong and transverse: the back fin has twenty-one rays; the pectoral ones have each fourteen; the ventral ones have six, and the pinna ani has fourteen: the tail is moderately large, and is even at the extremity: the whole body is covered with large scales, and is elegantly variegated with shades of several colours.

The antients were acquainted with this species. Aristotle, Ælian, Oppian, and Galen call it Ἰουλὶς; and from them most of the Latin writers have called it Julis; Gaza and Salvian, Julia; the Italians call it Donzellina and Zigorella; the Venetians, Donzella; and the people of Marseilles, Dovella; the Portuguese call it Zigorella; the people of Crete, Aldelins; the Neapolitans, Monchina di Re; and the people of Rhodes, Zillo.

Labrus

Labrus maxilla inferiore longiore, cauda bifurca, lineis utrinque transversis nigris.
The Labrus, with the lower jaw longest, the tail forked, and transverse black lines on the sides.

The Channa.

This is a less elegant species, in regard to colour, than some of the former: the head is large; the mouth is closed by a pair of very thick lips; the eyes are large, and stand at a distance; the nostrils have each a double aperture; the lower is round, the upper oblong and transverse; the body is thick, and somewhat broad: the sides are variegated with several transverse black lines: the back fin has twenty-one rays, ten of which are prickly: the pectoral fins have each of them thirteen rays; the belly fins have six rays each, and the pinna ani has nine: the scales are large and soft, and the whole body is covered with an unctuous fluid.

This species is frequent in the Mediterranean; and, as it varies in size, and sometimes in colour, it has been described by authors in general, two or three times over, under as many names, and as if of different species. In it's more usual form, Willughby calls it Sachettus Venetorum, or Cannadella Bellonii et Rondeletii; Ray calls it also Sachettus Venetorum; and Gesner, Sacheto; Bellonius, Aldrovand, and Jonston call it Hepatus piscis; Aldrovand, Channadella, Sachetus Venetis. In it's other varied state Gesner calls it Chanadella, vel Channadella potius, a similitudine Channæ; the Greek writers, [illegible], [illegible], and [illegible]; and the Latins from them Channa, Channe, and Channus; Salvian calls it Hatula sive Canna; the Italians call it Sopracielo; and the Portuguese, Serran.

Labrus ex purpuro, viridi-cæruleo, et nigro varius.
The Labrus variegated with purple, greenish-blue, and black.

The variegated Scarus.

This is a very beautiful species: the head is large, and the mouth closed with a pair of lips not so thick as in some of the others, but fleshy and wrinkled: the eyes are large, and stand at a distance from one another; their iris is red, and the pupil are round and black: the nostrils have each a double aperture, and stand nearer the eyes than the extremity of the rostrum: the back is elated; the belly is flat; the sides are rounded and fleshy: the scales are large and beautiful, and the whole body of the fish is elegantly variegated with a deep coppery purple, a bluish-green, and a black.

This species is frequent in the Mediterranean; all the writers on fishes have called it Scarus varius.

The other species of the Labrus are, 1. The forked-tailed Labrus, with the body all over red, called Anthias, Anthias, and Sacer by authors 2. The Labrus, called by the Ichthyological writers the common Scarus. 3. The yellow Labrus, with the back purple and the back fin continued from the head to the tail. This is described by authors, under the names of Alphestus and Cynaedus.

SPARUS.

THE coverings of the gills in the Sparus are scaly; there are lips which cover the teeth in the same manner as in quadrupeds, and the teeth themselves are either like those of the human head, or else like those of a dog; the grinders or molares are like those of quadrupeds; the teeth stand only in the jaws and fauces; the palate and the tongue are smooth: there is only one back fin; the tail is usually forked, and the eyes are covered with a lax skin.

The species of this genus are numerous, and may therefore be conveniently arranged under different divisions, according to the shape and figure of the teeth in the jaws.

SPARI.

Division the First.

Those which have the teeth in the jaws sharp and subcylindric.

Sparus dorso acutissimo, linea arcuata aurea inter oculos.
The sharp-backed Sparus, with a crooked, gold-coloured line between the eyes.

The Gilt Head.

THIS is a very beautiful fish: the head is large, and the body is somewhat compressed: the eyes are large; their iris is whitish, but sometimes tinged with reddish or yellowish, and the pupil is black: the mouth is large, and the teeth in front are tall, slender, rounded, and acute: the lips fall over them, and cover them compleatly, when the mouth is shut; the nostrils have each a double aperture, and stand somewhat nearer the eyes than the extremity of the head, and in the middle, between the eyes, there runs a beautiful crooked, gold-coloured line, which makes the top of the head look as if gilt: the back is elevated into a very sharp ridge; the belly also is ridged: there is but one fin on the back, and that has twenty-four rays; the pectoral fins have seventeen rays each, and the ventral fins have each six: the pinna ani has fourteen rays, and the tail is forked: the ground colour of the body is an olive brown, but it is elegantly variegated with a number of different colours.

This species is frequent in the Mediterranean, and in many other places; the antients were very well acquainted with it. Aristotle calls it Χρυσόφρυς; Ælian and Oppian also give it the same name; Varro and Ovid call it, from the Greeks, Chrysophorus; all the other Latin writers, Aurata: the Italians call it Aurata; the Venetians, Ora; the Spaniards the Dorada; we call it the Gilt Head, the Golden Head, and the Gilt Poll.

Sparus lineis utrinque luteis, longitudinalibus, parallelis, iride argentea.
The silver-eyed Sparus, with yellow, longitudinal, parallel lines.

This is also a very beautiful fish: the head is moderately large; the eyes are large, and their iris of a fine bright silvery hue; the pupil is large, and of a deep bluish-black; the mouth is large, and the teeth sharp, but there are no bony tubercles in it, but in their place only a few asperities or flat roughnesses: there is but one back fin, which is not very tall; the back is elevated, the belly is somewhat flatted; the sides are fleshy and prominent: the ground colour of the fish is a yellowish-brown, but it is elegantly varied with longitudinal lines of a gold yellow running parallel to one another.

This is sometimes caught in the Mediterranean, but is not very common. Authors have, from the earliest times, thought there was some resemblance between the lineations of it's sides and those of the upper wings of the beetle-kind, and have thence given it it's name. Aristotle and Oppian call it Κάνθαρος; Ælian, Κάνθαρος θαλάττιος, and the generality of the Latin writers, Cantharus; but Gaza and some others, Scarabeus.

Sparus totus rubens iride argentea.
The silvery-eyed, red Sparus,

This is also a very beautiful fish: the head is moderately large; the mouth is not very wide; the teeth are sharp and rounded, and the lips fleshy; they fall over them, and perfectly cover them, when the mouth is shut: the eyes are beautiful; their iris is of a fine silvery white, the pupil black; the body is moderately broad, in proportion to it's length, and is throughout of a strong and elegant red; there is but one fin on the back; the pectoral fins are short and obtuse; the ventral fins are narrow, acute, and each furnished with six rays: the pinna ani is long and low, and the tail is very forked.

This is very frequent in the Mediterranean. Aristotle, Oppian, and Ælian call it Ἐρυθρῖνος; and Oppian, in some parts of his work also, [illegible]. The Latin writers in general call it Erythrinus sive Rubellio; Paulus Jovius, Fragolinus and Pagrus sive Phagrus; the Venetians call it Albero and Arboro; the Portuguese, Pagro; and the Italians, Fragolino and Frangolino; the Spaniards, Pagel.

The Sea-bream.

Sparus rubescens cute ad radicem pinnarum dorsi et ani in sinum producta.
The reddish Sparus, with the skin at the base of the back fin, and pinna ani forming a sinus.

This is a very broad fish, in proportion to it's length: the head is not large, but the mouth opens wide, and is well furnished with teeth: the eyes are large and beautiful; the nostrils are conspicuous, and have each a double aperture; they stand at about a middle distance between the eyes and the extremity of the head: the back is elevated into a high ridge; the belly also is ridged; and at the base of the back fin, and of the pinna ani, the skin forms a kind of sinus: the pectoral fins are small and obtuse; the pinna ani is long and low; the tail is forked, and there are granulous tubercles in the mouth.

This species is frequent about the coasts of France, and is sometimes caught about those of England; the antients were well acquainted with it. Aristotle, Ælian, and Athenæus call it [illegible]; and from hence the Latin writers have also called it Phagrus and Pagrus; the French call it Pagre; and we the Sea-bream.

The Dentex.

Sparus varius dorso acuto, dentibus quatuor majoribus.
The variegated Sparus, with a ridged back, and with four large teeth.

This is a very singular species: the head is not very large; the rostrum is obtuse, and the body somewhat compressed and broad, in proportion to it's length: the mouth is not large, but there are four very large teeth in it, which make a singular figure: the back is raised into an acute ridge; the belly is also ridged; the colour is a deep olive brown, but it is elegantly variegated with darker and paler spots; the back fin is very long, and has ninety rays in it; the pinna ani has only ten rays: the tail is forked.

This species is not unfrequent in the Mediterranean; the antients were well acquainted with it. Aristotle calls it [illegible]; and Ælian, [illegible]; Athenæus uses both these names; Ovid and Charleton after him call it Synodon; others, Pagrus, Dentex, Dentix, and Dentrix; Bellonius, Dentalis.

The Boops.

Sparus lineis utrinque quatuor argenteis et aureis parallelis longitudinalibus.
The Sparus, with four gold and silver-coloured, longitudinal, parallel lines on each side.

This is a very beautiful fish; the head is large, and the body is not flatted, but rounded, and considerably thick, in proportion to it's length: the mouth is moderately large; the teeth are sharp and slender, yet very strong; they are defended by lips which fall over, and perfectly cover them, when the mouth is shut: the eyes are large and beautiful; the back is convex, and the belly somewhat flatted; the sides are rounded, fleshy, and prominent: the body of the fish is of a pale colour, but below the lateral line there run on each side four very elegant, longitudinal, parallel lines of a gold and silver colour: the back fin is long, and has thirty rays; the pinna ani has nineteen rays; the pectoral fins are red.

This species is frequent in the Mediterranean; the antients were very well acquainted with it. Aristotle calls it Βωψ; and Oppian and Athenæus, Βὸξ; Pliny calls it Box and Boca; Paulus Jovius, Boca; and Gaza, Vora. Rondelet, Gesner, and Charleton call it Boops; Aldrovand, Boops Bellonii; the Italians and French call it Boga and Bogue.

Sparus varius macula nigricante in medio latere, dentibus quatuor majoribus.
The variegated Sparus, with a black spot in the middle of the side, and with four large teeth.

The Mæna.

The head of this species is large, and the coverings of the gills are beset with small but firm scales: the mouth is moderately large, and there are four remarkably large teeth; the lips fall over them however, and perfectly cover them, when the mouth is closed: there are other lower teeth in the fauces, but the palate and the tongue are smooth: the eyes are large; the apertures of the nostrils are double, and very conspicuous: the back is elevated pretty high; the belly is flat; the sides have each a large black spot in their middle: the back fin has twenty-three rays; the pectoral fins have each fifteen rays; the ventral fins have six rays each; the pinna ani has twelve, and the tail is a little forked: the whole body of the fish is variegated with an elegant admixture of colours, but the spots on the sides are very distinct.

This species is frequent in the Mediterranean, and is caught also, sometimes, on the coasts of France; the antients were very well acquainted with it. Aristotle, Oppian, and Athenæus call it [illegible], Ovid calls it Menerela; almost all the other Latin writers, Mæna and Manula; Gaza, Alec. The Italians call it Menola.

Sparus macula nigricante in utroque latere medio, pinnis pectoralibus caudaque rubris.
The Sparus, with a black spot in the middle of each side, and with the pectoral fins and tail red.

The Smaris.

The head is not very large; it is of a compressed figure and obtuse; the eyes are large; the teeth are acute and rounded, and the lips fleshy and thick: the nostrils have each a double aperture; the lower one is larger and rounded, the upper oblong and narrow: the back is elevated, and the belly somewhat flatted; the sides are prominent and fleshy, and have on the middle of each a large black spot: the pectoral fins are short and obtuse, and of a bright red colour; the tail is also of the same bright red, and is forked.

This species is frequent in the Mediterranean; the antients were well acquainted with it. Aristotle and Oppian call it Smaris; and from them the Latin writers in general call it also Smaris and Maris. Charleton calls it Smaris et Maris Leucomænides; Gesner calls it Mæna candida sive Smaris; and Pliny and Martial, and with them some other of the Romans, Gerres; the Venetians call it Geroli or Gerroli; the Massilians, Gerret.

Sparus maxilla superiore longiore, lineis utrinque duodecim nigris, transversis, parallelis.
The Sparus, with the upper jaw longest, and with twelve parallel, transverse black lines on each side.

The Mormylus.

This is a very singularly marked species: the head is large and somewhat compressed; the rostrum is acute, and the upper jaw is somewhat longer than the under: the teeth are sharp and rounded; the lips are thick and fleshy, and compleatly cover them, when the mouth is closed: the back is elevated; the belly is somewhat flat; the sides are prominent, and are beautifully variegated each with twelve black lines, running parallel to one another in a transverse direction: the back fin has twenty-three rays; the pectoral fins have fourteen rays each; the ventral fins have each six rays, and the pinna ani has thirteen: the tail is forked.

This species is frequent in the Mediterranean, as well as in the Archipelago, and many other seas; the antients were well acquainted with it; they not only knew it at their tables, but used it's figure in their ornaments of many kinds. Aristotle, Athenæus, and Eustathius call it [illegible]; Oppian, [illegible]; Salvian calls it Mormylus; and Ovid, Mormyr. Gaza calls it Mormur, and most of the other Latin writers, Mormyrus; the Italians call it Mormillo; the Venetians, Mormiro; and the French, Morme and Morme.

Sparus officulo secundo pinnarum ventralium in longam setam producto.
The Sparus, with the second ray of the belly fins extended in in form of a bristle.

The Chromis.

The head of this species is of a compressed form, acute at the rostrum, and not very large; the mouth is moderately wide, and the teeth small; the lips are fleshy and firm; the eyes are large, and stand very distant; the nostrils have a double aperture; the anterior on each side round, the other oblong and transverse; the back is elevated; the belly somewhat flatted, and the sides prominent and fleshy: there is on each side a black spot, at the root of the pectoral fins: the lateral line is faint, and reaches no farther than to the back fin; the tail is forked: the back fin has twenty-three rays; the pinna ani has twelve; the pectoral fins have seventeen each; the ventral ones have only six.

This species is frequently caught in the Mediterranean, and sometimes in our own seas, but that more rarely; the antients were very well acquainted with it. Aristotle calls it Χρόμις, Χρέμις, and Χρέμψ; Ælian and Athenæus call it Χρόμις; and from them all the Latin writers also call it Chromis: Charleton calls it Chromis sive Chremis; the Italians and Genoese call it Castagnole; the Sicilians, Monachelle.

Sparus varius macula nigra ad caudam, in extremo æqualem.
The Sparus, with the tail not forked, and with a black spot near it.

The Orphus.

This is a very beautiful species: the head is large, compressed, acute at the rostrum, and of an elegant red colour: the mouth is moderately large, and the teeth are slender, rounded, and sharp; the lips are large and thick; the eyes are large; the apertures of the nostrils double, and the back part of the top of the head flatted: the body is elegantly variegated with red, brown, yellowish, and a fine purple; the back fin has ten prickly rays, and the tail is not forked.

This species is more frequent in the Archipelago than in any other part of the world; the Greeks were very well acquainted with it. Aristotle, Ælian, and Oppian call it Ὀρφὸς; Athenæus, Ὀρφώς; Ovid, Pliny, and most other of the Latin writers, Orphus; Willughby and Ray call it Orphus veterum, and Gaza, Cernua.

SPARI.

Division the Second.

Those which have the teeth in the jaws broad, especially at the base.

Sparus unicolor flavescens, macula nigra annulari ad caudam.
The yellow Sparus, with a black annular mark at the tail.

The Sargus.

THIS is a very elegant species: the head is large and compressed, and the rostrum acute; the mouth is large, and the teeth are broad and flatted; the lips are large and fleshy, and the eyes stand at a considerable distance: the back is elevated and ridged; the belly is somewhat flat; the whole body is of an elegant yellow colour, without any variegation, except that, near the origin of the tail, there is a beautiful, annular spot of a deep black: the back fin has twenty-six rays; the pectoral fins have fourteen rays each; the ventral fins have only six each, and the pinna ani seventeen: the tail is large, and is deeply forked.

This species is frequent in the gulph of Venice; the antients were very well acquainted with it. Aristotle, Ælian, Oppian, and Athenæus call it Σαργὸς; and from them all the Latin writers Sargus; the Venetians, Sargo.

Sparus lineis longitudinalibus varius, macula nigra utrinque ad caudam.
The Sparus, with longitudinal lines, and with a black spot on each side at the tail.

The Melanurus.

The head of this species is large and compressed; the opening of the mouth is large; the teeth are large, and somewhat broad: the lips are thick, fleshy, and fall over them; the palate and tongue are perfectly smooth, but, beside those in the jaws, there are also teeth in the fauces: the eyes are large, and are covered with a lax skin: the nostrils are conspicuous; they have each a double aperture, and they stand nearly at an equal distance between the eyes and the extremity of the rostrum: the iris of the eye is yellow, and the pupil is round and black: the back is somewhat acute; the belly is a little flatted; the sides are prominent: there is only one back fin, and the tail is forked: the sides of the fish are elegantly variegated with a number of longitudinal lines, and the back is duskier than the rest, or has a blackish tinge in it.

This species is frequently caught about the coasts of Italy; the Greeks were very well acquainted with it. Aristotle, Ælian, Oppian, and Athenæus call it Μελάνουρος; the Latins, in general, Melanurus; Gaza, in his comments on Aristotle, Oculata; the Italians, Occhiata; and the Massilians, Oblada.

Sparus lineis utrinque undecim aureis parallelis longitudinalibus.
The Sparus, with eleven parallel, longitudinal, yellow lines on each side.

The Salpa.

The head of the Salpa is large and compressed; the eyes are not large; the opening of the mouth is wide: the lips are thick, and perfectly close over it: the teeth are broad, and those of the upper jaw are bicuspidate, or have a forked apex; the palate and tongue are smooth, but there are other teeth in the fauces: the nostrils have each a double aperture; the anterior one round; the hinder one oblong, in form of a fissure; they are situate somewhat nearer to the eyes, than to the extremity of the rostrum: the back is elevated; the belly is somewhat flat; the back fin is long, and has twelve prickly rays: the skin, at the extremity of the back fin, and of the pinna ani, is continued into an eminence which is sinuous in the middle: the belly fin is long, and the tail is forked.

This species is frequent in the Mediterranean. Aristotle, Ælian, Oppian, and Athenæus call it Σάλπη; and from them all the Latin writers Salpa; the Italians call it Sarpa; the French, Sarpe.

SCIÆNA.

THE whole head of the Sciæna and the coverings of the gills are scaly, and one of the laminæ of these coverings is serrated at the edges: the body is compressed and broad; the back is acute; there are teeth in the jaws and fauces: the palate and tongue are smooth; there is only one fin on the back, but it has the appearance of a bifid one, being divided in the middle to the very base; the tail is equal at the extremity, not forked, lunated, or any way divided.

Sciæna maxilla superiore longiore, inferiore cirrosa.
The Sciæna, with the upper jaw longest, and the under one bearded.

The Umbra.

This is a very large fish: it's length is frequently five feet, and it's weight sixty pounds: the ground colour is a dusky olive, with a tinge of bluish, but it is all over variegated with an admixture of other colours: the scales are moderately large; they stand in an imbricated manner, and are of a roundish figure: the lateral line is crooked; the mouth is small, and there is a black edge to the coverings of the gills: the teeth are small and slender; there are a great many of them in the fauces, beside those in the jaws, but the palate and tongue are smooth; there is in the angle of the lower jaw

jaw a single short cirrus or beard; the upper jaw falls over this at the end of the nostrils, and have each a double aperture: the lamina of the gills is deeply serrated at the edges, and in the extremity of the lower jaw there are three large foramina: the tail is nearly even at the end: the first division of the back fin has nine or ten prickles; the other division has twenty-four rays, but only the first of them is prickly: the pectoral fins have each fifteen rays; the belly fins have only six; the first of them short and aculeated: the pinna ani has eight rays, and the first of those also is aculeated.

This species is frequently caught in the gulph of Venice; the antients were very well acquainted with it. Aristotle calls it [illegible]; Ælian, Oppian, and Athenæus, [illegible]; Bellonius calls it Chromis; the generality of the Latin writers, Umbra; Gesner and Aldrovand call it Umbra marina; Paulus Jovius, Umbrina; Salvian, Coracinus; Jonston, Sciæna sive umbra; Pliny, Sciena; and the Venetians call it Corvo.

Sciæna ex nigro varia pinnis ventralibus nigerrimis.
The blackish, variegated Sciæna, with the belly fins black.

The Umbrino.

The head of this species is large, and the opening of the mouth much larger than in any other of this genus: both jaws are furnished with strong teeth, but the palate and the tongue are smooth; the nostrils have each a double aperture; the anterior one is round, and the posterior oblong, and they are situated about at a middle distance between the eyes and the extremity of the head: the back is elevated, and the belly is somewhat flatted; there are a number of oblique lines of various colours on the body; the general colour, however, is an olive so deep, that it approaches to black: the lateral line runs parallel to the ridge of the back: the first division of the back fin has eleven rays in it; the second has twenty-four; the pinna ani has nine rays, two of which are prickly; the pectoral fins have each seventeen rays, and the ventral ones only six, the first of these only is aculeated: the tail is nearly equal at the extremity, and is turned naturally upwards.

This species is frequent in the Mediterranean; the Greeks were well acquainted with it. Aristotle, Ælian, Oppian, and Athenæus call it [illegible]; Rondelet and Gesner call it, Coracinus niger and Coracinus albus; the Italians call it Umbrino and Corvedi fortiera.

PERCA.

THE scales of the Perca are hard and rough to the touch, or are armed with a kind of little points at their extremities: the upper lamina of the coverings of the gills is serrated at the edge; there are sometimes two fins on the back, and sometimes there is only one: the branchiostege membrane on each side has seven bones.

The difference of the back fins, in this genus, is so considerable, that it parts the species into two divisions.

PERCÆ.

Division the First.

Those which have two fins on the back.

Perca lineis sex utrinque nigris transversis, pinnis ventralibus rubris.
The Perca, with six black transverse lines on each side, and with the belly fins red.

The Pearch.

THIS grows, in some places, to a foot and a half in length, but it's more usual size is about eight or nine inches in length; it is considerably thick, in proportion, and the sides are fleshy and prominent: the back is elevated, and somewhat sharp; the belly, all the way from the head to the anus, is flat: the head is compressed; the two jaws are equal in length, but the upper one, when the mouth is shut, falls a little way over the other: the opening of the mouth is immoderately large, and both the jaws are very thick set with small teeth; there are also in the palate three little spaces, which are covered in the same manner with teeth: the middle one is small, and is of a triangular figure; the two others are oblong: in the fauces there are four bones, which are, in the same manner, rough with numerous teeth: the two upper ones are larger; the under ones smaller, and these unite or grow into one as it were: the tongue is smooth, and is somewhat loose in the under part: the nostrils are very large, patulous, and nearer the eyes than the extremity of the rostrum; they have each a double aperture, and are placed at a great distance from one another; the anterior aperture has a little valve to cover it: there are on each side of the head, between the eyes and the rostrum, four ducts, the business of which seems to be the secreting a mucous fluid: the eyes are large; the iris is of a deep yellow, or of a yellow with an admixture of black: the pupil is oval and greenish; the coverings of the gills are formed of four laminæ, of a bony texture, and seven broad and crooked spines conjoined by a membrane: the upper lamina is serrated all the way round it's edge, and the under one is terminated by a bony apophysis; there are some very small scales on these laminæ, but they do not appear, unless on a close examination: the clavicles consist each of four bones; they adhere to the body just above the pectoral fins, and the first and third of these are slightly serrated round the edges; the others are smooth: the bones of this fish is not naked, as the generality of writers on these subjects have asserted, but is covered with very small scales: the lateral line is crooked; it runs near the back, and is bent toward it, and runs all the way much above the line, formed by the interstices of the muscles; this is straight, and easily visible; it runs at about an equal distance from the back and the belly, along the middle of the fish: there are on the sides six transverse lines or zones; these are all black, and that which is nearest the tail is least of all.

The scales of the Pearch are very hard; they are moderately large; they adhere firmly to the skin: those on the belly are white; those on the lower part of the sides of a pale yellowish, and those toward the back are darker; they are somewhat semicircular in figure, and they are armed at the edges with little uncinulæ turning backwards, whence it is that the fish seems so rough, on drawing the hand over it: there are two fins on the back; the former has fourteen, sometimes, though rarely, fifteen, rays; these are all of them prickly, and the last or hindermost ray is much the smallest: the membrane of this fin is of a dusky bluish-brown, but toward the end there is a large black spot: the hinder back fin has sixteen rays; the first of these is very small, and is aculeated; the others are longer, and are somewhat ramose at their extremities: the pectoral fins stand not on the belly but on the sides, they are of a greyish-brown colour, and have each fourteen rays; the two first and the three last of these are small and undivided; the intermediate ones are longer, and ramose at the extremities; these are connected together by a very weak membrane: the belly fins are of a bright and elegant red; they have each six rays; the first of these is simple and aculeated, but all the others are ramose; they are all of them robust at their origin, and the hinder one is connected by a membrane to the belly, so that the fins cannot be elevated into a perpendicular direction: the pinna ani is also of a strong red, and has twelve rays, sometimes only eleven; the two first of these are prickly; the rest are ramose; the hinder ones are very small, and the third and fourth are the longest: the tail is somewhat forked, and is reddish at the end and along the sides; there are seventeen long rays in it: the ovary in the female Pearch is long, simple, and undivided, and of a cylindric figure, filling up almost the whole cavity of the abdomen: the vesica seminalis in the male is double, or is composed of two portions conjoined at the bottom: the liver is of a pale red, and is divided into two lobes; the left is the larger, and the gall-bladder is situated in the middle between them, and hangs low down: the peritonæum is of a silvery white: the stomach is very large, and is very distinct from the intestines, and below the pylorus it has three large vermiform appendages: the intestine makes but one volution, and is usually found deeply covered with fat: the spleen is oblong and red; the air-bladder is oblong and simple; it is large, and is continued along the whole upper part of the cavity of the abdomen: the ribs are on each side nineteen; these separate very easily from the vertebræ; the vertebræ are forty-one.

This species is frequent in all our rivers and fresh waters; it is very voracious, and will seize upon roach and other fish of a considerable size: in our rivers, the Pearch usually grows to about eight inches, but in some places it greatly exceeds that

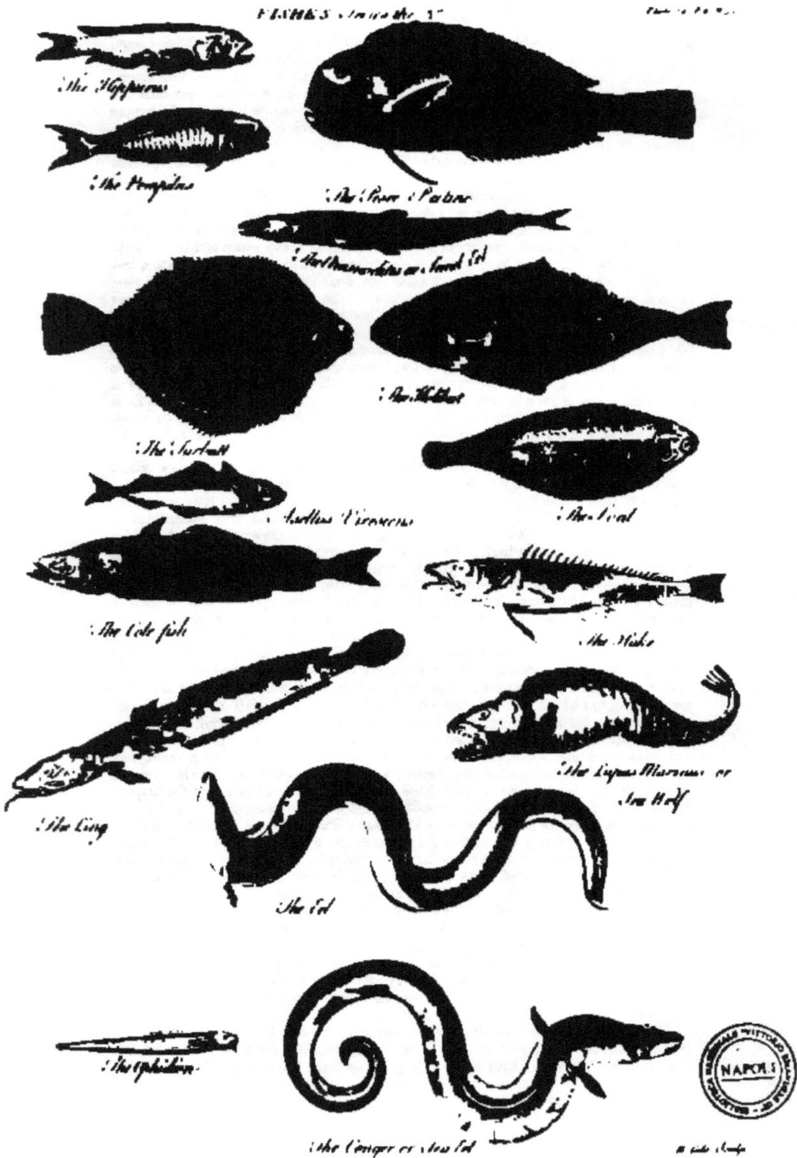
FISHES
The Cole fish
The Lupus Marinus or
Sea Wolf
The Eel
The Conger or Sea Eel

ſize. I was the laſt ſummer fiſhing for pike on Whittleſea-meer in Cambridgeſhire; we had only large baits that we might not take ſmall fiſh, and yet the greateſt part of theſe were taken by Pearch inſtead of pike; few of theſe were leſs than a foot in length, and ſome of them conſiderably more. In Germany they have them ſtill larger.

The ancients were very well acquainted with this fiſh. Ariſtotle, Ælian, and Athenæus call it πέρκη; the Latin writers in general Perca; Wotton, Bellonius, &c. Perca fluviatilis; Aldrovand, Perca fluviatilis major; and Schoneveldt, Jonſton, and others, Perca major: the Italians call it Perſego; the Germans, Barſch.

Perca pallida maculoſa dentibus duobus utrinque majoribus.

The pale ſpotted Perca, with two large teeth on each ſide.

The Lucioperca, or Sandre.

This is a conſiderably large ſpecies; it grows frequently to two feet in length, and ſometimes to conſiderably more than that: the head is compreſſed and large, and the opening of the mouth is very large, but not ſo enormouſly great, in proportion to the ſize of the fiſh, as in the common pearch: the breaſt is flat, but the belly is convex; the head is narrower and longer, in proportion, than in the common pearch: the back is convex, but it is not elevated as in that ſpecies, and the body toward the tail is more long and ſlender: the noſtrils are large, and have each a double aperture; the anterior is round, and has a little valve: the eyes are very large, and are of a ſomewhat elliptic figure; they ſtand at a great diſtance from one another: the pupil is roundiſh; the iris is of a mixt ſilvery, yellowiſh, and blackiſh-blue: the upper jaw, when the mouth is ſhut, falls a little way over the under, but it is very inconſiderable: the coverings of the gills are long; they are compoſed of four bones, ſeeming to form only two larger, as is the caſe in the common pearch, and of ſeven crooked oſſicula connected by a membrane: the lamina, which is neareſt the eye, is large, and is ſerrated round it's lower edges, and aculeated; and the lower lamina is alſo ſerrated a little, and does not terminate in a bony apophyſis, as in the common pearch: three of the laminæ are covered with ſmall ſcales, but this is moſt diſtinctly viſible in the full-grown fiſh; the clavicles conſiſt each of four bones, the loweſt or largeſt of which has it's origin below that of the pectoral fin, and the firſt and third are ſomewhat acute.

The head is blackiſh on the upper part, and has there a long depreſſion or furrow, reaching from the noſtrils nearly to the back; from the extremity of this to the back fin there runs alſo a narrower furrow, but this is not ſo conſpicuous: the lateral line is ſtraight, and runs much nearer the back than the belly: the ſcales are hard and rough, and are ſtriated in their hinder part; thoſe on the belly are white, and on the ſides they are of a white colour alſo, but tinged in an elegant manner, with a caſt of a gold yellow and a ſilvery white; on the back the colour is a duſky greyiſh-brown, and there are in this part a great many oblong black ſpots, partly placed tranſverſely, and partly quite without order.

The teeth in both the jaws are of an equal ſize: in the upper jaw there are two very large and long ones near the extremity, but with a ſpace between them; behind theſe the jaw is furniſhed each way with a ſingle ſeries of ſmaller teeth, and between the two great ones in the front there are two ſmaller, which, when the mouth is cloſed, ſtand out beyond the lower jaw: the lower jaw has alſo, near it's extremity, ſometimes two large teeth on a ſide, ſometimes only one; theſe form to themſelves a kind of cavity in the upper jaw, and between them there is alſo a vacant ſpace, and behind them, each way, there runs a ſingle row of teeth, ſmaller but unequal: on the anterior part of the palate there alſo ſtand two large teeth, with two, three, or more ſmaller ones between them; on the extreme verge of the palate, each way, there is a ſingle row of ſmall teeth; the reſt of the palate is ſmooth; in the fauces alſo there are ſeveral bones furniſhed with teeth: there are two large and roundiſh ones on the upper part, and on the lower there are ſix ſmaller, all covered with teeth, and ſo ſituated, with regard to one another, that they ſeem to form only two larger bones of an oblong form.

The tongue is white, ſmooth, and looſe, not faſtened down to the jaw, as in many other ſpecies: the pectoral fins are whitiſh; they have ſixteen rays each; they

are

are joined by a very thin and weak membrane, and are small, in proportion to the size of the fish: the belly fins are whitish, with some spots of a blackish-grey, and with the admixture also of a little red; they have each six rays, and the first of these is prickly; they are all thick and robust at their origin: the first back fin is elegantly variegated with spots of a deep black, and of a greyish-yellow; it has fourteen rays, and all of them are robust and prickly: the hinder back fin is of a greyish-colour, with smaller spots of black and of yellow; this has twenty-three rays; the two first of these are prickly, the others are ramose: the pinna ani is white, but it is variegated with very minute black spots, and has fourteen rays, the two first of which are prickly: the tail is forked.

The breast of this fish has been said, by some, to be naked, but it is covered with scales: the liver is red; it is principally in the left side, and is divided into two large lobes, and some smaller apophyses: the gall-bladder is small and yellow; the stomach is moderately large and oblong, and has six appendages below the pylorus; these are large and red, and both these and the stomach are usually found full of a thick, white chyle: the intestine makes but one turn, but it has a great variety of convolutions in that course, and is usually covered very thick with fat: the spleen is of an oblong, triangular form, and a dusky red colour, and is placed near the intestine: the ovaries in the females seem two distinct ones placed one on each side, but, when more closely examined, they are found to coalesce at the bottom; these are, at the season, distended with a vast number of yellowish, white ovula, and the vulva is situated just below the anus: the air-bladder is long and thick, and is extended along the back-bone; it is divided into two parts, as it were, at top, in the manner of a heart at cards: the pneumatic duct is situated at it's anterior and upper part; the peritonæum is of a silvery colour, and the flesh of the fish is white, firm, and well tasted.

It is an extremely voracious fish, and will seize almost on any thing: it is very frequent in the Northern seas. Gesner call it, Piscis quem Schilum Germani vocant, alii Nagemulum; Aldrovand retains the same name for it; Ray, Willughby, and several others call it Lucio-perca, or the Pike-pearch; Johnston, Schilus; and Charleton, Schilus sive Nagemulus. The Swedes call it Gios; and the Germans, in different places, Sandat, Schindel, and Nagemaul.

Perca lineis utrinque octo vel novem transversis nigris.
The Perca, with eight or nine black transverse lines on each side.

This is a small species: the head is large, and of a somewhat depressed form at the top; the eyes are large, and stand at a distance: the nostrils are situated at a small distance below them, and have each a double aperture: the mouth is very large, in proportion to the bulk of the fish, and is furnished with a great number of sharp teeth: the back is somewhat ridged; the sides are prominent, and the belly, from the gills to the anus, is somewhat flatted: the colour on the back is a deep blackish-brown; the sides are of the same colour, only much paler, and with a cast of yellowish, and have each eight or nine transverse black streaks on them: the first of the two back fins has eight rays; the second has thirteen: the pectoral fins have each fourteen rays, and the pinna ani has twelve: the belly fins have only five each.

This is very frequent in France, and in some parts of Germany. Ray, Willoghby, Jonston, and many other writers have called it Asper pisciculus, a name which they have also given to the common Stittle-back or Gasterosteus, with some confusion; Gesner and Aldrovand call it Asper pisciculus Gobioni similis, which distinguishes it from the other Asper pisciculus. Gesner, in some of his works also, calls it Gobius asper.

Perca radiis pinnæ dorsalis secundæ tredecim, ani quatuordecim.
The Perca, with thirteen rays in the second back fin, and fourteen in the pinna ani.

The Basse.

This is a very singular and a very large species; it grows to three feet in length, and is moderately thick, in proportion, and it's general figure somewhat resembles the salmon: the scales are large, and adhere very firmly: the back is of a deep blackish colour,

colour, and the belly is of a ſilvery white; the opening of the mouth is very large; there are teeth in the jaws, and on the palate, and in the fauces, and a rough bone runs all the way along the middle of the tongue: the noſtrils are conſpicuous, and have each a double aperture: the eyes are large, and ſtand at a diſtance; they are covered with a ſkin that adheres very firmly to them, and their iris is of a ſilvery white, but with a little circle of yellow: the tail is ſomewhat forked, but not much ſo: there are two ſhort ſpines at the pectoral fins, and at the angles of the branchiæ there is, on each ſide, a ſingle one larger than the others: the middle lamina of the coverings of the gills is ſerrated round the edges, and has three large denticles beyond the angle: the firſt back fin has nine rays; the ſecond has thirteen: the pectoral fins have each fifteen; the ventral ones have ſix each, and the pinna ani has fourteen, of which the three former are prickly: the pylorus has five or ſix appendages; the air-bladder is affixed to the back.

This ſpecies is frequent in the Mediterranean, and in ſome other ſeas; we have it about our own coaſts: it is a very voracious fiſh; it's fleſh is firm, and extremely well taſted: the antients were well acquainted with it. Ariſtotle, Ælian, Athenæus, and Oppian call it Λάβραξ; and Ovid, Varro, Pliny, and from theſe Ray, Willughby, and almoſt all the modern writers, have called it Lupus; ſome Lupus vulgaris; others, Lupus marinus; Paulus Jovius, Spigola ſive Lupus. The Italians call it Spigola; the Venetians, Branchini; the Spaniards, Lupo; the French, Lubin; and we, the Baſſe. The giving this ſpecies of Perca the name of Lupus has, however, occaſioned ſome confuſion, as that name is alſo given to the Anarrhicas.

P E R C Æ.

Diviſion the Second.

Theſe which have only one fin on the back.

Perca dorſo monopterygio, capite cavernoſo.
The Perca, with a cavernous head, and with only one fin on the back.

The Ruff.

THE uſual ſize to which this ſpecies arrives is four or five inches, though ſometimes it will grow conſiderably larger: the head is, in it's general form, compreſſed, but it is flatted a little between the eyes, and thence to the mouth it is ſomewhat declivous, and it has a great number of cavities or cavernulæ depreſſed on it's upper part, and on both ſides, and ſome on the lower; it is theſe hollows that have given riſe to it's being called the Hollow-head in ſome places: the back is ſomewhat acute; the breaſt and belly are flatted: the breaſt appears to be wholly naked, or without ſcales; the jaws are both of the ſame length, and the opening of the mouth is conſiderably large: the eyes are large, and of a variety of colours; the noſtrils are alſo large, and ſtand about the midway between the eyes and the extremity of the head; they are always full of a mucous humour, and the whole body of the fiſh is, indeed, covered with the ſame matter.

The teeth are ſmall, but numerous; there is a row of them in each jaw, and on the anterior part of the palate, toward the roſtrum, there are a number of teeth ſo minute, that they can ſcarce be ſeen, but they are very eaſily felt; and the reſt of the palate is all ſmooth; in the upper part of the fauces there are two bones, the ſurfaces of which are alſo covered with teeth; and on the lower part, near the gills, there are two others, which are alſo covered with teeth; theſe, as well as the former, ſtand ſo cloſe, that they ſeem to form only one bone of the two.

The lateral line is ſomewhat crooked. The colour of the fiſh is a browniſh-yellow, and there are, on the back and ſides, a number of black ſpots of different ſhapes and ſizes; theſe, though very conſpicuous, are not in the ſcales, but in the ſkin which is under them.

There is only one fin on the back, and this has twenty-eight rays, ſometimes one or two leſs; the fourteen or fifteen anterior ones of theſe are prickly: the pectoral fins have each fifteen rays, and the belly fins have each ſix; the pinna ani has eight rays, of which the two firſt are ſtrong, firm, and prickly: the tail is ſomewhat forked, but it is not deeply so; and the membranes of all the fins have black spots in them, especially in the back fin; those on the others are smaller as well as fewer.

This species is frequent in the rivers and lakes in many parts of Germany, and in France and England; and it is caught also sometimes at sea: the antients were well acquainted with it. Athenæus calls it Κέρκη ποταμία; and Ælian, Κέρκη; Pliny calls it Acerina; Bellonius, Gesner, Ray, and Willughby, Cernua fluviatilis; Charleton, Cernua fluviatilis, aliis Perca minor; Gesner, Percæ fluviatilis genus minus; and, in other parts of his writings, Porcus fluviatilis and Cernua fluviatilis. Tragus calls it Aurata fluviatilis; J. Caius, Aspredo; and Schoneveldt adds to it's other names those of Porculus and Porcellus. The Swedes call it Giers; the Danes, Horch; the Germans, Kaulbrass, Steurbrass, and Storer; the Dutch, Posch and Post; we call it a Ruff or Rough.

The writers on fishes have described what they suppose another species, under the name of Piscis Danubii quem Ichrel Germani vocant; but it is the same, in all respects, with the common Ruff, only it grows larger in the Danube.

Perca lineis utrinque transversis septem nigris, ductibus cæruleis et rubris in capite et ventre.
The Perca, with seven black transverse lines on each side, and red and blue marks on the head and belly.

The Sea-Pearch.

This is a larger species; it grows to a foot, or more, in length, and the body is of the general form of the common pearch, but not quite so thick: the mouth is very large, and almost always open: the back is ridged; and the belly, from the gills to the anus, is somewhat flatted: the head is large, and toward the back part, and from thence, a part of the body is of an elegant reddish colour, and is variegated with a great number of beautiful bluish and reddish lines; and there are also lines and streaks of both these elegant colours on the anterior part of the belly, and on the sides there are seven large black lines.

The back fin has twenty-five rays, ten of which are short and prickly, the other fifteen longer and weak: the pinna ani has ten rays, three of which are prickly, the others not: the pectoral fins are yellowish, and the belly fins rather of a whitish-grey, but still with some tinge of yellow in it: the tail is not forked or lunated, but plain and undivided: the apertures of the gills are extremely large, and are yellow on the inside: the eyes are large, and stand at a considerable distance; and the nostrils have each a double aperture, and are situated near them; there are sharp teeth in the palate, and in the fauces as well as in the jaws: the tongue is long and smooth; the middle lamina of the coverings of the gills is denticulated or serrated round the edge, and in the exterior angle of the lamina there are two large spines: the scales are moderately large; the lower part of the belly is prominent; the stomach is very large, and the air-bladder is of an oblong form, and is considerably large, and it's membrane firm and tough.

This species is frequent in the Mediterranean: it was very well known to the antients, and by them distinguished by the simple name of Perca. Aristotle, Athenæus, Ælian, Oppian, and Galen call it simply Πέρκη; Ovid, Pliny, Paulus Jovius, and some of the later writers, Perca; but Ray, Willughby, and the rest, Perca marina and Perca pelagia; the Italians call it Percia. Aldrovand, and after him some others, have figured and described two or three other species, as they call them, of the Perca marina; but the only differences they mention are, in point of colour, and those will happen by accidents in the same species. The other parts of their descriptions prove that they are all the same species as the common Sea-pearch.

Perca dorso monopterygio, lineis utrinque longitudinalibus nigris.
The Perca, with one fin on the back, and with longitudinal black lines on the sides.

The Schraitser.

This is the thickest, in proportion to it's length, of any of the species: the head is large, and the opening of the mouth is wide; the eyes are large, and their pupil round, and

and of a bluish-black; the iris is usually yellow: the nostrils have each a double aperture, and stand a little below the eyes; the back is ridged; the belly is flattish, but somewhat prominent toward the lower part, and the sides are very rounded and prominent: the colour is a dusky brown, with an admixture of yellowish, and on the sides there are longitudinal black lines: the back fin has thirty rays, eighteen of which are prickly: the pectoral fins have sixteen rays each; the belly fins have each six rays, and the pinna ani has eight; the two anterior ones prickly, the rest soft and weak.

This species is frequent in Germany, and some other parts of Europe, but we have it not in England. Authors in general have described it, but that so variously, that the greater part of them seem not to have seen it. Ray, Willughby, and others call it Schraitser Ratisbonensibus.

TRACHINUS.

THE head of the Trachinus has always certain rough tubercles on it, and there is either a single spine, or more, at the upper angle of the coverings of the gills: the eyes are placed near one another, in the upper part of the head: there are two back fins, and the first of them is very short. When the fish is opened, there are always found eight, ten, or twelve appendiculæ at the pylorus.

Trachinus maxilla inferiore longiore cirris destituta. **The**
The Trachinus, with the lower jaw longest, and without beards. **Weever.**

This grows to six or eight inches in length, sometimes to considerably more, and is moderately thick, in proportion: the head is large, and somewhat compressed; the eyes stand near one another at the top of it, and the iris is of a gold yellow: the nostrils are conspicuous, and have each a double aperture: the body is compressed, and the lateral line is straight; there is on each side, at the opercula or coverings of the gills, a large and robust spine; the tail is scarce at all forked: the first back fin has five prickly rays, which inflict an invenomed wound; the second has thirty-one rays: the pectoral fins have each sixteen rays, and the pinna ani has thirty-two: the three anterior rays of the pectoral fins are ramose.

This species is frequent in the Mediterranean, and in our own and other seas; the antients were very well acquainted with it. Aristotle, Ælian, and Oppian call it Δράκων; Ælian, Δράκων θαλάττιος; Pliny, Bellonius, Salvian, and others call it Draco marinus; Rondelet, Jonston, and some others, simply, Draco; others, Araneus piscis; Gesner, Draco sive Araneus Plinii; and Schoneveldt, Araneus vel Draco marinus; Isidore calls it Araneа; and Paulus Jovius, Trachurus, Trachina, and Trachinus. The Swedes and Danes call it Fjarsing; the Italians, Pesce Ragno; the French, Viver; and we, the Weever. Gesner, Aldrovand, and others describe what they call other species of this fish, under the names of Draco major, Draco minor, and Aranei sive Draconis marini species altera; but they have led one another into this mistake: there is no other species but this, and what they then figure and describe are only rieties from the different periods of growth and other accidents.

Trachinus cirris multis in maxilla inferiore. **The Star-**
The Trachinus, with numerous cirri on the under jaw. **gazer.**

This is a fish of an extremely singular figure; the body is rounded, or, if any thing, a little depressed: the back is broad; the sides are prominent, and the belly is somewhat flatted: the head is large and depressed; the mouth is wide, and of a very singular figure, being divided, as it were, into three spaces under the tongue, and the lower jaw turning upwards: the eyes are large, and stand near one another, not on the sides, but on the top of the head, so that the fish naturally looks straight upward; their iris is of a gold yellow; the pupil is of a bluish-black: the nostrils are conspicuous; they have each a double aperture, and are placed at some distance under the eyes: the whole head and the coverings of the gills are beset with a great number of rough and sharp tubercles: there are two back fins, the first has three prickly rays; the second has

has fourteen: the pectoral fins have each sixteen rays; the ventral ones have each five, and the pinna ani has thirteen.

This species was well known to the antients. Aristotle and Ælian call it Callyonymus; Athenæus, Ὀλιγονωνυμος; Athenæus, Ἁγνὸς; and Oppian, Ἡμεροκοιτης; Pliny and many of the Latin writers call it Uranoscopus; others, Callyonymus; Gesner and others, Callyonymus sive Uranoscopus; and Gaza, Pulcher piscis. It is caught in the Mediterranean, and is called upon the coasts there Mesoro, Locerne, and Pesce prete, Bocca, Tapecon, and Raspecon.

TRIGLA.

THE head of the Trigla is very declivous from the vertex to the rostrum: it is large, and of a figure approaching to square, and is usually beset with tubercles or prickles: the body grows gradually smaller from the head, till it terminates in a very slender tail. In many of the species of this genus, there are two or three articulated appendages under the pectoral fins: the eyes stand near together, on the top of the head, and are covered with a cuticle: there are two fins on the back, the anterior one of which is aculeated: the pectoral fins, in some of the species, are very large; in some, indeed, they are so long, that the fish can fly above the surface of the water, by means of them, in the manner of what is usually called the Flying-fish, or Exocætus; and many of the species have a power of making an odd noise.

Trigla capite glabro, cirris geminis in maxilla inferiore. **The bearded Mullet.**
The Trigla, with a smooth head, and with two cirri on the lower jaw.

The head of this species is remarkably large, and of a strangely irregular figure, angulated, but not prickly: the opening of the mouth is large, and there are rows of teeth, not only in the jaws, but deep in the fauces, and on the palate: the eyes are large, and stand near one another on the top of the head; the colour of the whole fish is a dusky yellow: the body is large toward the head, but it diminishes gradually in size to the other extremity, where it is very small; the back is convex; the sides are prominent and fleshy, and the belly is somewhat flat: the scales are large and beautiful, but they adhere so loosely, that the least force wipes them off: the first of the two back fins has seven prickly rays, the first of which is the tallest; the second fin of the back has nine rays, which are all soft and ramose: the pectoral fins have sixteen rays; the ventral fins stand at the same distance from the extremity of the head with these, and have six rays each: the pinna ani has seven rays; the tail is forked, and has seventeen rays; the appendages of the pylorus are twenty-six.

This species is frequent in the Mediterranean, and is caught in a particular abundance about the coast of Italy: the antients were well acquainted with it. Aristotle, Ælian, Oppian, and Athenæus call it Τρίγλη; Ovid, Columella, Pliny, and all the old Latin writers, Mullus; Ray, Willoughby, and many of the moderns have also retained the same name; others call it Mulus; and some, Mullus minor and Mullus barbatus; Salvian, Mullus sive Mulus minor.

Trigla capite glabro, lineis utrinque quatuor longitudinalibus luteis. **The Sur-mullet.**
The Trigla, with a smooth head, and with four yellow longitudinal lines.

This is a very beautiful species; it grows to more than a foot in length, and is very elegantly variegated, and, as it were, painted: the head is large, somewhat depressed, and of a kind of square figure: the mouth is wide, and there are teeth in the fauces and on the palate, as well as in the jaws: the eyes are large, and very beautiful; their iris is of a fine scarlet; they stand near one another on the top of the head, and at some distance below them stand the nostrils, which have each a double aperture: the body is thick near the head, but it grows taper to the tail, where it is very small; the back is somewhat depressed; the sides are prominent, and the belly is flatted: the general

general colour is a dusky brownish olive, and along the sides there run four parallel longitudinal lines, of a fine bright and golden yellow: the fins and tail are also of an elegant yellow, stained irregularly with a very bright scarlet: the second fin of the back has eight rays; the lower jaw is furnished with two oblong, white cirri or beards.

This species is very frequent in the Mediterranean, and it is also caught frequently about the coasts of the West of England. Salvian calls it Mullus major; Aldrovand and Jonston, Mullus major ex Hispania missus; Ray and Willughby, Mullus major noster et Salviani. Salvian is, indeed, the first author who has described it.

Trigla capite glabro, tota rubens, cirris carens.
The red, smooth-headed Trigla, without any beards.

The King of the Mullets.

This is a very singular and a very beautiful fish: the head is large, and of a very odd figure, angulated and approaching to square: the eyes are large; they stand near together on the top of the head, and their iris is of a gold yellow: the nostrils stand a little below them, and have each a double aperture: the mouth is monstrously large, and there are a great number of teeth in the jaws, in the palate, and in the fauces, but the tongue is smooth: the whole fish is of a beautiful red colour: the back is convex, and the sides distended so as to give it a bellied look: the scales are very large, and adhere but loosely.

The first back fin has six rays, of which the second is the tallest; the second has ten rays, of which the first is prickly: the pectoral fins have each twelve rays, and the ventral ones have six each; the first of these is aculeated: and the pinna ani has ten rays, of which the two first are aculeated; the tail is somewhat forked, and has about twenty rays: there are but four appendages to the pylorus, and the air-bladder is simple.

This is frequent in the Mediterranean. Ray and Willughby call it Mullus imberbis sive Rex Mullorum.

Trigla capite parum aculeato, pinnula singulari, ad pinnas pectorales.
The Trigla, with a somewhat prickly head, and with a singular pinnule at the pectoral fins.

The Hirundo Piscis.

This is one of the Triglæ which have the pectoral fins so long, that they are of use in flying; and it has thence, by some inaccurate writers, been confounded with the Exocœtus or Flying-fish, distinctively so called: the head is large, of a somewhat depressed figure, angulated and approaching to square; it's surface is, in several parts, irregular, but behind it's covering is continued into two long spines, which lie extended on the back; the coverings of the gills also terminate each in a long and very sharp spine: the eyes are large, and stand near one another on the top of the head; the nostrils have each a double aperture, and stand a little below them: the mouth is large, and is very well furnished with teeth, both in the jaws, in the palate, and in the fauces, but the tongue is smooth: the body is large toward the head, but it grows extreamly small at the tail: the back is convex, the belly flatted, and the sides rounded and prominent: the scales are hard and rough to the touch, and are elevated in the middle into a kind of eminence: the first of the two back fins has five rays, the second has eight, and the pinna ani has only six: the pectoral fins have twenty-seven or twenty-eight rays each; these are immoderately long, and are connected by a membrane sufficiently firm and tough, and of a brownish colour, spotted with black: the ventral fins stand in the middle of the belly, and they also are very long, and have five rays each: in the middle of the back there stands a short spine: the tail is forked, and there are two large scales at it's origin: near each of the pectoral fins there is a singular pinnule of six rays; the pectoral fins serve it to fly with, but it never flies far or high.

This species is frequent in the Mediterranean, and in some other seas. Aristotle, Ælian, and Oppian call it Χελιδὼν; Athenæus, Ἱέραξ; and Oppian, in another place, Ἱέραξ; Pliny calls it Milvago; and many of the old Latin writers, Milvus; Rondelet and Gesner call it Hirundo Piscis; Gillius, Accipiter; and some, Lucerna and Pora-

Y y y bele.

bele. The Spaniards call it Volador; the Italians, Rondine; the Sicilians, Falcone; and the Swedes, Flygande Fiske.

Trigla capite aculeato, appendicibus utrinque tribus ad pinnas pectorales.
The Trigla, with a prickly head, and with three appendages at each of the pectoral fins.

The Tub-fish.

This is a very singular species; the head is large, and of an irregularly square figure, and is prickly: the orbits of the eyes are surrounded with a series of short spines, and there runs a series of the same kind of spines also round the verge of the upper jaw: the sides of the head also are sharp, and, on the back part, the bony covering of it is protended backwards into the form of two horns, each terminating in a very sharp spine: the eyes are large, and stand very near one another; the nostrils have a double aperture, and stand at some distance below them: the mouth is large and well-furnished with teeth in the jaws, on the palate, and the fauces, and there are several little irregularities about the verge of it.

The body of the fish is large near the head, but it becomes very small toward the tail; it's colour is a dusky purplish-red, variegated with a silvery white, and with a bright green: the scales are very small and smooth, and they adhere very firmly; there is a peculiar squamous over each of the pectoral fins, which has a strong prickle: the lateral line runs simple to the tail, and is not forked, as in some of the species: the first of the two back fins has nine rays, the second has eighteen; the pectoral fins have each ten rays; the ventral ones have six each, and the first of these is prickly: the pinna ani is very long, and has nineteen rays; the tail is a little forked, and has only ten rays: there is a kind of sulcus in the middle of the back, and in this stand twenty-five prickles: the air-bladder in this fish is divided, as it were, into three lobes.

This species is frequent in the Mediterranean, and in some other seas; we not unfrequently meet with it about the coasts of Cornwall, and elsewhere on the British shores: the antients were acquainted with it: Athenæus calls it Κόκκυξ; Aldrovand has figured it under the name of Hirundo piscis prior; Jonston, Willughby, and Ray call it Hirundo; and Pliny, Isidore, and Salvian, Corvus. The Italians call it Capone; and our common people of Cornwall, the Tub-fish.

Trigla rostro parum bifido, linea laterali ad caudam bifurca.
The Trigla, with the rostrum a little bifid, and the lateral line forked toward the tail.

This is a very beautiful species: the head is large, and of a square figure; the top is depressed, and the rostrum somewhat bifid at the extremity, but the division not deep: the eyes are large, and stand near one another on the top of the head, and there are a number of short but sharp spines about them, and on several other parts of the head: the nostrils are conspicuous, and have each a double aperture; the mouth is large, and is well furnished with sharp but short teeth; beside those in the jaws, there are several on some oblong bones in the fauces, and the palate also has many on it, but the tongue is smooth.

The body is biggest, just at the head; it grows very suddenly less from thence, and is extreamly small at the tail: the back is convex; the belly is somewhat flatted, and the sides are prominent; the general colour is a mixture of olive brown and yellowish, but on the sides there is a beautiful tinge of red: the scales are small and smooth, and there runs a furrow along the middle of the back, in which there are placed twenty-five spines; at the pectoral fins, on each side, there are three articulated appendages: the lateral line is not simple all the way, as in the former, but toward the tail it is forked: the first of the two back fins has ten prickly rays; the second has seventeen rays: the pectoral fins have each ten rays; the ventral ones have six each, and the pinna ani has fifteen.

This species is frequent in the Mediterranean; Pliny calls it Milvus and Milvago; Rondelet and Aldrovand call it also Milvus; Gesner, Lucerna sive Milvus; Willughby and

and Ray, Lucerna Venetorum; and Salvian, Cuculus. The people of Naples call it Cocco; the Genoese, Organo; and the Massillans, Galline.

Trigla tota rubens rostro parum bicorni, operculis branchiarum striatis.

The red Trigla, with the rostrum somewhat bifid, and the coverings of the gills striated.

The red Gurnard.

This is a very singular fish, and is not without it's beauty: it is all over of a fine strong red, between a crimson and scarlet: the head is large, depressed, and of a figure approaching to square: the mouth is large, and well furnished with teeth in the jaws, and on the palate, and in the fauces; the rostrum is bifid at the extremity, but the slit is not deep: the eyes are moderately large, and stand at a small distance from one another on the top of the head: the nostrils have each a double aperture; the anterior is round, and the other a transverse slit; they stand nearly in the middle between the eyes and the rostrum: the coverings of the gills are deeply striated, and each of them is armed with three spines; two of these stand near one another on the lower part, and the other, which is larger than either of them, on the upper: the extremity of the rostrum, where it is forked, terminates also in two sharp spines: the body is largest toward the head, and thence grows immediately smaller, and toward the tail is very slender: the back is convex, and of a deeper red than the sides; the belly is of a pale flesh colour.

This species is frequent in the Northern Seas; it makes a singular noise, when out of the water: some have supposed it like the grunting of a hog, others have compared it to the singing of a cuckow; it is not easy to suppose that both accounts are right. The antients were very well acquainted with this fish. Aristotle calls it Κόκκυξ; Ælian, Oppian, and Athenæus also retain the same name for it; Rondelet, Gesner, and several others call it Cuculus; Bellonius and Gesner, Cuculus minor; Schoneveldt, Cuculus Lyræ species; Charleton calls it, simply, Lyra; and Paulus Jovius, Capo. The French call it Marrude; the Dutch, Hunchem; and we, the red Gurnard; and, in some places, the Rotchet. We have it in great abundance about the Yorkshire shore.

Trigla varia rostro diacantho, aculeis geminis ad utrumque oculum.

The variegated Trigla, with a bifid and spinose rostrum, and with two spines at each eye.

The grey Gurnard.

The head of this species is very large, and of an angulated figure, nearly square: the mouth is very wide; the eyes are large, and are placed at a small distance from one another, at the top of the head: the extremity of the rostrum is formed into two spines, each of which is armed with five or six little spinules; the back part of the head is also extended into two horns, and the coverings of the gills have each two robust spines: the iris of the eyes is of a silvery white, and the pupil is bluish: the body of the fish is large toward the head, but extreamly small, as it comes toward the tail: the back is convex, but it has a furrow in it, with several denticles arising from it, as in some of the other species: the sides are prominent, and the belly also is somewhat prominent; the lateral line is rough and eminent, and there is a bony squamose near the pectoral fins, terminating backwards in a very robust spine: the pectoral fins are small and obtuse; there are two back fins which are moderately long; the pinna ani also is long, and the tail is broad, and supported by nineteen rays.

We have this species in abundance in our own seas: the fishermen on the coast of Cornwall take it in great quantities; it makes a singular noise, when taken out of the water: the antients seem not to have known it. Charleton calls it Piscis Cuculus quem grey Gurnard vocant; Willoughby and Ray, Gornatus sive Gurnardus griseus, names formed from the vulgar English one, grey Gurnard.

Trigla

Trigla rostro diacantho, naribus tubulosis.
The Trigla, with a bifid rostrum, and tubulose nostrils. **The Piper.**

The head of this species is very large, in proportion to the body: the mouth is remarkably wide; the eyes are large; they stand at a very small distance from each other at the top of the head, and are covered with a skin: the bony covering of the head is angulated, and terminates in two horns at the hinder part; the rostrum is formed into two spines, and at the upper part of the orbits of the eyes there is also a robust and crooked spine: the body is somewhat rounded, and of a conic figure; very large toward the head, and extremely small at the tail: over each of the pectoral fins there stands a very robust and sharp thorn, and there are, on each side, three articulated appendages: the back is rounded or convex, but it has a furrow in it with twenty-six prickles on each side: the lateral line is scarce at all rough, and there are about seventy puncta on each side; the fish grows to more than a foot in length: the first of the two back fins has ten prickly rays, the second has eighteen; the pectoral fins have twelve each; the ventral ones have each six, and the first of them is aculeated.

This species is frequent in the Mediterranean, nor less so in the Ocean; we have it in great abundance about our own shores: when caught, it makes a very singular and loud noise. Aristotle and the rest of the Greek writers call it Λύρα; Paolus Jovius calls it Capo; Rondelet, Gesner, and Jonston, Lyra; Ray and Willughby, Lyra prior. The Genoese call it, Organo; and we, the Piper.

Trigla cirris plurimis, corpore octagono.
The Trigla, with numerous beards, and with an octagonal body.

This is an extremely singular species: the head is very large, and of an angulated figure: the mouth is wide; the eyes are large, and stand near one another on the top of the head; their iris is of a golden yellow, and the pupil black: the rostrum or anterior extremity of the head is protended into two long and sharp spines: the scales are hard, elevated, and of a rhomboidal figure; there are six series of them, and these give the body an octagonal figure: there are two articulated appendages. The colour of the whole fish is a pale red: the orbits of the eyes are surrounded with a double series of spines, and the nostrils are not conspicuous: the back fin is supported by twenty-six rays, which are not prickly: the pectoral fins have each twelve rays; the ventral ones have six each, and the pinna ani has nineteen: the fins all are of a paler colour than the body, as is also the tail.

This species is frequent in the Mediterranean, and is not less so in the Ocean about our own coasts. Rondelet, Willughby, and Ray call it Lyra altera; Salvian, simply, Lyra; Gesner calls it, Lyra cornuta sive Lyra altera; Bellonius, Coccyx altera sive major. The Italians call it Pesce capone and Pesce forca: the Genoese call it Malarmat; and we, the Piper, confounding it by that name with the former species.

SCORPÆNA.

THE head of the Scorpæna is very large and very prickly; there is only one fin on the back, and that is low in the middle: the body grows very slender, as it comes toward the tail: the eyes stand very near one another, and are covered with a skin, and there are teeth in the fauces and on the palate, as well as in the jaws: the branchiostege membrane contains seven bones.

Scorpæna pinnulis ad oculos et nares.
The Scorpæna, with pinnules at the eyes and nostrils.

This species very much resembles the common pearch in it's general figure: the head is large, and oddly figured, and is ornamented with four pinnules; two of these are placed at the eyes, and the other two at the nostrils: the eyes are large, and stand almost close to one another on the top of the head: their iris is of a gold yellow, and the pupil black: the mouth is large, and furnished with a great number of teeth; the body is large toward the head, but it grows very small, as it approaches the tail: the

 back

back is arcuated; the sides are prominent, and the belly is flatted; the colour is variegated, in an elegant manner, with blackish, olive, and yellowish; there is a remarkable depression or hollow on the head between the eyes: the belly is of a flesh colour, and the ventral fins are of a deep red: the back fin has twenty-one rays, twelve of which are prickly; the pectoral fins have fifteen rays, and the ventral ones only six; the first of these is prickly: the pinna ani has eight rays, and three of them are prickly: the tail is large and rounded at the extremity; the scales are very small, and cover one another, in a manner, standing like those of the serpent-kind: the urinary bladder in this fish is remarkably large.

It is frequent in the Mediterranean: the ancients were very well acquainted with it. Aristotle calls it Scorpios, and the female Σκορπὶς; Athenæus and Oppian also call it Σκορπίς, and after them Rondelet calls it Scorpios; Jonston, Scorpios and Scorpis; Willughby and Ray, Scorpius minor et Scorpæna; Paulus Jovius, Scorpæna; and Pliny, Salvian, and others, Scorpæna; Isidore, Caba, and Wotton call it Scorpio; Gesner, Scorpio vel Scorpæna, id est, Scorpius minor; and Charleton, Scorpides sive Scorpæna. The Italians call it Scrofanello and Pesce Scorpione.

Scorpæna tota rubens cirris plurimis ad os.
The red Scorpæna, with numerous beards.

This is of three times the size of the former species: the head is large, and of an irregular figure, and is armed with a number of spines: the eyes are large, and stand at a small distance from one another on the top of the head; the pupil is black, and the iris is variegated with a number of red spots: the body is large toward the head, and grows very small, as it approaches the tail; the colour of the whole body and head is a strong, but somewhat dusky, red; there is a tinge of blackish with it on the back, which gives the whole a dusky purple hue, and the belly, on the contrary, is pale and flesh-coloured: the back is rounded, and the sides are prominent; the belly is flat; the fins are large: the pectoral ones have each nineteen rays, and are obtuse or rounded at the extremities: the anus in this species is situated at a considerable distance from the pinna ani.

This is caught, with the former, in the Mediterranean, but it is less frequent. Salvian calls this, simply, Scorpius; Gesner, Scorpius simpliciter vel major; Ray and Willughby, Scorpius major; Charleton, Scorpio. The Italians call it Scrofano; and the people of Marseilles, Scorpena and Escorpena.

COTTUS.

THE head of the Cottus is of a depressed figure and prickly, and is broader than any part of the body: there are two fins on the back; the anterior one has flexible prickles: the belly fins are very small, and have only four rays each, and those soft and flexile: the skin is smooth, or not covered with scales; and the branchiostege membrane contains six distinct bones.

Cottus alepidotus, glaber, capite diacantho. **The Miller's Thumb.**
The smooth Cottus, with two spines on the head.

This species grows to about three inches in length: the head is broad, flatted, and irregular in it's figure; it is more rounded on the upper than on the under side, and it's surface is uneven: the upper and under jaws are nearly of the same length, but the extremity of the rostrum turns a little upwards: the nostrils are not very conspicuous, but there is a kind of valve or cirrus between the eyes, and the extremity of the rostrum, in the place where the nostrils of other fishes stand: the eyes stand very close to one another on the top of the head; they are large, and their pupil is greenish, and the iris of a dusky yellow: the head is covered with a kind of bony laminæ, and one of them on each side terminates in a spine, which is crooked, and turned backwards: the branchiostege membrane is very convex, and seems as if it were inflated with air: the mouth is very wide, and has a great number of teeth; there are more than a single series of them in each of the jaws: there is a bone in the anterior part of the palate, which is covered also with a great many teeth; and in the fauces there are two bones in the upper part, which are thick-set with teeth; these are of a roundish figure and large, and in the lower part there are two others smaller, and of an oblong figure, which are also covered with teeth: the middle of the palate and the tongue is smooth.

The body, by degrees, grows smaller from the head to the tail; it is rounded, and approaches to a conic figure, only that toward the tail it is a little depressed: the skin is smooth and lubricous, being covered with an unctuous fluid, and having no scales; the lateral line is very conspicuous, and runs nearly straight, only that toward the anterior part it is a little bent downward: the back and sides are of a dusky brownish-yellow colour, variegated with a number of large black spots: the upper part of the head is blackish, and the belly is whitish; the fins are of a dusky yellow, with an admixture of black.

There are two fins on the back, and they are nearly contiguous to one another; the anterior one is small, and has only seven short rays; the edge of the membrane of this fin is reddish: the second back fin is almost continuous to this, and is variegated with black spots; it has seventeen rays, and the middle ones are the longest of these: the pectoral fins are spotted with black; they are large and rounded at the ends, and they have fourteen rays, the extreme ones of which are the shortest: the membrane of these fins does not reach up to the summits of the rays, whence they appear serrated at the edge: the belly fins are very small, and of a whitish colour; they stand low, and have only four rays, the two middle ones of which are the longest: the pinna ani is spotted; it has thirteen, or sometimes fourteen, rays, the middle ones the longest as in the other fins: the tail is variegated with black and brown spots, and has eight long rays, beside a number of short and inconsiderable ones; it is not quite plane, but somewhat rounded, at the extremity.

The branchiæ are four on each side: the liver is large and undivided; the stomach is very large and round, and has the appearance of a bag; there are four appendages to the pylorus; the intestine has but one inflexion, and then runs straight to the anus: the vesiculæ seminales and ovaries in the female appear double, but they are united at the base, and are included in a black membrane: the kidnies and the urinary bladder are large, and stand at the lower part of the abdomen; the peritonæum is black; the vertebræ are thirty-one, and are compressed at the sides: the ribs are ten on a side, and are but slightly affixed to the vertebræ.

This little fish is very common in our shallow brooks, though it's structure and form of it's parts have not been observed: the antients were acquainted with it. Aristotle call it [illegible] and [illegible]; Gaza, Bellonius, Rondelet, and Aldrovand, simply, Cottus; Gesner, Cottus sive Gobio fluviatilis capitatus; Cuba, Capitatus; Salvian, Citus; Jonston and Charleton, Gobio capitatus. The Italians call it Messore and Capo Grosso; the French, Chabot; and we, the Bull-head, the Great-head, and the Miller's Thumb; the Swedes call it Erra Kaukh.

Cottus scaber tuberibus quatuor carniformibus in medio capite.
The rough Cottus, with four horn-like protuberances in the middle of the head.

The head of this species is very large and depressed; it is very irregularly figured, and has a great many prickles about it, especially at the sides; on the very middle also there stand four very singular, large, and remarkable tubercles, having the appearance of so many horns; these form a square figure, and the two anterior are the larger; these are sometimes round; the anterior ones are always smaller and oblong, and the superficies of them all is rough, porous, and unequal: beside these there are more than twenty bony and prickly apophyses arising from the jaw-bones, and the laminæ about them, and covered only by a thin membrane or cuticle: there are two of these considerably large at the upper part of the branchiostege membrane, and three above these on each side near the horns, two at the nostrils, and one a little above the origin of each of the pectoral fins; there are also two broader behind the horns, at the origin of the back, beside several small, obtuse, and scarce observable ones.

The opening of the mouth is very wide, and the upper jaw is a little longer than the under: the nostrils stand somewhat nearer to the eyes than to the extremity of the rostrum; they are at a distance from each other, and each of them has two apertures, but they are not very conspicuous: the eyes stand in the upper part of the head, near one

one another: the iris is ſmall, and of a reddiſh yellow; the pupil is ſometimes green, ſometimes bluiſh or yellowiſh, and is not perfectly round, but of a ſomewhat oval figure: the body is largeſt near the head, and thence gradually becomes ſmaller to the tail; it is rounded in the anterior part, but from the anus to the tail it is ſomewhat compreſſed: the lateral line runs ſtraight, and is nearer the back than the belly, and above this line near the back there is a ſeries of tubercles reaching from the head to the tail; they are ſmall, rough, of a roundiſh figure, and ſomewhat depreſſed, and form two lines; there are more than forty in the upper line, which is the longeſt; and in the under one, which is ſhort, there are no more than fourteen: theſe are, in their ſubſtance and ſtructure, very like the tubercles on the head, but they are ſmaller than theſe: under the lateral line there alſo are a number of tubercles, but they are ſmaller than theſe, and of an oblong or lanated figure; they are about forty in number, but that is not determinate, any more than in thoſe above the lateral line, which are ſometimes more, ſometimes fewer: the ſkin between theſe tubercles is perfectly ſmooth, or without ſcales: the colour is greyiſh or heavy, ſometimes a very duſky blackiſh, but there always are a number of black, tranſverſe lines on the ſides; the belly is white; the fins are often variegated with ſpots of black: there are two fins on the back; the firſt of theſe is lower and ſmaller, and has uſually nine rays; ſometimes however it has eight, or only ſeven, ſometimes ten; theſe are ſomewhat ridged, but not prickly: the hinder back fin is longer and taller; it has fourteen, and ſometimes fifteen, rays; theſe are two inches long, and are undivided at the top; the firſt and the laſt are the ſmalleſt; the middle ones are the longeſt, and they are all ſcabrous or rough on the outſide: the pectoral fins are very large and broad, and have each ſixteen or ſeventeen rays; theſe are two inches long in the full-grown fiſh, and are undivided at the extremity: the ventral fins are placed directly under the pectoral ones, but a little nearer to the anus; they have four undivided rays: the firſt of theſe is the ſmalleſt, and adheres to the ſecond in ſuch a manner, that they form, as it were, only one; the pinna ani has fourteen rays: the tail is nearly even at the extremity, and has twelve long rays; the gills are four on each ſide, the three upper ones larger than the fourth; the ſtomach is large and ſtrong, and there are ſeven appendages to the pylorus: the vertebræ are forty.

It is a native of the Northern Seas; it is very frequent about the coaſts of Denmark, Norway, and Sweden. I have alſo ſeen it caught near Scarborough, but it is not common in our ſeas; it feeds on ſea-inſects and ſmall fiſh.

Cottus alepidotus capite polyacantho, maxilla ſuperiore paulo longiore.
The Cottus without ſcales, with many ſpines on the head, and with the upper jaw longeſt.

The Father-laſher.

The head of this ſpecies is very large, appearing much diſproportioned to the body; the ſurface of it is very unequal, and there are a number of ſpines on different parts of it: ſome of theſe are ſhorter, and reſemble only tubercles; others are longer, ſlenderer, and very ſharp-pointed; the whole head is of a depreſſed form, broad, in proportion to it's perpendicular meaſure: the body is large toward the head, but becomes gradually ſmaller, all the way to the tail; the anterior part of the body is depreſſed, but the hinder part, from the anus to the tail, is compreſſed: the belly is broad, and ſomewhat prominent: the back is of a deep brown; the ſides are of an olive brown, with ſome tranſverſe ſtreaks of black, and the belly is much paler: the lateral line runs ſtraight, and is nearer the belly than the back.

We have this ſpecies in conſiderable plenty on our own coaſts. Jonſton calls it Scorpius marinus; Schonevelde, Scorpius marinus vel Scorpius aculeatus; Ray, Willughby, and Aldrovand, Scorpæna Bellonii ſimilis; and we, the Father-laſher, a ſtrange unmeaning name, given it by the children who play about the ſhores in Cornwall, but the only one our language has for it.

Cottus

The Sea-dragon.

Cottus pinna ſecunda dorſi alba.
The Cottus, with the ſecond back fin white.

This is a very ſingular, but not a very beautiful, ſpecies: the head is large, flatted, or depreſſed, irregular in ſhape, and armed with a great number of tubercles and ſpines, eſpecially toward the ſides; the eyes are large, and ſtand near one another on the top of the head, and are covered with the common ſkin of the head: their iris is of a duſky orange colour, and the pupil is greeniſh: the noſtrils are ſituated a little below the eyes, and have each a double aperture; there are ſeveral ſeries of teeth in both the jaws, and in the anterior part of the palate there is alſo ſemi-lunar bone, on which there ſtand a multitude of ſmall teeth; and in the fauces alſo there are two bones above, and two below, which are all covered with teeth: the body is broader, depreſſed toward the head; but toward the tail it is compreſſed, and it is throughout ſmooth and free from ſcales, but covered with a mucilaginous fluid: the firſt of the two back fins is the ſmaller, and is of a duſky olive colour, with an admixture of grey: the hinder one is larger, and is white; the lateral line runs tolerably ſtraight, and is much nearer to the belly than to the back: the ſides are of a yellowiſh olive, variegated with tranſverſe ſtreaks of black; the belly is paler, and the back ſtill darker than the ſides; the pinna ani is long and low, and the tail large and equal.

This is frequent in the Mediterranean; and is ſometimes, though not ſo often, met with in the Northern Seas. Rondelet, Aldrovand, Ray, &c. call it Dracunculus; Geſner, Dracunculus, Aranei ſpecies altera; and in another place he deſcribes it again under the name of Exocœti tertium genus.

The Cataphractus, or Pogge.

Cottus cirris plurimis corpore octagono.
The Cottus, with numerous cirri and an octagonal body.

This, as well as all the other ſpecies of this genus, is a very ſingularly ſhaped fiſh: the head is depreſſed, but it is convex and angulated on the upper ſide, and perfectly flat below: the body is very large toward the head, but is grown extremely ſmall, as it approaches the tail, it is angulated all the way, and, when ſtrictly examined, proves to be of an octagonal form; though in the whole, from the little elevation of the angles, it at a diſtance looks ſomewhat cylindric: the back, from the head to the extremity of the ſecond back fin, is hollowed, or has a depreſſion in form of a furrow, but from this part to the tail it riſes into a kind of ridge, or is ſubacute: the belly is of the ſame figure, or a little more plane and even; the anus is in the middle of the belly, or much nearer to the ventral fins than to the pinna ani: the mouth is ſmall, and is not ſituated at the extremity of the head, but ſomewhat underneath the tip of the roſtrum, and is of a lunated or ſemicircular figure: the upper jaw is hence much longer than the under; there are two cirri on the upper jaw, and on the lower there are ſeveral, both at the mouth and under the branchioſtege membrane: the noſtrils have each two apertures, but the hinder are very ſmall; the anterior one is tubuloſe, and is ſituated at a great diſtance from the other, and very near the extremity of the roſtrum: the eyes are round, and placed on the ſides of the head, not on it's top, as in ſome others of this genus; they ſtand ſo high, however, that they are very near one another: the apertures of the gills are moderately large; there are a great number of teeth in both the jaws, and in the fauces alſo there are bones covered with ſmall and fine teeth, but the palate is ſmooth: the head is bony, hard, and irregular in figure; it has eight very conſiderable ſpines on it; four of theſe ſtand on the roſtrum, two on each ſide, of which the hinder ones are bent backwards: the ſpace between theſe ſpines is hollowed, and the apex is bifurcated: the other four ſpines are ſituated on the ſides of the head, two on each ſide; they are ſhort, robuſt, and bent a little backwards: the hinder part of the head is bifid, and is terminated by two tubercles, but they are not prickly.

The back and ſides of this fiſh are of a duſky greyiſh, variegated with four or five tranſverſe blotches on the upper part; the belly is of a yellowiſh white: the body is octagonal, only till the hinder end of the back fin; from thence to the tail it is ſexangular: the whole body is covered with a kind of bony laminæ in the place of ſcales;

they

they are very hard, and have in their middle a prominence which turns backward; this is most eminently observable in the laminæ on the back, and it is to these that the body owes it's angular figure; these laminæ are, in the anterior part, disposed in eight longitudinal series: the lateral line runs nearly straight along the middle of the body, but it bends a little toward the back, as it passes the pectoral fins.

There is only one back fin, but this is cut in so deeply toward the middle, that it has the appearance of two fins; the anterior part of it has five simple and somewhat prickly rays; the hinder division has seven rays, but they are soft, and not all prickly: the pectoral fins are very large, and rounded at the extremities; they stand low, or nearer the belly than the breast of the fish, and are variegated with black spots; they have each fifteen rays; these are all undivided, and the middle ones are largest: the ventral fins are very small; they are placed almost close to one another, and stand but a little below the pectoral ones: these have each only three rays; the first of these is very short, and is prickly at it's point, but it adheres firmly to the second; the pinna ani stands just over-against the hinder division of the back fin; it has only six rays, and they are all undivided at the extremities: the tail is rounded at the extremity, and has eleven or twelve rays, which are not at all prickly, nor ramose.

This species grows to about six inches in length, and is considerably thick, in proportion. Ray, Willughby, and almost all the writers on fishes have called it Cataphractus; and we, in Cornwall, call it the Pogge: it is very frequent in the Baltic, and is not uncommon about our own shores in many places.

The other species of the Cottus are, 1. The very small, flat-headed Cottus. 2. The larger, thick-headed Cataphractus. 3. The larger, slender-bodied Cataphractus; and, 4. The smaller, gibbose-backed, thick-bodied Cataphractus.

ZEUS.

THE body of the Zeus is very broad, thin, and compressed; the head also is compressed and thin: the scales are rough; there is only one fin on the back, but it is very long, and is cut in so deeply near the anterior part, that it appears to be two fins: the branchiostege membrane does not consist of parallel bones, as in other fish, but has a number of ossicles of various figures; some of them placed longitudinally, some transversely, and some obliquely.

Zeus ventre aculeato, cauda in extremo circinnata.
The Zeus, with an aculeated belly, and the tail rounded at the end.

The Doree.

This is a very singularly shaped fish, but it is very far from a beautiful one; it is extremely broad, in proportion to it's length: it's head is large, compressed, and of a monstrous form, and the mouth is enormously wide, and strangely cut: the eyes are large, and the nostrils have each a double aperture: the body is very thin, though so very broad; it's colour is a dusky olive, with a strong admixture of a fine gold yellow, and in the middle of each side there is a very large, round, black spot: the lateral line is very crooked, and is turned up toward the back, in the greater part of it's course.

The anterior division of the back fin has ten prickles, under which there are as many rays: the second, or hinder, division of the back fin, has twenty-four rays; the pectoral fins have each fourteen rays; the ventral fins have each seven rays, and the first of them is prickly: the pinna ani has twenty-six rays, the four first of which are prickly and long; there are also series of prickles at the base of the back fin, and of the pinna ani, and all along the belly: the tail is rounded.

This species is frequent in the Mediterranean, and is sometimes caught in our own seas: the antients were acquainted with it. The Greeks call it [illegible] and [illegible]; Pliny calls it Zeus and Faber; Paulus Jovius calls it Citula sive sancti Petri piscis; and, in another part of his work, Corvus; Rondelet, Gesner, Willughby, and Ray call it Faber sive Gallus marinus; and most of the other writers, simply, Faber. The Spaniards calls it Gal; the Italian, Citula and Pesce san Pietro; the Genoese, Rotula; the French, Doree; and we, the Doree, or John Doree, a corruption of the Jaune Doree of the French, a name they give it to express the gilded yellow colour it has intermixed with the brown, and sometimes singly.

Zeus cauda bifurca.
The Zeus, with a forked tail. **The Indian Dorée.**

This also is a very singular fish; the body is extremely flat and thin, and more resembles a flounder than any other fish in shape: the head is moderately large and compressed; the eyes are large, and their iris is of a silvery white; the pupil is of a dusky greenish: the mouth opens very wide, and is cut in an odd manner; the nostrils have each a double aperture, and are situated at some little distance below the eyes; the whole inside of the mouth is perfectly smooth; there is not the least asperity or resemblance of a tooth in either of the jaws, or on the palate: the back is of a dusky colour, mixed, of an olive and a blackish grey; the sides have more of the olive, with a tinge of yellow, and the belly is more yellow than any other part: there is but one back fin, but it is very long, and extends to the tail; the pinna ani also is long, and reaches in the same manner also to the tail; the tail itself is large and forked: from the place of origin of the pectoral fins, there runs on each side a kind of long filament, and there is such another arising just before the back fin.

This is a native of the American Seas, but it has also been sometimes met with in Europe. Willoghby calls it Gallus marinus seu Faber Indicus. The natives of the Brasils, where it is extreamly common, call it Abacatuaia; and the Portuguese there, Peixe Gallo. Jonston calls it Abacatuaja Lusitanis Peixe; and Ruysch, Ican Kapelle. We frequently have it in the collections of the curious, sent over preserved from the Brasils.

Zeus totus rubens, cauda aequali, rostro sursum reflexo.
The red Zeus, with an even tail, and the rostrum turned upward. **The Aper, or Riondo.**

This is a small species; it rarely grows to more than three inches in length, but it is very broad, in proportion to that length, and is thin and sharp at the back and belly: the head is large and compressed; the eyes are large, and stand at a moderate distance; their iris is of a silvery white, and the pupil of a bluish-grey: the mouth is large, but there are no teeth in it: the nostrils have each a double aperture, and the rostrum is turned up at it's extremity: the scales are rough and small, and the lateral line is turned up toward the back.

The anterior division of the back fin, or, as the generality of authors have called it, the anterior back fin, has nine prickly rays of which the third is the tallest; the hinder division has twenty-three rays; the pectoral fins have each fourteen rays, and the ventral ones have six each; the first of these is prickly: the pinna ani has twenty-six rays, of which the three former are short and prickly: the intestines of this fish have frequent convolutions, and there are two or three appendages to the pylorus.

The colour of the whole fish is a deep and strong red; it is dusky or purplish on the back, and degenerates into a pale flesh colour about the belly; but on the sides it is an elegant crimson, especially about their middle.

The fish is frequent in the Mediterranean; the antients were acquainted with it. Aristotle and Athenaeus call it Κάπρος; and all the Latin writers, in general, Aper. The Italians call it Riondo; and the Genoese, Suivale.

CHÆTODON.

THE body of the Chaetodon is compressed, and is broad, thin, and short: there is only one fin on the back, which is extended the whole length of it: the tail is large, and there are in all six fins, exclusively of the tail: the mouth is small, and has a pair of lips which are moveable at pleasure, but which, when shut, cover the teeth: the teeth are oblong, contiguous, and flexile; the scales are rough, and the eyes are covered with the common skin of the head: the branchiostege membrane on each side contains four or five very slender bones.

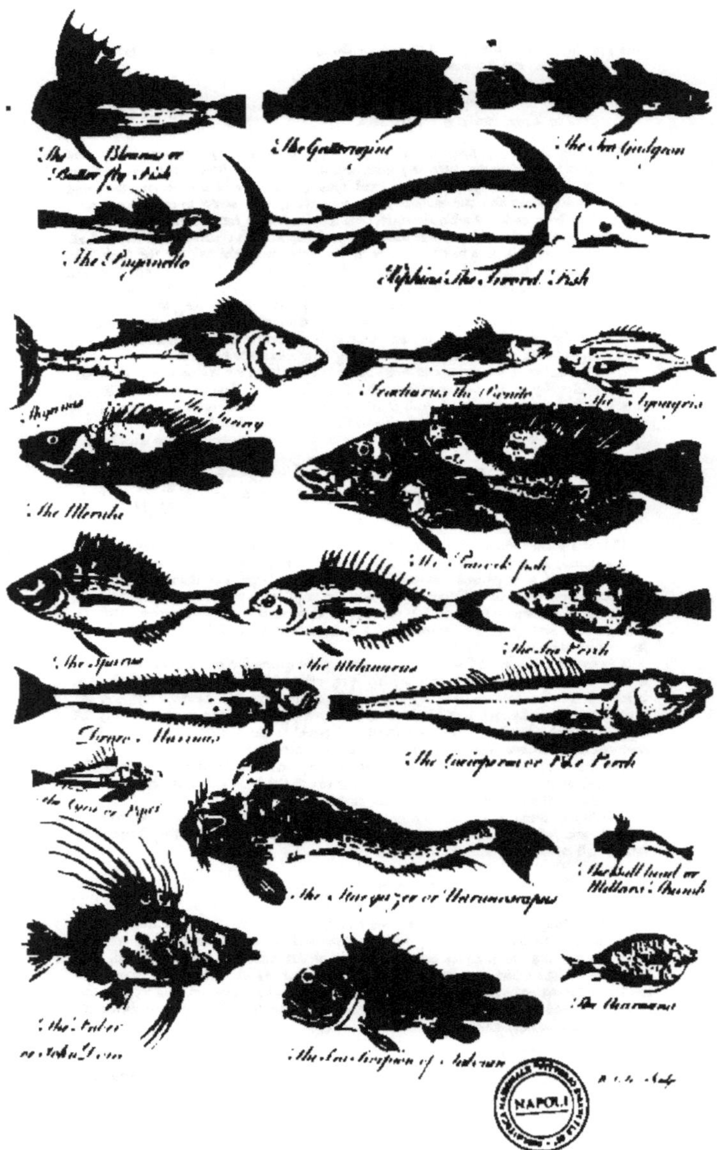
The Gattorugine
The Sea Gudgeon
Xiphias the Sword Fish
The Peacock fish
The Sparus
The Melanurus
The Sea Perch
Draco Marinus
The Faber or John Dory

Chætodon lineis longitudinalibus varius, cauda bifurca.
The Chætodon, variegated with longitudinal lines, and with a forked tail.

The head and body of this species are both greatly compressed; the body is very broad and thin; the back is acute, but the belly, in the anterior part, is somewhat plane; and the anus is situated in the middle, between the ventral fins and the pinna ani; the mouth is very small, and, when it is shut, the upper jaw is a little longer than the under; there are no beards: the head, from the eyes to the rostrum, is very declivous; and the lips may be drawn backward and forward, from the extremity of the jaws, with a finger: the nostrils are situated very near the eyes, and have each a double aperture, of which the anterior is the larger: the eyes stand at the sides, not on the top of the head, but they stand high, and are very near one another; they are large, round, and are covered with the common skin of the head; the iris is of a bluish-white: the openings of the gills are small and narrow; there is in each jaw a single row of white, slender, subulated teeth; they stand contiguous to one another, and are serrated at the point: the palate and the tongue are smooth.

The scales are obtusely quadrangular in figure, and have, at their lower verge, a number of small uncinulæ scarce perceptible; at the upper part they are very lightly striated: the whole body of the fish is elegantly variegated with longitudinal lines; there are on each side nine of these lines of a bluish colour, with a brown edge on each side of them, so that there are, as it were, eighteen of these brown lines accompanying the nine blue ones, and between these there are six broad white ones, sometimes seven: on the top of the head there are also some of these longitudinal bluish lines: the lateral line is distinct from all these; it is very crooked, and runs near the back.

The back fin is of a dusky grey colour, and runs along the whole length of the back; it has thirty-six rays; nine of these are prickly, the rest are soft, and divided at the extremities: the pectoral fins stand nearer the belly than the back, and have each fourteen rays: the ventral fins are oblong and black at the extremity; they stand lower down, or nearer to the anus, than the pectoral ones; they are fixed down to the belly by a membrane, by which they are prevented from being raised into a perpendicular; they have each six rays, of which the first is prickly: the pinna ani is whitish at the base, but it is black all the way; from this it has twenty-nine rays, of which the first three are prickly: the tail is large and forked, and has sixteen long bones or rays, and on the extremity of it there is a large, white, lunated spot.

There is on each side, near the tail, a long and robust spine; this is affixed to the body of the fish, and there is a furrow in the side, into which it is received.

This is a native of the American Ocean; but we meet with specimens of it dried frequently in the cabinets of our collection.

Chætodon nigrescens cauda albescente æquali, utrinque aculeata.
The blackish Chætodon, with a white, undivided tail, aculeated on each side.

The body of this species is broad, in proportion to it's length: the back is elevated and acute; the belly is flatted, and the anus stands at a middle distance between the pinna ani and the belly fins: the head is of a depressed form; the mouth is very small, and the upper jaw is a little longer than the lower: the lips are moveable, and completely cover the teeth: the eyes are large and round; they stand high, and at no great distance from one another; they are round, and their iris is small, and of a silvery white; the pupil is very large, round, and bluish: the nostrils have each a double aperture; the anterior one larger and round, the hinder a narrow slit; they stand a little below the eyes: the teeth are numerous, slender, and contiguous; they are serrated at the top, and there is a single row of them in each jaw; the palate and the tongue are smooth; the whole head is of a depressed form, and very declivous from the eyes to the rostrum.

The scales are smaller on this fish than on the other species, and the colour of the whole fish is one uniform, blackish brown, without any variegations of whatever kind: the pectoral fins are grey or hoary; the back fin and the pinna ani are black, only that they have ſome whiteneſs about the baſe; the ventral fins are alſo black; the pectoral fins have each ſixteen rays, and the belly fins have ſix rays each; the back fin is very long, and has thirty-eight rays, of which the nine anterior ones are prickly, the others are ſoft, and divided at their ends: the pinna ani has twenty-nine rays, and the three firſt of theſe are prickly; the tail is large, broad, and even, not divided at the end; it is white, and has ſixteen long bones or rays; when extended, it is of a ſomewhat triangular figure: there is on each ſide a ſtrong prickle near the tail, which has a furrow formed in the ſide of the fiſh to receive it: the whole fiſh is of a depreſſed form, and is conſiderably broad, in proportion to it's length.

It is a native of the South-Seas; none of the writers of fiſhes have deſcribed it, with any degree of accuracy.

Chætodon niger capite diacantho, lineis utrinque quatuor tranſverſis curvis.

The black Chætodon, with two ſpines on the head, and four tranſverſe, crooked lines on each ſide.

The Guarenta.

The head of this ſpecies is moderately large, and is of a depreſſed form, and very declivous from the eyes to the extremity of the roſtrum: the eyes are moderately large, and are ſituated very forward on the head, much nearer the roſtrum than in either of the preceding ſpecies: the roſtrum is ſhort and obtuſe, and the lower jaw is a little longer than the upper; in each jaw there ſtand ſeveral ſeries of oblong, ſlender, and moveable teeth; they ſtand cloſe together: the palate is ſmooth, but in the fauces, both on the upper and under parts, there are a great number of villi or hairs, which are flexible, and are ſcarce at all rough: the coverings of the gills are lightly ſerrated at their edges toward the middle, but at the lower part they terminate in a large and robuſt ſpine, which turns backwards; the apertures of the gills are larger in this than in any other ſpecies: the whole body of the fiſh and all the fins are of a dark blackiſh-brown colour, but there run down the ſides four yellow, or ſometimes whitiſh, lines, in a tranſverſe direction, but lunated or crooked: the firſt of theſe is leſs crooked than the others, and runs through the middle of the coverings of the gills; the ſecond and third ſurround the middle of the body, and are lunated; the convex part or back is turned toward the head, the points or horns toward the tail: the fourth or laſt line ſurrounds the beginning of the tail; this is extended from the ſides, and ſurrounds, as it were, the tail: beſide theſe lines on the body, there is alſo on the head a tranſverſe line on each ſide near the mouth, and another longitudinal and ſtraight one on the top of the head.

The ſcales on this ſpecies are ſmall and rough; the lateral line is crooked, and runs more to the back than the belly; the back is elevated, and ſomewhat acute, and the belly is flatted; the anus is at a ſmall diſtance above the pinna ani.

The back fin is very tall, and in the middle forms a kind of horn; it has forty-one rays, and the nine or ten anterior ones are prickly: the pectoral fins are blackiſh, and have each nineteen or twenty rays; the ventral fins are extremely black, and have ſix rays each, the firſt of which is prickly: the pinna ani is very large, and is raiſed into an eminence in the middle; it has twenty-ſeven rays, and the three firſt of them are robuſt and prickly: the membrane of the back and belly fins are very robuſt and firm, and are covered with ſmall, rough ſcales, up to the very tops of the rays: the tail is nearly even at the extremity, but a little convex; it has ſeventeen long rays.

This ſpecies grows to ſix inches in length, and a little more than three in breadth; it is a native of the Braſilian Seas. Marcgrave, and Ray after him, call it, by it's Braſilian name, Guarenta; Liſter has named it Acarauna exigua nigra tæniis aliquot luteis eleganter depicta.

Chætodon aculeis duobus brevibus ſupra oculos, oſſiculo tertio pinnæ dorſalis altiſſimo.

The Chætodon, with two ſhort ſpines above the eyes, and with the third ray of the back fin talleſt.

The body of this ſpecies is depreſſed, broad, and ſhort; the back ſomewhat acute, and the belly flatted: the roſtrum is very long and ſlender, and the head is very decli-

vous

bout from the eyes downwards: the mouth is ſmall, and the teeth are numerous; both jaws are very well furniſhed with them, but the palate and tongue are ſmooth: the lower jaw is a little longer than the upper, but the teeth in the upper one are longer than thoſe below; the lips are fleſhy and moveable: above the eyes there is, on each ſide of the head, a ſerrated bone, and above this there ſtands a ſmall, but thick, bony ſpine, which turns a little backwards: the ſcales are very ſmall, hard, and rough to the touch, and they are placed very cloſe together: the lateral line is crooked, and runs conſiderably nearer the back than the belly; the whole body of the fiſh is variegated with black and white: the anterior part of the head is white, with a black line on the top, and another below, at the lower jaw: there is a very broad black line, which runs between the belly fins and the back, and divides the body tranſverſely into two parts; and at the anterior part of this there is on each ſide a narrow, crooked, tranſverſe line; there is, behind this black line, a broad, tranſverſe, white one, which again divides the body; and below this there is alſo another broad, tranſverſe, black line: about the beginning of the tail there is another white line; and, finally, the tail itſelf is variegated in the ſame manner: there is, near the baſe, a broad, ſtraight, black line, and farther toward the extremity, a broad lunated one; and, finally, the extremity of the tail itſelf is white.

The back fin is very tall in the anterior part, but in the hinder part it is low; it has forty-ſix rays, of which the ſeven anterior ones are prickly, but two of them much more ſo than the other five: the two firſt are very ſhort, but the third is extremely long, and is extended in form of a briſtle beyond the tail: the pectoral fins ſtand in the middle of the ſides, or a little nearer the belly than the back; they have each eighteen rays; the belly fins have ſix rays each: the pinna ani is broad at it's origin, but very narrow toward the extremity; it has thirty-ſix rays, of which the three anterior ones are prickly; the tail is large, and a little forked, and has fifteen long rays: the body of this ſpecies is ſo broad, that in the middle it meaſures very nearly as much acroſs as in length; the fauces are ſmooth, but the orifice of the ſtomach is furniſhed at it's upper part with a ſharp bone.

This alſo is a native of the American Ocean, and has not been accurately deſcribed by any author.

Chætodon caneſcens aculeo utrinque ad os, oſſiculo tertio pinnæ dorſalis ſetiformi longiſſimo.

The hoary Chætodon, with a prickle on each ſide at the mouth, and with the third ray of the back fin long.

The head of this ſpecies is large, and the roſtrum long and obtuſe at the end: the eyes are large and round, and ſtand but at a ſmall diſtance from one another; their iris is of a ſilvery grey, and the pupil large and greeniſh; the noſtrils ſtand at a ſmall diſtance before the eyes, much nearer to them than to the extremity of the roſtrum; they have each a double aperture, the anterior one round, and the hinder one a tranſverſe fiſſure: the mouth is ſmall, and the lower jaw a little longer than the upper; both are furniſhed with a great number of ſlender, ſubulated teeth, and have fleſhy lips which cover them; theſe are moveable on touching, and are very tough in their ſtructure.

The body is very broad and depreſſed; the ſcales are all ſmall, hard, and rough, and the lateral line is crooked, and runs much nearer the back than the belly: at the angles of the mouth, on the upper jaw, there ſtands a robuſt ſpine, ſerrated at the baſe, and bent backwards, and behind the eyes there run three very viſible longitudinal bones, but there is no ſpine there as in the former: the colour of the whole fiſh is a greyiſh; in the middle it is a fine ſilvery grey, with ſcarce any duſkineſs in it, but there runs a black tranſverſe line down from the middle of the back fin, and another behind the eyes: the back fin has forty-ſix rays; the two firſt of them are a little aculeated, and the third is very long, and of the figure of a briſtle: the pinna ani has thirty-ſix rays; the pectoral fins have ſeventeen rays each, and the belly fins have each ſix rays: the tail is ſomewhat forked at the extremity, and has fifteen long rays, and, beſide theſe, ſome ſhort ones.

This is a native of the American Ocean, and is caught among the other already described species; but, though we meet with it frequently in the cabinets of our collectors, none of the writers on fishes have accurately described it.

Chætodon macrolepidotus, lineis utrinque duobus nigris, officulo quarto pinnæ dorsalis longissimo setiformi.

The Chætodon, with oblong scales, and with two black lines on each side, and the fourth ray of the back fin long.

The head of this species is longer than that of the last, and the rostrum obtuse: the lower jaw is somewhat longer than the upper, and the whole head is narrower than in any of the other species: the eyes are large, and stand but at a small distance from one another; their iris is white, and the pupil of a greenish black; the orbits are somewhat acute in the upper part: the apertures of the gills are larger than in almost any other species, and the middle lamina of their coverings is somewhat serrated on the lower part: there are in each jaw several series of teeth, placed very close to one another, and flexible; and there are two prominences in the anterior part of the palate; the rest of the palate is smooth, and the tongue also is smooth.

The scales are very large and thin; they are of an oblong figure, somewhat striated, rectilinear at the base, but rounded, and somewhat rough at the hinder edge: the head is of a silvery white, but there is a black line between the eyes, and a black spot at the rostrum: the body is variegated with black and white, disposed in transverse lines.

The back fin has thirty-seven rays, eleven of which are robust and prickly; the others are soft, and are ramose at the extremities; the fourth ray is so very long, that it resembles a bristle, and is extended beyond the tail; the rest are very short: the pectoral fins are white; they have each eighteen rays, and they are situated much nearer the back than the belly: the belly fins are black, and have each of them six rays, of which the first is prickly, broad, and robust; the rest are weak and ramose: the pinna ani is broad, and rises highest toward the middle; it has twenty-three rays, the three anterior of which are robust and prickly, the others are soft and ramose, and the middle ones are tallest: the tail is white; it is nearly even at the extremity, and has seventeen rays: the anus is much nearer to the pinna ani than to the ventral fins.

This fish is very broad, in proportion to it's length, though less so than the preceding species. It is usually about four inches long, and three inches and a quarter in breadth. It is caught on the Brasilian shores.

Chætodon macrolepidotus, lineis utrinque tribus nigris, latis, linea quarta in cauda.

The long-scaled Chætodon, with three black lines on each side, and a fourth on the tail.

This is also very broad, in proportion to it's length, and is of a more depressed figure than almost any of the other species: the head is oblong, slender, and obtuse at the extremity, and is very declivous from the eyes to the rostrum: the eyes are large, and placed near one another, and their iris is of a silvery grey, the pupil black: the nostrils have each a double aperture; the anterior one is a round foramen; the posterior a transverse slit; they stand nearer to the eyes than to the extremity of the rostrum: the openings of the gills are large, and the middle lamina is serrated on the lower edge: the jaws are both of the same length, but, when the mouth is open, the under one appears somewhat the longer: there are in each jaw several rows of teeth, and in the anterior part of the palate there is a very large protuberance.

The whole fish is variegated in a very elegant manner with black and white: the whole anterior part of the head is white, and there is a black line on each side near the eyes; behind this there runs another very broad white line: there goes a broad black line again, from the anus to the beginning of the back fin, and in the middle of the body there is a very broad white line; and in the middle, between the back and belly fins, there is also a black line: the hinder part of the body, and one half of the tail, are of a bright and elegant white; and the exterior edge of the back fin, and

 the

the pinna ani, are black: in the middle of the tail there is a black transverse line, but the extremity of the tail again is white: the back fin is not so high as in some of the other species; it has thirty-three rays, and twelve of them are prickly and very robust, the rest are soft and ramose; the middle ones are of the same general length, and there is not any thing of that hair-like appendage as in the others: the pectoral fins are yellow, and have each fifteen rays; the belly fins are very black half way, and have each of them six rays: the pinna ani has twenty-one rays, covered with scales; the three first are prickly, and very robust: the tail is convex, or rounded at the extremity, and has seventeen long rays.

This also is caught about the Brasilian shores, and is sometimes met with dried among our collections.

GASTEROSTEUS.

THE belly of the Gasterosteus is almost entirely covered with oblong bony laminæ: the ventral fins consist only of two rays each, of which the one is much larger than the other, and prickly: the branchiostege membrane contains three slender bones.

Gasterosteus aculeis in dorso tribus. **The common**
The Gosterosteus, with three prickles on the back. **Stickle-back.**

Common and disregarded as this little fish is, it has that in it's structure which sufficiently merits observation. The body is throughout, in some degree, compressed, but more particularly so in the hinder part: the head is large, long, compressed, and very declivous from the eyes to the rostrum: the rostrum itself is subacute, and the opening of the mouth is considerably large; the head is of a very firm structure, bony, striated, and of a greyish colour on the upper part: the under jaw is somewhat prominent beyond the upper; the eyes are large, and situated at the sides, not on the top of the head, and their iris is of a silvery colour: the apertures of the nostrils are extreamly small; they are situated at a middle distance between the eyes and the rostrum, and each has only a single aperture instead of the double one in most fishes: the coverings of the gills are large; they are composed each of two large bony laminæ and the branchiostege membrane: there are a great number of teeth in the verge of each jaw, but the palate and tongue are smooth.

The lateral line is somewhat crooked, and runs near the back, and parallel to it; toward the tail it rises above the surface of the body, and forms a kind of fin-like prominence, whence that part of the body has a quadrangular figure in some degree: the head and back of this fish are of a dusky colour, or an admixture of grey and brown; the rest of the body is of a silvery white: the breast is covered by two oblong bones, which coalesce at the top, and form, as it were one; over this armature of the breast there is a large muscle covered with a smooth skin; this stands before the pectoral fins, and in it's anterior part is the bone of the clavicle, which is large and simple: the belly is covered with a double scutiform bone, of an oblong figure, very firm and hard, and reaching to the anus; and at the anterior part of that bone there is situated another bone, at right angles, on either side: this serves to strengthen the sides, and is on each side continuous with the bone of the belly already mentioned: beside these, the whole lower part of the body, almost to the tail, is covered with transverse bones, twenty-six or twenty-seven in number; the middle ones of these are the largest, but those next the head and tail are remarkably small.

There is only one back fin, but this is large, and is extended almost the whole length of the back; it has fifteen rays, of which the three anterior ones are prickly, and stand at a great distance from one another; the others are soft, and somewhat forked at their extremities: of the prickly ones, the two anterior are large, long, equal, and serrated at their sides; the third is scarce equal to one third of the length of these, but all three are in part connected by the membrane, not distinct and separate, as authors, who have described this fish, have generally said.

The pectoral, or, as they may be called from their situation, the lateral, fins, are oblong, whitish, and even at the extremities; they have each ten rays, which are all of the same length, and all soft and undivided at the ends: the belly fins consist only of of two rays each; the anterior of these is large, aculeated, and serrated at the edges; and the other, which has escaped the observation of those who have examined this fish, is small, short, white, and soft, and is, indeed, not very conspicuous: the pinna ani is whitish, and has nine rays; the first of these is very short, and is aculeated; the others are soft, long, and a little bifid at their extremities: the tail is oblong, and even at the end; it has twelve long rays, all which, except the two extream ones, are bifid at the ends.

This little fish is very frequent in our shallow brooks and rivulets, and often in fish-ponds; it's smallness makes it overlooked, but it is often the cause of great mischief; it is the most voracious fish in the world, and devours the young of valuable fish, when first hatched from the spawn, in incredible quantities; it also feeds on the spawn itself, and many a good breeding pond is spoiled by it. All the writers on fishes have described it. The older Latin writers have called it Pugnitius and Spinacha; the later ones, in general, Pisciculus aculeatus; Aldrovand calls it Pugnitius piscis, Rondelet, Pisciculi aculeati primum genus; Gesner, Pisciculus aculeatus, primum genus; Ray and Willughby, simply, Pisciculus aculeatus. The Swedes call it Skittpig and Skittar; and we, the Stickle-back and Bansticle.

Gasterosteus aculeis in dorso decem.
The Gasterosteus, with ten prickles on the back. — **The lesser Stickle-back.**

People, in general, have confounded this with the former species, and call both by the common name of the Stickle-back, supposing all of one kind; but this is not only constantly smaller than the other, but has it's distinctive characters as evident as any species in the world.

The head is large, in proportion to the body, and is of a somewhat compressed figure: the lower jaw is a little longer than the upper; the jaws are both furnished with a number of very minute teeth, but the palate and the tongue are smooth: the eyes are large, and their iris is of a greyish colour: the nostrils have each only a single aperture; the openings of the gills are large, and their coverings are composed each of two large bones and the branchiostege membrane: the body is of a compressed form, and the lateral line runs in a direction parallel to the back, and is much nearer that than the belly; toward the tail this line rises into a kind of fin-like prominence on each side, as in the former species, and gives the body a quadrangular figure.

The breast is covered with two bones, as in the former; and these are covered also with a muscle, as in that species, and the structure of the armature of the belly is the same as in that: the head of this is, however, longer than that of the other, and has nothing of that declivity from the eyes to the rostrum: the back of this species is of a dusky grey, approaching to black, and is spotted with large blotches of a deep black; the sides are of a silvery white, and are also variegated with very minute spots of the same black: the sides and belly appear perfectly smooth, the naked eye being not able to distinguish the least appearance of scales on them: the prickles on the back and belly of this fish, when erected, are very difficultly depressed, and they arise out of a kind of furrow: the exterior prickle of the ventral fins is lightly serrated at the sides, and the membrane, which connects these, is of a silvery white: there is but one fin on the back, and that has twenty rays, the nine or ten first of which are prickly, robust, and short, and are placed in an alternate order; four on each side turn outward, and the ninth is erect, and they are all of them affixed down to the back by a thin membrane in their lower part; the other eleven rays are soft and longer than these; the middle ones are bifid at their tops: the pinna ani has eleven rays, of which the first only is prickly; this is short and robust, the rest are soft, and the middle ones are bifid at their extremities: the liver in this species is very large, and of a flesh colour, and is divided into three lobes; two of these are short, but the third is very large, and reaches down on the right side, almost to the anus: the gall-bladder is small, and is affixed to the upper part of the longest lobe: the heart is not so large as a hempseed; it is of a triquetrous figure, with a plane base: the aorta is large and white; the spleen is small, red, and obtusely triangular, and is situated at the left side of the stomach: the stomach is long and thick, and has no appendages; the intestine is a little elevated above the pylorus, but from thence it tends straight down to the anus, and is all the way covered with fat: the air-bladder is large, thick, and

simple,

ſimple, and is affixed to the back-bone; and in the lower part of the body there is alſo affixed to the ſpine an oblong, red body, which unqueſtionably does the office of the kidnies.

This ſpecies grows to about an inch and a half in length, and is very frequent with us in ſhallow waters, little brooks, rivulets, and ponds, and alſo in the ſea, among the ſtones that lie about the ſhores. Though the vulgar confound this ſpecies with the common one, the generality of authors, who have written on theſe ſubjects, have diſtinguiſhed it. Rondelet and Geſner deſcribe it under the name of Piſciculi aculeati alterum genus; Ray and Willughby call it Piſciculus aculeatus minor; Aldrovand and Jonſton, Alterum genus Pungitii; Belloniu, Spinarella puſilla; Schoneveldt, Aculeatus minor levis et glaber; and we, the leſſer Stittle-back, Banſtickle, or Sharpling.

Gaſteroſteus aculeis in dorſo quindecim.
The Gaſteroſteus, with fifteen prickles on the back.

The great ſea Stittle-back.

This is greatly larger than either of the other ſpecies; it grows to ſix or ſeven inches in length, and is moderately thick, in proportion: the head is oblong, narrow, and ſomewhat compreſſed; the eyes are large, and their iris is of a ſilvery grey; the pupil is round, and of a greeniſh-black; the noſtrils have each only a ſingle aperture, and are placed at nearly a middle diſtance between the eyes and the roſtrum: the back and ſides are of a deep brown, with a tinge of olive: the lower parts of the ſides and the belly are of a ſilvery white; the lateral line is large, and the whole body is of a figure approaching to ſquare: the back fin is long, and has fifteen prickly rays; the ventral fins conſiſt only of two rays each, one of them very large, and prickly; the other ſmall and inconſiderable: the breaſt and belly are covered with large, oblong bones, and the ſides are ſupported in the ſame manner.

This ſpecies is frequent in the Northern Seas; we have it on our own coaſts, but not ſo frequent as it is in Sweden, Norway, and Denmark. It is a very nimble and lively fiſh, and the ſpines about it ſeldom fail to wound thoſe who take it up, though that be done ever ſo carefully. It is extremely voracious, and uſually is found behind a ſtone near the ſhore, or under the cover of a tuft of the fucus or common ſea-wrack, watching for whatever comes by that is ſmaller than itſelf. The writers on fiſhes have moſt of them deſcribed it. Schoneveldt calls it Aculeatus ſive Pungitius marinus longus; Willughby and Ray give it the ſame name. The Germans call it Steinbicker and Erſtkruper; and our people in Yorkſhire, the ſea Stittle-back and great Stittle-back.

F I S H E S.

Claſs the Third.

BRANCHIOSTEGI.

Fiſhes which have the tail placed perpendicularly, the rays of the fins bony, and the branchiæ deſtitute of thoſe oſſicles which defend them in the fiſh of the two former claſſes.

BALISTES.

THE Baliſtes has only one belly fin, which is ſimple, aculeated, and is ſituated perpendicularly and longitudinally, exactly in the ſame manner as the pinna ani: on the back there are two, three, four, or more, robuſt ſpines: the jaws are furniſhed with very large teeth, which are placed contiguous to each other, and are protended forwards, and have much the appearance of thoſe of the human mouth; and in other ſpecies of thoſe of the hog: the body and the head are compreſſed and broad: there is no branchioſtege membrane, as in the former claſſes of the acanthopterygious fiſhes.

Balistes aculeis dorsi tribus, cauda bifurca.
The Balistes, with three spines on the back, and a forked tail.

The Old-wife Fish.

The head of this species is large and compressed; the rostrum is oblong, and is protended beyond the lower jaw: the eyes are moderately large, and their iris is silvery; their pupil is black; the nostrils have each a double aperture, and stand at a small distance below the eyes: the mouth is moderately large; there are eight teeth in the lower jaw; they are extremely large, and of a square figure, and there are fourteen more in the upper jaw: the scales are very large and very hard, and are of a square figure.

The first back fin has three very robust and thick spines; the hinder back fin has thirty rays, but they are all soft and flexile: the pectoral fins have each fifteen; the belly fin, for there is but one, has fifteen, sixteen, or seventeen prickly rays: the pinna ani has twenty-seven rays, all of them soft, and the anterior ones longest: the tail is extremely forked, and has twelve rays.

This species grows to a foot, or more, in length, and to about five inches in breadth, but it is usually met with much smaller: there is on the belly, near the ventral fin, an extremely large and robust spine, which points backwards.

This fish is frequent in the Brasilian Seas. Marcgrave calls it Guaperva; and Jonston, and most of the other writers, have called it by the same name. We have it among our collectors, who call it the Old-wife.

Balistes aculeis dorsi duobus cauda quadrata.
The Balistes, with only two spines on the back, and a square tail.

The long Guaperva.

This is a very singular and a very beautiful species: it grows from ten to fifteen inches in length, but it is more frequently met with of three or four: the head is large, oblong, somewhat compressed, but flatted at the top; the eyes are large, and the nostrils are placed a little below them; they have a double aperture each, and are very conspicuous in the living fish, though much less so in the dead: the mouth is moderately large, and is furnished with a number of extremely large teeth, standing contiguous to one another, and protended forward: the body is longer, in proportion to it's breadth, than in the former species, but it is compressed in the same manner: the scales are very large, very firm and hard, and of an irregularly angulated figure, approaching to square; the anterior back fin has only two prickly rays, but they are very robust and strong; the hinder back fin has several soft rays; there is no spine at the belly: the tail is not at all forked, but approaches to a square figure.

This species is caught with the former, in the Brasilian Seas. Lister calls it Guaperva longa, cauda fere quadrata minime forcipata, capitis vertice latiusculo.

Balistes cauda bifurca, pinna dorsi maculosa.
The Balistes, with a forked tail, and with the back fin spotted.

The broad Guaperva.

This is a very singularly-shaped species: the body is compressed, and very broad, in proportion to the length; the head also is compressed: the eyes are small; their iris is silvery, and the pupil round and greenish: the nostrils are not very conspicuous, but there is an appearance of a double aperture to each; the mouth is not very large, but it is well furnished with teeth, which are large, and placed close to one another; and the anterior ones somewhat protended: the upper part of the head is somewhat flatted, but it is not broad: the scales are large and hard, and of a singular figure; on the back there stands a single spine, which is large, and has the appearance of a horn; it is on every part tuberculose, or armed with a number of obtuse teeth; the second back fin is long, and moderately tall, and is of a pale brownish-grey colour, variegated with several spots of black: the belly fin is single, and stands in the same direction with the pinna ani; the tail is forked, and is of an olive colour.

This

This ſpecies is frequent in the American Ocean, and we meet with preſerved ſpecimens of it in ſome of our collectors cabinets. Liſter calls it Guaperva lata cauda forcipata, pinna dorſi maculis quibuſdam diſtincta.

Baliſtes lineis ſtriatis cauda bifurca.
The forked-tailed Baliſtes; with ſtriated lines.

The ſtriated Guaperva.

This is broad, in proportion to it's length, but leſs ſo than the preceding ſpecies: the head is compreſſed, and narrow at the bottom, but it has ſomething of a flatneſs on the top: the eyes are ſmall; the noſtrils have each a double aperture, but not very conſpicuous, placed at about the midway between the eyes and the extremity of the roſtrum; the teeth are large, and are like the dentes inciſores of the human ſpecies, and are protended a little forward: the body alſo is compreſſed, and only marked on the anterior part, but toward the tail it is regularly ſtriated in a longitudinal direction, with a number of fine ſmall lines: on the back there ſtands a ſingle ſpine, and that but a ſmall one; it is ſerrated or denticulated on the anterior part, and it's edge is ſmooth behind: the belly fin is large, and the tail is deeply forked.

This ſpecies is a native alſo of the American Ocean, and is preſerved dry in ſome of our muſeums. Liſter call it Guaperva lata corpore ad caudam ſtriata.

Baliſtes aculeis quinque in utroque latere.
The Baliſtes, with five prickles on each ſide.

The Porcupine Guaperva.

This is a ſpecies conſiderably different from all the others: it is of a compreſſed form, and is conſiderably broad, in proportion to it's length, but it is thicker than moſt of the others: the head is compreſſed; the eyes are moderately large; the noſtrils are ſcarce at all conſpicuous: the mouth is moderately wide, and is furniſhed with broad and large teeth, protended a little forwards: the belly is narrow and rounded; the ſides are but little prominent, in proportion to the thickneſs of the fiſh; there are, on each, five robuſt and firm ſpines, and toward the tail there is a remarkable hairineſs.

This ſpecies is alſo caught in the American Seas, and is preſerved in ſome of our collections. Liſter calls it Guaperva Hiſtrix.

Baliſtes aculeis binis loco pinnæ ventralis, ſolitario infra uno.
The Baliſtes, with two ſpines in place of the belly fin, and a ſingle one below them.

The Scolopax.

This is, in ſome degree, like the former, but differs in many of the eſſential characters, as a ſpecies: it grows to about four inches in length, very rarely to any thing more: the body is compreſſed and very thin, and it's breadth is not more than a fourth of it's length; the ſcales are ſmall, but they are very rough; the head alſo is compreſſed, and the roſtrum is bony, very long and ſlender, and ſtraight: the mouth is ſmall, and is at the extremity of the roſtrum, and has a kind of operculum formed for the cloſing it: the eyes are large, and their iris is of a greyiſh-white, with ſome faint tinge of the ſilvery hue; but the noſtrils are not very conſpicuous, but each has a double aperture, and they are ſituated at a ſmall diſtance below the eyes: the pectoral fins in this ſpecies have each fourteen rays, all undivided at the ends, and in the place of the belly fins there are only two ſmall ſpines like teeth, and, at a ſmall diſtance behind theſe, there ſtands another ſmall one near the anus: the pinna ani has eighteen rays; there are two back fins, and both of them ſtand near the tail: the anterior one has five prickles, of which the ſecond is longeſt, and has an articulation; the others are very ſhort; the other has twelve rays: the tail is but little forked, and the belly is very acute on the anterior part.

This ſpecies is frequent in the Mediterranean, and is brought to market at Rome. Many of the writers on fiſhes have deſcribed it. Rondelet, Aldrovand, Willughby, and Ray, all call it Scolopax; Geſner calls it by the ſame name, but he alſo believes it to be the Serra of Pliny. The Genoeſe call it the Trumbetta, and the Italians, Soffietto.

OSTRACION.

THE body of the Oſtracion is of an odd figure; it is either globoſe, ſpheric, or roundiſh, or elſe it is oval, or ovato-oblong; or, finally, ovato-quadrangular, or approaching to cubic: the ſkin is always very firm and hard; in ſome of the ſpecies it is ſmooth, but in moſt it is covered either entirely with ſpines, or it has them in particular places; there are no belly fins: the number of fins in the whole is five; there are two pectoral or lateral ones, one on the back, the pinna ani, and the tail: the mouth is ſmall; the teeth, however, are large: the eyes are covered with the common cuticle: the apertures of the noſtrils are two on each ſide, and ſtand a little way before the eyes; the lips are moveable, and in part cover the teeth; there is no branchioſtege membrane.

The ſpecies of this genus are numerous, and may conveniently be arranged under ſeveral diviſions.

OSTRACIONES.

Diviſion the Firſt.

Thoſe which have numerous teeth, and are of an oblong and ſquare form.

Oſtracion oblongo-quadrangulus tuberculis quatuor majoribus in dorſo.
The oblongo-quadrangular Oſtracion, with four large tubercles on the back.

The great ſquare Fiſh.

THIS is a very ſingularly-ſhaped fiſh: the head is ſmall, and the roſtrum obtuſe; the eyes are large, and are covered with a thin ſkin; their iris is of a greyiſh-white, and the pupil round, and of a greeniſh-black: the noſtrils have each a double aperture, very conſpicuous, and ſituated at a ſmall diſtance before the eyes: the mouth is ſmall, and has a pair of moveable lips, which in part cover the teeth; the teeth themſelves are large, and ſeem quite diſproportioned to the ſize of the mouth in which they ſtand: the body is of a ſtrange figure, approaching to ſquare; the back is broad and flat; the belly alſo is broad and flat, and the ſides are very little prominent, ſo that the whole figure is tolerably regularly quadrangular: on the anterior part of the back, or but a little way behind the head, there ſtand four very remarkable protuberances: the pectoral fins are ſmall and obtuſe at the ends; the pinna ani is oblong; the back fin is ſhort, and the tail is large.

This is caught in the Eaſtern Seas, and in the mouths of the larger rivers in that part of the world: moſt of the ichthyological writers have deſcribed it. Ray calls it Piſcis maximus quadrangularis ſive dorſo plano; Willughby, Piſcis quadrangularis tuberculis quatuor in dorſo non longe a capite inſignitus; Aldrovand, Jonſton, and Willughby call it Oſtracion prior; Geſner, Oſtracion Nili; and Willughby, in another place, Oſtracion Nili quem Bellonius Holoſteum appellat; Aldrovand and Jonſton alſo, in other places, call it Holoſteum Bellonii ſive Oſtracion Geſneri; and Charleton uſes the ſame name. It greatly reſembles the ſpecies deſcribed, under the name of Orbis ſcutatus, by the ſame authors.

Oſtracion oblongo-quadrangulus gibboſus.
The oblongo-quadrangular and gibboſe Oſtracion.

The gibboſe Oſtracion.

This is a very ſingular, but alſo a very beautiful, fiſh: the head is ſmall; the eyes are large, and are covered with a common membrane; the apertures of the noſtrils are two on each ſide, and are placed at a ſmall diſtance below them: the mouth is ſmall, and has a pair of ſomewhat fleſhy and moveable lips: the teeth are large, and are always partly in ſight, even when the lips are cloſed, for they do not meet in all parts: the body is of an extremely ſingular figure; it is oblong, and in ſome degree quadrangular, the back being flat, and the belly flat, and the ſides parted each way from both

by

by a ridge, and not very prominent. Notwithſtanding the flatneſs of the ſuperficies of the back, however, it riſes into a kind of globoſity in the whole: the ſkin is hard, and is marked all over, in a very elegant manner, with hexagonal figures: the back fin is ſhort and moderately broad; the pinna ani is longer: the lateral fins are oblong and broad, and the tail is large; they are all of a pale greyiſh-olive colour, the tail ſomewhat darker than the reſt.

Moſt of the writers on fiſhes have deſcribed this ſpecies. Aldrovand and Jonſton call it Oſtracion alter; Willughby, Ray, and Liſter, Oſtracion alter gibboſus; and the ſame authors, Piſcis quadrangularis gibboſus. We have it in plenty in the muſeums of our collectors of natural curioſities.

Oſtracion oblongo-quadrangulus roſtro acuto, maculis in dorſo et capite.
The oblongo-quadrangular Oſtracion, with an acute roſtrum, and ſpots on the back and head.

This is a moderately large ſpecies: but the head is ſmall, and the roſtrum oblong and acute: the eyes are large, and are covered with a membrane; the apertures of the noſtrils are two on each ſide, and are ſituated at a conſiderable diſtance below the eyes, toward the roſtrum: the body is of an oblong figure, and is quadrangular in ſome degree, the back and belly being both flatted, and parted from the ſides by a kind of ridge: the mouth, in this ſpecies, is remarkably ſmall; the teeth are large and diſproportioned: the general colour of the fiſh is a pale brown, with a tinge of a greyiſh-olive, but it is elegantly ſpotted on the back and on the head, with a deep colour approaching to black.

The pectoral fins are large and obtuſe at the extremity; the pinna ani is oblong; the back fin is ſhorter, and they are all of them of a pale olive-grey: the whole ſkin is hard, and there are no tubercles, nor any ſpines on any part of it.

Some of the writers on fiſhes have deſcribed it, but the generality ſeem not to have been acquainted with it. Liſter calls it Oſtracion tertius roſtratus. We meet with it in the cabinets of ſome of our collectors of curioſities.

Oſtracion quadrangulus maculis variis plurimis.
The quadrangular Oſtracion, with numerous differently ſhaped ſpots.

This is a very ſingular and a very beautiful ſpecies: the head is ſmall, and the roſtrum ſubacute; the eyes are remarkably large; they are covered with a membrane, and their pupil is round: the noſtrils have each a double aperture; they ſtand at a moderate diſtance below the eyes, and are very conſpicuous: the mouth is ſmall, but the teeth in it are numerous and large: the body is broad and thick, in proportion to it's length, and is of a ſquare figure: the back is broad, and the belly alſo is ſomewhat broad and flatted, and the ſides are ſeparated by a ridge each way, ſo that there is a ſtrong approach to a quadrangular figure: the ſkin is very hard, and the whole fiſh is tolerably ſmooth; the ground colour is a deep brown, with an admixture of yellowiſh, but it is very elegantly variegated all over with ſpots of a rounded figure, and of different magnitudes, and different colours: the ſcales are of a hexagonal figure, and have ſeveral lines and little prominences on them; on the ſides of the fiſh there is a ſingle ſpot on every ſcale, but on the belly, where the ſpots are much ſmaller, there are two or three of them on each ſcale or hexagonal figure: the upper part of the ſides is more ſpotted than the lower, and the back moſt of all.

The pectoral or lateral fins have ten rays each: the back fin has nine rays, and the pinna ani eight; the tail has ten large rays. Few of the writers on fiſhes have known any thing of this ſpecies. Liſter, in his Appendix to Willughby, calls it Piſcis mediocris quadrangularis maculoſus.

OSTRACIONES.

Division the Second.

Those which have numerous teeth, and are of an oblong and somewhat triangular form.

Ostracion triangulatus duobus aculeis in fronte, et totidem in imo ventre.
The triangular Ostracion, with two spines on the forehead, and two on the lower part of the belly.

The horned, triangular Fish.

THIS is about eight inches in length, and is moderately broad and thick, in proportion: the head is small, and the rostrum acute; the mouth is small, and has a pair of moveable lips surrounding, but not entirely closing, it: the teeth are numerous and large; the apertures of the nostrils are two on each side, and are very conspicuous; they stand at about the midway between the eyes and the extremity of the rostrum: the eyes are large, and are covered by a membrane; the body is of a triangulated form, and is covered with a number of very beautiful angulated bodies in the place of scales; these are not all of exactly the same form, but the greater number of them are hexagonal; they are radiated, and make a very beautiful appearance: the pectoral fins have each eleven rays; the back fin and the pinna ani have each ten rays; the tail is not forked, but equal at the extremity, and has from eight to ten long rays.

This species has not been described by the generality of writers on these subjects, but we have it in some of the collections of our naturalists. I bought a fine specimen of it among the late Duke of Richmond's specimens.

Ostracion triangulatus aculeis duobus in capite, et unico longiore superne ad caudam.
The triangulated Ostracion, with two spines on the head, and a single one at the tail.

This is also a very singular and a very beautiful species: the head is small, and the rostrum obtuse; the mouth is small, and is surrounded by a kind of fleshy lips: the teeth are numerous and large; the nostrils have each a double aperture, but they are not very conspicuous; they are situated at a considerable distance above the extremity of the rostrum; the eyes are large, and covered with a membrane, and not far from them there arise two irregular spines resembling horns: the body is of a figure approaching to triangular, and is covered with a hard and firm skin, elegantly configurated and spotted: the pectoral fins are oblong and obtuse, and have each ten rays; the pinna ani has eight rays, and the back fin nine; the tail is large, and not forked, it has eleven long rays, and is of a dusky greyish colour.

Many of our collectors have dried specimens of this species. Lister, in his Appendix to Willughby, calls it Piscis triangularis capite cornutus, cui aculeus longus ad caudam erigitur.

Ostracion triangularis limbis figurarum hexagonarum eminentibus, aculeis duobus in imo ventre.
The triangular Ostracion, with the verges of the hexagonal figures eminent, and with two spines on the belly.

This is a very singular species in it's form, and is the largest of all the triangular kinds: the head is irregularly shaped; the rostrum is obtuse; the mouth is very small, and the teeth in it disproportionately large: the eyes are large, and are covered with the common skin of the head: the nostrils are conspicuous; they have each a double aperture, and are situated at a middle distance between the eyes and the extremity of the rostrum;

roſtrum: the body is of an extremely ſingular figure, approaching to triangular; the ſkin is hard and firm, and is every-where formed into a ſingular kind of hexagonal figures, which are ſtriated, and have their edges prominent; it has no ſpots, but is uniformly of the ſame browniſh colour: the teeth in the lower jaw are about eight in number, and thoſe in the upper are eleven or twelve; the lips are moveable, and, in part, cover them; the pectoral or lateral fins are ſhort and ſomewhat broad; the back fin ſtands very backward; the pinna ani is long and low, and the tail is long.

This ſpecies is a native of the American ſeas, but we meet with dried ſpecimens of it very frequently. Cluſius calls it, ſimply, Piſcis triangularis; Ray and Willughby, Piſcis triangularis Cluſii cornubus carens.

Oſtracion triangulatus totus maculatus ac tuberculoſus, aculeis duobus in imo ventre.

The ſpotted and tuberculoſe triangular Oſtracion, with two ſpines on the belly.

This nearly approaches to the ſize of the former ſpecies: the head is large and extremely irregularly ſhaped; the eyes are large, and are covered with the cuticle; the covering of the head about the eyes, and principally above them, riſes into a kind of eminences reſembling eye-brows; the noſtrils have each a double aperture; they ſtand at about a middle diſtance between the eyes and the end of the roſtrum: the mouth is ſmall; there are eight very large teeth in the lower jaw, and twelve in the upper, and the whole mouth is ſurrounded with a kind of lips, that in a great meaſure, but not entirely, cover them: the body is of a figure approaching to triangular; the ſkin is hard and firm, and is all over beſet with tubercles at ſmall diſtances, and has, between theſe, brown ſpots larger than the tubercles; theſe are round, and all of nearly the ſame ſize: the back of this ſpecies is more gibboſe and acute than that of any other; the pectoral fins are broad, and have each eleven or twelve rays; the back fin has ten rays; the pinna ani alſo and the tail have each the ſame number: the whole fiſh is of a duſky brown, with a tinge of yellowiſh; the tubercles are paler than the reſt, and the ſpots between them are of a reddiſh tinge with the brown.

It is a native of the American, and ſome of the Eaſtern, Seas, and is ſometimes ſent over to us preſerved as a curioſity. Liſter, Ray, and Willughby call it Piſcis mediocris triangularis ad imum ventrem prope caudam cornutus, ex toto maculis æqualibus ſubroſis denſe inſignitus.

Oſtracion triangulatus tuberculis hexagonis radiatis, aculeis duobus in imo ventre.

The triangular Oſtracion, with hexagonal, radiated tubercles, and with two ſpines on the belly.

This is one of the ſmalleſt ſpecies of this genus: the head is ſmall; the mouth is very ſmall, but the teeth are large and numerous; there are no leſs than eight in the lower jaw, and twelve in the upper: the noſtrils are very conſpicuous; they have each a double aperture, and ſtand at about a middle diſtance between the eyes and the extremity of the roſtrum: the eyes are large, and are covered with the cuticle; the body is of an irregularly triangular figure; the back is gibboſe and very acute, and the belly is broader than in moſt other ſpecies: the ſkin is very hard and firm; it riſes into a kind of eminences reſembling eye-brows on the upper part of the head, and there are on each ſide two large ſpots: on the body it is formed into hexagonal figures, which are radiated, and the radiations formed of ſmall tubercles, and there are two tranſverſe lines on them: the pectoral fins have each twelve rays; the dorſal fin, the pinna ani, and the tail have ten rays each.

This ſpecies is brought to us dried, with the others, from the American Seas. Liſter, Ray, and Willughby have called it Piſcis triangularis parvus, non niſi imo ventre cornutus.

Oſtracion

Ostracion triangulus tuberculis exiguis innumeris, aculeis carens.
The triangular Ostracion, with numerous small tubercles, and no spines.

The head of this species is small, the rostrum obtuse, and the mouth extremely small: the teeth are large, and there are about eight in the lower, and eleven in the upper, jaw; the eyes are large, and covered with the cuticle: the nostrils are not easily seen; they have each a double aperture, but very small, and placed at a little distance below the eyes: the upper part of the head is rounded, and the skin above the eyes somewhat elevated: the back is gibbose and acute; the belly is very broad, and the whole body very much of a triangular figure: the skin is hard, and the scales, if they may be called so, are of a hexagonal, or perhaps, more properly speaking, of a triangular, figure: the other angles being scarce perceptible, they rise each into an eminence in the middle, and are ornamented with a vast number of little tubercles, which are disposed regularly in lines, and render them striated: the general colour of the fish is a pale brown, with a very slight tinge of yellowish in it; but it is variegated with a great number of white spots, especially toward the back; there are from one to four of these on each scale: the pectoral fins have twelve rays each; the dorsal or back fin has ten rays; the tail is rounded at the extremity, and has eight or ten rays.

We have this brought over dried, with the others, from the Eastern and American Seas. Lister calls it Piscis triangularis ex toto cornibus carens.

OSTRACIONES.

Division the Third.

Those which have only four teeth.

Ostracion sphaericus tetraodon aculeis undique exiguis.
The spheric, four-toothed Ostracion, with short spines all over the body.

The Globe-fish.

THIS is one of the smaller species; it rarely exceeds seven inches in length, and is about as much in diameter, the body being nearly of a spherical figure, excepting for the part toward the tail: the head is scarce at all distinguished from the rest of the body; the mouth is small, and has only four teeth in it: the eyes are large, and covered with a cuticle; they stand high upon the head: the apertures of the nostrils are two on each side, and are very conspicuous; the whole body is covered with short spines of a whitish colour and firm texture, placed at a small distance from one another: the pectoral fins are broad and short; the back fin and pinna ani stand opposite to one another, and at a great distance from these the tail is oblong; the colour of the whole fish is a dusky brown, with an admixture of yellowish and of olive colour.

It is frequent about the mouth of the Nile, and of several of the larger rivers in Africa and the East Indies; all the writers on fishes have described it. Pliny, Bellonius, Rondelet, Isidore, Salvian, Gesner, and Jonston, all call it, simply, Orbis; Ray and Willughby, Orbis primus; Charleton, Orbis vulgaris; Aldrovand, Orbis species ex Gesnero. The Venetians call it Pesce Columba. Dried specimens of it are very frequent in the museums of our collectors.

Ostracion maculosus aculeis undique densis exiguis.
The spotted Ostracion, with frequent small spines.

The rough Globe-fish.

This grows to about twelve inches in length, and to eight or nine in diameter: the anterior part of the body approaches to a spherical figure, but it is lengthened at the tail: the eyes are large and covered with the cuticle; the nostrils are conspicuous, and stand at about an equal distance from the eyes, and at the extremity of the rostrum there are two apertures to each: the mouth is small, and the teeth are only four: the whole body is beset with small, whitish, bony spines, placed at small distances, and the tail is of an oblong form: the ground colour of the fish is a pale whitish-brown, but

but it is all over spotted with small, round, or less regularly figured dots of black, or of a very deep reddish-brown.

This is brought to us dried from the Eastern Seas, and has sometimes been caught in the Mediterranean. Lister, in his Appendix to Willughby, calls it Orbis asper maculosus.

Ostracion cathetoplateo-oblongus ventre tantum aculeato, et subrotundo.
The oblong and compressed Ostracion, with only the belly rounded and prickly.

This is the most singular in it's figure of all the species of this remarkable genus: it grows to about a foot in length, but does not exceed five inches in diameter in the largest part: the mouth is small, and has only four teeth, which are but imperfectly covered by the lips: the eyes are large, and stand high; they are covered with the cuticle: the nostrils have each a double aperture, placed at some distance below the eyes, and not very conspicuous: the body is of an extremely singular figure; it is oblong, and the back is not elevated but straight, and the hinder part of the body, toward the tail, is slender: what gives it the resemblance of the rest of this genus is, that the belly is rounded and prominent, in the manner of that of the others of this kind: the back and the sides are smooth; the belly, in this prominent part, is covered with spines: the pectoral fins are broad and short; the dorsal fin and the pinna ani are small; the tail is large, and is somewhat forked, but not very deeply.

We have this dried from the East, and from America, and it is not uncommon in the collections of our naturalists. Willughby calls it Orbis cauda productiore, dorso lævi, ventre spinoso; and Grew, in his catalogue of the musæum of the Royal Society, Orbis lagocephalus.

OSTRACIONES.

Division the Fourth.

Those which have bony jaws, without any distinct teeth.

Ostracion sphericus aculeis undique densis triquetris.
The spherical Ostracion, with numerous triquetrous spines.

THIS is a small but a singular and beautiful species: the general size is about that of a goose's egg, to which it in some degree approaches in form: the head is small, and not at all erect; the eyes are small, and covered with the cuticle: the mouth is small, and the jaws are bony; there is a division, which gives the appearance of two teeth, but in reality there is no tooth at all: the nostrils are scarce at all conspicuous; there are, toward the eyes, two small apertures; the lower one round and small, the upper larger, oblong, and transverse; but they are not distinguishable, except on a very close attention: the whole body is covered with spines of a bony structure, short and triangular at the base; those on the back and toward the head are longer than the others, the rest are about equal in length: the general colour of the fish is a pale brown; the spines are whiter than the rest, and the fins, which are nearly of the colour of the skin, are spotted with a deeper brown, approaching to black.

We have dried specimens of this, as of the others, from the East, and from America. Lister calls it Atinga alter minor orbicularis.

Ostracion subrotundus aculeis undique brevibus planis, ventre glabro.
The roundish Ostracion, with short plane spines, and with the belly smooth.

The head of this species is small and erect; the eyes are large, and covered with the cuticle; the mouth is small, and the jaws are bony and hard, but there are no teeth: the lips are fleshy and moveable, and, when closed, they cover the jaws almost entirely: the nostrils are very conspicuous; they have each a double aperture, and stand at about an equal distance from the eyes, and the extremity of the rostrum: the body approaches to a spherical figure; the back and sides are furnished with a great number

 of

of short and flat spines, and the belly is smooth: the general colour is a pale yellowish-brown, but behind each of the pectoral fins there stands a black spot; at the root of the back fin there is also, on each side, a black spot, and there are some other small ones of the same colour about the head.

This is a native of the Brasilian Seas. Marcgrave calls it Guamaiacu atinga; Willughby and Ray, Orbis ranæ rictu; and Clusius, in his Exotics, Orbis spinosus. The mouth, when opened, is somewhat like that of a frog in shape, not round, as in the other species, but flat.

Ostracion subrotundus aculeis undique brevibus triquetris raris.
The roundish Ostracion, with flat, triquetrous, and rare spines.

This is of a figure considerably different from the preceding species: the back and belly are both a little flatted; the sides are rounded and prominent; the mouth is small, and the jaws are naked and bony, but there are no teeth: the eyes are small, and are covered with the skin of the head; the nostrils have each a double aperture, and are fairly conspicuous: the covering of the head, just over the eyes, is raised into an oblong eminence on each side; the mouth stands high in the anterior part of the head, in a very singular direction: the body is of a figure, in general, approaching to spherical, but depressed: the spines on it are not large; they are broad and triangular at the base, and they do not stand so close as in many of the other species: the general colour of the body is a pale brown, but there are on several parts of it spots of a deep black.

We have specimens of this from the American Seas, but it is not so frequent in our collections, as many of the other species. Lister calls it, Orbis muricatus et reticulatus; others, Orbis muricatus.

Ostracion rotundo-oblongus tuberculis utrinque, pinna dorsi longissima.
The oblong Ostracion, with tubercles on both sides, and with the back fin very long.

This is an extremely singular species, and carries the general appearance, in many respects, of a fish of some other genus: the head is small; the opening of the mouth is wider than in most of the others, but it is not round but flat: the body is of an oblong and rounded figure, and the fish grows to a foot and a half in length: there run on each side of the body two longitudinal series of spines or tubercles: the back fin, quite contrary to all the preceding species, is extremely long, and runs from the back part of the head quite to the tail.

We have specimens of this from the American Seas, but it is more rare than many of the other species. This is properly the species which Clusius calls Orbis ranæ rictu, the frog-mouthed Orbis, though the authors who have followed him, have quoted him as using that name for a very different species before described. Indeed all these toothless species have the mouth of this flat form, when opened, not round as the others.

Ostracion subrotundus aculeis undique densis, basi triquetris.
The roundish Ostracion, with numerous spines on all parts, triquetrous at the base.

The head of this species is small; the eyes are very large, and covered with the cuticle: the mouth is small, and has fleshy lips; there are no teeth, but the jaws are naked and bony: the nostrils have each a double aperture, and stand at a small distance below the eyes: the whole fish is about fifteen inches in length; the body is rounded, and is all over covered with spines set very close and thick; they are not very long, but are robust, and are of a triquetrous figure at the base: there are no belly fins; the pectoral ones are broad and short; the back fin and pinna ani are oblong and narrow; the tail is oblong: the colour of the whole fish is a pale brown; the spines are whiter than the rest of the skin, but there are no spots nor variegations of any kind on it.

We

We have specimens of this brought dried and stuffed from the East, where it is frequent about the mouths of large rivers. Authors seem to have imperfectly described it under the name of Orbis muricatus, but they have given the same name also to other species, which they have confounded with it. Willughby and Ray suppose it the species Clusius meant by this name, and call it Orbis muricatus Clusii.

Ostracion conico-oblongus aculeis undique longis teretibus, imprimis in lateribus.
The conico-oblong Ostracion, with long rounded spines all over it, especially on the sides.

The Porcupine-Fish.

The head of this singular species is small, plane, and erected; the eyes are very large, and covered with the cuticle; the mouth is small; the lips are fleshy and moveable, and the jaws are bony and naked, but there are no distinct teeth; the nostrils are not very conspicuous, but they have each a double aperture, and are placed at a little distance below the eyes: the body is of an oblong figure, approaching to conic, and is covered in all parts with very long spines; those at the sides are often near three inches in length; there are four very sharp and robust ones just over the eyes, and all those about the mouth point forwards; they stand very close on the middle of the body, but more thin toward the tail: the general colour of the skin is a tawny brown, but there are a number of black spots on several parts of it; and the general colour is deeper toward the head and tail than in the middle.

This species is caught about the mouths of large rivers in Africa and the East Indies, and is described by most of the authors who have written on fishes. Clusius calls it Hystrix piscis, and most of the succeeding authors have borrowed the same name.

Ostracion oblongus holocanthus aculeis longissimis ad caput.
The oblong Ostracion, with very large spines about the head.

The small-headed Porcupine-fish.

The head is small and slender; the eyes are large, and covered with a cuticle: the nostrils are conspicuous; they stand high on the head, and have each a double aperture: the mouth is small, and the jaws are naked and bony, but there are no regular teeth: the body is of an oblong figure, inflated or rounded in the middle, but small both at the head and toward the tail: the whole body is covered with spines of an oblong form, and tolerably robust; but the longest and most formidable are on and about the head; toward the tail they are short, and not so numerous, but they are very robust: the general colour is a dusky olive-brown, but the spines are much paler than the skin.

It is brought to us stuffed from the East. Lister calls it Hystrix alter capite angusto, subrotundo, et admodum spinoso, scilicet spinis praelongis in ipso capite et scapulis donatus.

Ostracion oblongus glaber capite oblongo, corpore figuris variis ornato.
The oblong Ostracion, with a smooth body, ornamented with figures.

The tortoise-headed Porcupine-Fish.

This is a large species; we sometimes see it of eighteen or twenty inches long, but those of about a foot are more frequent: the head is short, broad, and somewhat flatted; the body is of an oblong figure, but very much distended and rounded; the eyes are large, and are covered with a cuticle: the nostrils are sufficiently conspicuous; they have each a double aperture, and are situated at a small distance before the eyes; the mouth is small, and the jaws are naked and bony, but there are no regular teeth: the pectoral fins are short and broad; the back fin and the pinna ani are oblong and obtuse; the tail is oblong: the skin of the whole body is firm, tough, and smooth; there

there are no spines on any part of it: the ground colour is a pale tawny-brown, but it is elegantly variegated with spots in different figures.

It is frequent about the shores of the American islands. Clusius, Wilughby, Ray, and others call it Orbis oblongus testudinis capite; Sir Hans Sloane, Orbis levis oblongus cinereis et fuscis maculis notatus.

Ostracion cathetoplateus subrotundus inermis asper, foraminibus quatuor in capite. **The Sun-fish.**
The compressed, roundish, rough-skinned Ostracion, with four holes on the head.

This is one of the most singularly-shaped fish in the world: it is of a very considerable size, often weighing more than a hundred pounds; and it's figure, at first sight, more resembles that of the head of some large fish cut off from the body, than of a compleat animal of this kind: the head is small, and not much distinguished from the body; the eyes are small, and are covered with a cuticle; the whole body, as well as the head, are of a compressed form: the skin is very rough: the colour on the back is blackish, and on the belly it is a silvery white: the belly and back are both sharp and edged: the nostrils are conspicuous; they have each two small apertures, but beside these there are four large apertures or foramina on the head: the pectoral fins are small, of a roundish figure, and are placed horizontally; they have each twelve rays: the apertures of the gills are of an elliptic figure, and are situated on each side at the roots of the pectoral fins; there is one fin on the summit of the back, and the pinna ani stands exactly over-against it on the belly: the tail is semicircular and perpendicular; it surrounds the extremity of the body.

The liver is undivided; the gall-bladder is large; the kidnies are very large, and the ureters run in a very conspicuous manner to the urinary bladder: the intestines are thick, and full of circumvolutions; it is said to be viviparous, but this is not yet well attested: the bones are cartilaginous; the flesh is soft, and is more than that of all other fish liable to shine in the dark.

It is caught frequently in the Mediterranean, and sometimes in other places. It has been taken about the coasts of Cornwall, and some other parts of England.

This singular fish was well known to the antients. Pliny calls it Orthagoriscus; and Gesner, and several others follow him, in calling it, simply, by that name; Rondelet calls it Orthagoriscus sive Luna piscis; Salvian, Jonston, Wilughby, and Ray, Mola. We call it the Sun-fish; and in some places, from it's flat figure and rough skin, the Millstone-fish.

CYCLOPTERUS.

THE body of the Cyclopterus is thick, oblong, and approaching to a rounded figure: the ventral fins coalesce at their extremities, and form a single, oblong hollow, and, in some degree, infundibuliform fin: the fins are six, and the branchiostege membrane on each side contains six oblong and slender bones.

Of this genus there is but one known species.

CYCLOPTERUS. **The Lump-fish.**

This is a very singularly-shaped fish, extreamly thick, in proportion to it's length, and of a remarkably clumsy figure: the head is moderately large; the eyes are small, and are covered with a cuticle; the mouth is moderately large, and there are a great number of small but sharp teeth in the jaws and fauces: the nostrils have each only a simple aperture, and that is oblong and tubulous: the back of the fish is elevated into a ridge; the belly is broad and flat: there are no scales, but on several parts of the body there are series of low but sharp tubercles: the colour is variegated of black and an elegant pale red and whitish; the tubercles are all of them black, and the back is blackish: on each side there are three series of acute and crooked tubercles turning backward, and on the back there is a single series, consisting of eight of these tubercles: the upper series on the sides has twenty-six of them, the middle series has sixteen, and the lower one has eight tubercles: before the back fin there stands a cutaneous protuberance, emulating also the form of a fin, but supported by by no rays: the back fin has nine rays; the pectoral fins stand very near the mouth of the fish, and have each twenty rays: the pinna ani has seven or eight rays, and the tail has ten: the kidnies and the urinary bladder in this species are remarkably large.

It is frequent about our own coasts, and in most of the Northern seas, but no where so very plentiful as in the Baltic; it's flesh is very well tasted. Turner, Gesner, Aldrovand, Jonston, Charleton, Willughby, and Ray, all call it Lumpus Anglorum; Schelhammer, Lumpus Anglorum piscis; Gesner, Orbis Britannici sive oceani species; and Schonevelde, Lepus marinus nostras, orbis species. The Swedes call it Syurigg-fish; the Dutch, Snottolf; the Scots, the Cock-paddle; and we, the Lump-fish, or Sea-owl.

LOPHIUS.

THE figure of the Lophius is monstrous and irregular, unlike that of all other fishes: the head is equal to all the rest of the body in size; the head and body are both of a depressed form: there are a number of fleshy pinnules or appendages surrounding the whole body of the fish.

Lophius ore cirroso.
The Lophius, with a bearded mouth.

The Rana Piscatrix, or Frog-fish.

This is one of the strangest fish in the world; it's figure and characters are indeed so singular, that one of them accidentally caught on our coasts, and brought to London, proved, for many years, a livelihood to the person who singly shewed it: it grows to five or six feet in length, and it's head, in this case, is near two feet in diameter: the head is of a rounded but depressed figure, and is more than equal to the whole body in size; the opening of the mouth is enormously wide, and the apparatus of teeth very terrible: there are several series of them in each jaw, and in the fauces also, and at the root of the tongue; they are oblong, slender, and sharp, and all turn inwards, so that it is impossible for any thing, once seized on, to make it's escape: the lower jaw is longer than the upper, and the mouth seems incapable of closing perfectly. The disproportion of the head to the body gives the whole fish somewhat of the figure of the tadpole: the eyes are large; their iris is grey, and the pupil round and blackish: there are no conspicuous nostrils, but several little protuberances in the place where they might be expected: there are many elegant lines disposed in meanders on the head, and continued to the lateral line; the eyes are placed not in the sides of the head, but on it's top, yet at a considerable distance from one another; and it is singular that there runs a transverse line of white, or of the colour of the iris, through the pupil: near the extremity of the rostrum there stand two bodies of an oblong, slender form, resembling pieces of whale-bone, and on the summit of each is a whitish, fleshy protuberance; these it vibrates about at pleasure, and there are three other shorter, of the same form and structure, on the back: the eyes have several spines about them, and there are others toward the verge of the jaw, and on different parts of the head beside: the body of the fish is very inconsiderable, in proportion to it's head; it is oblong, small, and depressed; the back is flatted, and of a dusky greyish-colour, with some variegations: there is but one back fin; this stands near the tail, and has ten rays: the tail is large, and is not forked at the end; the pectoral fins stand at the sides of the fish, and have each twenty rays: the pinna ani stands over-against the back fin, and has nine rays, but the most singular fins are the ventral ones; they stand very high, and are placed contiguous to each other; they have each five rays, and are divided into so many parts very deep toward the base, and, being thick, fleshy, and of a pale reddish-colour, they have greatly the appearance of hands with their fingers: the appendages, surrounding the body of the fish, are of the nature of these ventral fins, only very small: the apertures of the gills are very large; they are under the roots of the pectoral fins, and the gills are three on each side: the intestines are very long, and have many circumvolutions; there are two appendages to the pylorus; the urinary bladder is very large. It is a very sluggish fish, and usually lies flat on the bottom, in which posture it's eyes are turned full upward, and distinguish every thing above it.

We have it in our own seas, but not common; that which was shewn in London was caught near Dover. It had seized upon the body of a drowned woman, and could not disengage it's jaws, but, on taking the body up, was brought on shore with it.

This fish has been known at all times to the naturalists. Aristotle calls it Ἁλιεὺς βάτραχος; and, in other places, βάτραχος ἁλιεύς; Ælian calls it βάτραχος ἁλιεύς; Athenæus and Oppian, simply, βάτραχος; Ovid, Pliny, Cicero, and the rest of the old Latin writers call it Rana and Rana marina; Bellonius, Salvian, and most of the moderns, Rana piscatrix; Gesner calls it Rana piscatrix sive marina; and Aldrovand, Rana piscatrix vulgaris. We call it the Toad-fish, the Frog-fish, and the Sea-devil.

There is but one other known species of this singular genus, that is, the American frog-fish, with a single horn on it's head. Marcgrave and others after him have called this, by it's Brasilian name, Guacucuja; Linnæus calls it, Lophius fronte unicorni.

FISHES.

Class the Fourth.

CHONDROPTERYGII.

Fishes, with the tail placed perpendicularly, and with the rays of the fins not bony, but of a cartilaginous substance.

PETROMYZON.

THE foramina or apertures of the gills, or of the lungs, are in the Petromyzon seven on each side, situated longitudinally; and there is, beside these, one in the middle of the head, between the eyes: the body is long and slender, and nearly cylindric, and is smooth; there are only two fins, and those are both on the back of the fish.

Petromyzon unico ordine denticulorum exiguorum in limbo oris, præter inferiores majores.
The Petromyzon, with a single row of little teeth in the verge of the mouth, beside the lower large ones.

The Lam-prey-eel.

This grows to about a foot long, but what are usually caught are under that standard: the body is slender, almost cylindric in figure, but a little compressed, especially toward the tail; it is all over perfectly smooth, or without the least appearance of scales, and is covered with a lubricous, tough matter in the manner of the eel; the whole back and the belly are convex: the colour on the back is a simple bluish-green, without any variegations; on the belly it is white and silvery: the head is but very little distinguished from the body; it is only somewhat smaller and slenderer; it's shape is rounded, and there are no distinguishable jaws: the mouth is not at the extremity of the rostrum, but on it's under part, and is round, large, and may be rendered wider or narrower by the creature's power of sucking, and drawing it more or less in: there is a row of extremely minute teeth in the verge or rim of the mouth; and in the upper part of it, within this limb, there are about six other series of teeth of the same size and form: at the sides of the mouth there are also three bones or teeth, of which the upper one is bifid, the second is sometimes trifid, sometimes only bifid also, and the lowest is always bifid; in the lower part, just at the entrance of the mouth, there is also a bone of an oblong figure, and serrated, so as to represent the divisions of seven teeth; and above this there is a semilunar and considerably thick bone: and, finally, deeper in the mouth there is another bone, armed with seven teeth, which is connected so loosely, that it may be easily pulled out; and at the lower part of the fauces there is a bifollate cartilaginous officle of a very odd figure: such is the singular armature of the mouth of this little fish.

The

The eyes are round, ſmall, and ſituated very diſtant from the roſtrum, and covered with a cuticle: their iris is of a ſilvery white, but obſcured by a number of little black ſpots; it eaſily becomes red, on preſſing the fiſh, in handling it, but that is not it's natural colour; near the eyes, and in other parts of the head, there are ſome little punctula, having the appearance of ducts, but nothing can be preſſed out of them: there are no viſible noſtrils, but there are two foramina or openings on the middle of the head; a little before the eyes theſe run backwards, and have not their opening, as is uſually ſuppoſed, into the mouth, but are continued to the lungs, and communicate with their anterior part, after paſſing in form of a duct all along, between the two bodies of which they are formed. There run in a ſtraight line a ſeries of ſeven roundiſh or oval apertures on each ſide, from the head downwards; theſe anſwer the purpoſes of the gills in other fiſhes.

In the middle of the head of this fiſh, between the eyes, there is uſually a round, large, red ſpot, but this is ſometimes faint, ſometimes almoſt entirely wanting; ſometimes it has no redneſs, but is merely white: the under part of the fiſh is wholly naked; there are no pectoral fins, no ventral ones, nor any pinna ani: there are on the back two fins; they are membranaceous and ſoft, and are both ſituated near the hinder part: the anterior of theſe is the ſmaller and ſhorter, the hinder one is taller and longer; it is higheſt in the middle, and decreaſes to the extremity: juſt as it approaches the tail, it becomes extremely low, and after this is again a little broader, and ſurrounds the extremity of the tail; ſo that what is called the tail in this ſpecies is not properly ſo, but is only a continuation of the hinder back fin. From the termination of the hinder part of this tail fin, there runs a ſlender red eminence, in form of a thread, to the anus: both the fins of the back are ſupported by cartilaginous inſtead of bony rays; they are numerous, and are ſimple at the bottom, but from the middle upwards they are bifid; this, however, is not to be diſtinguiſhed, unleſs the ſkin of the fiſh be pulled off.

There is no regular lateral line, but there ſtand certain punctula, at a diſtance from one another, in the anterior part of the body; but theſe are not diſtinguiſhable, unleſs when the fiſh is freſh caught: the inteſtine is ſimple, and has no volutions, but runs ſtraight from the ſtomach to the anus: the ovary is ſingle, but it is divided into a number of lobes; the liver is large and undivided; the heart is conic, there is no diſtinguiſhable air-bladder: the ſtomach is large and red; there are no gills, but in the place of them there are ſeven bodies on each ſide, which ſupply the place of lungs, and have ſomewhat of the form of thoſe of ſome animals; each of theſe has it's aperture in the anterior part, communicating with the external aperture which runs obliquely to it.

This fiſh is frequent with us in our brooks and rivulets in ſome places, and is common alſo to moſt of the northern parts of Europe: all the ichthyological writers have deſcribed it. Rondelet, Geſner, Willughby, and Ray call it Lampetra parva et fluviatilis; Aldrovand, Jonſton, Schonevelde, and Charleton call it Lampetra fluviatilis Geſneri; Salvian, Lampetra ſubcinerea maculis carens, and, in other places, Lampetra minor; Aldrovand has alſo called it Lampetra minima; Pliny and moſt of the old Latin writers mean this ſpecies by their Muſtela: Bellonius alſo calls it Muſtela fluviatilis, and Auſonius, Muſtela; Caſſiodore calls it Exormiſton, and Albin, Muræna.

Many of the writers on fiſhes have alſo deſcribed it over again, as if of a different ſpecies, under the name of Alterum genus Lampetræ and Lampredæ genus minus; Salvian alſo calls it, in this ſtate, Pryk; and Jonſton and Aldrovand, Fricka. We call it a *Lampern*; Plot calls it the Pride of the Iſis; the Swedes call it Natting and Nuenoges; and the Germans, Nuenaugen.

Petromyzon maculoſus ordinibus dentium circiter viginti. **The Lamprey.**
The ſpotted Petromyzon, with about twenty rows of teeth.

This grows to two feet and a half, or more, in length, but the greater number that are caught are much leſs: the head is ſmall, rounded, and obtuſe, and ſcarce at all diſtinguiſhed from the body, except by it's want of thickneſs: the eyes are ſmall; they ſtand on the ſides, not at the top of the head, and are placed very high, or at a great diſtance from the extremity of the roſtrum; the noſtrils are not conſpicuous:

the mouth is large and rounded, and is situated on the under part of the head: the teeth are small, but extremely numerous; there are not less than twenty rows of them: the body is considerably thicker than that of the former species; the back is of a dusky greyish-blue, with a number of spots, some of which are yellow, but the greater part black; there is also a white spot on the top of the head: the belly is of a silvery white, and the sides of a gradual decrease of the olive tinge to this absolute whiteness: the whole under surface of the fish is smooth; there are no fins, except on the back; there are two there, the anterior short, the posterior longer.

We have this species in some of our rivers toward the sea, but it is less frequent than the former. All the Ichthyological writers have described it, and the greater part of them have called it, simply, Lampetra; Gesner calls it, Lampreda marina; Salvian, Lampetra maculosa et bicolor; Aldrovand and Jonston, Lampetra major; Bellonius and Gesner, Mustela five Lampetra. The Italians call it Lampreda, and we the Lamprey, or the Lamprey-eel. This species seems to be the [illegible] of Oppian, and [illegible] of Galen. The other Greek writers call it Μύραινα ποταμία; but most of the authors of that nation have not been at all acquainted with it, as Galen, who saw it at Rome, observes that it was never found in the seas about Greece.

Petromyzon corpore annuloso, appendicibus utrinque duobus ad os.
The annulated Petromyzon, with two appendages on each side at the mouth.

This is a very small species; it rarely exceeds eight inches in length, and is very slender, in proportion: the head is small, rounded, and obtuse; the eyes are extremely small, and scarce visible; the mouth is very small, and formed like the round opening of a purse under the head: the nostrils are not distinguishable; the body is very slender and rounded, and not equal and smooth as in the other species, but annulated in the manner of that of the insect-kind, only not so deeply: the back is of a dusky olive colour; the sides have a tinge of yellowness; the belly is white and silvery; there are apertures to the lungs, as in the other species, seven on each side, and two at the top of the head: the belly is smooth, and without fins, but toward the end of the back there are two; the anterior one short, the hinder oblong, and both supported by cartilaginous rays. What distinguishes this from both the other species the most obviously, however, is, that it has on each side, at the verge of the mouth, two appendages.

This is not unfrequent with us, but it is much more common in many other of the northern parts of Europe: all the Ichthyologists have described it. Swenkfeld calls it Lampetra minima caeca. Willughby, Lampetra caeca oculis carens argentina Ein Blinder Neunogen, id est, Enneophthalmus caecus; and Ray calls it Lampetra caeca sen oculis carens.

ACCIPENSER.

THERE is only one foramen or aperture for the gills on each side in the Accipenser: the mouth is situated on the under part of the head, and has the appearance of a tube; there are no teeth in it: the body is of an oblong form, and the number of fins is seven.

Accipenser corpore tuberculis spinosis aspero.
The Accipenser, with the body armed with rough tubercles. The Sturgeon.

This is a very large fish; it grows to fourteen, fifteen, or eighteen feet in length, though the greater number are caught much smaller: the head is large, oblong, and acute at the rostrum; on the extremity of the under jaw there are four cirri or beards: the eyes are large, and stand at a great distance from the extremity of the rostrum: their iris is of a silvery white; the pupil is of a greenish black; the nostrils have each a double aperture: the head is of a somewhat angulated figure: the skin is rough; the belly is flat and smooth, but on the upper surface there are five series of spinose tubercles, one on the middle of the back, and two others on each side of that: these consist of different numbers of tubercles, and there are, beside these, two tubercles on the

the hinder part of the body; on it's under surface they are situated behind the anus: the pectoral fins have a very strong bone at their anterior part: the ventral fins are situated low, and near the anus; the pinna ani is situated in the midway between the anus and the beginning of the tail: there is only one fin on the back; the tail is bifid, and it's upper segment is longer than the under: the appendages to the pylorus are numerous; the air-bladder is simple; the intestines have only one volution, and the spina dorsi is continuous.

It is caught in the Adriatic, and in some other seas, and in most of the large rivers in Europe and Asia; we sometimes take it in the Thames, but it is not so frequent there as in many other larger rivers in Europe: all the Ichthyological writers have described it. Athenæus calls it [illegible] and Ἀντακαῖος; Plautus, Cicero, Martial, and Pliny, and most of the old Latin writers, Accipenser; Gesner, Acceptenser, Aquipenser, and Sturio; and Aldrovand, Jonston, Willughby, and Ray, Sturio; Salvian, Sturio sive Silurus; Athenæus, Gallus Rhodius, Isidore, Sus. The Italians call it Surione; and we, the Sturgeon.

The Greek writers have mentioned a fish under the name Ἔλοψ and Ἔλλοψ, from whom Ovid, Pliny, and many other of the Latins have mentioned Elops and Helops, as the name of another fish; but it is certain, from Pliny, that this was the same fish, and that Aristotle and Ælian call our Sturgeon by this name; though we had lost the connection between the name and the fish for many ages.

Accipenser tuberculis carens.
The smooth-bodied Accipenser.

The Huso, or Isinglass fish.

This is a larger fish than the sturgeon; it grows to twenty-four feet in length, and is thicker, in proportion, than that species: the head is large; the rostrum is extremely long, and there are eight beards or cirri under it: the mouth is very small, in proportion to the size of the fish, and there are no teeth in it: the eyes are large; their iris is grey, and their pupil of a deep black; the nostrils have each a double aperture, and are very conspicuous: the body is somewhat depressed; the back is less elevated than in the former species, and is rounded; the belly is flat; the whole body is smooth; there is no appearance of those tubercles, which are so conspicuous on the common sturgeon: the back is of a deep black, and the sides grow paler, till they approach the belly, which is yellow: there is only one fin on the back, and that stands very backward toward the tail; there are, beside this, the two pectoral fins, the two ventral fins, and the pinna ani, all supported by cartilages instead of bony rays.

This species is more frequent in the Danube than in any other part of the world, so far as is yet known; but even the accounts, given from such as have been caught there, have divided it, as it were, into three species: it was not unknown to the old Latin writers. Pliny calls it Mario; Gesner, Jonston, and Charleton, Huso. Willughby and Ray, Huso Germanorum; Rondelet, and some others, Exos piscis sive Ichthyocolla piscis; other authors, as Jonston, Rondelet, Willughby, Ray, and the rest who copy from these, have described it a second time as a distinct fish, under the name of Ichthyocolla; and a third time under that of Antacæus and Antacæus Borissthenis; the Greeks have also described it under this name as, Ἀντακαῖος. Ælian and Strabo both mention it; but the sole difference, as is evident from their own accounts, between the Huso and the Antacæus is, that one came from the Danube, and the other from the Borissthenes. The Ichthyocolla or Isinglass of the shops, famous as an agglutinant, and used also in the fining of wines, is the produce of this fish: it is made by boiling down the membranaceous parts of it to a jelly.

SQUALUS.

THE foramina of the gills are five on a side, and are situated in a longitudinal direction from the sides of the head down to the pectoral fins: the head is of a depressed form; the body is oblong, and is rounded or angulated, and the skin is rough: the eyes stand not on the top, but at the sides of the head; the tail is bifid, and the upper part of it is longer than the under: the mouth is usually transverse, and in the under part of the rostrum, not at it's extremity.

 The

The species of this genus are numerous, and may be conveniently arranged under several divisions.

SQUALI.

Division the First.

Those which have granulous teeth.

Squalus rostro longo, cuspidato, osseo, plano, utrinque dentato.
The Squalus, with the rostrum very long, flat, and dentated on both sides.

The Saw-fish.

THIS is one of the most singular of the whole fish-kind; it grows to a very considerable size, often, including the saw or rostrum, to twelve or more feet, and it is very thick, in proportion. The head is large, and terminates in this singular prominence, which is three or four feet, or more, in length, and is of a bony structure, flat, of a brownish colour, and furnished all along, on both sides, with very long, robust, and sharp teeth, or denticulations: the mouth is placed in the under part of the head, and is moderately large, and not round, but transverse and wide: the teeth are large, but not long; the apertures of the gills are five long slits on each side toward the head; the body is of a rounded figure, and the skin is very firm, hard, and rough: the tail is divided into two parts, and the upper division is much longer than the other.

This singular fish was well known to the antients. Aristotle call it Πρίστις, Athenæus and Oppian also give it the same name; Ælian calls it Πρίστις, Pliny and most of the old Latin writers, as well as most of the moderns, call it Pristis; Charleton, Pristis, Serra; and Clusius, Rondelet, and others, Pristis sive Serra piscis; some, simply, Serra; and others, Serra marina. We call it the Saw-fish.

Squalus levis dentibus obtusis.
The smooth-skinned Squalus, with obtuse teeth.

The smooth Hound-fish.

This is a large fish, though much inferior to the former in size: the head is moderately thick, and of a depressed form; the rostrum is obtuse; the mouth is large, and the teeth are numerous, but they are short, thick, obtuse, and granulous: the nostrils are very conspicuous, and have each two apertures; the eyes are large, and stand pretty high on the head, but not on it's top, but at the sides; the body is oblong, and of a rounded form; the skin is smooth: toward the head there are five apertures to the gills on each side; they stand in a line running from the head to the pectoral fins; there are two back fins; the pinna ani is but one: the tail is forked or divided into two parts, and the upper portion is much longer than the other.

We have this species in our own seas, but it is more frequent in those of Denmark and Norway: the Greeks were well acquainted with it. Aristotle, Athenæus, and Oppian call it Γαλεὸς λεῖος; Bellonius, Rondelet, Gesner, and Charleton, Galeus Lævis; Aldrovand, Galei lævis species ex Gesnero; Salvian and Jonston, Mustelus lævis; and Willughby and Ray, Mustelus lævis primus; and we, the smooth or unprickly Hound. And, beside these, we have a whole series of other names for it, under it's occasional varieties. Aristotle, under it's stellated form, calls it Ἀστερίας; Oppian and Athenæus, Γαλεὸς ἀστερίας; Bellonius and Gesner, Galeus Asterias; and Rondelet, Aldrovand, and Willughby, Galeus stellatus; Salvian, Mustelus stellaris sive varius; and Jonston, Mustelus Asterias.

SQUALI.

Diviſion the Second.

Thoſe which have acute teeth, and prickles on the back.

Squalus pinna ani nulla, ambitu corporis ſubrotunda. The
The Squalus, with a rounded body, and with no pinna ani. round-fiſh.

THE head is large; it is of a depreſſed figure and ſubacute, and the roſtrum, toward the extremity, is pellucid: the body is of an oblong form and rounded, or from the pectoral fins to the anus a little compreſſed, and from thence to the tail ſomewhat depreſſed, but neither in any conſiderable degree: the noſtrils are ſituated at the middle between the eyes and the extremity of the roſtrum, but they are not on the top or ſides, but toward the under part of the head; each has a double aperture, and a mucous matter may at all times be expreſſed from them: the eyes are of an oblong figure; they are placed on the ſides of the head; the iris is very large and white; the pupil is extremely ſmall, black, and tranſverſe: on each ſide of the head, behind the eyes, there is a ſemi-lunar hole; the convex part of theſe apertures is toward the eyes: the mouth is a tranſverſe cut, and is ſituated on the lower part of the head, and at a greater diſtance from the roſtrum than the eyes are: it is large, and of a ſomewhat lunated figure; there are in it three rows of ſharp teeth, affixed to a kind of common bone.

The apertures of the gills are five on each ſide, reaching from the head to the pectoral fins: they are placed a little obliquely, and, beſide the two apertures over the eyes, there are alſo two other very ſmall ones between them: the lateral line is ſtraight, and runs much nearer to the back than to the belly: the whole body and the fins alſo are rough; this is moſt plainly perceived on drawing the hand upwards, from any part of the body of the fiſh toward the head: the back and upper part of the ſides are of a greyiſh-blue colour; the lower parts of the ſides and the belly are white, and there are on the back about eight or nine roundiſh white ſpots, placed at a diſtance from one another: the pectoral fins are large, and are placed horizontally, and are broader at the extremity than at the origin: they are thin, but tough, and have no cartilaginous rays: the ventral fins are diſtinctly two; they are alſo placed horizontally, one on each ſide toward the anus: each of theſe terminates on it's lower ſide in an oblong body, complicated at the extremity, and terminating in two ſpines, the one ſtraight, and the other formed into a kind of hook; the anus is ſituated in the middle between theſe, and in the males the penis uſually alſo is found hanging out: there are two back fins; the firſt of theſe ſtands nearer to the pectoral than the ventral fins, and is perpendicularly erected, and has a ſpine adjoined to it's anterior part: the hinder one ſtands at about an equal diſtance from the anus and the tail; it is ſmaller than the other, but has a larger ſpine in it's anterior part: theſe ſpines in colour, texture, and ſubſtance, reſemble the claws of birds, and they both turn ſomewhat backwards: the tail is large, and of a compreſſed form; it is divided into two parts, and the upper portion is more than twice as long as the other, and is erected in a perpendicular direction.

This fiſh grows to about two yards in length, though it is more uſually caught of four or five feet; when it is ſkinned, the body has the appearance of that of an eel; there is a redneſs at each of the muſcles externally, but the fleſh within is perfectly white, and inſtead of bones it is ſupported by hard cartilages.

It is frequent in the Mediterranean, and is often caught alſo in our ſeas, and elſewhere: the antients were well acquainted with it. Ariſtotle, Athenæus, and Oppian call it Ἀκανθίας γαλεός; Rondelet, Geſner, and Charleton, Galeus Acanthias; Aldrovand, Willughby, and Ray, Galeus Acanthias ſive Spinax; Gaza, Bellonius, Salvian, and Geſner, Muſtelus Spinax; Scaliger, Muſtelus Spinus; Schonevelt, Canis Acanthias. The Germans call it Dornhund; the Venetians, Asio; the Italians, Scazone; and we call it the Hound-fiſh, or the prickled or pickled Dog.

Squalus

Squalus pinna ani carens, naribus in extremo rostro.
The Squalus, with no pinna ani, and with the nostrils at the extremity of the rostrum.

The Magree.

This grows to about five feet in length: the head is large, and of a depressed form; the rostrum is sebaceous, and the nostrils are situated almost at it's extremity, each having two apertures: the eyes are large, and the mouth stands transversely on the under part of the rostrum: the body is of an oblong and somewhat depressed form; the back is almost flat, and is less rough than the rest of the body: the sides are prominent, and the belly is somewhat flatted, and it's skin is extremely rough: the general colour of the fish is brown, but the belly, instead of being paler, is of a deeper colour than any other part: there are two back fins, and each has a spine at it's anterior part; there are two belly fins moderately large, but there is no pinna ani: the tail is large, compressed, and divided into two parts, of which the upper is the longer.

This is very frequent in the Mediterranean, and is caught in different parts of the Ocean also, but more rarely. Willughby and Ray call it Galeus acanthias five spinax fuscus, the brown, prickly Hound-fish.

Squalus pinna ani carens, ambitu corporis triangulato.
The triangulated-bodied Squalus, with no pinna ani.

The Centrine.

This is a large species: the head is large, and of a depressed figure; the rostrum is obtuse; the nostrils have each a double aperture, and are situated low; the eyes are large, and their iris is of a greyish-white; the pupil is small, oblong, and blackish; the mouth is a transverse fissure on the under part of the head, not at the extremity of it. There are in the upper part three rows of sharp teeth, but in the lower part there is only one row: the body is oblong, thick, and of a somewhat triangulated form; the apertures of the gills are five on each side, running obliquely, and placed in a line from the head to the pectoral fins: there are two fins on the back, and the prickle or spine at the anterior one is turned somewhat forward: the ventral fins are moderately large, and there is no pinna ani: the tail is divided into two parts, and of these the upper is much the longer.

This species is frequent in the Mediterranean, and in some other seas: it was well known to the antients. Ælian, Athenæus, and Oppian call it Κεντρίνη; Rondelet, Salvian, Gesner, and Aldrovand, Centrine; Jonston, Willughby, and Ray, Centrine; Gesner, Galeus Centrites; Bellonius and Gesner, in some others parts of his works, Vulpecula. The Italians call it Centrine and Pesce Porco.

Squalus pinna ani carens, ore in apice capitis.
The Squalus, with the mouth at the extremity of the head.

The Monk-fish, or Angel-fish.

This grows to about six feet in length, and is considerably thick, in proportion; and, though it has all the other external characters of the Squali, differs greatly from them in the situation of the mouth, which is not, as in them, situated on the under part of the head, but at the extremity, as in the generality of other fishes: the nostrils are conspicuous, and have each a double aperture; the eyes are moderately large, and the apertures of the gills are five on each side, situated between the back part of the head and the pectoral fins: there are three rows of teeth in the upper part of the mouth, and as many in the under; the whole number amounts to a hundred and eight about the region of the eyes, as also near the nostrils there are several spinose tubercles, and on the extream verge of the pectoral fins there are also some short spines, and some others on the ventral fins; and on the middle of the back there are two back fins, but they stand extremely backward, both behind the anus: the pectoral fins are moderately large; the belly fins are also large and broad, they are placed contiguous to one another, and in an horizontal direction: the tail is forked, and the upper portion is longer than the under.

This species is caught in our own seas, but more frequently in the Mediterranean: the antients were well acquainted with it. The Greeks call it Ῥίνη; Pliny, and the other older Latin authors, Rhina five Squatus; Isidore, Squatus; and almost all the modern authors, simply, Squatina; Willoghby and Ray call it Squatina Græcis Rhina; Gesner, Squatina five Angelus marinus; Albertus and Coba, Squamis. The Genoese call it Pesce Angelo; and we, the Monk-fish, and in some places the Angel-fish.

SQUALI.

Division the Third.

Those which have acute teeth, and have no spines or prickles on the back.

Squalus capite latissimo, transverso, mallei instar.
The Squalus, with a very broad, transverse hammer-like head.

The Zygæna, or Ballance-fish.

THIS is one of the most extraordinary fish in the world in it's form: the general size is five or six feet, but it grows to be much larger: the head is of the most extraordinary figure of that of any fish; it is not oblong, and running in a line with the body, as in the generality of fish, but is placed transversely, and has the appearance of the head of a hammer fastened to it's handle: the eyes are large, and placed at the two extremities; the mouth is a transverse cut on the lower part of the head, and is furnished with three or four rows of sharp teeth: the nostrils are small, and not very conspicuous, and the foramina at the eyes are oblong and large; the body is oblong, and is moderately thick; there are two back fins, and a pinna ani, which many of the preceding species have not: the apertures of the gills are ten oblong slits, five on each side, running from just below the head toward the roots of the pectoral fins: the tail is divided into two parts, and the upper of these is much longer than the under.

This species is sometimes caught in the Mediterranean, and sometimes in different parts of the Ocean. It is so singular in it's figure, to have escaped the notice of the naturalists of any period. Aristotle calls it Ζύγαινα; Ælian and Oppian, Ζύγαινα; the Latin writers, as well antient as modern, have taken Aristotle's name, and called it Zygæna; Ambrosius calls it Zigena; and Gaza and Salvian, Libella. The Italians call it Cambetta; and we, the Ballance-fish, and the Hammer-headed Shark.

Squalus cauda longiore quam ipsum corpus.
The Squalus, with the tail longer than the body.

The Sea-Fox.

This has obtained it's common names of Vulpecula and the Sea-fox, from the extraordinary length of it's tail: it is a large fish, it's weight often more than a hundred pounds; it's head is large, and of a somewhat depressed form, and acute; the mouth is small, and is situated on the under part of the head, but at a small distance only from the rostrum: the eyes are moderately large, and the nostrils are sufficiently conspicuous: the body is thick, short, and rounded; the back is of a dusky greyish colour, with an admixture of bluish and of brown; the belly is white; the apertures of the gills are five on each side, and reach in a series from the back part of the head, nearly to the root of the pectoral fins: the tail is of a very singular figure; it is falcated, and, as it were, of the shape of a sword-blade, only not straight: it is immoderately long and narrow, in proportion.

This species is not unfrequent in the Mediterranean, but there has been no instance of it's being caught in the Ocean: the antients were very well acquainted with it. Aristotle and Oppian call it Ἀλώπηξ; Athenæus, Ἀλωπεκίας; and Ælian, Θαλάττιος ἀλώπηξ; Pliny, Gesner, Willoghby, and Ray call it Vulpes marina; Gesner, Vulpes galeus; Rondelet, simply, Vulpes; Salvian calls it Vulpecula; Aldrovand, Jonston, and Charleton, Vulpecula marina; Bellonius, Simia marina. We call it the Sea-fox, and the Sea-ape.

Squalus naribus ori vicinis, foraminibus exiguis ad oculos.
The Squalus, with the nostrils near the mouth, and small holes near the eyes.

The Tope.

This grows to five feet in length, and it's weight is usually about a hundred pounds: the head is large, and of a depressed figure; the extremity of the rostrum is pellucid: the mouth is large, and of a semicircular figure, and is situated on the under part of the head: the nostrils are conspicuous; they are placed near the mouth, and have each a double aperture; there are three rows of sharp teeth in the upper, and as many in the under, part of the mouth: the body is of a rounded figure, and the lateral line runs very conspicuous from the head to the tail: there are two fins on the back; the first of them is nearer to the pectoral than to the ventral fins; the hinder one is placed a great deal more backwards: the apertures of the gills are five on each side, and run from the head nearly to the pectoral fins.

This is frequent in many parts of the Ocean, and in the Mediterranean: the antients were very well acquainted with it. Aristotle calls it Γαλεὸς ...; Pliny, Canicula; Rondelet, Galeus canis; Gesner, Galeus canis vel Canicula Plinii; Salvian, Willughby, and Ray, Canis Galeus; Aldrovand and Jonston, Canis Galeus vulgaris. The Italians call it Lamiola and Canosa; and we, in Cornwall, the Tope.

SQUALI.

Division the Fourth.

Those which have the rostrum shorter.

Squalus ex rufo varius, pinna ani media inter anum et caudam.
The brownish, variegated Squalus, with the pinna ani in the middle between the anus and tail.

The Bounce.

THIS is one of the smaller Squali; it rarely grows to more than three feet in length, and is but moderately thick, in proportion: the head is large, and of a depressed figure; the rostrum is obtuse, and the mouth is situated on the under part of the head, but at a small distance from the extremity of the rostrum: the teeth are numerous and sharp, and they bend a little inward: the nostrils are remarkably large and conspicuous: the body is of a somewhat depressed form; the back is broad, and the belly is somewhat flatted: there are two fins on the back, and they stand but at a small distance from one another; they both are placed on the hinder part of the back; the pectoral fins are large, and of a triangular figure; they have their origin about the third aperture of the branchiæ.

This species is frequent in the Mediterranean, and is also caught in our seas: the antients were very well acquainted with it. Aristotle calls it Σκύλιον; and Athenæus and Oppian, Σκύμνος; Rondelet and Gesner call it Canicula Aristotelis; Salvian, Aldrovand, and Jonston, Catulus major; Willughby, Catulus major vulgaris; and Ray, Catulus major vulgaris et Salviani. The Italians call it Scorzone; the Venetians, Pesce Gatto; and our common people in Cornwal, a Bounce.

Squalus dorso vario, pinnis ventralibus concretis.
The Squalus, with a variegated back, and with the belly fins concreted.

The Morgay.

The head is large, and of a depressed form; the eyes are large, and stand pretty high on the sides of the head: the nostrils are very conspicuous; they have each a double aperture, and are situated not very far from the extremity of the rostrum; the mouth is not at the extremity of the rostrum, but on the under part of it, and opens transversely: the body is long, rounded, and slenderer than in most of the other species: the apertures of the gills are five oblique or nearly transverse openings on each

 side

side below the head, and reaching to the pectoral fins; the teeth are numerous and sharp: the lower or under part of the rostrum has a multitude of foramina in it: the belly of this species is of a fine silvery white; the back is of a dusky greenish-black: the pectoral fins are somewhat broad; the back fins stand very low; the tail is divided into two portions, and the upper is much longer than the under.

We have this in abundance in our own seas, especially on the western coasts; it is frequent also in the Mediterranean. Salvian calls it, simply, Catulus; Aldrovand and Willughby, Catulus minor; Ray, Catulus minor vulgaris; Bellonius, Mustelus stellaris tertius. The Venetians confound it with the former species, under the name of Pesce Gatto; and we, in Cornwall, call it the rough Hound, or Morgay.

Squalus cinereus pinnulis ventralibus discretis.
The grey Squalus, with the belly fins separate.

This grows to five or six feet in length, and is considerably thick, in proportion: the head is large, and of a somewhat depressed form; the rostrum is subacute; the mouth is large, and is not at the extremity of the rostrum, but is a transverse opening underneath it; the under part of the rostrum is full of numerous foramina; the nostrils are conspicuous, and stand at a considerable distance from the extremity of the rostrum; the eyes are large, and stand high: the back is of a dusky colour, variegated with a few large spots; the sides are paler, and have also a few spots on them; the belly is white: the pectoral fins are oblong, and approach to a triangular figure; the pinna ani is much nearer to the anus than to the tail; the back fins stand very low: the tail is divided into two parts, and the upper one is much the larger: the belly fins are moderately large, and they stand separate, and are not united as in the preceding species.

This is frequent in the Mediterranean, but we do not meet with it in our seas. Rondelet, Gesner, and Willughby call it Canicula saxatilis; Ray, Catulus maximus; Bellonius and Gesner, Mustelus stellaris primus.

Squalus fossula triangulari in extremo dorso, foraminibus nullis ad oculos.
The Squalus, with a triangular fossula on the back, and no foramina at the eyes.

The blue Shark.

This is one of the most known and most terrible of the Squali: it grows to six, seven, or eight feet in length, and is considerably thick, in proportion: the head is oblong and large; the rostrum is long, and of a depressed form, and has a number of foramina both on the upper and under parts: the mouth is large, and stands not at the extremity of the rostrum, but on it's under part opening transversely: the teeth are not very numerous, but they are large, broad, and some of them serrated at the edges; the eyes are large, and the nostrils are very conspicuous, and stand low: the skin of this species is less scabrous or rough than that of most of the other Squali: the back is of a deep blue colour; the sides are of a paler greyish-blue, and the belly is of a silvery white: the pectoral fins are very long and acute: there are two fins on the back; the anterior stands about the middle, and the other near the beginning of the tail; the pinna ani stands low, and over-against the hinder fin of the back: the tail is large, and is divided into two parts, of which the upper is much the larger.

This species is frequent in the Ocean, but is rarely met with in the Mediterranean; we have it sometimes about our own coasts, but it is much more common in many other places: the antients were very well acquainted with it. Ælian calls it [illegible]; Rondelet, Gesner, Aldrovand, Willughby, and Ray, Galeus Glaucus; Charleton, simply, Glaucus. We distinguish it by the name of the blue Shark, from the colour of it's back.

Squalus dorso plano, dentibus plurimis ad latera serratis.
The flat-backed Squalus, with numerous teeth serrated at their edges.

The White Shark.

This is a very large and very terrible fish, it is the largest of all the species; it's weight, when full grown, being not less than a thousand pounds: the head is large, and

and somewhat depressed; the rostrum is oblong: the eyes are large; the nostrils are very conspicuous: the mouth is enormously wide, and the teeth are very numerous and terrible; there are from four to six rows of them, and they are broad and triangular, and many of them serrated in the upper jaw; those in the lower are more rounded and smooth, or but a few of them are serrated, in proportion to the number of the others: the fish is very thick, in proportion to it's length; and the back is very broad, and, as it were, flatted: the pectoral fins are large; the first of the two dorsal fins stands near the middle of the back, the hinder one toward the origin of the tail: the tail itself is large and compressed, and is divided into two parts, of which the upper is the larger: the general colour of the fish is whitish; the skin is very rough to the touch: the fish is very bold and voracious, pursuing and swallowing almost any thing smaller than itself.

It is a native both of the Mediterranean and the Ocean, and in some places is vastly frequent: the antients were very well acquainted with it. Aristotle calls it Λάμια; Oppian, Λάμια; Athenæus, Καρχαρίας; and Ælian, Κύων Θαλάττιος; Pliny, Gaza, Rondelet, and Gesner call it, Lamia; Willughby and Ray, Canis Carcharias sive Lamia; Aldrovand and Charleton, Canis Carcharias; Ray calls it also Tiburonus recentiorum; and some others, Tiburo and Tiburone; some call it Piscis Jonæ. The Swedes call it Haa; and we, the white Shark.

RAIA.

THE apertures of the gills in the Raia are five on each side, and they are situated on the breast at a little distance below the mouth: the head is depressed, and the whole body also is very depressed or flat: the sides are terminated by broad fins, which supply the place of the pectoral fins in other fishes: the eyes stand in the upper part of the head, and the mouth, in most of the species, in the lower; and, behind each of the eyes, there is a single foramen. The tail, in this genus, is usually long and slender.

Raia oblonga unico tantum aculeorum ordine in medio dorso.
The oblong Raia, with only a single row of prickles in the middle of the back.

This is a moderately large species; it grows to about the weight of twelve pounds, and is very broad and thin, in proportion to it's length: the rostrum is oblong and acute, and the mouth situated on the under part of it: the body is depressed and flat; the back is of a deep dusky brown: the belly is white, with a tinge of reddish; the skin of the whole fish is very rough, and there is a single row of spines running all down the back; there are also two simple spines placed just at the foramina or holes which are behind the eyes: the eyes themselves are large, and their iris is yellow; the nostrils are large and oblong; there runs all round the sides of the fish a kind of fimbria, which seems a commissure of the upper and lower parts.

This species is frequent in the Mediterranean; it is met with in other seas, but more rarely: the antients were well acquainted with it. Aristotle calls it Ῥινοβάτος; Gaza, Aldrovand, Charleton, and Jonston, Squatino-Raia; Bellonius, Salvian, and Gesner, Squatino-Raia sive Rhinobatos; Paulus Jovius, Rhinobatus sive Squali Raia; and Rondelet, simply, Rhinobatos.

Raia aculeata dentibus tuberculosis, cartilagine transversa in ventre. The Thornback.
The prickly Raia, with tuberculose teeth, and a transverse cartilage in the belly.

The head and the whole body are very flat and depressed; the figure of the body, exclusively of the tail, is nearly square; the tail is long and slender, but it is also a little depressed or flatted: the belly or lower part is altogether plane; the back is, in general, plane, but it rises a little in the middle into a convexity: the eyes stand on the upper part of the body, at a very considerable distance from the rostrum; they are somewhat

ſomewhat protuberant, and are covered with a ſimple and naked ſkin: the pupil and iris are not turned upwards, but directly to one ſide, ſo that they are formed for looking horizontally: the pupil is of a greeniſh-black, and the iris of a ſilvery white; there is, in the upper part of the pupil, a whitiſh operculum, and about the edge there is an elegant fringe under the cornea; this, by preſſing of the finger, may be forced upward and downward, till it hide the whole pupil: the cryſtalline humour in the eyes of this fiſh is round, pellucid, and is hard before boiling. Immediately behind each eye there is a ſingle, oblong foramen; this runs tranſverſely, and it's anterior ſide is ſtriated, and ſerves as a valve to cloſe up almoſt the whole aperture; but the hinder part of the aperture is ſmooth within: theſe foramina have each a double duct; they are opened in the anterior part directly into the mouth by a large foramen, and in the hinder part they have another which runs to the gills: the uſe of theſe apertures is the ſame with that of thoſe of the gills, to take in and let out water, and to aſſiſt in keeping the gills in motion, and promoting thus the circulation of the blood: the noſtrils are very conſpicuous; they are ſituated on the under part of the head, a little before the opening of the mouth, and have each only a ſingle aperture, but that very large; theſe communicate with the mouth, and have their aperture within, covered by a fine thin membrane; they are quite open before, except that on the outer ſide there is a little valve: theſe noſtrils have a long and large cavity within, and it's bottom is ſtriated or perforated with oblong and narrow holes, through which a mucous matter is forced.

The mouth is ſituated on the under ſide of the body, and ſtands in a tranſverſe direction: it is very large, and ſtands at the ſame diſtance from the extremity of the roſtrum that the eyes do. It is very well furniſhed, in both jaws, with a kind of granulous and rhomboidal tubercles by way of teeth: the palate is ſmooth, and the tongue alſo is ſhort, broad, and ſmooth.

The foramina or apertures of the gills are very ſmall; there are five of them on each ſide; they ſtand a great way below the mouth, at the ſides of the breaſt, in a ſtraight line, and the loweſt of theſe comes into immediate contact with the tranſverſe cartilage of the breaſt; theſe apertures are not round but oblong, and they run tranſverſely: the back, or upper part of the fiſh, is of a duſky brown, with a ſmall admixture of a greyiſh tinge, and is elegantly variegated with large, white, roundiſh ſpots, placed in no regular direction: the belly is white; the back is alſo ſometimes white, and is ſpotted with brown or with black, but that is not the uſual ſtate of the fiſh: the back is all over beſet with innumerable ſmall ſpines, the points of which turn backwards; the belly is ſmooth.

The ſides of the fiſh terminate each in a broad fin: theſe are very large, and ſtand in the place of the pectoral fins in other fiſhes; the extremity of each of theſe lateral fins is terminated in an acute angle: the cartilaginous rays of theſe fins are numerous, and are eaſily diſtinguiſhed in the entire fiſh, but much more beautifully when the ſkin is peeled off, in which caſe they are found to be elegantly variegated with tranſverſe knots: the ventral fins ſtand one on each ſide near the anus, or at the beginning of the tail; they are placed horizontally, in the manner of the pectoral fins, and their ſhape is very ſingular: in the open part, where they come near the lateral fins, each terminates in a large cartilaginous apophyſis, and at the other end, or at the beginning of the tail, they have two other apophyſes, but theſe laſt have no cartilages: the interior part of both theſe fins is joined by a thin membrane to the tail.

The tail is longer than the whole body; it is ſlender, but not rounded in figure, but depreſſed, and the lower ſide of it is plane or flat: there are two little fins on the upper ſide of the tail, near it's extremity, between which there are alſo uſually two little ſpines; at the extremity of the roſtrum there are alſo, both on the upper and under ſides, two ſpines, but theſe are not found diſtinct in the young fiſh, nor are they very conſpicuous at any time in the females.

There runs one row of large ſpines ſometimes from the beginning of the back, but ſometimes only from it's middle to the extremity of the tail; the ſpines on this ſeries are about thirty, and they run in a ſtraight line: there are two ſpines of this kind alſo at the anterior part of each of the eyes, and three behind each of them; and in the anterior part of the back alſo there are frequently four large ſpines; when the whole four are there, they ſtand in a ſquare form, but, in the not fully-grown fiſh, the two hinder ones only are viſible: and, finally, on each ſide of the tail alſo there is a ſingle ſeries of ſmaller ſpines, but theſe are often very faintly perceptible in the ſmaller fiſh:

the roſtrum itſelf, in fine, on it's under part, at the ſides, and in the middle, is rough, with innumerable very minute ſpines; all theſe are very like the prickles of the ſtalks of the roſe-buſh in form and direction.

The belly, or under part of the fiſh, is divided, as it were, into two parts, each of a ſemicircular figure; one of theſe is the region of the breaſt, the other of the belly or abdomen, and the edges of theſe two circles are often beſet with ſmall ſpines, but it is not ſo in all the fiſh; between theſe two regions there is placed a tranſverſe cartilage; this divides the breaſt from the belly, and is very conſpicuous externally: the anus ſtands between the ventral fins, a little above the beginning of the tail; it is of an oval figure, and is placed longitudinally: there are two ſmall foramina or apertures immediately at the lower part of the anus; theſe in the males communicate with the veſiculæ ſeminales, and in the females with the ovaries.

The gills are five on each ſide; and their apertures, on the external part of the body, are ſmaller, and communicate with much larger apertures within; theſe are oblong and tranſverſe: the heart of this fiſh is ſmall, and of a flatted figure; it is ſituated in the lower part of the breaſt: the diaphragm is robuſt and ſtrong; the liver is large, and is ſituated in the upper part of the abdomen, and is divided into three lobes: the ſtomach is large and oblong.

This is very frequent about our own coaſts. Rondeletius calls it Raia clavata; Bellonius, ſimply, Raia; Geſner, Raia proprie dicta. We, in Engliſh, call it the Thornback and the Maid. Many of the writers on fiſh have figured and deſcribed it alſo a ſecond time, as if another ſpecies, under the names of Raia clavata altera, Raiæ clavatæ ſpecies altera, and Raia clavata altera Rondeletii.

Raia corpore glabro, aculeo longo anterius ſerrato in cauda apterygia.
The ſmooth Raia, with a long ſpine ſerrated before in the tail.

The Paſtinaca marina, or Fire-flaire.

This is a moderately large ſpecies, very broad and thin, and the tail is remarkably long: the head is as flat and depreſſed as the body; the eyes ſtand on the upper ſide of the body, the mouth on the lower; there are two apertures or holes behind the eyes, one to each; the eyes themſelves are large: their iris is of a ſtrong yellow: the pupil is blackiſh; the mouth is large and tranſverſe; the teeth are ſtrong, ſhort, numerous, and granuloſe; the roſtrum is long, and very acute: the apertures of the gills are five on each ſide, and they are ſituated on the breaſt, beginning a little below the mouth: the ſides are all the way terminated by a broad fin: the whole body is very thin, and the tail is remarkably long, and is rounded; it has no fin on it, but terminates in a fine ſmall point: there is a large ſpine about it's middle; this is ſerrated on the anterior part, and is long, bony, and ſharp at the point: the back of the fiſh is ſomewhat gibboſe, but the belly is flat.

We have this ſpecies in our own and ſome other of the Northern Seas, but it is more frequent in the Mediterranean: the antients were well acquainted with it. Ariſtotle calls it Τρυγών; Ælian, Athenæus, and Oppian, Τρυγών; Rondeletius and Salvian, Paſtinaca; Geſner and Jonſton, Paſtinaca marina; Aldrovand, Paſtinaca marina altera; Schoneveldt, Paſtinaca marina Oxyrynchos; Bellonius, Paſtinaca marina lævis. The Italians call it Brucho or Bruco; and we the Fire-flaire, Fire-flaw, or Fierce-flaw: it gives a very ſevere and invenomed wound, with the ſerrated ſpine in the tail.

Raia corpore glabro, aculeis ſæpe duobus poſtice ſerratis in cauda apterygia.
The ſmooth-bodied Raia, with two ſpines on the tail, ſerrated behind.

The double ſtinged Fire-flaire.

This is a moderately large ſpecies: the roſtrum is long and acute; the head is depreſſed; the back is ſomewhat gibboſe, and the belly is flat: the eyes are large, and ſomewhat prominent; their iris is yellow, and their pupil black; the mouth is large and tranſverſe; the teeth are numerous, ſhort, and obtuſe; there are two large apertures or holes behind the eyes: the foramina of the gills are small; they are five on each side, and run lengthwise down the breast from the mouth: the lateral fins are very large and broad; the tail is but about half as long as the body, but is is usually armed with a double spine, serrated on the hinder part: the points of the aculei, in this species, turn backwards, and the aculei themselves have long furrows.

We have this species also sometimes in our own seas, but it is less frequent than the other: Columna call it, Pastinaca marina altera [illegible] aliacula dicta; and Ray has taken the same name.

Raia corpore glabro, aculeo longo serrato in cauda pinnata.
The smooth-bodied Raia, with a long serrated spine on a finny tail. **The Sea-eagle.**

This is a large and a very singular species; the rostrum is oblong, obtuse at the extremity, and has somewhat of the figure of the head of a toad: the eyes are large, prominent, and set at a considerable distance from one another; their iris is of a pale orange colour, and the pupil of a greenish-black: the back of the fish is somewhat gibbose, and the whole body is thicker than in many of the other species: the mouth is large and transverse, and the teeth are numerous; the apertures of the gills are small; they are placed five on each side, running down each way of the breast from the mouth, and behind the eyes there are also two large apertures, one to each eye: the side fins are remarkably large; they are pointed at the summit, and are easily moveable in any direction: the tail is very long and slender, and has, at it's origin, two little fins, and it is furnished with a long, sharp, and serrated spine, of a white colour, and bony structure.

This species is frequently caught in the Mediterranean, and sometimes, though rarely, in the Northern Seas: the antients were well acquainted with it. Aristotle calls it 'Αετὸς; and Athenæus and Oppian give it the same name; Pliny, Jonston, Salvian, Willughby, and Ray, all call it, simply, Aquila; Bellonius and Gesner, Aquila marina; Aldrovand, Aquila prior; and Columna, Pastinaca marina levis altera [illegible] Aquilone dicta. The Italians call it Aquilone; and we, the Eagle-fish, or the Sea-eagle.

RAIÆ.

Division the Second.

Those which have oblong and acute teeth.

Raia toto dorso aculeato, duplici ordine aculeorum in cauda, simplicique ad oculos.
The prickly-backed Raia, with two series of prickles on the tail, and one series over each eye. **The white Horse.**

THIS is a singular species: the body is considerably broad, in proportion to it's length, but it is also thick; the back is somewhat gibbose, but the belly is more flat; the rostrum is oblong and acute: the eyes are not very large, but they are prominent, and there is an aperture behind each: the mouth is transverse and large, and it is furnished with a number of sharp teeth: the apertures of the gills run down from it on each side along the breast; they are small, and there are five of them on each side: the back is of a greyish-yellow colour, with some spots of black, and is covered all over with short spines, standing very close to one another: the belly is white, and has no spots; the tail is long and slender, and has two rows of spines on it, and there is also a single series of them on each of the eyes.

This species is frequent in our own and other of the Northern Seas. Willughby calls it Raia aspera nostras; Rondelet, Gesner, Charleton, and others, Raia fullonica; and our common people call it the white Horse.

Raia

The Barracol.

Raia dorso utrinque glabris, aculeis ad oculos, ternaque eorum ordine in cauda.

The smooth Raia, with spines about the eyes, and three rows of them at the tail.

This is a small species; it rarely grows to more than a foot in length, and the far greater number met with are not more than eight inches; it's breadth is equal to somewhat more than half it's length, and it is moderately thick, in proportion: the rostrum is oblong and acute: the eyes are large and prominent; there are some spines placed about them, and just behind each there is a large aperture: the mouth is large and transverse, and is furnished with sharp teeth; the apertures of the gills are small; they run down the breast in two rows, five in each row: the back is of a pale reddish colour, and on the middle of it there are two large purple spots, black at the edges, and below these there are several white, transverse lines.

This species is frequent in the Mediterranean, but it is not met with in our seas. Rondelet, Gesner, and Aldrovand call it Raia oculata, levis; Jonston and Charleton, Raia oculata; Willoughby and Ray, Raia levis oculata; Salvian, Raia stellaris; and Bellonius, Miraletus. The Venetians, in whose markets it is frequent, call it Barracol; the Italians, Arzilla; and the Massilians call it Miralet.

Raia varia tuberculis decem aculeatis in medio dorso.

The variegated Raia, with ten prickly tubercles in the middle of the back.

This is a large species, the weight of a full-grown one is ten or twelve pounds: the rostrum is very long, and is more acute or sharp at the point, than in any of the preceding species; the eyes are large and prominent, and there is a large aperture behind each of them: the mouth is transverse, and the teeth are sharp; the apertures of the gills are small; they are five on each side, and run down in straight lines from the mouth, along the breast: the back is somewhat prominent; the belly is more flat, and there runs along the middle of the back a series of ten tubercles, which are prickly on the summit: there are two thick and fleshy pinnae ani, or rather ventral fins, with two thick appendages, beside what are called the penes; and toward the extremity of the tail there is always one, and sometimes there are two fins: in the full-grown fish there are always certain spines also at the sides of the tail, but these are scarce perceptible, unless in such.

This is frequent in the Mediterranean, and has been caught in some other seas: the antients were well acquainted with it. Aristotle and Oppian call it Βοῦς; Athenaeus, Βοῦς θαλάττιος; Ovid and Pliny call it Bos; and Paulus Jovius, Bos Bellonii. Authors have also described it a second time, as if a different species. The antients, as well as the moderns, have fallen into this error. Aristotle calls it Λειόβατος; Bellonius, Levis Raia. Among the moderns also, many have described it even a third time, as if a still different fish. Aldrovand calls it Bos Bellonii, Oxyrynchus major Rondeletii; Jonston, Oxyrynchus major; Rondelet, Willoughby, and Ray, Raia Oxyrynchos major; and Gesner, Raia Oxyrynchos major quem aliqui Bovem antiquorum esse putant. The Italians call it Mucosa and Bovosa; and the Massilians, Flassade.

The Skate.

Raia varia, dorso medio glabro, unico aculeorum ordine in cauda.

The variegated Raia, with the middle of the back smooth, and one row of spines on the tail.

This is one of the largest of the Raiae; it grows to more than a yard in length; it's breadth is equal to about three fourths of it's length, and it's thickness is so considerable, that it often weighs a hundred pounds: the back is somewhat gibbose; the belly is more flat; the colour is a pale grey, variegated with irregular spots of black: the rostrum is long and subacute; the eyes are large and prominent, there are two apertures, one behind each: the mouth is large and transverse, the apertures of the gills are small, and run in two series, five in each down the breast: the lateral fins in the

the male fish have a great number of little spines on them, both on the upper and under sides; these are not found in the female, and hence Rondelet and some others have described the male Skaite as a distinct species, under the name of the Raia spinosa: the back, as well as the belly, in this species, is smooth, but there runs a single row of spines along the tail.

This species is very frequent in the Northern Seas, and was well known to the antients. Aristotle, Ælian, Oppian, and Athenæus call it Βάτος; Cuba, Auctor, and Albertus call it Rayte, Rayche, and Rubus; Rondelet, Gesner, and Aldrovand, Raia undulata sive cinerea; Gesner, Jonston, and Charleton, Raia undulata; and Schonevelde, Raia levis. We call it the Skaite and the Flaire. Aristotle has distinguished the sexes in this species; he calls the male Skaite, Βάτος, and the female, which has no spines on the fins, Λεία.

Raia tota levis. *The wholly smooth Raia.* The Torpedo, or Cramp-fish.

This is a large species; it is considerably broad, in proportion to it's length, and it is very thick: the whole body, head, and tail are all perfectly smooth; there is not the least sign of any spine or tubercle for the origin of a spine in any part: it has also two little fins or pinnules in the middle of the back: by these characters it is sufficiently distinguished at sight from all the other species of this genus. The rostrum is oblong and subacute; the eyes are large and prominent, and there is an aperture behind each: the mouth is large, transverse, and well furnished with teeth: the apertures of the gills are larger than in many of the other species, though they are not immoderately so: the back is somewhat gibbose; the belly is flat, and the sides are terminated by broad fins, which stand in the place of the lateral or pectoral fins of other fishes: the colour on the back is a dusky greyish, and the belly is white. The singular property of this fish is, that, when out of the water, it has a power of affecting the hand that touches it, in a very remarkable manner: the shock is instantaneous, and resembles that given by electricity, only that the effect lasts longer: the part of the limb affected, which is usually from the fingers ends up to the elbow, if the hand has touched it, is affected with a sensation much like that which we call the cramp: this is all that is in the power of the fish, but those who have related it, have raised the effects into almost miracles. Monsieur de Reaumur has given a long memoir in the Paris Transactions, in which he has endeavoured to account for the manner in which the effect is produced: he resolves it into the instantaneous action of a vast multitude of small muscles on the surface of the body of the fish, which it brings into action in great numbers at once; but there seems something more required to the perfectly explaining so odd a phænomenon.

The fish has been known from the earliest times. Aristotle, Ælian, Oppian, and Athenæus call it Νάρκη; Cuba, Narkos; the same Cuba also and some others call it Rabus; the greater number of the Latin writers call it, simply, Torpedo; Ray calls it, Torpedo Græcis Narce; Willoughby, Torpedo Græcis Νάρκη Genuensibus Batta potta; Gesner, Aldrovand, and Jonston, Torpedo non maculosa; and the same Gesner, in other places, Torpedo maculosa and Torpedo maculosa supina, making two other imaginary species from it's varieties; Aldrovand calls it Torpedo Salviani maculosa; and Bellonius, Torpedo oculata; Rondelet also, as well as some others, has described it under the different circumstances of growth and variegations in colour, three or four times over, under the names of Torpedo, and Torpedinis species secunda tertia and quarta. It is from these accidental variations, and some little distinction between the male and the female fish, that it is not easy to give a description of it, that shall be accurate in the minute particulars. The Italians call it Occhiatella; and we, the Cramp-fish, or the Numb-fish.

F I S H E S.

Class the Fifth.

P L A G I U R I.

CETACEOUS FISHES.

Those which have the tail placed horizontally, not perpendicularly, as in all the former classes.

THESE fish respire by means of lungs; they are viviparous, and the males have a penis and testicles, the females a vulva, ovaries, and paps.

P H Y S E T E R.

THE Physeter has teeth only in the lower jaw, and they are crooked: on the back there is a fin, or a large and tall spine; and the opening or fistula for the discharge of the water is in the front part of the head.

The crooked-toothed Whale.

Physeter maxilla superiore longiore, spina longa in dorso.
The Physeter, with the upper jaw longest, and with a long spine on the back.

This is a very large whale, and it's figure is very singular: the head is monstrously big, it is equal in length to the whole body without the tail: the body is very thick, and of a figure approaching to rounded, only the back is elevated; the head, however, is thicker than the body, even in it's thickest part: the upper jaw or rostrum is at least five feet longer than the under, and, even in the under, the rostrum is continued two feet beyond the extremity of the jaw-bone: the shape of the whole head is irregular, and there are some large depressions on it's upper part: the eyes are remarkably minute in a fish of this enormous magnitude; they are scarce larger than those of a common whiting: the pipe or fistula for the discharging water is in the middle, or a little higher than the middle of the rostrum; it is divided into two passages within, but it is covered by one common operculum: there are no teeth in the upper jaw, but in the under there are no less than forty-two; they are large, long, of a nearly rounded figure, but somewhat compressed, and are not straight, but bent in the manner of a sickle; they are thickest and most bent in the middle; they terminate at the top in an acute cone, the point of which turns inward, and, what is more singular, they diminish in thickness also toward the bottom, and there terminate in a much smaller and slenderer root: the tail is large, and placed horizontally, and on the back there is a large and long spine, which occupies the place of a fin.

This species is a native of the Northern Seas, but not of the very remotest; we have it sometimes about our own northern coasts. The naturalists were not acquainted with it till the time of our Sibbald; he has described it under the name of Balæna major in inferiore maxilla tantum dentata, dentibus arcuatis falciformibus, pinnam seu spinam in dorso trahens; Ray has borrowed the same name.

The plane-toothed Whale.

Physeter pinna dorsi altissima, apice dentium plano.
The Physeter, with the back fin very tall, and the summit of the teeth plane.

This is a larger species than the former, and somewhat resembles it in it's general form: the head is enormously big; it is equal in length to three fourths of the body, and toward the hinder part is thicker than the largest part of the body: there are several oblong and irregular cavities, or depressions, before the eyes; the eyes themselves are small, but not so extremely minute as in the preceding species: the fistula or aperture for discharging water is in the middle of the head, and has a single valve, but it is divided into two parts or passages within.

There are no teeth in the upper jaw, but in the under there are a considerable number; they are long, tall, rounded, and a little bent inwards; they are thicker at the roots than those of the preceding species, and are less bent or falcated, though their extremities turn inwards; these also are not conic, as in the others, but plane: there are three fins on this fish; that on the back is the most singular; it is large and very tall, simple, and erect, and looks like the missen-mast of a ship: the tail also is very large and broad.

This species has been thrown on shores about the northermost part of Scotland; we owe our knowledge of it to the same author who described first the other. He calls it Balæna macrocephala tripinnis quæ in mandibula inferiore dentes habet minus inflexos et in planum desinentes; Ray also has continued this name.

DELPHINUS.

THE Delphinus has teeth in both the jaws, the fistula, or opening for the discharge of water is in the middle of the head, and the back is pinnated or finned.

Delphinus corpore fere coniformi, dorso lato, rostro subacuto.
The Delphinus, with a coniform body, a broad back, and a subacute rostrum.

The Porpess.

This is a very large and not a very beautiful fish: the head is large, and the body is very thick toward the head, but it grows gradually smaller to the tail; the whole thence has somewhat of a conic form: the back is broad, and the diameter in the largest part is more than equal to half the length of the fish: the rostrum is protruded; the head is not very large; the opening of the mouth is wide, and there are teeth in considerable numbers in both jaws; there are about forty-eight in the full-grown fish in each jaw, and they are acute and somewhat moveable: the fistula or aperture for the discharge of water is nearly in the middle of the head, and beside this there are six other smaller foramina about the rostrum: the apertures for the ears are so small, that they are scarce perceptible: the body is, in general, of a rounded figure, but, as it approaches the tail, it becomes a little compressed: the tail is large; it is placed horizontally, and is a little divided or forked: the stomach is triple: the intestines are extremely long; the kidnies are composed of a great number of lobules, and the ribs are thirteen on each side.

This species is not unfrequent in the Northern Seas, and frequently comes a great way up the large rivers; we have it sometimes in the Thames: the writers of all times have been acquainted with it. Aristotle calls it Φώκαινα; Gaza, Tursio; Pliny, Bellonius, and Rondelet, Tursio; Jonston, Charleton, Willughby, and Ray, Phocæna; Gesner and Aldrovand, Phocæna sive Tursio; Gesner, Phocæna sive Thursio; and Schonevelde, Parvus Delphinus vel Delphinus septentrionalium aut orientalium. The Swedes and Danes call it Marsuin; and we, the Porpess: this, however, is not so determinate a name as it ought to be, for we call the following species or proper dolphin also the Porpess.

Delphinus corpore oblongo subtereti, rostro longo acuto.
The Delphinus, with an oblong, subcylindric body, and a long, acute rostrum.

The Dolphin.

This is a considerable longer fish than the former, but it's body is not thick, in proportion: the head is not immoderately large; the rostrum is very long and acute; the opening of the mouth is vastly wide, reaching on each side to the breast: both jaws are furnished with numerous teeth, and the opening them has a very terrible appearance; there are two longitudinal depressions before the eyes, and in the middle of the head stands the fistula or aperture for the discharging water; it is double, or has two passages within, but it terminates singly: the body is of a rounded figure, and does not diminish so much in thickness toward the tail, as the former species: the parts of generation in both the sexes are very visible, and stand at a middle distance between the anus and the navel.

This

This species is frequent in the Northern Seas, and often makes it's way up rivers: it was very well known among the antients. Aristotle and Athenæus call it Delphis; Ælian, [illegible]; and Oppian, [illegible]; Pliny, and after him most of the other Latin writers, call it Delphinus; Aldrovand, Delphinus prior; and Ray, very properly, Delphinus antiquorum. We call it sometimes the Dolphin, but more usually the Porpess. It is extreamly wrong, however, to confound two fish so perfectly distinct, under the same name; and the more so, as we have two distinct names for them.

Delphinus rostro sursum repando, dentibus latis serratis.
The Delphinus, with the rostrum repandous upwards, and with serrated teeth.

The Grampus.

The size of this fish has made it generally taken for a whale, though it is properly of the same genus with the porpess and dolphin: the head is large, and somewhat depressed toward the rostrum; the lower jaw is larger and more robust than the upper: the opening of the mouth is very wide, and both the jaws are armed with strong teeth: the fistula or aperture for the discharge of water is in the middle of the head; the eyes are large, and the apertures of the ears are very small: the body is very thick, it's diameter is more than equal to half it's length; it's figure is somewhat rounded, but the back is broad, and the belly flatted: the tail is large and horizontal.

This is frequent in the Northern Seas, and is apt to play about upon the surface of the water, so that it is oftener seen than any of the others, even in places where they are equally frequent. All the old Latin writers call it Orca, and most of the moderns have taken the same name; Sibbald and Ray call it Balæna minor utraque maxilla dentata; Paulus Jovius calls it Capidolius; and Marten, Butz kopf. We call it the Grampus; and others the Springer, the North Caper, the Tandshye, and Loper.

BALÆNA.

THERE are in the upper jaw of the Balenæ certain laminæ of a horny matter, which supply the place of teeth, but there are none such in the lower jaw: the fistula or aperture for the discharge of water is double, and is situated either on the forehead, in the middle of the head, or in the rostrum: the back has in general no fin upon it.

The horny laminæ, in the upper jaw of this fish, are the substance which we call whalebone.

Balæna fistula in medio capite, dorso caudam versus acuminato.
The Balæna, with the fistula in the middle of the head, and the back ridged toward the tail.

The Whale.

This is the fish determinately and properly called the Whale, the principal object of the Greenland fishery, and the first known species. It grows to a monstrous size: the head is extreamly large, and of an irregular figure: the lower jaw is much larger than the upper, and covers it at the sides; the upper is narrow and oblong: the fistula is double, or has two distinct apertures, and is situated in the middle of the head between the eyes: the eyes are very small, in proportion to the enormous bulk of the head, and are placed at a great distance from one another; they are a little higher up in the head than the fistula: the whole head is somewhat depressed, and has several irregularities on it's surface: the body is very thick, and somewhat rounded, but toward the extremity of the back there is a subacute angle, extending itself longitudinally to the tail: the tail is somewhat forked, but not deeply so; it is very large, and in it's horizontal situation makes a very singular figure.

This is an inhabitant of the most Northern Seas; the Greenland fishery has this species for it's principal object, though it occasionally takes in any thing that can yield the materials for oil. Aristotle calls it [illegible], and Pliny just mentions it under the name Musculus; but these authors had but very imperfect ideas of it, all they knew being from hearsay and relation. Sibbald, who has entered into the distinctions of of these fish better than most others, calls it Balæna major laminis corneis in superiore maxilla habens, fistula donata, bipinnis; Ray, Balæna vulgaris edentula dorso non pinnato; Rondelet, Balæna vulgo dicta, sive Musculus; Gesner, Balæna vulgo dicta, sive Mysticetus Aristotelis et Musculus Plinii; Aldrovand and Jonston call it Balæna vulgi; Charleton, Balæna vulgaris; and Bellonius, Charleton, and many others, simply, Balæna. Cuba calls it Musculus piscis, and Marten and some others, Balæna Spitsbergensis; we, properly and simply, the Whale.

Sibbald describes another Balæna agreeing with this species in all respects, except that it has no fistula; but this is only owing to an error. There is no Balæna which has not a fistula; and this account of Sibbald's is formed from Faber's bad description of the common Greenland whale, for it is evident that he means no other, though he had but a very imperfect idea of this.

Balæna fistula in medio capite, tubere pinniformi in extremo dorso.

The Balæna, with the fistula in the middle of the head, and a pinniform tuberosity on the back.

The Physeter.

This is a very singularly-shaped fish for one of this genus; it is equal in length to the Greenland whale, but it's thickness is not one third so much: the head is very large, and the lower jaw is larger than the upper; the opening of the mouth is wide, but there are no teeth in it: the upper jaw has, in their place, a quantity of horny laminæ, as in the former species; and in the lower there is nothing: the head is somewhat depressed, and has several irregularities in it's figure, particularly some oblong depressions in the front, and toward the sides: the eyes are very small, in proportion, but they are placed at a less distance than in the former, though in this not at all near one another: the fistula is single at it's aperture, but it is double in it's passage; it stands nearly in the middle of the head, and is larger than in any other species: the fish also makes a more violent use of it than any other, tossing the water to an immoderate height, and with vast violence from it.

The head in this species is long, and, in proportion, slender; it is round at the sides, but somewhat ridged on the back, especially on it's hinder part, and the belly is flatted: toward the tail there stands, on the very ridge of the back, a tuberosity resembling a fin: the tail is very large, and a little forked, but it is but very little so; the fish has a power of moving it with vast violence.

This species was known to the Greeks. Ælian and Oppian call it φυσητήρ; Pliny, and after him most of the Latin writers, have called it Physeter; Gesner calls it Physalis Bellon sive Physeter; and Charleton, Physeter, Physalus. Our people, on the Greenland fishery, call it the Fin-fish.

Balæna fistula duplici in rostro, protuberantia corniformi in extremo dorso.

The Balæna, with a double fistula in the rostrum, and a corniform protuberance on the back.

This is a considerably large species, tho' inferior in size to the Greenland whale, and the former: it grows to fifty feet, or more, in length, and the body is considerably thick, in proportion: the head is large, and the rostrum, in comparison of the others, is acute: the opening of the mouth is large, and the upper jaw is furnished with the horny laminæ, in the manner of the Greenland whale; the eyes are small, and they stand very distant: the fistula is double, or has a double aperture, and is placed more forward on the rostrum than in any of the others: the body is rounded, but the back rises into a ridge, and has on it's hinder part a protuberance shaped like a horn: the belly is wrinkled; the tail is very large, and somewhat divided at the middle, but not deeply so, and the fish has a surprising force with it.

This species is sometimes thrown on shore on our northern coasts, and is also met with in the remotest Northern seas. Sibbald has called it Balæna tripinnis maxillam inferiorem rotundam, et plicis in ventre; Ray has taken this name; the rest of the Ichthyologists have not known it.

Balæna fiſtula duplici in fronte, maxilla inferiore multo latiore.
The Balæna, with a double fiſtule in the forehead, and with the lower jaw broadeſt.

This is a conſiderably large ſpecies; it has been met with thrown on ſhore on our own coaſts, of ſeventy-eight feet long: the head is remarkably large, and of an irregular figure, having ſeveral depreſſions on it: the opening of the mouth is very wide; the upper jaw is furniſhed with the horny laminæ in conſiderable abundance, in the place of teeth; the lower jaw has none of them: the lower jaw is vaſtly broader than the upper, and ſtands all the way prominent beyond it: the eyes are ſmall, and are ſituated high on the head, at a conſiderable diſtance from one another; the fiſtule is of a pyramidal figure, and is divided into two parts by a ſeptum: the body is thick; the back is elevated; the belly is ſomewhat flatted, and the tail is very large and moveable with a ſurpriſing force.

It is a native of the moſt Northern Seas, but it is ſometimes alſo thrown up on our ſhores, eſpecially in the moſt northern parts of Scotland. Sibbald, who had met with it there, calls it Balæna tripinnis maxilla inferiore rotunda et ſuperiore multum latiore; Ray alſo has borrowed the ſame name.

MONODON.

THE Monodon has only one tooth; this is remarkably long, and is fixed in the upper jaw, and runs parallel with the length of the fiſh, ſo that it has more the appearance of a horn than a tooth. There is no fin upon the back, and the fiſtule is in the vertex, or uppermoſt part of the head.

Of this ſingular genus there is but one known ſpecies.

MONODON. The Unicorn-fiſh, or Nar-whal.

This is an extreamly ſingular fiſh; the length of a full-grown one is about five and twenty feet, but from fifteen to twenty is more common: the body is extreamly thick; it's diameter equals at leaſt half it's length, and it is very unwieldy: the head is ſmall, and is ſhaped like that of a roach or carp: the mouth is very ſmall, and there are no teeth in either jaw, except the ſingular one, called a horn in the upper; this grows from the left ſide of the jaw, and is protended forward. It is fixed in a gomphoſis in the jaw, altogether in the manner of other teeth, whence, and from it's ſtructure, appears the impropriety of calling it a horn. This tooth grows to ten feet, or more, in length; it is about the thickneſs of a man's wriſt toward the baſe, and thence becomes gradually ſmaller all the way to the point; it is a little flatted, where it is let into the jaw; in the other part it is rounded: it is of a bony texture, and of a fine ivory white colour: the ſurface is wreathed in a very elegant, ſpiral manner, and the tooth is hollow almoſt all the way up. The eyes of this ſpecies are very ſmall, and ſituated at a great diſtance from one another; the fiſtule ſtands very high up in the head, and is large; the whole body is covered with a very tough ſkin: the back is greatly elevated; the ſides very prominent and rounded, and the belly ſomewhat flatted: the tail is horizontal and very large, and is a little divided in the middle.

This ſingular fiſh is frequent in the Northern Seas, and ſometimes comes as far towards us as Denmark and Sweden, but rarely; the people employed in the whale fiſhery frequently kill it, and procure oil, &c. from it, as from the common whale. One of theſe people brought the ſkeleton of one to London, about two years ſince, and ſhewed it for money, under the name of an unicorn. I had an opportunity of examining the tooth in this to great advantage, and found it to be abſolutely and entirely ſuch, with no title to the name of a horn at all.

This ſingular fiſh was not known to the antients, nor well, indeed, to any, till ſince the eſtabliſhment of the Greenland whale-fiſhery. Charleton calls it Monoceros, unicornu marinum; Ray, Monoceros piſcis de genere cetaceo; and Willoughby, Monoceros piſcis qui de genere cetaceo eſſe fertur. The tooth of this fiſh was known among the collection of curioſities, long before the fiſh itſelf was ſo. When the creature had been

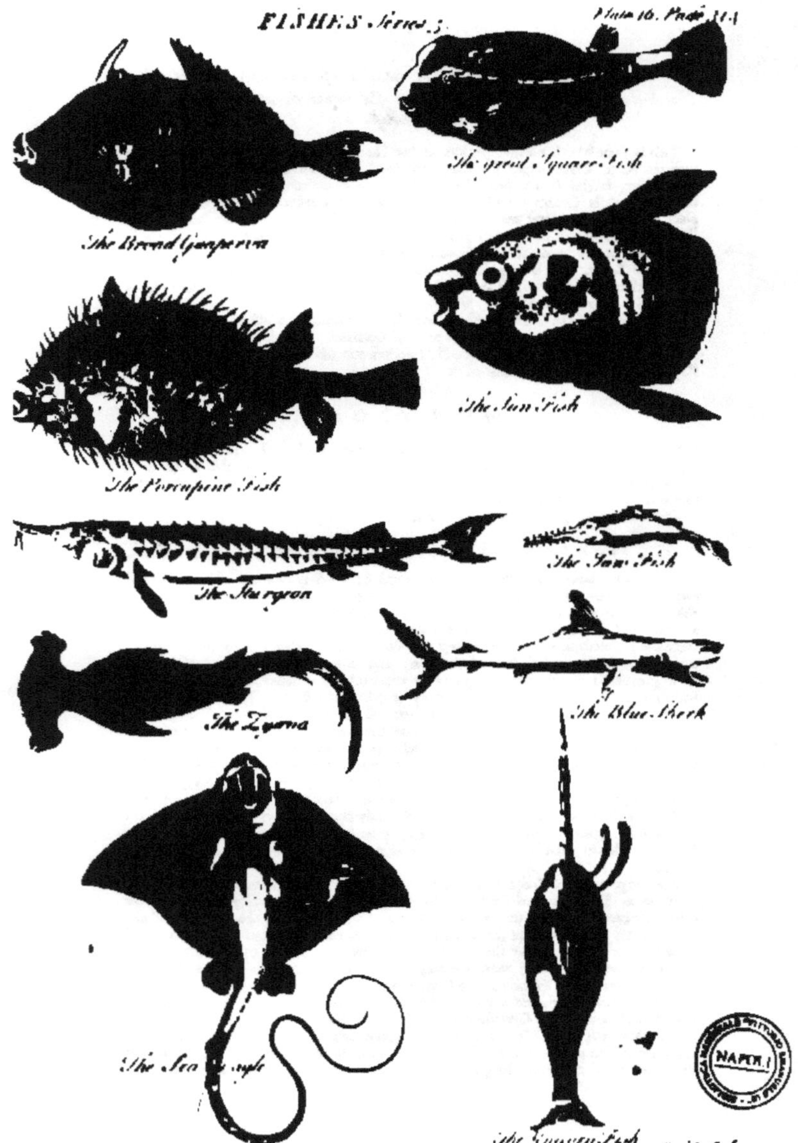
FISHES Series 3.
Plate 16.
The great Square Fish
The Broad Guaperva
The Sun Fish
The Porcupine Fish
The Sturgeon
The Saw Fish
The Zygæna
The Blue Shark
The Sea
The Unicorn Fish

been washed on shore, and it's body destroyed by the waves, the tooth still remained, and it's singularity induced any body that came in sight of it to pick it up. It was long known, before the fish was heard of; and the first conjecture about it was, that it belonged to some land-animal; it was immediately after declared to be the horn of the unicorn. Medicinal virtues were then ascribed to it, and it's scarcity made it sell for a vast price. It is but a century ago that there were only two or three of them known of, and one of those, in the cabinet of a German Prince, had been purchased at the expence of two thousand pounds: at present we have scarce any of our collectors without it; and it is so common, as to be made a kind of sign at the doors of many of our druggists shops.

CATODON.

THE Catodon has teeth only in the lower jaw: there is no fin on the back, and the fistule is either in the head, or very forward on the rostrum.

Catodon fistula in rostro.
The Catodon, with the fistule in the rostrum.

This does not grow to the size of many of the other whales, thirty-five or forty feet being it's usual length, when full-grown, and it being more frequently met with much smaller: the head is small, short, thick, and of a roundish figure: the mouth is small, and the upper jaw is smooth, but in the lower there are teeth; the eyes are very small, and they stand high up in the head, and at a considerable distance from one another: the fistule is prominent, and stands very forward on the rostrum; the body is thick and unwieldy; it's diameter is nearly equal to half it's length, and it grows gradually smaller to the tail: the back is ridged; the sides are rounded and very prominent, and the belly is somewhat flatted.

This is not met with in the Greenland Seas, but has been thrown on shore at the Orkney islands, and on some other parts of Scotland. Sibbald calls it Balæna minor in inferiore maxilla tantum dentata, sine spina aut pinna in dorso; Ray has borrowed this name from Sibbald, and none of the other writers have mentioned it.

Catodon fistula in cervice.
The Catodon, with the fistule in the neck.

The Sperma-ceti Whale.

This is a considerably large whale; it grows to forty feet in length, and to more than half that in diameter: the body is very thick, bulky, and unwieldy; the head is very large; it is elevated at the vertex, and broad toward the rostrum, and has several irregular depressions on it: the eyes are larger than in most of the whale-kind, and yet they are very small, in proportion to the bulk of the fish: the mouth is large; there are no teeth in the upper jaw, but there are about two and forty in the lower; these are placed in two series, and are each about the bigness of a man's thumb: the fistule is large, and is situated in the very hindermost part of the head, so that it seems in the neck: the back is very much elevated, and the belly is flatted; the sides are prominent and rounded, and the whole fish is covered with a tough and firm skin: the tail is very broad and horizontal.

This species is frequent in the most Northern Seas, and sometimes has been thrown on shore in Holland, and on the northern coasts of Scotland: it has been very well known to the generality of authors. Ray and Sibbald call it Balæna major in inferiore maxilla tantum dentata, macrocephala, bipinnis; Clusius calls it, simply, Cete; Willughby, Cete Clusio descriptum Pot-walfish Batavis oræ accolis dictum; Charleton calls it Cetus dentatus; and Purchas and some of our English writers, a Trump.

The drug so much in use at this time, under the improper name of Sperma-ceti, was originally made only from the oil procured from the head of this particular species of whale, and hence the fish obtained the name of the Sperma-ceti Whale: but at present our druggists purchase it of people, who make a trade of preparing it, and who make it from the sediment of any kind of fish-oil; they generally purchase, for this purpose, the bottoms of the casks at our oil-shops, and from this coarse matter make the finest sperma-ceti we are acquainted with. While it was really made from the oil

oil obtained from the brain and out of the diploe of the cranium of this species of whale, which was the case at first, it is no wonder that it was kept up at a considerable price; but, when it was discovered that any oil would do, the price soon became moderate. The method of making this pure substance out of so coarse a matter, is, by boiling the oil a considerable time, with a solution of any fixed alkaline salt; the German pot-ash or pearl-ashes is generally used, where it has been boiled with this liquor, till it become white and firm; it is melted over-again singly a great many times, and at length, after several washings to get out the saline particles that might remain in it, it becomes the white, firm substance we see, and is cut out into flakes, with knives made for that purpose.

It was proper to be the more large on the origin and preparation of this medicine, because people have run at all times into great errors about it; nor are there wanting, even at this time, some who, from the idle name that has been given it, suppose it to be the real sperm of the whale. What our repeated boilings and bathings do to this oil, the motion of the salt-water will sometimes do for it, without any farther trouble; and to this accident has been owing the first knowledge of the drug. It was found sometimes floating in large irregular masses, on the surface of the water, in the Northern Seas; it was afterwards observed, that this matter was principally found in places where the carcasses of dead whales had been broken up, and washed to pieces by the water; and succeeding observation shewed, that the oil about the head of this particular whale was what most generally afforded it. On the first discovery that it belonged to the whale-kind, there prevailed a random opinion of it's being the semen or sperm of that fish; and, before so much was known as that it belonged to the whale or to any fish at all, it was supposed to be a mineral substance, a kind of bitumen thrown up from the earth at the bottom of the sea, and floating on it's surface. We find this opinion strongly maintained by Schroder and others of his time; and it is, indeed, but very lately that we have been let into the secret of what it truly is, and in what manner prepared.

When we trace in this manner the several steps to truth in a recent instance, we shall not wonder at the many errors we find in the old writers, in regard to the drugs, and other productions of nature or art, in different climates, while they were very well acquainted with their external surfaces, their qualities and effects.

THRICHECHUS.

THE Thrichechus has teeth in both jaws: there is no fin upon the back, and the skin is very tough, firm, and hairy.

Of this singular genus there is but one known species.

THRICHECHUS. — The Sea-cow.

This singular creature seems to be the link in the great chain of beings, uniting the fish and the quadruped tribes, as it is the only fish whose skin is hairy, and like the quadrupeds; it's whole appearance also has something in it different from the fish-kind, and approaching to that of the land-animals: it grows to fifteen feet, or more, in length, but it's more frequent size is about ten or twelve feet, and it is considerably thick, in proportion: the head is moderately large, and is oblong, rounded, and has much the appearance of that of some of the quadruped kind; it has been supposed to resemble that of a cow; but, on a strict observation, it will be found more like that of a hog: the eyes are very small, and there are two small apertures in the place of external ears: the mouth is not very large, but it has fleshy and thick lips, resembling those of quadrupeds, and there are each way two long teeth which shew themselves: these are six inches long in a well-grown fish, and of the thickness of a man's thumb.

There is no fin on the back; the pectoral fins are of an extremely singular figure; they stand on the thorax, and resemble the feet of quadrupeds: each of them is formed of five toes, as it were, connected by a membrane; each toe has three articulations, but the whole fin is not capable of contraction.

The tail is large, and is placed horizontally, in the manner of that of the cetaceous fishes: the females have between the pectoral fins two large, round, and fair

breasts

breaſts; and both ſexes have the parts of generation and the navel perfectly reſembling thoſe of the human ſpecies: the ſkin is hard, firm, and almoſt impenetrable, and has ſhort, browniſh, or greyiſh, hairs on it, but they do not ſtand thick, or cover it in the manner of thoſe of land-animals.

It is a native of Africa, and of the Eaſt and Weſt Indies, and generally lives about the mouths of large rivers; it is found, however, ſometimes out at ſea, at great diſtances from land. It is common in the Red Sea, in ſome of the great rivers of Africa, and in the Amazons river in America, and is often ſeen ſleeping on the ſurface of the water, about the American and Eaſtern Iſlands.

It feeds on vegetables, and it's fleſh is white, firm, wholeſome, and of a very agreeable taſte; it has a power of making an odd noiſe with it's breath: it is extremely ſhy, when out at ſea, and has ſuch a quick ſenſe of hearing, that the people who go out in ſearch of it, though they ſee many in an hour, find it very difficult to get within reach of one.

Geſner calls it Manatus; Aldrovand and Jonſton, Manatus Indorum; Charleton and others, Manati; Ray, in his Synopſis of Quadrupeds, among which he reckons it, calls it Manati ſive Vacca marina; Cluſius and Hale, Manathi; Herrera, Taurus marinus, Zuchelli La Donna; and Rochefort, Sec-koejen. The Spaniards call it Manathi and Manati; the French, Lamantin and Namantin; the Portugueſe, Peixe Moulier and Mugor; and we, the Sea-cow, and the Manatee.

It is probably from an imperfect view of this fiſh, that the opinion of mermaids, mermen, and ſyrens firſt aroſe. This creature has a way of raiſing itſelf upright, and ſtanding, for ſome minutes together, half out of water; a perſon who ſhould look at it from a diſtance in front in ſuch a poſture, would ſee ſomething like hands and breaſts, and this ſeems to be all that has give origin to the reports of ſeeing mermaids, &c. We are told alſo of ſome of them having been caught and taught to' ſpin, but we have no exact deſcriptions of theſe mermaids. The manatee and the ſeals are all wonderful tractable creatures, and will be taught a thouſand odd things: we had lately, in London, a ſeal ſhewn publickly, which was more obedient, and had more tricks than a ſpaniel. From ſome ſuch inſtance in a tame manati, it is very poſſible the ſtory of a mermaid, and her being taught to ſpin, may have got footing in the world; for it is very certain, that there never was any ſuch creature as is deſcribed under that name.

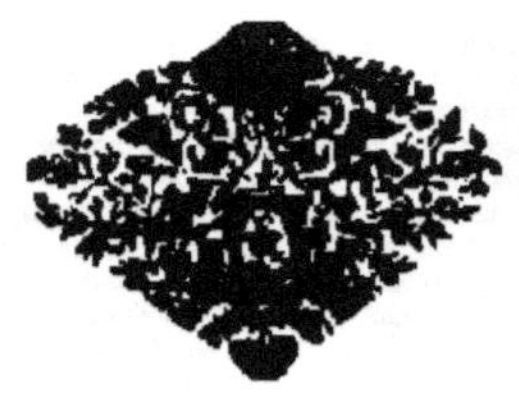

THE HISTORY OF ANIMALS.

PART V.

Of BIRDS.

BIRDS are animals, whose bodies are covered with feathers, and which have two legs, two wings, and a beak of a firm, bony, or horny texture. They lay eggs also, which are covered with a shell of a firm, crustaceous substance.

As the genera, as well as the species of birds, are considerably numerous, it will be proper to take the most obvious and distinctive characters for the arranging them into classes, before we descend to the genera. The beaks alone afford sufficient matter for such a general distinction, and happily leave all the other characters to be used in the generical and specific distinctions.

The whole family of birds will be conveniently arranged, according to the shape of their beaks, into six classes, under the names of, 1. Accipitres; 2. Picæ; 3. Anseres; 4. Scolopaces; 5. Gallinæ; and, 6. Passeres.

The Class of Accipitres comprehends all those which have the beak uncinated or hooked.

The Class of Picæ comprehends those which have the beak convex and compressed.

The Class of Anseres comprehends those which have the beak dentated or serrated.

The Class of Scolopaces comprehends those which have the beak subcylindric and obtuse.

The Class of Gallinæ comprehends those which have the beak of a conic form, but crooked, and the upper maxilla imbricated.

The Class of Passeres comprehends those which have the beak conic and attenuated.

After the classical names have been thus generally laid down, it will be necessary, before we proceed to the enumeration of the species under their several genera, to explain what are the principal parts from which the specific distinctions are drawn, and by what words they are expressed: these will occur more frequently in the names than in the descriptions of the species; but as they are some of them such as the earlier authors in ornithology have not used at all, and others, such as they have employed in a very vague and indeterminate manner; it will be necessary to ascertain their meaning according to the writings of the naturalists of these improved times.

The term Cera *expresses that membrane or naked tunic, which is propagated from the anterior part of the head, and surrounds and extends itself over more or less of the base of the beak.*

The term Urhopygium *expresses that part of the body, which is between what we call the lower extremity of the back and the insertion of the tail. This we call the Rump in English.*

The

The term Papillosa *expresses a surface naked, or not covered with feathers, but on which there are a number of smaller or larger papillæ.*

The term Artus *is used to express the legs, the wings, and the tail.*

The feet of birds, when the toes are otherwise than naked, are expressed by the two terms Palmati *and* Lobati. *The former of these* Palmati *signifies, that the several toes are connected by membranes one to another. The term* Lobati *expresses the toes having a membrane on each side to increase their breadth, but this not connecting them one to another.*

The other terms used, in regard to the feet, are expressive only of the number of toes on each; these are tridactyli for those which have three toes, tetradactyli for those which have four, and so of the other numbers.

The feathers which distinguish the several species are to be carefully attended to, and nothing could be more necessary than the distinguishing the several series of them by determinate names.

The term Rectrices *expresses those large feathers which constitute the tail of the bird.*

The term Remiges *expresses the large feathers of the wing, by means of which the bird flies; these are separated under two distinct terms of primary and secondary.*

The primary remiges are those placed from the flexure of the wing to it's extremity: the secondary Remiges are those which are situated between the body and the flexure of the wing: the feathers, which serve as a covering to the wings, as the others do to the body, are distinguished by the name of Tectrices.

So few terms as these are all that are necessary to be understood, in order to the being familiar with the names given to the several species. The rest will be easy from the descriptions.

B I R D S.

Class the First.

ACCIPITRES

Birds which have the beak of an uncinated or hooked figure.

STRIX.

THE feet of the Strix have each of them four toes: three of these stand forward, and one backward; but the exterior one of the three anterior is capable of turning backwards.

Strix capite aurito, corpore rufo.
The Strix, with an aurited head, and a reddish-brown body.

The great horn-owl, or Eagle-owl.

This is an extremely singular and beautiful bird, and is very worthy to appear first in the enumeration of the species: it is of the bigness of a goose; the wings are very large, and the principal feathers extremely long and strong: the head is short, round, and large; it is of the size, and much of the figure, of that of a cat, and at the auricles there stand two remarkable series of black feathers; these rise to the height of three fingers breadth above the surface of the head, and very perfectly resemble ears: the eyes are very large and look remarkably fierce; the feathers on the rump are long, extremely close set, and soft: the beak is short and hooked, and is black; the apertures of the ears, when the feathers are put away, are found to be very large and patent: on each side of the face, about the place of the nostrils, there are a great number of a kind of whiskers resembling hairs.

The feathers covering the whole body are much alike in the several parts; the belly is paler than the back, but those elsewhere are variegated in a tolerably uniform manner, with black and white in spots and clouds, upon a reddish-brown ground; they are large and elegant, and lie in a remarkably regular and beautiful manner on the body.

The

The legs are long and very robust; the claws are large, black, sharp, and crooked; when fully exerted, they make a very formidable appearance: the feet are feathered down to the extremities of the toes, and the covering plumes are of a mixed whitish and reddish-brown colour.

The tail is longer than the wings, and is variegated with several dusky brown fasciæ; the wings are not fasciated, but they are variegated with black spots, and there are some black, transverse, undulated lines, and some larger, longitudinal, black ones also on the body: the iris of the eye is of a fine orange yellow, and there is no cera or membrane at the base of the beak, but the whiskers already described supply it's place: the plumage that covers the legs is rather like the down of some quadrupeds, than of the regular figure of feathers.

This is a native of the Northern Countries. It is frequent in Norway, Denmark, and Sweden, and has been seen wild in England about the Suffex coast, and near some of the high cliffs in Scotland: it generally lives in thick woods, or among rocks, and sometimes builds among ruins, though more frequently in the high rocky cliffs. It is a very bold bird, and seizes not only lesser birds, but hares, rabbits, and any kind of small quadrupeds. It preys like the other owls by night, and in the the time of breeding it's young is so voracious, as to do infinite mischief in the neighbourhood of the places where it builds. Gesner calls it Bubo primus; Bellonius, Bubo; and most of the other writers have taken the one or the other of these names.

Strix capite aurito, corpore albido.
The white, aurited Strix.

The black and white Horn-owl.

This also is an extremely beautiful bird, but it is much more fierce than the former. It is equal to a turky in size, and is of a beautiful snow-white colour, elegantly variegated with spots and lines of black: the head is large, rounded, short, and decorated in a very beautiful manner, with a pair of ears or horns, as they are usually called; they are tufts of erect feathers, having their origin from the verges of the apertures of the ears: the large feathers of the wings are of a snow-white, with a few black variegations; the tail also is white, and has a few oblong, black variegations: the variegations of black on the body are principally also oblong, some of them are transverse, and others longitudinal; and some are plane, others undulated.

The legs are very robust and long; the claws are very sharp, long, and formidable.

This species is a native only of the coldest climates: there was one of them shewn alive some years ago in London; it had been brought over by the people concerned in the Greenland fishery. Few of the Ornithologists have described it. Rudbeck calls it Noctua Scandiana marina et albo et cinereo variegata, but the spots on that I saw were black.

Strix capite aurito pennis sex.
The Strix, with the head aurited with six feathers.

The lesser Horn-owl.

This is a very beautiful bird, but vastly inferior to either of the preceding in size: it is equal to a large pigeon, and it's weight is about three quarters of a pound: the wings are very large, and, when fully extended, measure, from tip to tip, more than three times the length of the body: the beak is black, moderately large and hooked; the tongue is thick, and somewhat bifid; the head is short and rounded, and the whole face has an agreeable, though a very fierce aspect: the eyes are large, and their iris is of a very beautiful yellow: the ears are long and beautiful; there runs a circle or ring round the face, marking it's outline; this consists of a double series of feathers: the exterior consists of feathers which have very elegant variegations, in little lines of yellowish, brownish, blackish, and white; the interior is almost entirely of a reddish-brown, but, where they meet, they are both black at the edges.

The feathers on the lower part of the belly are of a beautiful reddish-brown, and a plumage of the same colour is continued down the legs, quite to the toes: those on the breast and the middle of the throat are black, but they have variegations of white, and of a reddish-brown about their edges: those on the under part of the wings are reddish, and toward the bases of the primary remiges or wing-feathers there are black spots, one to each feather; these are oblong and large: there is also another very large black ſpot juſt at the flexure of the wing: the reſt of the feathers which cover the wings are very beautifully variegated with black, reddiſh-brown, and white.

The back feathers are like thoſe of the wings; they are black in the middle, and yellow and white toward the edges, and they are laid with an uncommon regularity and order over one another.

The horns or ears conſiſt each of only ſix feathers; they are very elegant and moderately long; the middle of each feather is black, but the edges are of a reddiſh-brown, and ſometimes whitiſh.

The tail is long, and is variegated with ſix or ſeven black, narrow, and tranſverſe areolæ; the intermediate ſpaces are greyiſh on the upper ſide of the tail, and yellow underneath: the principal or long feathers of the wings are of the ſame colours with thoſe of the tail, but in ſome of the exterior ones there is a broad reddiſh-brown faſcia toward the baſe; and, on the largeſt of theſe feathers, the black areolæ are alſo conſiderably bigger than thoſe on the others.

The legs are very robuſt and moderately long; the claws are large, black, and ſharp, and that of the middle toe is, on the interior part, compreſſed into a kind of edge.

On opening the body, the gall-bladder is found to be remarkably large; the appendages of the inteſtines alſo are very large and tumid; and the ſtomach large, and uſually filled with the bones of mice, ſometimes with the cruſtaceous remains of large beetles, and other inſects.

It is a native of moſt of the warmer parts of Europe; Italy, and the South of France, abound with it. It was long ſuppoſed not to be a native of our own country, but it has been diſcovered wild, and breeding with us, but that very rarely. It is ſo like the great horn-owl in every thing but ſize, that many of the earlier writers on theſe ſubjects were in debate, whether it were properly a diſtinct ſpecies, or only the young of that. All the writers on this part of natural hiſtory have deſcribed it. Bellonius call it Aſio and Otus; Willughby, Otus ſive Noctua aurita; Ray, Otus ſive Noctua aurita, Aſio Latinis. It ſometimes inhabits ruins and deſerted buildings, but more frequently the hollow trunks of trees in large foreſts, and flies abroad in the evening.

Strix capite lævi, corpore ferrugineo oculorum iridibus atris,
remigibus prioribus ſerratis.
The brown Strix, with a ſmooth head, with black eyes, and
the primary wing-feathers ſerrated.

The grey Owl.

This is a large and beautiful bird; it is equal in ſize to a well-grown pullet: the general colour is grey, with a ſlight tinge of a ferrugineous brown in it: the head, the back, and the wings are of this colour; the tail alſo is of the ſame colour, only paler than the reſt, and all theſe parts are variegated with ſpots of white and of black; the breaſt is of a paler grey than any other part, and the belly is white, but variegated with a number of elegant black ſpots: the head is remarkably large, thick, and full of feathers; it is of a rounded figure, and has no ears or horns: the eyes are large and beautiful; their iris is of a deep black, and they are ſurrounded in a beautiful manner, with each a circle of ſhort white feathers; and within theſe, at the verge of the eye-lids, there is a circle of red: the beak is crooked, ſtrong, and of a greeniſh colour; the noſtrils are very wide: the wings are very large; they are, when cloſed, ſo long, as to reach to the extremity of the tail: the legs are moderately long, and very robuſt; they are feathered all the way down to the very ends of the toes; the claws are large, ſtrong, very ſharp and grey.

Behind the ears, in this ſpecies, there is a ſegment of a circle formed of extremely odd feathers; and, toward the ſides of the back, the white ſpots are large and frequent. The three firſt of the remiges or wing-feathers are ſerrated at the exterior edge, and are gradually ſhorter, and all of them are of a pale grey colour, with a number of broad, brown faſciæ: the tail-feathers are greyiſh, and have ſive brown faſciæ, excepting only the two intermediate ones; and the feathering of the legs is white, with black ſpots.

This ſpecies is frequent in many parts of Europe; it principally inhabits thick woods, and lives in the trunks of hollow trees. Geſner calls it Ulula; Aldrovand alſo calls it, ſimply, by the ſame name; he has given a tolerably accurate deſcription of it, from one which he kept in his houſe alive: Willughby calls it Ulula Aldrovandi; Ray, Ulula Latinis; and in another place, Noctua. The Engliſh names of Owl and Howlet, and the French of Hulot, ſeem to be derived from this Ulula; but the bird, commonly expreſſed by thoſe names, is not this ſpecies, but the other, deſcribed by authors under the name of Aluco. This is a bold feeder; it will ſeize upon almoſt any kind of poultry, and feeds alſo on the leſſer quadrupeds.

The yellow Owl.

Strix capite levi, corpore luteo.
The yellow Strix, with a ſmooth head.

This is a moderately large ſpecies, it is equal in ſize to a well-grown chicken: the head is large, thick, rounded, and elegantly covered with feathers, but there are no erect ones to form what are called ears or horns: there are circles of ſhort and elegant feathers about the eyes and ſegments of circles, formed of feathers of the ſame kind behind the ears: the back is of an elegant yellow; the breaſt is of a paler yellow; and the belly is whitiſh: the legs are very robuſt and ſtrong, and the claws are large, long, and ſharp, and are of a blaiſh-grey colour.

This is a native of ſome of the northern parts of Europe, but it is not common; we have it not in England. It lives in thick woods, hiding itſelf by day in the trunks of old trees; it comes out in the evening, and feeds on mice, ſmall birds, and almoſt any living thing ſmaller than itſelf that it can lay it's claws on. None of the authors who have written on birds have deſcribed it, except the Swediſh. Rudbeckius calls it Strix tota flammea; and Linnæus, Strix capite levi, corpore luteo. They meet with it in Sweden occaſionally, but it is far from frequent there; it is oftener ſeen in the foreſts in Norway, and is met with alſo in Denmark; but it is every-where a very ſhy bird, ſo that it is rarely that it comes into any body's hands, even in the places where it is moſt frequent.

The hazel-eyed Owl.

Strix capite levi, corpore fuſco, oculorum iridibus fulvis.
The brown Strix, with a ſmooth head, and the eyes brown.

This is of the ſize of a pigeon: the head is large, rounded, and elegantly ornamented with feathers, but they are not elevated in any particular part, nor is there the leaſt approach toward an appearance of ears or horns: there is a circle of ſoft, regularly diſpoſed, and pale brown feathers round each eye, and a ſegment of a circle formed of the ſame kind of feathers about each ear, but theſe lie flat: the general colour of the feathers of the head is a duſky brown, but the ſhort and ſoft ones are paler than the others: the back and ſides are of a deep duſky brown; the breaſt is of a paler brown, and the belly is much paler than this, and has an admixture of white: the eyes are large, and their iris is of a duſky brown, with an admixture of a reddiſh-yellow, or orange colour: the beak is broad, ſhort, thick, hooked, and ſharp; the legs are ſtrong and thick; the claws are long and robuſt; they are of a bluiſh-black, and are extremely ſharp at the points.

This is a native of the northern parts of Europe; we have it in England, but not common. I ſhot one, about four years ago, in Charleton foreſt in Suſſex. It was the only one I have ſeen in England, nor do any of the writers on birds mention the ſpecies, except the Swediſh ones, Olaus, Rudbeckius, and Linnæus; the former of theſe call it Noctua major oculorum iridibus croceis; the latter, Strix oculorum iridibus fulvis, corpore fuſco.

The yellow-eyed Owl.

Strix capite levi, corpore fuſco, oculorum iridibus luteis.
The brown, ſmooth-headed Strix, with the iris of the eye yellow.

This is a larger bird than the former, but it reſembles it in many particulars: it is of the ſize of a raven: it's head is very large and full of feathers; the eyes are beautifully ſurrounded with each a circle of ſoft and regularly diſpoſed plumes: the whole head alſo has a verge of the ſame kind of feathers, diſtinguiſhing it from the body, and there is a ſegment of a circle of the ſame at the ears: the body is of a greyiſh-brown colour, elegantly variegated with large, round ſpots of white; the circle ſurrounding

rounding the head is dusky and blackish, and the head itself is somewhat paler than the body, otherwise it is of the same kind of colour: the beak is broad, short, hooked, and of a whitish colour; the eyes are very large and beautiful, they are of a pale yellow: the breast is grey, with but little tincture of the brown; the belly is still paler: the legs are long, but not so thick as in the former; and the claws are long, sharp, and formidable.

This is a native of Sweden, Denmark, and Norway; it is sometimes seen in England, but has not been known to breed here. It lives principally in ruined buildings; with us, it is chiefly seen about the high cliffs on some of our sea-coasts, so that it is probable it only comes to us occasionally. None of the naturalists have mentioned it, except Rudbeck and Linnæus; the former of these call it Noctua major oculorum iridibus pallido-luteis; the other, Strix capite levi, oculis flavis.

The variegated-tailed Owl.

Strix capite levi, corpore supra fusco, albo maculato, rectricibus fasciis albidis.

The smooth-headed, brown Strix, spotted with white, and with white fasciæ on the tail.

This is a very beautiful species; it grows to the size of a common crow: the head is large, short, rounded, and elegantly feathered: the back and sides are of a dusky brown, variegated with moderately large round spots of white; the breast and belly are of a fine silvery grey, approaching to white, and are variegated with short and somewhat undulated lines of brown; every feather has five or six of these fasciæ: the wings are brown, and have only a few spots of white on them; the long wing-feathers are brown also, and have only a few, small, white spots about their edges: the tail is very remarkable; it is near twelve inches in length, and is very narrow, in proportion: it's colour is brown, and it has from nine to twelve white fasciæ on it: the head is grey, and is variegated with a great number of spots and undulated lines: the regions of the temples is of a greyish-white; both the canthi of the eyes are black, and from the ears there runs on each side an oblong black spot to the base of the beak; this becomes narrower by degrees, and there is an oblong, grey line, variegated with black, which unites with it at the end: the throat is ornamented also with a black spot; the legs are very robust and strong; they are feathered down to the feet, and the claws are remarkably large, long, and sharp, and are of a bluish-black colour.

This has been sometimes killed in England, but very rarely; I was favoured with a specimen, shot not far from Scarborough, last summer: this is the only one I have seen. Linnæus mentions it as a native of Sweden, and is in some doubt about it's being a distinct species, but to me it appears evidently so.

The little Owl.

Strix capite levi, corpore fusco, remigibus albis maculis quinque ordinum.

The brown Strix, with a smooth head, and five series of spots on the wing-feathers.

This is an extremely pretty little bird, and is so small, indeed, that it appears very singular to see the marks of this genus on it. It is about the bigness of a black-bird: it's wings are long, and, when extended, measure at least double the length of the whole bird: the beak is short, broad, hooked, and white; the tongue is a little bifid; the lower part of the palate is black; the head is large, round, and covered thick with feathers, and has all the marks and characters of that of the owl-kind: it is separated from the body by a circle of peculiar feathers, and this is broad toward the lower part, but behind the ears it becomes so narrow, as to be scarce distinguishable.

The back and sides are of a chesnut brown, with some admixture of a ferruginous red, and are variegated with oblong, white, transverse spots, and there are five or six distinct transverse lines on the tail: the tail itself is not very long; it consists of twelve feathers, all of the same length: the circles of feathers about the ears, and on the other parts of the face, are variegated with white and brown; the breast is variegated with oblong spots of a dusky brown, but the belly and the throat are of a bright and beautiful white.

The

The remiges or long, wing-feathers are very beautifully variegated at their edges, with five ſeries of white round ſpots.

The ears of this ſpecies are large; the eyes, on the contrary, are ſmall, but they are extremely bright and beautiful, more ſo indeed than thoſe of almoſt any other ſpecies: the iris is yellow; the legs are ſhort and robuſt: they are feathered almoſt to the ends of the toes, only that there are two or three naked annules: the lower part of the foot is yellow; the claws are black and ſharp; the head is elegantly ſpotted with numerous white dots, and the baſe of the beak is ornamented with a kind black whiſkers.

This elegant ſpecies is frequent in Germany, and in many other parts of Europe, but we have it not in England: moſt of the naturaliſts, who have written on birds, have deſcribed it. Geſner and Aldrovand call it, ſimply, Noctua; Ray, Noctua minima; Willughby, Noctua minor et minima. It lives in woods, in the hollow trunks of trees, and feeds on mice, ſmall birds, and every living thing that is ſmaller or weaker than itſelf.

Strix capite levi, corpore albido.
The white Strix, with a ſmooth head.

The great, White Owl.

This is a very large and a very ſingular ſpecies: it is nearly equal to a gooſe in bigneſs, and it's colour is a beautiful bright white all over: the head is very large and rounded; it is ſeparated from the body by an elegant and regular circle of feathers, ſofter than the reſt, and beautifully arranged; and there are two other ſuch circles about the eyes, and parts of ſuch as the ears: the reſt of the head is of the ſame white with the body, but it is oddly and irregularly ſprinkled over, as it were, with little ſpots of a blackiſh colour: the tail alſo is white, and is ſpotted with the ſame manner with black: the breaſt and belly are of the ſame white with the back and ſides, but they are variegated with ſome tranſverſe, undulated lines of a greyiſh-brown, and the covering feathers of the wings are ornamented alſo in the ſame manner.

The long feathers of the wings, when expanded, are found to be of a ſnow-white, but variegated with a double ſeries of ſmall, black ſpots, and ſometimes there is a third ſeries of them.

The legs are very robuſt, and the claws long and terrible; they are of a bluiſh-colour, and very ſharp.

This ſpecies is a native of many of the northern parts of Europe, but it has not been ſeen in England. It lives in foreſts, and is a very bold and voracious bird, feeding on hares, rabbets, and other of the ſmaller quadrupeds, and ſometimes attacking lambs, when very young. The authors who have written on birds have been in general unacquainted with it. Rudbeckius and Linnæus have both deſcribed it, however, as a native of Sweden. Rudbeck calls it, Noctua Scandiana maxima ex albo et cinereo variegata.

Strix capite levi, corpore ferrugineo, remige tertia longiore.
The brown, ſmooth-headed Strix, with the third of the wing-feathers longeſt.

The common brown Owl, or Jay-owl.

This is the moſt frequent with us of all the owl-kind: it is of the ſize of a pigeon, but the body is ſhorter, in proportion to it's thickneſs; it's weight is about three quarters of a pound: it's wings are large and ſo long, that, when extended, they meaſure two feet and a half from tip to tip: the beak is large, broad, and hooked, and is of a bluiſh or horniſh colour: the opening of the mouth is extremely large, the fiſſure reaching beyond the immediate root of the beak: the tongue is not large, but it is ſomewhat biſid at the extremity, and on the palate there is a cavity fit to receive the tongue: the eyes are vaſtly large and protuberant; the eye-lids have a broad, reddiſh edge: the apertures of the ears are alſo very large, but they are covered with a membrane: the face is marked by a circle of feathers peculiarly arranged; this circle conſiſts of two ſeries of feathers; the exterior one is more rigid, and is variegated with black, white, and brown; the inner is compoſed of ſofter feathers, and they are only variegated with white and flame-colour: the ears are ſurrounded by this circle, or are within it's verge, and indeed furnish feathers to it: the eyes stand nearer to the ears in this bird, than in any other known animal: between the nostrils and the eyes there are a number of rigid setæ, forming a kind of beard or whiskers.

The back and sides of this species are of a mixed, ferrugineous, and black colour; every feather appearing black in the middle, and ferrugineous at the edges; but if any one feather be separated, and attentively examined, it will be found elegantly variegated with transverse lines of grey and brown: the belly is of the same general colour with the back, but there is a mixture of white that gives it a paler hue; the bottom parts of all the feathers are black, and all the feathers are longer, larger, and more downy than in almost any other bird. All the owls have this singularity in their plumage, and it is owing to this, that they appear much larger-bodied birds than they really are; but this has it more than any other species of the whole genus.

The legs are moderately long and very robust, and they are covered down to the very toes with a thick plumage of a downy structure; this is of a whitish colour, variegated with dusky, undulated lines, and interrupted by two or three naked annules.

The great feathers of the wings are twenty-four in each, and both these and the tail-feathers have each six or seven areolæ of a dusky white, variegated with brown, and a dusky, ferrugineous tinge; and the covering feathers of the wings, and those of the upper part of the back, which are longer than the others, are variegated with spots of white: the tail consists of twelve feathers; the middle ones are longest, and the others gradually decrease to the verge, but all of them are sharp-pointed, which is very different from their form in the common screech-owl.

The under part of the foot is hard, callous, and of a pale whitish-brown, with a tinge of yellow; the claws are long, sharp, and blackish; the intestines are very long, and make a great number of circumvolutions: the liver is divided into two lobes; the gall-bladder is large: the testicles are large and black, and the stomach is remarkably robust and thick.

This is one of the most frequent of all the owl-kind; it is common to all parts of Europe, and to most other places: all the Ornithologists have described it. Willughby, Ray, Aldrovand, and the generality of other writers call it, simply, Strix; we, the common Owl, the brown Owl, and the Joy-owl. It is this species which utters that cheerful and agreeable hooting, which we hear on evenings: this seems to be a note of exultation and joy in the creature, and is erroneously, though very poetically, described by the author of a late anonymous poem *, one of the best in the English language:

> *Save that, from yonder ivy-mantled tow'r*
> *The moping Owl does to the moon complain,*
> *Of those who, wand'ring near her secret bow'r,*
> *Molest her antient, solitary reign.*

It is pity so elegant a description should not be just; but it is certain, that the bird has no title to the epithet moping, while she thus lengthens out her even song; nor is it complaint, or the effect of molestation.

Strix capite pinna singulari aurita.
The Strix, with the head aurited by a single feather. **The Scops.**

This is an extreamly elegant little species; it is of the bigness of a fieldfare, but has all the characters of the owl-kind in the strongest manner about it: the head is large, short, rounded, and very thickly covered with feathers; they are short, but very downy, and are of a dusky blackish-grey, or what we properly express by the term lead colour: the ears or horns, as they are called, are short, but very erect; they consist each only of a single feather, but that is very well plumed, and makes an extreamly pretty appearance: the back and sides, and the covering feathers of the wings, are of a colour approaching to that of the head, but not exactly the same; they are paler, and are of a simple grey, without that admixture of the blue tinge, which gives the other what we call the lead colour. This, however, is not the sole simple colour,

* An elegy written in a country church-yard.

every feather has on it a number of elegant, round, white ſpots, and the whole back of the bird is by theſe variegated in an extremely beautiful manner: the larger feathers of the wings and tail are of the ſame grey with thoſe of the back, only paler; and they are variegated with almoſt innumerable little white ſpots, diſpoſed in ſeveral tranſverſe ſeries, and there is beſide a long black line on each: the ſhoulders and upper part of the wings, in the male birds, have a beautiful caſt of reddiſh-brown in the grey, which gives the white a yet greater advantage: on the belly the feathers are paler than any where elſe, and toward it's lower part they are, indeed, whitiſh; but in theſe, as in all the others, the roots are black.

The eyes of this ſpecies are very large, and remarkably bright; their iris is of a flame colour: the beak is ſhort, black, and crooked; the legs are not very robuſt, but they are feathered down to the toes, which makes them look thick; the plumage upon them is grey, with a tinge of reddiſh-brown: the toes are ſmall and ſlender, they are of a greyiſh or lead colour; the feathering does not reach over them, but they are ſquammoſe.

This is frequent in Italy, but is not ſo common in any other part of Europe: it lives in woods, and ſometimes about old buildings, and comes abroad only in the night, then making a loud beating noiſe. It feeds on fi:ld-mice and inſects; moſt of the naturaliſts have deſcribed it. Aldrovand, Ray, Willughby, and others call it Scops; others, Scops Aldrovandi; the Italians call it Chiovino.

Strix capite aurito, albida, cauda longiore.
The Horn-owl, with a long tail.

The German Horn-owl.

This is of the ſize of a jack-daw, and of a beautiful ſilvery grey colour, approaching to white: the head is round, large, and very full of feathers: the ears or horns are ſhort, but very erect, and compoſed each of about ſix feathers: the beak is very broad at the baſe, but it is ſhort, crooked, and of a pale bluiſh colour: the eyes are very large and bright; their iris is of a fine colour, and the pupil black: the head is of a ſomewhat greyer colour than the body, and both are variegated in a very elegant manner with minute ſpots of a ſnow-white; there are many of theſe ſpots on every feather, but they are not diſtinguiſhable, unleſs on a cloſe inſpection.

The wing-feathers, and thoſe of the tail, are of a very beautiful pearly grey; thoſe of the tail are remarkably long; they are twelve in number, and are all of the ſame length, and obtuſe at the ends: the legs are ſhort, thick, and plumoſe down to the toes; the plumage upon them is fine and downy, and is of the ſame pearly grey with the wing-feathers, but it has a little tinge of brown in it.

The bottoms of the feet are of a lead colour, and the claws, which are leſs formidable than in many other ſpecies, are black.

This ſpecies is frequent in the South of France; I have had three ſeveral times ſpecimens of it from thence, yet it has been unknown to the generality of the writers on this ſubject. Aldrovand is the only one who ſeems but to have named it; and he barely mentions it, at the end of his deſcription of the Scops. He ſays, it is a native of Germany, and is like the Scops, but of a whitiſh colour, and that it has longer ears, and a much longer tail.

It lives in foreſts, in the ſtumps of decayed trees, and comes abroad only in evenings. It feeds on field-mice and inſects.

Strix dorſo variegato, capite lævi.
The variegated-backed, ſmooth-headed Strix.

The Church-owl, or leſſer Barn-owl.

This is a very beautiful bird, and, if it were leſs common among us, would not fail to be admired extremely; it is of the ſize of a pigeon: the head is large, ſhort, and rounded, and is ſurrounded by a circle of white feathers like a ruff, ſupported behind by another circle of ſhorter, but more rigid, feathers, of a duſky yellow; this takes it's origin from behind the eyes, and runs down on each ſide to the throat, and has much the appearance of a woman's head-dreſs ſurrounding the face: the back part of the head is of a pale brown, ſprinkled over with little whitiſh ſpots: the eyes are large, but they ſeem, on account of this circle of feathers, to be ſunk deep into the head: the noſtrils are oblong; the apertures of the ears are broad and patulous, but there

z   are

are no erect feathers ariſing from them; they are covered, however, by a kind of valve, which has it's origin from the upper and interior part, near the eyes.

The breaſt and belly of this bird are covered with elegantly variegated feathers; they are white, and the ſpots which form the variegations are brown, and of a quadrangular figure: but 'tis on the back that the beautiful variegations ſhew themſelves; the feathers which cover this part of the bird, are beyond that of any other of this genus in beauty; they are each of the ſame yellowiſh-brown colour at the baſe, but toward the tip they are variegated with undulated lines of black and white, which are diſpoſed in a regular and elegant manner, and form, by their aſſemblage, an elegant grey or hoary colour, when viewed at a little diſtance; and along the body of the feather there run alſo lines formed of alternate black and white ſpots, ſome of them placed ſingly, ſome in orders of four, three, or two together.

The long feathers of the wings are twenty-four in number, and the largeſt of theſe are variegated with brown areolæ, uſually four on each feather, and the ſmaller have uſually three: the intermediate ſpaces are of a yellowiſh-brown, and are ſprinkled over with a kind of irroration, as it were, of minute black ſpots: the exterior feathers are terminated at the ſides by disjunct pinnules, ſo that they appear pectinated, or deeply denticulated: the w..k wings are ſo long, that, when cloſed, they reach a little beyond the extremity of the tail.

The tail is compoſed of twelve fe..ers, all very nearly of the ſame length, and in colour the ſame with thoſe of the ..ngs; the interior edges of all theſe, as well as of thoſe of the wings, are white: the legs are robuſt and long; they are covered to the toes with a thick plumage of a whitiſh colour, but the toes themſelves have only a kind of hairineſs in the place of this plumage: the claws are large and formidable, and that of the middle toe, in particular, is ſerrated on the interior ſide, in the manner of that of the heron, and ſome other birds of that kind, but not ſo viſibly: the ſerrature, however, in the full-grown bird, may always be diſtinguiſhed.

This ſpecies is extreamly common with us, and in almoſt all other parts of Europe; nor indeed is it peculiar to this, or, as it ſhould ſeem, to any part of the world. I have received a ſpecimen of it, as a great curioſity, from Sardinia; and Marcgrave figures and deſcribes it as a native of ſo remote a part of the world as the Braſils, under the name of Tuidara. Aldrovand, and many of the other writers on birds, have called it Aluco minor; ſome, Ulula variegata. We call it the Church-owl, and the leſſer Barn-owl; it hoots in the ſame manner as the common joy-owl. It is frequent among ruined buildings, and in barns and other places where there is food for mice and the other creatures on which it preys: it is ſeldom ſeen abroad, but often heard through a great part of the evening.

Strix capite lævi, dorſo plumbeo, roſtro albeſcente.
The ſmooth-headed Strix, with a bluiſh-grey back, and a whitiſh beak.

The White-beaked Owl.

This is a very beautiful ſpecies; it is ſomewhat larger than our common joy-owl, and it's head ſeems, in a remarkable manner, too big for the body: it is round, and is incircled by a very beautiful ſeries of pale-coloured feathers, approaching to whitiſh, and behind theſe there run a ſeries of ſhorter black ones: the back part of the head is of a deep and duſky bluiſh-grey, and is variegated with a few ſpots of white: the the eyes are very large, and altogether black; they ſeem all iris, and that of this deep colour; and they make a very ſingular figure, as they appear ſunk in the head, by means of the height of the circle of ſurrounding feathers: the apertures of the ears are wide, but they have no erect ſeries of feathers riſing from them: the beak is broad, crooked, and whitiſh.

The back is of a deep duſky lead colour, with a great many ſpots of a bright white, which form very elegant variegations; and the breaſt and belly are white, and have many ſpots of black on them; theſe are not large, nor diſpoſed with any ſort of regularity.

The legs are robuſt and long; they are covered with a white and tolerably thick plumage down to the toes, and the toes themſelves, with a number of ſcattered hairs of the ſame colour, reſembling a kind of rarer plumage: the tail is long, and the wings are ſo long, that, when cloſed, they reach to the end of it.

This

This is a native of the warmer parts of Europe: it is very frequent in Spain and Portugal; it is also frequently met with in Italy, and in the South of France, but we have it not with us. Aldrovand calls it, Aluco prior; and Willoghby and others have borrowed the same name.

Strix corpore fusco albis maculis variegato, rostro flavescente.
The yellow-beaked Strix, with a brown body, variegated with white.

The yellow-beaked, American Owl.

This is a very small, but an extreamly singular, species: the head is large, round, and incircled with a border of beautiful feathers: the whole bird is not much larger than a thrush. The head is almost equal to the body in size, when the creature erects the feathers on it, as it does, when angry or surprized; but, without this extraordinary circumstance, it is very large, in proportion to the body: the beak is broad, short, crooked, and yellow; the foramina of the nostrils are two, and are very conspicuous, the eyes are very large, beautiful, and of a bright yellow; the apertures of the ears are patulous, and have a membrane taking it's origin from the interior upper part, and extending over them: the upper part of the beak is remarkably broad, and about it's base, as also under the eyes, there are placed a number of black, rigid hairs, making a kind of whiskers.

The back is of a dusky brown or amber colour, and is elegantly variegated with spots of white; these on the head and neck are very small, but on the back they are much larger: the breast and belly are white, and are variegated with spots of the same brown, which is the principal colour of the back.

The legs are robust and short; they are covered down to the toes with a very elegant and beautiful plumage, of a yellowish-white: the inside of the feet is yellow, and the claws are long, sharp, and black: the tail is remarkably broad.

This is a native of the Brasils, nor do we know of it in any other part of the world. It is described by Marcgrave under the name of Noctua Brasiliensis Cabure dicta; and he tells us it is very tractable, and will be learned to play tricks in the manner of a monkey.

Strix dorso flavescente maculis nigris variegato.
The yellow and black Strix.

The Jacurutu.

This is a very elegant species; it is of the size of a small pigeon: the head is very large, and of a roundish figure; the eyes are large, and seem somewhat sunk in among the feathers; the circle that surrounds the whole head, and separates it from the body, is formed of a double series of fine, soft feathers; the anterior one paler, the hinder darker-coloured, and shorter: the ears are patulous, but there are no erect feathers growing from them, like those which give an appearance of horns to some of the others: the beak is broad at the base, and is short and crooked, and of a bluish colour: the wings are long; the back is beautifully variegated; the ground colour is a bright yellow, with a tinge of orange colour, and the variegations are of a deep black, and are disposed in form of spots: the legs are robust, but not long; they are feathered down to the toes: this feathering is like a kind of fine down, and is of a pale whitish hue, as is also the belly of the bird; the breast is also pale and spotted.

This beautiful bird is a native of the Brasils; we have it sent over to us sometimes as a curiosity dried. Linnæus calls it Feliceps flavescens maculis nigris, a name first given to it by Barrilierus; and Marcgrave, in his account of the Brasilian birds, calls it Jacurutu.

These are all the known species of owls properly so called; authors have, indeed, included some very different birds, under the same name. The Caprimulgus, or Churn-owl, of authors, though almost universally ranked among these, is of a very different genus; it is properly a swallow.

FALCO.

THE beak of the Falco is strong and uncinated, and the feet have always three toes before, and one behind, no one of the anterior ones being capable of turning backwards.

Falco cera lutea, pedibus lanatis, corpore rufo.
The ferrugineous Falco, with feathered legs, and the cera yellow. The Eagle.

This is a very large and beautiful, as well as a very terrible, bird: the size is that of a turky, and the weight not less than ten or twelve pounds: the head is large, and the beak is remarkably large, and very thick and convex toward the base, and at the point the upper part of it is three quarters of an inch longer than the under, and bends over the other; it is of a blackish colour, and very firm substance: it's base is covered with a yellow cera or membrane, in which the nostrils are placed in a transverse direction; the edges of this upper part of the beak are sharp, and the tip is blacker than the rest, which is often rather bluish than quite black: the opening of the mouth is very wide; the tongue is in shape like that of the human species; it is rounded at the anterior extremity, and has two appendages of a hooked figure on each side at the base: the palate is perforated; the eyes are very large, and the part of the head, immediately over them, is prominent, so that they seem sunk in a kind of hollow; they are very bright and piercing: the iris is greenish, with a cast of fire colour in it; the pupil is black: nature has taken surprizing care to defend them from injuries, for, besides the nictitating membrane which serves other birds in the place of our eye-lids, this has four eye-lids, two above and two below, which it closes over the eyes at pleasure.

The feathers of the neck are rigid, and of a chesnut brown: the wings are very large, and extend to a surprizing breadth; they are of a brown colour, as is also the tail: the whole body is of a dusky, ferrugineous brown, with a few spots of white on the back, and more on the belly: the large feathers of the wings, though extreamly long, have that part, which we call the barrel, shorter than in the goose-quill; but it is very firm and elastic, and they make the finest pens in the world for writing: the feathers which cover the body are not large, and toward the lower part of the back they are of a darker colour than in any other part, and have fewer spots.

The legs are robust, and are feathered down to the toes; the plumage on them is brown: the feet are yellow, and the claws are very long, sharp, and terrible, and of a bluish-black; they are equal to the toes themselves in length, and are very thick at the base: the anterior part of the body is of a brighter, the posterior of a duskier, brown, and the tail is nearly of the colour of the anterior part of the body, and has several white spots on it.

This is frequent in many parts of Europe; the forests in Germany abound with it, and it is met with in some plenty in Denmark, Norway, and Sweden. I have shot one in Charleton forest in Sussex, and once found a nest there, built in a strange wild manner, and with four eggs in it. It frequently builds in the high rocky cliffs about the Scotch shores, and in some of the rocky precipices in Ireland.

It is a very capacious and bold bird; it will seize on lambs, and, in the time of it's having young, scarce any thing is safe for it. All the writers on birds have described it. Gesner calls it Aquila Germana; Jonston, Aquila Gesneri sive Chrysaetos aquila; Willughby, Chrysaetos; Ray calls it also by the same name. The Swedes call it Ornand; we, the Eagle; and some of our writers, the golden Eagle.

Falco corpore castaneo, cera cærulescente, cruribus plumosis.
The chesnut-coloured Falco, with feathered legs, and a bluish cera. The brown Eagle.

This is a very bold and fierce bird: the head is large, and the feathers which cover that and the neck are rigid and long, but narrow: the beak is short, but very

 robust;

robuſt; the baſe is convex, and the point of the upper part is very hooked and ſharp, and hangs a good way over the lower; it is of a bluiſh-black colour, and ſpotted with a darker hue; there is a hollow or channel along the top of the under part, which receives on each ſide the ridge of the upper: the extremity is very ſharp, and the baſe is covered with a cera, or membrane of a variegated blue or whitiſh colour, and in this the noſtrils are viſible, opening in a tranſverſe direction: the cera runs up very high in this ſpecies, and the face is bald almoſt up to the eyes: the opening of the mouth is wide, and the palate and tongue are fleſh-coloured; the ſize of the bird is that of a large dunghil cock; the colour is a duſky cheſnut-brown, with a few whitiſh ſpots, and in ſome with ſcarce any; the tail is long, and is beautifully variegated with a tranſverſe, annular mark of white: the legs are robuſt and yellow; they are feathered down below the joint, but not quite to the toes: the feet are yellow, and the claws are remarkably large and ſtrong.

This ſpecies is frequent in ſome of the large foreſts in Germany; we have had one of them lately ſhewn alive in London. Willughby and Ray call it Chryſaetos cauda annulo albo cincta.

Falco pedibus ceraque cæruleis, corpore ſupra fuſco, capite albo.

The Falco, with the cera and the legs bluiſh, the head white, and the body brown.

The bald Buzzard.

This is a bird of that ſize and figure, that Aldrovand, and many of the other writers on birds, have ranked it among the eagles. It is of the ſize of a large cock; it's weight is between three and four pounds: the head is large and white, whence it has obtained the name of bald, as at a diſtance it appears as if there were no feathers on it; but it is as thick ſet as any other part with them, only that this ſingularity of colour gives ſo odd an appearance: the beak is large and ſtrong, and is very crooked; the colour is a deep bluiſh-black, and the extremity is hooked, and very ſharp: the baſe of it is covered with a membrane which is of a bluiſh colour, and in which the noſtrils are viſible; they are oblong and oblique: the angle of the lower jaw is rounded; the tongue is broad, ſoft, and rounded at the extremity: the eyes are large and very bright; the iris is yellow, and the pupil black; there are upper as well as lower eye-lids to the eyes, but the lower are much the larger: the eyes are not ſunk in under a kind of prominent eye-brows, as in the eagle, but they are prominent.

The back and wings are of a deep duſky colour, between ferruginous and black; the breaſt and belly are white: the legs are long and robuſt, and are covered with ſhort and ſoft feathers: the feathers of the wings are very large, and they are pointed at the extremities; the largeſt of them are the darkeſt-coloured, and the interior edges of them all are variegated with white and brown: the feathers under the ſcapulæ are white, but they have ſpots of a ferruginous black toward their tops. The third and fourth order of thoſe which cover the roots of the large wing-feathers underneath are beautifully variegated with ſpots of a ferruginous colour, and are brown at the edges; over theſe there ſtand other ſmall white ones, and over theſe larger brown ones: the tail conſiſts of twelve feathers, nearly equal in ſize, and variegated at the edges with white and brown.

The legs are long and robuſt; the feet are of a bluiſh colour, and very ſtrong; the toes are thick, and the middle one is longeſt; the outer one of the three anterior is a little longer than the inner, and the hinder toe is much ſmaller than any of the others; the claws are black; they are very large, of a ſemicircular figure, and ſharp: the feet are ſquammoſe or ſcaly, and their under part is ſcabrous, or furniſhed with little punctules, by means of which it is enabled to hold any thing the more firmly.

The heart is large; the liver alſo is large, and the gall-bladder proportioned to it: the ſpleen is round, and of a blackiſh colour, and the ſtomach is large and robuſt: the inteſtines are long and ſlender, and are very much convoluted.

It is a native of England, and of moſt parts of Europe; it lives principally in damp places, among the reeds and ſedge by the ſides of rivers and large ponds, and by the ſea; it builds on the ground, among the tall graſſes and flags, and lays three or four large eggs of an elliptic figure, and altogether white in colour. It feeds on fiſh: this is a very ſingular circumſtance, as it ſeems by no means qualified by nature for catch-

ing

ing such prey, having neither webbed feet, nor a long neck or legs; that it does catch them, however, is evident, from it's stomach being always found loaded with the bones of them. This I found in one, which I killed, last summer, in an osier bed near Peterborough; and all the authors who have described it, mention the same circumstance.

Gesner calls it Cyanopoda and Falco cui pedes cærulei. Aldrovand ranks it, as before observed, among the eagles, and describes it, twice over, under the names of Haliætus and Morphnos. Willughby calls it Balbusardus, a strange name formed from the English one; and Ray calls it Balbusardus Anglorum.

Falco cera flava, rectricibus albis, versus apices nigris.
The Falco, with a yellow cera, and with the tail-feathers white, and black at the end.

The Pygargus.

This is a very large bird; it at least equals a peacock in size, and it's weight is about eight or nine pounds: it's wings are very large, and, when fully extended, measure, from tip to tip, very near three times the length of the whole body: the head is not very large, in proportion to the body: the beak is short, but it is very robust and hooked; and the bent part of the upper mandible reaches a finger's breadth beyond the lower; the whole beak is of a deep yellowish colour, and this extream part is nearly black: the cera, or membrane covering the base of the beak, is yellow, and the nostrils are very conspicuous in it; they are oblique, and are near half an inch long: the tongue is broad, fleshy, and black toward the extremity, and the palate is hollowed to receive it: the eyes are very large, but they do not stand prominent; they are rather a little sunk in, under the swelling of the part of the head above them: their iris is hazel, or in some of a reddish yellow, or deep orange colour: the legs are very long and robust, and the feet are strong, and have a callous tubercle on the under part; they are yellow; the claws are black, and are very large; the hinder one is not less than an inch in length, and is very sharp.

The head is white; the feathers are long and narrow, and their scapi are blackish: in the space between the eyes and the nostrils there are no feathers, but there are in their place a kind of setæ, with the bases lanuginous: the beginning of the neck is of a somewhat reddish-brown, and the neck is all the way down covered with oblong and narrow feathers: the whole body is covered with feathers of a dusky, ferrugineous colour, but the rump is black, and the tail is in part white: it consists of twelve feathers, and they are all of them white from the top to half way of their length, and black in the other half: the long wing-feathers are twenty-seven in number, and the third and fourth of them are longer than the others; the wings, when closed, do not reach quite to the extremity of the tail: the barrels of the large quills are short, but they are very firm and elastic, and make excellent writing-pens; these large feathers are all black, and the edges of some of the smaller covering feathers of the wings are grey.

The liver in this bird is very large, and the gall-bladder is in proportion: the testicles in the males are large and oblong and the intestines are slender, but very long, and convoluted; the throat is extreamly wide, but the stomach is but small: the female, in this species, is somewhat paler-coloured than the male; and the belly, in both sexes, is paler than the back.

It is frequent in Italy, and many of the warmer parts of Europe, but has not been met with in England, and but rarely in Germany. It lives in forests, and is a very bold feeder; hares, and other of the smaller quadrupeds, are it's usual food, but it will seize on almost any thing. Gesner calls it Pygargus; Ray also gives it the same name; Willughby calls it Pygargus sive Albicilla, quibusdam Hinnularia. The name Pygargus has, however, been used by some of the ornithologists in a confused manner, and we are not always to understand this species by it in their writings. Aldrovand's Pygargus is the same with our second species; the brown Eagle, and the Pygargus prior of Bellonius, is no other than the male of that species of hawk to be described hereafter, and which we call the Hen-Harrow.

Falco

Falco cera flava, cauda forcipata, corpore ferrugineo, capite albidiore.
The forked-tailed Falco, with a yellow cera, a brown body, and a whitiſh head.

The Kite.

This is a common bird with us, but it is a very ſingular one in it's form and manner of flying, and is not without it's beauty: it is about equal, in the bulk of the body, to a large tame pigeon, but the wings are immoderately long and large; they meaſure nearly three times the length of the body, when fully extended: it's weight is about two pounds.

The head is ſmall, and is of a pale, or ſomewhat whitiſh, colour; the throat is alſo of the ſame greyiſh-white, and the beak is ſhort, but broad, and the hooked part of the upper mandible turns a good way over the under: both the head and the throat have variegations of a blackiſh hue among the grey, formed of oblong lines, deſcending down the ſcapi of the feathers: the neck is of a reddiſh-brown, but the middle parts of the feathers are black; the back is of a dark brown, but the feathers near the tail are of a paler brown, nearly of the ſame colour with thoſe of the tail itſelf, but the ſcapi of theſe alſo are black in the middle: the ſmaller feathers which cover the wings are variegated with a reddiſh-brown, with black and with white; the black always occupying the middle part along the ſcapi: the long feathers on the ſhoulders have black lines on them, like thoſe of the long wing-feathers; and thoſe which cover the under part of the wings, are of a reddiſh hue, with black ſcapi: the feathers on the throat are of a greyiſh colour, and red at the edges, and black in the middle; thoſe on the breaſt and belly have leſs and leſs black on them, all the way down to the tail; and thoſe under the tail have only the ſcapi black.

The long wing-feathers are four and twenty in number; the five outer ones are black, and the ſix next to theſe are of a greyiſh-black, and then all the reſt are black, to the very hinder ones of all, which are variegated with brown, white, and reddiſh: all theſe, except the five or ſix exterior ones, are variegated with tranſverſe black lines on the outer edges, and the middle are paler: the wings, when ſhut, are of the length of the tail; they reach as far as the middle or forked part of the tail, but not to the ſides.

The tail is forked, and that very deeply; the feathers in the middle being more than two inches ſhorter than thoſe at the ſides: the colour is a brown, not very deep, and with a caſt of reddiſh in it; the exterior feathers are blackiſh, and all, indeed, except the middle ones, have black tranſverſe lines on their interior edge; the tips or extremities of all are white.

The beak is black, broad, ſhort, and hooked; the tongue is broad and thick, and there is a cavity formed in the palate for the receiving it; the inſide of the mouth is yellow, and the membrane on the top of the beak is alſo yellow: the noſtrils are very conſpicuous in this: the eyes are very large and fierce, and their iris is of a whitiſh colour, with a fine tinge of yellow.

The legs are yellow; the outer toe is connected to the middle one, almoſt half way of it's length, by a membrane; the claws are large, black, and ſharp; that of the hinder toe is much longer than any of the others; that of the middle one is formed into an edge on it's anterior part.

The mouth opens very wide in this bird, and the ſwallow is very great: the gall-bladder is large, and the inteſtine below the appendages is greatly dilated, as is the caſe alſo in ſome other birds of prey: the forked ſhape of the tail diſtinguiſhes this at ſight from all the other birds of prey, yet known.

The

The buſineſs of flying ſeems in greater perfection in this bird than in any other; it has a way of lying up on the air with it's wings expanded, and unmoved for a long time together; and, when it changes place, oftener does it by a kind of ſliding, than by the vibration of wings, practiſed by other birds. The tail is the great inſtrument of direction to this bird, while it's wings lie expanded: we often ſee this moved, while they are perfectly at reſt; and Pliny delivers it as the opinion of the antients, that the uſe of the rudder in ſhips was found out, by means of the uſe this bird was obſerved to make of it's tail.

The kite uſes it's power of flying to a very proper purpoſe; it traverſes whole ſeas, and, though uſually ſeen ſingle with us, is, in ſome places, and on certain occaſions, ſeen in flights like thoſe of ſwallows. Bellonius, from his own obſervation, mentions flights ſo large on the hills, about the ſhores of the Euxine, in April, that people were aſtoniſhed to conceive how or where it was poſſible they ſhould find food. There are many parts of Europe in which the kite is a bird of paſſage, but they breed and ſtay all the year with us.

It is a general opinion that the kite will feed only on fleſh; but Bellonius aſſerts, that he has ſeen them in Ægypt ſettle on the palm-trees, and feed voraciouſly on the dates: but, though this be a fact, there is reaſon to believe that hunger, and an abſence of all other proviſion, could only be the cauſe of it. It's moſt favourite food ſeems to be the young of domeſtic fowls, chickens, turkies, and geeſe. It has a ſurpriſingly piercing eye, and, when it ſeems diverting itſelf in the air with it's indolent wavings, it generally has it's looks about, from that height, on ſome brood of one or other of theſe fowls: it does not chuſe to fall upon them while together, or under the protection of the parent, but will be whole hours upon the watch to ſee one ſtraggling from the reſt, and, when that happens, it is down in an inſtant, and carries off the prey. The parent animals underſtand the buſineſs of it's hovering, though at ſo vaſt a diſtance in the air, and make ſignals to their young, whenever they ſee it.

All the writers on birds have mentioned this ſpecies, and all under the ſame ſimple name Milvus. The Swedes call it Glada; and we, the Kite; and, in ſome places, the Glead.

Falco pedibus flavis, corpore cinereo, maculis fuſcis,
cauda faſciis quatuor.
The yellow-legged Falco, with a grey body ſpotted with
brown, and with four faſciæ on the tail.

The gentle Falcon.

This is a very beautiful bird; it is of the ſize of a raven, or ſomewhat larger: the head is ſmall, and ſomewhat flatted on the crown; the beak is broad, ſtrong, and hooked, but not very long; the membrane on it's baſe is of a fine bright yellow, and the noſtrils are very conſpicuous in it: the eyes are large, and very bright and piercing; their iris is yellow.

The upper part of the head is of pale ſilvery grey, with a few ſmall, browniſh ſpots on it; the greater part of the body is grey, and has alſo ſeveral ſpots of brown on it; theſe are much larger than thoſe of the head, and are of an irregular figure; the tail is grey, and has four brown faſciæ on it; the under feathers of the head are black, and their edges are ferrugineous: the head, the neck, and the upper part of the wings are brown, but the tips of the feathers are ferrugineous: the breaſt and belly are of a yellowiſh hue, but there are a few oblong, brown ſpots under the throat, as alſo on the breaſt and belly; and there are ſome of the ſame kind on the thighs, only they are narrower.

The long feathers of the wings are brown on the outer ſide or edge, and white on the inner, but they have ſome whitiſh or greyiſh ſpots, and four or five brown, tranſverſe lines: the tail-feathers are all brown, both on the upper and under ſides, and they have each five or ſix broad and black faſciæ on them; the ſmaller feathers which cover the under part of the wings are of a yellowish-white, with some brown, longitudinal lines.

This species is a native of Germany, Sweden, and some other parts of Europe, and is sometimes seen with us but rarely, nor are we assured that it breeds here. This is the kind that was in principal favour, at the time when hawking was a favourite diversion. All the authors on natural history have mentioned it. Willughby calls it Falco gentilis, id est, nobilis; and he also describes it in another place, as if a different species, under the name of Falco Montanus; and Ray, and most of the writers since his time, have used the same name. It is a very bold and voracious bird, but is easily made tame and tractable.

Falco pedibus rostroque caeruleis, maculis albis nigrisque longitudinalibus.

The blue-legged Falco, with oblong, black and white spots.

The Lannar, or Lannaret.

This is a very beautiful bird; it's size is that of a common crow; it's colours are beautifully variegated, and it's whole figure elegant, and yet formed in an uncommon manner for strength: the head is small; the beak is broad at the base, not very long, hooked, and extremely sharp at the point, and of a deep blue colour: the eyes are large, and very bold and piercing: the nostrils are small and transverse.

The head and neck are pale, and elegantly variegated; the colours are black and white, but the white is in greatly the larger proportion; the variegations are not in form of spots or transverse lines, as is most usual, but in longitudinal ones: the back and the upper part of the wings are much darker than the head and neck; the colours are indeed the same, but the black is predominant, and the white lines are faint, as well as few in number.

When the wings are extended, their under part is seen variegated in a very uncommon manner with white spots; these are round, and resemble so many small pieces of money; they are scattered irregularly over the surface; the round figure of these, as the other marks on the body, are oblong, have a particularly odd appearance.

The tail is moderately long, and the feathers are spotted, and very elegantly arranged, and are obtuse at the ends.

The legs are short and blue; the claws are very sharp and black.

The male of this species is smaller than the female, and is paler-coloured, and both of them immediately, after the moulting time, have a yellowness diffused over the other colours, which might deceive an unwary eye into the opinion of the species being different; but this tinge wears off after a few days.

It is a very bold bird, when wild, but it is very easily tamed, and is usually kept in many parts of Europe for the diversion of hawking. Though smaller than some of the others used to that purpose, it will seize on any thing, the heron and cranes not excepted.

It is a native of some of the warmer parts of Europe, and is seen occasionally in many places where it is not so; it breeds in France, and lives there the whole year: all the writers on these subjects have described it. Gesner calls it Lanarius; Aldrovand, Lanarius Gallorum; Willughby, Lanarius cujus mas tercinarius. It is a perfectly distinct species from the Italian one next to be described, though too many of the writers on these subjects have confounded them together.

The Italian Lannar.

Falco dorso subcaeruleo, pectore flavescente.
The blue-backed Falco, with a yellow breast.

This is an extremely beautiful bird; the size is that of a rook, and the colours are very bright and beautiful: the head is small and flat at the top; the colour is a pale yellow, with an admixture of bluish in some parts, toward the edges of the feathers: the eyes are large and black; the beak is very robust, short, and blue; it's point is hooked and very sharp; the back is of a deep and very beautiful iron grey, with a fine tinge of blue, so that in some lights it appears wholly blue; there are some black transverse lines on it, and some few yellowish ones: the breast and belly are yellowish, and have a multitude of ferruginous, irregular spots, and the extremities of the wings are ornamented with a number of beautiful, round, white spots, in form of eyes.

The wings are long, and the tail also is long: the legs are short, and not very robust; they are blue, and quite naked; the claws are very long, sharp, and black.

This is a native of Italy, and some other of the warmer parts of Europe. Most of the writers on birds have confounded it with the French or common lannar; but some of them have described the bird, which they call by that name, in such a manner, as to shew that they have meant this species by their accounts, and not that intended by the authors from whom they have borrowed the name. It is a very bold bird, and is very troublesome about the houses of people who breed up the domestic fowls.

The Gyr-falcon.

Falco rostro nigro pedibus luteis, corpore supra fusco, subtus albo-cinereo, maculis transversis.
The yellow-legged Falco, with a black beak, and with a brown back, and a spotted breast.

This is a large and very bold bird; it is indeed the largest of all the birds of this genus: it's size is that of a common dunghil cock: the head is small, flatted on the top, and grey in colour: the beak is short, but it is extremely thick and strong, and is sharp, and very much hooked at the point: the eyes are large, and very fierce and piercing; the iris is bluish, the pupil black.

The back is of a brownish colour, though with a tinge of grey among it: the breast and belly are of a very pale grey, approaching to white, but variegated with transverse spots of brown; the feathers of the tail and wings are of a greyish-white, but they are ornamented with spots of a deep black, of a heart-like shape, and in some measure resembling the eyes in a peacock's tail: The largest of the wing-feathers are terminated by one of these black spots bigger than the others, and edged with a verge of white: the tail is long, and the wings also are very long; when expanded, they reach to a vast breadth, and, when closed, they come very nearly to the tip of the tail: the throat is of a pure and elegant white, variegated with brown spots: the tail has some transverse black marks on it; the legs are yellow; they are very thick and strong, and the claws are very sharp and strong.

The size of this bird of prey makes it a match for any thing of the winged kind, and it is bold and fierce to a surprising degree; it will seize herons, and the larger birds of all kinds.

It is a native of some of the northern parts of Europe; most of the writers on birds have described it. Rudbeck calls it Falco peregrinus albo-cinereus viridis, pedibus flavis; Willughby, Ray, and others, Gyrfalco. We call it the Gyrfalcon, or the Jerfalcon, and the male the Jerkin. The term Gyrfalco is of German origin, and is given it on account of it's size and boldness, Gyr being the German name of the vulture.

Falco

Falco cera luteo-viridi, pedibus luteis, corpore ferrugineo, vertice fulvo. **The Moor-buzzard.**

The yellow-legged Falco, with a ferrugineous body, and yellow head.

This is of the size of a common crow: the head is small, and not so flat as in many other birds of prey at the top, but narrower than in most, and more rounded: the beak is robust and moderately long, and is of a deep bluish-black colour: the membrane of it's base is of a greenish-yellow: the nostrils are very conspicuous; they are oblong, broad, and of a somewhat kidney-like shape: the inside of the mouth is partly bluish, and partly black, and the tongue is broad and thick: the rima or fissure in the palate is broad and papulous: the eyes are large, bright, and fierce; their iris is yellowish, and the pupil is black.

The upper part of the head is of a pale and whitish brown, sometimes deeper, and with a tinge of reddish, but in either case it is variegated with transverse lines of a deep black; the upper part of the throat also is of the same colour: the whole body is of a dusky, ferrugineous colour, only the breast and belly have more of a tinge of yellowish than the back; the wings are of a ferrugineous colour also, but there is, on the middle of each, a spot of a yellowish-white: the feathers, at the origin of the tail, are of a yellowish-brown.

The wings are very large, and so long, that, when closed, they reach to the extremity of the tail: the long feathers are twenty-four in each, and the exterior one of these is a great deal shorter than that which is immediately next to it: the short feathers, which cover the under side of the wings, are of a mixt yellowish and brownish colour: the tail is about seven inches long; it is composed of twelve feathers, all of the same length; and, when the bird expands it, it forms a kind of semicircle at the extremity: the colour is a bright glossy brown, with some admixture of a darker brown, but both colours are elegant, and the whole surface is very glossy.

The legs are long, and they are feathered a little lower than the knees: the legs and feet are yellow; the claws are very long and strong, and are of a deep black: the outer toe of each foot is connected above half way down to the middle one, by a membrane; and the claw of the middle toe is sharpened into an edge, on it's inner side.

The gall-bladder in this bird is remarkably large; the appendices to the intestines or czca are small and short: the stomach is large and membranous.

The feathers which cover the upper parts of the legs are long, narrow, and of a greyish-brown, and are variegated with oblong and irregular spots of black: the beak is of a bluish-black, often quite black.

This species is a native of most of the northern parts of Europe; we have it in some parts of England in considerable abundance. It lives frequently in the midst of great heaths, and other tracts of barren and unfrequented lands, and generally is seen sitting on the stumps of low oaks. It builds in moorish and boggy places: most of the writers on birds have described it under the name of Milvus æruginosus, and the Moor-buzzard. Bellonius calls it Circus; and Aldrovand and some others, not perceiving this, have described the Circus of Bellonius over again, as if a different species. It is to such errors, which indeed are much too common in the writings of the naturalists, that we owe the seeming multiplicity of the species. Nature has limited the number in moderate bounds; but when one of the old writers gives one name, and another another to the same bird, their careless followers, not attending to the similarity of their descriptions, give two in imitation of them, as if they were speaking of two different species; and too often, not knowing to what they belonged, have made a third species, by describing afterwards the creature under a name of their own.

Falco

Falco pedibus, cera palpebrisque flavis, capite fusco, nucha et abdomine albis.
The yellow-legged Falco, with the head brown, and the shoulders and belly white.

This is a large bird: the head is small, and flatted at the top; the beak is broad, short, strong, very hooked, and of a blue colour: the membrane, at it's base, is yellow; the eye-lids also are yellow: the eyes are large, and very bright and piercing; their iris is red: the upper part of the head is of a dusky blackish-brown, and the hinder part, toward the shoulders, is white: the sides of the head also are white, and so is the throat.

The belly is white, variegated with spots of black; there is a spot of black of an oblong form on every feather: the legs are of a dusky grey, with a tinge of brown, and the tail is, on the under part, of a mixt, ferruginous, and whitish colour.

The wings are very large, and cover a great space, when they are expanded; when closed, they reach beyond the end of the tail.

The long feathers of the wings are of a bluish tinge, with a mixture of black and grey, except the first of them, which are absolutely black, only that the verge of the extremities of them is white, and the others have from five to ten spots of white on the inner edges.

The tail is long and acuminated; the two intermediate feathers are longer than the rest, and have no variegations, but all the others are spotted; they have each about nine spots on them, and these are more distinguishable on the inner edge than elsewhere: all the feathers on the body are of a dusky iron grey toward the base.

The legs are long, naked, and yellow; the claws are very robust and sharp, and are of a deep horn colour.

This species is a native of some of the northern parts of Europe; I have met with it in Yorkshire and in Sussex, but it is not common with us: wherever I have seen it, it has been in the midst of large and undisturbed woods. It is a bold bird; it usually sits toward the top of some high tree, under cover of the boughs, and will fly at almost any thing that comes in sight: the wood-pigeon seems it's favourite prey. It has been overlooked by many of the naturalists, though a very singular bird, but it has not often fallen in the way of people in these studies; those who have described it, have called it Subbuteo and Dendro falco.

Falco cera pedibusque luteis, dorso fusco, pectore pallido, maculis longitudinalibus fuscis.
The yellow-legged Falco, with a brown back, and a pale breast spotted with brown.

The Buzzard.

This species is of the size of a pheasant; it's weight is about thirty ounces: the head is large, which is contrary to the most of this genus: the crown or summit of the head is depressed, broad, and flat; the beak is short, but it is very thick and strong, and is of a bluish-black colour, and very hooked and sharp at the point: the cera or membrane at the base of the beak is yellow, and in this are placed the nostrils, which are oblong, and very conspicuous; the opening of the mouth is wide, and it is yellow within: the tongue is large, thick, fleshy, and obtuse at the end; the palate has a furrow in it, equal to the size of the tongue: the angle of the lower jaw is circular; the eyes are large, and have a bold, fierce, and piercing look; their iris is usually whitish, but sometimes it has with the white a tinge of red or yellow: the lower eye-brow is lanuginous, and the nictitating membrane is blue.

 The

The whole upper part of the body, the back, shoulders, neck, rump, &c. are of a dusky brown, with a cast of reddish in it, but so deep, that it tends to blackness; it may not improperly be called a ferrugineous black: the feathers covering the upper part of the wings are sometimes variegated with white spots, and these spots are sometimes disposed quite irregularly, sometimes so as to form a line; but this is very uncertain, and in many of the birds there is not the least appearance of any white at all; in some also, the feathers on the shoulders have a few of these white spots, but this is equally uncertain and irregular; these feathers on the shoulders, and those on the upper parts of the wings, however, are usually yellowish about the edges.

The breast and the belly are of a pale whitish colour, with a faint tinge of yellow, and are variegated with oblong spots, of a dusky, ferrugineous brown; these are not disposed transversely, but longitudinally running down the feathers: the feathers which cover the under parts of the wings are also variegated in the same manner; but toward the sides, and on the thighs, they are variegated with oblong spots and dashes, which are disposed transversely, not longitudinally: the throat is of a ferrugineous colour, but the stems of the feathers are black; and between the eyes and the nostrils there are a number of setæ or whiskers of a black colour: the feathers growing on the upper part of the back are long, and reach down so far, as to cover the whole back; on the middle of it there grow no feathers, there is only a downy matter.

The long wing-feathers are about twenty-four in each wing; the exterior one of these is much shorter than the others, and the third and fourth are the longest: the extremities of the four exterior ones are also narrower, and blacker than those of the others, though most of the others have their extremities indeed quite white, and they are all variegated on the inner part with transverse lines of white and brown, in the manner of those of the snipe: the rest of the under surface of the wings is white, variegated in a regular and beautiful manner, with transverse lines of black running parallel to one another: the wings are so long, that, when closed, they reach very nearly to the extremity of the tail; when open, they expand over a great surface.

The tail is seven or eight inches long: it is composed of twelve feathers, and, when expanded, is rounded at the extremity; the extremity of each feather is grey; beyond this grey part there runs a transverse black line, of the breadth of a finger: the rest of the feather is variegated with transverse lines, alternately black and grey, and the base is white.

The thighs are long, very robust, and muscular; the legs are short, thick, and robust, and they are feathered a little lower than the middle joint: the lower parts of the legs and the feet are yellow and scaly; they are very hard and harsh to the touch, and the claws are long, robust, and sharp: the outer toe is connected a little way by a membrane to the middle one, but it is not so much as in many other species: the claw on the outer toe is the smallest of all, and that on the hinder one is the largest.

The liver is large, and is divided into two lobes; the gall-bladder is very large: the spleen is of an oval figure, and the testicles in the male are large and oblong: the stomach is large, and is not musculous as in the others, but membranous, and much like that of some quadrupeds.

When this bird grows old, the upper parts of the head and the back become grey, and, at last, almost white; the males and females also differ in the having or wanting white spots and variegations on these parts, so that it would be easy for a person, not upon his guard as to these varieties, to make two or three species from this one.

This species is a native of most of the northern parts of Europe; it breeds with us, and is, in some parts of the kingdom, very frequent, particularly where there are thick woods to screen it. It is a very ravenous bird, and, when it has not such food as it likes, will take up with almost any thing. It will seize upon rabbits, and even hares: it's swallow is so large, that whole sparrows have been found in the stomach; but, in

scarcity

scarcity of other food, it will eat worms and beetles, and is often seen rooting up the dung of oxen, in search of the common black beetle. It builds in woods, and lays three or four large eggs; these are sometimes wholly white, and sometimes they are variegated with large red spots.

Most of the naturalists have described it. Gesner calls it Buteo; Willughby, Buteo vulgaris; and Ray, Buteo vulgaris sive triorchis. This latter name is borrowed from the old authors, who had an opinion, that the buzzard had three testicles. Pliny tells us, that the Greeks called by this name, and for this reason, the bird which the Romans named Buteo; but there is no better foundation, than the mistake of some ignorant person in the dissection of the bird, for this opinion: the testicles in this are two, as in all the others; and they are so like to those of the other birds of prey in their figure and situation, that it is a wonder any thing particular should ever have been said about them.

Falco pedibus seminudis flavis, cera nigra, capite cæruleo, caudæ fascia cinerea apice albo.
The yellow-legged Falco, with a black cera, a blue head, and a variegated tail.

The honey-buzzard.

This is a large species: it is of the size of a full-grown hen, and of the shape of the former species or common buzzard, except that it is longer, in proportion to it's thickness: the beak is large, and somewhat long; it is of a black colour, very hooked, sharp at the point, and protuberant between the nostrils and the forehead: the cera or membrane at it's base is large, thick, wrinkled, and black; the nostrils are very conspicuous in this; they are of a figure not rounded, but somewhat oblong and bent: the opening of the mouth is extremely wide, and is yellow within; the angle of the lower mandible is semicircular: the eyes are large, and have a very fierce and piercing aspect: the iris is of a deep yellow, approaching to orange colour.

The head is of a grey colour; the crown or top of it is broad, flatted, and grows narrow toward the base of the beak: the bottoms of the feathers, in the hinder part of the head, and on the top of the back, are white: the back is of a dusky brown, somewhat between the ferrugineous and the mouse-colour; and the extremities of the long wing-feathers, and of some of the investient ones, are white.

The wings are not so long as in some of the other species; they are, when closed, considerably shorter than the tail: there is on each of the wings, and also on the tail, a broad transverse areola of grey, and over this another, though narrower, of black; and the tips of the wing-feathers are grey, and of those of the tail white: the long wing-feathers are twenty-four in each; the tail-feathers are twelve; they are ten or twelve inches long, and are variegated with zones, or broad transverse lines of blackish and white: the extremities are white, and just at the edge of the white part there is a narrow black line, and this is succeeded by a space of three fingers breadth, which is grey.

The throat, and the lower part of the belly, near the origin of the tail, are white as snow, without any variegations: the breast and belly are white, variegated with black spots.

The legs are short, robust, and thick; they are feathered below the joint; the naked part and the feet are yellow; the claws are long, black, and very sharp.

The intestines in this bird are shorter, and have fewer convolutions than in the others: their appendiculæ are but few, thick and short: the liver is divided into two lobes, and the gall-bladder is remarkably large: the spleen is oblong, and the testicles are large and oblong.

This

This a native of most parts of Europe, but is not very frequent in England, though it builds with us. It forms it's nest of twigs of trees, and lines it with wool: often it takes the deserted nest of some other large bird, as the kite or common buzzard; it lays four eggs, which are large, of an elliptic figure, and spotted irregularly with a dusky crimson. It builds with us in woods, and sometimes on single trees on heaths. It is a very singular bird in it's manner of feeding; for though so large, and able to seize on almost any thing that the others do, it feeds principally on reptiles. As the other birds of this kind are principally on the wing, this is frequently seen searching it's prey on the ground; it runs very swiftly in the manner of our domestic fowls, and feeds on worms, beetles, and caterpillars. It is frequently seen hunting out the nests of wild bees and wasps, feeding on the creatures themselves, and on their produce: it usually destroys the combs of these insects in great quantity at it's breeding time, the nymphae of the bee and wasp being the principal food it provides it's young. All the writers on birds have described it. Willughby calls it Buteo apivorus, sive vespivorus; and Ray, Buteo apivorus et vespivorus. Some have called it Accipiter Palumbarius; but this is a name which confounds it with another of the birds of prey.

Falco pedibus ceraque flavis, dorso rufescente, pectore maculato, cauda rotundata.

The yellow-legged Falco, with a brown back, a spotted breast, and a rounded tail.

The Kastril.

This is a very beautiful bird; it is of the size of a pigeon: the head is small, flatted on the crown, of a greyish colour, and elegantly variegated with long and narrow streaks of black, the middle part of every feather being of that colour: the beak is short, and broad at the base, and very hooked, and sharp at the point; it is of a deep bluish-black colour, and it's upper mandible has on each side a denticulation, and the lower, on each side, a sinus: the membrane covering the base of it is yellow, and somewhat wrinkled; in this the nostrils are conspicuous, though they are not very large, the naked part about the eyes also is yellow.

The throat is white; the back and the feathers which cover the wings are of a deep reddish-brown, or ferrugineous colour, and every feather is ornamented with a single acute spot near the extremity: the breast, the belly, and the thighs, are of a paler, ferrugineous colour than the back, and they are variegated with longitudinal black spots; these are narrow and linear on the breast, but they are broader, and approach to an oval figure on the belly, and on the thighs they are more rounded, though still oblong, and are very numerous; under the tail there is a space of a pale colour, without a single spot of any kind; and the space between the lower part of the back is of a greyish colour.

The legs are long and robust, and they are, in great part, naked and yellow.

The long feathers of the wings are twenty-two in number, and of these the second is the largest, and from this, to the end, all the others become gradually shorter: they are all of them brown on the upper part, and of a hoary grey underneath; they have each seven or eight white marks on their hinder part, the anterior part is not variegated.

The tail-feathers are very long, but the lateral ones are somewhat shorter than the others; they are of a greyish or hoary colour, and each of them has a broad, black fascia toward the end; and each has also, in the hinder part inward, seven or eight lines of black, excepting only the two intermediate ones, which have none of these spots, but are simply and entirely grey, except that they have the same fascia of black at the extremity with the rest.

The feathers which cover the under part of the wings are white, and are variegated in an elegant manner, by a number of black spots.

The

The males and females in many of the birds of prey differ, in some respects, from one another; all the variation in these is, that in the female the tail is somewhat different in colour: the twelve long feathers which compose it are of a ferruginous colour, but each is terminated by a large black spot at the extremity, and each has also nine or ten transverse spots of the same black colour.

This species is frequent in many parts of Europe, but less so in England than elsewhere. It does not live in forests, as most of the others, but, like the owl, is fond of ruined buildings, and sometimes builds in them, but more frequently in the high stony cliffs on the sea-coasts. It lays four roundish eggs of a beautiful white, spotted with purple. It is a very bold bird, considering it's size; there is scarce any fowl that escapes it. Most of the writers on birds have described it. Gesner calls it Tinnunculus accipiter; Aldrovand, Tinnunculus sive Cenchris; Willughby and Ray, Tinnunculus.

Falco cera viridi, pedibus flavis, pectore albo undulato, cauda fusca fasciata.

The yellow-legged Falco, with a white, undulated breast, and a fasciated brown tail.

The Sparrow-hawk.

This is about the bigness of a pigeon, but it is considerably longer-bodied, in proportion to it's thickness: it's wings are long, and, when expanded, measure to twice the length of the body and tail; the tail is short, but very strong; it is broad and thick at the base, very sharp at the point, and considerably hooked; it is of a bluish colour, except at the point, where it is black: the cera or membrane covering it's base is of a greenish tinge, and in this the bird is very singular, it's legs being yellow, whereas, in almost all the others, the legs and the cera are of the same colour; there is, indeed, some faint yellowness among the green in this membrane, but the green is evidently the prevailing colour, and it has on each side an appendage under the nostrils, of an angulated figure: the nostrils are of an oblong figure; the palate is blue; the tongue is large, thick, and blackish, and the tip of it is somewhat bifid: the eyes are moderately large, and their iris is yellow and bright; they are, as it were, sunk in the head, or defended by prominent eye-brows.

The head is small, and somewhat flatted; it is of a brown colour on the upper part, only that over the eyes, and toward the hinder part, there is some whiteness, and the bottoms of all the feathers on the top, and on the hinder part of the head, are white: the back, the sides, and the wings are of the same dusky brown colour, except that some of the covering-feathers of the wings, which are nearest the back, are variegated with spots of white; sometimes also there is a tinge of greyish all over the back, and, in some particular birds, the grey is the prevailing colour, but this is less common. As the whole upper part of the bird is of this simple brown colour, the whole under part, the throat, breast, belly, and covering-feathers of the under part of the wings, are variegated with white and brown: there run alternate, undulated, transverse lines of white and a ferruginous brown, so deep, that it in some places approaches to black all over the breast, throat, and belly: the white lines are much broader than the brown ones, so that the white seems the ground colour: the feathers, immediately under the base of the belly, and at it's angles, are wholly white, except that in their middle, and that principally toward the extremity, they have some faint tinge of brown.

The wings, though they expand to a considerable breadth, yet, when closed, do not reach farther than to the middle of the tail: their long feathers are twenty-four in each, and they are variegated, on the lower side toward the middle rib, with several transverse lines of brown.

The tail is about seven inches long; it is composed of twelve feathers, and each of these is variegated with five or six bands of black, and their tips are whitish.

The legs are long, and very robuſt; the claws long, ſharp, and black: the legs and feet are yellow, and the toes are bony; the outer toe in this, as well as in ſome other ſpecies, is connected by a membrane to the next, almoſt half way it's length.

This ſpecies is very frequent with us in woods, and ſometimes about ruined buildings. It is a very bold feeder, ſeizing upon almoſt any thing of the feathered kind. It lays five moderately large eggs; they are white, but ornamented, toward the obtuſe end, with a kind of crown, formed of ſmall ſpots of a blood-red colour: all the writers on birds have deſcribed this ſpecies. Geſner calls it Falco Fringillarius; Aldrovand, Accipiter Fringillarius; and Willughby, Accipiter Fringillarius ſive recentiorum Niſus et Sparverius. Some of the writers miſtaking the male and female of this kind as diſtinct ſpecies, Aldrovand calls the male Moſchetus; and Geſner, Accipiter minor mas quem vulgo Niſum ſive Sparverium appellant.

Falco cera flava, dorſo ferrugineo, cauda ruffo et nigro variegata.
The Falco, with a brown back, and a variegated black and brown tail.

The Hen-harrier.

This is a conſiderably large ſpecies; it is equal to a well-grown pullet in ſize; the head is ſmall, and ſomewhat flatted at the top: the beak is large, and very robuſt; it is broad and thick at the baſe, very hooked, and extremely ſharp at the point: the baſe of it is covered with a thick, yellow membrane, in which are ſituated the noſtrils, and toward the angles of it, above the noſtrils, there are a kind of black hairs, reſembling whiſkers; theſe are very rigid and hard, and they all turn forwards: from the hinder part of the head there runs a circle of ſomewhat erect feathers round the ears, and the baſe of the back; the middle part of theſe feathers is of a reddiſh-brown, and their edges are variegated with a paler brown and whitiſh. This circle of feathers is a very ſingular thing, and gives an appearance of a kind of crown to the head of the bird; from this circle of feathers there runs a naked ſkin, which ſurrounds the region of the ears: the back is of a duſky, ferruginous colour; the edges of the feathers which cover the neck are of a reddiſh tinge; thoſe on the top of the head are of a more ſimple brown, and thoſe on the back part are white at the bottom: there is a white ſpot under the eyes, and the breaſt and belly are of a whitiſh colour, with ſome tinge of brown, and are variegated with long and narrow lines of a pale brown, which run down the middles of the ſeveral feathers.

The middle of the throat is brown, or of a duſky, ferruginous tinge; the edges of the feathers, however, have all a reddiſh caſt: there are ſome white feathers on the rump, which are all of them variegated toward their middle with a number of round, ferruginous ſpots.

The wings have each twenty-four long feathers; the exterior verge of theſe is of the ſame colour with the feathers on the back, but their inner part is variegated with tranſverſe black and white lines: in the larger of theſe feathers, the white lines are largeſt; in the more anterior ones, the black are the broadeſt and moſt conſpicuous: the edges of the innermoſt are wholly brown; there is ſome white on thoſe that are next them, but in theſe it degenerates by degrees into brown, till there is no trace of it to be ſeen: the extremities of the exterior feathers of the ſecond order are white, and thoſe of the inner ſeries are reddiſh; the reſt of the feathers are of the ſame colour with thoſe of the back of the bird.

The tail is about eight inches long, and is compoſed of twelve feathers; the extremities of theſe are of a browniſh or ferruginous colour, with ſome tinge of red in it, and all the reſt of their length they are variegated with tranſverſe lines of this reddiſh-brown and black; but the black lines are much larger than the others, and in the two middle feathers the brown wholly diſappears.

The

The legs are long, very robuſt and yellow, and the claws are very long, ſharp, ſtrong, and black: the middle toe is longeſt, the inner one the ſhorteſt, and the outer one is connected, for nearly half it's length, to the middle one by a membrane: the opening of the mouth is wide, and the ſwallow very large; the tongue is large, broad, fleſhy, and undivided at the extremity, and there is a cavity of the ſame dimenſions in the palate.

We have this ſpecies in ſome of our large woods; I killed two ſome years ago in Charleton foreſt in Suſſex: they feed on all kinds of fowls, and on the leſſer quadrupeds; they will ſeize on young rabbets and hares. It builds in high trees, and lays four, or ſometimes five, eggs; they are large and blotted all over, as it were, with a duſky purple, the white hardly appearing any where through it.

The female of this ſpecies is what we uſually meet with, and is the firſt deſcribed here; the male is ſmaller, and differs much in colour: the head, neck, and back are not of the ferrugineous brown of the female, but of a duſky lead colour, or bluiſh-grey, ſomewhat reſembling that of the wild pigeon, only the feathers on the ſhoulders, which are conſiderably long, have ſomething of this ferrugineous brown: the breaſt is white, but is variegated with a number of tranſverſe lines of brown: the exterior wing-feathers are black, but their tips are grey, and their bottoms white, and the exterior covering-feathers of the wings are of a grey, approaching to that of the back, only paler; and the firſt ſeries of thoſe which cover the under part of the wings have ſome ſpots of brown.

It is not a wonder that the male in this ſpecies, ſo very different from the female, ſhould have been miſtaken for another bird. Aldrovand's bird of prey, which he deſcribes under the title of Palumbo ſemilis, is evidently this; and, indeed, there is an appearance of that author's having made two or three ſpecies out of this, from it's differences in age, ſex, and other accidents: other authors call this ſpecies Pygargus Accipiter and Subbuteo, but both theſe names confound it with others birds. We have two names for it, as if thoſe who gave them had taken the two ſexes for two diſtinct ſpecies. We call the female the *Ringtail*, and the male the *Hen-harrow*, or *Hen-harrier*.

Falco dorſo variegato, cauda albo et fuſco faſciata.
The Falco, with a variegated back, and a brown and white tail.

The Caracara.

This is one of the moſt beautiful of the hawk-kind; it is about the bigneſs of a tame pigeon: the head is ſmall and flatted; the beak is broad at the baſe, but ſhort and conſiderably hooked; it is very ſharp at the point; it is of a duſky bluiſh-brown colour, except juſt at the tip, where it is black: the cera or membrane at the baſe is yellow, and ſomewhat wrinkled, and the noſtrils in it are oblong, and ſtand obliquely: the wings, when expanded, reach to a great breadth, and, when cloſed, they extend to the tip of the tail.

The back is of a pale brown colour, variegated in an extremely elegant manner in the ſpots of white, and of a gold yellow: the tail is long, and is beautifully faſciated with tranſverſe, broad lines of white and brown, placed in an alternate order.

The legs are long and very robuſt, and are of a bright yellow colour: the toes are long, and the outer one is connected by a membrane, nearly half way of it's length to the middle one; the claws are black, long, and ſharp.

This ſpecies is a native of the Braſils; Marcgrave has deſcribed it under the name of a Milvus. He gives it's Braſilian name Caracara, and it's Portugueſe one, which is Gaviano. Willoughby and other writers have copied his name, but we have not had an opportunity of ſeeing any ſpecimens of the bird in Europe; from which more perfect deſcriptions might be formed, than that which we at preſent have from Marcgrave and Piſo.

Falco pedibus caeruleis, dorso ferrugineo nigricante. *The blue-legged Falco, with a dusky, ferrugineous back.* — The Sacer.

This is a very large, but not a very beautiful, species; it is of the size of a full-grown hen. It is a very swift flier, and so bold, that there is scarce any bird it will not seize upon: the head is large and rounded, not like those of the other birds of prey, which have it small and flatted on the crown: the beak is short, but it is very strong; it is broad at the base, and hooked at the point; the opening of the mouth is very wide, and the swallow remarkably large: the body is long, in proportion to it's thickness; and the wings and tail are longer also than in most other species: the back is of a deep, ferrugineous colour, with a tinge of black: the breast is paler, and the wings and tail are of a middle colour, between that of the back and that of the breast: the legs are short, but they are very robust; they are of a bluish colour: and the toes are long, and the claws very long, black, and sharp.

It is frequent in some parts of Germany, but is not a native of England; it is very mischievous to the neighbourhoods, particularly where it builds: it will at any time seize upon the largest domestic fowls, and will often kill four or five, before it carries one off; but in the breeding time it will attack young kids and fawns, and, in short, almost any thing that comes in it's way; and, if the whole of it's conquest be too heavy to be carried, it will take it piece-meal. Most of the people who have written on birds have described it under the name of Falco Sacer.

Falco dorso livido, pedibus flavo-viridescentibus. *The greenish-legged Falco, with a livid back.* — The haggard-falcon.

This is a large species; it equals a full-grown hen in size: the head is moderately large, but it is somewhat flatted at the top, and is narrower toward the beak, and broader from thence to the hinder part: the beak is short, but it is very strong; it is of a bluish colour, but very deep, and with somewhat of a tinge of olive; it is broad at the base, and pointed at the extremity, and is very hooked: the cere or membrane covering it's base is of a dusky olive colour; the nostrils are large and open; the eyes are large, and have a very fierce look, but the part of the head immediately over them is swelled out in such a manner, that they seem sunk in their sockets, or hid under very prominent eye-brows: the neck is short, and of a dusky livid colour, with some admixture of brown in it: the shoulders are large, and, with the prominence of the tops of the wings, they have an appearance of gibbosity: the back is of a livid or deep greyish colour; the breast and belly are of a paler grey, and are somewhat variegated; the wings are very long; when closed, they reach to the tip of the tail; and, when expanded, they measure in extent, at least, three times the length of the body: there are usually a few black spots on the wings, and more on the back of this species, but this is not universal.

The legs are long and very robust; they are of a dusky greenish colour, with an admixture of blue and of yellow, in shades: the toes are long, and the claws are very sharp and formidable.

This species is caught sometimes in Germany, and in many other parts of Europe; but it is a bird of passage, and we do not know where it breeds. It varies in colour, whence the writers on birds have divided it into several species. They have called it Falco gibbosus and Falco peregrinus. It is described twice in Willughby, and three times in Aldrovand.

Falco

Falco pedibus flavis, corpore albo-flavo variegato.
The yellow-legged Falco, with a white body variegated with yellow.

The variegated Falcon.

This is one of the most singular and beautiful birds of the Falcon-kind; it is of the size of a moderately-grown pullet: it's head is small, narrow toward the beak, broader behind, and flatted on the crown: the beak is short, but very robust; it is broad at the base, very much hooked, and terminated by a very sharp point: it is of an extreamly pale colour, a whitish, with a cast of bluish and of yellow, except that at the point it is black: the back, and the breast, and belly also are of a whitish colour, but they are all over spotted with a pale and very faint yellow: the tail is composed of twelve large feathers; they are white, and are spotted with the same pale yellow with the body; but the wings, which are very long, and reach nearly to the extremity of the tail, when closed, are of a perfect beautiful white, without the least appearance of any spot at all.

The legs are moderately long, and very robust; they are of a strong and deep yellow, and the toes are long, and the claws very long and strong, and black: the cera or membrane covering the base of the beak is yellow, and the nostrils are oblong and transverse; the iris of the eye also is yellow, and the look is very bold and piercing.

This is a native of Italy, and is a very bold and mischievous bird: it lives principally in thick woods, where it feeds on young hares, rabbets, and other the smaller quadrupeds, as well as on birds of all kinds: it sometimes takes up it's habitation near the villages, and is then very troublesome to the inhabitants, sparing hardly any thing that is incapable of defending itself. Aldrovand, Jonston, Willoghby, and others have described it under the name of Falco albus.

Falco pedibus flavo-virescentibus, dorso nigrescente.
The black Falco, with yellowish-green legs.

The Tree-falcon.

This is a smaller species than many of the former, but it is as bold and desperate as any of them; it is about the size of a large tame pigeon: the head is small and flatted; the beak is broad at the base, moderately hooked, and very sharp at the point, and the upper part of it is much longer than the under: the cera or membrane covering it's base is of a mixt colour of yellow and green, which shew themselves variously in different lights, but the beak itself is bluish.

The head is of a deep iron grey, and the back is almost black, but the edges of many of the feathers have a circular verge of a ferruginous tinge round their extremities.

The tail is composed of twelve large feathers, they are almost black, and a little spotted, except the two middle ones; and the wings are also black, only that on that part nearest the body the feathers have some of them a number of ferrugineous spots: the breast is beautifully variegated with black and white spots; and on the sides of the head, as also toward the upper part of the breast, there are some yellowish feathers, which add very singularly to the variegations.

The eyes are large and black, and have a remarkably bold and fierce look; the bird itself has, indeed, in it's whole appearance, a peculiar boldness of aspect: it stands more firm and erect than any other species, and is usually seen on the topmost bough of some tall tree, taking a survey of every thing about it, and ready to drop on, or rise to it's prey.

This is a native of Germany, but it is less frequent than most of the other species; it is a very bold bird in it's manner of feeding. Most of the ornithologists have described it. Ray and Willoghby call it Litho-falco, Dendro-falco, seu Falco Lapidarius et Arborarius. These are the names under which Gesner, Aldrovand, and others have described it; but most of these writers have made two species of it, not considering the constant difference in size between the males and females in the birds of prey, and the slighter variations in colour. This species builds in great abundance in the Hartz-forest, where it is common to see the male and female together on that occasion; and any one who does so, will be convinced of the error of those who have made them two species. It lays four eggs of a dusky greyish-white, variegated, principally toward the larger end, with irregular blotches of purple. It is at all times a bold feeder, but in the time of breeding is more than ordinarily so, and will seize on the largest birds, and on the young of quadrupeds.

Falco pedibus flavis, corpore fusco, capite majore. *The large-headed Falco, with a brown body, and yellow legs.*

The Barbary-falcon.

This is of the size of a pullet, but it's body is slender, in proportion to it's length, and it stands very erect, and has at once an elegant and majestic appearance: the head is large and rounded: the beak is large, robust, and very much hooked; it is of a dusky horn colour, with a tinge of bluish all over, except at the point which hangs over the under chap, and is very hard, sharp, and black: the membrane that covers it's base is yellow and wrinkled, and the nostrils in this are large, oblong, and placed obliquely: the eyes are very bold and fierce in their aspect; the iris is yellow, the pupil black, and the part of the head immediately over them is prominent, so as to make them appear sunk in their sockets: the colour of the head is a dusky brown of the ferruginous tinge, but with some admixture of a deep grey or lead colour, especially in the hinder part; for toward the beak, and especially about the region of the ears, it is paler: the back is of the same dusky brown, with a tinge of the ferruginous hue, and an admixture of grey: these three colours are not separate in variegations, but are all blended in every feather, so as to form a different tint from all three in the whole, and they are rather distinguishable in different lights than any other way: the tail is composed of twelve large feathers; they are darker, or have more of the unmixed grey, than the feathers of the back, and the two middle ones are darker than the other ten.

The wings are very long; when extended, they measure, at least, three times the length of the body, and, when closed, they reach very nearly to the tip of the tail: the long feathers in these are almost black, and have no variegations, except that the inner ones, or those next the body, have somewhat of a ferruginous cast on their inner side: the feathers that cover the upper side of the wings are of the same colour with those of the back; and those which cover the under part of them are paler, and have some variegations of whitish and yellowish, especially toward the edges.

The breast is paler than the back, but is of the same general colour, only it has some dark variegations: the belly is yet paler than the breast; and, just at the origin of the tail, there are some pale brown feathers, variegated with white.

The legs are very robust, though not very long; they are of a dusky yellow, and toward the feet they have some admixture of greenish or olive colour: the toes are long, and the claws are very strong, and remarkably long and sharp; they are black, and that of the hinder toe is much longest.

This is a native of Africa, but it is a bird of passage, and is sometimes seen in Europe. Most of the writers on birds have described it under the name of Falco Tunetanus. The male is somewhat different in colour from the female here described, and is smaller; but the difference is less than in most other of the birds of prey.

Falco

Falco ſupine rubeſcens pedibus flavis, capite depreſſo.
The yellow-legged, reddiſh Falco, with a flatted head.

The red Indian Falcon.

This is a very large and an extremely beautiful ſpecies; it is equal to a well-grown pullet in ſize, but it's body is ſlender, in proportion to it's length, and it has a way of ſtanding in a remarkably erect manner.

The head is very ſmall, in proportion to the ſize of the bird, and it is remarkably flatted on the crown: moſt of the birds of prey have the head ſomewhat flatted, but this more than all; the hinder part is very broad, but there is not the leaſt appearance of any eminence or riſing in it: the feathers which cover the head are of a greyiſh colour, with a faint admixture of brown, and thoſe on the back of the neck are of the ſame tinge: the beak is remarkably large and ſtrong; the baſe is broad, and on the upper part it is ſomewhat prominent: the point is extremely hard, firm, and ſharp, and bends a conſiderable way over the lower chap: the colour of the beak is a greyiſh-blue, with a tinge of horn colour, and the cera or membrane that ſurrounds it's baſe is yellow, wrinkled, and has the noſtrils in it; they are large, patulous, and ſtand obliquely: the eyes are large, and look very fierce; their iris is brown, and the pupil black, and the membranes of the eye-lids are yellow at the edges; and, from the leſſer canthus of each eye, there is continued a beautiful oblong red ſpot.

The back and the upper part of the wings are of the ſame colour with the head, only a little darker; but the breaſt, the belly, and the upper part of the under ſurface of the wings are of a beautiful red; the rump alſo is of this elegant colour, as are alſo the feathers which cover the thighs.

The throat is of the ſame elegant red colour, but it is variegated with a beautiful, oblong ſpot, of a dark grey; and in ſome birds the breaſt has ſome variegations of the ſame colour, in form of irregularly oblong ſpots; but this is not univerſal, many having not the leaſt ſpot there. The ſides where they are covered by the wings are not of this high colour, but of a duſky grey.

The wings are very long. When cloſed, they reach to the extremity of the tail, and, when expanded, they meaſure three times the length of the body: the tail is long, and is compoſed of twelve feathers; theſe are very large and firm, and are beautifully variegated with alternate circles, or parts of circles of black and grey.

The legs are very robuſt, but not remarkably long; they are yellow: the toes are long and ſtrong, and the claws are black, and are remarkably ſharp and formidable.

The male of this ſpecies is conſiderably ſmaller than the female; the colours are fainter, and it has not that ſpot on the throat, which is ſo ſingular in the female, and the legs are of a paler yellow.

This beautiful ſpecies is a native only of the Eaſt Indies; it builds in the thickeſt woods, and is very ſhy, but it is a bold feeder both on birds, and on the ſmaller quadrupeds. Aldrovand has deſcribed it under the name of Falco ruber Indicus; and moſt of the writers who have ſucceeded him, have borrowed his name, and his deſcription. We ſometimes ſee rude delineations of it in the Chineſe pictures; and ſpecimens of the bird are ſometimes, though rarely, brought over, as curioſities, by the Captains of our Eaſt India ſhip.

Falco capite cirrato.
The Falco, with a cirrated head.

The creſted Falcon.

This is an extremely beautiful ſpecies; it is of the ſize of a well-grown pullet: the head is ſmall and flatted, and of a deep black, and is ornamented with a creſt of long feathers, which hang down behind it; the outer ones of theſe are longer than thoſe

those in the middle, so that it appears forked: the breast and belly are variegated, and extremely beautiful; the colours are black and white, and they are disposed in alternate, transverse lines, or oblong spots, and both are extremely bright and glossy colours: the back is black.

The eyes are large, and have an extremely piercing aspect; their iris is yellow, and the pupil black: the beak is of a deep bluish-black; it is very broad at the base, extremely hooked, and the point or extremity of the upper chap or portion, which hangs to a considerable distance over the other, is black: the membrane which surrounds the base is yellow, and somewhat wrinkled; the nostrils are conspicuous on it; they are oblong, and placed transversely.

The legs are moderately long, and very robust; they are feathered down almost to the toes: the feet are yellow, and the toes are very long and large: the claws are black, long, and sharp; the lesser wing-feathers are fringed, as it were, with white at their edges: the tail is long and broad, and is elegantly variegated with alternate rows of grey and black; all the rest of the bird is black.

This elegant bird is a native of the East Indies; the forests in China, and some parts of Tartary, also afford it: we have sometimes had it brought over alive to us, but rarely. Ray mentions one of them kept alive in London; and I remember to have seen one at the late Duke of Richmond's, among his collection of living animals, at Goodwood. Ray, in his edition of Willughby's Ornithology, calls it Falco Indicus cirratus.

Falco cera lutea, dorso nigricante, ventre rufo.
The black-backed, brown-bellied Falco, with a yellow cera. **The Hobby.**

This is a large and a very bold bird; it's size is about that of the pheasant, but it's wings are so long, that, when extended to the full, they measure more than two feet and a half.

The head is small, and somewhat flatted on the crown; the beak is short, and so extremely hooked, that it appears of a kind of semicircular figure: it is of a bluish colour, except toward the upper part, where it becomes whitish: it's base is covered with a yellow membrane, which is tolerably smooth, and in which the nostrils are conspicuous; they are large, oblong, and transverse; the extremity of the upper chap falls a great way over the lower, and is very hard, sharp, and black: on each side also, at the angle of the beak, there is on the upper chap an appendage or denticle, which is received into a hollow in the under one: the opening of the mouth is large; the tongue is large, thick, fleshy, and blackish: the palate also is black, and there is a hollow formed in it for the reception of the tongue: the tip of the tongue is somewhat bifid; the eyes are large, and their iris is hazel; the eye-lids are black.

Immediately over the eyes there runs a line of a whitish-brown, and the rest of the head is variegated with black, and a bright chesnut colour; the scapi of the feathers being all of the former, and their edges of the latter colour: the feathers on the neck are of the same whitish-brown with those which make the two lines over the eyes: the back and the upper part of the wings are of a very deep iron grey, approaching nearly to black, and in some are absolutely black; the middle part of the back and the largest of the wing-feathers are the more intensely black; the rump and the smaller wing-feathers are more grey: the throat is of a yellowish-white, and there run from the head to this part two oblong, white spots on each side, one from the aperture of the mouth, and the other from the hinder part of the head: the lower part of the belly is of a reddish-brown, and all the intermediate part is elegantly variegated; the feathers being black in their middle, and white at the edges: the feathers which cover the thighs or upper part of the legs are of a reddish-brown, spotted with small, irregular, and not very numerous spots of black.

The long feathers in the wings are twenty-four to each; of theſe the ſecond is the longeſt; the larger feathers have their extremities blackiſh, and all the others have variegations of black and white along the ſcapi: theſe are more numerous and conſpicuous in the inner feathers, and in ſome birds they are almoſt wholly wanting in all: the ſhort feathers that cover the under parts of the wings are black, and are beautifully variegated with little round white ſpots.

The tail is not very long; it conſiſts of twelve feathers; the middle two are the longeſt, the others grow gradually ſhorter to the edges, but the decreaſe in length is but inconſiderable: the middle ones are of the ſame colour on each ſide of the ſcapus or ſtem; but the others are variegated with oblong ſpots of a reddiſh-brown toward the inner part, and are white at their extremities.

The legs are long and very robuſt, but they are yellow: the toes are long, and the claws very ſharp: the outer toe is connected to the middle one, by a membrane toward their baſe.

The liver is large, and the gall-bladder, as in all other birds of prey, is remarkably large alſo: the appendages to the inteſtines are ſhort.

This is a bird of paſſage; it comes over to us in April, and builds with us, but it leaves us again about September. It builds with us in high trees, uſually in the midſt of thick woods, but ſometimes in ſingle ones, on hills, and in hedges. It is a bold feeder, and will at times ſeize on almoſt any thing, but ſmall birds and particularly larks ſeem it's favourite prey. We have people who make a trade of catching larks by means of this bird; they call it daring of larks: the ſight of the hawk makes them lie ſo cloſe, that the nets are eaſily drawn over them.

The antients were acquainted with this ſpecies. The Greeks called it Ὑποτριόρχης, Hypotriorchis; and the Latins, Buteo minor. Some of our writers, as Ray, Willughby, and others call it Subbuteo, but this confounds it with another ſpecies already deſcribed.

Falco pedibus flavis, dorſo variegato, pectore albo ferrugineo. The yellow-legged Falco, with a variegated back, and a brown and white belly. **The Merlin.**

This is a very ſmall and a very beautiful hawk: Bellonius calls it the leaſt of all the birds uſed in hawking, and, excepting the lanius, it is ſo: it is not much larger than a thruſh.

The beak is ſhort, broad at the baſe, and very hooked and ſharp at the point: it is of a bluiſh colour all the way, except at the extremity of the upper chap, where it falls over the other, and there it is black and very ſharp: it has alſo, in this portion, an appendage or denticle toward the baſe on each ſide, received into a hollow in the lower chap.

The head is ſmall, and flatted on the crown; the eyes are large, and have a very piercing look; they ſtand forward; their iris is hazel, and the pupil black: under the back part of the head there runs a kind of chain of a yellowiſh or reddiſh-brown; the throat is white.

The back and whole upper part of the body are of a duſky hue, but, when nearly viewed, it is found to be variegated with three diſtinct colours; theſe are a deep, duſky, ferrugineous brown, a deep greyiſh-blue, and an abſolute black; and the middle of all the feathers, both on the head and back, are black, that colour following the courſe of the ſcapi or ſtems in all; the external edges in moſt are fringed, as it were, with a very roſly orange colour, or a ferrugineous brown: the long feathers of the wings are black, but they are variegated with ferrugineous ſpots: the tail is long, and is elegantly variegated with zones, or broad transverse lines of black and a brownish-white; there are sometimes fourteen of these in the whole tail.

The breast and belly, or indeed the whole under part of the bird, from the throat to the insertion of the tail, is of a pale whitish-brown, with a slight tinge of the ferrugineous reddish, and is slightly variegated with spots of a blackish-brown; these are not very numerous, and they stand in a longitudinal, not a transverse, direction, seeming to follow the course of the feathers.

The legs are less robust than in most of the birds of prey; they are long, slender, and of a yellow colour: the toes also are long and slender, and the claws are long, black, and sharp.

The male in this species, as in all the other birds of prey, is considerably smaller than the female, and it is also distinguished by some blue feathers on the rump, which are wanting in the female: the back of this bird loses all it's variegations by age, and becomes solely of a dusky bluish colour; this is an incident that happens more or less to all the Falco kinds, as they grow old, but it is in none of them so conspicuous as in this little species.

Though this is one of the smallest of the birds of prey, it is as bold as any of the larger: it will attack almost any thing, but the partridge is it's favourite prey: it will fly at a whole covey of these, and destroy quicker than any other hawk.

It is a native of England; it builds with us in woods, and sometimes in hedges. All the writers on birds have described it; Aldrovand, and after him most of the modern writers, call it Æsalon and Æsalo, supposing, though with no great certainty, that it was the Æsalon of the old Romans.

Falco pedibus flavis, dorso fusco, pectore albo lineolis nigris variegato.

The yellow-legged Falco, with a brown back, and a white variegated breast.

The Gos-hawk.

This is a large and a very beautiful bird; it is bigger than the common buzzard, and is very like it in shape: the head is small and flatted: the beak is short, but very broad at the base, and extremely hooked; the point of the upper chap hanging over the lower, and very sharp: the eyes are extremely sharp and piercing, and the nostrils are roundish.

The upper part of the head and neck, the back, and, in short, the whole upper part of the body, is of a deep dusky brown, not unlike that of the buzzard; but the breast and belly are very beautiful: they are of a snow-white, and are elegantly variegated with narrow, oblong, and undulated lines of black, disposed in a regular and a very beautiful manner.

The legs are very robust, moderately long and yellow: the toes are long, and the claws are very long, black, and sharp: the beak is of a blackish colour, and the membrane that covers it at the base is of a yellowish-green: the wings are short; they do not reach, when closed, to more than half the length of the tail: this is a singularity which distinguishes the Gos-hawk, at first sight, from all the other species.

The tail is long; it is of a greyish-brown colour, and is variegated with three obscure zones, or transverse dusts, placed at a great distance from one another.

This is a very bold bird; it feeds on pheasants and partridges, but it will fly at any thing: the heron makes admirable diversion for the hawkers, under an attack from this bird: it may also be taught to attack wild geese and hares. The French call this species Autour, and their Latin writers, Astur; Aldrovand, however, declares that

that the Astur of the French writers is the same with the Asterias of the old Greeks, mentioned by Aristotle, and different from our Gos-hawk: there is not much to be built upon this, however, when we consider that Aldrovand, as is evident from his description, knew little or nothing of the Gos-hawk. It breeds with us in woods, and is extremely bold in the defence of it's young. I remember to have seen a servant, whom I employed to take a nest of them in Rockingham forest, attacked with the utmost fury by both the old ones, and wounded in the face, while he was up in the tree, and had but little use of his hands for his defence. All the writers on birds have named it. Willoughby, Ray, and the rest call it Accipiter Palumbarius.

Falco dorso cinereo, ventre albo variegato, pedibus nigris. The black-legged Falco, with a grey back, and a white variegated belly.

The Butcher-bird.

This is the smallest of all the birds of prey, used by the falconers; it is not quite equal to the black-bird in size, and is smaller than the merlin, though Bellonius calls that the smallest species: the head is small and flatted; the beak is moderately long, and very broad at the base; it is moderately hooked at the point, and the extremity of the upper chap hangs over the lower, and is hard and black, and very sharp: toward the base of the beak also there are two appendages, one on each side, and there is a hollow in the lower chap for the reception of each of these: the tongue is slender, fleshy, and bifid at the end; the palate has a cavity formed for the receiving it: the nostrils are roundish, and over them there are placed a number of short and rigid setæ or bristles, in manner of whiskers: on each side, from the angles of the mouth to the hinder part of the head, there runs an oblong black spot; the rest of the head is a pale grey.

The back, the upper part of the wings, and the rump, are of a darker grey than the head: the belly is white, and the breast and throat are also white, and are variegated with oblong, dusky spots, running in a transverse direction.

The long wing-feathers are eighteen in each; all these, except the four exterior ones, are white at the extremities, and the second and third have their exterior edges white: the bottom of the very first feather also has a little white on it, and from this the white is more and more considerable on all the others to the tenth, in which it occupies more than half the feathers; from the tenth feather the white part decreases again in all the others, but on the interior ones it is extended along the edges, quite to the extremities: on those which are nearest the body there is not any white at all; the large wing-feathers, as well as those of the tail, are elsewhere black.

The tail is moderately long, and is composed of twelve feathers; the two middle ones are the longest of these, and the others grow gradually smaller to the exterior ones; but the decrease in length is but very little, till toward the very outer ones, but these are only a fourth part shorter than the middle two: the two exterior feathers are entirely white; the two middle ones are white at the tips, and the others are black throughout, but the black is deeper in the middle than in the exterior ones.

The legs are moderately long; they are slender and black: the toes are long and black, and the claws, though not very robust, are long, and extremely sharp and black: the gall-bladder is large in this bird, and the testicles are very small and round.

This species is frequent in Germany, and some other of the northern parts of Europe; it has been doubted whether it was a native of England, but that is a question I can answer in the affirmative, having shot more than one of them in Rockingham forest in Northamptonshire. I have seen it also in Yorkshire; but it is no where very plentiful, and is always very shy, so that it is not a wonder many have overlooked it. It may be known sitting from almost all the other hawks; for it generally holds it's tail erect, and, as they are fond of the tallest trees, whence they can look down on the country for a great way round, and keep themselves concealed all the time under the

the boughs, this, on the contrary, usually takes it's post on the tops of a shrub or furz-bush.

It builds in the midst of thickets, in the most inner recesses of our thick woods; it's nest is formed of moss, wool, and small stems of hoary herbs, and it lines it first with the softer branches of the common erica or heath, and over these with the leaves of soft and tender plants. It lays six eggs, and the young ones, while in the nest, carry very little resemblance of the parents, except in the beak and legs: their feathers, which indeed are properly no more than the rudiments of future feathers, are green toward the base, and the whole birds have a dusky olive tinge.

It feeds on beetles and other large insects, but not on these only; it will seize on any bird that is not greatly larger than itself; the thrush forms it's most capital prey, but from this downward nothing escapes it: there is a peculiar cruelty also that it is guilty of, which is the taking the young out of the nests of other birds. Most of the writers on birds have described it. Ray, Willughby, and others call it Lanius cinereus major; and we, in the North of England, the Wierangle, a name borrowed from the German one Werkangel or Warkangel. The Germans also, about the Hart's-forest, where it is very frequent, call it Neghen-doer, a word expressing it's great fierceness, that it will kill nine birds, before it begins to eat of one of them, but with us it is not so terribly mischievous.

Gesner mentions what he calls a larger species of this bird, and names it Lanius cinereus maximus; he says, it is in all things like this, only that it is larger: probably, this was no more than the difference between male and female, or some other as trivial error.

Falco pedibus cæruleis, capite nigrescente, dorso ferrugineo.
The blue-legged Falco, with a black head, and a ferruginous back.

The lesser Butcher-bird.

This is a very singular and a very beautiful little bird, though a perfect hawk in all it's characters; it is not larger than a lark: the head is large, rounded, and not at all like those of the generality of these birds: the beak is large, in proportion to the size of the bird; it is broad at the base, and very hooked at the point: the upper chap is much longer than the under; it's hooked extremity turns over that, and it has near this part two pointed appendages or denticles, which are not received into cavities, or hollows formed for that purpose in the lower chap, but hang over it: the nostrils are round; the inside of the mouth is yellow: the tongue is divided into many parts at the extremity, and the palate has a cavity to receive it, which is hairy on each side, as well as the tip of the tongue: there are about the angles of the beak certain rigid bristles or hairs, which serve as whiskers.

The middle of the back, and the middle series of the feathers which cover the wings, are of a deep dusky, ferrugineous colour: the head and the rump are black; from the angles of the beak there runs beyond the ears a black line on each side, of a deeper colour, and more glossy than any other part; this is considerably broad, and is terminated at the edge by a white line: the throat and the breast are of a very pale, ferrugineous colour, almost white; the lower part of the belly is absolutely white.

The long wing-feathers are eighteen in each; the first or most exterior of these is very short and little; the third is longer than any of the others, and from this they gradually become shorter and shorter to the innermost, but the diminution in length is very small in each: the whole wing is not very long; it spreads to a considerable extent, in proportion to the bulk of the bird, when opened; but, when closed, it is not so long as the tail: the larger feathers are brown; the smaller, that are nearest the body, are of a reddish-brown, and the others have their middle black.

The

The tail is moderately long, and is compoſed of twelve feathers; the exterior of theſe are the ſhorteſt, and the others grow gradually longer to the middle; the two middle ones are almoſt entirely black, and the next to theſe have ſome whiteneſs toward the baſe, eſpecially at the interior edge; four of the others on each ſide are white at their extremities, and even half way up, and the interior pinnules of the farther ones are alſo entirely white.

The legs are moderately long, but very ſlender; they are of a deep greyiſh-blue, approaching to black: the toes are long, and the outer one of each foot is connected a little way to the middle one by a membrane: the claws are not very thick, but they are long, in proportion to the ſize of the bird, and are very ſharp.

The gall-bladder in the bird is particularly large; the inteſtines are ſhort; the teſticles in the male are large, round, and white.

This ſpecies is not uncommon with us in England, though it's ſmallneſs makes us ſeldom diſtinguiſh it as of the hawk-kind. It frequents woods, and is ſometimes ſeen among the ſingle thickets on heaths: it builds in the crab, the white-thorn, or the holly, at about ſix or eight feet from the ground: it builds it's neſt with graſs, moſs, the ſtalks of tender plants, and feathers. It lays ſix eggs; they are of an oblong ſhape, and are large, in proportion to the ſize of the bird; they are white toward the ſmaller end, but toward the larger they have a kind of circle or crown of browniſh or purpliſh ſpots.

Moſt of the writers on birds have deſcribed this ſpecies. Ray and Willughby call it Lanius minor, and the leſſer Butcher-bird; Aldrovand, Lanius tertius; and others, Lanius tertius Aldrovandi. We call it, in the North of England, where it is more frequent than in any other part, the Flaſher.

Falco pedibus nigris, capite rufo, dorſo variegato.
The black-legged Falco, with a brown head, and variegated back.

The least Butcher bird.

This is the ſmalleſt of all the hawk-kind; it is not much larger than a ſparrow, but it is a bold looking bird, and even a bold feeder: the head is ſmall, but it is not flatted, as in many of the hawk-kind, but rounded: the beak is large, in proportion to the ſize of the bird; it is broad at the baſe, and prominent on the upper part; the extremity of it is very hooked: the upper chap is much longer than the under, and the hooked extremity, which turns over it, is very ſharp; it is of a dark bluiſh-black toward the baſe, and at the extremity it is quite black: the cera or membrane ſurrounding the baſe of it is of a deep gloſſy, lead colour; and the noſtrils, which are ſeparated in this, are ſmall and round: the tongue is divided at the end, and there is a cavity in the palate fitted to receive it.

The head is of a duſky brown, with a tinge of the ferruginous in it; the back is alſo of a reddiſh-brown, and is elegantly variegated with tranſverſe lines of black: the feathers of the rump are of a more reddiſh tinge than any of the others, and they are variegated with ſemicircular ſpots of yellowiſh: the throat and breaſt are of a pale colour, but they alſo are variegated in an extremely elegant manner, with ſemicircular lines or ſpots of a deep black: the lower part of the belly is entirely white.

The long feathers of the wings are brown; but thoſe which are neareſt the body, and all the covering feathers, have their edges of a reddiſh-brown.

The tail is of a very deep, ferruginous brown, approaching to black, but the exterior feathers have the pinnules on the outer ſide of the rib entirely white; the four ſucceeding ones on each ſide are white at the extremities, and the two middle ones are entirely of a reddiſh-brown.

This species is not uncommon with us in Yorkshire, and is easily known from the other small birds, by it's being always single, and usually avoided; sometimes hunted, and followed by them with a great noise. The male is a little smaller than the female, but the difference is inconsiderable: the nest is formed of dry stalks of grass and feathers: the eggs are five or six; they are of the shape of a hen's egg, and of a dusky brownish-white, spotted all over it in an irregular manner with purple.

This has hardly been exactly distinguished by any of the writers on birds. The Lanius secundus of Aldrovand comes nearest it; but either it is not the same, or the author has not been over accurate in the description. Ray and Willoughby have described a female of it, which they call Lanius fœmina similis secundo Aldrovandi.

Falco pedibus subcæruleis, dorso cinereo et russo, pectore pallide fusco.

The blue-legged Falco, with a grey and brown back, and a pale, brown breast.

The Italian Butcher-bird.

This also is an elegant little hawk; it is about the size of a lark, but is longer-headed: the head is small and flatted on the crown: the beak is large, in proportion to the size of the bird; it is broad at the base, and very sharp at the point: the cera or membrane covering it next the head is of a greyish-blue; the nostrils are somewhat oblong, and stand obliquely: the upper part of the beak is somewhat gibbose, and the hooked part at the extremity is formed of an elongation of the upper chap, falling over the extremity of the under one.

The head is of a pale reddish-brown: the shoulders and upper part of the back are of a ferruginous brown, and the lower part, toward the rump, is greyish; it becomes paler and paler, as it goes lower, and the rump itself is white; there are some brown transverse lines under the throat, and all the rest of the under part of the whole breast and belly are of a whitish-brown, the breast somewhat darker, and the belly paler.

The feathers of the wings are only eleven in each; they are all of them of a dusky colour, but the exterior ones have the half next the body white: the tail is moderately long, and is of a deep brown, with a tinge of grey. These are marks, by which this little hawk may be distinguished with great readiness and certainty from all the other kinds; but there is another yet more striking; this is, that it has a large and more beautiful spot of snow-white on the shoulders. The male and female in this species differ considerably, but this always remains in both, and is an essential and obvious mark of the species.

It is a native of Italy, and of the South of France, but is is not known in England; I have been favoured with two or three specimens of it from Italy. Aldrovand seems to have meant this species by his Lanius minor primus, but his description is imperfect. Ray and Willoughby had both met with it: Ray mentions some variation in the colour of that which he saw, from that described by Willoughby; but this was only that Ray's was a male, and Willoughby's a female; they make a question whether it be the same with that described by Aldrovand, and have named it Lanius, an minor primus Aldrovandi.

In the male, according to Mr Ray's description, the head and neck had more of the reddish-brown about them, and were of a deeper colour; and Willoughby mentions one shot on the banks of the Rhine, the upper part of the head of which was perfectly reddish, with very little brown, and the tail surrounded with a line of white. All this is no more than the variations of the male from the female of the same species, and that of the different seasons nearer to, and more distant from, the time of moulting.

This bird, though very small, is a very bold feeder, and flies very swiftly. It is often seen perched on ruined buildings, or on shrubs, in less frequented places; and frequently

frequently is observed on the wing many hours together, about the shores of rivers and lakes, preying on the small birds that frequent those places, for the sake of the insects about them.

VULTUR.

THE Vultur has four toes on each foot, and three of these are placed forward, only one backward: the neck is long and almost bare of feathers, and the legs are covered with feathers down to the feet, or nearly so, and under the throat there is a space covered with hair instead of feathers: the head also, in many species, is naked, or has, at the utmost, only a downy matter on it, instead of feathers, and the under part of the wings is downy.

Vultur cinereo-nigrescens cauda brevi. **The black Vulture.**
The greyish-black Vultur, with a short tail.

This is a very large, but by no means a beautiful, bird; it's aspect is at once terrible and distasteful; it is of the size of a full-grown turkey: the head is large, and of a pale colour, approaching to whitish; there are no feathers on it, but in their place there is a kind of downy or woolly matter: the eyes are very large and piercing: the beak is extremely large, long, and formidable; it is not arched all the way as in most birds, but it runs straight from the head, almost to the extremity, where it turns round, and is very hooked; it is of a dusky, olive colour.

The opening of the mouth is large, and the nostrils are oblong and transverse: the neck is long, robust, and thick, and it is also naked: at it's lower part there is a large tuft or fringe of a woolly matter of a hand's breadth, and of a pale colour.

The back, and the upper part of the wings, are of a deep colour, approaching to black; the breast is of a paler hue, approaching to a deep iron grey, or of a mixt appearance between that and a mouse-colour; the belly is still paler: the long feathers of the wings are of a deep mouse colour, almost black: their under part is somewhat paler, especially toward the interior edge; the under surface of the upper part is woolly, and of the colour of the tuft at the throat.

The legs are very robust and strong; they are feathered in a beautiful manner very low, almost to the toes: the feet have an appearance of peculiar strength, and the claws are very sharp and terrible: the tail is composed of very large feathers, but it is not long; the wings are of a very great length; when closed, they reach to the extremity of the tail, or very near it; and, when opened, they expand to a surprising breadth: the colour is the same dark one with that of the long feathers of the wings, or rather more black than those.

This is a very bold feeder, and is so strong, that it is able to seize upon the young of the larger quadrupeds: it had been an opinion of the old writers, that the Vultur never killed any thing, but fed only on the carcasses of what it found dead, but that is found to be an error: it seizes on birds of almost all kinds, and on fawns, young kids, and many other animals. It was also an opinion of the old writers, that no birds of prey were gregarious. Aristotle declares them all solitary, but this also is an error: I have already observed, that the common kite is in some places such; and the Vulturs of this species are scarce ever seen, except in large flights.

It is a native of many parts of the East, but it is not known in the colder countries: most of the writers on birds have described it. Bellonius and Gesner call it Vultur cinereus; and Aldrovand, Vultur niger; we very rarely see it in Europe. The Duke of Richmond had two in great perfection among his living curiosities at Goodwood.

Vultur corpore caſtaneo, cauda brevi.
The cheſnut-coloured Vultur, with a ſhort tail.

The Baie Vultur.

This is a large and noble bird, but it is ſomewhat inferior to the former in ſize; it is about equal, in that reſpect, to a full-grown hen: the head is large, and rounded at the top, not flatted, as in moſt of the birds of prey: the neck is long, and moderately thick; it is not abſolutely naked, as in the former ſpecies, but has, eſpecially on the hinder part, ſome feathers; theſe are not ſhort and broad, as thoſe of the body, but they are long, and very narrow and ſlender, like thoſe on the neck of a cock: they are of a bright brown, with a tinge of reddiſh in it, and are very gloſſy and beautiful: the back is of a fine bright cheſnut colour, and the feathers are remarkably ſhort, broad, and elegantly arranged in the manner of ſcales; the breaſt and belly are covered with larger feathers of the ſame colour and ſhape; but the long feathers of the wings and tail are of a very dark colour, almoſt black: the upper part of the wings is a little paler than the back, and the rump is ſomewhat darker than the upper part.

The tail is ſhort; the wings are very long, and expand to a ſurpriſing breadth; they are of a paler colour on the under ſide than on the upper, and are not ſo even and regular at the ends, as thoſe of the birds of prey in general, but have more of the rough appearance of thoſe of the wood-pecker.

The legs are ſhort, but they are extremely robuſt: they are covered with feathers down to the extremities of the toes, and the whole foot has an appearance of great ſtrength: the claws are long, black, and very terrible: the beak is long, and very robuſt; it is ſtraight all the way, from the head to near the extremity, but there it is turned down and very ſharp.

This ſpecies is a native of ſome parts of Africa, but it is more rarely ſeen among us than the other. None of the Vulturs are wild in this part of the world, but this is ſo tender, that it is not to be kept alive by art. I ſaw one expoſed for a ſhow in London, about four years ſince, but it was then dying. It is a very bold feeder, but it's prey is principally the larger birds, not any thing of the quadruped kind; it is uſually ſeen ſingle. Moſt of the writers on birds have named it: they call it Vultur fuſcus, and Vultur Baeticus, from it's colour, this being a term very frequently uſed among the naturaliſts, to expreſs that kind of brown which we uſually diſtinguiſh by the word cheſnut colour.

Vultur capite criſtato, pedibus flavis.
The yellow-legged, creſted Vultur.

The Vultur Leporarius, or Hare-catcher.

This is a very noble and majeſtic bird; it is of the bigneſs of a gooſe, and is very erect and ſtately in it's appearance: the head is large, and almoſt naked, except that it has a ſeries of not very long feathers on the crown, toward the hinder part, which it erects at pleaſure, and which, in that ſtate, form a very beautiful creſt: as it flies, they are uſually laid flat to the head; but, when it ſtands or ſits, they are uſually erect; and when it is preparing to ſeize it's prey, they ſtand up in a very peculiar manner: the beak is very large and ſtrong; it is ſtraight and thick all the way, till near the point, but there it turns down, and becomes hooked and very ſharp; it is of a blackiſh colour: the tip is quite black and extremely hard: the membrane that covers it's baſe is yellow, and the noſtrils are placed tranſverſely in it, and are very large and conſpicuous: the eyes are bold and piercing, but they are not beautiful, and the whole aſpect is indeed rather diſguſtful and forbidding.

The back and upper part of the wings are of a beautiful gloſſy colour, between a reddiſh-brown and black: the breaſt is paler, and approaches to an orange colour, with a great predominance of the yellow: the wings are darker, and their long feathers are almoſt black, as is alſo the tail: the wings are extremely long, and ſpread to a ſurpriſing extent, as the birds fly, and the tail alſo is long; there is both in the wings and

and tail a glowing tinge of red along with the dark colour, seen in some particular lights very beautiful.

The legs are very robust and long; they are feathered down to a considerable depth, but not quite to the toes: the naked part is yellow, and the claws are very long, sharp, and black.

This species is a native of the East, and flies usually in considerable numbers together; it is an extremely beautiful, but a very terrible, bird: it's wings are so long, that it makes a great noise in flying; and it's legs are so well formed for running also, that it pursues it's prey on foot, and will often overtake it that way. It is very voracious; it feeds on birds and quadrupeds, and does not decline feasting on carcasses of any kind. Most of the authors who have written on this part of natural history, have described it. Aldrovand calls it Vultur Leporarius, from it's feeding on hares; and many of the later writers have borrowed that name from him. The natives of the countries, where it is frequent, are not only afraid for the young of their cattle, but even for their children.

Vultur pedibus caeruleis, dorso nigricante.
The blue-legged Vultur, with a blackish back.

The golden-breasted Vultur.

This is a very large, and a very beautiful, bird; it's size is equal to that of a turkey-cock, and it is remarkably stately in it's port and manner of standing: the head is large, and rising on the crown, not flatted: the eyes are very large and piercing; their iris is a deep hazel, but the whole aspect of the face is disagreeable: the beak is very long for a bird of prey, and very strong: it is rounded on the upper part, and runs straight from the head, almost to the extremity, where it is very hooked and sharp: it is throughout of a horn colour, and the membrane which covers the base is bluish-black; the nostrils are very conspicuous in this, and placed transversely.

The back and shoulders are of a very deep colour, approaching to black, but in some places, especially toward the neck, they are variegated with a few spots of a ferruginous brown, and in some parts with white ones, but these are few and small: the breast and belly are of a kind of pale and very bright orange colour: the yellow in this is predominant, but there is every-where a tinge of the reddish: there is more of this toward the neck, and on the upper part of the breast, than elsewhere, and the belly is almost simply yellow.

The long feathers of the wings are of a deep dusky brown, somewhat paler on the upper side than on the under; the tail is moderately long, and is of the same colour with the wings: the legs are robust and remarkably long; they are very thick covered with feathers on the upper part, but this does not reach quite to the feet; the naked part is of a deep bluish colour: the feet are large; the toes long and thick, and the claws very formidable.

This is a native of the East, and we have sometimes the skin of it stuffed, brought over by people of curiosity, who have been up the Levant; but the living animal has not been seen in Europe. Some of the writers on birds have described it but imperfectly. Gesner, from the colour of it's breast, and more especially of it's belly, calls it Vultur Aureus, and others have borrowed the same name; whence those who have known nothing of it farther than the name, which is the case of the generality of those who at this time call themselves naturalists, have supposed it to be all over of a gold yellow. It is a very rapacious bird, and seizes on almost every thing smaller than itself, but it feeds more on quadrupeds than birds.

Vultur pedibus flavis, dorſo fulvo.
The tawny Vultur, with yellow legs.

The brown Vulture.

This is of the bigneſs of a large capon: the head is large, and elevated, not ſtatted, as in the falcon-kind: the beak is long for a bird of prey; it is very robuſt and black; it runs ſtraight a conſiderable way from the head, but toward the point it bends downward, and becomes hooked: the membrane which covers it's baſe is yellow, and the noſtrils are very conſpicuous in it; they are large, and ſomewhat oblong: the eyes are large, and have a very fierce and cruel look; their iris is of an orange colour.

The head, the whole neck, and the upper part of the breaſt are wholly without feathers; they are covered, in the ſtead of theſe, with a white, ſhort, and ſoft woolly or downy matter, and on the hinder part of the neck, at it's bottom, there are a number of very beautiful, oblong, and very narrow feathers; theſe are longer than any of thoſe on the back or breaſt, and of a paler colour, and more gloſſy: the back is of a tawny or yellowiſh-brown; the breaſt and belly are of the ſame kind of colour, but ſomewhat paler; and in the lower part of the belly, toward the inſertion of the tail, there is a caſt of whitiſh: the upper part of the wings are of the ſame tawny colour with the back, and the rump is alſo of the ſame, only a little paler.

The long feathers of the wings are of a deep brown, approaching to blackiſh: they are very long, and the plumage remarkably broad; their quills make the fineſt of all pens for writing: the tail is compoſed of feathers reſembling thoſe of the wings, but not quite ſo dark in colour.

The legs are covered down, below the middle joint, with feathers of the ſame brown colour with thoſe of the body, except that they are a little paler: the naked part of the legs is yellow, and the toes are very ſtrong, but not remarkably long; the claws, however, make ſome amends for this, and are remarkably long and ſharp.

This is a native of Paleſtine and other parts of the Eaſt; there are uſually ſeen large flights of them together, and they are leſs miſchievous than moſt other of the large birds of prey: they take long flights, and rarely ſtop to do any miſchief: they feed on carcaſſes, and ſeem to have a very remarkable ſcent, for they will aſſemble in companies, and go a great way to find them. We have ſometimes had this ſpecies alive with us, among the people who live by ſhewing foreign creatures, but rarely: none of the authors who have written on birds have perfectly deſcribed it. It comes the neareſt to the Boetic Vultur of Geſner before deſcribed, but differs in many ſo eſſential particulars, that it can by no means be allowed the ſame ſpecies.

Vultur pedibus albeſcentibus, dorſo nigricante variegato.
The white-legged Vultur, with a black, variegated back.

The Braſilian Vulture.

This is the ſmalleſt of all the Vulturs properly ſo called, but it is not the leaſt beautiful: the bigneſs is about that of a kite, but it is longer, in proportion to it's breadth, and it's legs are longer than in that bird: it alſo ſtands more erect, and in the whole makes a very elegant appearance: the head is ſmall, and not flatted; the beak is moderately long and very robuſt; it is ſtraight a great way from the baſe, but toward the point it turns down, and is very ſharp and hooked: the colour is a deep blackiſh, but it is covered, above half way from the baſe, with a yellowiſh membrane, in which are placed the noſtrils: theſe are long, and ſtand tranſverſely: the eyes are very large and beautiful; their iris is of a bright red, and the pupil is round and black.

The head is naked, but the ſkin which covers it is wrinkled, and is divided, as it were, along the top: the upper part of this ſkin is blue, that below the eye yellow, and that on the top of the head is of a deep browniſh-red; the colours on the reſt of the head are varied in the ſame manner; the variations are quite irregular, and often the two ſides of the head are quite unlike one another.

The

The neck is in part naked, but the lower part of it has before some fine downy matter, and behind a few long feathers: the back is of a dark dusky brown: the breast and belly are of a paler brown; the wings and tail are darker; the wings are remarkably long, and the legs, where they are not covered with feathers, are white.

This is a native of the Brasils. Marcgrave calls it Urubu, and Nieremberg, Aura; Ray and Willughby call it Vultur Brasiliensis Urubu dictus.

PSITTACUS.

THE beak of the Psittacus is of a hooked or uncinated figure, and the toes are four in number, two of which are situated before, and two behind the foot.

Division the First.

The larger Psittaci, called Macao's.

Psittacus cauda cuneiformi, temporibus nudis rugosis.
The Psittacus, with a cuneiform tail, and with naked, rugose temples.

The Macao.

This is a large and an extreamly beautiful bird: it is equal to a well-grown pullet in size, and it's tail is so long, that, when full-grown and in perfection, it measures, from the top of the head to the extremity of it, more than two feet and a half: the head is moderately large, and rises on the crown; the beak is very large, thick, and strong, and is so hooked, that it is of a semicircular figure: it is more than two inches and a half long, and the upper chap of it is two fingers breadth longer than the under, and the whole is black: the eyes are large; they are white and black, and they are surrounded by three long black lines arising from the base of the beak, and continued in a crooked form, so that they represent the letter S quite to the neck: the top of the head is a little flatted, though the sides rise into a convexity; it is green: the throat is ornamented with a black line, which surrounds it in the manner of a necklace: the breast, the belly, the thighs, the rump, and the under part of the tail, are of a fine, strong, and elegant saffron colour: the upper part of the neck, the back, the wings, and the upper side of the tail, are all of an extreamly beautiful blue: the tail is very long, and of a cuneiform shape.

The legs are very short, but they are robust and strong; they are brown, and the toes are long, and are armed with very long, sharp, and strong black claws.

This is a native of the southern parts of America, and of some parts of the East; almost all the writers on birds have described it. Gesner calls it Psittacus Cyanocroceus; Aldrovand, Psittacus maximus Cyanocroceus; and Ray, Willughby, and others have borrowed the same name. We have it frequently brought over to us alive, on account of it's beauty, and, when in perfection, it is extreamly elegant. It feeds on fruits, and builds in the hollow trunks of trees; it lays two or three eggs, which are roundish, white, and large; when kept with us, it naturally learns to imitate the human voice, and the noises of many of our domestic and other animals.

Psittacus cauda cuneiformi, temporibus nudis albentibus, dorso puniceo.
The scarlet-backed Psittacus, with naked, whitish temples.

The red Macao, or Cockatoon.

This also is an extreamly elegant bird: it is of the size of our raven, and in colour is equal to any of the species in elegance and splendor: the head is very large, and flatted at the top: the beak is very large also, and is so extreamly hooked, that the whole forms a kind of semicircle: the upper chap is white, and the lower one black; and the upper one hangs over the lower to a considerable length: the nostrils are small

and

and round; they stand near one another at the upper part of the beak, near the feathers: the region of the temples is white, naked, and wrinkled.

The back, the beginning of the wings, the throat, the breast, belly, thighs, and the upper part of the tail, are all of the most strong and beautiful scarlet imaginable: the large wing-feathers are also of the same bright colour on their inside, but the second series of the feathers covering the wings are yellow, with purple edges, and they have each a little and beautiful eye, or round spot of blue near the extremity: the upper surface of the long feathers of the wings, and the lower part of the rump, are of a beautiful blue, and the tinge is very elegant.

The legs are short, but they are very robust and strong; they are of a dusky greyish colour: the toes are long and strong, and are armed with very long, sharp, and black claws.

This is very common in the East Indies, and in some parts of America. It is sometimes brought over alive to us, but is not easily kept so, unless great care be taken to preserve it from the cold in winter, and at best it loses a great deal of it's spirit and vivacity with us: it is, when native, one of the nimblest birds in the world, but, with us, there hardly is a slower creature in all it's motions: it indeed appears torpid and frozen, to a numbness of it's limbs. It will be taught to imitate the human voice, and is often very docile. It lives, in it's wild state, in thick woods, and builds in the decayed trunks of trees; it lays four eggs perfectly white, and of the shape and size of those of the pigeon, and it's young are a great while before they arrive at their beauty. Most of the writers on birds have described it. Aldrovand calls it Psittacus maximus alter. Gesner, Psittacus maximus puniceo-cæruleus; and Ray, Willughby, and others have borrowed one or other of these names. We call it the Macao and Cockatoon.

Psittacus temporibus nudis levibus, capite miniaceo.
The scarlet-headed Psittacus, with naked, smooth temples.

The red-headed Macao.

This also is a very beautiful bird; it is of the size of the raven, and very elegantly variegated with colours of an extream brightness: the head is large, and of a beautiful bright scarlet: the beak is very large, and is not so extremely bent as in the other species; it is very considerably hooked, indeed, but it does not form a semicircle: the upper chap is white, and the lower one black, and the upper one is considerably longer than the other: the nostrils are small and round; they stand near one another, and are situated at the base of the beak, near it's insertion at the head; and from the base of the beak to beyond the eyes there are no feathers, but the head is so far covered only with a whitish skin, which is thick and tough, but not wrinkled.

The breast and belly are of the same elegant scarlet colour with the head: the wings and the tail are very beautifully variegated with red, yellow, and blue: the tail is very long and very beautiful; it is of a fine sky-blue, and the two principal or middle feathers are acuminated at the end.

This is very frequent in the East, and is also a native of many parts of America: it is easily kept alive with us, and learns to imitate the human voice very readily. It bears the cold of our climate better than any of the other species; so well indeed, that the late Duke of Richmond had a number of them wild in his garden, where they kept on the trees in a remote part of the grove, near a house at which they were fed, and in which there was a German stove for warmth; they came into this, when they pleased, and were so happy and so healthful, that they built and laid their eggs there, but they never hatched any young. They did not build, as in America, in the trees, but probably, on account of the warmth, kept in the house: the nest was in some corner of the ground; the eggs were only two, at the most three, and were white, and much like pigeons eggs. While the hen sat upon these, the cock would walk before

before her to guard her, and would scarce suffer even the person who fed them to approach within three or four yards of her.

Most of the writers on birds have described this species. Ray, Willughby, and others have called it Psittacus major diversicolor; Gesner, Psittacus Erythrocyaneus; and Marcgrave, by it's Brasilian name, Araracanga. We call it by the general name of the Macao.

Psittacus cauda cuneiformi, temporibus nudis, lineis plumosis.
The cuneiform tailed Psittacus, with naked temples, and plumose lines.

The Ararauna.

This is another very beautiful species; it is of the size of a full-grown pullet: the head is large and flatted at the top; the beak is very large, and extremely hooked: the upper chap is a finger's breadth and a half longer than the under one, and turns down over it: the colour of this is a dirty brownish-white, and that of the lower chap a dusky brown, almost black, sometimes entirely black: the nostrils stand very high near the insertion of the beak at the head; they are small, round, and almost contiguous: the region of the temples is naked, and variegated with a dusky, blackish colour, on a whitish-ground, and decorated with plumose lines: the feathers which form these are black and very small, and the whole variegation caused by these lines looks like needle-work: the anterior part of the head, a little above the insertion of the beak, is ornamented with a kind of crown or mitre of green feathers: the throat on the upper part, just under the beak, has a circle of elegant black feathers; and the sides of the neck, the rest of the throat, the whole breast also and belly, are yellow: the head, the neck behind, the whole back, and the upper part of the wings, are all of a strong and elegant blue.

Toward the extremities of the wings there are some yellow feathers, and there are also some of the same colour about the tail: the tail itself is very long and blue, but these feathers, and some of the others, are only blue on their upper surface, being black underneath, and having often somewhat of blackness about the edges also.

The legs are very robust and thick, but they are short; they are of a dusky greyish colour, and the toes are long, strong, and armed with very sharp claws.

This is a native of many parts of the East Indies, and also of South America. Most of the writers on birds have named it. Marcgrave has described it under it's Brasilian appellation of Ararauna; and most of the writers since his time have called it by the same name. We have it sometimes brought over to us alive, but it is miserable in our winters.

Psittacus maximus albus capite cristato.
The great white Psittacus, with a crested head.

The crested Cockatoo.

This is a very stately and fine bird; the size is that of a common hen: the colour is all over a bright and beautiful white: the head is large, and is ornamented with a very elegant crest, which it erects at pleasure, and which is formed of a great number of moderately long feathers: the beak is very large and hooked; the upper chap is whitish, and much larger and longer than the other, which is black: the eyes are small; and the nostrils, which stand at the top of the beak, are also very small, roundish, and almost contiguous: the wings are moderately long, and the tail is very long and white.

It is a native of America, and is frequently brought over to us; it lives with us more comfortably than many of the other species, and very familiarly learns to imitate our voices.

Psittacus cauda cuneiformi cinereo-plumbeus.
The lead-coloured Psittacus, with a cuneiform tail.

The grey Cockatoo.

This is of the size of our common crow, and is less beautiful than most of the other species of this genus: the head is large, and flatted on the crown; it is well covered with feathers, but it has not the least appearance of a crest or crown: the beak is very large; the upper chap is very much hooked, and is two fingers breadth longer than the under: the nostrils stand high, and they are small, contiguous, and round: the eyes are small, but very bright, and the whole front of the head feathered, not in part naked, as in some species.

The neck is thick and short, and is of the same grey with the rest of the bird, only on the throat it is a little paler, and at the sides a little darker than elsewhere: the whole body, the wings, and the tail are of a dusky grey, with some admixture of bluish in it: the tail is long, and the two principal or middle feathers are pointed at the extremities.

The legs are short and thick; they are scaly, and of a bluish-grey colour, deeper than that of the body of the bird; and the toes are long and very robust, and armed with long and strong claws.

This is very frequent in the warmer parts of America, but it is not known in the East Indies; it is very clamorous in the woods. Most of the late writers on birds have described it. Ray, Willoughby, and others call it by the Brasilian name Maracana, mentioned first by Maregrave. It is sometimes brought over to us as a curiosity, but, having no great beauty, it is less regarded than the others, whose gaudy colours strike the eye.

Psittacus cauda cuneiformi, capite viridi, alis intus rubris.
The green-headed Parrot, with a cuneiform tail, and the wings red within.

The green-headed Cockatoo.

This is of the bigness of a large tame pigeon, and is a very beautiful bird: the head is large, and flatted on the crown; the beak is very large, and so hooked, that it approaches to a semicircular figure; the upper chap is much larger and longer than the under, but both are black: the nostrils are small, round, and situated near one another at the top of the beak; the skin about the eyes is naked, and of a whitish colour, but it is beautifully variegated, as if with needle-work, with black lines: the eyes are small, but they are very bright; their iris is of a fine deep orange, or saffron colour, and the pupil is round and black.

The head is of a very bright and elegant green colour: the neck, back, and wings are also of a green, but the colour there is deeper than that of the head: the top of the head, indeed, is somewhat bluish, and this shews the reason of the difference of the rest of it from the green of the body, there being a faint tinge of the bluish all over it, though scarce distinguishable, unless by the comparison with the more perfect green of the body.

The tail is very long and very elegant; the feathers are green on their upper side, except that toward the tips they become blue, and of a fine strong scarlet on the under side: the wings also are of the same high scarlet underneath, and the tips of the wings, like the extremity of the tail, are also bluish.

The legs are very short, but they are thick and strong, as in the other species; they are of a dusky brown: the toes are long and robust, and the claws are also very long and sharp. There is frequently, but not universally, in this species, a spot of brown on the front of the head, a little above the insertion of the beak.

This

This is frequent in the Brasils, and in some parts of the East Indies; it is very clamorous and noisy: it lives principally in the thickest woods, and feeds on fruits: it builds in hollow stumps of trees, and lays two or three white roundish eggs. We have it sometimes brought over to us from the East Indies, but rarely: it is very noisy with it's own wild notes, when we have it here, but it seldom learns any thing. Margrave has described it under it's Brasilian name Maracana, by which, however, he confounds it with another very different species. Most of the later writers have mentioned it also by the same name, and have contented themselves with such a description as they could form upon his account, though the bird itself might have been seen.

PSITTACI.

Division the Second.

The smaller kind, called Parrots.

Psittacus pedibus flavis capite cristato albus.
The white-crested Psittacus, with yellow legs.

The White Parrot.

THIS is a beautiful and a very singular species; it is about the bigness of a large tame pigeon: the head is large, and well-covered with feathers; the beak is large, hooked, and of a grey colour, tending to black: the nostrils are small, round, and situated near one another on the top of the beak, near the head; and between them there is a prominent tubercle: the mouth opens very wide; the tongue is broad and red: the eyes are large, and very bright and beautiful; their iris is yellow, and the pupil black: the whole body is all over white, and the crest on the head is of the same colour. It is composed of a considerable number of large feathers, which are two inches long, and stand nearly erect, only that their middle turns a little backward, and their extremities, which are sharp or pointed, turn again forwards; there are about ten of these, and they are of a brighter and more elegant white than the rest of the body.

The tail is of a very singular figure; it is about five inches long, and is not carried longitudinally, as in the other species, but is turned up as our hens carry it: it is composed of a considerable number of feathers, and seems to answer some way, by it's port and form, to the crest on the head.

The legs are short, robust, and yellow; the feet are strong; the toes are thick and scaly, but the claws short and inconsiderable.

This is a native of some parts of America, but it is not common any where. The late Duke of Richmond had one of them alive many years; and we have seen two others shewn about London as curiosities among other foreign birds. Most of the writers on this part of natural history have described it. Aldrovand calls it Psittacus albus cristatus; and Ray and Willoughby have borrowed his name.

Psittacus viridis alis rubro variegatis.
The green Psittacus, with the wings variegated with red.

The common Green-parrot.

This is of the size of a tame pigeon: the head is large, and well-covered with feathers; the beak is large, prominent, or gibbose on the upper part, and considerably hooked: the upper chap is longer than the under, and is of three different colours, when the bird is in health and perfection; it is black at the extremity, thence it is a little way bluish, and toward the base it has a tinge of reddish; the lower chap is of a pale, simple, brownish-white.

The nostrils are situated toward the base of the beak; they are two small, round apertures, and stand very near one another: the eyes are large and very beautiful; they are of a deep saffron colour in the iris, and the pupil is black: the upper part of the head is yellow.

The body is throughout of a very elegant green: the back is deeper, and the belly is paler, and has often a cast of yellowish; the wings are of the same deep green with the back, but the largest of the long feathers in them is bluish, and the upper edge of each wing is of a very beautiful red.

The tail is short, and has a great deal of red in it on the upper side, and has some yellow, and a spot or two of the same red on the under part: the legs are grey; they are short, but very robust: the feet are strong, and the toes long and scaly; the claws are black, but they are not very sharp or long, though of sufficient thickness.

There is usually a circle of a whitish colour about the eyes, but this, though very common, is not universal: the colour of the back is, in the male, of a brighter, in the female, of a duskier, green, but in both it is very elegant.

This species is a native of many parts of America, and of the East Indies: all the writers on birds have described it. Aldrovand calls it Psittacus viridis alarum costa superne rubente; and most of the late writers have mentioned it under the same name. It is the species most frequently kept in our houses, and learns to imitate the human voice with great facility.

Psittacus viridis pedibus plumbeis, alis variegatis.
The green Psittacus, with variegated wings, and bluish legs.

The painted-winged Parrot.

This is of the size of a crow: the head is large, and well-covered with large feathers; the beak is remarkably large and strong: the upper chap is a great deal larger than the under, and it's hooked point bends over it, and is extended singly for half a finger's breadth in a hooked form; the colour of the upper part of it is a bluish-green, and at the sides it is yellow; toward the extremity also it has a transverse white spot: the lower chap is of a lead colour, but somewhat yellow in the middle.

The upper part of the head is ornamented with a number of yellow feathers: the sides of the head and the whole body are green: the back and upper parts of the wings are of a darker, the breast and belly of a paler, green, and, indeed, toward the belly, there appears a little yellowness in the green; and the roots of the feathers are every-where grey.

The large feathers of the wings are, at their origin, green, but this by degrees turns lower down the feather into a bluish, and thence into an elegant purplish, hue, and toward the extremities there are evident tints of a fine purple, and some black. Those of the second order are entirely yellow, and make a very beautiful variegation in the middle of the wing; and the feathers on the interior part toward the belly are beautifully stained with purple, black, green, and scarlet.

The tail is large, and very beautiful; it is composed of twelve long and very large feathers: the four outer ones each way are green at their base, but lower down they are yellow on the inner side, and of the most elegant scarlet on the outer: and the others have, beside these colours, variegations of yellow and green: the four middle feathers have much less variegation; they are of a simple green all the way to the extremities, where they are tipped, as it were, with yellow.

The legs are robust and short; the feet are strong, and are of a dusky lead colour: the toes are thick, but not long; the claws are black, and they are longer and sharper than in almost any other of the Parrot-kind.

The

This species is a native of the East Indies, and is frequent also in some parts of America, and in other warm regions; it is not unfrequently brought over into Europe, and learns as readily as any of them to imitate the human voice. The writers on birds have all described this species. Aldrovand and Gesner call it Psittacus Poicilorynchos; and Ray, Willoughby, Charleton, and others have borrowed the same name, though it would not have been difficult to have devised a better.

Psittacus viridis capite et pectore flavescentibus, alis coccineo variegatis.
The green Psittacus, with the head and breast yellow, and the wings variegated with scarlet.

The yellow-breasted Parrot.

This is of the size of a common pigeon, and is a very beautiful bird: the head is large and rounded; it is covered thick with broad, though short, feathers: the beak is large, very strong, much hooked at the extremity, and throughout of a deep black: the nostrils are situated near it's base, and are two round apertures near one another: the eyes are very bright and beautiful; their iris is of a deep saffron colour, and the pupil black.

The head is of a bright and beautiful yellow, except just on the crown, where it is of a bluish-green: the throat is also of the same bluish-green, and the breast and belly are of the same yellow with the rest of the head; but the belly, toward the lower part, has some faint tinge of greenish among the yellow: the back is green, and the upper part of the tail is partly yellow, and partly green: the neck and the upper part of the wings are of the same full green with the back, but the upper verge of the wings, toward the body, is tinged with a beautiful scarlet; and the feathers in each, next to the body, are black, and the others tinged with scarlet at the extremities: the rump is also tinged with a fine scarlet at the hinder edge: the legs are brown; they are very short, but very robust: the feet are large and strong, and the claws are black and strong, but not very long or sharp.

This is a native of the West Indies, and of some parts of the East; it is very noisy in the woods, but, when kept in a cage, it is not very docile. Most of the authors who have written on birds have described it. Gesner and Aldrovand call it Psittacus viridis melanorynchos; and Ray, Willoughby, and others who have written since them, have copied the name they found in them, to save the trouble of forming a better.

Psittacus viridis fronte albescente, gula cinnabarina.
The green Psittacus, with a white front, and red throat.

The red-throated Parrot.

This is an extremely beautiful species; it is of the size of a large pigeon: the head is large and rounded; the beak is very large, thick, and white; the nostrils are small, round, and situated very near one another, and very high: the eyes are large, but they are less bright than that of many other species; their iris is of a dusky, ferrugineous brown, with some tinge of red: the pupil is black.

The fore-part of the head is white, variegated with small spots of black, and the hinder part, as also the neck, back, wings, and rump, are all of a full deep green: the throat is of a very bright and beautiful scarlet, and the tops of the wings are also of the same colour: the breast is green, and so are also the feathers which cover the thighs, but the belly, between the thighs, is of a reddish-brown: the upper parts of the wings have some of the smaller feathers blue, variegated with white; and the very extream part of the belly, near the insertion of the tail, is yellow.

The tail is short, but very beautiful; it is red in the middle, and yellow at the sides, and in some parts towards the edges variegated with blue and green among the yellow; and almost all the feathers are black at their extremities.

The legs are of a pale grey colour; they are short, but robust, and the toes are long, and not so thick and clumsy as in many others of this species: the claws also are longer, slenderer, and sharper at the extremities than in almost any other species: they are of a deep black.

This is a native of the American Islands, but it is less frequent there than most of the other species; we have it sometimes brought over to Europe, but it does not succeed so well with us as the others, generally becoming sickly, and losing it's feathers. Many of the writers on these subjects have mentioned it. Ray, Willughby, and others of the late ones have retained Aldrovand's name of Psittacus leucocephalos, the white-headed Parrot, though that expresses too much, as the whole head is not white, but only the fore-part of it.

Psittacus dorso caeruleo, ventre virescente.
The Psittacus, with a blue back, and green belly.

The blue Parrot.

This is an extremely beautiful bird: our English name but very imperfectly expresses it's singularity in colouring; it is smaller than the common green Parrot: it does not indeed exceed a common wild pigeon in size, but it is thicker, in proportion to it's length: the head is moderately large and round; the beak is not so large as in the generality of these birds, but it is considerably hooked, and is all over of a black colour: the nostrils are small, round, and situated near it's base; the eyes are bright, and their iris is of a pale brown, their pupil black.

The head is of a bright and beautiful blue, but the region of the eyes is whitish; and there is an elegant spot of yellow on the crown: the neck and breast are blue, but the belly is green, and the feathers which cover the thighs are of a beautiful whitish-green: the rump is yellow; the back is of a beautiful strong blue, sometimes paler, but always elegant: the wings are variegated in an extremely pretty manner with green, yellow, and red: the extream part of the back is yellowish.

The tail is short, and composed of twelve robust feathers, and there are a few small ones that cover the bases of the others; they are all green: the legs are robust, short, and grey, and the toes are short and clumsy; they are of the same grey with the legs, or a little paler, and the claws are long, black, and sharp.

This elegant species is very frequent in the Island of Madagascar; we have it also from some other places, but it does not live so long, or so healthfully, with us as most of the others: it feels our winter very severely, and usually is in a bad condition, as to it's feathers. Most of the authors who have written on birds have described it. Ray and Willughby borrow the name of Aldrovand, and most others after them have also called it Psittacus versicolor sive Erythrocyaneus.

Psittacus cinereo-subcaeruleus.
The bluish-grey Psittacus.

The grey Parrot.

This is one of the least beautiful species of the Parrot-kind that we are acquainted with, but it is valued for it's great docility, and it's articulate voice; it is of the bigness of a large tame pigeon: the head is large, round, and covered with broad feathers: the beak is large, moderately hooked, and black; the nostrils are round, and stand near one another in a white membrane, that covers the base of the beak.

The head and body are of the same colour, which is a simple, uniform, bluish-grey; but the lowest part of the back, the belly, and the rump, are paler than any other part, and have less of the bluish-cast than the rest.

2 The

The wings are long, and are of the same colour with the rest of the bird; the covering feathers are paler than those of the back, and the under part of the wings is also yet paler. The bird is not a swift nor an easy flier, notwithstanding the advantage of this breadth of wing, but it's body is bulky, and it's motions none of them quick.

The tail is very short; it scarce reaches at all beyond the tips of the wings, when they are closed; but, though not conspicuous for it's length, it is sufficiently so in it's colour, being of a bright and elegant red.

The legs are of a bluish-grey; they are short, but robust: the toes are long and strong, and the claws are remarkably long, large, and sharp in the wild state; though, with us, the treading on the dung in a cage injures them greatly.

This is a native of the East and West Indies, and of many parts of Africa. It is very common with us in houses, and, though the least beautiful of all the Parrot-kind, is valued for it's docility, and the clearness of it's voice: it is the happiest mimic of all the race, not only of the human voice, but of all other sounds. All the writers on birds have described it. Gesner, Aldrovand, Charleton, Willughby, and Ray, all call it Psittacus cinereus sive subcaeruleus; we, the grey Parrot.

The red-rumped Parrot.

Psittacus cinereus alis et uropygio rubro tinctis. The grey Psittacus, with the rump and wings tinged with red.

This is a very beautiful bird; it is one of the largest of the Parrot-kind: it's size is about that of a well-grown pullet: the head is large and rounded; the beak is remarkably large; it is very hooked and black all over: the upper chap is much longer than the under, and is very prominent or convex toward the middle: the nostrils are round, and stand near one another at the base of the beak: the eyes are small, but very bright; and the tongue is large, thick, and fleshy.

The head and neck are covered with short, broad, and very thick-set feathers, the body with longer; the whole bird is of a very pale and beautiful grey: it has nothing of the dusky lead colour or bluish tinge of the common grey Parrot, but the grey is pale, silvery, and almost white, and the hinder part of the back, and the white rump, are of a strong scarlet, very bright and beautiful.

The wings are long and large; their covering feathers are of the same whitish-grey with the rest of the bird, but the long feathers are of a beautiful scarlet, the same with that of the rump; these make an elegant variegation in the colour, as the bird sits; and still more so, when they are expanded for flying, as much more of the scarlet is then seen.

The legs are short, thick, and of a dusky lead colour; the toes are long, thick, and scaly, and the claws are robust, black, and sharp, but not very long.

The tail is short, it hardly reaches beyond the tips of the wings; it is of the same colour with the body: it's size would naturally have led authors to the ranking it among the macao's, but the shortness and form of the tail forbid.

This is frequent in the woods in the American islands, and is sometimes brought also from the island of Madagascar, and some other places; we have it often brought over alive, but it does not bear our winter so well as some of the other species. The late Lord Petre had one which lived several years, and spoke very articulately. Gesner and Aldrovand have described it under the name of Psittacus Erythroleucos; and Ray, Willughby, and many of the late writers have borrowed the same name, though it is not quite proper, the general colour being not an absolute white, but a fine pale, silvery grey.

Psittacus

Pſittacus viridis capite flaveſcente, vertice cæruleo.
The green Pſittacus, with a yellow head, and blue crown.

The painted Parrot.

This is one of the beautifulleſt of the whole Parrot-kind; it is of the ſize of a tame pigeon: the head is large and round, and is remarkably beautiful in it's form, as well as colouring; the beak is very large, long, hooked, and black; the upper chap is much longer than the under, and the noſtrils are round, and ſtand near it's baſe, but not ſo cloſe to one another as in the other ſpecies: the eyes are very bright and beautiful; the iris is of a gold yellow, and the pupil black, and they have a peculiarly briſk and ſprightly aſpect.

The head is of a fine gold yellow, only on it's crown there is a beautiful and regular oval ſpot of a bright blue: the throat alſo is of the ſame bright yellow with the head.

The body is of a very beautiful graſs-green; the upper part of the back is darker than any other part; the wings are long, and their principal feathers are variegated in an extremely elegant manner; they are one half yellow, and the other half black, except at the very tips, where they are of the ſame beautiful blue with the top of the head, and they have alſo ſome green about them, paler, but not leſs bright, than that of the body, and very happily intermixed with the other variegations.

The tail is moderately long; when cloſed, it appears ſimply green like the body; but, when the bird ſpreads and expands it, the feathers are found to be elegantly variegated toward their edges with black, red, and blue, as if ornamented with fringes of thoſe colours: the legs are robuſt, but ſhort, and the toes are ſtrong and ſcaly; they are of a duſky grey, with a caſt of bluiſh, and the claws are long, black, and ſharp.

This is a native only of the warmer parts of America; it is frequent in the Braſils, and in the woods, about the mines of Potoſi. *Marcgrave*, and other of the writers on the natural hiſtory of thoſe countries, have deſcribed it. They have called it, by it's Braſilian name, *Ajurucura*, and our authors, who have written on the ſame ſubjects, have ſaved themſelves the trouble of forming an intelligible one by keeping this, as they have done alſo in too many other of the Braſilian animals.

Pſittacus viridis capite flavo viridi et albo variegato.
The green Pſittacus, with the head variegated with yellow, green, and white.

This is alſo an extremely elegant bird; it is ſmaller than the former ſpecies, it's ſize not exceeding that of a jack-daw: the head is ſmaller than in moſt of the Parrot-kind: the beak, however, is large; it is of a mixed, browniſh, and bluiſh-grey colour, and the upper chap is conſiderably longer than the under: the noſtrils are very large, and ſtand at ſome diſtance from one another, but very near the baſe of the beak; they are round, and of a paler colour than the reſt of the membrane in which they ſtand: the tongue is large, thick, and fleſhy; and the eyes are ſmall, but very bright and piercing: their iris is yellow, and the pupil black.

The head is of a very elegant blue, but about the center of the crown there is an elegant ſpot of yellow, with ſome white intermixed among it; between that ſpot and the origin of the beak there is alſo a very beautiful ſpot of a ſea-green; and over the eyes there is a broad ſpace of an elegant and pure yellow: the neck is green on the hinder part, but the throat is of the ſame gold yellow with the upper part of the ſides of the head: the body is of a fine graſs-green; the back is darker, and the breaſt and belly paler: the wings are long, and the tail ſcarce exceeds them by half an inch, when cloſed.

The

The legs are short, but they are less robust than in many other species: the toes are slender, and of a bluish colour, and the claws are long and black; the whole bird has a more delicate appearance than almost any other of the Parrot-kind; it stands more erect, and holds it's head in a stately and elegant manner.

This species is frequent in the woods of almost all parts of South America, but it has never been brought alive into England. Sir Hans Sloane has a stuffed one, and I believe it is the only specimen in Europe. Marcgrave and Piso have described it but imperfectly.

Psittacus virescens capite nigro cæruleo et flavo variegato. *The green Psittacus, with the head variegated with blue, black, and yellow.*

This is another of the elegant species known only in South America; it is of the bigness of a common pigeon: the head is large and round, and is close covered with short feathers, which the bird frequently raises into an almost erect position, and by that means makes it appear much larger: the beak is large, very hooked, and of a dusky brown colour: the upper chap is much longer than the under, and the extremity of it, which hangs over the other, is quite black: the base is covered with a thick and wrinkled bluish membrane, and in this are situated the nostrils; they are large, and stand but at a small distance from one another: the tongue is thick, fleshy, and black: the eyes are moderately large, and very piercing; they are yellow in the iris, and the pupil is black.

The ground colour of the head is green, but the crown is ornamented with a large spot of a mixed blue and black colour, and in the center of this there is a spot so yellow: the sides of the head below the eyes have also a great deal of yellow, and the throat is of the same beautiful gold colour, but in the midst of it there is a large and elegant spot of blue.

The hinder part of the neck, the back, the wings, and the rump, are all of a fine strong grass-green; the breast and belly are of an elegant green, but paler: the wings toward their lower part also are paler, as is also the tail; but the extremities of the long feathers in the wings are beautifully variegated with yellow, a strong crimson, and a deep blue: the tail is but short, and, though solely green on the upper part, it is variegated with green and yellow on the under.

The legs are short and robust; the toes are moderately long; the claws are very long and sharp, but they are not thick; they are of a glossy black.

This is a native of the Brasils, and of the woods about the gold mines of Peru. Marcgrave and Piso have described it; they have only given it's Brasilian name Ajurucuraca; and Ray, Willughby, and other of the modern ornithologists, have casually mentioned it only under the same name; we have had two or three very well-preserved specimens of it lately in England.

Psittacus niger pectore rubro. *The black Psittacus, with a scarlet breast.*

The black Parrot.

This is an extreamly singular bird; it's form, and the structure of it's beak and feet, sufficiently distinguish it to be a Parrot, though it's colours, so different from those of the generality of those birds, would induce a person not so well acquainted with this part of natural history, at first sight, to doubt it: the head is moderately large and round, and is thick, covered with short and broad feathers, which it can erect at pleasure, and make it look much larger; the beak is large, only prominent or convex on the upper side; the hooked part of the upper chap hangs a good way over the extremity of the under, and the whole of a dusky grey colour: it is covered at the base by a brownish membrane, with a tinge of blue, and in this stand the nostrils; they are situated very near one another, just at the base of the beak, almost under the front feathers of the head: the eyes are very sharp and piercing, but not very large; their iris is of a bright gold yellow, and the pupil black.

The head is of a deep black, without any variegation: the neck and back in the female are also of the same deep black, but, in the male, the upper part of the back is red: the breast and belly are of a very strong and deep scarlet, but the lower part of the belly, toward the tail, is black: the wings and tail are black, and the legs are of a bluish-grey.

It is a less clumsy bird than most of the Parrot-kind: it stands very erect, and has at once a sprightly and a stately appearance: the legs are longer and less robust than many of the species; and the toes, though they stand two each way, as in the rest, are smaller than in any other: the claws are long, black, and sharp.

This is a native of the forests in Paraguay, and some other parts of South America. It is sometimes carried alive into Portugal, where it is taught to imitate the human voice, as the other Parrots do, and is much valued for the singularity of it's colouring.

Psittacus viridis, capite, pectore, et summitatibus alarum coccineis.
The green Psittacus, with the head, breast, and top of the wings red.

The red-headed Parrot.

This is a very singular and a very beautiful species; the head is small, and less rounded than in most of the others, being compressed at the sides, and a little flatted on the top: the beak is large, very hooked, and it's upper chap prominent a great way beyond the under; the nostrils are small and round, and they stand so near the base of the beak, that they are not easily seen, unless looked for: the eyes are small, but they are very bright and piercing; the iris is of a saffron colour, the pupil black.

The head is of an elegant crimson; the body is green, but the breast and the tops of the wings are of the same strong and elegant red with the head: the wings are long, and the tail reaches about three quarters of an inch beyond their tips, when they are closed: the legs are short, but robust, and the feet are also thick and strong; the claws are sharp and black.

This is a native of South America. Marcgrave has described it under it's Brasilian name Tuiete; and Ray, Willughby, and others have borrowed it.

Psittacus cruribus albis corpore toto viridi.
The white-legged, green Psittacus.

The red-eyed Parrot.

This is a very singular species, though in beauty it is much inferior to the former, and to many of the others: it is of the size of a moderately grown pullet; the head is large, round, and covered thick with short and broad feathers: the beak is very large and very hooked; it is of a dusky brown colour; the upper chap is much longer than the under, and it's upper part is gibbose, and it's point which hangs over the other very sharp: the base of it is covered for a considerable way with a thick, white membrane; in this the nostrils are very conspicuous: they are two roundish apertures, placed very near one another, and toward the base of the beak.

The eyes are small, but very bright and piercing; their iris is of a beautiful red.

The feathers which cover the head are all of the same uniform green: the neck is of a somewhat darker green than the head; and the body, especially the back and upper parts of the wings, are of a green yet darker than that: the breast and belly are somewhat paler, but the whole is without variegation.

The wings are moderately long, and their long feathers very robust; the tail is short; it does not reach more than an inch beyond the tips of the wings, when they are

are closed: the legs are robust, but they are very short and white; the feet are also white, and the toes less thick and clumsy than in many of the others: the claws, however, are very long, black, and sharp.

This species is a native of South America, and we have sometimes had it brought from the island of Madagascar, but it is less valued than the other species, from it's wanting their variegations; a very remarkable one for it's articulation of voice in the city was of this species. Few of the writers on birds have described it. Marcgrave mentions it under the name of Ajurucatinga; and Ray, Willughby, and others have preserved the name, though they have troubled themselves about very little more. Marcgrave has also mentioned it a second time, under the name of Ajurupara, describing a smaller bird of the same species, as if a different one.

PSITTACI

Division the Third.

The lesser Parrots, commonly called Parroquets.

Psittacus viridis torquatus ventre subflavescente.
The green, torquated Psittacus, with the belly yellowish.

The common Parroquet.

THIS is a very pretty little bird, though not so much variegated, in point of colour, as some of the other species; it is smaller than any species of the common Parrot: the head is moderately large and round; the feathers which cover it are short, broad, and well plumed: the beak is remarkably large, and is of a bright red colour; the upper chap is much longer than the under, and the form of the whole is gibbose and hooked: the eyes are not large, but they are very bright and piercing in their aspect; the iris is yellow, and the pupil black.

The head is of a fine grass-green, and the whole body also is green, but the back and upper parts of the wings are darker, and the breast and belly paler, and the belly has some tinge of yellowish: from the beak there runs a black line on each side, which is continued under the chin to the beginning of the breast, and thence to the sides of the neck again, and the joins torquis or circle of red which is extended across the back part of the neck, and makes what is called the necklace of Parroquets. This elegant mark is of the breadth of a man's little finger in the middle, and grows each way somewhat narrower to the sides.

The wings are long, and the large feathers are, some of them, of a duskier green than the others, and these have each a single small round spot of red on them: the tail is very long, and very beautiful; it is not less than seven inches in length, when the bird is in perfection, and it is of a fine pale yellowish-green.

The legs are short, robust, and of a dusky greyish colour, and the membrane which covers the base of the beak is also greyish: the feet are clumsy, the toes short and thick, and the claws less sharp than in some of the other species.

It is a native of the East Indies, and was the first bird of the Parrot-kind known in Europe; we meet with accounts of it in the oldest writers, and we find, by those accounts, that they knew no other. The authors who have written on birds have all described it. Ray, Willughby, Aldrovand, and others call it Psittacus torquatus, Macrourus antiquorum; others, Psittacus torquatus.

Psittacus

Psittacus totus viridis macrouros pedibus rubentibus.
The green, long-tailed Psittacus, with red legs.

The red-legged Parroquet.

This species is distinguished at sight from the former, by it's want of the scarlet collar on the back of the neck, and by the redness of it's legs, as well as by it's size; yet so it has happened, that, in despite of these obvious characters, they have been confounded together: it is but little bigger than a thrush; the head is larger, in proportion to the body, and is round, and covered thick with short and somewhat broad feathers, which, though they usually lie very close, it can erect at pleasure, so as to make the head seem very nearly equal to the body in diameter: the beak is large, and of a fine bright red colour: the upper chap is longer than the under, and is hooked at the point; it is very thick at the base, and is there covered with a thick and tolerably smooth scarlet membrane, in which are placed the nostrils; they stand very high, and are round and situated over one another: the eyes are not very large, but they are very bright, and have a pleasing and beautiful, as well as a piercing, aspect; their iris is of a saffron colour, and their pupil is black.

The head, as well as the whole body, are of a strong and beautiful green, but the back and tops of the wings are of the deepest colour: the head, throat, and breast are somewhat paler, and the belly is yet paler than those, and has a tinge of yellowish.

The tail is very long and narrow; it is pointed at the extremity, and is of the same green with the wings, which is a duskier than that of the rest of the body.

The legs are short, robust, and of a bright red; the toes are short but strong, and the claws are short, not greatly hooked and blackish.

This species is a native of many places both in the East and West Indies, and is frequently brought over to us, though less esteemed than the former species: all the writers on birds have described it. Aldrovand calls it Psittacus minor macrouros totus viridis; and Ray, Willughby, and most of those who have written since his time, have continued the name.

Psittacus macrourus viridi rubro caeruleo et albo variegatus.
The long-tailed Psittacus, variegated with green, red, blue, and white.

The variegated Parroquet.

This is one of the most beautiful birds we are acquainted with: the Parrot-kind are many of them elegantly variegated, but this has more variety of colouring, and the colours themselves are brighter than in any other: it is about the bigness of a common pigeon, and, though in it's general form it approaches to the Parrot-kind, it is, in many particulars, extremely different from them.

The head is small and not rounded, as in most of the others, but flatted on the crown, and somewhat compressed at the sides: the beak also is small, and not only differs in this from that of all the other Parrot-kind, but also in it's shape; it is, indeed, hooked at the extremity, as they are; but whereas in theirs the hooked extremity is formed only of the end of the upper chap, continued beyond the under, and turned over it; on the other hand, this has the under as well as the upper chap turned down, to form the hooked point.

The body is variegated with four colours, all of them very bright and glossy; the principal are green and red: the whole back, the neck, the top of the head, and the covering feathers of the wings are green, except that about the shoulders there are some few blue feathers: the two outer long feathers of the wings are green, but all the

the rest are of an extreamly beautiful blue; and all the feathers of the wings, as well the long ones as the covering ones, have the scapi white.

The variegations about the head are elegant: the beak itself is red, but the membrane which covers it's base is of a deep black colour: the eyes are large, and their iris is of a strong scarlet, and the pupil black: the throat is of a dusky colour, but with a considerable tinge of red: the sides of the head are variegated each with blue spots, one before, and the other behind, the eyes: the breast and belly are altogether red; the tail is very long, and is green, only the scapi of the feathers are white: the legs and feet are of the same deep and glossy black, with the membrane that invests the base of the beak: the feet are clumsy; the toes short and thick, and the claws not very long or crooked.

This is a native of Japan. The first account we had of it was from Aldrovand, who had never seen the bird, but had formed a full description from a painting which had been done from the life: the singularity of the colouring, and of the form of the beak, made Ray and Willughby doubt of the truth and accuracy of the figure; but the bird has been since sent over preserved, and answers perfectly to it.

Most of the writers since Aldrovand have mentioned it, and they have all borrowed his name; he calls it Psittacus Erythroleucus macrouros.

The crested Parroquet.

Psittacus cristatus macrouros crista et cauda rubentibus.
The long-tailed, crested Psittacus, with the crest and tail red.

This is a very singular and a very beautiful species; it has much the appearance of the great, long-tailed Parrot, or Macao-kind, but that it is so much smaller; it hardly exceeds a jackdaw in size: the head is moderately large, rounded, and very thick covered with short and broad feathers, and on it's top has a beautiful crest: the beak is large, protuberant on the back of the upper part, and hooked at the end: the eyes are moderately large, and very bright and piercing in their aspect; their iris is of a fine strong red, and the pupil black: the back and shoulders are of a fine deep grass-green; the breast is of a paler green, but still of the same pure tinge; the belly is yet paler, and is somewhat yellowish: the head also is of an elegant grass-green, but the crest, which is long and very beautiful, is of a fine bright scarlet: this crest is composed of six feathers, three of them longer, and three shorter, and is usually carried erect, though the bird can lay it down flat at pleasure, in which state it gives the appearance of a fine, long spot of red, carried from the hinder part of the head down to the neck, toward the shoulders: the tail is long and narrow; it is formed but of a few feathers, and is of the same elegant scarlet colour with the crest.

The legs are short, robust, and of a bluish-grey: the toes are short and scaly, and the claws are long, black, and sharp.

It is a native of the East Indies; they say that it has been also sometimes brought over from South America, but this seems less attested. Aldrovand calls it Psittacus erythrochlorus cristatus, and most of the writers, since his time, have continued the same name to it.

Psittacus minimus macrouros totus virescens, rostro nigricante.
The very small, long-tailed Psittacus, all over green, but with a black beak.

This is an extreamly pretty little bird, it is not larger than a sparrow: the head is small, not round, as in most of the Parrots, but compressed at the sides, and a little flatted on the crown: the beak is moderately large; the upper chap is longest, and is hooked

hooked at the point; the base of it is covered with a membrane, in which are situated the nostrils very near to one another; and, toward the very base, the upper part is protuberant, and the whole is of a deep black.

The eyes are small, but very sharp and piercing; their iris is of a saffron colour, and their pupil is black: the whole head is of a fine, deep, grass-green.

The body is slender, and is throughout of the same bright green with the head: the wings are short, but the tail is very long, narrow, and pointed at the extremity; this, as well as the whole bird, is of a fine strong green.

The legs are slender, but short; the toes are long, and also very slender, and the claws long, sharp, hooked, and black.

This species is very frequent in the woods of South America, but has not yet been met with any where else. Marcgrave has described it under it's Brasilian name Tui, with the distinction of prima species; and Ray, Willughby, and others have continued the same name, though they seem never to have seen the bird. We have it at this time sometimes sent over dried, as a curiosity.

Psittacus macrourus virescens, capite macula crocea insignito.
The long-tailed, green Psittacus, with a saffron-coloured spot on the head.

This species is about the size of a lark: the head is small, but very elegantly variegated; the beak is moderately large, protuberant on the upper part, pointed at the end, and all over of a deep shining black: the eyes are large; their iris is of a dusky, ferruginous colour; and the pupil black; the form of the head is somewhat compressed and flatted on the crown, but less so than in some others.

The ground colour of the head is a fine grass-green: on the summit or crown there is an elegant and large saffron-coloured spot, and round the eyes there runs a circle of yellowish-feathers: the green of the lower part of the back has a cast of a dusky bluish among it; and, on the upper part of the tail, there is yet more of this blueness: in some lights, indeed, the tail appears rather blue than green; the legs are short, but they are not very robust; the feet are large; the toes long and scaly, and the claws very long, black, and sharp: the legs and the membrane that covers the base of the beak are of a dusky bluish-grey.

This is a very common species on the continent of South America, and in other of the warm climates. Marcgrave and Piso have described it under the name of Tuipararojuba; and Ray, Willughby, and others have given it no other name, though they have all mentioned it, and transcribed as much as these imperfect writers have said about it. We sometimes have specimens of it brought dried from the warmer parts of America.

Psittacus viridis cauda brevi, rostro rubente, et pedibus caeruleis.
The short-tailed, green Psittacus, with a red beak, and blue legs.

The Tuiririca.

The name of this species expresses a singularity, that scarce any other of the Parrot-kind has; it's beak, and the membrane at the base of it, are of a colour remarkably different from that of the legs; whereas it is almost universal, that the legs and that membrane are of the same colour.

This is of the size of a pigeon: the head is large and round, and is covered very thick with short and broad feathers, which it erects at pleasure; the beak is large and very

very hooked; it is of a pale red colour: the upper chap is considerably longer than the under one, and turns over it, and is of a deep blood colour at the point, which is very sharp and solid: the membrane investing the base of it is a strong red, deeper than the generality of the beak, but not so deep as the tip of it: this is somewhat wrinkled, and, toward the part where it is connected to the forehead, stand the nostrils: they are large, round, and situated very near to each other.

The head is of a strong green, but when the feathers are raised, as the bird frequently carries them, there appears a tinge of yellowish throughout them: the back and the upper parts of the wings are of a deep green, and the breast and belly are much paler, and have more yellowness among the green, than there is on the head.

The wings are long, and the principal feathers have, in some lights, a good deal of a bluish tinge mixed with the green: the tail is short; it scarce at all appears beyond the tips of the wings, when they are closed: the colour is the same green with that of the back, but on the under part it is somewhat pale and yellowish.

The legs are short, robust, and of a dusky bluish colour: the feet are large; the toes both long, and thick, and scaly; the claws are very strong, but they are shorter, and less sharp at the point than in some other species.

This is a native of the Brasils, and is often carried over alive to Portugal, where it is kept in cages; it easily grows tame and familiar, and will learn to imitate the human voice with great readiness. Marcgrave and Piso have described it under it's Brasilian name of Tuitirica; and Ray, Willoughby, and the generality of the later writers on birds have preserved that name, and some imperfect descriptions of it.

Psittacus viridis cauda brevi, pectore subflavescente.
The short-tailed, green Psittacus, with a yellowish breast.

This is one of the smaller Psittaci, and is a very beautiful bird, though of no great variety of colouring; it is about the bigness of a lark: the head is large and round, and the feathers which cover it are short, but they are broad and well plumed; they always lie very close, and in a beautiful imbricated order, resembling scales; and the bird never erects them, as is usual with many others of the species: the eyes are large and black, and there is a naked and rough skin of a pale grey about them: the beak is large and thick; it is very hooked, and is all over of a black colour: the upper chap is much longer than the under, turning over it at the end, and the very point is black: the membrane which covers the base of it is of a deep bluish-grey, and the nostrils stand toward the top of this; they are roundish, and are somewhat oval, and are smaller than in almost any other species; but they stand almost close together, as in the generality of the others.

The whole head is of a strong and bright green, only that on the crown there is a tendency to yellowishness, which is not at all seen on the sides: the back and the upper part of the wings are of a very beautiful strong green: the breast is yellowish, and the belly still paler and more yellow: the whole upper surface appears of this strong and elegant green, but in the wings there is a blackishness mixed with it in some degree, that is not at all distinguishable on the back.

The tail is short and broad, it scarce at all extends beyond the tips of the wings, when closed; it is of the same green with the back on the upper side, and on the under it is of a paler green, but there is not much of the yellow tinge of the breast or belly to be distinguished in it.

The legs are moderately long, and are not so thick as in many of the Parrot-kind; they are of a dusky bluish colour and scaly: the toes are short, thick, and scaly, and the claws are long, black, and very sharp.

This

This is a native of Africa, and seems also, from the accounts of Marcgrave and Piso, to be common to the Brasils, and some other parts of South America. We have had it alive in London. Marcgrave mentions it, but he does not give it any peculiar name; he only says, that it in some degree resembles the Tuipararejuba, the species before described.

Psittacus glaucus cauda brevi, capite et pectore luteo.
The bluish-green Psittacus, with a yellow head and breast, and a short tail.

The Jendaya.

This is a very singular species; it is of the bigness of our common thrush: the head is large, round, and very beautifully covered with feathers; the beak is large, very robust and hooked, and is throughout of a deep glossy black; the membrane which covers it's base is also black; it is extended over a third part of the beak, and is of an irregular surface: the nostrils stand toward the base of this, and are of an oval figure, and placed very near one another: the eyes are large, and they are remarkably bright and piercing; the iris is yellow, but it is not broad; the pupil is very large and black.

The whole head is of a yellow colour, deeper at the sides, and paler on the crown: the breast and the belly are also yellow, but of a paler tinge than the head: the shoulders, back, and wings are of a very beautiful bluish-green, a colour not met with in this perfection in any other species, and the whole bird is not only tinged equally on the upper part with this colour, but the feathers are remarkably even, glossy, and shining.

The wings are large, and of the same blue green colour throughout, except at their tips, where they are black: the tail is of the same blue green in the back; the legs are short, robust, and of a deep shining black; the toes are robust and short, and the claws short but thick, and very sharp at the points.

This is a native of Peru, Mexico, and the Brasils: the Portuguese there are very fond of it for it's docility, and sometimes send it over into Europe, but it seldom lives long out of it's native country. Marcgrave and Piso mentioned it under it's Brasilian name of Jendaya; and Ray, Willughby, and others have preserved the name, and copied their imperfect descriptions. The specimens which have been sent over dried to England, and other parts of Europe, since their time, have enabled us to make much more full descriptions of these birds; and to these are, in a great measure, owing the improvements made within the last forty or fifty years in natural history.

Psittacus viridis caeruleo variegatus rostro rubente, pedibus cinereis.
The green Psittacus, variegated with blue, with a red beak, and grey legs.

The Tuiete.

This is one of the small species, but it has an extremely elegant and clean appearance; it is about the bigness of our lark, but it is thicker, in proportion to it's length, and stands remarkably erect and firm: the head is large, in proportion to the body, and is round, and very regularly and beautifully covered with short and broad feathers: the eyes are large, and very piercing; their iris is of a dusky saffron colour; the pupil large and black: there is a small naked space about the eyes, the skin of which is of a pale grey colour, and granulated surface: the beak is very large, in proportion to the size of the bird, and is of a beautiful pale red colour: the upper chap is longer than the under, and it's point turns over the extremity of the other, as in the rest of the Parrots: the membrane at the base is of a bluish-grey; the nostrils are very conspicuous toward the upper part of this; they are perfectly round, and stand very close to one another; the whole membrane is of a wrinkled surface.

The

The head is of a beautiful and ſtrong green: the back, ſhoulders, rump, and tail are alſo of the ſame green, and the ground colour of the wings is alſo the ſame; but the beginnings of the wings are tipped with a beautiful blue, and the extremities of ſome of the other feathers are alſo blue, and theſe ſtand in ſuch a regular order, that, when the wing is cloſed, their tips form a very beautiful and regular oblong blue ſpot, running down the length of the inner ſide of the wing: at the lower part of the back alſo, near the origin of the tail, there is a very elegant blue ſpot.

The tail is ſhort; it is ſcarce at all ſeen beyond the tips of the wings, when they are cloſed; it is of the ſame green with the back: the legs are ſhort, robuſt, and of a pale bluiſh-grey colour: the toes are ſhort, robuſt, and ſcaly, and are of a yet paler grey than the legs; the claws are long, ſharp, and black.

This is a native of the Braſils, and is deſcribed by Marcgrave and Piſo. They call it by it's Braſilian name Tuiete; and Ray, and others who have ſince mentioned it, have continued the ſame name.

The Tui-para.

Pſittacus viridis cauda brevi, macula miniata in fronte.
The ſhort-tailed, green Pſittacus, with a ſcarlet mark on the head.

This is an extremely beautiful little ſpecies: it is of the ſize of a lark, and is very erect and majeſtic in it's poſition, when on foot: the head is round and moderately large; the beak is ſmaller, in proportion even to the ſize of the bird, than in moſt of the Parrot-kind, but it is as much hooked as in any: the whole beak is of a deep blackiſh-grey colour, and the membrane which covers it's baſe is ſo little different from the reſt, that it is ſcarce perceived, but, on a cloſe inſpection, it is of a granulated ſurface, and the noſtrils ſtand near one another almoſt at it's baſe; they are of an oval figure, and are duſkier-coloured within, than any other part of the membrane.

The head is green, but juſt in the front it has an extremely bright and beautiful ſpot of ſcarlet conſiderably large, and forming a kind of crown: the back and wings are of a bright ſhining green, but in the middle of the wing there are ſome beautiful variegations of yellow: the tail is ſhort, and ſcarce appears at all beyond the tips of the wings, when they are cloſed: the legs are of a greyiſh colour, and they are robuſt, though ſhort: the toes are ſhort, thick, and ſcaly; and the claws are very long, black, and ſharp.

This is a native of many parts of South America, and is ſometimes ſent over dried to Europe, among the other curious birds of that country. We had two ſpecimens lately ſold among ſome American birds in glaſs-caſes, at the auction of the late Duke of Richmond's curioſities; they were a male and female, and the only viſible difference was, that in the male the beak was of a fine bright fleſh colour. Marcgrave has mentioned it among his Braſilian birds, under the name of Tuipara; and Ray, Willughby, and others have preſerved the name, and ſome of the more obvious characters.

The Anaca.

Pſittacus variegatus pectore ferrugineo.
The variegated Pſittacus, with a ferrugineous breaſt.

This is one of the moſt ſingular, and alſo of the moſt beautiful, of the Parrot-kind; it is exactly of the form and figure of the common green Parrot, but it is not much bigger than a ſparrow: the head is large and round; the beak is not very large, but it is conſiderably hooked; the upper chap is longer than the under, and in all reſpects formed like that of the other Parrots; the whole beak is of a bright brown, but the membrane which covers it's baſe is greyiſh, and the noſtrils ſtand near one another, and are round: the eyes are large, and very bright and piercing; their iris is yellow, and the pupil black.

The head is of a brownish colour, with a tinge of ferrugineous at the sides, but on the top it is of a bright liver colour: the upper part of the neck and the sides are green; the throat is grey, and the breast and belly are of a mixt colour, the principal tinge in which is a ferrugineous brown: the back is green, but in it's middle there stands also a large brown spot, like the colour of the breast, but not so much ferrugineous: the tail also is of a pale brown.

The wings are long; when closed, their extremities reach to the end of the tail; toward their top there is a very beautiful spot, which is of a bright blood colour in the male, and brown in the female; the middle part of the wings is green, but toward the extremity there is somewhat of a bluish tinge with it: the feathers which cover the thighs are of a bright pale green.

The legs are robust and moderately long; they are of a greyish colour; the toes are long and slender, and the claws are long, black, and sharp: the whole foot is of a paler grey than the legs.

This is a native of South America; specimens of it are often sent over to us, and to France, among those of other birds of that climate, preserved in spirits, or by stuffing. Marcgrave has mentioned it under the name of Anaca; and Ray, Willughby, and others have followed him in it.

Psittacus totus aureus cauda longa.
The long-tailed Psittacus, all of a gold yellow.

The Quijubatui.

This is another extreamly singular, and extreamly elegant, bird: it is of the exact shape of the common Parrot, but it is not larger than a lark: the head is small and rounded; the beak is large, in proportion to the size of the head, and is very much hooked, and of a grey colour: the membrane which invests it's base is of a fine bright red, and in this the nostrils are very conspicuous; they are round, and stand near one another at the base of the beak.

The eyes are large; their iris is saffron-coloured but narrow; the pupil is very large and black, and there is a naked, granulated skin about them.

The head and the whole body, back, breast, and wings, are all of a fine gold yellow, extreamly bright and glossy, except that the extremities of the wings only are of a dusky greenish: the tail is long, and is of a bright and very beautiful yellow.

The legs are robust and short, and are of a beautiful flesh colour; the toes are long and scaly, and the claws are very long, sharp, and black.

This is a native of many parts of South America, and is frequently brought alive into Europe: it becomes very tame and familiar, and easily learns to imitate the human voice, and other sounds. Marcgrave has described it under it's Brasilian name Quijubatui; and most of the late writers on birds have continued it under the same denomination, though few of them seem to have known any thing more of it than what they learned from that author's short description.

Psittacus ruber alis viridi et nigro variegatis.
The red Psittacus, with the wings variegated with black and green.

The scarlet Parrot.

This is an extreamly elegant but small species; it is not larger than a blackbird: it's head is small, round, and very prominent on the crown: the eyes are large and very bright, their *iris* is yellow: the beak is large, and very much hooked; it is of a yellow colour, and the membrane at it's base is black; in this stand the nostrils; they are round, large, and placed very near one another, and just under the forehead, at the very base of the membrane.

The

The head is of a very beautiful and bright scarlet, except that about the eyes there is a naked whitish skin: the back and breast, indeed the whole bird, excepting only it's wings, is of the same scarlet with the head, only that the colour is somewhat deeper on the back, and paler on the belly.

The wings are large and long, and are very elegantly variegated with green, black, and yellow: the long feathers are black, and the covering ones green; some of these are green throughout, others are only of this colour on their upper side, the under part being scarlet: the edge of each wing is of a bright yellow, and there is sometimes also a little of the same yellow on the inner side of the wing, next the body.

The tail is about four inches long, and is composed of twelve robust feathers; they are all of them half red, and half of a greenish yellow, so that the upper part of the tail is of the greenish colour, and the lower half red: the legs are very short, robust, and black, and above the knee they have some few green feathers on them.

This is a native of the island of Madagascar; our East India ships sometimes bring it home, and we have them in London: no species of Parrots more readily learn to imitate the human voice. None of the ornithologists, except Willughby, have so much as mentioned this species. That author calls it Psittacus coccineus alis ex viridi et nigro variegatis.

Psittacus ruber alis coccineo viridi et aureo variegatis.
The red Psittacus, with green, red, and yellow wings.

The changeable-coloured Parrot.

This is another extreamly beautiful, but small, species of Parrots: the head is round and large; the beak is not so large, in proportion, as in most of the Parrots, but it is sufficiently hooked; it is of a dark grey colour: the membrane which surrounds it's base is of a pale or ash-coloured grey, and the nostrils are very conspicuous; they are two oval depressions standing near one another, at the very top of the membrane: the eyes are large; their iris is of a silvery white, but the pupil is large and black.

The head is of a very beautiful rose colour, with some shade of blue, as it is viewed in some lights; and the feathers on the crown are long, and very beautiful, and the bird has a power of erecting them at pleasure, so as to form a kind of crest: the back and the upper part of the neck are of the same elegant red with the head, and are shaded in the same manner, with a flying tinge of blue disposed, as we see it in the changeable silks: the breast and belly are of the same colour, only paler; and the belly has more of the blue than is to be seen any where else.

The wings are long, and extreamly beautiful; their ground colour is of a fine grass-green, but they are variegated in the most elegant manner with red and yellow; and these colours are so disposed and blended, that, on looking at the wing against a good light, there are an infinite variety of fine colours seen, such as neither the pencil nor words can describe.

The tail is very long; the bird is not larger in the body than a lark, but the tail is more than eight inches long: it is of the same elegant rose colour with the body, but has a tinge not only of the blue, which is visible on the body, but also of greenish and whitish.

The legs are short, robust, and of an ashy grey; the toes are long and scaly, and the claws are thick, not very long or sharp, and black.

This species is very frequent in the woods in the inland parts of China, and elsewhere in the East. They fly together in large companies, and make a loud noise, as they

they fly: when tamed, they learn to imitate the voice very readily, and do it very agreeably. Many of the writers on birds have described this species. Bontius was the first; he has only called it Psittacus parvus, and from him Ray, Willoghby, and others have called it Psittacus parvus Bontii.

Psittacus totus niger.
The wholly black Psittacus. **The black Parrot.**

This is a very singular species, and, though of one simple colour, is not without it's beauty; it is of the size of our thrush: the head is large, round, and prominent on the crown; the beak is very large; the under chap is straight, but the upper one is much larger and longer than that, and is rounded on the back, and very hooked at the point, turning over the other a great way, so that in the whole it is almost of a semilunar figure: it is of a deep black colour, but the membrane which covers it's base is of a dark grey: in this are placed the nostrils; they are two small round apertures near one another, and situated at the base of the membrane.

The head and the whole body, as also the wings and tail, are all of a deep and very glossy black, without any the least variegation; but the feathers are so glossy, and lie in such regular and perfect order, that there is great beauty in the form and disposition of them, and a peculiar look of cleanness about the bird.

The tail is very long and narrow, and is of the same deep black with the rest of the bird: the legs are moderately long, and not so robust and clumsy as in the generality of the Parrots; they are of a deep iron grey, almost approaching to black: the toes are long and slender, and the claws are long, black, and sharp.

This is a native of South America; it is very common in the woods there, and makes a very clamorous noise. Marcgrave has described it; he calls it Psittaco congener avi Brasiliensium; and others have borrowed the same name.

B I R D S.

Class the Second.

P I C Æ.

Birds which have the beak convex and compressed.

RAMPHASTOS.

THE beak of the Ramphastos is very remarkably large. In most species it is equal to the whole body in magnitude; there are no visible nostrils: the feet have each four toes, two of which stand forward, and the other two backward, as in the parrot.

Ramphastos uropigio coccineo.
The Ramphastos, with a red rump. **The Toucan.**

This is an extremely singular bird; it is of the bigness of our magpy, or hardly quite so large: the head is large, and is rounded at the sides, but somewhat depressed on the crown: the eyes are very large, and stand in the middle of the head; they have a very piercing aspect, and are of a singular structure; the iris, properly so called, is yellow, but within this there is a circle of a silvery white surrounding the pupil, which is coal black, large, and round.

The

The beak is the most singular part of the bird; there requires, indeed, a head of an extraordinary size to support it: it is about seven inches long, and in the largest part is three inches broad; the under chap is much smaller than the upper: the beak is nearly straight all the way, to near the extremity, but there it is somewhat bent: it has the appearance of an intolerable burthen to the bird, and it would indeed be such, were it not provided by nature that all this size should have but little weight. Though of a firm and bony texture, it's substance is as thin as a fine parchment, and it is all the way hollow; the great cavity within, and the thinness of the bony laminæ, letting the air through into that cavity, render nostrils unnecessary to this singular bird: the apertures formed for nostrils, in the usual way, would have weakened the tender substance of this enormous beak near it's base, and might have made it liable to break there, to the utter destruction of the bird; the air, however, finds a sufficiently free passage into the cavity of the mouth, for the beak is denticulated all the way along at each side, and by this means is rendered incapable of shutting closely: the colour of this surprising beak is not uniform throughout; the upper chap is of a pale lemon colour, approaching to white; the under one is of a stronger yellow, and the tip of both is of a high red: the whole inside of the beak is also red.

The head, the back, and the wings of this bird are black, but there is a light cast of whitish diffused over them, more visible in some lights than in others: the rump is of a very fine high crimson: the breast is of an elegant and splendid yellow, with an admixture of a fiery red; the belly is wholly red, and the feathers that cover the thighs are of the same colour.

The wings are long, and have very little variegations; and the tail is all the way black, except at the extremity, where it is of a bright and elegant red.

The legs are robust and short; the toes are long; they stand two forward, and two backward, and have long, sharp, and black claws.

This singular bird is a native of South America, and of some parts of the East Indies; all the late writers on birds have described it. Thevet has figured it, and most who have followed him have copied his figure; Aldrovand calls it Pica Brasilica; Marcgrave, Tucana sive Tucana Brasiliensibus; Nieremberg gives it it's Mexican name Xochitenacatl; we call it the Brasilian Magpy. We have specimens of it often brought over, by way of curiosity, dried; and, by some very singular accident, a single bird of this species was some years ago shot by a country sportsman, on our own coasts. It feeds on the fruits of the West Indies, and is so tame and familiar with the natives, that it will breed in their houses.

Ramphastos uropygio luteo.
The Ramphastos, with a yellow rump.

The Pepper-bird.

All four of the species of this singular genus are fond of pepper, but this eats it the most voraciously; the name of Pepper-bird has been thence applied to them all, but it is more appropriated to this; and, where any distinction is meant, this is the species intended by it; it is of the bigness of our jack-daws: the head is large and round; the eyes also are large; their pupil is round and black, and their iris of a strong yellow: the beak is six inches, or more, in length, and is three in diameter at the base; it is somewhat hooked at the end, but very little so, and is denticulated along the sides, and covered with a kind of scales that are easily raised by the finger: it is of a firm but very thin substance, the whole being only a hollow case, as it were, with a thin covering of a bony substance; it is of a yellowish colour throughout, deepest at the point, and palest on the upper part of the upper chap: there are no nostrils visible on it.

The head, back, and wings are black, only at the rump there is a transverse band, as it were, of yellow: the tail is moderately long, and is black, but with a shade of purplish and yellowish.

The legs are short, robust, and clumsy; the toes are long and scaly, and the claws black and sharp.

This is very frequent in the South American islands, and in some parts of the East Indies. It feeds on the fruits in general, but on none so voraciously as on the ripe pepper; and the use of it's beak is seen on this occasion, as it draws off the fruit from the clusters, by means of it's length and denticulations, in a surprisingly ready manner: few of the modern writers have been acquainted with this species. Barrilier describes it under the name of Piperivora nigra urrhopigio luteo; Linnæus, in his systema Naturæ, has borrowed the same name.

Ramphastos urrhopigio albo.
The Ramphastos, with a white rump.

This is smaller than either of the former species; it is indeed but about equal in size to the common field-fare: the head is large, and somewhat depressed on the crown; the beak is of the same immoderate size with those of the former two species, and is of a pale yellow throughout: it is five inches in length, two and a half in diameter at the base, and very sharp at the point: it is serrated all along the edges, and is of a fine gold yellow within.

The head, back, and wings are black, but with a shade of a silvery grey thrown over that colour: the breast is of an iron grey; the rump is white, and the belly also, toward the sides, has some tendency to whitish: the legs are short, thick, and scaly; the toes are long, and the claws sharp.

It is a native of South America, and feeds on the spices and other fruits there; few of the authors who have treated of these subjects have named it. Barrelier has figured it under the name of Rostrata nigra urrhopigio albo; and Linnæus has continued the name.

Ramphastos viridans rostro variegato.
The green Ramphastos, with a party-coloured beak.

This species is somewhat smaller than a jack-daw, but it's body is thicker, in proportion to it's length: the head is large, and of an odd figure, much compressed at the sides, somewhat depressed on the crown, and very broad at the front: the eyes are large, and their iris of a brown ash-colour; the pupil is large and black; the beak is equal in length to the whole bird, head and body; it is but little crooked all the way to the point, where the upper chap hangs over the lower, and is there bent; it is somewhat prominent on the upper part of the back, and is all the way denticulated along the edges: it's colour is partly a bright but not very deep red, and partly a fine shining, glossy black; the red usually takes it's place in the part next the head and the back, at the other extremity, but this is uncertain: sometimes the whole lower chap is black, and the whole upper one of a coral-like red, except the tips; and sometimes the colours form a clouded appearance, in the manner of a tortoise-shell. We have the beaks brought over by way of curiosity, and preserved in our museums, and find all these varieties in them.

The body is, on the upper part, of a deep green, with a few small and indistinct spots of white in it, which, when not seen separately, give a general paleness to the whole, that makes it greyish; the breast and belly are of a somewhat paler greenish-grey; the wings are long, and the tail short: the legs are short, robust, and of a bluish colour: the toes stand two forward and two backward, and are short and thick, and the claws black, sharp, and long.

It is a native of the Molucca islands, and some other places in the warm climates, and is a great destroyer of the spices, and other valuable dry fruits. Barrilier has figured it under

under the name of Roſtrum viridans roſtro partim nigro, partim rubro; and Linnæus has borrowed the ſame name. Scarce any other of the naturaliſts have ſo much as mentioned it.

BUCEROS.

THE beak of the Buceros has, towards it's baſe, a large gibboſity riſing above the reſt of it's ſurface, and turning backwards at the point; and the upper chap of the beak is, in this genus, conſiderably longer than the under.

Buceros niger capite majore.
The black Buceros, with a great head. **The Indian Raven.**

This is an extremely ſingular bird; it is of the bigneſs of a well-grown pullet, and in ſhape ſomewhat reſembles the crow-kind, whence, and from it's ſize, it has been called the Indian Raven: the head is remarkably large, as indeed it had need to be, for the ſupporting ſo enormous a beak; it is rounded at the ſides, but depreſſed on the crown: the eyes are not large; their iris is bluiſh, with a tinge of brown, and the pupil black: the beak is more ſurpriſing in ſize and ſhape than even that of the ſeveral ſpecies of the former genus. It is about nine inches long, and, in the largeſt part, at leaſt three and a half in diameter: it is nearly ſtraight, or but little bent; the upper chap is conſiderably the largeſt, and is very ſharp and firm at the point, where it is protended over the other, and is not bent much down in that part: on the upper part, toward the baſe, there grows a very large gibboſity or protuberance: it is not leſs than three inches in length, or an inch and a half in diameter; it is affixed all the way down to the back of the beak at it's baſe, but at the point it turns up, and bends backward, forming in the whole ſomewhat of the appearance of a horn growing down to the beak to it's extremity, and there turning up it's obtuſe point: the colour of this remarkable beak is a pale red, variegated with white; theſe colours are diſpoſed without any regularity, but the red takes up the greater ſhare.

The body of the bird, as well as it's head, neck, rump, and tail, are of a deep and very gloſſy black, without the leaſt tinge of any other colour: the wings are very long, and the tail is but ſhort: the legs are long, robuſt, and of a deep bluiſh-black; the toes are long, and the claws are remarkably long and ſharp.

It is a native of the Eaſt Indies, and of ſome parts of China and Tartary, but it is a very ſhy bird. It lives remote from cities, and uſually in the thick parts of foreſts, where there run brooks or large rivers through them; it is moſt frequently ſeen among the ſedge on the banks of theſe feeding on frogs and other water inſects. Few of the writers on birds have tolerably deſcribed it. Barrelier has figured it under the name of Hydrocorax niger. We have the beak frequently brought over to us as a curioſity, and kept in the muſeums of our collectors, under the name of the beak of the Indian Raven; and ſome years ſince the ſtuffed ſkin of one was brought to us, though much injured by accidents in the voyage.

Buceros cæruleo nigreſcens capite minore.
The bluiſh-black Buceros, with a ſmaller head.

It is not to be underſtood by the name of this ſpecies, but it's head is ſmall, in proportion to the body; ſuch an immenſity of beak as all the ſpecies of this genus have, could not be carried, without a head ſufficiently big for the ſupport of it; but as the head in the other is remarkably large, in proportion to the body, that of this ſpecies is no more than proportioned to it; the bird is of the bigneſs of our raven: the head is rounded, and but very little depreſſed on the crown; the eyes are ſmall, and their iris is bluiſh: the beak is about ſeven inches long; and, in the largeſt part that is about the middle of the gibboſity, it is near three inches in diameter: the up-

per

per chap of it is much longer than the under, and is sharp and strong; the gibbosity runs about one third of the length of the whole, beginning near the base; it is highest in the middle, though not much so, and is turns up at the point; the whole beak is elegantly variegated with a pale red, and a dusky white; the colours are in such equal proportion, and so irregularly dispersed, that it is not easy to say which is the ground colour, and which the variegation.

The body of this species is thick, in proportion to it's length; the tail is moderately long, as the wings are also; when closed, they reach within an inch of the tip of it: the whole body is of the same uniform colour, which is a very beautiful one, and can no way be so well described, as by comparing it to that of a common black-beetle, on every part of which there is a fine and elegant tinge of blue shining over the black; the same elegant shade, but of a deeper blue, is cast over the whole body of this bird: the legs are long and robust, and the toes long, and armed with sharp claws.

This is a native of Tartary and China. Barrelier calls it Hydrocorax cæruleo-nigricans; no other writer has taken any notice of it: the beak is often brought over to us as a curiosity, and sometimes, though very rarely, the whole bird.

Buceros subvirescens.
The greenish Buceros.

This is an extreamly beautiful, as well as singular, bird; the head is larger than in the former, but not so remarkably large as in the first described species: the whole bird is not bigger than our common crow, but the neck is somewhat slenderer, and the legs longer: the head is depressed at the crown, but not at all flatted at the sides; it has a very large quantity of feathers on it, and those on the hinder part are long and slender, hanging a little way down the neck: the others are broad and short, and both are capable of being elevated, though not quite erected at pleasure, and in this state the long ones form a kind of crest: the beak is about five inches and a half long, and in the largest part is near three inches in diameter, the protuberance being larger, in proportion, than in any of the others: this is fixed down all along the upper part of the beak, except at it's point, where it turns up, and points a little backward; it is at least three inches long, and considerably elevated above the rest of the beak, but is not higher in the middle than elsewhere: the whole beak is somewhat curvated, but not much; the point is sharp, and is formed only of the upper chap, the under terminating half an inch within it's extremity: the colour of the beak is white, only that the ridge of the upper chap, and the point of the horn, are a little reddish.

The head is of a deep greenish-cast, but with an admixture of black: the body also is of a green colour, as viewed in almost any light; and yet this seems only a very strong tinge of the changeable kind, like the blue in the former a black, though much less distinguishable, forming, in reality, the ground colour, though this elegant tinge is so universally and strongly diffused over it.

The tail is about three inches long; the wings are moderately long; when expanded, they spread to a great extent, but, when closed, they reach very nearly to the end of the tail; both the wings and tail are of the same green glossy hue with the body, but the black underneath is more distinguishable in these than elsewhere, and the green would be almost universally allowed in these parts to be a superadded colour.

The legs are very robust, and moderately long; they are of a deep bluish-black, and the toes are long, scaly, and of the same colour, and are furnished with very long and sharp claws.

This species is more rare than any of the former; it is only found in China, and no where there, except in the inland counties: we had a stuffed skin of it sent over to to England, about eight years ſince, a very great curioſity. It fell into the hands of a perſon in the city, who had a large collection of foreign birds, as well as thoſe of our own kingdom, dried.

None of the writers on theſe ſubjects ſeem to have been acquainted with it, unleſs Barrelier means it by the ſpecies which he has figured under the name of Hydrocorax viridiſcens, but his figure does not perfectly agree with the ſpecimens ſent over to us. It frequents ponds, and, it is ſaid, feeds on fiſh; but probably it haunts them, for the ſake of the frog and newt-kinds, and other creatures which inhabit their ſhores, like the other of it's kind. The beak ſeems formed for ſtriking through the bodies of ſoft animals of this ſort.

CORVUS.

THE beak of the Corvus is of a convex and cultrated form; the upper and the under chaps are nearly equal in ſize, and the baſe is ornamented with a kind of ſetæ or briſtles.

This genus comprehends the jay and magpy-kind, beſide thoſe uſually underſtood by the name Corvus, the crow, raven, and jack-daw tribe. Theſe ſometimes ſimply called Picæ, are carefully to be diſtinguiſhed from the Picus, a genus heretofore too nearly approaching to them in name, though quite different in form and manners.

Corvus ater dorſo cæruleſcente.
The black Corvus, with a blue back. **The Raven.**

This is a bird of no great beauty, but it's ſize diſtinguiſhes it among the crow-kind; it is of the bigneſs of a common hen: the head is ſmall, in proportion to the body, and is ſomewhat depreſſed on the crown, and flatted at the ſides: the eyes are large, and very bright and piercing; the beak is conſiderably large: it is two inches long, moderately thick, ſomewhat ridged on the back, and ſharp at the point, and the under chap is very little either ſhorter or narrower than the upper: it is of a duſky blackiſh colour.

The head is black, with a ſlight tinge of a duſky blue, viſible only in ſome lights; in others it is totally and invariably black: the neck, back, and rump are alſo of a deep black, but they have a very ſtrong and elegant tinge, of the ſame blue with that of the head, and is on theſe parts viſible in all lights and all directions: the breaſt and belly are perfectly black: the wings are long, the tail is but ſhort; when the wings are cloſed, their tips reach nearly to it's extremity: the legs are robuſt, and of a bluiſh-black, and the toes are long and ſlender; the claws are black, and are not very long, nor very ſharp.

The long feathers of the wings are twenty in number in each, and the two exterior ones in each wing are ſomewhat ſhorter than the others: the large feathers of the tail are all of a length.

This is a very common bird with us; it builds in high trees, and makes it's neſt in a careleſs manner, but it is very bold in the defence of it's young. It has a remarkably quick ſcent; it's food is principally carcaſſes, and it will ſmell them at a great diſtance. When the air is clear, we often ſee it entertaining itſelf with flying to a ſurpriſing height. It's voice is naturally hoarſe but loud; it will be kept tame about houſes, and will learn to imitate the human voice in the manner of the parrot, and does it very articulately: it will alſo mimic the notes of other animals more nearly than any other of the talking birds.

Corvus ater
The wholly black Corvus. **The Rook.**

This is somewhat smaller than the raven, though a large bird; it is equal in size to the biggest of our tame pigeons: the head is small, and somewhat flatted on the crown, but rounded at the sides: the eyes are large, but not very bright and piercing; the beak is moderately long and pointed; it is of a brownish-black colour, but toward the base it is often whitish; it is naturally covered there with a kind of short bristles, but, in thrusting the beak into the ground in search of worms and other food, these are frequently rubbed off, and the naked scull appears whitish: the nostrils are round, and stand at some little distance from one another; the tongue is long, thick, and whitish, and is bifid at the end.

The head, neck, and the whole body are of a deep glossy black, without any tinge of bluish: the long feathers in each wing are twenty; of these, the fourth from the verge is the longest: the scapi of the smaller remiges, in the middle of the wing, terminate in a kind of spines or short bristles; the tail is moderately long, and is composed of twelve feathers, the outer ones of which are somewhat smaller than the middle ones: the whole bird is of a deep black colour.

The legs are robust, and moderately long; the toes also are long, and armed with strong and sharp claws; but the claw of the hinder toe is much larger and stronger than those of the others, and the outer of the three anterior toes adheres for some space to the middle one.

This is very frequent in all the northern parts of Europe; with us, it builds in high trees, and that frequently about houses, where it was once a custom to encourage them, as the noise in the building-time was thought very pleasing. It feeds promiscuously on animal and vegetable substances; fruits are the principal food, but it eats also worms and other insects.

The business of building the nest and hatching the young is carried on mutually by the male and female; the one sits on the eggs as well as the other, and, in the fabricating of the nest, the one keeps possession, while the other goes to get the materials for finishing it, otherwise the neighbouring ones steal away every thing that is left to assist in their own. This bird, though encouraged in it's produce by the gentlemen, is mischievous in a very great degree to the farmers, and they are at a great expence to drive them from their fields.

Corvus ater oculis magnis, naribus setis reflexis obtectis.
The black Corvus, with large eyes, and reflex bristles at the nostrils. **The Carrion-crow.**

This is of the size of the largest tame pigeon, and is all over of a deep and fine black colour: the head is small, and somewhat flatted on the crown; the eyes are large, and of a very piercing aspect: the beak is long and robust, and of a blackish colour; and the tongue is hard, firm, and divided at the extremity.

The wings are very large; they measure more than two feet from tip to tip, when extended: the long feathers of each are twenty; the three first of these are gradually shorter than the fourth; this is the longest of all, and from it the others gradually shorten again: the legs are long, and moderately robust; the toes are long and strong, and the claws are long, black, and sharp: the nostrils are small and oval; they are placed in the upper part of the membrane, which covers the base of the beak, and are covered with short and crooked bristles.

This

This species is very common with us; it builds in high trees, and lays four or five eggs. It feeds like the raven on carcasses, and smells them from a very considerable distance. It may be kept tame in the manner of the raven, and will learn to imitate the human voice: the knowledge of this is as old as Pliny, who mentions one famous for it in his time. Most of the writers on birds have described it. Ray, Willughby, and others call it Cornix.

Corvus capite, gula, alis, caudaque nigris, trunco cinerascente.
The Corvus, with the body grey, and the head, throat, wings, and tail black.

The Royston-crow.

This species is somewhat larger than the common rook: the head is small, and is somewhat flatted on the crown; the eyes are large, and their iris is of a hazel brown, with some admixture of grey: the beak is large, strong, and smooth; it is of a shining glossy black throughout, except at the tip, where it is white.

The head and throat, down to the breast, are of a deep and beautiful black; the wings also are black, of a glossy hue, and with somewhat of a bluish cast: the breast and belly, and also the back, and the upper side of the neck, are grey; but the scapi of the feathers, even on these parts, are black, and the back is of a somewhat duskier or deeper grey than the belly.

The wings are long and large; the long feathers in them are twenty to each; of these the first is the shortest, and the third and fourth are the longest: the tail consists of twelve large feathers; the two middle ones of these are the longest, and the others become gradually shorter to the edges, but the diminution in length is but inconsiderable.

The legs are long, robust, and black; the toes are long and strong, and the hinder one is much larger than any of the others: the outer of the three fore-toes is equal in length to the middle one, and is connected to the middle one near it's base: the nostrils are round, and are covered with bristles.

This species is less frequent with us than the others, but it breeds annually in some parts of the kingdom in great abundance: it is fond of the tops of hills and thick woods, but in the hard weather, in winter, it comes down into the low grounds. They are very common on the high downs in Sussex, and on Newmarket-heath, and particularly so about the town of Royston, from which they have obtained their English name; most of the writers on birds have described the species. Aldrovand calls it Cornix cinerea frugilega; Willughby and Ray continue the same name to it; and some call it simply Cornix cinerea. It is a very coarse feeder; it eats carrion and insects of all kinds, and, like the rook, will also feed on fruits; it often frequents the sea-shores, and will there prey upon such shell-fish as it can get open.

Corvus fronte nigra, occipite incano, corpore nigro fusco alis caudaque nigris.
The Corvus, with a black and grey head, a brownish-black body, and black wings and tail.

The Jack-daw.

This is one of the smallest of the crow-kind, but it is a very erect and well-shaped bird: the head is large, and flatted on the crown; the eyes are large, and have a piercing aspect; their iris is whitish, and the pupil large and black; the ears are large also, and more conspicuous than in most of the others: the beak is large, long, and robust; it is of a blackish colour, and the base is covered half-way down, with plumules bent forwards; among these are situated the nostrils, which are round, and stand high: the front of the head is black, but the hinder part is grey, and that colour extends itself down to the middle of the neck: the breast and belly are greyish; the

back

back has a tinge of browniſh, and the wings and tail are black: the black in this kind is very bright and gloſſy, and has ſomewhat of a bluiſh tinge diffuſed over it.

The long feathers in each wing are twenty; of theſe the outer one is not half ſo long as the ſecond, and the third and fourth are longeſt of all: the tail conſiſts of twelve long feathers; the two middle ones of theſe are a little longer than the others, but the difference is inconſiderable.

The legs are moderately long and ſlender; the toes are long, and the hinder one is much longer and larger than any of the others: the claws are large and black, but not very ſharp; and the claw of the hinder toe is alſo conſiderably longer than that of any of the others.

This ſpecies is very common with us; they are continually ſeen flying about old churches, and other high ſtone buildings, making a loud noiſe; they build in theſe places, and ſometimes, though more rarely, in trees. Almoſt all the writers on birds have deſcribed the ſpecies. Willughby and Ray call it Monedula ſive lupus; others, ſimply, Monedula. It feeds on inſects and fruits, and it may be kept tame, and learned to imitate the human voice.

Corvus dorſo ſanguineo, remigibus nigris, rectricibus viridibus. *The Corvus, with a blood-red back, a green tail, and black wings.*

The Roller.

This ſpecies, though very beautiful in it's colouring, is the leaſt conſiderable in it's ſize of any of the Corvus-kind; it is ſmaller than the jack-daw, but it is very like it in ſhape: the head is large, and is ſomewhat flatted on the crown: the eyes are large and piercing; their iris is whitiſh, and the pupil black; there is a naked ſpace about the eyes, and above them there ſtand alſo two large and naked tubercles: the beak is large, long, ſtrong, and black; the tongue is black, and lacerated at the end, and the palate is green.

The head is of a deep bluiſh-black, with a caſt alſo of green in it; the middle and upper part of the back are of a blue colour alſo, with a very ſtrong tinge of a blood-red; and the rump, as alſo the ſmaller and inner covering feathers of the wings, are of a fine ultramarine blue: the breaſt is of the ſame greeniſh-blue with the head; there is a deep tint of black at the bottom of this gloſſy and varied colour, and there are ſome variegations of white in tranſverſe lines, on the ſcapi of the feathers on the breaſt: the belly is of a pale and not very elegant bluiſh-grey.

The long feathers of each wing are twenty in number, and the lower half of all theſe, except the firſt, is blue, and the upper half black; the exterior ones are wholly black, except that there is a little bluiſh toward the edges: the tail is formed of twelve feathers; the ten middle ones of theſe are all of the ſame length, but the two exterior ones are about a finger's breadth longer than theſe: the colour of the tail is variable, but in moſt lights it appears green: when the feathers are examined ſeparately, the two middle ones appear of a greyiſh-blue, and thoſe next to theſe are of a whitiſh-blue toward their extremities, and the ſucceeding ones have more of this pale colour: the reſt of the feathers appear bluiſh or greyiſh in the interior, and blackiſh in the exterior ones; the extremities of the two outer ones are black, and, in the general, whenever the feathers are black on the upper ſide, they are of an elegant blue on the under.

The legs are ſhort, and reſemble thoſe of the pigeon in form, but they are of a duſky yellow colour: the toes are long, the middle one conſiderably the longeſt, and the claws are black; that of the middle toe is on the inner part ſharpened into a kind of edge, and all the toes are divided quite to the baſe.

This

This is a native of the East, and of many other parts of the world; also they have it in Italy, but not common; and it is met with in some parts of Germany, particularly about Strasburg. It feeds on insects; almost all the writers on birds have described it, though many, who have done so, seem never to have seen it. Gesner calls it Cornix cærulea; Aldrovand, Cornix cærulea and Garrulus Argentoratensis; Willughby, Wormius, and others, Cornix cærulea. I once saw one of this species in Charlton-forest in Sussex, but was not able to shoot it. I think it the most beautiful of all the European birds.

Corvus variegatus, tectricibus alarum cæruleis, lineis transversis albis nigrisque.

The variegated Corvus, with the covering feathers of the wings blue, variegated with black and white.

The Jay.

This is a very beautiful species, but it comes in no degree of competition with the former: the head is moderately large; the whole bird is about the size of a common pigeon, but the head is considerably bigger than in that bird: the eyes are not very large, but they have a very piercing and a very sprightly aspect; their iris is whitish, the pupil black and large: the beak is large, strong, and black: the nostrils stand toward the upper part of the membrane that invests the base of the beak, and are oval and moderately large; the tongue is black, and is bifid at the extremity.

The whole plumage of this bird is very delicate, and the feathers have a kind of crestiness, which is not usual in any of the species: the head is of a very beautiful greyish-brown, and there are two black spots near the base of the lower chap of the mouth: the throat and the lower part of the belly are white; the shoulders and the breast are of a ferrugineous brown, but pale, and with somewhat of a blush of a pale red, or flesh colour, intermixed, which makes the whole very beautiful: the forepart of the head is paler than the hinder, and has some longitudinal black streaks on it: the middle part of the back is of the same reddish brown with the shoulders, but with something of blueness in it, which is not observable in that part; and the rump is white.

The wings are moderately large; the long feathers in each are twenty: of these, the first is not more than half as long as the second, and the fourth is the longest of all: the first of them is black throughout, except that the lower part is white, but this is the case only in that; the six succeeding ones are greyish on the outer side, and the three succeeding ones have some tinge of blue, and have several transverse streaks of blue and of black on the lower part; the others are in part black and in part white, and the sixteenth feather is elegantly variegated with transverse lines of blue, black, and white; the seventeenth is black, and has only a few spots of blue; and the eighteenth is black, with only a little variegation, which is reddish; the nineteenth is reddish, and it's tip is black: the underside of all these is brown, but that of the last is of the same colour with the upper side; the first fifteen of these long feathers are covered by a number of very beautiful short feathers, which are of the most bright and splendid blue, variegated with lines of black and white; the surface of this part of the wing is so glossy, and the colours are so bright, that it has the appearance of enamel; the other covering feathers are black.

The tail is moderately long, and is composed of twelve feathers: the legs are short, and not very robust; they are of a dusky ferrugineous colour; the claws are long, but not very sharp; that of the middle toe is longer than any of the others; the hinder one is as long as the outer one of the anterior three, and the claw of it is longer than that of any of the others: the outer toe is connected to about a third of it's length to the under one.

The male and female in this species differ so very little, either in size or colouring, that it is hard to know them asunder.

It is a native of most parts of Europe, and has been described by all the authors who have written on birds. Gesner calls it Pica glandaria sive garrulus avis; Ray, Willughby, Aldrovand, and others Pica glandaria: it feeds on vegetable fruits. It has it's name from it's eating of acorns, but the whole race of our wild fruit-trees supply it: it is particularly fond of the black-berries, and other of the softer fruits. It is frequently kept in cages, and will learn to imitate the human voice.

Corvus cinereus cauda alisque nigris.
The grey Corvus, with the wings and tail black. **The Caryocactes.**

This species is somewhat smaller than the jay: the head is small; the eyes are moderately large, and their iris is of a hazel colour: the beak is long, strong, and obtusely pointed, and the upper chap is somewhat longer than the under: the tongue is short; it scarce reaches indeed beyond the angle of the mouth, and it is bifid at the end, and the division is more deep than in any other bird; from the extremity of the tongue to the very end of the beak, there runs a kind of ligament of a wrinkled surface, which fills up the channel in which the tongue would otherwise lie, so that it is evident that the tongue never can be extended farther than it's common dimensions: the lower part of the palate, and the sides of the fissure, are hairy; and the nostrils which stand in the upper part of the membrane, that covers the base of the beak, are round; they are placed at a distance from one another, and are surrounded, and, in a manner, hid, by a quantity of short and reflex white bristles.

The whole body, as well on the upper as the under part, is of a dusky reddish-brown, very elegantly variegated all over with moderately large spots of a snow-white, and of a regular triangular figure: there are none of these spots, nor indeed any variegations at all, on the head: the breast and belly have more of the red mixed with the brown than the back, and the spots on the breast are larger than those on any other part: there is a space between the eyes and the insertion of the beak, which is white, and the feathers under the base of the tail are also of a snow-white.

The long feathers of the wings are black; the tail is moderately long, and is composed of twelve feathers; the two outer ones are half white, but in the others the white has a smaller and smaller share, till at length it wholly disappears in the middle ones: the part of the feathers, which is not white, is of a fine deep, glossy black; and, in some birds of this species, the tail has been observed wholly of this shining black.

The legs are moderately long, and very robust; they are of a very deep bluish-black, and of a glossy surface: the toes are long and slender; the outer one is fixed to the middle one, by a membrane toward it's base: the claws are long and black, and that of the hinder toe is longer than that of any other.

This species is a native of Germany, and many other parts of Europe, but it has not been seen in England: it feeds on fruits, and it's voice is like that of our magpy. Most of the authors who have treated of birds have written on it. Gesner calls it Caryocactes, and Ray gives it no other name; Willughby calls it Caryocactes Gesneri et Turneri.

Corvus cauda cuneiformi.
The Corvus, with a cuneiform tail. **The Magpy.**

This is a very well-known bird, and when in full feather, and in it's wild state, has a great deal of beauty; we see it to much disadvantage, when kept in cages, where it is always dirty, and usually out of health; it's weight is about eight ounces; the size of it's body is about that of the jack-daw, but it's variegated wings and length of tail make it seem larger: the head is moderately large; the eyes are bright and piercing, their iris is of a pale hazel; the pupil is black, and there is a yellow spot vi-

sible

sible on the nictitating membrane; the beak is moderately long, very strong, black in colour, and pointed; the upper chap is prominent on the back, and a little bent: the tongue is black, and is divided into two parts at the extremity; the sides of the fissure of the palate are hairy.

The head is of a fine deep black, but with a shade of a changeable blue, with a mixture of green thrown over it: the neck, the throat, the back, and the lower part of the belly, are also of the same deep black, with the same variegating colour diffused over them, only the lower part of the back, toward the rump, is greyish: the breast is of a fine and delicate snow-white; the sides are also snow-white, and the feathers which cover the first joint of the wings are also of the same bright colour.

The wings are shorter than in almost any other bird of the same size, but their long feathers, as also those of the tail, are of the same black with the head, shaded over with the changeable green or blue tinge, in more strength than any other part; and there is often, when the bird is in high vigour, a kind of deep purple, diffused among the green and blue, in this shading variation.

The long feathers in the wings are twenty; to each of these the outer one is not half so long as the second, this also is considerably shorter than the third, and the third than the fourth; but the difference in the length is not so much in these, as between the first and second; the fourth and fifth are longest of all: the eleven first feathers are all stained with white; this is in a larger proportion in the foremost, and becomes less and less in the others, till in the eleventh, instead of occupying the greater part of the feathers, the white is no more than a spot.

The tail is very long, and of a singular shape and structure; it is composed of the usual number of feathers, that is, twelve; the two middle ones are greatly longer than the other ten; these became gradually shorter, till the exterior ones are very inconsiderable, in proportion to the others: the whole tail is of a cuneiform figure; and the two principal or middle feathers are of a mixed greenish, purplish, and bluish hue, with a deep black for the ground colour under all. The lower part of each is purplish, and the very tips are blue.

The legs are slender, not very long, and black; the toes are long, and the outer one is joined, toward the base, to the middle one; the claws are long and black, and that of the hinder toe is considerably longer than any of the others.

This is a very common bird with us, but it is subject to some variations in colour, under which it makes a very singular appearance: it is sometimes met with of an uniform brown throughout, and sometimes quite white, but these variations are rare. It builds with us in trees, and sometimes in tall hedges. The nest is of a very singular structure; it is not open at the top, as those of most other birds, but is covered close on every part, only a narrow entrance being left in one place for the bird to go in at; and the whole is guarded from attacks, by being covered with sharp thorns, the points of which are turned outwards. It lays five, six, or sometimes seven eggs; they are somewhat smaller than those of the rook, and are of a pale colour, spotted with red. It feeds on insects and on fruits, but it is a bold bird, and sometimes invades the properties of those which are expressly carnivorous: it is frequently seen to attack sparrows; and Ray, an author of undoubted veracity, declares, that he saw one seize on a thrush, and eat it. All the writers on birds have mentioned it. Most of them, simply, under the name of Pica; some under that of Pica varia and Pica caudata.

Corvus niger roſtro longiore rubente.
The black Corvus, with a long red beak.

The Corniſh Chough.

This ſpecies is about the ſize of the jack-daw, and is not unlike it in figure: the feathers ſtand more looſely and irregularly on it than on that ſpecies, and it thence looks larger, but their weight is about equal: the head is moderately large, and is flated on the crown; the eyes are large, and of a very piercing aſpect; the beak is very ſingular, both in it's figure and colour; the upper chap is ſomewhat longer than the under; the whole beak is ſomewhat longer, in proportion, than in any of the Corvus-kind, and is ſtrong, and not very ſharp at the point: the colour is throughout a bright and beautiful red; the membrane which inveſts the baſe of it is of the ſame colour, but deeper, and in this ſtand the noſtrils; they are of a roundiſh figure, but approaching ſomewhat to oval, and are ſurrounded by ſhort and curled briſtles inſtead of feathers.

The body is ſlender, in proportion to it's length; the feathers, however, ſtand ſo looſe, that this is not ſo readily perceived; the whole is a deep black, without any variegation: there is, indeed, a ſhade of a deep purpliſh, changeable hue, diffuſed over the back and breaſt, but the abſolute black is the ground colour throughout: the long feathers of the wings are twenty in each, and they are of a deeper black than thoſe of the body: the tail is moderately long, and is compoſed of twelve feathers, and theſe alſo are of the ſame deep black, but with a ſingle tinge of the purple diffuſed over them: the lower part of the belly is darker than any other part.

The legs are moderately long and ſlender; they are of a fine bright ſcarlet, and the feet are of the ſame colour: the toes are long and ſcaly; the claws are not very long, but they are ſharp, and of a deep black, and that of the hinder toe is longer than any other.

It is a native of all the northern parts of Europe; with us, it is very common about the rocky cliffs of our ſhores, and about high buildings that are near the ſea. It is no-where ſo frequent as in Cornwall, whence it obtained it's name of the Corniſh Chough. Authors have deſcribed it under the different names of Coracias and Pyrhocorax. It's voice is like that of the jack-daw, and it makes the ſame inceſſant noiſe. The children about our ſea-coaſts ſometimes breed it up tame, and learn it to imitate the human voice, which it does very happily. All the Corvus-kind, indeed, have organs for the doing this, and many other birds, which nobody has ever thought of inſtructing.

Corvus vireſcens capite variegato.
The greeniſh Corvus, with a variegated head.

The green Magpy.

This is an extremely ſingular, and alſo an extremely beautiful, bird; it is of the ſize of the common jay, and greatly reſembles it in ſhape, but nothing can be more different than it's colours from thoſe of that bird: the head is moderately large, a little compreſſed, and conſiderably flated on the crown: the eyes are not large, but very piercing in their aſpect; their iris is of a gold yellow, and their pupil black: the beak is about three quarters of an inch long, not very thick, ſtraight, and pointed at the extremity, and both chaps of it are very nearly equal: the noſtrils ſtand toward the top of the membrane that inveſts it's baſe; they are round, and are placed at a ſmall diſtance from one another, and are ſurrounded by ſome ſhort, curled feathers.

The head is of a deep cheſnut-brown on the crown, and the ſame colour extends itſelf down ſome part of the neck behind, but, at the ſides near the eyes, it is of a very elegant gold yellow: the reſt of the neck is of a pale brown; and the whole body, back, breaſt, and belly, are of a very elegant and ſhining green, ſomewhat like

like that on the body of the common parrot: the wings are long, and they have each twenty long feathers; the two first of these are shortest, the third is longer considerably, and the fourth is longest of all: these have some brown about them, but the covering feathers of the wings are green, and that a colour not at all inferior to the most elegant tint of the back.

The tail is moderately long; it consists of twelve feathers, and of these the two middle ones are the longest; they are all of a deep and fine green, only that toward their extremities; the middle ones, especially, are of a brownish tinge, like that of the crown of the head.

The legs are moderately long and brown; the toes are long, slender, and scaly; the outer of the three anterior ones of each foot is connected for about a third of it's length to the middle one, but the claws of all of them are black, long, and obtuse; but that of the hinder toe is longer than any of the others.

It is a native of Italy, and frequents the sea-coasts, but it is not very common. It is not certain that any of the ornithologists have described it. Aldrovand mentions a species of Pica, according to his distinctions, which he calls *Pica marina*; and his description, so far as it goes, seems in most things to agree with this bird, but it is too imperfect to form any certain judgment on. When wild, it flies about very swiftly, making a loud and hoarse noise, but it may be kept tame; and we are told, that some which have been so have very happily imitated the human voice. There is a stuffed skin of this in Sir Hans Sloane's Museum.

Corvus nigro et flavo variegatus rostro albicante.
The black and yellow Corvus, with a white beak.

The Persian Magpy.

This is a larger bird than either our magpy or jay, but it approaches considerably to the jay in figure: it's head is small and rounded, but a little compressed: the eyes are large, and very bright and piercing; their iris is yellow, and the pupil black: the beak is moderately long, but not very robust; it is pointed at the extremity, and the upper chap is a little longer than the under: the membrane which covers the base of it is of a greyish-brown, and in that stand the nostrils; they are of a roundish but somewhat ovated figure, and are covered by a number of short and curled plumes.

The head is altogether black; the back also is black, and the breast and belly, though somewhat brownish, approach to the same colour: the rump is of a fine gold yellow; and the feathers on the wings, which answer to the variegated blue ones on the wings of our jay, are in this species of the same gold yellow: the wings are short, and have each twenty principal feathers, of which the fourth and fifth are longest: the tail is moderately long, and is of a deep black, only some of the feathers toward the base have a little yellowishness about them.

The legs are short and slender; they are of a greyish-brown: the toes are long and scaly, and the claws of all of them are long and black, but that of the hinder toe is longer and sharper than the others.

This species is a native of Egypt, and some other parts of the East. It has not been described by any of the authors, except the *Pica Persica* of Aldrovand be it; but his description is very imperfect.

Corvus cauda longa, roſtro rubente, capite cæruleo. *The long-tailed, blue-headed Corvus, with a red beak.* — The Chineſe Magpy.

This is a very beautiful bird, and, though perfectly unlike to our magpy in colour, yet it ſo greatly reſembles it in ſhape, that nobody would fail at ſight to call it by that name; it is of the bigneſs of a common pigeon: the head is moderately large; the eyes are ſmall, and their iris is hazel; there is a ſmall naked ſpace of a fleſh colour round about them, and a little naked tubercle alſo, at ſome little diſtance above each: the beak is three quarters of an inch long; it is ſtraight and tolerably ſtrong; both chaps are nearly equal in ſize, and it is pointed, but not very acutely at the end: it is throughout of a bright and ſtrong red, but the tip is deeper than any other part: the membrane which inveſts it's baſe is alſo red, and the noſtrils which ſtand in the upper part of this are ſmall and roundiſh, and are covered by little twiſted plumules.

The head and neck are of the moſt beautiful ſky blue; the back wings and tail are of a very deep colour, approaching to black, but with ſome blue, and alſo a caſt of greeniſh in it: the breaſt and belly are of a ſnow white.

The wings are ſhort, ſo that the bird flies in a noiſy, fluttering manner, but yet very ſwiftly: the long feathers are twenty in each, and of theſe the third and fourth are the longeſt; they are all of a deep black, but there is ſome flying tinge of bluiſh thrown over the whole wing.

The tail is very long, and is compoſed of twelve feathers, and in all reſpects exactly reſembles that of the magpy: it is throughout of a blackiſh colour, but with a flying tinge of a mixed blue and green over it.

The legs are moderately long and ſlender; they are of a fine bright ſcarlet, and the feet are of the ſame colour, only a little paler, or more dead: the toes are long and ſlender; the claws are very long and ſharp, and are of a coal black; that of the hinder toe is much longer than any of the others.

This ſpecies is a native of China, and of ſome other parts of the Eaſt; it is much eſteemed by the natives, and is taught to imitate the human voice; the wild notes are not diſagreeable: ſcarce any of the writers on theſe ſubjects have mentioned it. Charleton, in his Onomaſticon Zoicon, mentions a bird under the name of Pica caudata Indica ſeu Japonenſis, which may probably be the ſame with this ſpecies; but there is not enough ſaid to aſſure us of it. No other writer ſeems to have had any notion of ſuch a ſpecies.

Corvus fuſco ſubrubens capite albido. *The browniſh Corvus, with a whitiſh head.* — The Indian Jay.

This ſpecies as remarkably reſembles the common jay in it's figure and ſhape, as the former does the magpy, but it is ſmaller; it is of the ſize of a fieldfare: the head is large, in proportion to the body, and is rounded at the ſides, but ſomewhat depreſſed on the crown: the eyes are ſmall, and their iris is of a fiery red; they are ſurrounded with a naked ſkin, and there are a couple of naked tubercles, one on each ſide the head, a little above them: the beak is ſhort but ſtrong, and is obtuſe at the point, and both the chaps are of the ſame length: the whole beak is of a pale whitiſh-brown, and the membrane which covers the baſe of it is of a bluiſh-grey: in this ſtand the noſtrils; they are large, round, and there grow a great number of ſhort and ſtiff briſtles about them; the tongue is ſhort and biſid at the end, and the inſide of the mouth is of a pale yellowiſh or lemon colour.

The head is of a very pale whitiſh-brown: the neck is of a ſomewhat darker brown, and the back and ſhoulders are of a deeper brown, with a tinge of red in it:

the

the breaſt and belly are nearly white; they have much the ſame colour indeed with the head, but that there is ſomewhat of reddiſh in it: the rump is paler than the reſt of the body, and the lower part of the belly, between the thighs, and under the root of the tail, is white.

The wings are ſhort, and yet the bird is a ſwift flier; the long feathers in each are twenty in number, and of theſe the fourth is the longeſt: the firſt is very ſhort; they are all of a deeper brown than the reſt of the body, and are tipped with a yet duſkier colour at the ends: the covering feathers of the wings have ſome variegations of yellow and brown about the part where, in our jay, there are the blue feathers.

The tail is ſhort; it is compoſed of twelve feathers, all of a deep brown, but ſomewhat duſkier at the tips than elſewhere: the under part of the tail is ſomewhat paler than the upper.

The legs are moderately long; they are ſlender, and of a deep duſky, bluiſh-grey: the toes are long, and the claws are alſo long, black, and ſharp.

The bird is a native of the Eaſt Indies, and has ſometimes been brought over ſtuffed from thence. It is very noiſy and chattering; it's wild notes are not very agreeable, but when kept tame, as it frequently is by the Indians, it learns to imitate the human voice. None of the writers on birds have named it, unleſs it be Charleton; he mentions a ſpecies of Picus, as he expreſſes it, under the name of Picus garrulus ſive Minus Indicus minor, which may poſſibly be the ſame with this; but, unleſs he had ſaid more, there is no being certain of it.

Corvus caſtanei coloris pectore ferrugineo.
The cheſnut-brown Corvus, with a ferrugineous breaſt. — **The Ampelis.**

This is a very ſingular and a very elegant bird; it is like our jay in ſhape, but it is greatly inferior to it in ſize; the utmoſt bigneſs of it is that of our common thruſh: it's head is moderately large, rounded at the ſides, but conſiderably depreſſed on the crown: the eyes are large, and very beautiful; their iris is of the higheſt ſcarlet, and the pupil black, and they have a peculiar piercing aſpect: the beak is ſtrong, but very ſhort; it is not longer than that of a ſparrow; both chaps are nearly of the ſame length; it is pointed at the extremity, and is throughout of a deep black: the membrane which covers the baſe of it is alſo black, and the noſtrils which ſtand in this are roundiſh: there are a number of curled briſtles growing about theſe, and from them to the beginning of the head.

The head is of a pale cheſnut colour on the forepart, but behind it is of a much deeper brown, approaching to that which our painters call an umber colour: at the ſides there are two black ſpots, formed, as it were, of a continuation of the briſtly covering of the noſtrils, and in theſe ſtand the eyes: the neck is ſhort; it is black before and behind, but of a ſtrong, though not very deep brown, at the ſides, and, as it approaches the head, it becomes whitiſh: the ſhoulders and back are of a deep reddiſh-brown, a very ſingular colour, and very elegant; the brown is bright, though ſtrong; and the reddiſh that is blended with it is not of the ſcarlet or crimſon kind, but of the pale bloſſom hue, or fleſh coloured tint: the breaſt and belly are of two different colours; the breaſt a ferrugineous brown, deep and ſhining; and the belly a pale, but not bright, grey, with no tinge of either brown or red in it: there is a ſhade of cheſnut colour viſible in certain lights, though not in all; and this is moſt diſtinguiſhable about the middle of the back; the feathers about the rump are yellowiſh.

The wings are ſhort, but they are very well feathered: the remiges or long feathers in each are twenty; the two firſt of theſe are ſhort, the third longer, the fourth longeſt of all, and the reſt from this gradually declining in length to the extremity or brown part: the long feathers are of a darker colour than any other part of the bird; in the male they are blackiſh; the covering feathers are of a bright and elegant cheſ-

nut

nut colour: part of these long feathers are very beautifully variegated with white and with scarlet spots, and part of them with yellow; the inner ones become paler than the outer, and some of the most hidden ones are greyish.

The tail is moderately long, and has this singularity, that in the male it consists of only ten, and in the female of twelve feathers; they are nearly of the same length, and are all of a deep mouse colour toward the base, but yellow at their extremities: the yellow is brighter in the male than in the female, but in both it is very elegant, and gives great beauty to the bird: and toward the base of the tail there are a number of other considerable long feathers, which form, as it were, another tail; these are of an elegant chesnut colour, and are not rigid and stiff like the rectrices, or long feathers of the real tail, but flexible, and soft as those which cover the body.

The legs are slender, moderately long, and of a deep bluish-black colour: the toes are long, slender, and scaly, and the claws are long, sharp, and very crooked, and are of a deep black: the claw of the hinder toe is longer and sharper than any of the others.

This is a native of Bohemia, and is said, by some authors, to be peculiar to that kingdom, but this is erroneous. It is frequent in many parts of Germany, and in Italy, and is seen, though not so abundantly, in the South of France. Most of the authors who have written on birds have named it. Aldrovand, and after him many others, call it Garrulus Bohemicus; and many, from it's peculiar fondness for grapes, Ampelis. It feeds on fruits of all kinds, but principally on the soft ones: it is naturally very noisy, but it's wild notes are hoarse and disagreeable; when kept tame, it is easily learnt to imitate the human voice.

PICUS.

THE beak of the Picus is straight, of a polyhedral or many-sided figure, and has it's point formed in manner of a wedge: the tongue is rounded, and very long; it resembles in form a worm, or some other such insect: the toes, in all but one species, stand two before, and two behind, as in the parrot.

Picus niger vertice coccineo.
The black Picus, with a scarlet head.

The great, black Wood-pecker.

This is a very singular bird; it's weight is about ten ounces; it's body, in size, is somewhat larger than that of a fieldfare: the head is moderately large, and very much depressed or flatted on the crown; the eyes are large and piercing; the beak is very strong, about two fingers breadth long, and of a deep blackish-blue colour; it is of a figure approaching to triangular, and is covered at the base by a membrane of a pale colour: the nostrils are large and round, they stand very forward, and are surrounded with reflex hairs.

The head is black at the sides, and toward the lower part, but of a beautiful red at the top; properly and distinctly speaking, there is a spot of this fine coral red extending from the base of the beak, longitudinally, to the very back part of the head: the neck, shoulders, back, and wings, in short, the whole bird, besides, is of a deep and glossy black.

The wings are not very long; the long feathers in them are nineteen in each: of these the first is very short and inconsiderable, in proportion to the others; these, however, are of a yet deeper black than any of the others: the tail consists of ten feathers, of which the exterior two are very short, and the others become gradually longer to the very middle ones; these are between six and seven fingers breadth long, which

which is therefore the extent of the measure of the tail; all the feathers, except the two inconsiderable exterior ones, are acute, rigid, and bent inwards: the provision of nature is very remarkable in this, for, while the bird runs up trees, or sustains itself on the surface of the trunk, as is frequently necessary in it's seeking it's food, these feathers sustain a great part of the weight of the body, relieving the feet, and giving it more opportunity of moving the anterior part of it's body without falling.

The legs in this species are robust, but short; they are covered with feathers on the anterior part, half-way down below the knee, but on the hinder part they are not covered below that joint; they are of a deep bluish-black: the toes are moderately long, and stand two before, and two behind; the claws are large, strong, and sharp, only that of the lesser of the two hinder toes is very inconsiderable.

The tongue of this bird, when extended to the utmost, is very long, and it is capable of a very swift motion of lengthening and retraction at pleasure.

This species is very frequent in Germany, Sweden, Denmark, and other of the northern parts of Europe, but it has not been met with, that I know of, in England: all the writers on birds have described it. Gesner calls it Picus maximus vel niger; Aldrovand, Picus maximus; Willughby, Picus niger maximus; and others, by the same or similar names. It feeds on the cossi or hexapode worms, hatched from the eggs of beetles; and, as many kinds of these are lodged on the decayed trunks of trees, it is most usually seen about them: it makes a loud and disagreeable noise, usually, when on the wing.

Picus viridis vertice coccineo. **The green**
The green Picus, with a scarlet crown. **Wood-pecker.**

This is an extremely beautiful species, but it is smaller than the former; it's weight is about three ounces; it's body is about the size of that of a lark: the head is moderately large, and is very much depressed on the crown; the eyes are bright and piercing, and of a singular structure: the pupil is large and black, and it is surrounded by a double circle, in the place of an iris: the interior circle is of a brownish-red, and the outer one is white: the tongue is very long, when thrust out to it's full extent, and is of a bony hardness, and very sharp at the point; it uses this to transfix the insects, on which it is about to feed.

The head is of a beautiful bright red colour, variegated with small and irregular spots of black; and on the lower part of each chap there is also a fine red spot: the throat, the breast, and the belly are of a beautiful pale green; the back, the upper part of the neck, and the covering feathers of the wings, are of a much deeper and more elegant green: the rump is yellow, very pale, and approaching to what we call straw colour; the under part of the tail is variegated with transverse lines of a deep brown.

The long feathers in each wing are twenty, though the first is so small, that it is an easy mistake to count but nineteen; and those nearest the body have only one side green, and the other of a dusky brown, variegated with spots of white, of a semi-circular figure; some of the outer ones have also the same variegations, and the short feathers which cover the roots of the long ones, on the underside of the wing, are of a whitish-green, and variegated with transverse spots of brown.

The tail is about four fingers breadth long, and is composed of ten feathers; they are rigid and stiff, and are turned inwards at the points, where they seem forked; this is owing to the scapus of the feather not being continued, as is usually the case, quite to the extremity: the two middle ones, and sometimes two or three of a side next them, have their ends tipped with black; the rest of the feathers is variegated on the upper surface, with spots of a greyish-white, and on the under with white.

The legs are short, and are not determinate in their colour; in some they are of a greenish-white, in others they are of a bluish-grey, or lead colour: the toes are moderately long; they stand two before and two behind, and the two anterior ones are connected at their bases.

This species is not unfrequent with us; we meet with it principally in woods, where it feeds on the insects living in the rotten trunks of trees, and on the ants, and their chrysalis, usually called ant's eggs. It is more frequently seen on the ground than any of the other species. It builds in hollow trees, and lays five or six eggs: all the writers on birds have described it. Gesner calls it *Picus viridis*; Aldrovand, *Picus martius viridis*; and Ray, *Picus viridis major*.

The male in this species differs so much from the female, that it might be mistaken for a different kind; that which is here described is the female: it is more beautiful, and somewhat larger than the male. In the male, the temples and space about the eyes are black; the throat is white, and the breast and belly are of a pale green, and undulated.

Picus albo nigroque variegatus, vertice nigro, rectricibus tribus lateralibus utrinque albescentibus.
The black and white variegated Picus, with a black head, and some of the tail feathers white.

The great, spotted Wood-pecker.

This is a very singular and beautiful bird; it is of the of the size common blackbird: the head is large and flatted on the crown; the beak is long, for a bird of this genus; it is nearly half an inch in length, and of a triangular figure: it is broad and thick at the base, and thence grows gradually taper to the point, and is all the way of a black colour, and perfectly straight; the point is sharp, and the angles are acute, and there are three or four furrows marked lengthwise on it: the nostrils are round, and there grow about them a number of curled, short, black setæ or bristles.

The eyes are large, and they are very bright and piercing; their iris is of a fiery red, and the pupil large and black: the tongue is very long, and is formed into a hard and bony point at the end: the bird darts it out to it's full length at pleasure, and it is of use in killing the insects on which it feeds, by transpiercing them, before they are taken up by the beak.

There is a beautiful white space at the base of each chap, and in the male there runs a transverse duct of a fine beautiful red behind the crown of the head, reaching quite across from the middle each way to that white: in the female, the throat and the breast are both of a pale yellowish-white: the lower part of the belly, near the insertion of the tail, is of a fine bright and beautiful scarlet: the head and back of the neck are black, and there runs a black streak along the sides of the neck, from the angles of the beak, down almost to the body.

The long feathers of the wings are twenty in each; the first of them is very small and inconsiderable, and they are all of a deep black, variegated with spots of white, of a semicircular figure: the covering feathers of the inner or under part of the wings are part white, and the remainder black; and those of the outer side are black, but they have each one, or sometimes two, spots of white, of a semicircular figure: the top of the wings is white: the tail is about three fingers breadth long; it is formed of ten rigid feathers: the two middle ones of these are longer than the others, and more rigid, and are turned downwards at their extremities; and both these, and all the other feathers of the tail appear bifid, at the end; which is owing to the rib not reaching to the extremity of the feather, as it does in most other birds: the two exterior ones, that is, one on each side, are black, only they have a single spot of white on the outer side; the two next are black only at the tips, and the rest of the feathers white, except that it is variegated with two transverse lines of black; the upper one going

going quite acroſs the feather, the under only reaching over the inner pinnules: the fourth pair have the black of the tips carried a great way farther up the feather, and the white part above in theſe has only one black ſpot, by way of variegation: the fifth pair are uſually altogether black, but they have each a white ſemicircular ſpot near the baſe, and the top or extremity of each is of a browniſh-white: the two middle feathers, which are larger and longer than the others, are all over perfectly black: the courſe of nature, in the variegations of every ſingle feather of the tail, is uſually ſo very exact, that it was not improper to give the ſingle variegations here; but it is not ſo univerſal, but that it may ſometimes vary.

The legs are ſhort, robuſt, and of a deep bluiſh-grey or lead colour: the toes ſtand two before and two behind, and the two anterior ones are connected to one another at the baſe: the liver is ſmall, and the gall-bladder is alſo ſmall, and hangs to it by a ſlender neck: the ſternum is remarkably long, it is extended almoſt to the anus: the ſtomach is ſmall; the guts have many convolutions, and are laid deep in the hinder part of the body; which is the proviſion of nature, that they may not be eternally diſordered by the violent motions which the bird gives the head and anterior part of the body continually, as it ſtrikes the trunks of trees with it's beak, in taking it's food.

This ſpecies is frequent with us; we meet with it in greateſt abundance in the northern parts of England, and it is principally ſeen about the tall trees in hedges, and makes a very loud and diſagreeable noiſe, as it flies acroſs the fields, from one to another. All the authors who have written on the birds of this part of the world have deſcribed it. Geſner calls it Picus varius major; and Ray, Willughby, and many others who have written ſince, have copied his name. Aldrovand and ſome others call it Picus major, but that is leſs expreſſive, and leſs appropriated. It's food is the hexapode worms of the beetle-kind; it will feed on other inſects, but theſe are it's principal and favourite ſupport.

The leſſer, ſpotted Wood-pecker.

Picus albo nigroque variegatus, rectricibus tribus lateralibus apice albo variegatis.
The black and white Picus, with the three lateral rectrices variegated with white at the top.

This ſpecies is extremely like the preceding in ſhape, colouring, and every other particular, but it is vaſtly ſmaller; it's weight is hardly more than an ounce: the head is ſmall, flatted on the crown, and very beautifully variegated; the hinder part of it is wholly black: there is an elegant white ſpot upon the crown, and the anterior part, toward the baſe of the beak, is browniſh: the eyes are ſmall, but of a very piercing aſpect: there runs from behind each eye a white line, which is continued down the ſides of the neck, turning inward all the way in it's courſe, till both meet at the bottom: the beak is ſhort, but it is very robuſt; it is of a pyramidal figure, and has two or three longitudinal furrows on it.

The back, and the upper feathers of thoſe which cover the outſide of the wings, are entirely black; the other covering feathers of the wings are black alſo, but they are very beautifully variegated with ſpots of white: the middle part of the back is white, but it is elegantly variegated with tranſverſe lines of black: the breaſt, belly, and throat, are all of a duſky white.

The wings, when extended, meaſure about eleven fingers breadth from tip to tip, and the tail conſiſts of ten feathers: the two middle ones of theſe are longeſt, and the others all the way gradually ſhorter on each ſide to the extream ones: the four middle ones are altogether black; they are much larger than any of the others, and they are turned inward at the extremities, and ſerve to ſupport the weight of the body in a great meaſure, while the bird is running up the trunks of trees.

The

The legs are feathered a great way down, but not quite to the feet; the naked part is of a deep bluish or lead colour, and the toes are placed two before and two behind: they are moderately long and strong, and the claws are very long, black, and sharp.

This species is also very frequent with us; it flies swiftly but irregularly, and is seen with great rapidity up the trunks of trees. It feeds on the worms hatched from the eggs of beetles, and on other insects found on the surface, and under the barks of trees: it is very swift in it's motions, and is incessant in the search of it's prey. All the authors, who have written on the birds of this part of the world, have described it. Gesner, Ray, Willughby, and almost all the others have joined in calling it Picus varius minor. We, in English, call it the lesser-spotted Wood-pecker, or the Hickwall; and the large, spotted kind last described we call, by a name somewhat like this in sound, the Witwall.

Picus albo nigroque varius rectricibus tribus lateralibus seminigris.
The variegated Picus, with the three lateral rectrices half black.

The middle spotted Wood-pecker.

This is a very pretty species; it's variegations are less striking at first sight than those of the first or largest of the spotted ones, but, on examination, they appear to be no less beautiful: the head is moderately large, and is flatted on the crown; the hinder part of the head is black, but the crown and the anterior part, almost to the base of the beak, is sometimes whitish, and sometimes red; the space just at the base of the beak only is grey; the temples also are grey: the eyes are small, but they are bright and piercing; the beak is short, but it is very robust and triangular in figure, and is marked with two or three longitudinal furrows: the nostrils are round, and they are covered with a few short and curled bristles.

The whole bird is of about the bigness of a sparrow: the throat, the breast, and the belly are all of a pale grey; the back is variegated with black and white, and the wings are black, and have six series of white spots on them: the tail is short, and is composed of ten feathers: the fourth and fifth are black, the others are about half black, and that is variegated with several spots of white.

The legs are short, but they are robust, and of a bluish colour: the toes are moderately long; they stand two before and two behind, as in the parrot, and the two anterior ones are connected to one another at the base; the claws of all of them are long, black, and sharp.

This is a native of England, but it is much less frequent than the others; it lives principally in thick woods. I have met with it in sufficient plenty in Charleton forest in Sussex, and in Rockingham forest in Northamptonshire, and some other places. Many of the writers on birds have omitted to name it, either from their not having seen it, or their confounding it with one or other of the preceding species. Ray has distinguished it under the name of Picus varius tertius; and our Albin calls it, but improperly, Picus varius minor.

Picus pedibus tridactylis.
The Picus, with only three toes.

The three-toed Wood-pecker.

This is a bird extremely singular in the construction of it's feet, in that particular, differing from all the others of it's genus; it is of the size of a common linnet: the head is small and depressed, or flatted on the crown, and is black but spotted: behind the angle of the mouth, on each side, there runs a white line; these, from the two sides of the head, join at the neck, and are thence continued, only broader all the way down the back, to the insertion of the tail: the breast and belly are variegated with

BIRDS. Series 1.
Plate 17
The Eagle Owl
The Scops
The Osprey
The Little Owl
The Vulture
The Kestrel
The Buzzard
The Moor Buzzard
The Green Parrot
The Parroquette

with black and white: the long feathers of the wings are black on the upper part, but with three or else five series of very minute spots of white; on the under side they are grey, and there are seven or eight rows of larger white spots: the feathers which cover the wings on the upper side are black; the tail is short, but firm and rigid: the long feathers are all of them black, but the two outer ones have some spots of white on them: the top of the head is of a fine strong saffron colour.

The legs are short, and of a deep bluish-black colour: the feet are also bluish, and the toes are only three, in which it differs from all the other birds of this genus: two of these stand forward, and one behind; of the two anterior ones the inner is the smaller and shorter; the hinder one is larger than the longest of the others.

This is a native of Sweden, Norway, Denmark, and some parts of Germany; it frequents the forests, and especially the mountainous ones. It has not been known to any of the ornithologists. We meet with it first described in the Stockholm Acts, under the name of Picus pedibus tridactylis, a name it retains every-where since.

Picus vertice coccineo cristato.
The Picus, with a scarlet-crested head.

The Brasilian Wood-pecker, or Ipecu.

This is a very singular and a very beautiful bird; it is the largest of all the wood-pecker-kind; it is equal to a common pigeon in size. The head is moderately large and flatted; the whole crown or depressed part is of an elegant scarlet; there is also a crest of long and very beautiful feathers, of the same colour, which has it's origin from the hinder part, and hangs down on the back: the eyes are small, but very bright and piercing; the iris is of a deep saffron colour, and the pupil of a bluish-black, with a shade of a changeable green also cast over it: the beak is about three quarters of an inch long, and of a bluish-grey colour; it is of an angulated form, and obtuse at the point, and is there of a darker colour than any where else: the membrane which invests the beak is of a deep bluish-grey, and in this stand the nostrils, which are round and small.

The neck is black, but there runs down on each side of it a longitudinal streak of white: the back is of an extremely dark iron grey: the rump is quite of a deep black; the wings and tail are also of a deep black on the upper side, but white underneath; and the breast, belly, and feathers which cover the thighs are of a pale grey, or a black, spotted with so numerous and small spots of white, that the whole appears grey and uniform.

The legs are short, and they are covered down to the joint with pale grey feathers: the naked or lower part of them is blue, and the toes also are blue, and are placed, two before and two behind: the claws are short and obtuse; the tail is short, but the feathers which compose it are rigid, and the middle ones, which are a little longer than the rest, are turned down at the points, so that they assist the bird in keeping it's place on the trunks of trees.

This is a native of the Brasils, and is frequent also in the inland parts of South America; we sometimes have it sent over to us dried. Maregrave, Piso, and the other writers who have mentioned the Brasilian and South American animals, have all given it the name Ipecu, by which it is known to the natives; and Ray, Willughby, and the rest have followed them. There is a very perfect specimen in Sir Hans Sloane's Museum.

Picus corpore flavescente.
The yellow Picus.

The golden Wood pecker.

This is an elegant bird, and, though extreamly different from all the other species of this genus in colour, is so perfectly like them in shape, in the structure of parts and manner of living, that even the peasants of the countries where it is native, distinguish it as such.

The head is small, depressed at the top, but rounded at the sides: the beak is about half an inch long, and is of a triangular figure, very thick at the base, and, though sharp, yet not small at the point; it has three longitudinal furrows on it, and is of a pale bluish colour: the membrane which covers it's base is of a somewhat greyer hue, otherwise nearly like it; and the nostrils, which stand pretty forward in this, are patulous and round: the eyes are large; their iris is of a dusky colour, and the pupil large and black, but with a changeable shade of bluish and greenish thrown over it.

The top of the head is of a fine gold yellow; the breast and the long feathers of the wings are also of the same elegant colour: the whole bird is indeed yellow, but the belly is pale and whitish, as is also the rump; and the back and the covering feathers of the wings are of a somewhat duskier yellow, but still a very agreeable colour.

The legs are short; they are covered down to the joint with short feathers, of a pale whitish-yellow, approaching to what we express by the term straw or lemon colour: the toes are long; there are two placed before, and two behind, and their claws are black, but not very long or very sharp.

This is a native of the South of France, and of some parts of Italy, but it is more rare than any of the other European species. It feeds, like the others, on insects that live upon, and under, the barks of trees, and has, like the rest, a long tongue, very sharp at the point, and capable of being darted out at pleasure. None of the ornithologists have described it. Barrelier mentions a bird under the name of Picus Citrinus, but the description does not perfectly agree with the preserved specimen I have received of this.

The other species of the Picus are, 1. The brown Picus, spotted with yellow. 2. The black Picus, with the wings and tail yellow. 3. The cirrated, black Picus, with the tail yellow. 4. The brown Picus, with a gold yellow crest. 5. The smallcoal-black Picus, with the scarlet crest. 6. The larger-crested, American Picus. 7. The smaller-crested, American Picus. 8. The elegantly-coloured, variegated, American Picus. 9. The larger, spotted, American Picus. 10. The lesser, spotted and variegated, American Picus.

J Y N X.

THE beak of the Jynx is smooth, and the nostrils are very conspicuous, and depressed and hollowed: the tongue is very long, of a rounded form, and resembles a worm: the toes are four, and they are placed two before and two behind.

Of this genus there is only one known species. Linnæus, in some of his works, has joined this and the cuculus; but 'tis better to preserve them distinct, as he has done in his Systema Naturæ.

JYNX.

The Wry-neck.

This is a bird of a very singular appearance; it is about equal to a lark in size: the head is small, and somewhat depressed on the crown; the beak is smooth, and not so robust as in the wood-pecker-kind, though somewhat approaching to them in shape;

it

It is not more than a third of an inch in length, and is of an obtusely angulated form, but much less angular than in the generality of the species of those birds: it is of a deep and dusky lead colour or bluish, and the membrane at the base is of a pale red in the male bird, and greyish in the female: the nostrils stand somewhat forward, and are of a rounded form, and hollowed; the tongue is very long, and terminates in a bony and sharp spine: the eyes are large and piercing; their iris is of a hazel colour, but with a tinge of yellow; and the pupil is very bright, and of a bluish-black.

The colouring of the whole bird is very elegant; the head is of a very bright grey, variegated with transverse lines and spots of black, white, and a reddish-brown, and it has a power of raising the feathers of the crown into a kind of crest: from the top of the head to the middle of the back there run a series of black feathers: the rump is grey, but it is variegated with transverse lines, and spots of black: the back and the covering feathers of the wings are variegated with black, brown, reddish, and grey: the throat and the lower part of the belly are of a pale yellowish colour, and there are several transverse lines of black on them: the breast is of a yet paler colour, and with less of the yellow; it is, indeed, almost whitish, and there are fewer spots on it than on the belly or throat.

The wings are moderately long, and very well feathered; the short plumes which immediately cover the roots of the long feathers are yellow, and have some variegations of black: the long feathers are nineteen on each wing; they are black, spotted with a reddish-brown, and some of these spots, especially those on the feathers, which are nearest to the body, are themselves again spotted with black, in very minute dots.

The tail is moderately long; it consists of ten feathers, and these are slender and weak, not rigid or formed as those of the wood-pecker-kind; they are of a grey colour, and each of them has four or five transverse black lines on it: the part of the feather which is nearest these lines is of a mixture of brownish with the grey; the rest of the feather is a mere unmixed grey, and is, when closely examined, found to be sprinkled over with small dots of black.

The legs are short, and not very robust; they are of a pale lead colour in the one sex, and of a flesh colour, or very pale red in the other: the toes are moderately long; they stand two before and two behind, and the anterior ones are connected at the base: the two outer toes on each foot are equal in size, and they are on each twice as long as the two inner ones.

We have this in England, but it is not very frequent. It inhabits large forests, and is seldom seen out of them; I shot two a few years since in Charleton-forest in Sussex. It feeds on ants and other small animals; it pierces these with it's tongue, which it darts out upon them, and never bites them with the beak, but to prepare them for going down the throat. It has a very singular way of twisting it's head about, and bending it's neck: it thence obtained of the Latins the name Torquilla; and, with us, that of the Wry-neck. All the writers on birds have described it. Bellonius calls it Jynx, Torquilla, Turbo, and Sisopigis; Gesner, simply, Jynx; Aldrovand, and since him Ray, Willughby, and the rest since his time, Jynx sive Torquilla; Linnæus, where he ranks it with the cuculus, calls it Cuculus subgriseo maculatus rectricibus nigris, fasciis undulatis; Charleton calls it Sisopigis Torquilla.

CUCULUS.

THE beak of the Cuculus is smooth: the nostrils are not hollow, as in the jynx, but protuberant: the tongue is entire, and of a sagittated figure: the toes are four, and they stand two before and two behind, as in the parrot and wood-pecker-kinds.

Of this genus there is only one known species.

Cuculus

CUCULUS. **The Cuckow.**

The cuckow is a bird of considerable beauty, when closely examined; and the male and female differ so much in their colour, that it would be easy to mistake them for distinct species. The female is about the size of a sparrow-hawk, or somewhat smaller. It is a very well-made bird, and usually stands erect and firm: the head is moderately large, and somewhat depressed: the eyes are large, and piercing in their aspect; the beak is a little hooked, but not much so; the upper chap is somewhat longer than the under, and is black; the under one is green: the nostrils are large, prominent, patulous, and naked: the tongue is flatted, and of a sagittated figure.

The head, the neck, and back are all of a hoary colour, with some dark grey feathers of a peculiar gloss, shewing themselves in some places: the throat, and indeed all the under part of the neck, is of an undulated flesh colour, or very pale red, with an admixture of yellow: the rump is of the same colour with the back, only paler; and the belly is whitish, but variegated with elegant, transverse lines, of a shining, brownish-black; the under part of the tail is also of the same hue, only whiter.

The tail is long; is composed of ten feathers; these are black, but with the tips white, and there are also some other white spots on them, both at the sides and in the middle: the wings are long; they are of a brownish-black on the upper sides, and on the under they are of a hoary greyish, with transverse lines of white: the long feathers are black.

The legs are moderately long, and are covered pretty low on the anterior part with feathers: their naked part is yellow; the toes are long, and they stand two before and two behind; the claws are long, black, and sharp.

In the male, the head is of a pale silvery grey, without any admixture of spots or variegations; the back is of a somewhat duskier grey than the head, but neither in this is there any variegation, or any feathers darker, or otherwise different from the rest: the under part of the throat also is grey, without any spot: the belly is of a darker colour than in the female, and the under part of the tail is yellowish: both male and female are sometimes darker, and sometimes paler coloured, and have more or less variegations of the different shades of grey, but there is no foundation for the dividing them from this accident into different species.

This breeds with us, but does not remain the whole year. All the writers on birds have described it, and all under the same simple name of Cuculus; only Linnæus, who makes the jynx a species of the same genus, calls this Cuculus rectricibus nigricantibus punctis albidis.

PARADISÆA.

THE beak of the Paradisea is of a cultrato-subulated form, and acute; and the forehead is gibbous: the two middle feathers of the tail are extremely long, and very firm.

Paradisæa subrubens alis coccineo et flavo variegatis.
The reddish Paradisæa, with the wings variegated with red and yellow. **The common bird of Paradise.**

This elegant species is about the bigness of a swallow, and it somewhat resembles that bird in shape, if the body only be examined; though, taking in the whole bird, it has not the least resemblance to any thing, but the other species of it's own genus; the whole

whole bird appears of variable colouring, as seen in different lights, but in all very beautiful; the head is large: the beak is in shape like that of the cuckow; it is moderately long, and a little bent or hooked: the nostrils are small, depressed, and of an oval figure: the eyes are small, but of a very piercing aspect, and their iris is of a fine orange colour.

The whole crown of the head, from the naked part at the base of the beak to the beginning of the neck, is covered with short, thick, and rigid feathers, of a fine gold yellow, and extremely bright and glossy: the lower part of the head and the throat are covered with feathers of a softer texture, and of a most elegant changeable deep blue and green colour, such as we see on the neck of some of the duck kind: the back and rump are of a beautiful colour; a red, but not bright or fiery, but tinged with a mixture of a kind of brown: the breast is of the same colour, only something paler; and the belly is still paler than that.

The wings are very well feathered; they are long, and pointed at the ends: the principal feathers are of a very deep colour, seeming formed of an admixture of black and red, and over these there stand some others of a singular structure, considerably large, and most elegantly variegated with scarlet and yellow: the tail is moderately long, but the two long feathers of all are of more than three times the length of the others; these also are of the same mixed, dark, but glowing colour with the principal feathers of the wings, but there are none of the variegated feathers toward their bases.

The breast and belly of this species are covered with thick-set and considerably broad feathers; the back with narrower and more rare ones: the legs are short, and of a pale flesh colour, the toes are long, and the claws are long, black, and sharp.

This, and indeed all the other species of this genus, are the natives only of the hotter countries; they keep almost continually on the wing, in the manner of our swallows, their wings and tail being extremely long, and their bodies very light. There grew from hence an opinion among the vulgar, that they had no legs; and those who have them ever preserved to us, used to favour the deceit, by pulling off the legs of such, before they dried them. Many of the writers on birds have described this species. It was one of the first brought over into Europe, and thence acquired among us the name of the common bird of Paradise, or, simply, the bird of Paradise, and among authors that of Avis Paradisi primus; Aldrovand calls it Manucodiata prima; and Ray, Willughby, and others have copied the same name.

Paradisea flavo-ferruginea capite pallidiore maculato.
The yellowish-brown Paradisea, with a pale spotted head.

The yellow bird of Paradise.

This is also a very beautiful species, though in the whole less so than the former; it is of the bigness of a lark: the head is large, and scarce at all flatted on the crown: the beak is moderately long, somewhat hooked, and of a pale brown: the upper chap is a little longer than the under, and the inside of the mouth yellow: the nostrils are small, hollowed, and oval; the eyes are very small, and their iris is of a fiery red, with a tinge of yellow diffused throughout: the head is of a very pale colour, approaching to white, and of a silvery gloss, and is elegantly spotted all over with a bright yellow; the spots in some lights having the appearance of burnished gold.

The back is of a beautiful colour, formed of a mixture of yellow and a ferruginous brown: the upper part, near the neck, is palest; it has very little of the brown, but in the place of it a greyish-white, mixed with the yellow and predominating. Toward the middle of the back there is a good deal of brown in the yellow, but it is yet very bright and glittering, and at the rump it is perfectly brown: the breast and

 belly

belly are of the same mixed colour, only paler; and the under part of the tail is of a silvery grey: the wings are very long, and their principal feathers are of a fine gold yellow, with a slight tinge of brown: the two long feathers of the tail are also of the same variegated colour, and very beautiful: the legs are short, and of a pale flesh colour, and the claws are long and black.

This is a native of Arabia: it has not been described by any author. We had some specimens of it brought over, about four years since, in very great perfection.

Paradisæa subalbida flavo et fusco variegata.
The whitish Paradisæa, variegated with yellow and brown.

The white bird of Paradise.

This is of the size of our lark, and is a very beautiful bird, though less so than the two preceding species: the head is moderately large, and a little depressed or flatted on the crown: the beak is two fingers breadth long, and moderately thick; it is of a mixed greenish and yellowish colour: the upper chap is a little longer than the under, and it is somewhat hooked at the extremity: the inside of the mouth is reddish, and the tongue is long, red, and sharp, and hard at the end, in the manner of that of the wood-pecker-kind: the nostrils are oblong and hollowed: the eyes are moderately large, and their iris is yellow: the verge of the eye-lids is red, and there is a redness also in the naked part, at the base of the beak.

The head is of a silvery white, variegated with spots of yellow: the back is of three colours; it's upper part is whitish, like the head, with scarce any variegation; a little lower there is a tinge of yellow in the white, and toward the rump of brown: the breast is of the same whitish colour, but with a blush of flesh colour diffused over it: the wings are very long, and their principal colour is the same whitish, but there is a mixture of ferrugineous at the tips: the tail is white toward the body, but afterwards it is also tinged with a ferrugineous hue, and the two long feathers are very beautiful; they have a fine gold yellow mixed with the brown.

The legs are short, feeble, and reddish; the toes are long, and the claws sharp and black.

This is a native of the same part of the world with the former; it is sometimes brought into Europe dried, and seems to be the species described, under the name of the Manucodiata secunda, by Aldrovand; and after him, by Ray, Willoughby, and others.

Paradisæa albo pectore castaneo.
The white Paradisæa, with a brown breast.

This species is considerably larger than either of the former; it's body is equal to that of our jay in size, and it's wings are more than ten inches in length: the head is large and flatted on the crown; the eyes are large, and very bright and piercing; their iris is of a bright gold yellow; the pupil of a greenish-black: the beak is three fingers breadth long, and considerably thick; it is very much hooked, and at the point is very sharp; the upper chap is longer than the under, and the nostrils which stand toward the base of it are large, patulous, and of an oval figure.

The sides of the head are of an elegant silvery white, without any spot or variegation: the crown is of a ferrugineous brown, and, toward the hinder part, this brightens with an elegant yellow: the back and rump are of a clean and beautiful white; the coverings of the wings are also white, and their long feathers are somewhat greyish: the throat and breast are of a fine deep chesnut brown; the belly is of a pale brown, and the under part of the tail is again whitish: the wings, when expanded, spread

to

to a great extent; the tail is moderately long, and the two long feathers are very narrow and flexible.

The legs are ſhort, but robuſt; they are of a duſky browniſh-yellow: the toes are long and ſlender, and the claws are long, black, and ſharp.

This is brought to us dried with the others, and it is but rarely that the legs are preſerved with it; moſt of the authors who have written of birds have mentioned it. Aldrovand calls it Manucodiata cirrata and Hippomanucodiata; Ray, Willoughby, and others have borrowed the ſame name. Aldrovand mentions it as much larger than the ſize given here; but this deſcription is from a ſpecimen ſent over to England; it is poſſible larger of the ſame ſpecies may have been ſeen.

Paradiſea griſea capite nigreſcente criſtata.
The creſted Paradiſea, with a black head.

This is of the ſize of our thruſh in it's body, but the wings are ſo long, that it has the appearance of being much larger: the head is large and flatted on the crown: the eyes are ſmall, and their iris brown; the beak is two fingers breadth long, and ſomewhat hooked: the upper chap is longer than the other, and is turned over it at the end: the whole beak is of a deep black; the noſtrils are large, open, and of a figure approaching to oval.

There is a ſpot of yellow near the baſe of the beak, but the reſt of the head and the neck are black: the back is of a duſky grey, and the breaſt and belly of a pale grey; the lower part of the belly is whitiſh; there is a creſt on the neck formed rather of a kind of ſcce than of feathers; this is of a yellow colour: the long feathers of the wings are black; the tail alſo is black, and ſo are the two oblong feathers of it.

The legs are moderately long, ſlender, and yellow: the toes are long, and the claws are long, black, and ſharp.

Moſt writers on birds have deſcribed it. Aldrovand, Ray, Willoughby, and others call it Manucodiata quarta ſive cirrata. We have ſpecimens of it ſometimes brought over to England.

Paradiſea nigreſcens capite minore.
The black Paradiſea, with a ſmaller head.

The black bird of Paradiſe.

This was one of the firſt ſpecies brought into Europe, and is conſequently deſcribed by ſome authors as the common kind; it is, in the body, about as big as our thruſh, but the wings are immoderately long: the head is ſmaller than in moſt of the others, but it is ſomewhat broad, and flatted on the crown: the beak is a finger's breadth and a half long, and is hooked at the end: the upper chap is longer than the other, and is black; the under one is of a kind of olive colour: the noſtrils are large and roundiſh: the eyes are ſmall, and their pupil black, with a tinge of greeniſh; their iris grey: the anterior part of the head is of a pale, ſilvery grey, but the reſt of it is black: the back is of a dark grey, the breaſt of a paler, and the belly whitiſh: the wings are very long, and their feathers extremely thin, light, and delicate, and thoſe of the tail are alſo very elegant, but ſomewhat more rigid; all theſe are black, and the two long feathers of the tail are blacker than all the reſt.

The legs are ſhort but robuſt, and of a deep bluiſh-black colour: the toes are moderately long and ſcaly; the claws are very long, black, and ſharp.

This has been long known in Europe by the ſpecimens ſent over of it; but as the firſt people, who made us in this part of the world acquainted with it, told of it's being always on the wing, and, to credit the ſtory, pulled off it's legs; the authors who

who have mentioned it, are found to have swallowed the falsity, and describe it under the name of Apos, or the bird without feet. Gesner has given a figure of it, and submits to it's want of feet with great readiness; but, as he would have it have some way of resting, he tells us, that, when inclined to do so, it twists the two long feathers of it's tail round the bough of a tree, and so hangs with it's head downward. It is at this time not so often brought over, as those of more beauty. Aldrovand calls it Manucodiata quinta sive vulgaris; and Ray, Willughby, and others have borrowed his name; Gesner and others call it Paradisea, sive Apos Indica; others, Paradisea vulgaris.

Paradisea fusca fronte viridescente, collo superiori flavo.
The brown Paradisea, with a greenish front, and the neck yellow.

The king of birds of Paradise.

This appears of the size of the jack-daw, but, when examined, the greater part of it's bulk is found to be owing to it's feathers, the body being, in reality, very small: the head is small, flatted on the crown, and rounded at the sides: the eyes are large, and of a piercing aspect; the beak is three quarters of an inch long, and of a blackish colour, with a tinge of olive: the upper chap is longer, and the under shorter, and the upper is almost entirely black, the other brownish: the nostrils are of an oval figure, and tolerably large: the head is of a beautiful brown, except that toward the base of the beak, and about the angles of the mouth, there are some beautiful feathers of a glossy green: the back-part of the neck is of a beautiful gold yellow, and the fore-part of the neck, from the beak to the breast, is of an elegant, glossy green, with a tinge of gold shining through it, much like the colour on the backs of some of our beetles: the back, the wings, the tail, and indeed the whole upper part of the bird, are of a deep brown: the breast is also of a strong brown, approaching to chesnut; a very elegant colour, and extremely glossy: the long feathers having their origin from under the wings, and hanging over the tail, are of an elegant gold yellow toward the base, and of a variegated brown and yellow in the rest; these are about a foot long, and of an elegant structure.

The two long feathers have their origin among these; they are more than two feet in length, and are of a gold yellow toward the base, but they have more brown toward their extremities, and they are turned or curled at the ends.

The legs are moderately long, robust, and of a brown colour: the toes are long; they stand, as in all the others of this kind, three before and one behind, and they are armed with long and sharp whitish claws.

We frequently meet with specimens of this in the collections of the curious. Margrave calls it Manucodiata Rex; and most of the authors who have written since, have continued the same name to it. Charleton calls it Paradisearum Rex.

Paradisea capite nigro viridescente, collo flavo, dorso aureo, cauda fusca.
The greenish, black-headed Paradisea, with a yellow back, and a brown tail.

This is a small species, but it is one of the most beautiful of the whole genus: the head is small, and is depressed on the crown, and compressed in some degree at the sides: the whole bird is not bigger than our common swallow, only the wings are much longer: the eyes are extremely minute; they hardly exceed a grain of millet in bigness, and are coal black: the beak is three quarters of an inch long; it appears very large, in proportion to the head: the upper chap is longer than the under, and is somewhat hooked: the nostrils are of an oval figure and small.

The

The front of the head is ornamented with a quantity of fine, soft, and short feathers, of a deep glossy black; these surround the whole base of the beak, both above and below, and have the appearance of velvet: the rest of the head is of a deep greenish-black, with a changeable shade of blue and purple thrown over it, as we see in the changeable colours on the neck of a peacock; the neck is of a strong yellow, with some tinge of brownish with it, but the back and the covering feathers of the wings are of a pure, unmixed, gold yellow; the breast is of a paler yellow, and the belly is whitish, but with a tinge of straw colour.

The long feathers of the wings are of a deep chesnut-brown; the tail also is of the same colour, but there is an appearance of it's being of a silvery white: there are a multitude of very long feathers, of an elegant and a singular structure, which have their origin from the sides of the body under the wings, and about the thighs; these are from six to ten inches in length, and of a singularly delicate structure: they are of a silvery white, and they hang all over the tail, so that they have been mistaken by inaccurate observers for the tail itself: the two long feathers of the tail exceed these considerably in length, and are brown, very glossy, and shining.

This is sent over dried, among the others, as a curiosity to Europe; and many of the writers on birds have described it, though most of them imperfectly. Marcgrave, having before described a larger kind, calls this Manucodiata, sive avis Paradisi altera; and Ray, Willughby, and others have continued the same name to it.

Paradisæa gutture viridescente, pectore nigro purpureo.
The Paradisea, with a greenish throat, and a blackish-purple breast.

This species is about equal to the sparrow in the real size of it's body, but it's feathers stand so loose, and it's wings are so large and long, that, when they are closed, it has the appearance of a much larger bird; most of the birds of Paradise are handsome, but this is more so than almost any of them: the head is moderately large, and is flatted on the crown: the eyes are very small and black; the beak is two fingers breadth long, and is almost straight, only the upper chap is somewhat longer than the under one, and at it's point bends a little over it: the whole beak is of a dusky olive colour, approaching to black; but the under chap is somewhat paler than the other, and the nostrils are small, depressed, and of an oval figure.

The head is covered with a peculiar kind of feathers; they are extreamly narrow, and resemble rough filaments of silk; they are of a gold yellow, except at the base, where they are brownish: the sides of the head, from the eyes down to the neck, are covered with the same sort of fine narrow feathers, but they are of a deep blackish colour, with a shade of changeable green and blue, such as is seen on the neck of a peacock: the breast is covered with elegant long and narrow feathers, of a beautiful glossy appearance, and of a colour made up of a deep brown, approaching to black, and a strong red: the belly is of the same colour, only considerably paler; and the back, the wings, and the tail, when viewed in a full light, appear only of a dusky velvety-brown; but, viewed in different directions, it is seen that this colour is composed of a mixture of brown and scarlet.

The long feathers of the wings are extreamly elegant; they are from five to fourteen inches in length, these also in some lights have a shade of brown upon them, in such manner, that they seem simply of that colour; but, viewed in other directions, they shew a most elegant gold yellow, a high scarlet, and a flaming orange colour: the tail has the same variegations, but this is in great part covered by the long and narrow feathers, which grow from the sides of the body: the two long feathers of the tail are, like those of the wings, of a mixed colour, part yellow, part red, but with a shade of brown over all.

The legs are moderately long, and very slender; they are of a dusky olive colour: the toes are long and slender, and the claws very long, sharp, and of a pale brown.

This is a species rarely sent over to us; Sir Hans Sloane has one in fine perfection. Ray and Willughby have mentioned it, under the name of Paradisea avis majoris generis.

Paradisea rubens cauda et alis ferrugineis.
The red Paradisea, with a brown tail and wings.

This is a species sometimes also honoured with the name of Rex avium Paradisearum. Authors had heard the Dutch merchants, who were the first that brought the knowledge of these birds into Europe, talk of one species more beautiful than the rest, which was called the King of the birds of Paradise, and they have occasionally honoured different species with that appellation.

This is about the bigness of a nightingale, but it's wings are extreamly large and long; the head is smaller than in many other species, and flatted on the crown; the eyes are small, and their iris is of an orange colour: the beak is more than an inch long; it is slender, weak, and somewhat hooked at the end: it is white throughout, only the nostrils, which are large, depressed, and oval, have some tinge of brown.

The upper part of the head is of an extreamly bright and fine scarlet: the back, the rump, and sides are of the same bright scarlet, but the sides of the head, and the whole under-part of the body, are a little paler: the wings have the covering feathers scarlet also, but the long ones have a tinge of brownish, and the long feathers of the tail have also the same brown colour: the long and slender feathers however, which grow from the sides of the bird and fall over the tail, and the two long feathers of all, are scarlet, only there is a little tinge of brownish toward the bases of the latter.

The legs are robust but short; they are white, and the toes are long, and armed with sharp but slender whitish claws.

This species is brought over with the rest into Europe. Ray and Willughby have described it, under the name of Rex avium Paradisearum majoris generis; and many of the other writers seem to have mentioned it, under the general name of Rex avium Paradisi.

UPUPA.

THE beak of the Upupa is arcuated, convex, compressed, and equal, and it has a furrow running along each side of it: there is a crest on the head, which is capable of folding back.

Of this genus there is but one known species.

UPUPA. **The Hoopoe.**

This is an extreamly singular bird; it's weight is about three ounces, but it is so thick covered with feathers, that it appears large, in proportion to that weight: the head is large, and is ornamented with an elegant crest: the eyes are small, but very bright and piercing: the beak is of a very singular figure; it is an inch and a half long, somewhat bent into the form of a bow, pointed at the end, very slender, and all over of a black colour: the nostrils are large and oval, and stand toward the base, and there runs on each side, all the way down, a longitudinal furrow: the tongue is short, and lies deep in the mouth; it is broad at the base, and pointed at the extremity, and is, upon the whole, of a figure approaching to triangular: the figure of the whole bird approaches to that of the plover.

The crest on the top of the head is extreamly elegant; it is composed of a double series of feathers, two fingers breadth high, and continued from the base of the beak to the very back-part of the head: the bird has a power of raising or depressing this at pleasure; it is composed of twenty-four or twenty-six feathers, and some of these are much longer than the others; these are tipped with black, and some of the under part with white; the rest of the crest is yellow, but that not a fine bright yellow, but with a tinge of a chesnut-brown: the neck is of a beautiful reddish-brown; the breast is white, variegated with lines of black tending downwards; but, as the bird grows older, the middle of the breast becomes absolutely white, and these black variegations are only seen on the sides.

The tail is between four and five fingers breadth long, and is black; it is composed of ten feathers; in the middle of it there is an elegant spot of white, large, and of the figure of a new moon; the back or gibbose part looking towards the body, the horns toward the extremity of the tail: the wings are moderately large, when expanded; they measure about sixteen inches from tip to tip, and, when closed, they reach nearly to the tip of the tail: the long feathers in each are eighteen; the ten first of these are black, with a white area on each; the succeeding ones have also a white area on them, and the tips and edges of some of the last are reddish.

The back is very elegantly variegated with black and white, in little alternate spaces on every feather: the legs are short, and not very robust; the toes are moderately long: the outer toe is connected to the middle one, some part of the way down, without the help of a membrane: the claws are moderately long and sharp.

This bird is a native of the northern parts of Europe, but is no where very plentiful. It had been long known in Germany and Sweden, before it was supposed to be a native of England, and several of our old writers declare it not to be such. Ray produces the testimony of people who had seen it with us; and I have myself shot it on the vast heath, called Hind-Head, in Sussex: it is a very shy bird; I found it on the wet parts of the heath. It's food is insects, ants, small beetles, and the like. The antients were well acquainted with it. The Greeks called it Ἔποψ; and Aristotle tells us, that it used human dung instead of clay for the lining of it's nest, but there is no late observation to countenance this. The Latin writers have all described it under the same simple name Upupa; some have also continued the Greek one, and called it Epops sive Upupa. We call it the Hoop or the Hoopoe; the Germans, the Widehoppe; names which seem to have been all formed from it's note, which is very loud, and almost continually repeated.

ISPIDA.

THE beak of the Ispida is of a trigonal figure, somewhat arcuated, compressed, and equal: there are four toes to each foot, but only one of them behind.

Ispida dorso caeruleo, pectore rufescente.
The blue-backed Ispida, with a reddish breast. — **The King-fisher.**

This is one of the most beautiful of the European birds: the head is moderately large and flatted; the whole bird is about the bigness of a lark, but of an extreamly different shape, broader on the back, and flatter: the eyes are moderately large and piercing; the beak is near an inch and a quarter in length, and is almost straight, strong, acute, and all over black, except at the corners of the mouth, where it is whitish: the two chaps are about of the same length; the nostrils are large and oblong, and the inside of the mouth is yellow: the tongue is broad and short, and is pointed, and not divided at the extremity.

The throat is white, but with some faint admixture of a reddish-brown: the breast and belly are also of the same colour, but with more of the reddish tinge; and the lower part of the belly, toward the tail, is of a full and strong reddish tinge; the sides also,

also, and the feathers under the wings, are of the same reddish tinge: on the breast there is a tinge of bluish-green at the tips of the feathers.

The whole back, from the neck to the insertion of the tail, is of an elegant blue colour, scarce to be described, and on which it is painful to the eye to remain looking: it is not a deep or strong colour, but pale, and extreamly bright, excelling the lustre of all the works of art: it appears uniform at first sight, but, if carefully observed, there will be found some transverse lines in it of a darker colour.

The crown of the head is of a blackish-green, variegated with transverse spots of blue: between the nostrils and the eyes there is a reddish spot, and there is also another spot of the same colour behind each eye; and behind each of these reddish spots there is yet another of a pale whitish hue, with only a slight tinge of reddish-brown in it: under all these, there runs a longitudinal line of a bluish-green.

The wings are long, and well plumed; the long feathers in each are no less than twenty-three in number; the third of these, from the outer one, is the longest: as well these long feathers, as the shorter, which immediately cover them, are of a beautiful dazzling blue both on the upper side, and brown underneath; and all the lesser ones, except those quite at the verge of the wing, have the same elegant blue at their tips: from the shoulders there hang a number of very long and very narrow feathers down each side of the back; these are of a mixed blue and green colour, very bright and glossy; the tips of them are bluer, and the bodies greener.

The tail is very short, it hardly equals an inch in length; it is composed of twelve feathers, and is throughout of a deep blue, with some blackness.

The legs are extreamly short; they are black on the front side, and of a red colour on the hinder part, as are also the feet and toes: the structure of the feet is singular; the outer toe of each adheres to the middle ones for the space of three joints, and the inner one only one joint; the inner toe is the smallest of all; it hardly equals half the length of the middle one: the outer one is nearly equal to the middle one, and the hinder toe is about equal to the least of the three anterior: the third bone of the leg is shorter in this than in any other known bird; and the toes are deeply marked with a number of transverse lines or furrows, which give them the appearance of being articulated.

This is a native of England, and of many other parts of the world. It is usually seen about brooks and rivers, and feeds on fish: all the writers on birds have mentioned it. Ray, Willughby, and others call it Ispida, or verorum Alcyon; and others, Alcyon and Halcyon. It builds about the sides of rivers, but it is not common.

Ispida capite cæruleo pectore variegato. **The Indian King-fisher.**
The Ispida, with a blue head and variegated breast.

This is a very small but a wonderful elegant bird, it hardly exceeds our common titmouse in size; the head is small, oblong, flatted on the crown, and rounded at the sides: the eyes are small, and their iris is of a dark dusky hazel; the beak is large, in proportion to the size of the bird; it is more than half an inch long, considerably robust, a little crooked, and very sharp at the end, it is of a pale flesh colour throughout, except at the tip, and along the verge of the under jaw, where it is black: the nostrils are very minute, and of an oblong figure and narrow; and the angles of the mouth are yellow.

The head is throughout of a most elegant shining blue, without any spot or variegation: the throat is of a deep purple; the breast and belly are of a pale but elegant crimson, variegated with oblique lines of a greenish colour, not very distinguishable, unless on a near inspection: the back and shoulders are of a deep green, shaded with a tinge of the richest and deepest blue imaginable, if viewed in certain lights; in others it is less distinct; the rump is whitish.

The

The wings are long, and there are twenty-two large feathers in each; these are of the same deep colour with the back, but those which cover them, especially toward the top of the wing, are of the brightest blue, like that of the head: the tail is very short, and is blue; the colour of the back of this bird greatly resembles that of the neck of a drake, which is blue in some lights, and green in others; or it more exactly resembles that high tinge of the large humming bird, so remarkable for it's colour.

The legs are very short, and of a very beautiful pale red: the toes are long, and the three anterior ones are connected a great way together: the claws are black, but not sharp.

This is a native only of the East Indies; we had a specimen of it brought over about two years since, which is now in my possession, purchased at an auction. None of the authors who have written on birds have named it, unless it be Charleton; that author, in his Onomasticon Zoicon, talks of an Indian King-fisher not bigger than a wren: the size very well agrees with this, but he has not left us any farther proofs of his acquaintance with it.

Ispida corpore ferrugineo, collo lunula albo cincto.
The brown King-fisher, with a white collar.

The Brasilian King-fisher.

This has nothing of the elegant colouring, either of the East Indian, or of the European species, but it has very great beauty of another kind; it is about the bigness of a thrush: the head is large, and very much flatted; the eyes are large, and their iris is of a deep hazel: the beak is two inches and a half in length, nearly straight, strong, pointed at the end, and throughout of a black colour: the nostrils are oblong, narrow, depressed, and of a brown colour within: the head is somewhat elevated at the base of the beak.

The whole upper part of the bird, the head, neck, back, rump, and wings are brown, except that there runs a broad and elegant transverse streak of snow-white across the neck: the brown has some faint tinge of the reddish in it, and is what we understand by the term ferrugineus, when we express a very bright colour by that word: the feathers are all amazingly bright and glossy; and the eye is as much dazzled with them, as with the intolerable blue of the European kind, which it is always a pain to continue but a moment looking upon: the breast, the belly, the throat, and the under part of the wings and tail, are all of a snow-white.

The wings are moderately long, and the tail is very short, not exceeding three quarters of an inch in length; the long feathers of the wings are twenty-three in each, and they are of the same brown with the body, but there are a few spots of white distinguished on some of them, when the wing is expanded; and the tail also is brown, only with a few spots of white on some of the feathers.

The legs are remarkably short and black; the feet are formed exactly, as in our king-fisher, both in regard to the sizes and connexion of the toes: the claws are black, long, and moderately sharp.

This is a native of South America; we owe the first description of it to Marcgrave who gives it under it's American name Jaguacati guacu; the Portuguese, from it's frequenting ponds and rivers, and feeding on fish, call it Papa peixe; Ray, Willughby, and others call it Ispida Affinis, and add the Brasilian name of Marcgrave. We have, within these few years, had some good specimens of it sent over into Europe.

Ispida capite variegato, pectore cæruleo.
The blue-breasted Ispida, with a variegated head.

The Bee-eater.

This is also a very beautiful bird; it is larger than our common king-fisher, and indeed equals the black-bird in size: the head is very large, in proportion to the body, and is remarkably depressed; the eyes are large, and have a very bold and fierce aspect; their iris is of an extreamly bright and beautiful red: the beak is very long and large, a little crooked, black throughout, sharp at the point, and very like that of the common king-fisher in shape: the nostrils are oblong, and the tongue is small and lacerated.

The head is very elegantly variegated in it's colouring; the front of it, about the base of the beak, is of a most elegant bright blue, with an admixture of green, but, in the midst, just above the nostrils, there is some white: the crown of the head is of a reddish-brown, and in some there is a cast of greenish; this is the difference of the sexes: on each side of the head there runs a black line from the angles of the mouth by the eyes, and quite to the hinder part of the head; and from this, on each side downward to the top of the throat, the colour is yellow, but it is extreamly pale, about the colour of flour of brimstone, only not so bright: the neck and shoulders are of a very elegant hue; it is green, but with a considerable admixture of red blended in it: the whole under side of the body, the breast, under part of the neck and belly, are of a beautiful blue: this is deepest on the neck, and of all is the most deep, where it joins the yellow under the base of the beak; on the breast it is paler, and yet more diluted on the belly, and throughout it has an admixture of red in it; but this is very faint, and in some birds, probably the male, there is a tinge of green with the blue.

The wings are moderately long; the large feathers are all of them black at the extremities, and the ten first have behind the black a space of blue, with a cast of green: the middle ones are of a very bright and beautiful orange colour, and the larger orders of the covering feathers of the wings are of the same colour, but those on the very top are bluish: the feathers of the shoulders, which are long, and hang down on the sides, are of a pale whitish-yellow, and are deeper at the tip than elsewhere. This is the general colouring of this elegant bird; but the several individuals of the species, though they agree in the principal things, yet differ greatly in particulars; some have more of the blue, some more of the green; others have more, others less, of the reddish tinge; and, though it is always easy to know the species solely by the colouring, yet the variation is surprising.

The tail is moderately long; it is four or five inches in length, and is composed of twelve feathers: it is blue, but there is some greenness in the outer feathers; the two middle feathers run out farther than any of the rest, and terminate acutely.

The legs and feet are formed exactly, as those of the common king-fisher; the legs are very short but thick, and the toes are proportioned and connected, as in that species; they are of a pale brownish-red: the claws are long and black.

This elegant bird is a native of Italy, and of the islands of the Archipelago, and many other places, but it is not met with in England: at Rome it is brought to market; and Bellonius says, that it is as common in Crete as our ordinary small birds here: all the writers on birds have described it. Bellonius, Gesner, Aldrovand, and Charleton call it Merops; Willoghby and Ray, Merops and Apiaster. It feeds on beetles and other insects. Aristotle says on bees, and all the rest of the naturalists have followed him in this, and many of them seem to have supposed that it eat nothing else. Bellonius tells us too, that it eats wheat, and other corn, and the seeds of several plants; and, as it is frequently seen about the water, it is probable that, like the king-fisher, it feeds occasionally on fish. It catches bees, and other flying insects, while

while on the wing, as the ſwallow does flies. It does not uſually fly ſingle, but in flocks, and makes a loud but not diſagreeable noiſe, ſomewhat like that of a man whiſtling.

Iſpida dorſo caſtaneo, pectore flaveſcente.
The brown-backed Iſpida, with a yellow breaſt.

This, though greatly inferior to the former ſpecies in colouring, is yet a beautiful bird; the bigneſs is about that of our king-fiſher: the head is large, in proportion to the body, and is depreſſed on the crown: the eyes are large, and their iris is of a bright red: the beak is an inch and a quarter long, and is moderately robuſt and thick: the two chaps, of which it is formed, are nearly of the ſame length, and the whole is bent a little toward the form of a reaper's ſickle, but in a leſs degree: it is black throughout, and the noſtrils which are ſituated toward it's baſe are oblong: the mouth opens to a great width, and the tongue is ſhort and ſmall.

The head is of a duſky, but not diſagreeable, yellow: the back is throughout of a beautiful cheſnut-brown, but the rump is paler, and greeniſh or yellowiſh: the breaſt, belly, and throat are of a pale yellow; the breaſt is ſomewhat deeper than the belly, and toward the ſides, and about the thighs, the colour is almoſt whitiſh: but the great ſingularity and ſtriking character of the bird is, a long black ſtreak on the middle of the head; this has it's origin at the baſe of the beak, and is continued quite along the head, to the very beginning of the neck.

The wings are long, and they are variegated in an extremely elegant manner; the long feathers are ſome of them all over blue, ſome all over yellow; ſome part blue, and part yellow, and ſome entirely black, only tipped with a bright ſcarlet at the extremities; it is not eaſy, indeed, to conceive a greater variation of colouring, than there is in the wing of this bird, and all the colours are high and glowing: the tail is moderately long, and is yellow, but green for a conſiderable ſpace at the end, ſo that at firſt ſight it appears half green and half yellow; the legs are very ſhort and yellow: the claws long and black.

This is a native of Germany, but it is not known in England. It is called by ſome authors Hirundo Marina, the ſea-ſwallow, but it is truly of the king-fiſher-kind, and is more like the merops than any other ſpecies. Aldrovand has therefore properly called it Merops alter ſive Meropi Congener. It feeds on inſects.

Iſpida vireſcens capite criſtato, cauda longiſſima.
The green Iſpida, with a creſted head, and very long tail.

The Guira Guainumbi.

This is perhaps the moſt elegant bird of this whole claſs, notwithſtanding that there are ſo many ſingularly beautiful ones of it; it is indeed ſo particular in it's figure as well as colouring, that, if ſpecimens of it had not been ſent over into Europe, we ſhould hardly credit the relations of thoſe who wrote upon the ſpot, that there was ſuch a bird.

It is about equal in ſize to our thruſh; the head is large, and is ornamented with a moſt elegant creſt; the beak is an inch long, tolerably robuſt and thick, a little hooked, and ſharp at the point: the noſtrils are oblong; the eyes are large and beautiful; their iris is of a gold yellow, and the pupil black.

The head is ornamented with a creſt of erect feathers, of the moſt lively blue imaginable; they are indeed exactly of the colour of the turquoiſe-ſtone; and in the midſt of this creſt is a round black ſpot, of the bigneſs of a ſilver three-pence; there are alſo ſome black feathers and ſome blue ones, which form an elegant, variegated line under the eyes: the neck, back wings, and tail, that is, the whole upper part of the

the bird, are of an elegant, strong, and shining green, equal to the colouring of the brightest of the parrot-kind: the throat, breast, and belly are of a bright and beautiful yellow: on the wings and tail there is a shade of a deep blue, mixed with the green, in the manner of the shaded colours on the neck of a drake: the tail is very long and narrow; it is composed of but few feathers, and has at first view somewhat of the appearance of the tails of the Macao-kind, which, together with the colouring, might at first sight induce people to suppose it of the parrot-kind; but the structure of the legs and feet, and of the beak, and other the most essential characters, declare them different.

The legs are short, and are feathered down to the middle joint, with fine, soft, and short feathers of a pale but glossy green; the naked part of the legs is black; the toes stand three before and one behind, and are, in proportion and structure, perfectly like those of the king-fisher: all the three anterior toes are joined together at their bases; the exterior one a great way down, the interior only for the compass of one joint: the claws are long and black; beside the other singular variegations on this species, there is a spot of most remarkable kind on the front of the neck; this is composed of three or four black feathers, and a series of pale blue ones round about them; it stands in the very middle of the neck, and makes a very beautiful appearance.

This beautiful bird is a native only of the Brasils, and of some parts of South America. Most of the ornithologists mention it after Marcgrave, under the name of Ispidæ seu Meropi affinis Guara Guainombi Brasiliensibus Tapinambis. We have seen specimens of it sent over in tolerable condition to Europe.

Ispida dorso fulvo albo variegato, pectore albo.
The Ispida, with a yellowish brown back, variegated with white, and a white breast.

The Creeper.

This is a very singular little bird, and has puzzled the naturalists of many ages in what class, or among what other birds, to place it: the shape of it's beak, however, arranges it among these; and the structure of it's feet brings it into the genus of the Ispida, though very unlike the common king-fisher in size, form, colour, and every other obvious particular.

It is an extremely small bird; it hardly exceeds a wren in size: the head is large, in proportion to the body, and the crown is flatted: the eyes are small, but they are very bright and piercing, and their iris is of a blue colour: the beak is long, slender, and a little crooked; both chaps are nearly of the same length, and the upper one is black, the lower whitish: the nostrils are small and oblong.

The head is of a deep dusky brown, with somewhat of a tawny yellow in it; there is, on each side of it, a spot of white: the back and the upper parts of the wings are of a tawny brown, but variegated in the middle of the feathers with white: the rump is of the same brown, only a little paler; the throat, the breast, and the belly are all white: the wings are short; the long feathers in each are eighteen in number; of these the first is short, and the fourth is the longest of all; they are brown, and have variegations of white in lines, and spots in their middle, and at their extremities: the covering feathers of the wings are of a darker brown than any of the others, and approach toward blackness; but they have white variegations, like those of the back ones, only not so large or numerous.

The tail is moderately long, in proportion to the size of the bird; it is not less than an inch and three quarters, and it is composed of ten rigid feathers, of a dusky brown, but with some faint tinge of a reddish, as well as yellowish, in it.

The

The Rook
The Roller
The Green Woodpecker
The Spotted Woodpecker
The Wryneck
The King Fisher
The common Bird of Paradise
The Cuckow

The legs are short, but they are robust; they are of a pale brown colour, and the hinder one is armed with a very long and sharp black claw; those of the three anterior toes are also black and sharp, but they are much shorter.

This singular little bird is a native of most parts of Europe; with us it is very frequent. It runs up the trunks of trees, in the manner of the wood-pecker, and feeds on the insects that it finds on their bark. It builds in rotten wood, and the hollows of old trees, and lays a vast number of very minute eggs. All the writers on birds have mentioned it, and almost all under the name of Certhia. Bellonius calls it Certhia, Certhius, and Reptatrix. We, in England, the Creeper, and, in some places, the ox-eye Creeper.

B I R D S.

Class the Third.

ANSERES.

THE beak or opening of the jaws in the Anseres is dentato-serrated: the feet are formed for swimming.

The word Anser has been used in general to signify only the goose-kind; but it is in this sense made to comprehend the swan, duck, widgeon, pelican, and a number of others.

PHOENICOPTERUS.

THE beak of the Phoenicopterus is of a strange figure; it is bent in such a manner as to appear as if broken, and is dentated at the edges; and the lower chap is broader than the upper: this singular structure of the beak excludes all other birds, from the genus of which there is therefore only the single species, usually known by the name.

PHOENICOPTERUS. **The Flamingo.**

This is one of the most singular birds in the world; it is very large, and most remarkably tall: the body is but small, in proportion to the extream length of the neck and legs: the head is very large, and remarkably rounded or prominent, not flatted on the top: the eyes are large, and very bright and piercing in their aspect: the beak is of a very singular figure and structure; it is long, and very broad; it seems as if broken toward the top, and the upper chap is depressed and dentated, and the under one very broad and thick: it is of a dusky blue throughout, except at the tip, where it is black.

The head is white; the neck also and the body are white: the neck is of so surprising a length, that that of a swan appears short to it.

The wings are short but broad; the long feathers of them are black, but the covering ones are all of the highest scarlet, and make a most glowing appearance: the tail is short and inconsiderable; the legs are wonderfully long, they exceed even the neck in length, and are robust, and of a high scarlet colour: the toes are three before, and one behind; the three anterior ones are connected by a membrane, and are very long; the hinder toe is short and inconsiderable.

It is a native of many parts of the West Indies and of Africa; we have it sometimes in Europe about the sea-coasts, but rarely. The Flamingo feeds on fish, shell-fish, and a variety of animals that frequent the sides of waters, as well as those which live in them; it is usually seen wading up to the mid-leg, and the use of it's length of neck is evident on these occasions, as it takes up any thing with ease from the bottom, at these depths. The writers of all times have been acquainted with it, and we have an account from Pliny, that it's tongue was esteemed by Apitius as one of the greatest of all delicacies at table; they have all called it, as we do, Phoenikopterus.

ANAS.

THE beak of the Anas is convex; the point or extremity of it is obtuse, and the whole verge is furnished with transverse, lamellose teeth: the tongue is obtuse and ciliated.

Anas rostro plano, apice dilatato rotundoque.
The Anas, with a flat beak, broad and rounded at the end. **The Spoon-bill.**

This species is so extreamly different from the generality of the Anas kind in it's general figure, that it has been usually accounted one of the storks; some have made it a species of pelican, and others of heron; but the figure and structure of the beak, which is, though referable in every particular to this class, yet one of the most singular in the world, brings it evidently to this family, as do also it's other characteristic parts, though it's general form be different.

It is of the shape of our common heron, and is about the same size: the neck is very long, and the legs are also proportionably long: the head is large, rounded, and very convex on the front: the eyes are small; the beak is very large, and of the most singular structure imaginable: it is very long and flat; it is broad throughout it's whole length, but, contrary to the custom of nature in other birds, which have it broadest at the base, and smallest at the point, this is largest there; it is flat, and of the same breadth all the way till near the extremity, where it expands into a round figure, and forms somewhat of the resemblance of the bowl of a spoon, the rest serving in manner of a handle; it has from this obtained the name of the spoon-bill: the beak is white, when the bird is young; but, when it is full grown, it becomes of a dusky olive, approaching to black: there are twelve or fourteen furrows on the broad part; the rest is almost smooth.

The whole bird is white, and, in it's wild state, the white is extreamly clean and elegant, as in the swan; when kept, as we sometimes have it in confined places, it becomes nasty, and loses the beauty of it's colouring: the wings are moderately long, and are the only part of the bird, where there is a feather that departs from the whiteness of the body: the three or four first of the long feathers have some blackness, especially at their tips: the tail is very short and inconsiderable.

The legs are very long, strong, and white; they are naked half-way up, above the middle joint, which shews they are intended for wading in the shallow waters; but the structure of the feet shews, that they are also formed for swimming where it is deeper: the toes are connected together by a membrane, as in the feet of the swan, &c. but not in so conspicuous a manner, or so deeply; the outer toe of each foot is connected to the middle one, as far as the end of the second joint, but the membrane which connects the others reaches no farther than the first joint.

This singular bird is a native of many parts of Europe, but not of England. It lives principally about waters, and frequents the watery countries most: it is no where so common as in Holland; but, though it lives about waters, it does not build among the sedge, and other furniture of their banks, as most of the water-fowl do, but on the tops of the highest trees: this it does in some parts of the Low Countries as frequent, and in as vast number, as the rook with us, and makes more noise. It feeds on on frogs and other water animals, as well as on fish, and, like the duck, on all kind of offal and garbage: it's food is apt to pass through it very quick and undigested, for it swallows voraciously. I remember to have seen at once the most singular and nasty incident in the world in some of these birds, kept in the late Duke of Richmond's yard; they had been fed with chickens guts, and we saw them taking hold of the ends of the guts, as they hung undigested out of the fundament, and pulling them out to eat over again. Most of the ornithologists have mentioned this singular bird, but they have been distressed what to call, or where to arrange it; it's general shape seeming to refer it to one class, and it's characters evidently to another. Gesner calls it Pelicanus sive Platea; Aldrovand, Leucorodius sive Platea, Pelicanus Ornithologi; Wottonus, Platea Ciceroni; Olearius, Platea sive Cochlearia; Willughby, Platea Pelicanus leucorodius sive Albardeola. The Dutch call it Lepelaer; and we, the Spoon-bill, or Spoon-beaked Stork.

Anas rostro semicylindrico, cera flava, corpore albo.
The white Anas, with a semicylindric beak, and it's membrane yellow. The Swan.

This is the largest and most beautiful in it's appearance, though simply white in colour, of all the Anas kind: it is considerably larger in the body than the best fed turkey, and it's plumage is more soft, bright, and elegant, than in any other known bird: the neck is extremely long, but the legs are short; the use of the length of the neck being to take up, as in others, food from the bottoms of shallow waters; the legs serving to carry it thither, by swimming, not by wading: the head is large, rounded, and very convex on the crown: the eyes are large and bright; the beak is large, and of a semicylindric figure, and is rounded at the extremity: in young birds it is of a lead colour, except the verge of the end, which is black, and a black areola that runs behind the eyes; but, in the full-grown ones, the beak is of a fiery or orange scarlet colour, and it's verge black; and, at it's base, there is a remarkable large black fleshy tumour: the space under the eyes always remains black; the beak is serrated, and the tongue is hairy, as it were, or furnished with numerous short and very slender appendages.

The whole body is of a snow-white, as are also the wings; they are extremely long and well feathered: the tail is short and inconsiderable; the legs are of a very deep lead colour, almost black; the feet are very large; the toes long, and connected by a broad and tough web, which is black; the claws are short and obtuse: the feet are so broad, when fully expanded, that they serve the bird excellently for swimming; and the intent of nature being, that it should pass it's time on the water, when forced out upon land, it walks but very badly and awkwardly.

It is a native of many parts of Europe, and is preserved, for it's beauty, in most others. It breeds with us; the nest is very large, and is made in the sedge, or among reeds, at the sides of rivers: it lays five or six eggs, very large, and all over white. The male and female sit on these by turns, and, when the one is on the nest, the other is always swimming about near the place to guard her. I have known a swan, in the defence of it's mate, attempt to drive off a boat that has come too near, by all the fierceness of aspect and fluttering of wings imaginable; and at length, in disregard of it's own defenceless condition, come into the boat. And I have known them keep a dog, that has ventured too near, under water, by the buffeting of their wings, till it has been drowned.

All the writers on birds have described the swan: they have called it Cygnus domesticus and Cygnus ferus, distinguishing it in it's wild and familiar state into two species; but this is idle and unnecessary; the bird is wholly the same in both.

Anas rostro semicylindrico, basi gibbo.
The Anas, with a semicylindric beak, gibbose at the base.

The Swan goose.

This is a very singular bird, and has so much of the stately appearance of the swan, and at the same time also so much of the general figure of the goose, that people who have judged only by the general figure, in these cases, have not known to which of the two kinds they should refer it. It is of the size of the largest goose: the neck is very long, and the legs short: the head is round and large; the eyes are moderately large, and the beak is of a semicylindric figure, and remarkably long and large: the back is of a dusky brown colour; the belly is of a snow-white: the throat and breast are not absolutely white, but have a tinge of reddish-brown, that is extreamly beautiful, and on the back part of the neck there runs a brownish-black line all the way down, from the head to the very back: the beak is black, and the singular and obvious characteristic of the bird is, that there is a very large fleshy tubercle at it's base, which hangs over it: this is larger in the male than in the females, and is much larger also in birds of five or six years old, than in those of only a year or two.

The wings are long, and are of the same dark colour with the back, only that the tips of some of the long feathers are white: the tail is short and inconsiderable, and the tips of it's feathers are also white.

The legs are short, but very robust and thick, and the feet are very broad, all the toes being long and connected together to the ends with a membrane: the legs are red, and the beak is also sometimes red, but it is more usually black; and over it's base, or between that and the eyes, there always runs a beautiful streak of white: the hinder toe is short, and all the claws are obtuse.

This is a native of the East, and of some parts of America; we have it kept for it's beauty in many parts of Europe. It loves the water, but it is much better qualified for walking on land than the swan, and often does it on choice, the other never. Whether walking or swimming, it carries it's head and neck in the same stately, erect manner as the swan. Most of the writers on birds have described it. Ray calls it Anser Cygneus Guinensis; Willughby, Anser Bisanitus vel potius Guinensis. We call it the Swan-goose, and the Spanish goose, the Turkish goose, the Guinea-goose, and the Siberian goose; there are indeed so many parts of the world where it is native, that it may be very variously named from them.

Anas rostro semicylindrico, corpore supra cinereo, subtus albido, rectricibus margine albis.
The brown-backed, white-bellied Anas, with the edges of the wing feathers white.

The Goose.

This is too common a bird to need a long description: the head is large and rounded; the beak large, and of a semicylindric figure, as in the swan; the eyes are large, and the nostrils oblong: the back is naturally of a mixt colour, composed of brown and grey, but it varies extreamly in this respect, when kept tame, which in it's wild state it is very regular in it.

The legs and the beak are yellow, while the bird is young, but they become reddish, as it grows to maturity: the wings are considerably long, and so well feathered, that, though the body be very bulky, they can support it with great ease in very long flights: they are of the colour of the back, but with more of the grey and less of the brown in it: the long feathers are twenty-seven in each wing, and the edges of many of them are white: the tail is short and inconsiderable, but it is longer, in proportion, than in the swan; it is composed of eighteen feathers: these are shortest

ſhorteſt at the edges, and thence gradually longer to the middle, where they are longeſt of all. It is equally calculated for walking or ſwimming; it carries it's head erect, and, when provoked, makes an odd hiſſing noiſe.

All the writers on birds have deſcribed the gooſe, but they have run into the ſame error in this, as they do in regard to the ſwan. They divide it into two ſpecies, under the names of Anſer ferus and Anſer domeſticus; but the bird is entirely and abſolutely the ſame. We have the wild gooſe flying over our heads, in the fens of Lincolnſhire, in vaſt flocks. They uſually fly at a vaſt height, but are ſometimes within the rowls of a peculiar kind of long guns, which are ſupported by ſticks, while the people fire them. I have ſeen them frequently brought down this way, and found them to be wholly the ſame with the tame ones in our yards.

Anas capite colloque nigris.
The Anas, with the head and neck black.

The Bernacle, or Clakis.

This is a very ſingular bird; it is conſiderably ſmaller than the common gooſe, but larger than the duck: the head is large and rounded, neither at all compreſſed at the ſides, nor depreſſed on the crown: the eyes are large; the beak is black, and is much ſmaller than in the common gooſe, but it is rather broader, in proportion to it's length: from the very baſe to the extremity it hardly meaſures an inch and a half; the part about the angles of the mouth, and juſt under the lower part of the beak, and whatever elſe is contiguous to it, is white, except for a deep black ſpace between the top of the baſe of the beak and the eyes.

The whole neck, both on the upper and under ſides, is of a coal-black quite down to the breaſt and to the back: the under part of the body is white, but not a pure and clean white, but with ſomewhat of an admixture of greyiſh. The feathers that cover the lower part of the thighs, a little above the knees, are black: the back is variegated with black and grey, and the covering feathers of the tail are part white, and part black: the wings are long and well plumed; the long feathers are of a blackiſh-grey; the tail is black: the covering feathers of the wings are very elegantly variegated with black, grey, and white, and make an extremely beautiful appearance.

The legs are ſhort; the feet are broad and webbed, as in the common gooſe.

It is a native of England, and is frequent, in particular, about the coaſts of Lancaſhire, in winter. Moſt of the authors who have written on birds have deſcribed it. Geſner calls it Brenta fiye Bernicla; others, Bernicla or Brenta; and ſome, Anſer Scoticus ſive Bernicla; Bellonius calls it the Cravant and Oye nonette, and ſuppoſes it to be the Chenalopex of the ancients. Moſt of the other writers have ſuppoſed this bird, and that diſtinctively called Brenta, to differ only in ſex; and, while that was the opinion, there was no impropriety in joining the names: but later obſervations have proved, that they are two abſolutely diſtinct birds, and conſequently the name Bernicla ought to be kept appropriated to this ſpecies, to which it was originally given; and that of Brenta to the other, next to be deſcribed.

This is the bird which Gerard and ſome other authors have declared to be produced from a peculiar ſpecies of ſhell-fiſh, called the Bernacle-ſhell, found on decayed wood that lies about the ſea-ſhores. The love of wonderful obſervations raiſed this firſt account of the bird's being produced from a ſhell-fiſh, that uſually adhered to old wood, into the ſtory of that ſhell growing upon a tree, in manner of it's fruit. Thus were the animal and vegetable kingdoms confounded; and it was pretended, that ſuch of theſe fruits as fell on land came to nothing, but ſuch as fell into the ſea; for they aſſure us, the tree grew no where but on the ſhores, diſcloſed living animals, which at firſt were rude and imperfect, but by degrees grew to the perfect form and ſize of this bird. The whole matter that gave origin to this ſtory is, that the ſhell-fiſh, ſuppoſed to have this wonderful production, uſually adhere to old wood, and that they have a kind of fibrils hanging out of them, which in ſome degree reſemble feathers of ſome bird: from this slight origin arose a story that they contained real birds: what grew on trees people soon asserted to be the fruit of trees, and from step to step the story gained credit with the hearers, till our Gerard, a man, in the rest, of candour and veracity, declares, that what he speaks of them is what his eyes have seen, and his hands have handled; though not one syllable of his account is true.

Anas collo nigricante torque albo insignito.
The Anas, with a black neck, and a white collar round it.

The Brent-goose.

This has usually been confounded with the former, the difference having been supposed to consist only in the sex; but it is, in reality, a distinct species; it is a little larger than our common duck; the head is large and round; the eyes are large and beautiful; their iris is of a bright hazel, and their pupil of a deep black: the beak is small, in proportion to the size of the bird, and black; but it is exactly of the figure of that of the common duck, and is serrated in the same manner at the sides: the nostrils are very large, and of an oblong form; the whole beak is black.

The whole head, the neck, and the upper part of the breast are black, but only in the middle of the neck there arises from each side an oblong white spot, and these, joining in the middle, form a kind of chain or necklace round it: the back is of a greyish-brown, and toward the rump it is darker than elsewhere, and almost black: the feathers, however, which lie immediately over the root of the tail, are white; the lower part of the breast is of a greyish-brown colour, and the belly is white.

The wings are long and well-feathered, and the long feathers in them are black; the tail also is black: the covering feathers of the wings are brown.

The legs are short and black; the feet are webbed, and the hinder toe is short.

Many of the writers on birds have described this, but they have in general mistaken it for the female of the preceding species. Aldrovand has figured it, and Bellonius has described it under the name of Anas torquata. We have it about the sea-coasts, in some places, in great abundance. It is very common in Scotland, and in some parts of Yorkshire.

Anas cinerea fronte alba.
The grey Anas, with a white forehead.

This is a very beautiful bird, and it is somewhat singular, that, though it be very common in many of the northern parts of Europe, scarce any of the writers on these subjects have described, or even mentioned, it. It is somewhat larger than our common duck: the head is large and rounded; the eyes are large, and their iris is of a bright hazel, with a tinge of an orange scarlet in it: the beak is large and long; it is flatted, rounded at the end, and dentated all the way along the sides; it is of a dusky red colour: the head is of a bright grey colour, but there is a spot of white in the front: the back of the neck is also grey, but somewhat darker than the head: the back, the wings, and the tail are also grey, and the breast, throat, and belly white, but on the breast there are some beautiful variegations of black in form of spots.

The legs are short, and of a bright red colour; the feet are large, and are webbed; the hinder toe is short, and the claws are all short, black, and obtuse.

This species is common in many parts of Germany, and in Sweden, Denmark, and Norway; it is not a native with us, but is sometimes seen on our coasts. Rudbeck was the first who made the world acquainted with it; he calls it Anser cinereus ferus torque inter oculos et rostrum albo Erythropus; Linnæus calls it, Anas cinerea fronte alba.

Anas

Anas albo variegata abdomine longitudinaliter cinereo maculato. **The Tadorna.**

The Anas, variegated with white, and with a longitudinal spot of grey on the belly.

This also is a very beautiful bird, and a large one; it is very little inferior to the common goose in size: the head is large and rounded; the eyes are large, and their iris is of a bright hazel, with an admixture sometimes more, sometimes less, of yellowish and reddish: the beak is short, but broad, and a little crooked; it is flat, and large, and rounded at the extremity: it is serrated all round the edges, and at the point, and is of a blood-red colour, except that the verge at the extremity is black, and the nostrils also are black.

There is an oblong and prominent tubercle of a fleshy structure at the upper part of the beak, or between the base of the upper chap and the head: the head itself is of a very deep, changeable green, so dusky, that at a little distance it appears black: the upper part of the neck also is of the same colour; the rest of the neck is of a snow-white; the shoulders, and the upper part of the breast, are of a very beautiful orange colour; this forms a kind of broad ring, surrounding the whole anterior part of the body: the breast and belly are white, but there runs all along them a beautiful line of dark grey in the middle, from the breast, nearly to the insertion of the tail. In the lowest part of the belly, just at the origin of the tail, there is a faint tinge of the orange colour, like that of the breast and shoulders: the back is white, but there are a number of coal-black feathers on the sides; and the long feathers, which have their origin from the shoulders, are also black: the wings are long, and are black and white; many of the long feathers, and all the covering or short ones, except those in the extream joint, are white; these excepted ones are black, and they make a beautiful variegation: some of the long feathers, which are white in the middle, are tipped with black at the ends, or edged with it at the sides: the long feathers are about twenty-eight in each wing: the tail is short, and is composed of twelve feathers; these are white, and are all of them, except the two outer ones, tipped with black at the extremities.

The legs are short and robust; they are of a pale flesh colour, and the skin which covers them is so thin, that the course of the several vessels is easily seen through it: the feet are large and webbed; the hinder toe is short, and the claws are all short, black, and obtuse.

It is frequent with us about the coast of Lancashire, and in Wales: it's flesh is coarse and ill-tasted, but it is eaten by the common people. It builds with us, and usually makes it's nest in holes of the ground. If there be any deserted rabbet-burrows, or any other hollows of a like kind, it never fails to build in them, and has hence obtained, among our common people, the name of the Burrough-duck. All the writers of birds mention it. Ray and Willoghby call it Tadorna and Vulpanser; Bellonius, simply, Tadorna. It's elegant variegations of colour, which, in their brightness and regularity of disposition, in some degree resemble those on certain shells, has obtained it also with us the name of the Shell-drake, or Shell-duck.

Anas rostri ungue obtuso, cera superne bifida rugosa. **St Cuthbert's Duck.**

The Anas, with the unguis of the beak obtuse, and the cera bifid and rugose.

This is a very beautiful bird; but the male and female differ so extreamly, that they may easily be mistaken for distinct species: the male is in general white, almost the whole body is of this colour; the head is large and rounded; the eyes are large, and

and their aspect bright and piercing: the beak is very large and long; it's length is not less than three inches; it is not so flatted, as in many of the Anas kind, but of a regularly semicylindric figure: the end is obtuse, and the whole verge is pectinated: the base is covered with a rugose membrane, divided into two portions at the top: the nostrils are large and black.

The top of the head is black, and the large spot of this colour, in that part so different from the rest, has the appearance of a hat on the head of the bird: this large spot is divided into two parts behind, and terminates in three points before; the several angles being directed toward the fines in the membrane which covers the base of the beak: the back is white, but the lower parts of the breast and belly are black: the rump also is black, and so are the long feathers of the wings, and those which cover the thighs: the tail is short, and all the feathers which compose it are black: the hinder part of the head, close behind the spot which forms the appearance of a hat, is of a fine glossy, changeable green: in some the anterior part of the breast is simply white, in others, though all males, it has a tinge of reddish, and this *fleshy* tinge varies extreamly in those which have it.

The female is of a quite different appearance; her body is all over of the colour of that of a wood-cock, a dusky grey, with brown spots, and some black ones, and has a very remarkable and narrow line of white running across the wings.

We have this species in plenty among us. It builds on our coasts among the rocks, and in the hollows and crannies of the cliffs, and lays a number of large and very well-tasted eggs. The nests are so often plundered for these, that one would almost wonder at there remaining enough to continue the surprising plenty of the species, that we see about the same places the next season. The feathers of this species are also softer than those of almost any other bird, and are greatly valued for stuffing of beds, and other such purposes. Bartholine calls it Anas plumis mollissimis vulgo Eider; Wormius, Willughby, and most of the other writers have continued to it the name of Anas plumis mollissimis; and our authors have also called it Anas sancti Cuthberti and Anas Farinensis. The peasants, in some part of the kingdom, where it builds in the midst of high cliffs, venture themselves from the top with ropes about their middle, in a surprising and frightful manner, for the sake of their eggs.

Anas cauda cuneiformi forcipata. **The Sea-pheasant, or Craker.**
The Anas, with a forked, cuneiform tail.

This is a very beautiful bird, but it is much smaller than any of the preceding species; the size is about that of our widgeon: the male and female of the Anas kind differ extreamly, but in few more than in this; the male is somewhat the larger: the head is large and round; the eyes are large; the beak is also large, in proportion to the size of the bird: it is convex on the back of a semicylindric figure, and obtuse at the extremity; it's general colour is black, but in the middle it is of a bright red: the nostrils are large and oblong; the verge of both chaps is dentated, by means of a kind of patulous squammæ: the fauces also are dentated, and the tongue is divided into three obtuse portions at it's extremity: the two lateral ones are small, and the middle one large.

The head and neck are white, but there are two large black spots on the head, one on each side; the back is black; the breast is also black, and the wings are, on their upper side, black, without a single spot of any other colour, and on the under-part they are uniformly grey: the tail is of a cuneiform figure, and is black on the upper side, and white underneath: the feathers which compose the tail are sixteen in number, and their proportions, as well as number and arrangement, are different from those of any other species: the two middle ones are considerably longer than any of the others, and run out into slender and sharp points; 'tis from this singularity that the bird has obtained the name of the Sea-pheasant.

The

The female of this species has the back all over of a grey, clouded with blackish, and the belly white: the tail is of the same singular figure as in the male, but the middle feathers, though longer than the others, do not in this sex so much exceed them: the legs are short and slender, but the feet are large, broad, and webbed: the toes are slender, and the claws are obtuse and short.

All the writers on birds have named this species. Willughby calls it Anas cauda acuta; Wormius, Anas Islandica, and in another place, Anas cauda acuta Islandica Kaertlde ipsis dicta. We call it the Craker and the Sea-pheasant.

Anas caudæ rectricibus intermediis recurvis.
The Anas, with the intermediate rectrices of the tail crooked.

The common Wild Duck.

The male and female in this common kind differ as much as in many of the more rare ones: in the male the head is of a very elegant and changeable colour, a deep blue and a deep green distinguish themselves in the different lights in which it is viewed; the back is of a ferrugineous brown, and the rump has somewhat of fine silky green: the breast and belly are of a pale whitish-grey, very elegantly undulated, and variegated with five slender and short striæ, placed very close together: the wings are brown on the upper side, and white underneath; the long feathers are partly grey, and partly brown: the grey ones have little variegation, but the brown are some of them tipped with white, and a considerable number of the others are of a deep violet blue in the middle, with some blackness about it: these stand together, and the dark and beautiful colour in them gives a very singular beauty to the wing, forming a large and elegant spot in it: the covering feathers of this part of the wing are grey, at the edges whitish in the middle, and black at the extremities; those next the sides are simply of a greyish-brown.

The tail is short; the four middle feathers of it are black and curled, or turned; the others are of a greyish-white: the legs are of a dark yellowish-brown; they are covered to the knees with feathers: the female has nothing of the beautiful variegations of the male, but is of a simple dark brown.

This is the most plentiful of all the Anas kind with us, and we keep it also tame in vast abundance for the supply of our tables. It feeds on almost any thing, whether of the animal or vegetable kind, and, when kept, some vary extreamly in colour, and to nothing oftener than to a perfect white. All the writers on birds have described it. Gesner calls it Anas fera torquata; Aldrovand, Boschas major sive Anas torquata; and Ray, Willughby, and others have continued the same name. Many of the late writers, however, have described it again in it's tame state, as if another species, and called it Anas domestica vulgaris. This is the case also with those authors, in regard to the goose, the swan, and, in short, every other bird that is kept tame at our houses, and is wild also in our fields or rivers; the custom was, at one time, carried so far, that one Lovel, who has written a history of animals under the name of a Panzoologicon, when he comes to treat of the bull, has one chapter for that, another for the cow, a third for the calf, and a fourth for the ox.

Anas facie nuda papillosa.
The Anas, with a naked papillose face.

The Muscovy Duck.

This is a very beautiful bird; it is much larger than our common duck, and is nearer the size of the goose: the male is, as in many other species of this genus, extremely different from the female in colouring, and is particular in this, that he has about the middle of the hinder part of the neck a number of red, naked papillæ: the females are very various in their colouring, but the principal colours are black and white: the head is large and rounded; the eyes are large, and very bright and piercing; the beak is large and long, but it differs from that of most of the duck-kind, in that it is

 bent

bent at the extremity: between the eyes, but lower, and toward the top of the nostrils, there stands a red, fleshy protuberance, of the figure of a common red cherry: the temples have no feathers on them, but are covered with a naked, scarlet, fleshy matter, of a granulated surface.

The body is principally black, but it has more or less variegations of white, and differs also in the disposition of them.

We have this in most parts of Europe tame, but where it is wild is not certainly known. All the late writers on birds have mentioned it. Aldrovand calls it Anas Indica; Ray, Anas moschata; and Willoughby, Anas Lybica Bellonii. The whole bird has a perfumed smell, approaching to that of musk. It is a very quick breeder, and lays a great number of eggs, which are large and well tasted.

Anas crista dependente, corpore nigro, ventre maculaque alarum albis.

The Anas, with a hanging crest, a black body, and a white belly, and a white spot on the wings.

The tufted Duck.

This is a very beautiful bird, and is about the size of our common duck: the head is large and round; the eyes are moderately large and bright; their iris is of a fine gold yellow: the beak is moderately long, and very broad; it is serrated all the way at the edge, and rounded at the end, and is of a deep blue colour throughout, except at the extremity, where it is black: the nostrils are large, and between them there runs in upon the beak an angle of feathers: the ears are small, as is indeed the case in all the birds that are formed for diving.

The upper part of the head is of a dusky purple colour, tending to black, or perhaps, more properly speaking, it is black, with a tinge of purple, whence it is in Venice, and some other places, called Capo nigro, or the black head, and it has a cirrhus or crest growing on the back part of the head, and hanging down upon the neck, of an inch and a half in length: the neck is black; the shoulders, back, rump, and covering feathers of the wings are also in general black, or of a very deep colour, nearly approaching to that.

The wings are short, but they are well feathered; the four outer long feathers are of the same colour with the body, but the others are many of them variegated, more or less, with a bright snowy white: the tail is very short, and is composed of fourteen feathers, all black throughout.

The fore-part of the neck and the top of the breast are black, but the rest of the breast and the whole belly are of a silvery white, and the side feathers, and those which cover the thighs, are of the same colour: the legs are short, and of a deep dusky bluish-black: the feet are large and webbed thoroughly; the toes are long; the membrane which connects them is black; the claws are short and obtuse: the body is thick and short; the sides of the head are, in some of the birds of this species, decorated with a white spot, each way at the angles of the beak, but this is not universal.

We have this species kept in some places. It is a native of the sea-coasts of some of the northern parts of Europe. Most of the authors who have written on birds have described it. Gesner calls it Anas fuligula prima; Aldrovand, Querquedula cristata sive Colymbus Bellonii; Ray, and others, Anas cristata. We call it the crested or tufted Duck.

The Golden-eye.

Anas nigro alboque variegata, capite nigro vireſcente, ſinu oris albo macula.
The Anas, with a greeniſh-black head, a black and white body, and a white ſpot at the mouth.

This is a very beautiful ſpecies: the body is ſhort and thick; the head is large and rounded; the eyes are large, and their iris is of a fine bright yellow, approaching to that of burniſhed gold: the beak is black; it is not long, but very broad, and is much broader at the baſe than at the extremity; it is ſerrated all the way round.

The head is of a very deep colour, approaching to black, but with a fine tinge of a graſs-green thrown over it; and at the angles of the beak there is on each ſide a ſpot of white, which is round, and has in ſome degree the appearance of an eye, whence the Italians have called the bird Quattro occhii, or four eyes: the neck is ſhort; the whole of that, as alſo the ſhoulders, the breaſt, and the whole belly are of a fine ſilvery white: the lower part of the back is all over black: the wings are variegated with black and white, and that in a very particular manner; for the middle feathers are all white, and the outer and inner are black: it is not only the long feathers or remiges that are of theſe ſeparate colours, but the covering ones too, thoſe over the white long feathers being white, and thoſe over the black ones black; from the ſhoulders there hang a number of long feathers down on each ſide of the back; theſe alſo have a little white in them, by way of variegation: the tail is compoſed of ſixteen feathers, and is entirely black.

The legs are very ſhort, but they are robuſt and thick; the feet are large, and very deeply webbed: the toes are long, and they have a blackiſhneſs about the joints, otherwiſe the whole ſoot, as well as the leg, is of a deep yellow, with a tinge of orange: the membranes which connect the toes are of a duſky colour; the hinder toe is ſmall, but it is increaſed in breadth by a membrane on each ſide, and it's claw is black, but very ſhort.

The female differs conſiderably from the male in this ſpecies: the head is of a bluiſh-red; the back, the ſides, and the tail are brown: the breaſt is of a grey in the anterior part, and is variegated at the ſides with undulated lines of a pale or hoary colour, and the belly is white; the wings are ſpotted, and variegated with black and white, as in the male, but they have more white than in that ſex; the legs alſo are of a paler yellow.

We have this ſpecies about the ſea-coaſts of ſome parts of England, but it is far from common with us, though in ſome other of the northern parts of Europe it is very frequent. All the writers on birds have deſcribed it. Geſner calls it Clangula; Aldrovand, Clangula and Anas Platyrynchos mas; and moſt of the others have called it either Clangula, or Platyrynchos Anas. We, from the remarkable yellow of the iris of the eye, call it the Golden-eye.

The Gadwall, or Grey.

Anas macula alarum rufa, nigra, alba.
The Anas, with the wings variegated with brown, black, and white.

This is of a middle ſize, between the teal and the duck, though more the ſize of the duck than of the other, and it's body is long and ſlender: the head is moderately large and rounded; the eyes are large and piercing in their aſpect; the beak is formed like that of the teal, broad and flatted, and armed with it's unguis at the end: the under chap of it is yellow, and the ſides of the upper chap are alſo yellow, but it's middle is black: the noſtrils are large, and the yellow of the beak has a tinge of the orange in it.

The

The head is of a greyish-black, but the edges of the feathers which fall down toward the throat, have a tinge of bluish: the sides of the head near the beak, and just under it, are white, variegated with spots of brown: the back is of a deep and dusky brown, only the edges of the feathers are of a whitish-red: the rump is wholly black; the lower part of the neck, the shoulders, and the top of the breast are covered with very smooth and glossy feathers, variegated in an elegant manner with black and white: the lower part of the breast is white; the belly is of a pale brown, and is variegated with transverse streaks of black, and under the tail there are some transverse streaks of brown: the sides are very beautifully variegated with alternate lines of black and white: the tail is very short; it scarce appears beyond the feathers which cover it, and is composed of sixteen robust but short feathers terminating in sharp points: these are white at the tips, and some of them also at the edges, but the rest of the feather is brown.

The wings are long, and there are twenty-six feathers in each; these are variegated with black, white, and brown, and make a very pretty figure in the wing; some of them being simply of the one of these colours; some partly of the one, and partly of the other: the covering feathers also are variegated with white, and with a reddish-brown, upon a ground of a simpler brown.

The legs are short and robust; they are feathered as far as the joint of the knee, and their naked part is white: the hinder toe is very small, the others are connected by a broad and strong membrane; they are long, and consequently the foot itself is large, and admirably calculated for the bird's swimming: the colouring in the wings of this species sufficiently distinguishes it at first sight from all the others of the genus; there are marks of black of a reddish-brown and of white, distinct and separate, and are placed just over one another.

This is a native of our fen countries, and is much esteemed at table; most of the writers on birds have described it. Gesner calls it Anas strepera; Aldrovand, Anas Platyrynchos; and Ray, Anas Platyrynchos rostro nigro et plano. We call it the Gadwall or the Grey.

Anas rostri extremo dilatato rotundoque, ungue incurvo.
The Anas, with the extremity of the beak broad and round, and it's unguis bent. **The Shoveler.**

This is a very singular species; it is very nearly equal to our duck in size: the head is large and rounded, and the eyes large and bright, but the beak is of so singular a form, that it distinguishes it at sight from all the others; it is considerably long and large, but as the others are usually broadest at the base, and somewhat smaller at the point, this, on the contrary, is broadest at the extremity: it is there flatted, and even hollowed in some degree, and is terminated by an unguis, which is small but bent: the edges of both the upper and under chaps are serrated, or furnished with a kind of pectinated border: the tongue is large, thick, and fleshy, especially toward the extremity.

The head and the upper part of the neck are of a beautiful blue, and sometimes of a deep green, with a cast of brown in it, changeable in the different lights, like the colour of the common drake's neck: the lower part of the neck is white in the forepart, and on the hinder of a mixt white and brown: the shoulders also are of the same brown and white mixed colour; the breast and the belly are reddish and black; just under the tail there are some black feathers: the back is of a deep brown, but there is a changeable shade of a very beautiful purplish thrown over it, and some lights discover a tinge of deep green in this: the feathers which cover the thighs are variegated with transverse black lines.

The wings are long and large; the long feathers are twenty-four in each: these are some of them entirely brown, and others have the edges tinged with greenish, pur-

p'ish,

plish, or bluish, as seen in different lights, and the four that are next the body have some variegations of white: the tail is short, and is composed of fourteen feathers; it is variegated with black and white: the female is very different from the male in this species; it entirely resembles the common wild-duck, except in the wings, which are like those of the male.

The legs are short, and the feet smaller than in most of this genus; they are of a bright red colour, and the membrane which connects the toes is serrated. It is said by many, that this bird changes it's colour in the winter; but I am rather apt to believe, that they mistake some different species for it, which they do not examine, or which they only see at a distance, or on the wing.

All the writers on birds have named this. Gesner calls it Anas Platyrinchos major; Aldrovand, Anas latirostra major, and Anas latirostra fusca; for he describes it twice, as if two species; Willughby calls it Anas Platyrinchos major sive Clypeata Germanica. We have it in some parts of England, but it is not frequent with us; in other parts of Europe it is very common, and most so in the more northern.

Anas macula alarum purpurea, utrinque nigra alboque, pectore rufescente.

The Anas, with the spot of the wings purple, and black on each side, and the breast reddish.

This is a very pretty bird; it is nearly, but not altogether, of the size of the common duck: the head is large and rounded; the eyes are small, and their iris is yellow; the beak is large, and is very broad, in proportion to it's length; the nostrils are oblong, and stand high for one of this species: the whole beak is of a mixed brown and yellow colour.

In the male there is an extremely elegant spot on the wing, of a violet colour, edged with black, and on the outside of that with white; this is so singular, that it cannot fail to distinguish the bird, at sight, without further description.

The female has a spot of the same form and kind with the male on her wings, only in this the colour is paler: in the male it is a deep violet or purplish-blue, but in the female it is a clearer blue, with much less of the addition of the purple: the breast in both the female and male is of a pale reddish-brown; the tips of the wings, and also their furthemit, is grey, and the tail is short and white: the back is of a brown colour, with a tinge of ferrugineous towards the sides; and the under part is of a ferrugineous colour, with spots of a dusky brown: the throat is of a very pale ferrugineous tinge, and is not at all spotted: the body of the wings is of a greyish-brown, and their long feathers are partly simply brown, and partly brown, tipped with white; and some have white near the extremity, but the absolute tip is black.

The legs are short, and not very robust, nor are the feet large, but they are deeply webbed: the hinder toe is very short; the colour of the legs and feet is a yellow, approaching to reddish.

This species is frequent in many parts of Europe, but I have not met with it in England. I have been told of a species caught in the decoys of Lincolnshire, which by description answers to this, but have never seen it. All the authors who have written on birds have described this. Aldrovand calls it Anas Platyrinchos pedibus luteis; and Ray, Willughby, and most of the others have followed him in this.

Anas oculorum iridibus flavis, capite grisea, collari albo.
The grey-headed Anas, with yellow eyes.

The Glaucion.

This is nearly of the size of the common wild-duck, and is a very elegant bird: the head is large, and of a rounded, not a depressed, form; the eyes are bright, and very piercing in their aspect; their iris is of a fine gold yellow: the beak is large, broad, and serrated all round the edges: the nostrils are large and oblong, and the base of the beak is remarkably firm and robust.

The head and half the neck are of a deep, dusky, ferrugineous colour in this part, or at the bottom of the ferrugineous portion of the neck there is a circle of white carried quite round it, in form of a collar: the breast is of a fine silvery grey colour; the belly is perfectly white: the back is of a deep black, and the wings, while they are closed, appear to be also black entirely, but, when they are expanded, they shew seven white feathers: the tail is short and black; the legs are short, but robust; the feet are broad and webbed, and the hinder toe is very short.

This is a fresh-water fowl, and is very frequent in most of the northern parts of Europe, but less so in England than almost any where else. It dives almost continually, and will keep under water a vast while at a time. Most of the writers on birds have named it. Bellonius calls it Glaucium and Glaucus; and most of the writers who have followed him, have called it Glaucion, or Glaucium Bellonii.

Anas capite brunno, fronte alba, cauda subtus nigra.
The Anas, with a brown head, a white front, and a tail black underneath.

The Widgeon.

This is a smaller bird than the duck, but considerably larger than the teal: it's head is large and round, and it's beak large, but not so long, in proportion to the body, as that of the common duck: the eyes are small; the nostrils are oblong and narrow: the beak is serrated all the way round, and is obtuse, but not remarkably broad, as in some species, at the extremity.

The head and the upper part of the neck are of a reddish colour, sprinkled over with blue spots: the front of the head, toward the base of the beak, is of a paler red than the rest, and has a tinge of yellowish-white in it: the upper part of the breast and the sides, under covert of the wings, are of an extremely bright and beautiful purple, variegated, in an elegant manner, with transverse lines of black: the middle of the back is brown; the edges of the feathers, however, are very beautifully variegated with grey, especially toward the hinder part: the covering feathers of the tail are black; the lower part of the breast, and the whole belly, are of a whitish tinge, but there is something of yellow mixed in it: the feathers which cover the thighs are variegated with numerous, oblong spots of a reddish-brown, and the feathers which are under the tail are of a mixt black and white, and have some of the same spots in them.

The tail consists of fourteen feathers; the twelve outer ones, six on each side, are of a deep brown, with their edges of a greyish-white; the two middle ones are black, only with a slight admixture of grey.

The legs are short, but they are robust; their colour is a very pale bluish-grey, or lead colour, with some admixture of white in it. All the writers who have treated of birds have described this, and almost all under the same name Penelope; some few have called it Anas fistularis. We have it in great plenty in our fen countries, as about Crowland in Lincolnshire in vast abundance. It feeds on the herbage at the bottom of fresh waters, and on the water insects.

Anas

Anas corpore obscuro, macula pone oculos, lineaque alarum alba.
The dark-coloured Anas, with a white spot behind the eyes, and a white streak on the wings.

This is one of the larger kinds; it is considerably bigger than a duck, and indeed little inferior to the goose in size: the head is large and rounded; the eyes also are large: the beak is very remarkably large, and resembles that of the goose rather than of the duck; it is very thick as well as broad, and is serrated all the way round at the edges: it's colour is red, except at the base, where it is black and gibbous: the nostrils are oblong and wide.

The head as well as the body are of so deep a colour, that it appears black, but, when closely examined, it is rather an extreamly dusky brown, with a tinge of iron grey: behind the eyes, and a little below the level with them, there is on each side an elegant spot of snow-white: the back is of a deeper colour than any other part.

The wings are long and well-plumed; the long feathers are all black, but the first series of the investient ones are white: the tail is short and black; the legs are red, and so are the toes, but the membrane which connects them is black.

This is the description of the male of this singular species; but the female, as in many others, differs considerably from it: the head, in those of this sex, is smaller; the beak also is considerably different; it is not red but brown, and has not that gibbosity on the upper part which characters the other, and the edges of the jaws are lamellose or dentated: the tongue is also fimbriated or lobated at it's edges, and the body is not of that deep colour, but of a fainter brown, with the tips of the feathers whitish, or at least of a vastly paler brown: the long feathers are some of them black, and others white throughout, except at the tips, where they are also black: the legs are of a paler red than in the male, but the membrane which forms the web connecting the toes, is also black; there is the same white spot behind the eyes in this as in the male, and the temples also in this are white.

This is a native of many of the northern parts of Europe, but has not been met with in England. It is a favourite fowl at the table, wherever it is native. Many of the authors, who have written on birds, have described it. Jonston calls it Anas fera fusca; Rudbeck, Anas fera nigra; Willughby, Ray, and others, Anas niger.

Anas alis cinereis immaculatis, urrhopygio nigro.
The Anas, with grey wings, and a black rump. **The Pochard.**

This species resembles the widgeon in most particulars, but it is very nearly equal to the duck in size: the head is large and rounded, and the eyes are but small; the beak is long, but it is not so broad as in many of the duck-kind: the upper chap is of a lead colour, and has a black angule at it's edge; the lower is entirely black, and both are serrated round the edges; the iris of the eye is of a reddish colour.

The head and neck are of a singular brown colour, with some faint tinge of red, but more of yellow in it: the middle part of the breast is white, but the extremities of the feathers have a yellowish tinge; the lower part of the breast, toward the belly, is also white, but it is variegated with lines of brown: the wings are long, and their principal feathers are all of a simple blackish-grey, without any spot or variegation: the tail is short, and it's feathers are also of one simple colour, which is a greyish-brown: the lower part of the neck and the rump are black, and the back is variegated with undulated lines of grey and brown.

The legs are short and slender, and of a dusky lead colour: the membrane connecting the toes is black.

This

This is frequent with us in the fen countries, and is taken in the decoys with the common wild-duck, but it is not so much valued as the common widgeon. Our people distinguish it by the name of the Pochard, but it is often sold under the name of the widgeon. Most of the writers on birds have mentioned it. Gesner calls it Anas fera fusca vel media; others, simply, Anas fera fusca; but this confounds it with another species.

Anas macula alarum viridi, linea alba supra oculos.
The Anas, with a green spot on the wings, and a white line over the eyes.

The Garganey.

This is about the bigness of the common teal, and in many particulars is greatly resembles it: the head is small, and, though not depressed, is much less prominent or rounded than in any other of the species: the eyes are small; the beak is oblong, flatted, somewhat broad and black, serrated at the edges, and obtuse and rounded at the end: the top of the head is black, and there is a white line on each side, running from the eyes quite down to the neck, and reaching to it's middle: the throat is black; the breast is variegated in an undulatory manner, with black and grey; the back is of a very singular colour, a deep brown, with a tinge of purple thrown over it: the feathers which cover the thighs are also variegated with black and white.

The wings are long, and the principal feathers of them are variegated with a deep brown, and a kind of mouse colour, and some of them also with white: the more elegant ones have a part of their anterior surface of a bright and beautiful green, and that forms the elegant spot on the wing: the tail is short, and is composed of fourteen feathers; they are brown, and the outer ones variegated with whitish-brown spots.

The female of this species is very like the male, but has less variety in the colouring, and is known at sight by the want of the black throat of the male.

This species is frequent in our fen countries, and is much esteemed at table. The authors who have written on birds have almost all described it. Aldrovand calls it Querquedula prima; and most of the others have called it by the same name, or else, by as little expressive a one, Querquedula altera.

Anas macula alarum viridi, linea alba supra infraque oculos.
The Anas, with a green spot on the wing, and a white line both above and below the eyes.

The Teal.

This is the smallest of all the duck-kind, but it is a very elegant and valuable bird: the head is small, and not remarkably rounded or prominent; the eyes are large, and their iris is of a singular colour, a hazel, approaching to whitish; the beak is short, but it is broad, obtuse, and a little turned up at the end; the edges are serrated all round, and the colour of the whole beak is black: the nostrils are conspicuous, and of an oval figure: the head is of a reddish-brown, as are also the throat, and the whole upper part of the neck: from the eyes there runs each way, to the hinder part of the head, a beautiful broad line, of a fine silky gloss, and of a dusky green colour; and between these, on the under part of the back of the head, there is a black spot: under the eyes, on each side, there runs a very bright streak or oblong spot of white, separating the green from the general brown colour of the head: the lower part of the neck, as also the back and the sides, under covert of the wings, are all of a beautifully variegated hue, every feather being particoloured, and the lines transverse and undulated, partly black, partly white: the lower and anterior part of the neck is, in some birds of this species, of a yellowish hue, spotted with black, but this is not the case in all: the breast is of a dingy white, or grey, as is also the belly, only paler, or

nearer

nearer a genuine white than the breast; and below the rump there is an elegant black spot, surrounded with a verge of yellow.

The long feathers in each wing are twenty-five, some of these are simply brown, some have some white in their variegations, and some have not only white but black also; at the sixteenth feather, from the edge of the wing, the green begins to shew itself, and it is continued in the several succeeding ones in such manner, as to make that beautiful spot which we see on the wings: some of the inner feathers, or those next the body, have not only black, but yellow, in the variegations.

The legs are short, and not very robust; they are of a pale brown, as are also the toes, but the membrane which connects these, and forms what is called the web of the foot, is black: the inner of the three anterior toes is the shortest, and the hinder one is not augmented by any membrane. The female of this species differs very obviously from the male: the head has nothing of that elegant red or green, which are so elegant in the male; nor is there the black spot so singular in the male, at the rump of that sex; the feathers of the back also want that undulatory variegation, which is so beautiful in the male.

This species is frequent with us on our large fresh waters, and it's flesh is so well-tasted, that it claims the preference against the whole genus at table. All the writers on birds have described it. Gesner calls it, simply, Querquedula; Aldrovand, Phascas sive Querquedula minor; Willoghby, Querquedula secunda; and others, Querquedula francia and Querquedula francia altera, distinguishing the male and female as if two distinct species, an error too common among this set of writers.

Anas cauda longitudine pedum, vertice fusco, pectore exalbido nebuloso.
The longer-tailed Anas, with a brown head, and whitish-clouded breast.

This is an uncommon but a very beautiful species; it is of the size of the widgeon, and in it's make more resembles that particular species which we call the sea-pheasant, than any other of the Anas-kind: the head is small, but rounded; the eyes are small, and their iris of a bright hazel, with an admixture of an orange colour: the beak is moderately large, broad, flatted, thick at the base, and rounded at the point; it is serrated all the way along the edges, and the nostrils are large and oval: the colour of the whole beak is a very deep olive, approaching to black.

The head is of a dusky brown at the top, and paler elsewhere: the neck is of a deeper brown, and the back also is brown, but not so deep: the breast is of a whitish colour, but clouded in a very elegant manner with a greyish-brown: the wings are moderately long, and there is an elegant spot at their origin: the tail also is longer than in the generality of the duck-kind, but not equal, in this respect, to that of the sea-pheasant: the legs are short and tolerably robust; the toes are long, and consequently the foot is large: the legs and the toes are of a dusky or dingy colour, between an olive and a lead blue, but the membrane which connects the toes is black: the hinder toe is very short, and the outer one of the three anterior is considerably longer than the inner.

This is a native of some of the most northern parts of Europe, but it has been little known; scarce any of the writers on birds have named it. Linnæus mentions it as a native of Sweden, and quotes Olaus Rudbeck for his authority. We have had stuffed skins of it brought over by the people concerned in the whale-fishery from Greenland.

The summer Teal.

Anas testaceo-nebulosa, superciliis albidis, rostro pedibusque cinereis.
The clouded, brown Anas, with white eye-lids, and grey legs and beak.

This is a pretty and an extremely singular species; it does not at all exceed the common teal in size, and in colouring it excels it: the head is small and rounded; the eyes are small, and their iris of a pale hazel, with more or less of a tinge of orange in it: the beak is small, but flatted, broad throughout, and rounded at the end in the manner of all the duck-kind, and resembles that of the teal in particular, by it's turning a little up at the extremity: the colour is a dusky grey, and the nostrils are large, oval, and black.

The head is of a pale brown, with a tinge of reddish; the upper part of the neck also is of the same colour: the eye-lids are white, and there is some whiteness also all about the eyes, which gives a singularity in the general aspect: the back is of a brown colour, not pale, but very bright and glossy, and clouded with grey; the breast and belly are of a whitish colour, but with a tinge of yellowish-brown; but the upper part of the breast, and the lower of the belly, are elegantly variegated with close-set spots of black; the wings are moderately long, and the tail is short: the legs are short and slender; they are of a greyish colour, and the membrane which connects the toes is black; the hinder toe is very short and inconsiderable.

We have this species on our fresh waters, but it has been overlooked by many of the writers on this subject, it's size making it frequently mistaken for one of the Dobchick. Gesner has described it, but imperfectly, under the name of Anas Circia; and those who have since mentioned it, have in general borrowed his name.

Anas rufa rostro pedibusque cinereis.
The reddish-brown Anas, with grey legs and beak.

This is a singular, but not a very beautiful, species; it has less variegations or gradations of colour than most of the species, but it strikes the eye by the remarkable tinge that it has almost universally diffused over it. It is of the size of the widgeon, or not quite so large: the head is large, in proportion to the bulk of the body; the eyes are small, and their iris is of a dead or whitish hazel: the beak is long and moderately broad; it is flatted all the way, except at the base, and is rounded at the extremity, and serrated all along the edges: the colour of the whole is a pale bluish, with a tinge of grey, only that the nostrils which are large and oval are quite black.

The back of the neck, and the whole upper surface of the body, the back wings, sides, and rump, are of a very singular colour, a reddish-brown; the red seems of a deep hue, and is so perfectly blended with the ground colour, that it scarce any where discovers itself in any light; and the result of the perfect union is a peculiar colour, approaching to what we express by the term copper colour; it is unlike any thing that we see elsewhere in the colouring of the duck-kind, and always claims attention: the wings are moderately long; the tail is short; the breast and belly are of a pale colour, and the legs, which are short, and not very robust, are of the same greyish-blue with the beak; but the membrane which connects the toes, and forms the web of the foot, is black.

I saw three or four of these birds the last winter, on Whittle-sea-meer in Cambridgeshire, and was so lucky as to kill one of them. I find, by the people who reside at Yaxley, and other little towns in the neighbourhood of that vast extent of fresh water, that it is common there. It is singular, that a bird, so well known to the peasants, should be so perfectly overlooked by the writers of our own country, but not

one

one of them have so much as named it. I have the skin of that which I killed now preserved, and the colouring, though so simple and uniform, strikes every eye that sees it, by it's singularity. Linnaeus has evidently mentioned it, and gives the authority of his countryman Rudbeck, that it is a native of Sweden, and lives about fresh waters: but that author calls it a rare species; he calls it Anas fluviatilis rufa rostro pedibusque cinereis.

Anas viridi-nigrescens capite cristato.
The greenish-black Anas, with a crested head.

The Brasilian crested Duck.

The general figure of this species has occasioned those, who distinguish the duck and goose-kinds, to refer this to the former, though the bigness would have naturally enough referred it to the latter; it is at least equal to our common goose in size: the head is large, and of a very singular appearance; it is very rounded and protuberant; the eyes are large, and their iris is of an orange scarlet; their aspect is very fierce; the beak is long and large, it is thick at the base, but flatted in all other parts, and rounded at the end: it's colour is a dusky orange; the nostrils are oval; large, and black; at the base of the beak there is a large protuberance of a naked, fleshy matter, of a high and elegant red colour, and the eyes are also surrounded with a naked, fleshy matter, of the same colour, and of a corrugated or wrinkled surface.

The top of the head is of a deep black, but there is an elegant green diffused all through that colour, and in some lights shewing itself almost solely, without any appearance of the black: the top of the head is ornamented with a crest, formed of a very large tuft of black feathers: the back, wings, breast, and belly, in short, the whole bird, are of the same blackish-green with the head, and that very beautiful, only the tops of the wings are white.

The legs are long and thick, and of a beautiful orange scarlet; the toes are long, and the membrane which connects them, and forms the web of the foot, is serrated; the hinder toe is very short and inconsiderable; the inner one of the three anterior is much shorter than the outer: the legs are not situated so backward as in many species, and hence the bird can walk on land better than most.

It is a native of the South American islands, and lives on the large fresh water lakes there. It frequently leaves the water, and enjoys the serene air on shore. The writers on the Brasilian animals have described it; and we have lately had the stuffed skin sent us over as a curiosity: the mixture of green with the black in it is very beautiful.

Anas capite et collo albis, corpore nigricante.
The white-headed Anas, with a black body.

The Ipecati Apoa.

This is another of the Brasilian ducks, and is a very singular, as well as beautiful, bird; it is somewhat smaller than our common goose, but considerably larger than the duck: the head is large and rounded; the eyes are bright and piercing, but not large; their iris is of a pale hazel, with an admixture of yellow: the beak is large, and of a singular structure; it is thick at the base, flatted all the way of it's length, and rounded, but withal somewhat bent at the end, the unguis, which terminates it, evidently turning down.

The head is of a snow-white, but variegated, in an elegant manner, with a few black feathers: the base of the beak is ornamented with a fleshy protuberance, but this is not red, as in the former species, but black, spotted with white: the neck is of the same elegant white with the head, and is in the same manner variegated with a small number of black feathers: the breast and belly are white; the back and wings are of a deep black, but with an admixture of a deep glossy green: the tail is short; the legs are long and robust, and are of a greyish colour.

This

This is a native of the Brasils, and of some parts of the continent of South America. It has been described by Marcgrave and Piso, under the name of Anas sylvestris Brasiliensibus Ipecati Apoa dicta Pour, id est, Anser Lusitanis. We have sometimes had the stuffed skin sent us over as a curiosity, very well preserved.

Anas niveus rostro partim flavo, partim miniaceo.
The snow-white Anas, with a red and yellow beak. — The Ipeca Guacu.

This, though of one simple colour, is an extremely beautiful bird: the head is large, in proportion to the body, and it is rounded; the eyes are moderately large, and their iris is of a bright orange colour: the beak is long and large, and is remarkably beautiful: the basis of it is of a fine bright scarlet, which colour is continued all over a naked fleshy substance, that covers the anterior part of the head: the other part of the beak is of a fine gold yellow; it is very broad, and serrated all the way at the edges, and rounded at the extremity. The bird is of the bigness of our common duck, or a little larger; it's colour is throughout a fine and elegant white, like that of the swan, when in it's utmost beauty and perfection: the legs are short, but robust, and of an orange colour; the feet broad and webbed.

It is a native of the Brasils. Piso has described it under the name of Ipeca Guacu Brasiliensium; and most of the writers on birds, since his time, have communicated it under the same name.

Anas griseo-rufescens macula utrinque ad rostrum rubra.
The reddish-grey Anas, with a red spot on each side the head. — The Mareca.

This is of the size of our common duck, and is a very elegant bird: the head is large, and is flatted on the crown; the eyes are moderately large, and their iris is of a bright hazel, with an admixture of yellow: the beak is large and broad, but not very thick; it is obtuse, and rounded at the extremity, and is serrated all the way about the edges: the nostrils are round and patulous; the whole beak is of a deep brown colour, but it is somewhat paler in the middle than at the sides.

The upper part of the head is of a dusky greyish-brown colour, but at the sides, under the eyes, it is of a snow-white; and just at the origin of the beak, on each side, it has a bright red spot: the back is of a very beautiful colour, a mixed greyish-red: the grey is the principal ground colour, but the reddish tinge is diffused universally through it, and in many lights appears the principal colour: the breast and belly are throughout of a dusky brown, with a tinge of yellowish.

The wings are extremely beautiful; their ground colour is the same reddish-grey with that of the back, but they are variegated in an elegant manner along one side, of almost all the long feathers, with a bright and pale brown, and toward their middle they are tinged with a fine and elegant green: the tail is simply grey; the legs are short and black, and the feet are very large, and deeply webbed: the flesh is much esteemed at table, where it is a native.

It is frequent about several parts of South America. All the writers who have treated of the animals of that part of the world, have mentioned it; and those who have treated of the same subjects more generally, have continued their accounts. Ray, Willughby, and others have described this species, under the name of Anas sylvestris Brasiliensis Mareca dicta prima Marcgravii.

Anas fusca gutture albicante, pedibus sanguineis.
The brown Anas, with a white throat, and with red legs.

The red-legged Mareca.

This is also a very elegant bird, though it have less variegation in it's colouring than many of the others: the head is large, and flatted on the crown; the eyes are large and black; the beak is nearly as long, and full as broad, as in our duck, and is of a glossy black: the head is of a deep dusky brown, such as our painters express by the term umber colour; but on each side, near the eyes, there is a small roundish spot, of a pale lemon colour: the hinder part of the neck is of the same dusky colour with the head, but the throat is all the way down of a snow-white: the back is of a very elegant brown, not so deep as that of the head, and with some tinge of the olive along with it: the breast and belly are of a pale grey, with a colour of gold yellow: the tail is black; the wings are of the same brown with the back, only there is a tinge of greenish visible on every part of them: the tips of the feathers are many of them white, and in the middle of the wing there is a large spot of a bluish-green, very elegant: the legs are short, and of a fine deep red; the feet are very broad and webbed; the hinder toe is short and inconsiderable.

This is a native of South America. The authors who have written on the animals of the Brasils have all described it, and from them the rest of the ornithologists. Ray and Willughby call it Anas sylvestris Brasiliensis dicta Mareca secunda Marcgravii. The flesh is much esteemed in that part of the world.

MERGUS.

THE beak of the Mergus of is a cylindric figure, and hooked at the extremity, and it's denticulations are of a subulated form.

Mergus crista dependente, capite nigro-caerulescente collari albo.
The Mergus, with a hanging crest, a bluish-black head, and a circle of white round the neck.

The Merganser.

This is a moderately large and a very singular bird; it's weight, without the feathers, is about four pounds; it's body is of an oblong and slender form, and the back flat: the male and female are so very different, both in their form and colouring, that our common people call them by two distinct names, and many of the writers on birds have described them as distinct species.

In the male, which we call the Goosander, the head and the upper part of the neck are throughout of a dark green, with a shade of a deep violet blue, a very beautiful and a very glossy tint, but so deep, that in many lights it carries an appearance of black: the lower part of the neck is of a bright glossy white: the middle of the back, and the exterior of the feathers of the shoulders, are black; the others are white, but the lower or hinder part of the back is of a dark grey: the rump, and the side feathers near it, as also the feathers which cover the thighs, are variegated with transverse lines of a deep grey, with some tinge of brown: the tail is composed of eighteen feathers, and is entirely grey.

The wings are large; the long feathers in them are twenty-six in number: the ten outer ones are black, the four next to these are black also, but their tips are white; the five which succeed these are, on the contrary, white, with the bottoms black; the others, or the six or seven next the body, are also white, but with the outer edges black: the covering feathers are very beautifully variegated with black and white.

The whole breaſt and belly are of a livid or duſky bluiſh colour, but it is darker toward the top of the breaſt, and it becomes whitiſh toward the lower part of the belly.

The beak is very remarkable, both in figure and in bigneſs; it is of the length of the middle finger, and is conſiderably thick and robuſt; but it is not flatted, as in the duck, but rounded on the upper part, and, though it runs ſtraight nearly to the point, it is there hooked and turned down: the under chap is black throughout; the upper one is alſo black along the middle, but it is reddiſh at the edges, and black as jet at the hooked extremity, and is all the way denticulated along both ſides; the denticulations are of a ſubulated form, and pointed inward: the inſide of the mouth is yellow, as is alſo the tongue.

The feathers of the head are long, eſpecially toward the hinder part, and they ſtand looſe, and form a kind of creſt, which hangs down over the upper part of the neck, though but a little way: the noſtrils are large, and the iris of the eye is of a blood-red.

The legs are long and robuſt, they are of a ſtrong red: the feet are webbed, and the hinder toe is increaſed in breadth by a membrane.

The female of this ſpecies, which our people call the Dundiver, or Sparlin-fowl, has the head ſmaller than the male, and the feathers do not ſtand ſo looſe on it: it is of a brown colour, and has ſome approach toward a creſt, but not ſo much as in the male: the upper part of the throat is whitiſh; the back is altogether grey: the breaſt and belly are of a bluiſh or livid colour, as in the male, but paler: the long feathers of the wings are variegated, nearly in the ſame manner as in the male; the beak alſo is entirely of the ſame colour, figure, and ſize; the wings are rather ſhorter than in the male; and the tail alſo is ſhorter, and of a ſomewhat paler colour.

The bird is frequent in many parts of Europe, and has been deſcribed by almoſt all the authors who have written on this ſubject. It frequents freſh waters, and feeds on fiſh: it does not often take flight, but, by the ſwift motion of it's wings on the ſurface of the water, makes way with great rapidity. Aldrovand, Willughby, and Ray call it Merganſer; Geſner, Mergus cirratus longiroſter; Scheffer, who deſcribes it in his hiſtory of Lapland, Kaipa; and Bellonius, Harle.

Mergus criſta dependente, capite nigro, maculis ferrugineis.
The Mergus, with a hanging creſt, with a black head, ſpotted with brown.

The long-beaked Duck.

This is nearly of the ſize of the common duck, and is not unlike it in the form of it's body, but the beak is perfectly that of the Mergus-kind: the head is ſmall, but the feathers ſtand ſo looſely on it, that it appears conſiderably large: the eyes are ſmall but bright, and of a piercing aſpect; the beak is very long, narrow, convex on the upper part, and hooked at the extremity, and dentated all along the ſides: the head is of a deep gloſſy black, but ſpotted all over with a ferrugineous brown: the creſt is not long, but hangs from the hinder part of the head: the breaſt is of a bright grey, ſpotted all over with innumerable minute dots of black; the belly is white, and the ſides are undulated with white and black: the back is altogether black; the long feathers of the wings are alſo black, but the ſhorter or covering feathers are white, only that the lower ſeries, or thoſe which fall immediately over the long ones, have their extremities black; this diſpoſition of the white feathers forms together a ſingle, broad, white line on each wing: the tail is ſhort and brown; the legs are robuſt, but not long, and the feet are large and webbed.

The wings in this species are remarkably short, so that it seldom rises into the air, but it is continually skimming along the surface of the water, by their help, with great rapidity.

It is frequent in many parts of Europe, and almost all the writers on birds have described it. Gesner calls it Anas longirostra, the long-beaked Duck, a name which many others have copied; Aldrovand, Anas longirostra Gesneri; Willughby, Mergus cirratus fuscus; others have called it Merganser minor, and Mergus cirratus minor.

Mergus capite griseo, crista destitute.
The Mergus, with a grey head, without any crest.

This is a very singular species, and is not only much smaller, but, in many other essential circumstances, different from the others: the head is large, and flatted on the crown; but this is the less conspicuous, as the feathers stand very loose on it: the eyes are large, and very bright and piercing in their aspect: the beak is not long, as in the others, but short and small; but it is of the same general figure as in those, rounded or convex, and straight all the way to the point, where it is hooked: the upper part of the head is black; the back of the neck also is black, and the whole back of the bird is also black, but of a less glossy and shining hue: the wings also are black, but have a broad, transverse line of white on them, formed of the covering feathers, which are white, all the long ones being of a deep black. It is singular to this species, that it has no crest on the head, nor indeed the least approach toward any thing of that kind: the throat, or anterior part of the neck, as also the breast, are perfectly white; and the sides of the rump are also white: the tail is short and brown: the legs are short, and of a dusky olive colour, with a very strong tinge of the green, and the feet are webbed.

This is frequent about the sea-coasts, in almost all parts of Europe, and has been described by most of the writers on this subject. Ray and Willughby call it Mergus Melanoleucos rostro acuto brevi.

The other known species of the Mergus are, 1. The larger, brown Mergus, with a spotted back and greenish legs. 2. The larger, deep-brown Mergus, with a longer crest. 3. The lesser, olive-coloured Mergus, with a very long beak. 4. The lesser, dark-brown Mergus, variegated with black and white, and with a shorter beak.

PELICANUS.

THE beak of the Pelicanus is very long, and is crooked and unguiculated at the extremity: the sides of it are not denticulated, and the anterior part of the head, toward the throat, is naked.

Pelicanus gula saccata.
The Pelicanus, with a bag at the throat. **The Pelican.**

This is a very large and an extremely singular bird: the head is moderately large; the eyes are large, and their iris is of a dusky grey, with a slight tinge of yellow: the beak is enormously large and long, and is of a strangely singular figure and structure; it is near a foot in length, and toward the base is of the thickness of a man's wrist: it is in this part of a dusky lead colour, but with a faint tinge of yellow, especially toward the sides, and it is also yellowish again at the tip: the upper chap is broad and depressed; the under chap is formed of two horny ribs, separate at their base, and joined at their extremity, and connected by a thick membrane of a yellowish colour, which is continued not only all along the beak, but beyond it's base, down to the throat: there is a tubercle at the base of the beak, but it's extremity is pointed, and somewhat hooked: the nostrils are small, and of a roundish figure, and are situated near the base of the beak, just under the plumage which invests it.

The

The head is naked at the sides, from the angle of the beak quite up to the eyes; the feathers on it's crown are long and narrow, and somewhat emulate a crest; they are of a whitish-grey, and the whole bird is also of the same colour, except that there is some yellowishness about the neck, and that the scapi of the feathers on the back are black, and the tips of the long feathers of the wings are black.

The tail is about six inches long, and is composed of two and twenty feathers; these are nearly of an equal length, but the exterior ones, when closely examined, are found to be somewhat shorter than the middle ones: the wings are moderately long; the large feathers are twenty-eight in each.

The legs are short, but very robust: the feet are very broad and webbed; the colour of the legs is a dusky bluish-grey, and they are naked to the middle joint. The bird is, upon the whole, one of the most singular in the world; it has, at first appearance, much of the general figure of the swan, but the addition of so enormous and strange a beak, to a bird of that kind, is so perfectly out of the course of what might have been expected, that it startles and surprizes every body.

The pelican is a native of some parts of Asia, and, where it breeds, is usually very plentiful: the food of it is fish and water insects, and it frequents equally the sea-shores, and all the larger fresh waters. All the writers on birds have mentioned it. Aldrovand calls it Onocrotalus sive Pelicanus; and most of the others have taken one or both these names.

Pelicanus subtus albicans rectricibus quatuordecim.
The Pelicanus, with a white breast, and with fourteen long feathers in the tail.

The Cormorant.

This also is a very large bird; it is equal to a well-fed goose in size, and is not without it's beauty: the head, neck, back, wings, and rump, are all of one uniform colour, which is a deep olive brown, with an additional tinge of a changeable green thrown over it, visible, in a different degree, in different lights: the breast and belly are of an elegant white, and the contrast of these so very different colours gives a very great beauty to the bird: the wings are long, and very well feathered; the long feathers are about thirty in each, and the tips of the greater part of these, as also of some of those of the second series, are greyish: the tail is about seven inches long, and is composed of fourteen feathers; it is hollow within, and, when expanded, is of a rounded figure at the extremity.

The beak is about three inches long; it is robust and straight to near the extremity, where it is hooked: the upper chap is altogether black, and is sharp at the edges; the lower has the edges flatted: the tongue is very inconsiderable, and the opening of the mouth is enormously wide: the eyes are large, and stand at a very small distance from the angle of the mouth; but, what is most of all remarkable in this bird, and at first sight distinguishes it from all the others, to which it has any resemblance in general figure, is, that it has a naked yellow skin, investing the base of the under chap of the beak. The legs are moderately long and naked; they are covered with a firm and hard skin, divided into a kind of cancellated scales: the feet are very large; the toes are four, they all stand forward, and are connected together by a large and firm web or membrane: the outer toe of each foot is the longest, the others are all gradually shorter, and the claw of the middle one is serrated on the inside.

This is an extremely voracious bird; it feeds on fish, and it's swallow is so extremely large, that it takes such as would surprize any one to conceive, and lets them down whole. It frequents the sea, and sometimes the large fresh waters. It breeds in some of the more northern parts of England about the sea-coasts, and that in a very different manner, for it sometimes makes it's nest in the cracks and caverns of the rocks, and sometimes in the tallest trees. This is the more singular, as it is a bird of the web-footed kind, not any other of that structure of feet being ever known to perch upon trees. It is a native also of many other parts of Europe. All the writers on birds

birds have described it. Gesner calls it Carbo aquaticus; Aldrovand, Ray, Willughby, and others, Corvus aquaticus; and others have taken one or other of these names.

Pelicanus subtus fuscus, rectricibus duodecim.
The Pelicanus, brown underneath, with twelve feathers in the tail.

The Shag.

This is considerably like the former species in shape, but quite different in colour; it is of the size of a well-fed duck; the head is large, and flatted on the crown; the eyes are small, and they are very remarkably situated; they stand more forward, and lower down on the head, than in any known bird: the beak is straight, and not very robust; it is neither compressed on the top, nor flatted at the edges, but is of a rounded figure: the whole beak is about three inches and a quarter long; the upper chap is somewhat longer than the under, and turns over it in a little hook at the end, otherwise the whole beak is straight; the upper chap is black; the under one is of an olive colour with a tinge of yellow: the opening of the mouth is enormously wide: the tongue is very small and inconsiderable: the nostrils are also very small, and scarce conspicuous: several authors have indeed said it has none.

The body is of a broad and depressed figure; the back flatted, and the shoulders very little raised: the colour of the whole upper part of the bird is a deep and elegant black, with a tinge of glowing purple thrown over it, and of green in some lights; and the whole has a gloss and brightness on it, that is scarce to be conceived, otherwise than by seeing it: the breast is brown, and so is the greater portion of the under part of the body; but the middle of the belly is somewhat greyish, and, toward the tail, it is black. And the throat has some whiteness in it.

The tail is six inches in length, and is composed of twelve large and rigid feathers; the middle ones are the longest of these, but the exterior ones are very inconsiderable, so that, when expanded, it appears of a kind of hyperbolic figure: the wings are not very long; when closed, they reach no farther at the tips than to the base of the tail: the long feathers in each are thirty.

The legs are short and robust; they are not rounded, but compressed, or of a flatted figure, and are feathered below the joint; they are covered with a firm and tough skin, not divided into scales, but cancellated: the feet are very large and webbed: the toes are four; the outer one of each foot is longest, the others gradually shorter: the legs are black, as is also the sole of the foot; but the membrane which connects the toes is brown: the claw of the middle toe is serrated or pectinated.

This is a native of the coasts of England; it is a sea-bird, and is usually seen swimming with it's head erect, and it's body almost entirely under water. It is very shy, and very difficult to be shot; it dives to admiration, and never fails to get far off, at the sight of a gun. It builds with us, and usually does it in trees, not among the rocks; the eggs are oblong, and totally white. Most of the writers on birds have mentioned this species. Aldrovand calls it Graculus palmipes Aristotelis, sive Corvus aquaticus minor. Ray and Willughby, Corvus aquaticus minor, sive Graculus palmipes.

A L C A.

THE beak of the Alca is of a convex and compressed figure, and is intervated and furrowed, or sulcated in a transverse direction: the feet stand very backward, and have each three toes.

Alca roſtri ſulcis quatuor, oculorum regione temporibuſque albis.
The Alca, with four furrows on the beak, and with the ſides of the head white.

The Pope.

This is a very ſingular bird; it is about the ſize of our widgeon, or ſomewhat larger, but is not quite ſo large as the duck: the head is large and rounded; the eyes are ſmall, and ſtand forward on the head, and lower down than in the generality of birds: the beak is ſhort and broad, but it is flatted in a different direction to that of the beaks of the duck-kind, being compreſſed from ſide to ſide, and not depreſſed from top to bottom. It is large on the baſe, and pointed at the extremity, and on the whole approaches ſomewhat to a triangular figure: the upper chap is a little longer than the under, and ſomewhat arcuated or hooked at the extremity: it's baſe is ſurrounded with a naked, callous matter, in the ſame manner as the beak of the parrot-kind; juſt under this callous ſubſtance ſtand the noſtrils; they are very long and narrow, and of a darker colour than the reſt: the beak itſelf is of two colours, of a grey or bluiſh-livid hue toward the baſe, and of a bright red at the extremity: there are four very remarkable furrows in it; they run in a tranſverſe direction, and are conſiderably deep, but the extream one is fainter than the reſt; ſo much ſo indeed, that many of the authors who have deſcribed the bird, have called them only three: the mouth is yellow within: the eyes are large; their iris is grey, and the eye-lids are armed with a thick black cartilage, and in the under one there is a ſmall, livid, roundiſh protuberance, and on the upper a triangular excreſcence of the ſame fleſhy matter.

The legs are, while the bird is young, of a bright yellow, but, when it is full-grown, they are red; they are ſhort, but robuſt, and ſtand ſo extremely backward, that, when the bird ſtands upon them, it ſeems to raiſe itſelf erect on the tail: the feet are moderately large; the toes are three, all placed forward, and connected by a firm membrane: the middle toe of each foot is longeſt, and the inner one much ſhorter than the outer: the claws are of a bluiſh-black.

The head, neck, back, and upper ſurface of the wings, are all black: the whole breaſt and belly are white; there is a collar of black drawn acroſs the throat: the jaws, the upper part of the throat to this collar, and the ſides of the head, are of an elegant ſnow-white, or in others of a very bright ſilvery grey: the eyes are, as it were, ſurrounded by a circle of this colour.

The wings are very ſhort, and compoſed but of few feathers; they were not intended by nature for long or high flights, but they ſerve the bird for ſkimming along the ſurface of the water at a prodigious rate: the tail is very ſhort, ſcarce an inch and a half long; it is compoſed of twelve feathers, and is totally black.

This ſpecies breeds with us, and is very common about many of our coaſts, eſpecially the more northern. It is not at the trouble of making any neſt, but lays it's eggs in any little hollow on the ground; and, what is very ſingular, each lays only one egg, but this is very large, in proportion to the ſize of the bird. If this egg be taken away, the bird will lay another in the place of it, and ſo on to the fifth; but, if undiſturbed ſhe lays only one, nor will ſhe ever ſit on more than a ſingle one. The egg is equal to the largeſt hen's egg, or larger, and is of a duſky yellowiſh-brown colour all over; it is large at one end, and very ſmall and pointed at the other. We have them in vaſt plenty about the coaſts of the Iſle of Angleſea, Caldey Iſland, the Iſle of Man, and other places, where they come on ſhore in prodigious numbers to breed; but they are birds of paſſage, for they leave us at the approach of winter, and do not return till ſpring. A few of them are uſually ſeen at the end of March, or in the beginning of April, which ſeem as if ſent to reconnoitre, for they go off again, after a few days, and in three weeks, or a little more, return with the whole body. If the ſeaſon be favourable, and the weather mild, they come in ſurpriſing numbers,

and

The Wild Goose
The Spoon Bill
The Gadwall
The Shell Drake
The Swan Goose
The Sea Pheasant
The Pelican

and in good condition; but, if it be stormy, they are thrown up dead by thousands at a time on the shore, starved and mere skeletons. It is a long way that they travel; and they can only come on the surface of the water, and must feed all the way, which is not to be done in tempestuous weather. In August, or in the beginning of September, they go away in a body. Our fishermen tell us, that they sometimes find a few of them in a torpid state, in the cracks of rocks in winter; but the stay of these is accidental, and probably they all perish.

All the writers on natural history have named this singular bird, but they have distinguished it by so many different appellations, some of them common also to other species, that their accounts have great confusion. Gesner calls it Puphinus Anglus; Aldrovand, Fratercula marina sive Pica marina; Martin, Avis glacialis; Clusius, in his Exotics, Anas arctica; and Ray, Willughby, and many others have borrowed this last name; Wormius calls it Anas arctica Clusii, and Bartholine, Lunda avis.

Alca rostri sulcis octo, macula alba ante oculum.
The Alca, with eight furrows on the beak, and a white spot before the eye.

The Penguin.

This is a very large and singular bird; it is equal to the common goose in size, but in all respects resembles the species of Alca already described, except in the specific distinctions: the head is large, and flatted on the crown; the eyes are moderately large, and their iris is grey, with a tinge of yellow: the beak is of a kind of triangular figure, compressed at the sides, and a little hooked just at the extremity; it is between three and four inches in height, and has eight of the furrows which distinguish the birds of this genus on it, but they are not exactly transverse, but oblique; these are toward the extremity of the upper chap, and the eighth is so faint, that it is easy to overlook it: the lower chap has no less than ten such furrows, but they are fainter: the whole beak is black, and the opening of the mouth is enormously wide; there is a protuberance also at the angle of the lower chap.

The head is black, only that there runs a white line on each side, from the beak to the eyes: the back and wings, and indeed the whole upper part of the body are black, and the breast and belly, or whole under part are white: the wings are very short, and the tail also is short: the feet stand very backward: the legs are short and black, and the toes connected by a membrane.

This is a native of most of the northern parts of Europe, and has been described by all the writers on birds. Willughby calls it Penguin nautis nostratibus dicta quæ Grosfugel Hoieri esse videtur; Bartholine calls it Avis Garfahl; Clusius, Anser Magellanicus; Wormius, Anser Magellanicus Penguin; and Ray, simply, Penguin. It feeds on sea-fish, and on many of the insects and small animals that frequent the shores.

Alca rostri sulcis quatuor, linea utrinque alba a rostro ad oculos.
The Alca, with four furrows on the beak, and with a white line on each side the head.

The Razor-bill.

This is a much smaller species than any of the former. It is not equal to the common duck in size: the head is large, in proportion to the body, and is rounded at the sides, but a little flatted on the crown; the eyes are moderately large, and their aspect is very bright and piercing; the beak is of the nature of that of the others of this genus, but it is still more singular than in any of them; it is about an inch and a quarter in length; it is throughout of a black colour, and is narrow and compressed at the sides: a little above the nostrils there runs a very deep furrow, and the beak is, from it's base to this mark, covered with a fine downy and soft plumage; there are three

three other furrows running transversely, or a little obliquely, at some distance below this; but they are less considerable, especially the last of them, which, in some specimens I have seen of this bird, has been almost obliterated, so that a man might have easily mistaken the number of furrows on the beak for only three: the upper chap of the beak is somewhat longer and larger than the other, and is crooked and hollow at the extremity, and receives the under chap into it.

The head, as also the neck, shoulders, back wings, and indeed every other part of the upper surface of the bird is black: the belly, the breast, and half way up the throat, are of a fine snow-white, but the upper part of the throat is reddish; there are two very remarkable white lines on the head; they run on each side, one from the angle of the beak to the eyes: the mouth is of a beautiful bright yellow within, and the iris of the eyes is of a pale hazel colour.

The wings are short, and are rather formed for assisting it's motions along the surface of the water, than for high flights: the tail also is short; the legs are short, and very robust; the feet are webbed, and they have no hinder toe: the legs stand so extremely backward in this species, that the bird walks but very awkwardly, and, when doing so, seems to stand erect upon it's tail. Nature has calculated all the parts of animals for the purposes they are to serve, and the business of these is to assist the animal in swimming, not walking.

This is a native of our own coasts; it builds in the crevices of rocks, and on cliffs. It makes no nest, but lays a single egg on the bare rock, and there sits on it. It has been described by all the authors who have written on these subjects. Willughby and Wormius call it, simply, Alca; Ray, Alca Hoieri. Our people, in different parts of the kingdom, have three separate names for it. Those about the western coast call it the Razor-bill; in the northern it is called the Awk, and in Cornwall, and some few other particular places, the Murrel. The eggs are esteemed a delicacy; they are very large, and though the creature lays but one egg, they are to be had in great abundance, for multitudes usually breed together.

COLYMBUS.

THE beak of the Colymbus is of a subulated figure, and somewhat compressed; it is more than equal in length to the head of the bird, and it has no denticulations at it's edges: the legs are placed very backward, so that the bird walks awkwardly.

Colymbus pedibus palmatis indivisis.
The Colymbus, with palmated, undivided feet. **The Lumme.**

This is a very beautiful bird, and is extremely common in some particular parts of the North of Europe, though wholly unknown elsewhere; the bird is about equal to our common wild-duck in size: the head is large, in proportion to the body, and is rounded at the sides, but somewhat depressed on the crown: the eyes are large, sharp, and piercing, and their iris is of a fine pale hazel: the beak is about an inch and three quarters in length, and toward the base it is moderately thick; it is from thence all the way smaller to the very extremity, where it is sharp, and is all the way compressed or flatted at the sides; it is all over of a deep glossy black: the head and neck are grey, and have the appearance of being covered with a friar's hood, or some ornament of that kind: the back and wings are black, but they are beautifully variegated with large and square spots of white; these are smaller on the wings than on the back: under the neck there is a large black spot, of an oblong, quadrated figure; this has the appearance of a shield, and is four inches in length, and two in breadth: the sides of it are variegated with grey and white feathers, in manner of a zone; the breast and belly are white.

The

The legs are short, but robust; they are of a deep blackish colour, and stand so very backward, that it is plain they were intended for swimming, not for walking.

This is very frequent about Norway and Sweden, and some places much farther North. It dives to a wonderful degree, keeping under water so long, and making so considerable a progress under it, as surprizes every body that sees it. It lays but one egg, but that very large. The eggs and the bird itself are eaten, and extreamly valued in many places. The skin also, when the feathers are taken off, is dried, and used in many places as swan-skin, by way of warmth, but is said to excel it. Most of the writers on these subjects have described it. Clusius calls it Mergus maximus Farrensis sive arcticus. Wormius, Colymbus arcticus Lumme dictus; Willughby and Ray, Colymbus arcticus; and Bartholin, simply, Anas aquatica.

Colymbus pedibus lobato-divisis, capite nigro.
The Colymbus, with the toes lobated and divided, and with a black head.

The great Didapper.

This is a very pretty bird; it's size is about that of the widgeon; the head is large, and somewhat depressed on the crown; the eyes are large and bright; their iris is of a deep dusky hazel, and the pupil of a greenish-black: the beak is an inch and a quarter in length, black all over, and large at the base, but sharp at the point, and compressed all the way: the feathers on the top of the head, and those on the upper part of the neck also, are elevated in such a manner, that the bird has not only a crest, but a kind of horns also, though they are short.

The upper part of the head is of a deep black; the sides are of a reddish-brown: the upper part of the throat is white, and there is also a white space on each side of the head, about the eyes; and this white on each part has somewhat of the reddish-brown adjoining to it: the neck is moderately long, and is partly black, and partly of this reddish-brown: the breast and belly are of a white colour, with an admixture of the reddish-brown, but it is faint, and less considerable.

The back and the upper surface of the wings are in general black, but there are some feathers of a reddish-brown, intermixed with the black ones: the wings are short, but they are not so extreamly short as in some of the former species, in which they seem scarce designed for any thing more than skimming along the water: this species, on the contrary, flies very well; the long feathers are all of a dusky whitish colour, and so is the upper verge of the wing; the edges downward are of a rusty brown, but so deep, that it approaches to black.

The legs are short, but they are very robust; they are somewhat flatted in figure; the feet are large; the toes on each of them edged with a membrane on both sides, but they are not webbed; the membranes not conjoining into one, but leaving the toes a liberty of separating from one another to any distance.

This is a native of England, and of most other parts of the North of Europe, but it is not so common as many of the other species. It builds with us among sedge and reeds, and lays four or five moderately large eggs. Most of the writers on birds have described it. Gesner calls it Colymbus major; Aldrovand, Colymbus major cristatus; Willughby and Ray, Colymbus major cristatus et cornutus. We call it the horned Didapper.

Colymbus pedibus lobato-divisis, capite rufo.
The Colymbus, with the feet lobated and divided, and with a reddish-brown head.

The Dobchick.

This is a very singular and a very pretty bird; it is smaller than the common teal, and, as it is seen swimming on the waters, appears like the young of some of the duck-kind,

kind, not yet fledged or feathered: the head is small, and is somewhat depressed on the crown: the eyes are large, and very piercing in their aspect; the beak is about half an inch in length; it is largest at the base, and from thence all the way smaller to the extremity, where it is pointed, and it is flatted all along at the sides: the upper chap is black, but the tips and the sides of it are, for a little breadth, of a pale or whitish yellow: the whole under chap also is of this pale or whitish yellow: the nostrils are conspicuous, and stand at a little distance below the plumage that invests the base of the beak: the tongue is long, and bifid at the end; and the iris of the eye is hazel.

The plumage of this bird is very thick and soft, and at any distance it has more the appearance of a downy or woolly matter, than of feathers: the back, shoulders, and upper part of the wings are of an extreamly deep olive colour, approaching to black, and in many lights appearing quite black: the belly and breast are of an elegant silvery white; the very lowest part of the belly is brown, and there is some reddish-brown along the sides of the throat: the feathers which cover the thighs have also a little brown on them: the neck is long, in proportion to the size of the bird.

The wings are extreamly short and hollowed: the large feathers are twenty-six in each; and are some of them altogether brown, and others variegated with brown and white. This bird is singular, in that it has absolutely no tail: the glands found in the rumps of other fowls, however, are not wanting in this in their due place, and there arises a tuft of feathers from them, as in the other birds.

The legs are very short, and stand so extreamly backward, that, though very useful to the bird in swimming, it can hardly walk at all; when on land, it seems to stand erect upon the rump: they are robust, of a greenish-black colour, and of a flatted form, and naturally bend backwards: the toes are long, and are armed at their extremities with broad and short claws, much resembling the nails of a human hand: each toe is also increased in breadth, by a membrane on each side; but the feet are not webbed, or the membrane not connected.

This is very frequent about our rivers and ponds, and is the most nimble diver of all the water fowls. All the writers on birds have described it. Gesner calls it Trapazorola sive Mergulus; Aldrovand, Colymbus minor; Willughby and Ray, Colymbus sive Podiceps minor. It builds with us among sedge and rushes, and lays four or five moderately large eggs. We call it by several names, expressive of it's diving; the Didipper, the Dipper, the Dobchick, the Douker, and the lesser Loon; some, from the situation of it's legs, the Arse-foot.

Colymbus pedibus tridactylis palmatis.
The Colymbus, with webbed feet, and three toes to each.

The Sea-turtle.

This is of the bigness of a large tame pigeon, and is supposed, though not very judiciously, to resemble it in some degree in figure: the head is large, compressed at the sides, and a little flatted on the crown: the eyes are large, and their iris of a bright hazel: the beak is about three quarters of an inch long, and is shaped somewhat like that of the common hen, but broader at the base, and more compressed at the sides; it is sharp at the point, and all over of a deep black colour: the head and neck, as also the shoulders and back, and the generality of the upper surface of the wings also, is black; the breast, belly, and throat are white.

The wings are short, but they are very well feathered; the remiges or long feathers are all of them black, but the covering or shorter feathers are white, and thence there appears a large white spot on each of the wings: the breast and belly of the male, in this species, are of a deep colour, approaching to the black of the back, so that, in that sex, the whole bird is black, excepting for the white spot on each of the wings.

The legs are moderately long, and somewhat robust; they are of a bright and elegant red colour, and the toes are moderately long; they are only three on each foot, and

and all before, there being no mark of a hinder toe at all; they are not connected by a single membrane, in the manner of the web-footed fowls; but the toes have each a separate membrane on each side, increasing their breadth.

We have not this species native in England, but it is very common in many other parts of the North of Europe, and most of the writers on birds have described it. Martin calls it, simply, Columba; Willughby and Ray, Columba Groenlandica; Redbeck, Colymbus minor; and we, the Greenland Dove, or the Sea-turtle. It builds among the rocky cliffs in Norway, Denmark, and Sweden, and lays only two eggs, which are large, and much sought after by the people on the sea-coasts as food.

LARUS.

THE beak of the Larus is straight all the way, except just at the point, where it turns down; and it is obtuse, and not denticulated along the sides; and it's lower chap is gibbous or protuberant.

Larus albus dorso cano.
The white Larus, with a hoary back. **The White Gull.**

This is a pretty bird, though it have very little variegation in the colouring: the whole bird is of a clear and bright white, except that the back and the upper parts of the wings are grey, or of a hoary colour, and the long feathers of the wings are also grey, and that somewhat deeper than the others: the three or four first of them have also a black spot at their extremity, and a white dot within this; and all the others are white on their outer edge: the tail is moderately long, and undivided at the end, and it is throughout of a white colour, equal in brightness to that of any other part of the bird.

The legs are short, and of a bright yellow; the feet are large, broad, and webbed, and the hinder toe is very short and inconsiderable: the thighs are feathered at the top, but they are at least half the way naked: the eyes are large, and their iris is grey: the region of the eye-lids is naked, and their edge of a scarlet colour.

The upper chap of the beak somewhat resembles that of the raven; it is convex and cultrated, and arched, and the lower chap is gibbous; the corners of the mouth are naked, and of a bright scarlet colour; the tongue is sagittated, and bifid at the extremity: the beak is itself yellow, and the palate of the mouth is denticulated.

The size of the bird is that of a well-grown pullet; the female is somewhat smaller than the male. It lives on salt-water about the sea-shores, and in the ponds and ditches in the salt marshes. The coasts of Scarborough afford a great number of them at some times, but at others there is hardly such a bird to be seen: they sometimes build with us, and lay three large eggs. Many of the authors who have written on birds have described it, though some have omitted it. They call it, in general, Larus albus; our people, the white Gull.

Larus albus dorso cinereo-fusco.
The white Larus, with a greyish-brown back. **The great grey Gull.**

This is a somewhat larger bird than the preceding, but it is not so handsome: the head is large, and flatted on the crown, and the eyes large and grey: the upper chap of the beak is black, and the under one yellowish, but there is nothing of that scarlet colour in this, which is so conspicuous at the angles of the mouth; nor are the verges of the eye-lids scarlet, as they are in that species.

The size of the bird is that of a large hen: the head is white, but it is not so bright and clear a white as in the preceding species: the neck is whitish; the breast and

and belly are grey, and the throat has also some greyishness, but less than the others in it: the back is of a somewhat dusky brownish-grey, and the upper part of it is darker than the rest; the feathers of the wings are irregularly coloured with black, grey, and white; some of the feathers are wholly black, except their tips, which are white, and in others the colours are laid on more and more irregularly: the tail is moderately long, and the feathers which compose it are white, irregularly clouded with grey, brown, and black: the tips of many of them are black, and the upper edges of almost all are of a pure white. It is singular, that the birds of the first year's growth, in this species, are extremely different from those of a more advanced age; indeed, the variegation is so great, that it would be scarce possible for a person, not informed of it, to know them for the same: they are all over of the colour of the wood-cock, or variegated with clouded spots of grey, white, and brown.

The legs are short, and not yellow, as in the preceding species, but of a dusky grey, with a tinge of bluish and olive.

This species is very frequent about our coasts, and breeds with us. The eggs are large, and there are usually about four in the nest. All the writers on birds have described it. Ray calls it Larus cinereus maximus; and most of the others, Larus cinereus and Larus major. As it is often seen swimming and feeding promiscuously with the former species, the people about our sea-coasts have an opinion of their being the male and female of the same kind, but this is erroneous.

Larus dorso cinereo, collo maculato.
The Larus, with a grey back, and a spotted neck.

The lesser Gull, or Sea-mall.

This is of about the size of a common tame pigeon: the head is small, and depressed on the crown; the eyes are large, and their iris is of a pale hazel; the beak is considerably large; the upper chap is narrow, but arcuated and acute, and the eminence on the under chap, which is the distinguishing characteristic of the genus, is so small in this species, that it is scarce distinguishable, unless when nearly examined: the whole beak is of a whitish colour, except at the tip, where it is yellowish.

The head and the upper part of the neck are white, spotted with moderately large and irregular spots of brown, but the lower part of the neck, down quite to the bottom, is simply of a pure white: the back is of a dusky grey, down quite to the origin of the tail, but the feathers which cover the base of that are white: the throat, the breast, and belly are all over of the most pure and bright white: the tail is also of a pure snow-white; and the covering feathers of the under part of the wing are white, but those of the upper part are grey.

The wings are large, and extend to a great measure from tip to tip: the long feathers of them are variegated with black, white, and grey; the feathers which cover the upper part of the thighs are variegated with grey and white; the legs are moderately long, and of a pale greenish-grey colour: the toes are long; the middle one has the claw acute, or sharpened on the exterior side; the hinder toe is very small.

This is frequent about our coasts, and most of the writers on birds have described it. They have called it Larus cinereus vulgaris and Larus cinereus minor; the latter is the name given it by Willughby and others, the former by Aldrovand.

Larus corpore toto cinereo-fusco.
The brownish-grey Larus.

The deep, grey, Sea-mall.

This is of the size of a moderately grown pullet: the head is small, and flatted on the crown; the eyes are large, and very bright, and have a mixture of a pale hazel and a gold yellow in the iris: the beak is large, and very much arcuated, for one of this species: the upper chap is turned down for some length at the end, and the eminence

eminence on the under chap is conspicuous, though not so large as in some of the species: the upper part of the head is of a shining blackish-green, somewhat like that of the drake, but not quite so elegantly changeable; the sides are paler: the neck, shoulders, back, and upper part of the wings, are of a deep iron grey, with a strong admixture of brown: the breast, belly, and throat are of a paler grey, and with less of the brown, but still with some tinge of it.

The wings are very long. When closed, they reach beyond the extremity of the tail, and, when expanded, they reach to a great extent: their long feathers are elegantly variegated with grey, black, white, and brown: the tail is moderately long, and is of a pale grey, with some variegations of brown and of black, but with very little of the brown intermixed among the grey: the legs are moderately long, and of a pale fleshy colour, as is also the beak: the hinder toe is very short and inconsiderable.

This is frequent in our sea-marshes about Gravesend, and elsewhere; yet most of the writers on birds have, by some strange error, overlooked it. Belmer calls it Larus major cinereus; and our Willughby describes it, from his figure, under the same name; yet he is so uncertain about it, that he has a suspicion of it's being the same with our common Lapwing, a sufficient proof of his not having met with it.

Larus capite albo, nigra utrinque macula.
The Larus, with a white head, with a spot of black on each side.

The Torrock.

This is a very singular and a very pretty species; it is of the size of the common jack-daw: it's head is small, and of an elegant snow-white, but on each side there is a large and round spot of black: the neck also is of a pure snow-white, and has not the least spot on it: the bottom of the neck, where it joins the back, is of a deep black, and this colouring is so disposed, that it forms a kind of ring round the insertion of the neck, and with the contrast of the white of the neck is very beautiful: the back is of a dusky greyish colour; the breast and belly are of a snow-white: the tail is moderately long and white, only that most of the feathers are tipped with black: the wings are long, and their principal feathers are variegated, in an elegant manner, with black, white, and grey: the covering feathers are all of a greyish colour.

The beak is about three quarters of an inch long, and all over of a deep black: the upper chap is nearly straight, but hooked down at the end; and the prominence on the lower chap, which characterises this genus, is very considerable and conspicuous.

The legs are moderately long and red; the toes are long, and the claws black; the hinder toe is very short, and indeed imperfect; it rather is a protuberance of fleshy matter than a toe, there being no claw upon it.

It is a native of our coasts, and breeds with us in the West of England in great plenty. Ray, Willughby, and others call it Larus cinereus Bellonii, that author having first mentioned it. Our common people in Cornwall call it the Torrock. It builds on the ground, and lays from three to five eggs, which are large, in proportion to the size of the bird, and are accounted a delicacy at table.

Larus cinereus capite nigricante.
The grey Larus, with a black head.

The Pewit, or Black-Cap.

This is a very elegant species; it's size is about that of the common tame pigeon: the head is small, and the eyes are bright; they are large: their iris is of a pale hazel colour; the verges of the eye-lids are red, and they are surrounded with a fine white plumage: the beak is nearly an inch in length; it is of a fine bright scarlet, and is a little bent downwards.

The head is black, and the anterior part of the neck or throat is also black half way down: the middle of the back is grey; the neck, the breast, and the belly are white: the wings are considerably long; their covering feathers are grey, but the long feathers, which are twenty-nine in each wing, are variegated with black and white; the tail is moderately long, and is even at the end, not forked, and is throughout of a pure white: the legs are slender and delicate; they are not very long, and their colour is a blood-red: the claws are black; the toe behind is very short and inconsiderable.

This beautiful bird is very frequent about our sea-coasts, and in some places at a distance from the sea; it builds on the ground in any accidental hollows. All the writers on birds have described it. Aldrovand mentions it twice; once under the name of Larus cinereus Ornithologi, and another time under that of Larus cinereus tertius. Willughby and Ray continue these names, and call it also Cepphus Turneri et Gesneri.

Larus dorso ferrugineo, pectore pallide fusco. **The Gannet.**
The Larus, with a ferruginous back, and a pale brown belly.

This is an extreamly singular species; it's whole aspect, in some degree, approaching to that of some of the land birds of prey, as well as it's colour; it is equal to the common duck in size: the head is large, and flatted on the crown; the eyes are very bold and piercing in their aspect, and their iris is of a hazel colour: the beak is very large and strong; it is shorter than in the other species, but is turned down at the end, just like the beaks of the hawk-kind, and in the same manner has a yellow membrane investing it, quite down to the nostrils: the legs also approach to those of the hawk-kind; they are robust and short, and the claws are long, strong, and hooked, and sharp, quite in the manner of those birds of prey, and wholly unlike those of all the web-footed birds.

The whole upper part of the bird is of a deep, rusty, ferrugineous colour, much like that of the common buzzard: the breast and belly are also brown, but paler: the wings are very large, and their principal feathers are black; the tail also is moderately long and black: the upper part of some of the larger feathers of the wings is white, but there is nothing of this in the tail, for that is altogether black.

This singular bird, which, if it were not for the protuberance on the under part of the beak, one could scarce suppose of this genus, is frequent about the coasts of the western parts of England. Willughby, Ray, and some others have called it Cataractes noster. Many of the foreign writers on birds have omitted the mention of it. Our common people in Cornwal call it the Gannet. It follows the shoals of pilchards and some other fish, flying at a distance over them, and at times precipitating itself downward with great violence upon them.

Larus dorso cinereo, pectore albidiore. **The great grey Gull.**
The Larus, with a grey back, and a whiter breast.

This is a large and a very beautiful bird; it wholly resembles the little grey Gull in form, but in colouring, as well as bigness, it is extreamly different; the bigness is about that of our widgeon: the head is small, and flatted on the crown, and it is also a little compressed at the sides: the eyes are large, and they have a dull or dead look, not that piercing aspect which some of the others have: their iris is of a pale grey, and the pupil is scarce black, but rather of a deep olive: the beak is an inch and a half long, and is black throughout; the upper chap is somewhat arcuated in it's whole form, and is particularly hooked at the end more than in most of the other species: the lower chap has a considerable large protuberance between the angle and the apex: the nostrils are very conspicuous; they are of an oblong form, and stand very high toward the base of the beak.

The

The neck is short, but it appears much shorter than it really is in the creature's usual posture, for it has a way of sinking the head, as it were, between the shoulders; all the birds of this genus have more or less of this, but none of them nearly so much as this species: the back is of a very elegant grey colour, mottled with white, and with a tinge of brown diffused over it; the breast and belly are of the same colour, only it is paler; there is more of the white, and indeed very little of the brown in it: the wings are large, and their principal feathers are very beautifully variegated with white, grey, and brown, more distinctly placed than in the other parts: the rump is paler than the rest of the back, and the feathers which immediately cover the tail are white, only that they have a few spots of brown on them: the tail itself is grey, brown, and white.

The legs are moderately long, and they are somewhat robust; they are of a very pale colour, whitish only, with a faint tinge of brown or olive: the toes are long, and the feet are webbed, only the hinder toe is very short and inconsiderable.

This is a native of most of our sea-coasts, and of those also of most other parts of Europe. Many of the writers on birds have described it. Ray and Willughby call it Larus cinereus major; and Aldrovand seems to mean it by his Larus albo-cinereus torque cinereo. We call it the great, grey Gull, and, in Cornwal, the Waggel. The Veterians call it Marinoze, and the Dutch, by a name expressing the Burgomaster of Greenland.

The Coddy Moddy.

Larus fusco et griseo superne variegatus, inferne albus. The Larus, with a brown and grey back, and a white breast.

This is of the size of the common carrion-crow: the head is large, and is depressed on the crown, and considerably flatted or compressed at the sides: the eyes are large, and have a bright and piercing aspect; their iris is of a fine, elegant hazel, and the pupil of a deep black: the beak is about an inch long, and of a pale whitish-brown colour: the upper chap is arcuated, and moderately hooked; the lower chap has that protuberance, which is the characteristic of this genus, very large and conspicuous: the tongue is long, white in colour, and bifid at the extremity.

The head is white, variegated very elegantly with spots of brown: the neck is short and thick, and is brown: the middle of the back is of a dusky iron grey, but toward the sides it is of the same brown with the neck: the breast and belly are white as snow, the rump also is whitish, with only a little admixture of brown, and scarce any grey: the wings are large, and, when expanded, have a great deal of white in them: the principal feathers are white, variegated, in different proportions, with dark grey and pale brown: the tail is moderately long, and is principally white, but darker at the sides, and toward the end, than elsewhere.

The legs are moderately long and robust; they are of a pale whitish-grey, but with a tinge of greenish or olive colour thrown over them: the feet are webbed, and the toes are long, excepting the hinder one, which is short and inconsiderable.

This species is very frequent in our fen countries; we have it in great plenty in Lincolnshire, where the shooting them is a great diversion among the country gentlemen. They fly high, and it is sometimes difficult to kill the first; but, after that, it is but leaving the dead bird upon the ground, and they gather in numbers about it: in defect of this, they sometimes lay a white handkerchief carelessly folded together on the ground, and the birds are tempted to come near, under an opinion of it's being a hurt companion.

Many of the authors have omitted to describe this species, though so frequent not only with us, but in many other parts of Europe. Ray and Willughby however have mentioned it; they call it Larus fuscus sive hybernus, in agro Cantabrigiensi Coddy Moddy vocatus.

Larus

Larus griseo et albo variegatus, rostro flavicante.
The grey and white Larus, with a yellow beak.

The Winter Mew.

This is of the size of our widgeon, and at a distance appears to be all over white: the head is remarkably large and rounded; it is not at all either depressed or compressed in any part: the ears are very large and conspicuous; the eyes are also very large and bright: their iris is of a beautiful gold yellow, and the pupil black as jet: the beak is about an inch and a quarter long, considerably thick, very much arched and hooked, and pointed at the extremity: the upper chap is of a beautiful yellow, except just at the point where it is black; the under chap is entirely yellow, and has a large protuberance: the nostrils are oblong, large, and conspicuous.

The head is white, variegated with moderately large and irregular spots of grey: the back is of a very pale colour, almost white, only variegated, in a clouded manner, with a pale grey: the breast and belly are also of the same colour, only with less of the grey; the wings are very large, and spread to a great breadth; they also have a great deal of white in them, but the long feathers are elegantly variegated not only with grey, but a chesnut brown: the tail is considerably long, and is of a pale colour, almost white in the greater part, but dusky at the extremity.

The legs are moderately long, very slender, and yellow: the thighs are naked half the way up; the feet are webbed; the toes are long and yellow, and the claws short and crooked: the hinder toe in this species is not inconsiderable, as in many of the others, but is moderately large, and armed with a claw like the others.

This species is met with on some of our coasts, but it is not common with us, though it is very much so in many other parts of Europe. Aldrovand calls it Larus major; and Bakner, Larus hybernus. Willoghby has preserved both these names.

Larus albo et fusco variegatus, alis nigricantibus.
The brown and white Larus, with black wings.

The Cepphus.

This is an extremely beautiful bird; it is very different from all the rest of the Lari in it's general form and colouring, and, except that the singularity of the protuberance under the beak, and the other less conspicuous generical characters, referred it to them, would be easily mistaken for a species of some very different class: the head is moderately large, but not so remarkably so as in many of the Lari, nor so remarkably small as in some of the others. The bird appears of the bigness of a large tame pigeon, but this is owing to the feathers standing loose, for the body is, in reality, very small and inconsiderable: the figure of the head is somewhat singular; it is a little compressed at the sides, but is remarkably prominent on the crown: the eyes are small, but they are very bright and piercing in their aspect; their iris is of a fiery red, the pupil black: the beak is an inch in length, and moderately thick; it is of a deep glossy brown, or horn colour, in the body, but somewhat reddish at the sides, and black at the point.

The head, neck, back, and breast are all of a very elegantly variegated hue; the colours are white and a chesnut brown, but there are also some spots of yellow in different places.

The wings are moderately long, and their principal feathers are black, and their tips yellow. This occasions a very singular, as well as a very elegant, appearance, when they are closed, as well as when the bird is flying: the legs are slender, and of a yellowish-brown; the feet are webbed, and the hinder toe is small, and has no claw.

This species is not frequent with us, but I have had a preserved specimen of it sent up from Scarborough, shot somewhere in that neighbourhood. Aldrovand has de-

scribed

scribed it under the name of Cepphus; and many other of the ornithologists have followed him in his accounts of it, though few of them seem to have seen the bird.

The Brasilian Gull.

Larus cinereus capite, et medietate alarum, nigricantibus.
The grey Larus, with the head and one half of the wing black.

This is one of the larger species; the size is about equal to that of our common hen: the head is very large, and the neck short and thick: the eyes are small, and their iris is of a whitish-grey: the ears are wide and conspicuous; the beak is an inch and half long; it is nearly straight all the way to the extremity, but there it is crooked and sharp: the upper chap is yellow along the middle, but of a deep brown on the sides, and black at the extremity: the under chap is yellow throughout, and has a large protuberance on it's lower part.

The head is black on the upper part, the neck and back are of a deep grey, and the breast and belly are white: the wings are very large; their covering feathers are grey, but the long feathers are all black half the way up from their extremities, and white in the rest, where they are covered by the shorter: the tail is moderately long, and is also white toward the base or root, but the lower part of it is black.

The legs are long and robust; they are of a dusky orange yellow, and the feet are webbed; and the hinder toe is short, but it has a claw like the rest.

This is a bird of South American origin, but we sometimes have the skin stuffed, and sent over to us. The Portuguese there call it Gavitáa; the natives, Gaucuguacu; Marcgrave, and the others who have written on the Brasilian animals, have described it.

The scare-crow Gull.

Larus nigricans alis cinereis, pedibus rubentibus.
The black Larus, with grey wings, and red legs.

This is of the size of the common pigeon, and has very little the appearance of the Larus-kind, on a general view, though it has, when examined, all their characters: the head is large, and the feathers stand somewhat loose upon it: the eyes are long, and of a piercing aspect; their iris is of a yellowish-hazel, and the pupil is black; the ears are large, patulous, and conspicuous: the beak is three quarters of an inch long; it is moderately arched in form, and is very hooked at the point; it is all over of a deep black, except that the nostrils, which are oblong and narrow, are rather brown.

The neck is short and thick: the whole head, the neck, and also the back, shoulders, breast, and belly are all of a coal black: the wings are very long; when closed, they reach beyond the extremity of the tail; they are of a very beautiful dark iron-grey: the tail is moderately long, and is of the same iron-grey with the wings, only that it is black at the end.

The legs are very slender and short, they are of a beautiful scarlet; the hinder toe is very short and inconsiderable, but it has a little claw on it.

The male and female of this species are no way to be distinguished, but by an elegant white spot, which the male has under it's throat. This species has been supposed not a native of England, but I killed several, about four years since, among the Bognor rocks on the Sussex coast, and was informed, by the country people, that they bred there. Most of the writers on birds have named it. Gesner and Aldrovand call it Larus niger; and Ray, Willughby, and others have continued the same name.

STERNA.

THE beak of the Sterna is of a subulated figure, and acute: it is compressed at the extremity, and has no denticulations along the sides: the feet do not stand so backward as in many of the water-birds, but it walks conveniently with them.

Sterna rectricibus extimis maximis dimidiato-albis nigrisque.
The Sterna, with the largest outer tail feathers half black, and half white.

The Sea-swallow.

This is a very beautiful bird, and it's figure is very singular; it is equal in length nearly to the jack-daw, but it is very slender, and the body, when stripped of the feathers, is very small: the head is small, rounded at the top, but somewhat compressed at the sides: the beak is moderately long, and almost straight, but it is slender; it is red all the way, except at the tip or extremity, where it is black: the mouth is red on the inside, and the tongue is pointed.

The head is black on the upper part, but white on the sides, and on the under; the throat, the belly, and the feathers which cover the under part of the wings are white, but the middle of the breast has some grey in it: the rump is white; the back and the upper part of the wing are of a dusky grey: the wings are long, and their principal feathers have a great deal of whitish in them, but they are variegated with black: the tail is long and forked; it is from this that the bird has obtained with us the name of the Sea-swallow: the whole is composed of twelve feathers; the outer ones of these are six or seven inches in length; their outer edge is of a greyish-black, the middle ones are hardly three fingers long, and are totally white; the others have their outer edge grey, and the inner white.

This species is frequent about our sea-coasts in many places; they usually fly in vast companies, and, when they build their nests, are placed in vast numbers close to one another, and they seem to bring up their young almost in common. They lay three, or sometimes four, eggs, and their nests are very carelessly formed of a few pieces of reed, laid irregularly together; and sometimes there is no nest at all, but the eggs are laid on the naked ground. They fly very swiftly, and feed on the wing. They are continually hovering at a small distance over the surface of the water; and, when they see any little fish in reach, they throw themselves precipitately down upon it. They will plunge under water after their prey, and in this situation they often become themselves a prey to some larger fish; if they escape this fate, they are up again in an instant, and very seldom fail to bring up the fish with them.

The length of the wing, and shortness of the feet in this bird, added to it's forked tail, and it's being almost continually seen flying, have obtained it the name of the Sea-swallow with us, and of Hirundo marina among authors. Gesner calls it Sterna; Willughby and Ray, Hirundo marina major, and Hirundo marina Sterna Turneri; our people, in some places, call them Scrayes, and in others Terns.

Sterna supra cana, capite rostroque nigro, pedibus rubris.
The greyish Sterna, with a black head and beak, and red legs.

This is a very beautiful species; the general figure wholly represents that of the former kind, and it's size is about equal, or, if any thing, a little smaller: the head is large, and of a very singular figure; it is rounded at the summit, a little depressed at the sides, and of an acuminated figure toward the anterior extremity: the beak is moderately long, and is considerably convex; it is pointed at the extremity, and is all over black: the nostrils are oblong and large, and are pervious and naked, and of a somewhat paler colour about their edges than the rest.

The

BIRDS
The Dobchick
The Penguin
The Greatest Loon
The Razor Bill
The Guillemot
The Cepphus
The Great Grey Gull

The head is of a fine deep glossy black; the back, shoulders, and rump are of a rusty grey colour: the breast and belly are also grey, but it is paler, and has a bright silvery appearance on the whole under surface of the bird, and particularly about the belly and sides.

The wings are long; the covering feathers of their under surface are white, and so are also the covering feathers of the under part of the tail: the tail is a little forked, but not nearly so deeply as in the former species; it consists of twelve feathers, and the two exterior ones of these are not so remarkably long as in the other: all the long feathers of the wings and tail are of one uniform colour; they are grey on the upper side, and white underneath.

The legs are short and red; the feet are large, and are semi-palmated: the toes are long, and the claws black: the hinder toe is inconsiderable, and the claw of the middle toe is marginated on the inner side.

This is less frequent on our coasts than the former species, but I have seen it in Yorkshire. It is much on the wing, but not so continually as the former; at other times it lies down among the sedge, not seeming very fond of walking. Few of the writers on birds have described it. Linnæus mentions the having met with it near Upsal, and calls it sterna supra grisea capite rostroque nigro pedibus rubris. It is no where very common, so that there is the less wonder that it has been overlooked by many.

Sterna rectricibus maximis nigris.
The Sterna, with the largest of the tail feathers black.

This is a very beautiful species, and in it's general figure more resembles the common sea-swallow, than the last described species; it's bigness is about that of the common black-bird; it's head is large, somewhat depressed on the crown, and a little flatted on the sides; the eyes are large, and their iris is of an orange colour, with a strong admixture of a fiery red: the beak is short; it is thick at the base, moderately convex on the back, and very sharp at the point.

The upper part of the head is of a deep black behind the eyes, but of a paler or greyish colour toward the front: the upper part of the neck, as also the shoulders, back, and sides are brown: the breast and throat are white, and the belly also is whitish, but it has a tinge of brown; the wings are moderately long; they are of a greyish-brown, with a tinge of olive on the upper side, and whitish underneath; the tail is long, and very forked; and the two extream or outer feathers of it, which are much longer than any of the others, are black: the legs are short, and of a dusk colour, with a tinge of greenish diffused over them; the feet are large and semi-palmated; the toes stand three before, and one behind; the hinder one is small and inconsiderable: the claws are all black, and that of the middle toe is serrated and marginated on the inner part.

This is not a native of England, but it is occasionally met with on our coasts. It breeds in Sweden, Denmark, Norway, and many other parts of the North of Europe. Many of the ornithologists have described it, though there are many also that have omitted to mention it. Ray calls it Struntjager, id est, Caprotherus; Bartholine, Troen sive Fur; the Danes call it Kyuffwa sive Tjufva; and the Swedes, Swartloffe and Labben. We have no name for it.

BIRDS,

BIRDS.

Class the Fourth.

SCOLOPACES.

THE characters of the Scolopaces are, that the beak is of a cylindric figure, rounded and obtuse; and the legs are often naked about the middle of the thigh.

ARDEA.

THE beak of the Ardea is very long, and somewhat compressed: the upper and under chaps of it are both of a length, and there runs a furrow from the nostril; the legs are very long: the toes are four, and these also are long and connected.

Ardea vertice papilloso.
The Ardea, with the top of the head papillose. **The Crane.**

This is a very large, stately, and beautiful bird: it's body is not very bulky, however, in proportion to the length of the neck and of the legs; it's general weight is about eleven pounds, and if measured in an extended posture, from the tip of the beak to the extremities of the toes, it is more than five feet in length.

The head is moderately large; it is rounded and prominent on the crown, and somewhat, though but little, compressed at the sides: the eyes are large and piercing; the beak is long, straight, sharp at the point, and somewhat compressed at the sides: it's colour is a blackish-green; the tongue is broad, and is of a firm and, as it were, horny structure at the extremity: the crown of the head is black; this colour extends from the front of the head, at the origin of the beak, to the hinder part, and all this is covered rather with short hairs than feathers: in the hinder part of the head there is a small space of a lunulated figure which is naked, of a scarlet colour, and of a granulated surface; and below this there stand some grey feathers, which form a triangular spot on the neck; at it's top, on the upper or hinder part from each eye, there runs also a fine white line, which continues it's course straight backward; these join the triangular grey spot already mentioned, just below the hinder part of the head, and are continued from thence again, quite down to the breast: the throat and the sides of the neck are of a dusky blackish colour: the back, shoulders, breast, belly, and thighs, and the covering feathers of the wings, excepting only those on the extream joint, are grey.

The wings are very large; the long feathers in each are twenty-four in number, and they are black, only the smaller ones have something of a reddish-brown tinge in the black, as have also the primary covering feathers on the extream joint: the tail is small and short, and is composed of twelve feathers: when it is expanded, it is round; the feathers which compose it are grey, but their tips or extremities are black.

The legs are extreamly long, and they are naked to at least a hand's-breadth above the joint, commonly called the knee: they are black; the toes also are black, and are very remarkably long: the outer toe of each foot is connected to the joint of the middle one, by a very tough and strong membrane.

There is something very singular in the wind-pipe of this bird; it enters the sternum by an aperture, formed for that purpose, and after penetrating deep into it, and making

making some turns in it, it comes out again at the same passage, and then makes it's way to the lungs.

This beautiful and singular bird is a native of Holland, and of many other of the northern parts of Europe, as well as some of the more southern, but it does not breed with us. It comes over to us occasionally, however; we sometimes see a single crane straggling in one or other part of the kingdom, but in the fens of Lincolnshire it is otherwise: I have there met with large flocks of them, and there are people who pretend that they breed there. Though the crane frequents waters, it is not one of those birds which feed on fishes. It's flesh is very delicate, and this, as well as the musculous structure of the stomach, declare for it's feeding on vegetables. It is frequent in the markets of Italy, and some other places; but, in some countries where it is very common, it is a superstition that they should not be killed. All the authors who have written on birds have named it, and all under the same simple name Grus.

Ardea vertice toto nudo papilloso, rostro longiore.
The Ardea, with the whole upper part of the head papillose, and the beak long.

The Indian Crane.

This is a smaller bird than the European crane, otherwise it is very like it: the head is large, and is somewhat compressed at the sides, but it is rounded at the crown: the eyes are large, and their iris is of a beautiful yellow: the beak is considerably longer than in the common crane, but it is less robust; it is short, and pointed at the end: the nostrils are oblong, and very conspicuous, and are of a greyish-colour; the rest of the beak is of a blackish colour, but with a tinge of olive among it.

The head is all over naked on the crown, and the flesh rises into a kind of granulations or papillæ, with a few black, short, and stiff hairs among them: the naked part extends from the very base of the beak to the top of the neck: the sides of the head are blackish, but toward the eyes there is some whiteness; the neck is immoderately long, and is of a deep ash-colour; the throat indeed is almost black: the breast and belly are white, and the back, shoulders, and covering feathers of the wings are grey: the rump only is a little paler than the rest.

The wings are very large; their long feathers are of a deep iron-grey: the covering feathers of the extreme joint are whiter than the rest: the tail is very short and inconsiderable; it is scarce seen beyond the tips of the wings: the legs are black; they are very long, and are naked half way up the thighs: the toes are very robust, but much shorter than those of the common crane.

This is a native of the East and West Indies, but it is no where so common as on the island of Madagascar. It feeds on water plants, and is continually seen wading in shallow brooks, and on the sides of rivers; the common people therefore suppose it feeds on fish, but that is an error. Most of the writers on birds have described it. Willoughby, Ray, and others call it Grus Indica, and some Grus minor Indica.

Ardea capite cristato.
The Ardea, with a crested head.

The Balearic Crane.

This is an extremely singular and beautiful bird; it's body is of the bigness of that of a goose: it's neck and it's legs are immoderately long: the head is moderately large, but the singular arrangement of it's feathers makes it appear much more so than it really is: the eyes are large, bright, and piercing; the beak is robust and thick at the base, but it is shorter than that of the crane; it does not exceed, indeed, three inches in length; it is throughout of a deep olive colour, except that the tip or point is quite black, and the nostrils are grey and pale.

The head is a little flatted at the sides, and rounded, not at all depressed, on the top: on the crown there stands a most singular and elegant tuft of bristles, in form of a crest: they are very numerous, and spread every way so, as to form a round figure in the whole; they very much resemble the bristles of a hog in thickness, and are of a deep brown colour, with a slight tinge of yellowish: on each side of the head there is a spot of white, which terminates in a red line; and under the base of the beak there is a red fleshy appendage, in form of the wattles of the turkey-cock, only smaller.

The whole bird is of a coal-black colour, except that some of the smaller of the long feathers of the wings are white: the wings are very large; and the tail is very short and inconsiderable, and is totally black: the legs are of a deep olive colour, approaching to black, and are naked a great way above the knees.

This singular species is a native of the Cape Verde islands, and some other places; it is frequent there in all the watery places, and has a way of using it's wings to assist it in running, which it does with great rapidity. All the authors who have written on birds have described it. Aldrovand calls it Grus Balearica; and Ray, Willughby, Charleton, and most of the moderns retain the same name. I have met with it once shewn as a curiosity in London, but it is not frequently seen alive in Europe.

Ardea rostro longiore, capite nigricante, corpore albido.
The long-beaked Ardea, with a black head, and a white body.

The Jabiru.

This is a very large and very singular bird, as it stands erect; it is five feet in height; the body is of the bigness of that of a swan: the legs are more than two feet in length, and the neck is long, in proportion: the head is large and rounded; the eyes are large and black; the beak is very long, and considerably broad and thick; it is straight all the way to the extremity, where it turns up a little: it is of a deep black, and of a glossy surface, only that the nostrils, which are oblong, are grey and rough.

The head is black, and the upper part of the neck is also blackish; this part is often naked, and the skin is then seen to be perfectly black: when the feathers are on it, they are but few, but those on the neck are whitish: the lower part of the neck is of a snow-white, and so is also the whole body of the bird, as well the back as the breast and belly: the wings also are perfectly white, and their long feathers are of a brighter and purer white than any other part of the bird.

The legs are long and very robust; they are of a coal-black all the way to the joint, called the knee, and are naked also half way up the thigh, but they are there of a deep olive colour: the toes are very long and robust, and are also black; the tail is very short, and does not appear beyond the tips of the wings, when they are closed.

This species is a native of the Brasils, and of no other part of the world, so far as is yet known. The natives call it Jabiru; the Dutch give it the name of Negro. It lives about waters, and feeds on the water plants, and sometimes, though more rarely, on fish.

Ardea alba, capite cinereo crista ossea ornato.
The white Ardea, with a grey head, and a white bony crest.

The Jabiru Guacu.

This is an extreamly singular bird, much more so indeed than the former species; it is four feet and a half high, as it stands erect; it's body is of the bigness of that of a goose, and the legs are very long, as is also the neck: the head is large and round; the eyes are also large and black, and the apertures of the ears are very wide: the

the beak is six inches long, and is straight all the way, till near the extremity, where it is a little crooked: the nostrils are oblong and large, and the lower part of the beak is hollow: the base of it is of a deep black, and the rest is also blackish, but there runs a tinge of olive through it.

The head and the upper part of the neck are naked; there are no feathers in them, but the skin is squammose, and of a grey colour: the top of the head is ornamented with a crest of a very singular appearance; it is of a bony structure, and of a mixed grey and white colour: the whole body is white; the wings are large, and their long feathers are of a very beautiful deep and glossy black, but with a glow of purple shining through it: the tail is very short and black.

The legs are long and robust; they are naked half way above the joint of the knee, and are of a deep shining black: the toes are long, robust, and black.

This is a native of the Brasils, where it is esteemed very delicate at table. The natives call it Jabiru Guacu, and, in some places, Nhandu Apoa; the Dutch call it Scurvogel. Most of the modern writers on birds have described it after Marcgrave, who gives it under these names.

Ardea cinerea fusco variegata, crista plumosa variegata.
The grey Ardea, variegated with brown, in a feathered crest. **The Cariama.**

This is a beautiful bird, and is extreamly different from all the former; it is about three feet high, as it stands erect; it's body is about equal to that of a well-grown pullet, and the legs and neck are very long, and much less robust than in the several preceding species.

The head is large, and of a rounded figure; neither depressed on the crown, nor flatted at all at the sides: the eyes are large; their iris is yellow, and the pupil black: the beak is short, in comparison with that of the generality of the birds, and is straight all the way, to near the extremity, where it turns up a little: it is of a deep, dusky, yellow colour, and the nostrils are blackish, oblong, large, and very conspicuous.

The neck and body are elegantly coloured; the ground colour is a deep iron grey, but this is variegated with spots of brown, and of a bright yellow, in the manner of the backs of several of our hawk-kind: the head is grey, with a smaller admixture of the brown or yellow; and on the front, just beyond the base of the beak, there stands a large and elegant crest, not of a bony structure, and naked, as in the preceding species, but composed of a great number of elegant feathers; it carries this erect, and it's ground colour is the same grey with that of the body, but it is elegantly and very deeply variegated with black.

The wings are very large; the tail is short; it just appears beyond the tips of the wings, when they are closed: the covering feathers of the wings and tail are of the same variegated hue with those of the body, but the long feathers of each are simply brown, so that they appear both of a deep and unmixed brown at their extremities.

The legs are very long and yellow: the toes are long, but not very robust: the hinder one is small, and stands very high up, at a considerable distance from the rest.

This is a native of the Brasils, where it's flesh is esteemed a great delicacy. It frequents watery places, but it feeds only on vegetables. It's note is loud, and like that of the peacock. Marcgrave has described it under the name of Cariama, and most others have followed him in this.

Ardea capite cornuto.
The Ardea, with a horn on the head. **The Anhima.**

This is another of the Brasilian birds, so very different from all the European, that, if one had not seen the skin of it preserved and sent over in a very good condition to us, it would be very difficult for us to be brought to believe there could be such a creature.

It's body is of the bigness of that of the swan, and it's neck and legs long; but the legs bear no proportion to the neck in length, as in other species: the head is large and rounded; the eyes are large, and very bright and piercing; the iris is yellow, and the pupil black: the beak is long, large, and black; it is straight all the way to the extremity, but there it is bent a little downward: the nostrils are oblong, and very conspicuous.

On the front of the head, just above the base of the beak, there stands a bony protuberance, of an oblong and conic form, turning forwards, and very much resembling a horn: this is white, and very sharp-pointed at the end; it is about an inch and a quarter in length, and of the thickness of a skewer: the head is of a mixt white and black colour; at a distance, the combination of these forms a kind of iron-grey, but, when seen near, they are found to be distinct colours, and the variegations not small: the bottom of the bony protuberance or horn is surrounded with a very elegant series of short and slender feathers; they stand erect, and are, like the rest of the head, variegated with black and white: the sides of the neck, toward the top, are black; part of the throat also is black, but the lower part of the neck, and the whole breast and belly, are variegated with white, black, and grey feathers: the back is blackish; the sides also and the upper part of the wings are black, and the belly, toward it's lower part, is paler than elsewhere, and the covering feathers of the thighs are also whitish.

The wings are very long and large, and their long feathers are of a deep and glossy black, with a tinge of changeable green; the covering feathers are of a deep and perfect black; the tail also is black, but it is moderately long, and shews itself to some extent beyond the tips of the wings, when they are closed.

The legs are robust and long, but not proportionably to the length of the neck, and they are of a deep and dusky brown colour: the toes are long and robust, and the claws are large, black, and sharp.

This is a native of the Brasils; it frequents the lakes and other fresh waters, and feeds on fish. The Brasilians call it Anhima, and Marcgrave has described it under the same name; and all the rest have followed him in this, as well as in all the particulars of the description.

Ardea tota alba capite laevi.
The white Ardea, with a smooth head. **The Garza.**

This is a large and beautiful bird, when it stands erect; it is more than three feet in height, but it's body is not very bulky, the whole bird seldom weighing more than two pounds and a half; the head is tolerably large, and somewhat depressed on the crown, but rounded at the sides; the eyes are large, and very bright and beautiful; the iris is of a pale yellow, and the pupil of a deep black; the beak is long and yellow, and the nostrils are oblong.

The head is naked on the front; the space between the eyes and at the base of the beak is covered with a greenish skin, and the eye-lids also are naked, papillose, and greenish: the whole bird is of a beautiful snow-white; the wings are long and large; the

the tail is short; the long feathers in each wing are twenty-seven; the tail is composed of twelve, and these, as well as those of the wings, are of a very bright and pure white: the legs are long and robust, and the claws are long.

This species is a native of Italy, and many other of the warmest parts of Europe, and there are authors who mention it's being seen in England. Most of the writers on birds have described it. Ray, Willoughby, and others call it Ardea alba major; and the Venetians, who have it very frequent among them, call it Garza.

Ardea crista dependente.
The Ardea, with a hanging crest.

The common Heron.

This is a tall bird, though it's body be not very large; it measures more than four feet from the point of the beak to the tip of the toes, and it's weight is about four pounds: the head is small, and depressed a little on the crown: the eyes are large, and of a very bright and piercing aspect: the beak is long and sharp; it is compressed, and the upper chap is black, but the under one is of a pale flesh colour.

The anterior part of the head is covered with short and soft white feathers; from the hinder part there hangs over the neck a crest of very long feathers, these are black, and the crest, in it's usual situation, hangs over the neck; the throat, just under the base of the beak, is of a snow-white: there are two spots of white near the eyes, which, in the male, blend and confound themselves with the white of the front; in the female they are more distinct: the upper part of the neck, the shoulders, the back, and the covering feathers of the wings are of a dusky bluish-grey; the under part of the neck is white, and is variegated with longitudinal spots of black; the breast and belly are white, and the feathers which cover the thighs yellowish.

The long feathers in each wing are twenty-four, and they are black: the tail is short; it is composed of twelve feathers, and is of a pale greyish colour, with a tinge of blue.

The legs are long and blackish, but the naked part of the thighs is of a light flesh colour: the toes are long and black; and the claw of the middle one, which is longest, is serrated on the outer side: the outer toe is joined to the middle one.

This is very frequent with us, especially in watery places, and wades up to the mid thigh in our brooks and ponds in search of fish, of which it is a great destroyer. All the writers on birds have described it. Gesner calls it Ardea palla sive cinerea; Jonston, Ardea cinerea; and Aldrovand, Ardea cinerea major, which last name has been copied by Ray, Willoughby, and most of the other writers since his time. We call it the Heron and Heron's-haw.

Ardea vertice nigro, pectore pallide maculato.
The Ardea, with the top of the head black, and the breast pale and spotted.

The Bittern.

This is a very singular bird; it is about equal to the common heron in size, but very different in the colouring: the head is moderately large, and somewhat depressed on the crown: the eyes are very large, bright, and piercing: the beak is long and sharp; it is of a convex figure, and a dusky black colour: the nostrils are very conspicuous; they are oblong, and are covered with a kind of scale on the upper side; from these there runs on each side a long furrow to the very point of the beak: the tongue is narrow, acute, and triangular.

The crown of the head is black, and there is also a black spot on each side, at the angle of the mouth: the throat and the sides of the neck are of a reddish-brown, and have frequent variegations of oblong black spots, running in a transverse direction: the neck is covered, on the back part especially, with long and narrow feathers, and the

the longest of these, which are those toward the breast, are black in the middle: the lower part of the belly, and the thighs on the inside, are of a pale, whitish-red, but the feathers, on the outer part of the thighs, are of a deeper colour, and they are spotted with black.

The back and whole upper part of the bird are elegantly variegated with brown, black, and grey, in a beautiful arrangement, and irregular blotches.

The long feathers in the wings are some of them black, from the tip up to a third part of the feather; the rest are irregularly variegated with black and brown, in transverse lines: the smaller feathers of the wings are of a pale reddish-brown: the black spots are larger, between the shoulders, than in any other part, and they run downward, and the reddish-colour from this part degenerates into a kind of yellowish tawny: the legs are long and robust, and they are naked more than half way up the thighs: the toes are very long and thick, and the claws are robust, long, and sharp.

This species is a native of England, and is very common in our fen countries, but more rare elsewhere. We meet with it skulking among the reeds and sedge, and it's usual posture is with the head and neck erect, and the beak pointed directly upwards; it will suffer persons to come very near it sometimes without rising, and has been known to strike at boys, who have got into the closer retirements in it's way up from the cover. It will do the same by the sportsman, when wounded, and unable to make it's escape; it always aims the stroke at an eye, and has sometimes been fatal. It flies principally toward the dusk of the evening, and then rises in a very singular manner by a spiral ascent, till it is quite out of sight. It makes a very odd noise sometimes, as it is among the reeds; and a very different, though sufficiently singular one too, as it is rising on the wing in the night; these different notes have obtained it, among our country people, the two names of the *Butterbump* and the *Night-Raven*; it's genuine and proper English name is the *Bittern*. All the writers on birds have described it. Gesner calls it Ardea stellaris minor; Aldrovand, Ardea stellaris minor sive Ocnus; Willoghby, Ray, and others, Ardea stellaris.

Ardea nigra, pectore abdomineque albis.
The black Ardea, with the breast and belly white. **The black Stork.**

This is an erect and beautiful bird; it is larger than our heron in the body, but it's neck is not quite so long, in proportion; so that, when it stands erect, it is hardly quite so tall as that bird: the head is large and rounded in the general figure, only a little depressed on the summit: the eyes are large and bright; the beak is long, strong, sharp at the point, and all the way of a green colour, only paler on the upper part than below: the tongue is red, and is short, though the length of the beak is not less than eleven inches: the mouth opens to a vast extent, and the throat is enormously wide; the neck is about ten or eleven inches in length, and is moderately thick; it is black, as are also the back shoulders and wings; indeed, the whole upper surface of the body is of this colour, but the breast and belly are white.

The legs are long, and not very robust; they are of a deep green colour, and they are naked more than half way above the knee: the toes are long, robust, and armed with sharp claws.

This is a native of Italy, and most of the warmer parts of Europe, nor is it confined to those; it is seen in Denmark, Norway, and Sweden, but not that I know of in England. All the writers on birds have described it, and almost all of them under the same name, Ciconia nigra.

Ardea alba remigibus nigris.
The white Ardea, with the long wing feathers black.

The common Stork.

This is a leſs beautiful bird than the former, but it is ſomewhat larger, when it ſtands erect; it is between three and four feet high, and it's body is of about the ſize of that of a gooſe: the head is large, round, and ſomewhat depreſſed at the top; the neck is long, but it is conſiderably thicker than that of the heron, and therefore appears ſhorter: the eyes are large and piercing in their aſpect; the beak is long and robuſt, very ſharp at the point, and all the way of a beautiful red: the head is white, as is alſo the whole neck; the ſhoulders, back, and ſides are alſo white, and ſo are the breaſt and belly, but the rump and part of the wings are black.

The tail is white, but the long feathers of the wings are black; and this, whether the wings be cloſed or expanded, gives a pleaſing variegation to the bird: the legs are red, and they are very long, and naked a great way up: the feet are of a deeper red than the legs; the toes are long and robuſt, and the claws long, black, robuſt, and not ſharp, but broad, like the nails of the human fingers: the barrels of the quills, from the wings, are as large as thoſe of the ſwan, and they make the moſt laſting and uſeful pens in the world.

This bird is a native of many parts of Europe, and is preſerved in many places from all inſults and injuries by way of ſuperſtition; it does not breed with us, and is indeed but rarely ſeen here. It frequents the ſea-coaſts, and feeds on fiſh and reptiles: It has a way of making a very odd noiſe, by clapping forcibly the two ſides of the beak together. All the writers on birds have deſcribed it. Geſner and many others call it, ſimply, Ciconia; but Ray, Willughby, and ſome more, by way of diſtinguiſhing it from the ſpecies before deſcribed, call it Ciconia alba, the white Stork.

Ardea nigro et albo variegata.
The variegated, black, and white Ardea.

The Braſilian Stork.

This is of about the ſame ſize with our common heron, but the head is larger, and the neck, though not ſhorter, is conſiderably thicker, and therefore appears to be ſo: the head is of a rounded figure, ſcarce at all depreſſed on the crown; the eyes are large, and their aſpect piercing; their iris is yellow: the beak is near a foot long; it is of a bright red all the way, except at the tip, where it is blackiſh: the opening of the mouth is enormouſly wide, and the tongue of a pale fleſh colour, and very ſmall; the head and neck are white, except that in the male there are a few longitudinal black ſtreaks on the back of it: the feathers at the bottom of the neck form a kind of ruff; they are of a ſnow-white, and are remarkably long and narrow: the back is entirely white to the rump, which is variegated in an elegant manner, with black and white; the covering feathers of the wings are alſo variegated with black and white in large blotches, and the long feathers are entirely black.

The wings are conſiderably long, but the tail is very ſhort; when the wings are cloſed, their tips reach to the extremity of the tail, ſo that in this ſtate there is no tail at all ſeen: there is a fine ſcarlet ſpace about the eyes, where the ſkin is naked; and at the baſe of the beak there is alſo a naked ſpace, covered with the ſame ſcarlet ſkin; this it can ſwell out into a kind of bag, when is it angry: the legs are long and robuſt, they are of a fine red colour: the toes are thick and long, and their claws broad and flatted.

This ſpecies frequents the freſh as well as ſalt waters in South America, and it's fleſh is eſteemed a delicacy there; it has a way of making the ſame odd noiſe that the ſtork does, by the ſnapping together the two parts of the beak. It feeds on reptiles that frequent the waters, but not on fiſh. Marcgrave, in his account of the Braſils, has deſcribed it under the name of Ciconia Americana Maguari Braſilienſibus; and moſt of thoſe who have written ſince, on the ſame ſubject, have taken his name.

Arde

Ardea nivea crista brevi suberecta.
The snow-white Ardea, with a short and somewhat erect crest.

This is an extremely elegant bird, and is one of the least of this genus; when it stands erect, it is not more than a foot and half high, and it's body is not larger than that of a small pullet: the head is small, of an oblong figure, and rounded at the top, not at all compressed, as it is in some species: the eyes are small, and their iris is of pale yellow; their aspect not so bright and piercing as in many others of this genus: the beak is about three inches and a half long, very large at the base, and sharp at the point, and is throughout black.

There is a green space naked about the eyes, and the base of the beak, especially on the under part, has somewhat of a greenish about it: the whole bird is of a beautiful snow-white, and there grows on the hinder part of the head a crest, but it is composed of shorter feathers than that of our common heron, and does not droop so much, but often appears in a suberect posture, especially when the bird is alarmed at any thing.

The wings are moderately large and long, and their principal feathers are of a brighter white than that of any other part of the body: the barrels of the quills in these are very thick, in proportion to the length of the plumage, and are greatly esteemed for making pens: the tail is short; when the wings are closed, it does not appear, for the tips of the wings reach to it's extremity; it is of a fine bright white, but not so lucid as the wings.

The legs are long and slender, and are properly of a deep green colour, but they are often covered with a kind of loose skin of a black colour: the toes are long, and the claw of the middle one is denticulated: the feathers which fall down on each side of the body of this bird from the shoulders are long, narrow, and of a very elegant figure, and snow-white colour; they are sold at a considerable price in Italy, and are even carried to the Turks dominions, for the ornamenting of hats, and other parts of dress.

This elegant species is frequent about Venice, and in some parts of Italy, but it is not seen in the northern parts of Europe. Most of those who have written on birds have described it. Gesner and Aldrovand call it Ardea alba minor sive Garzetta; Bellonius, Aigretta Gallorum; Willughby and Ray call it Ardea alba minor Venetiis, Garza Giovanno, sed falso.

Ardea castanea capite flavo et nigro variegato.
The brown Ardea, with the head variegated with black and yellow. — The Cirris.

This species is extremely different in colour from the generality of the Ardeæ, but it's form is almost exactly the same with that of the common heron; when it stands erect, it is about sixteen inches high; it's body is of the bigness of a small pullet, and it's general colour is brown: the head is small, and a little depressed on the crown: the eyes are small, and their iris is of a pale yellow, approaching to that we express by the term lemon colour: the beak is very long and robust; it is toward the base of a bluish-green, but toward the point of a deep black; one half of the beak is of one these colours, and the other half of the other.

The ground colour of the head is a bright and elegant chesnut-brown, but it is variegated with black, and with gold-yellow; these colours run in a mixed manner down the back part of the head, and the top of the neck, and form the appearance of a crest: the back, shoulders, and rump, and also the upper part of the wings, are all of a deep but very glossy brown: the breast and belly are of the same colour, only paler, fpaler, and there is fome tinge of yellowifh in it: the whole bird, however, at a fmall diftance, appears of one uniform brown.

The neck is fhorter in this fpecies than in any other of the genus, but the legs are long; they are of a bright red colour, and are naked about the joint, commonly called the knee: the toes are very long, and the claws large and broad.

This fingular fpecies is a native of Italy, and of fome parts of the Eaft; many of the writers on thefe fubjects have defcribed it, though fome have omitted it. Aldrovand calls it Ardea Hæmatopus, five Ciris Virgilii Scaligero; others have called it Ardea Hæmatopus.

Ardea fulva capite et collo nigro, flavo, et albovariegatis.
The yellowifh Ardea, with the head and neck variegated with black, white, and yellow.

The Squacco.

This, when it ftands erect, is about fourteen inches in height; the body is moderately large, but the neck is fhorter than in moft others of this genus: the head is large and round; the eyes are large, and their iris is of a gold yellow: the beak is not more than three inches in length, but it is very robuft; it's colour throughout is blackifh, but with a mixture of a ferruginous brown.

The neck is very thick, and this, as well as the head, is very elegantly variegated with oblong and irregular blotches of black, white, and yellow: the back and fhoulders, as alfo the rump, are of a deep ferruginous brown, with fome obfcure variegations of blackifh and yellowifh, but no white: the wings are in great part white, but there is fome variegation of a yellowifh-brown in them: the breaft and belly are perfectly white, and fo is the tail, which is fhort, and does not appear beyond the tips of the wings, when they are clofed.

The legs are long and yet robuft; they are yellow, and are naked a good way above the knee: the toes are long, and the claws flatted.

This is a native of the coafts of the Levant in many places. Aldrovand is the firft author who has defcribed it, and from him moft of the others have borrowed their accounts of it. He calls it Ardea quam *Squacco* vocant; and Ray and Willughby have continued the name.

Ardea ferruginea, crifta capitis nigro et albo variegata.
The brown Ardea, with a black and white creft.

The Squaiotta.

This is about eighteen inches high, when it ftands erect: the head is fmall, and appears the more fo, as the neck is thick: the eyes are moderately large; their iris is of a deep orange colour, with an admixture of a fiery red: the beak is between four and five inches in length, of a yellowifh colour toward the bafe, and black at the extremity: the tongue is fmall and red.

The head in general is of a deep ferruginous brown, but it is ornamented on the hinder part with a moderately long creft of a mixed brown, black, and white colour; this hangs down upon the neck, and is a very great ornament to the bird: the neck is of a deep ferruginous colour: the back and the upper part of the wings are alfo of the fame colour, but with a greater admixture of reddifh: the long feathers of the wings, are black; the tail alfo is black, and is a little longer than the tips of the wings, when they are clofed. In the males, the middle feathers of the creft are of a pure white, and the outer ones black; but in the females the variegation is irregular, and there is an admixture of brown, like that of the body: the legs are long, robuft,

 and

and green; they are naked but a little way above the knee, and the toes are long, and armed with red claws.

This is a native of the East, and of some parts of Europe. Aldrovand calls it Ardea vulgo Squaiotta dicta, and most of the writers since have used the same name.

Ardea cinerea pectore et lateribus rufis.
The grey Ardea, with the breast and sides of a reddish-brown.

This is a singular and a beautiful bird; it is of the size of the common heron, but it's neck is not quite so long as in that species: the head is large and rounded; the beak is long, and very robust; it is of a very deep olive, or a mixt tinge of greenish and black: the eyes are very large, and their iris is of a deep orange colour: the crown of the head is of a deep black colour, and from the hinder part of it there hangs a considerably long crest of the same black, but sometimes toward the edges a little variegated with a deep iron grey: the neck is of a very deep grey on the upper or outer side, and of a much paler at the throat.

The back is of a bright ashy grey, but toward the rump it it is much darker than elsewhere; and in the male bird it is all over very dark, in comparison of what it is in the female: the shoulders give origin to a considerable number of long and slender feathers, which fall down to the sides, and are reddish, or of a ferruginous brown: the breast also is of the same reddish-brown, and the belly has the same tint but paler; the covering feathers also, of the under part of the wings, are of this reddish-brown.

The thighs are of a whitish brown, but still with some tinge of the ferruginous or reddish among it: the legs are robust and long, and are of the same dark olive colour with the beak.

We have this species in our marshy places. Aldrovand has described it under the name of Ardea stellaris major five rubra; and most authors, who have written since, have copied the name, though it very ill expresses the colouring of the bird.

Ardea ferruginea nigro variegata.
The ferruginous Ardea, variegated with black. **The Soco.**

This is nearly of the size of our common heron, but the body is smaller, in proportion to the length of the neck and the legs: the head is large, and of an oblong figure, narrowest at the front, and broader behind: the beak is five inches long, and of a greenish-olive colour; the eyes are large, and their iris yellow, and they have a very piercing aspect.

The head is of a very deep brown, approaching to black, and has no crest, and very little of the black variegation: the whole body is of a bright ferruginous brown, spotted in a beautiful manner with black, in small variegations; the back is darker, and has more of these; the belly is paler, and has fewer.

The wings are large, and their long feathers are edged with black: the covering ones are of the same brown with the body, and spotted in the same manner: the tail is short, and does not reach beyond the tips of the wings, when they are closed: there is a little whiteness toward it's base, and at the sides of the belly; otherwise the bird is throughout of the same brown.

The legs are long, robust, and of a blackish-olive colour: the toes are long, and the claw of the middle one is serrated.

This species is a native of the Brasils and South America. Marcgrave calls it Soco Brasiliensibus.

Ardea cinerea nigro variegata, crista erecta variegata.
The black and grey Ardea, with an erect, variegated crest.

The Cocoi.

This is one of the stateliest and most elegant of the whole genus; when it stands erect, it is three feet and a half from the feet to the top of the crest, and it walks in a remarkably erect manner, with the neck elevated to it's whole length: the head is large and round; the eyes are of a very beautiful gold yellow in the iris; the beak is five inches long, and is very robust, remarkably thick and large at the base, and sharp at the extremity; and is of a yellowish-green at the base, but darker toward the point.

The head is of a mixed black and grey; and the crest, which is large and very beautiful, and naturally stands erect, not drooping over the neck, as in our common heron, is of the same silvery grey, variegated in the most elegant manner with black: the neck is white, but it is variegated, on the anterior part, with some spots of black: the back is also of a beautiful grey, variegated in the same manner, but less elegantly; from the shoulders there hang down on each side a number of long, slender, and very elegant feathers, of a snow-white: the breast and belly are white.

The legs are long; the upper half of the thighs is covered with whitish feathers: the wings are long, and of a mixed black and grey, and the tail is simply grey, and is so short, that it is not seen beyond the tips of the wings, when they are closed.

This is a native of the Brasils, and many parts of South America. It frequents watery places, and it's flesh is esteemed pleasant, but it tastes coarse to an European palate. The natives call it Cocoi, and Marcgrave describes it under that name.

Ardea fusco-flava, nigro variegata, pectore albescente variegata.
The yellowish-brown Ardea, with black variegations, and a whitish-spotted breast.

This is, when it stands erect, at least three feet in height: the head is large, rounded, and very well covered with feathers, but it has no crest: the eyes are large, and their aspect piercing: the iris is of a gold yellow, the pupil black: the beak is five inches and a half long; the base is very thick and yellowish, the rest is brown; and, what is singular in a bird of this genus, it is serrated half way down on each side of either chap; the rest is smooth, and sharpened at the edge.

The head is of a dusky brown, variegated with black; the neck is also of the same colour on the outer part, but whitish at the throat: the back is of a yellowish-brown, elegantly spotted with black, and the breast and belly are of a whitish colour, variegated with yellowish: the tail is short, and does not reach beyond the tips of the wings, when they are closed: the legs are long, robust, and of a deep olive colour, but with a tinge of yellow.

This is a native of the Brasils, but is not common, nor has any name among the natives. Marcgrave, who found it there, has described it under the name of Ardea Brasiliensis rostro serrato stellari similis.

Ardea tota nivea rostro flavicante.
The snow-white Ardea, with a yellow beak.

The Guiratinga.

This is a very elegant bird, and perfectly resembles the heron-kind, though it is much smaller than any other known species: the head is oblong, rounded, and large, in proportion to the bulk of the bird; it's body is not larger than that of a pullet, but it's neck and legs are very long: the eyes are small, but very bright and piercing in their aspect; the iris is yellow, the pupil black: the beak is three inches and a half long.

long, and is throughout of a yellow colour; but brighter toward the base, and darker at the point.

The whole bird is of a beautiful snow-white, but the feathers are particularly elegant toward the base of the neck; they are there as soft and fine as those of the ostrich.

The legs are long and yellow; the toes are long, the claws all long and robust, and that of the middle toe serrated: the wings are long; the tail is short; when they are closed, it is not seen beyond their tips.

This elegant little bird, for such it is in comparison of the rest of the Ardea, is a native of the Brasils, and of several parts of South America. The Brasilians call it Guiratinga; the Portuguese there, Garsa; Marcgrave commemorates both these names; and those who have described it since, have copied them from him.

Ardea dorso nigro, pectore et ventre variegatis.
The Ardea, with a black back, and variegated breast and belly. — The dwarf Heron.

This is a bird so much out of the common size of it's genus, that, unless one had seen it, it would be hard to credit the writers who first described it. It is a compleat heron, with the body not larger than that of a common wild pigeon: the head is small and rounded; the eyes are small, but very bright and piercing, and their iris yellow: the beak is perfectly of the shape of that of the common heron, and is about an inch and three quarters in length; of a brownish colour on the upper part, and yellowish below.

The head is of a very dark and glossy hue, a colour and lustre much resembling that of polished iron: the neck is very slender, and is about five inches in length; it is of a whitish colour, variegated with brown and black, and the breast and belly are of the same colour and variegations: the back is perfectly black; the tail is short, and does not reach beyond the tips of the wings, when they are expanded: the wings are greyish, but so dark, that they appear at first sight absolutely black, and the long feathers in them entirely and perfectly so.

The legs are slender, and are about four inches and a half in length: their colour is of a brownish-yellow: the toes are long, and the claws oblong and sharp.

It is a native of the Brasils, but whether it be that it is taken for the young of some other species, or for whatever other reason, the natives have no name for it. Marcgrave, who found it there, describes it under the name of Ardeola Brasiliensis.

RECURVIROSTRA.

THE beak of the Recurvirostra is of a depressed or flatted figure, and is pointed at the extremity, and bent or turned upwards.

Recurvirostra albo nigroque varia.
The variegated, black and white Recurvirostra. — The Avosetta.

This is a most extreamly singular bird; it's size is about that of our common lapwing, or a little larger; the weight of a full-grown one is about ten ounces: the head is large and rounded; the eyes are small, but their aspect is bright and piercing: the beak is of a very strange form; it is very long, and in colour black throughout; it is narrow and flatted, or depressed, and appears to be of a coriaceous substance, rather than of that firm and solid matter, of which the beaks of other birds are formed;

it

it is turned up in a strange manner toward the extremity, bending into a part of a circle, and is extremely sharp at the point, and is there absolutely membranaceous, though it is no where rigid: the nostrils are oblong, and are pervious to the light: the tongue is short, and not bifid: the length of the beak, in the whole, is at least equal to three times that of the head.

The colour of the head is black, only the hinder part is paler or greyish: the upper part of the neck is also greyish; the whole body is white, except that the rump, the sides of the back, and part of the wings are black: the long feathers of the wings have some white in them, and this becomes gradually more and more conspicuous, as they shorten inward of the wing, till some of the shortest become properly white, only with a spot of black at the extremity: some of the covering feathers are totally white, and some are spotted with black: the wings are moderately long; the tail is very short and white.

The legs are moderately long, and of a bluish colour; the toes are oblong, and the feet are palmated; the thighs are naked half the way up.

This is a native of many parts of Europe; it is extremely common at Venice, and we have it in sufficient plenty about the western coasts of England. All the writers on birds have described it. Gesner calls it Avosetta, Recurvirostra; Aldrovand, Avosetta Italis dicta; Willughby and others, Recurvirostra, Avosetta Italorum.

Recurvirostra pectore croceo. **The Scotch Woodcock.**
The Recurvirostra, with a yellow breast.

This is of the size of the common pigeon: it's head is small and perfectly round; the eyes are moderately large, and their iris is of a deep yellow; the beak is long, slender, depressed, and pointed at the extremity, and toward that part it is a little turned up, but not at all like the other kind; the turning up in this, indeed, is very inconsiderable, but the form and structure of the beak is otherwise entirely the same.

The head is of a dark iron-grey; the upper part of the neck is also of the same colour: the whole back, shoulders, and upper part of the wings are also grey, but not so deep as the head and neck; and the breast, belly, and thighs of a strong yellow.

The tail is short, and scarce appears beyond the tips of the wings; when closed, it is of a paler grey than the back, but the rump is darker: the legs are moderately long, slender, and black, or of a very deep and dusky brown.

This is a native of England, though none of our writers on birds have chanced to meet with it. I shot one last year in Lincolnshire, and was told by the common people they bred there. Rudbeck has called it Scolopax rostro recurvo, pectore rufescente, pedibus nigris; and Linnæus, Recurvirostra pectore croceo. These are the only writers who have named it; and the latter of them doubts, whether it properly belongs to this genus; but it certainly does.

HÆMATOPUS.

THE beak of the Hæmatopus is of a compressed form, and terminates in a cuneiform or wedge-like figure; both the upper and the under chaps are equal in length.

Of this genus there is only one known species.

HAMATOPUS. **The Sea Pye.**

This is of the bigness of the common magpye; it's length, from the tip of the beak to the extremity of the tail, is about thirteen inches: the head is small and rounded; the eyes are large and bright; their iris is of a bright scarlet colour, as are also the edges of the eye-lids: the beak is more than an inch in length; it is short, compressed, and slender; the whole is of a bright and beautiful red colour; the nostrils are oblong and pervious: the head and neck are of a deep black; the back also, to it's middle, is black, and so are the wings on the upper part: the breast, the belly, and the hinder part of the back are of a bright white; it is this mixture of black and white, together with it's size, that has given it the name of the sea Magpye: the covering feathers of the tail, both above and below, are altogether white; the long feathers of it are white, but variegated with a great deal of black.

The wings are long and large, and the principal feathers are black, but not without some variegation of white in some part of them: the legs are long; the thighs are naked more than half the way up; the whole naked part of the leg is of a fine bright scarlet: the feet are moderately large; the toes long and red, but they are only three, there being no hinder toe.

This is a native of our sea-coasts, and in the western parts of England is very common. All the writers on birds have described it. Bellonius calls it Hæmatopus; Gesner, Himantopus; Bartholine, Pica marina; and most of the authors, who have written since, have adopted one or other of these names. Some have taken two, or all three of them.

ORTYGOMETRA.

THE beak of the Ortygometra is shorter than the toes; it is of a compressed figure, and terminates in a kind of point; both the chaps are equal in length.

Ortygometra alis fusco-ferrugineis.
The Ortygometra, with ferrugineous-brown wings.

This is of the size of our magpye, and is a singular and elegant bird: the head is small and oblong; the eyes are large, and their iris is reddish, the pupil black; the beak is short, and fashioned in some degree like that of the common hen: the colour of it is a dusky bluish or livid grey, and it is furrowed on each side toward the base; the nostrils are oblong and pervious; the opening of the mouth is wide; the swallow is large, and the tongue undivided.

The head, neck, back, and tail are of a bright and elegant brown, variegated in a very beautiful manner with spots of black: the real state of the variegation is, that the middle of every feather is black, and it's sides and end brown: the long feathers of the wings, as also those of the tail, are of a ferrugineous-brown both above and below: the throat is of a pale whitish-grey, and there is a spot of the same colour on each side of the head, behind the eyes: the breast is grey; the belly and sides are brown, variegated with spots of a dusky or brown, and white; the thighs are naked half way up; the upper part is covered with feathers of a mixed grey and brown; the variegations are undulated, and very beautiful: the legs are moderately long, and of a livid or bluish colour: the toes are robust, and moderately long.

It is not uncommon with us, and is a native also of most other parts of Europe. It frequents rich pastures, generally hiding itself among the tall grass, and is not easily raised. It feeds on worms and other insects, and makes an odd creaking sound with it's throat, that often discovers it. All the writers on birds have described it. Most of them under the names Ortygometra and Ortygometra alarum; Turner calls it Crex Aristotelis.

NUMENIUS.

NUMENIUS.

THE beak of the Numenius is of a figure approaching to cylindric; it is obtuse at the point, and is longer than the toes: the feet have each four toes connected together.

Numenius rostro arcuato, alis nigris maculis niveis.
The Numenius, with an arcuated beak, and black wings with white spots.

The Curlew.

This is a moderately large bird; the weight of a female, tolerably fleshy one, is near two pounds: the head is small and rounded; the eyes are large and bright; the beak is of a very singular figure; it is long, slender, bent or arched, and black: the extremity is obtuse; the nostrils are oblong; the tongue is very short.

The head, neck, and back are of a variegated mixture of black-grey and brown: the middle of each feather is black, and the edges are grey and brown: the throat and breast are variegated with black, white, and a reddish-brown: the middle of the feathers is always black, but the edges of those on the throat are of a pale reddish-brown, and of those of the breast white: the belly is of a pure white; the rump also is white, without any variegation: the wings are long, and are variegated with black and white; the covering feathers being many of them wholly white, and the principal of the long feathers black.

It is frequent in watery places with us, and in most other parts of Europe, and is every-where esteemed a delicacy at table. Most of the writers on birds have described it. Aldrovand calls it, simply, Numenius; Gesner, Numenius sive Arquata; Ray, Numenius sive Arquata major; and some, simply, Arquata. We call it the Curlew, or Curlieu.

Numenius rostro arcuato, dorso maculis fuscis rhomboidalibus variegato.
The arcuated-beaked Numenius, with rhomboidal brown spots on the back.

This is of the bigness of a tame pigeon, but a bird of a very different form: the head is small and rounded; the eyes are small, but of a very piercing aspect: the beak is black, and it is slender and arcuated in a very remarkable manner.

The head, neck, and breast are of a pale brown, marked with longitudinal spots of a darker colour, the lower of which are sinuated as it were: the back is marked with larger spots of a very deep brown, and of a rhomboidal figure: the belly is white; the thighs are covered with pale brown, variegated feathers, and behind them there are some regular spots: the legs are long, slender, and of a bluish colour; the toes are long and connected.

The wings are long, and spread to a great extent, in proportion to the size and genus of the bird: the long feathers of them are black; they are twenty-six in number, and have some variegations of white: the large feathers of the tail are fourteen; they have eight fasciated variegations of a paler and a deeper brown.

This is a native of England, but it is not common; I have only met with it in the northern counties, but I have two or three times shot it there, and the country people are well acquainted with it. They call it a Curlew, but are very well apprized of the difference between it, and the bird commonly called so. Linnæus, in his account of the animals of Sweden, has named it after Rudbeck, but no other author mentions it.

Numenius

Numenius capite utrinque linea nigra notato.
The Numenius, with a black line on each fide the head.

The Woodcock.

The woodcock is a very beautiful, as well as delicate, bird; it is fomewhat fmaller than the partridge: the head is moderately large and rounded; the beak is long and flender, of a cylindric figure, and obtufe at the point: it is brown toward the extremity, but paler, and often flefh-coloured, toward the head: the upper chop is a little longer than the under, but the difference is inconfiderable: the ears are large and open; the eyes ftand higher up the head than in any other known bird; this feems a provifion of nature, that they may not be injured, while the bird is plunging it's beak, as it frequently does, up to the bafe, in the foft and wet ground.

The head has on each fide no elegant and regular line of black: the upper part of the body is of a mixed colour, mottled with black, grey, and a reddifh-brown, in a very beautiful variegation: the front of the head is of a greyifh-brown; the breaft and belly are of a pale grey, with little tranfverfe lines of a bright brown: the upper part of the throat is of a whitifh-yellow, and the under furface of the tail is alfo fomewhat yellowifh: the hinder part of the head is principally black, but there are two or three tranfverfe lines of brown on it.

The long feathers of the wings are twenty-three in each; they are black, but have fome tranfverfe lines of brown, and the feathers under the wings are beautifully variegated with grey and brown: the tail is fhort, and confifts of twelve feathers, the tips of which are grey on the upper fide, and white underneath, and their edges are marked with a dentated line of brownifh, the reft in general black.

The legs are of a pale brown, and the claws are black; the hinder toe is very fmall: the male is fomewhat darker than the female in this fpecies, in it's general colouring.

We have the woodcock in confiderable plenty in our woods, efpecially where the ground is damp, the whole winter feafon; early in fpring they leave us. They fly away in pairs, in order to build; but we fometimes have feen them left here, and they have built and brought up their young. All the writers on birds have mentioned this fpecies. Gefner calls it Rufticola five Perdix ruftica major; Aldrovand, Scolopax five Perdix ruftica; Willughby, Scolopax; others, Gallinago major; the general name among the prefent writers is Scolopax.

Numenius uropygio albo pedibus virefcentibus.
The green-legged Numenius, with a white rump.

The great Plover.

This, though honoured with the name of the great Plover, is not a very large bird: the head is round and fmall; the eyes are fmall, but of a piercing afpect: the ears are large and patulous: the beak is more than two inches in length; it is flender, and of a cylindric figure, and is black, except toward the bafe, where it is red: the head, neck, fhoulders, and back are elegantly variegated with greyifh, brownifh, and whitifh, but the whole mixture feigns what we call a clouded grey: the edges of the feathers, which cover the top of the head, are white; but the middle part of the fame feathers is black as jet: the rump is of a fnow-white, and the whole breaft and belly are alfo of that elegant colour.

The wings are moderately large; their long feathers are twenty in each, and they are all brown; the five exterior ones are darker than the reft, and many of them have little fpots of white irregularly fcattered over them: the tail is about an inch and three quarters in length, and is very beautiful; it is compofed of twelve feathers, and is marked with alternate and undulated fafciæ, or tranfverfe bands of white and brown.

The

The legs are extreamly long, and very slender; they are naked for more than an inch above the knee, and their colour is greenish, but with a tinge of blue among it: the toes are long and slender, and the claws are black: the hinder toe is short and inconsiderable, and the outer one is connected by a membrane to the middle one a great way down.

This is a native of England, and of most other parts of Europe; and most of the authors, who have written on birds, have described it. Gesner calls it Glottis and Limosa Veterum; Aldrovand and Willughby, Pluvialis major; Sloane, Glottis sive Pluvialis major.

Numenius capite quatuor lineis fuscis notato.
The Numenius, with four brown streaks on the head. **The Snipe.**

This is a small but a beautiful and a delicate bird; it's weight is about four ounces; the male is somewhat smaller, and our sportsmen distinguish it under the name of the Jack Snipe, the Gid, and the Jaddock; nor are there wanting authors who have described it as a different species.

The head is small and round, and is elegantly variegated by the four longitudinal streaks, which run above and below the eyes, from the front to the back part of it: between the eyes and the rostrum there is a space of brown just under the beak; it is white.

The neck is variegated with brown, and a ferrugineous-reddish; the breast and belly are whitish: the feathers which fall on each side from the shoulders are very long; they reach almost to the insertion of the tail, and are very beautiful; their middle is black, and their tips and edges of a bright and glossy ferrugineous-brown: the feathers which grow on the back are brown, variegated with white, disposed in transverse lines: those which cover the base of the tail are brown, and have transverse lines of black.

The wings are not very large; the long feathers are twenty-four in each, and are brown, with more or less white on the larger: the tail is short, and it appears indeed shorter than it really is, being hid under the feathers which fall over it at the origin: it is composed of twelve feathers, variegated with a paler and a deeper brown in alternate bands, and with some white toward the tips of some of them: the beak is long and slender; the legs are greenish.

It is a native of most parts of Europe, but in many is a bird of passage only; it is, in general, such with us; but, as the woodcock will sometimes stay and breed with us, this does so much oftener. It is fond of wet places, and it's flesh is delicate, and much esteemed at table. All the writers on birds have described it. Gesner calls it Gallinago sive Rusticola minor; Aldrovand, Scolopax sive Gallinago minor; Willughby, Ray, and others, simply, Gallinago minor. When it breeds with us, it builds on the ground by the sides of brooks, and lays four or five eggs.

Numenius uropygio albo, rectricibus nigris basi albis.
The Numenius, with a white rump, and the tail-feathers black, but white at the base.

This is of the bigness of a common pigeon: the head is small, oblong, and depressed on the crown: the eyes are small, but their aspect is bright and piercing; the beak is very long, slender, and pointed obtusely; it's colour is black, and the nostrils are very conspicuous on it.

The head, neck, and breast are of a yellowish colour, but variegated in an elegant manner with transverse lines of brown: the belly is white, but it is variegated also in

 an

an elegant manner, with spots of a yellowish-grey: the back is grey, with spots of a deeper and a paler brown; the sides are grey, and are not spotted at all.

The wings are moderately large: the long feathers are black, variegated more or less with white, but none of them very much so: the tail feathers are of a fine, deep, shining black, only that they are white at the base, and the covering feathers, as well as the rump, are also white: the legs are long and slender, they are black; the toes are four, they are long and divided; the claws are robust.

This is a native of the northern parts of Europe; we have it not in England. Only Linnæus has described it.

TRINGA.

THE beak of the Tringa is of a cylindric figure, obtuse at the extremity, and in length about equal to the toes: the feet have each four toes, and they are connected.

Tringa facie granulata, rostro pedibusque rubris.
The Tringa, with a granulated face, and with a red beak and legs.

The Ruff.

This is about the bigness of the common Jackdaw: the head is moderately large, round, and well covered with feathers, except at the anterior part, where it is naked, but the skin is elegantly granulated with small red tubercles disposed regularly and closely over it: the eyes are large; their aspect bright and piercing, and their iris of a bright hazel colour: the beak is moderately long, and obtuse at the end; it is of a fine bright red at the base, sometimes all over, but usually it is black toward the extremity: the upper chap is a little longer than the under, and the tongue is extended to the very tip of the beak.

The male has such a variety of colouring, that it is scarce possible to give any regular description of it; when they are first arrived at their full growth, there are no two of them alike, but, after the first moulting or casting of their feathers, they become much more uniform.

The head is of a brownish-grey, spotted with black; the neck is grey; the back and the long feathers which grow from the shoulders, and fall down the sides, are also grey, and are variegated with spots and clouds of black and of white: the throat is variegated with white and grey; the breast and belly are altogether white.

The wings are moderately large; the principal feathers of them are, in general, black, but some of the inner ones are variegated with a little white: the covering feathers of the wings are variegated with grey, white, and black: the tail is about an inch an half in length, and is brown, but the extremities of the feathers are whitish. Thus far the males and females are alike, except that the females are somewhat smaller; but the male is distinguished by a kind of collar of long and narrow feathers round his neck; this in some is white, in some black, in some yellow, and in others grey, and sometimes is of a fine shining bluish-black, and is then most elegant of all.

The females of this species are as peaceable as those of other birds, but the males are eternally fighting; they are fatted up in places closed every way from the light, and, while kept thus in the dark, are peaceable enough; but, the moment the light is let in, they all get to fighting, and the battle never ends till one of them is dead. They are frequent in Lincolnshire, and some other of our fen countries; they fly in companies, and are as quarelsome in their wild state, as when kept for fatting. The sportsmen take the advantage of their fighting to draw their nets over them: they are a very delicious food, and are greatly esteemed. It is observed, that, after the breeding time, the

the males are much more numerous than the females, but that they fight till the additional number is destroyed. Most of the writers on birds have described this species, and all under the same name of Avis pugnax, the fighting bird. We call the male the Ruff, and the female the Reeve.

Tringa rostri apice punctato, dorso fusco viridi.
The Tringa, with the tip of the beak punctated, and the back greenish.

The Cinclus, or Tringa.

This is of the bigness of the common blackbird, or somewhat more: the head is round, and somewhat depressed on the crown; the eyes are large, very bright and piercing in their aspect, and their iris is of a bright hazel: the beak is almost an inch in length; it is straight, slender, compressed a little at the sides, and of a greenish colour toward the base, and black and punctated at the point: the upper chap is a little longer than the under, and the tongue is acute, and not bifid.

The colour of the whole back and covering feathers of the wings is an elegant shining green, with a tinge of brown; but the feathers about the shoulders, and some of those of the wings, are very elegantly dotted, as it were, with white toward their edges: the crown of the head has also a great number of these spots so disposed, that they form a kind of streaks: the part about the eyes is white; the throat is white, but there are some spots of brown on it: the feathers of the middle of the back are darker than the others in their middle, but white at the edges: those which cover the base of the tail are of a snow-white; the breast and belly are also throughout of a snow-white.

The wings are long, and their principal feathers in general are brown: the tail is short; it is composed of twelve feathers; these are all nearly of a length: the edges of some of them, and a yet greater part of many of the others, is white: the legs are moderately long and greenish.

This is a native of most of the northern parts of Europe, and usually flies singly, except in the breeding time, when they are always seen in pairs. Most of the writers on birds have described it. Bellonius calls it Cinclus; Gesner, Gallinae genus quod ignoto nomine Somethoombe nominoo; and, in other places, Rhodopus and Ochropus medius; Aldrovand calls it Gallina Rhodopus sive Phoenicopus and Tringa; Willoughby and others, simply, Tringa.

Tringa rostro levi, corpore cinereo, lituris nigris, subtus albo.
The grey Tringa, spotted with black, with a smooth beak, and a white belly.

The lesser Tringa.

This is a little but a very pretty bird; it's weight is no more than two ounces: the head is small, rounded, and a little depressed on the crown; the ears are very wide and patulous: the eyes are large and bright, their iris is of a beautiful hazel: the beak is moderately long, and is of a whitish colour toward the base, and black and smooth at the tip.

The head is of a pale brownish colour, with a tinge of green; the crown especially is variegated with many longitudinal lines of a darker colour: the neck is grey; the back, the shoulders, and the covering feathers of the wings are all of an elegant brownish-green, a very bright colour, with a fine silky gloss, and all variegated with short, transverse lines, of a darker colour: there runs a line of white on each side of the head, just over the eyes: the breast and belly are throughout perfectly white, but the throat is of a somewhat brownish-white, with the scapi of the feathers darker than the rest.

The wings are moderately long, and their principal feathers are of the colour of those of the back, but more or less variegated with white: the legs are moderately long,

long, and are slender, and of a pale olive colour; the toes are long, and the claws black, and not very short.

It is a native of most parts of the North of Europe; we have it in England, but it is not very frequent with us. It flies singly, except at the time of breeding, when they are usually seen in pairs. Most of the writers on birds have described it. Gesner calls it Gallinula aquatica hypoleucos, and, in another place, he makes it a Motacilla; Aldrovand calls it Gallinula hypoleucos; Willughby and Ray, Tringa minor; the Indians call it Becassine; and our people in Yorkshire, and some other places, the Sandpiper.

Tringa crista dependente, pectore nigro.
The black-breasted Tringa, with a hanging crest.

The Lapwing.

Were this species less common among us, it would be greatly esteemed for it's beauty; it is about the size of the common pigeon: the head is small, but very beautiful; a little depressed on the crown, but not at all at the sides: the eyes are bright and piercing, but they are not large; the beak is moderately long, and the nostrils are very conspicuous in it.

The head is very elegantly variegated; the sides of it are white, but there runs a black line along them by the eyes to the very hinder part of the head, and the hinder part of it is ornamented with a beautiful crest, composed of twenty long and slender feathers, and hanging over the hinder part of the neck: the whole throat, to the breast, is of a coal-black; the lower part of the breast, and the whole belly, are white: the back, and the long feathers of the shoulders, which fall down along the sides, are all of a beautiful shining green; and on each side, near the wings, there is an elegant spot of purple: the legs are long, slender, and red, sometimes brown; the hinder toe is very inconsiderable, the others are moderately long; the outer toe of each foot is connected by a membrane to the middle one, for a considerable way.

This is very frequent in our fen countries, and the wet places in most other parts of Europe. All the writers on birds have described it. Bellonius calls it Capra, Capella, and Vanellus; Gesner, Vanellus; Aldrovand, Capella sive Vanellus; and most others have continued those names.

Tringa rostro nigro, basi rubra, pedibus coccineis.
The red-legged Tringa, with a black beak red at the base.

The Totanus, or Godwit.

This is about the size of the fieldfare, and is esteemed a delicate bird at our tables: the head is small and flatted on the crown: the eyes are remarkably small, but they have a very piercing aspect; the ears are open and large, and the beak is moderately long, smooth, thickest at the base, and smaller all the way to the extremity; smooth all over, and red toward it's insertion at the head, but black every-where else.

The head is of a deep iron grey, spotted irregularly with black; the back, shoulders, sides, and upper part of the wings, are also of the same colour, and variegated in the same manner: the breast is white, but variegated with spots of black; the belly is of a pure white, without any variegation: the legs are moderately long; they are slender, and of a bright red: the toes are long, slender, and red, and the claws black.

We have this in England, and it is common also to most other parts of Europe. The authors who have written on birds have all described it, and almost universally, under the same name Totanus.

The Stork
The Heron
The Little Heron
The Black Stork
The Recurvirostra
The Crane
The Balearic Crane
The Bittern
The Woodcock
The Snipe
The Curlew
The Ruff
The Lapwing
NAPOLI

Tringa cinerea alis albo variegatis.
The grey Tringa, with the wings spotted with white. **Canuti avis.**

This is of about the size of the starling, or somewhat less: the head is small, oval, flatted, and compressed at the sides: the eyes are small, their iris hazel; the ears are patulous; the beak moderately long, and of a dusky colour; the tongue has eight denticulations near it's base, and there are also a number of others in the opening of the mouth, all pointing inwards: the head, the neck, and the back are of a very deep, dusky, and mixed colour, undulated with black, brown, and grey: the whole breast, belly, and throat are white, without any variegation: the wings are large; their long feathers are twenty in each, and they are of the colour of the back, but variegated elegantly with spots and lines of white.

The tail is moderately long, and is variegated with black and white.

We have this in the fens in the Isle of Ely, and it is common to many other parts of Europe, yet it has been little known to the writers on birds. It lives about waters, and runs very swiftly, usually moving it's tail all the time, in the manner of the wagtail.

Tringa remigibus fuscis prima rachi nivea.
The Tringa, with the wing feathers brown, the first rib white.

This is of the size of the common black-bird: the head is large, compressed at the sides, and rounded at the top: the eyes are moderately large; their iris is of a hazel colour, with some tinge of the orange: the beak is moderately long, largest at the base, obtuse at the extremity, and all the way black, except for some whitishness toward the base of the lower chap: the back is brown, with deep dusky spots of a ferrugineous hue, approaching to black: the rump is white, but the feathers which compose it have some blackness toward their base: the breast and belly are white.

The wings are in general of the same colour with the back, and are spotted in the same manner; but the principal feathers are brown, and the middle rib of the first white, and there are some other variegations of white, especially about their tips: the tail is moderately long, and is variegated with white and brown, disposed in an undulatory manner; the outer feathers are indeed principally white: the legs are long, slender, and brown.

This is frequent in Sweden, and in some of the northern parts of Europe; it sometimes is found in England, but not in any degree of plenty, nor does it stay the year.

Tringa nigra albo punctata, pedibus virescentibus.
The black Tringa, spotted with white and with green legs.

This is of the size of the common black-bird: the head is small, and compressed a little at the sides, but not flatted on the crown; the beak is moderately long and obtuse: the head, neck, back, and wings are black, spotted with white: the breast is of a whitish colour, but spotted with black; the belly is entirely white.

The wings are long, and the principal feathers black, but many of them white at the edges and tips: the tail is moderately long, and of a variegated colouring; the black and white being laid, in an undulatory manner, one over another: the legs are moderately long, very slender, and of a bright olive or greenish colour: the toes are slender, long, and green.

It is frequent in many of the northern parts of Europe, but we have it not in England.

Tringa luteo cinereoque maculata, pedibus rubris.
The red-legged Tringa, spotted with yellow and grey.

This is about the bigness of the fieldfare: the head is small and rounded; the ears are large and patulous; the eyes are small, but their aspect is bright and piercing, and their iris of a hazel colour: the beak is moderately long, obtuse, and throughout of a red colour: the base brighter; the extremity darker than the rest.

The head is of a bright grey, variegated with streaks and clouds of a tawny yellowish; the back, sides, shoulders, and coverings of the wings are mottled in a very agreeable manner, with grey and black; the breast and belly are white, but not perfectly or purely so: the anterior part of the breast, in particular, is variegated with tawny spots: the wings are large, and their long feathers are of a deep colour, but not without variegations; the tail is short and grey, but also clouded.

The legs are long, slender, and throughout of a bright red; the toes long, and the claws obtuse; the hinder toe is inconsiderable, but it has it's claws like the rest, and black.

This is a native of the northern parts of Europe, but not, that I have discovered, of England. I have received specimens of it from Denmark. Few of the writers on birds have described it.

Tringa nigro ferrugineo et albo variegata.
The black, brown, and white mottled Tringa. **The Tolk.**

This is of the bigness of the thrush: the head is small and rounded, but a little flatted on the crown: the ears are broad and patulous; the eyes small, piercing in their aspect, and their iris is orange-coloured: the beak is short, obtuse at the point, and black; the nostrils are oblong and pervious.

The head is white, variegated with a great many large spots of grey: the neck, shoulders, and back are mottled in a very elegant, though perfectly irregular, manner, with grey, white, and a ferruginous colour; the grey is so very deep, that it appears at a little distance quite black: the wings are long; the principal feathers are black, except that they have some whiteness toward the base: the tail is short, and it's feathers are white both at their base, and at the extremity, except only the two middle ones.

The legs are long, slender, and red; the thighs are naked half the way up; the toes are four; they are moderately long, divided, and each armed with an obtuse black claw.

This is a native of most of the northern parts of Europe. We have it in the fens of Lincolnshire, and some other parts of England; but it has been overlooked by the writers on birds, though the country people in these places are well acquainted with it.

Tringa nigro-fusca rostro nigro, pedibus virescentibus.
The blackish-brown Tringa, with a black beak, and green legs. **The grey Plover.**

This is a beautiful bird: the head is small, and somewhat depressed on the crown; the ears are not very patulous: the eyes are small, and their iris hazel; the beak is moderately long, large at the base, obtuse at the point, and black; the nostrils are oblong and conspicuous, and the opening of the mouth is large.

The head is of a deep iron grey, with a tinge of brown, and it is spotted with small, but very numerous, spots of black: the neck is of the same colour, with the same

same variegations; but the shoulders and back, as also the covering feathers of the rump and tail, are of a deep dusky brown, almost black, and scarce at all variegated: the breast and belly are white.

The long feathers of the wings are twenty-four in each; they are of a deep brown, approaching to black, but more or less variegated with white, especially about the edges, near the extremities; the covering feathers of the upper part of the wings have also some transverse variegations.

The tail is moderately long, and consists of twelve feathers; they are of a deep blackish-brown, but variegated with transverse streaks of white, narrow, and placed at a distance from one another, except on the two middle ones, where they are broader, and stand nearer: the lower part of the back is blacker than any other portion of the body, but the rump itself is white, like the breast and belly.

The legs are long, slender, and greenish; the toes are long and slender, and are transversely cut in form of the bodies of some of the insects; the claws are obtuse and black.

This is found in most of the northern parts of Europe, and has been described by many, though not all the writers on birds. Ray and others call it Pluvialis cinerea.

CHARADRIUS.

THE beak of the Charadrius is of a figure approaching to cylindric, and is obtuse at the extremity, and short: the feet have only three toes, and those are connected.

Charadrius nigro lutescenteque variegato. The green Plover.
The variegated black and yellowish Charadrius.

This is of the bigness of the common lapwing: the head is large and rounded; the eyes are large, and of a very lovely aspect; the beak is short, but it is robust, straight all the way, furrowed about the nostrils, and black: the neck is short, and the body slender: the head, neck, back, breast, and belly are all of the same colour, only that the breast is palest; this, at a distance, appears of a kind of green, and thence the bird has been named, but in reality it is a complex variegation of black and yellow; the ground colour is black, the other has the variegation; the middle of every feather is of the first colour, the other is sprinkled and lineated about the edges: the colour on the breast, where it is paler, shews more of the green: the wings are long, and their principal feathers are variegated more or less with white, though the ground colour is the same in them all: the tail is short; it consists of twelve feathers of the same colour with those of the back, and, when it is expanded, is round at the extremity: the legs are long and slender, and are naked above the knees, though they are black; the toes are three; they are moderately long, and the claws are black.

This is frequent in our fen countries, and is esteemed a very delicate bird at the table. All the writers on birds have described it. Ray calls it Pluvialis viridis, and almost all the other writers among the moderns use the same name.

Charadrius nigro lutescenteque variegatus pectore nigro.
The black-breasted, variegated Charadrius.

This is of the size of a pigeon: the head is large; the eyes are small, but piercing; the beak is short, obtuse, and robust: the head, neck, back, and wings are nearly of the same variegated colour with those of the grey plover, but the breast in this species is black; and on the back there is some white in the variegation, that is wanting in that of the other: the wings are large, and their long feathers are black, with but little

little variegation: the tail is short; it consists of ten feathers, and they are black, but variegated with transverse lines of yellow: the legs are long and slender; the thighs are covered more than half way down toward the knee, and the feathers of their upper part are white.

This is a native of many of the northern parts of Europe, but we have it not in England.

Charadrius pectore ferrugineo, torque albente.
The Charadrius, with a ferrugineous breast, and a white ring round the neck.

This is of the bigness of our jackdaw: the head is small, and flatted on the crown; the eyes are small, and the beak is short, straight, obtuse, and black: the head, neck, and back are of a pale grey; the breast is of a reddish-brown, a bright ferrugineous tinge: the belly is black; there is a sort of ring of white at the bottom of the neck, and a double line of white and black between the breast and belly: the legs are long, slender, and black.

It is common in many parts of Europe, but we have it not in England. Few of the writers on birds have mentioned it. Albin calls it Morinellus, as have also some others.

Charadrius fronte nigricante, lineola alba.
The Charadrius, with a black front, and a white line on it. **The Sea-lark.**

This is a very singular and a very pretty bird; it is about the size of the common lark, or, if any thing, a little larger: the head is large and rounded, and it very elegantly variegated with black and white; a black space appearing at the base of the beak, and continuing itself along on each side, and a white one running from the interior corner of one of the eyes to the interior corner of the other: the hinder part of the head is grey; the neck also is grey, but it has two streaks surrounding it in form of collars; the one white, and the other black; the back and the covering of the wings are of a pale grey; the breast and belly are white.

It is frequent about our coasts, and in most other parts of Europe. All the writers on birds have described it. Jonston calls it Charadrius; Aldrovand, Charadrius sive Hiaticula; Marcgrave, Matuiti Brasiliensibus, for it is common to that distant part of the world.

Charadrius abdomine rufescente pectore cinereo.
The grey-breasted and reddish-bellied Charadrius.

This is of the bigness of the common fieldfare: the head is large; the eyes small, and the beak short, obtuse, and black: the whole upper surface is of a dusky grey, spotted with a reddish-brown; the breast is of a bright and beautiful grey; the belly is reddish; the legs are long, slender, and grey.

It is a native of Sweden and Denmark, but we have it not in England, nor have any of the writers on birds mentioned it. Linnæus, in his Fauna Suecica, describes it; and we have had specimens sent over into England, which prove his accuracy in his description.

BIRDS.

Class the Fifth.

GALLINÆ.

THE beak of the Gallinaceous tribe is conic, and somewhat incurvated, and the upper chap is imbricated. This class comprehends the ostrich and cassowary.

STRUTHIO.

THE feet of the Struthio have only two toes to each, and those are both placed forward; and the head is simple, or not ornamented with the appendages, which are common to many of this class. Of this genus there is but one known species, which is the common ostrich.

STRUTHIO. The Ostrich.

This is the tallest of all the bird-kind; when it stands erect, and stretches it's head to the height, it measures between seven and eight feet from the ground: the head is small, in proportion to the body, and is flatted, and in form somewhat like that of the goose: the beak is also compressed, and of a somewhat triangulated form; it is small, in proportion to the size of the bird, and is of a horn colour, and black at the extremity: the skin of it terminates in a sort of semicircle at the nostrils; the opening of the mouth is large, and is extended nearly to the eyes: the eyes are large, and iris is of a hazel colour.

The head and neck, down to the breast, or nearly so, are in a manner naked; they are covered with a loose and scattered matter, of a downy or hairy nature, in the place of feathers: the legs, and the part of the body that is under the wings, are altogether naked: the lower part of the neck, where the plumage begins, is white: the wings are small, and serve only to assist the creature in it's running, for they are not calculated for flying, as those of other birds.

The body is not large, in proportion to the length of the neck and legs; the feathers which cover the back are of a darker colour in the male than in the female, in which sex they are only of a deep brown, in the male quite black: they are very soft, and resemble a kind of wool rather than feathers: the feathers of the wings are of the same colour, only on the upper part they are white as snow: the tail is of a clustered structure, and in shape round, not spread out with breadth as in all other birds; the feathers of which it is composed are white in the male, but in the female they are brownish, but with the tips white; these are the feathers so greatly esteemed.

The legs are very long, very robust, and naked; the structure of the foot, having only two toes, is very particular: the claws are very robust and large.

It is a native of Arabia, and other parts of the East, and is often seen in such numbers together in those places, that they have the appearance of an army drawn up in order of battle. All the authors who have written on this subject have described it, and all under the names of Struthio and Struthio-camelus.

CASUARIUS.

THE feet of the Casuarius have each three toes, and these are all placed forward: the head is ornamented with a kind of naked comb. Of this genus also there is only one known species.

CASUARIUS. **The Cassowary.**

This bird, when it stands erect, is about four feet and a half high, and is more singular in it's appearance even than the ostrich: it appears at first sight to be covered with hair rather than feathers; and, though it's legs and neck are long, they are less so than in that bird: the body also is larger, in proportion, than in the ostrich, and the whole bird more robust.

The head is small, and it ornamented with a kind of comb or crest, not fleshy as in the cock, but bony, hard, and of a reddish colour; two inches and a half in height, and considerably thick: the ears are large, and are surrounded with a kind of bristles; the eyes are very large, and of a remarkably fiery and fierce aspect: the beak is moderately large, broad at the base, and smaller, and somewhat hooked toward the point.

The head and neck are almost naked; they are beset with a number of coarse, large hairs, of a black colour, resembling bristles; but the colour of the skin is easily seen through these, and is bluish, excepting only on the hinder side of the lower part of the neck, where it is red: on the anterior part there hang two lobes of a fleshy substance over the breast: the opening of the mouth is very wide, extending in the manner of that of the ostrich, much beyond the base of the beak, it's angles running up almost to the eyes.

The body is large and thick; it is covered with a very strange and singular kind of plumage, of a very dark brown colour, approaching to black, and having, when seen at any distance, the appearance of hairs rather than of feathers; when nearly examined, they are found, however, to be genuine feathers, two growing always together, and their structure very elegant; they are very long, and extreamly narrow, and, when taken off from of the bird, do not support themselves erect.

The wings are still smaller, and more imperfect than in the ostrich; they have, indeed, more of the appearance of rudiments of wings, than of any thing that can be regularly called by that name: each of them has five quills only, which have more of the appearance of the armature of the porcupine, than of any part of the plumage of a bird.

The legs are long, and very robust; and the toes of each foot three, and very thick and strong.

It is a native both of the East and West Indies; and all the writers on birds have described it. Aldrovand calls it Emeu sive Eme; Boutius, Emeu vulgo Casoarius; and others, Cassuarius.

OTIS.

THE feet of the Otis are each composed of three toes, all turned forward: the head is naked, or has no comb. There is but one known species of this genus, and that has been confounded, by those who had no regard to the characteristic of the genera, with the common turky.

OTIS. **The Bustard.**

This is a bird more nearly allied to the ostrich and cassowary-kind than people are aware, and like them it runs at a prodigious rate, and but rarely rises on the wing; it is of the size of the common peacock: the head is large, and is well covered with feathers, but it has no comb: the eyes are large, and of a very piercing aspect: the beak is short, but very robust; is is exactly of the form of that of the common turky, thick at the base, and pointed and hooked at the extremity: the neck is of no remarkable length.

The

The head and neck are of a very bright and beautiful grey colour; the back is of the same colour for the ground, but it is very beautifully variegated with transverse streaks of brown, and of black: the wings are small, in proportion to the bulk of the body, and their principal use to the creature is to assist it in running: the legs are robust, but not remarkably long; the toes are three, and all placed forward; the claws are short, but they are thick and black.

We have this bird in many parts of England, where it feeds on vegetables, and on corn, when it can get at it. I have seen great numbers of them on the downs in Sussex; they run away at the approach of men, but rarely, and indeed difficultly, take wing. They are often taken by greyhounds in a fair course, in the manner of a hare. Their flesh is very well-tasted. All the writers on birds have named this under the names of Otis, Otus, and Avis tarda.

P A V O.

THE feet of the Pavo have each four toes: the head is ornamented with a crest of feathers.

Pavo cauda longa.
The Pavo, with a long tail. **The Peacock.**

The male of this species is the most specious and handsome of all the bird-kind; the body is large; the neck and legs are moderately, but not remarkably, long; the tail is more considerable, both in length and structure, than that of any other bird.

The head is small, and of a greenish colour, but variegated on each side with two white spots and a black one, and ornamented on the top with a crest of feathers erect, and of a most elegant as well as singular structure: the whole neck is of a beautiful deep changeable green, and is slender, in proportion to the size of the body: the back is of a pale bright grey, variegated with transverse spots of black: the wings are of a dark grey, approaching to black; the length and beauty of the tail, and the various forms in which the creature carries it, are sufficiently known and admired.

The legs are of a greyish colour, robust, and moderately long; the toes three before, and one behind: the flesh is coarse and ill-tasted, but it is eaten in some places, and is singular in keeping a long while without putrefaction, when it has been boiled.

It is a native of the East, but is common in all parts of Europe kept tame; and all the writers on birds have described it.

Pavo dorso griseo, pectore nigrescente.
The Pavo, with a grey back, and black breast.

This is of the size of the common peacock, and resembles it in form, but it wants the singular ornament of it's tail: the head is small, and of a pale grey colour, without any variegation, excepting a small spot under the eye on each side: the neck is slender, and moderately long; the back is of a pale grey, variegated with undulated, distant, and narrow lines of brown, in a transverse direction: the breast is black, the belly a reddish-brown.

The tail is short and inconsiderable; the wings are considerably long, and their principal feathers are black: the legs are robust and black; the toes are four, and they are long, thick, and armed with sharp claws.

This is a native of the East, but is kept as a curiosity in some places. Few of the writers on birds have been acquainted with it.

Pavo

Pavo griseus corona aurea.
The grey Pavo, with a golden crest.

This is a very beautiful species: the body is of the bulk of that of our ordinary peacock, but the neck is shorter and thicker: the head is small, and of an oblong form; the beak is short, robust, and a little hooked at the point: the crown of the head is ornamented with a fine downy crest, of an orange yellow, though with less of the reddish tinge in it, than we usually express by that phrase: the eyes are large and piercing, and the opening of the mouth is much larger than would be conceived from the length of the beak, but it's angles reach very nearly to the eyes: the back is of a dusky grey, with a tinge of brown: the wings are moderately long, and their principal feathers are black, with very little variegation: the tail is long and beautiful, the legs are robust and grey.

It is a native of the East; few of the writers on birds have described it. The Balearic crane, referred by Linnæus and some others to this genus, is described among the Ardeæ.

CRAX.

THE feet of the Crax have each four toes: the head is ornamented with a feathered crest.

Crax niger corolla albo nigroque varia.
The black Crax, with a black and white crest.

The Indian Cock.

This is a stately and elegant bird: the head is large, and well covered with feathers; the eyes are large, and very bright and piercing; the apertures of the ears patulous and broad: the beak is short, but very robust, thick at the base, and pointed at the extremity, and there a little hooked: on the top of the head there stands a very elegant crest of mottled black and white feathers, which is not erect but revolute: the neck is short, but thick; the body corpulent: the colour of the whole bird is black; the wings are moderately large, and their principal feathers are blackest of all: the legs are robust, but not very long; they are of a very deep grey, approaching to black: the toes are long and robust; the claws thick, but short and obtuse, and of a deep black.

This is a native of the East, and of many parts of America. Sir Hans Sloane calls it Gallus Indicus; and the French writers have named it, as if from him, Coq. Indein.

Crax niger corolla atra, rostro rubro.
The black Crax, with a black crest, and a red beak.

The black, crowned, Indian Cock.

This is a large and majestic bird, and is not without it's beauty, though of one simple colour: the head is large, and somewhat depressed on the crown: the eyes are large and bright; the ears patulous, and surrounded with a double series of short and semi-erect feathers: the beak is short, robust, formed exactly like that of the common cock, but of a scarlet colour: the crest is moderately large, revolute, and of a deep and glossy black, as is also the whole bird.

The wings are short, and the tail is not so long as in the former species: the legs are very robust, and of a deep colour; the toes long, and armed with large black claws.

This

This is a native of many parts of America. Few of the authors who have written of birds have been acquainted with it.

Crax punctulatus corolla atra, rostro fusco.
The spotted Crax, with a black corolla, and a brown beak.

This is a more beautiful species than either of the preceding: the head is large, and depressed at the front, but elevated toward the hinder part, and there ornamented with a large and beautiful crest of a deep glossy black, long and revolute: the eyes are large, and their iris is of a strong orange colour: the beak is short, robust, and of a brownish-yellow; the point of it is very sharp, and it is more hooked than in any of the other species: the neck, back, breast, and, indeed, every part of the bird is black, spotted in an irregular manner with little specks of white; these are more numerous on the breast than in any other part, and are no where so large as on the upper part of the wings: the legs are robust, moderately long, and of a dusky yellowish colour; the toes are long and robust, and the claws are long and black.

This is a native of Domingo, and some other of the American islands; it is not found on the continent. Few of the writers on birds seem acquainted with it.

MELEAGRIS.

THE anterior part of the head of the Meleagris is covered with a fleshy, pendulous substance: the sides of the head also and the throat are covered with a papillous fleshy matter; and there is also a longitudinal fleshy crest, of a reddish, bluish, or purplish colour, and a soft substance. Of this genus there is but one known species.

MELEAGRIS. **The Turkey.**

This is a large but an unweildy bird: the head is large, and is strangely covered and ornamented with a pendulous, soft, fleshy substance, of a scarlet colour, but varying with blue and purple, and many other changeable colours, and in some degree also altering it's form according to the pleasure, or as influenced by the passions, of the creature: the eyes are small, but bright and piercing; the apertures of the ears less patulous than in many others.

The wings are moderately long, though not at all formed for supporting so large a bulk in long flights; they have each twenty-eight long feathers: the tail is long and large, and the creature has a power of erecting and spreading it in a beautiful manner: the legs are moderately long, and very robust.

It is a native of North America, where it is always, without variation, black throughout; with us it is frequently grey, and of other colours: we keep it for the sake of it's flesh. Most of the writers on birds have described it. They call it Gallo-pavo, Meleagris, and Numidica avis.

GALLUS.

THE front of the head in the Gallus is ornamented with a longitudinal fleshy crest or comb: the wattles on the throat are two; they also are longitudinal and fleshy.

Gallus cauda compressa ascendente.
The Gallus, with a compressed ascendent tail. **The Cock.**

The common domestic cock, in his natural state, is a very robust and beautiful bird: the body is large and fleshy; the head small, and the legs remarkably robust;

 the

the head is compressed and oblong; the eyes are small, but lively and piercing in their aspect: the crest or comb on the crown is oblong, tall, thin, erect, and dentated at the edge: the back is of a pale brownish-yellow; the neck is of a somewhat paler colour, but of the same kind: the wings are variegated with the same tawny colour and black, and so is the tail: the legs are brown, and are armed behind with long, strong, bony spurs, turning up, and very sharp-pointed.

This is the form and colouring of the cock in his wild state; with us the variegations in colour are endless, and those in the form and disposition of the feathers, and in the structure of the comb are as various. These have, by the less scientific authors, been understood as specific distinctions; and hence in the common way, from this single species, have arisen those imaginary ones, distinguished by the names of the Feather-topped fowl, the Rumpless fowl, the Friesland hen; and from hence alone those of the Bantam fowl, and the like. These are barely varieties; the common dunghil cock is the single species, and these are produced and farther varied by accidents and mixed breeds, as our variety of auricula's from the seeds of the simple and original kind.

PHASIANUS.

THE area or space about the eyes in the Phasianus is naked; it has no wattles.

Phasianus pectore nigro purpurascente.
The blackish purple-breasted Phasianus. **The Pheasant.**

The pheasant differs more in bulk and weight, according to the plenty or scarcity of food which it has met with, than almost any bird; when in good condition, it is little less than a common unfatted fowl: the head is small; the ears are patulous, and very obviously distinguished: the eyes are remarkably bright and piercing; their iris is of a bright and elegant yellow, and they are surrounded with a naked space of a fine scarlet colour, which gives a great beauty to the bird: the beak is not very long, but robust, and resembles that of the common fowl; there are two fleshy tubercles toward it's base, under which the nostrils are hid: the sides of the head are blackish; the top of it, and part of the neck in the male, are of a fine glossy and changeable green; but this is paler on the head than on the neck: the feathers near the ears are long and elevated, and form what the old naturalists call the ears of the pheasant; a purple colour is seen on the throat, and at the sides of the neck; but all these tinges, though very bright, are changeable, and hardly inherent in the feathers.

The lower part of the neck, the shoulders, the middle of the back, and the breast are covered with elegant feathers of a blackish or purplish tinge, as seen in various lights: the tail is long, and composed of few feathers; the wings are but short, in proportion to the bulk of the body, but they are beautifully variegated; the legs are robust, and of a pale colour.

This valuable bird would be very common in our woods, if it were not so universally the delight of the sportsman and of the table. All the authors who have written on birds have described it, and all under the same general name Phasianus.

Phasianus pectore coccineo.
The scarlet-breasted Phasianus. **The red Pheasant.**

This is nearly of the size of the common pheasant: the head is larger, in proportion to the body; the eyes are very bright, and their iris is of a fine strong orange colour; they are surrounded by a naked skin, not scarlet, as in the former species, but of a deep fir colour: the top of the head is of a deep chesnut; the neck has a changeable glow of a deep blue, a deep green, and a very strong blackish-purple: the back is of a fine, bright, ferrugineous tinge, with a glow of purple; the feathers are all very

very gloſſy: the breaſt is of a high ſcarlet, and the belly red, but not ſo bright: the tail is long and compreſſed; the legs are robuſt and yellow; the toes are very long and ſtrong, and the claws ſharp: the ſpur behind the leg is often very long, pointed, and horny.

This is a native of many parts of Europe, but we have it not wild in England; in Italy it is common.

Phaſianus ſplendidiſſime variegatus cauda longiore.
The long-tailed, elegantly variegated Phaſianus.

The East Indian Pheaſant.

The two preceding ſpecies are extremely beautiful, but this is vaſtly more ſo than either. Indeed, words cannot deſcribe adequately the luſtre of it's colouring; nor, if they could, would any one believe them, who had not the atteſtation of his eye-ſight: the head is ſmall, depreſſed, and a little prominent at the ſides: the ears are elegantly plumed; the eyes are large, and their iris is of a fine fiery red colour: the beak is half an inch long, robuſt, a little hooked, pointed, and of a pale colour: the whole body, wings, tail, and head are variegated with a profuſion of the brighteſt colours, yellow, red, bluiſh-green, and almoſt every other tinge, varying in different lights, and all having the luſtre of gems: the wings are ſhort, but the tail is longer than in the common pheaſant: the legs are robuſt, and of a pale colour.

It is a native of the Eaſt Indies, and is ſometimes brought over to us.

TETRAO.

THE part of the head that is over the eyes in the Tetrao, is naked and papilloſe.

Tetrao alarum baſi macula alba inſignita.
The Tetrao, with a white ſpot on the baſe of the wing.

The Cock of the Wood.

This is a very large and noble bird; it is nearly of the bigneſs of the turkey, and much reſembles it in figure, only that it is not ſo unwieldy: the head is large and rounded; the beak is ſhort, robuſt, a little hooked, and ſharp at the point: the eyes are large, and their iris is hazel; there is a naked ſpace over them, by way of eye-brow, which is of a fine bright ſcarlet.

The breaſt is of a pale reddiſh-brown, variegated with tranſverſe lines of black, and ornamented with a little white, diſpoſed principally about the tips of the feathers; the belly is grey: the back is variegated with black, grey, and a reddiſh-brown; and the head is of a purer black, but with an admixture of a purpliſh gloſs.

The legs are ſhort and robuſt, and are naked behind; but on the anterior part they are feathered nearly to the toes: the long feathers of the wings are twenty-ſix in each: the wings are moderately long, but the tail is ſhort.

This ſpecies is a native of many parts of Europe; we have it not in England, but it is ſometimes met with in Scotland and Ireland. All the writers on birds have deſcribed it. Geſner calls it Urogallus and Grygallus, diſtinguiſhing the male and female; others Urogallus ſive Tetrao major.

Tetrao alis albo variegatis, cauda bifurca.
The forked-tailed Tetrao, with variegated wings.

The Grouſe.

This is a noble and valuable bird, though much inferior to the former; it is of the bigneſs of a well-grown fowl: the head is large, and the eyes bright and piercing; the ears

ears are pennated: the beak is three quarters of an inch long, and of a pale colour, robust, somewhat hooked, and pointed at the extremity; the scarlet protuberance over the eyes is very bright and beautiful.

The male and female differ greatly: the male, excepting for the little variegation of white in his wings, is totally black, but there is a fine changeable tinge of a deep blue thrown over the feathers of the neck, and in some degree over those also of the back; the female is brown and mottled, and not a little resembles the woodcock in colour; but in her, as well as in the male, there is a great deal of white discovered, when the wings are expanded; the principal feathers in these are twenty-six in each; they are robust, but the whole wing is not long or large, in proportion to the bulk of the bird.

The tail consists of sixteen feathers, and is forked, the exterior ones being greatly longer than the interior: the long feathers in the male turn back, but in the female they are straight.

This is a native of England, but is not very frequent; it lives on large mountainous heaths. The male and female in the former species, as well as in this, are so different, that they have been described by Gesner and others as separate birds. All the writers on birds have named this. Gesner calls it Urogallus minor, and Grygallus minor; others, Tetrao, sive Urogallus minor.

Tetrao cauda bifurca subtus albo punctata.
The Tetrao, with a forked tail, spotted with white underneath.

The Moor-cock.

This is as big as our large Dorking fowl; and the male and female differ so extreamly, that they may easily be mistaken for distinct species: the female is grey, variegated with transverse lines of black, and the tail is moderately long, forked, and the outer feathers straight: the male is throughout of a very deep iron-grey, without any variegation, except that his head and neck have a beautiful tinge of a changeable blue thrown over them: the head is large, and the beak very robust and black, somewhat gibbous on the upper part, and pointed at the end: the eyes are large; their iris is hazel, and the scarlet eye-brows are very bright and beautiful: the legs are very robust, but short; and the tail in the male has the long exterior feathers turned back or curled.

This is a native in England, but very rare. I did not know it was at all found here, till I killed two brace on Hindhead, a vast mountainous heath in the Portsmouth road. The writers on birds have not mentioned it; they either have not seen it, or have confounded it with the others; only Linnæus says something of it, from the account of one Leech.

Tetrao rectricibus albis, intermediis nigris, apice albis.
The Tetrao, with the tail feathers white, or tipped with white.

The Lagopus.

This is of the size of our largest tame pigeon: the head is large; the ears are pennated; the eyes are bright, and their iris is of a hazel colour, and over them there is a naked space by way of eye-brow, of a fine scarlet colour, and granulated figure: the beak is very short and black; it is gibbous on the upper part, hooked at the extremity, and very sharp-pointed.

The whole bird, excepting only the tail, is as white as snow, only that in the male there is a longitudinal streak of black on the head, which distinguishes it from the female: the wings are moderately long, and there are twenty-six of the principal feathers in each: the tail is four inches, or more, in length, and is composed of sixteen feathers, which are in general white, but some of them are black toward the base, and only white at the extremity: the legs, and even the toes, down to the very claws, are

are covered with a fine white plumage: the claws are very long, and of a lead colour.

This is a native of many of the northern parts of Europe, but we have it not in England; it lives principally on the tops of mountains, covered with snow. All the writers on birds have described it under the names of Lagopus and Perdix alba.

Tetrao cauda cinerea, punctis et fascia nigra variegata.
The Tetrao, with a grey tail spotted, and fasciated with black.

The hazel Hen.

This is of the size of a moderately grown pullet: the head is large and well-feathered, except over the eyes, where there is a naked space of a strong scarlet in the male, and of a flesh colour in the female: the eyes are large and bright; the beak is about half an inch in length, black, gibbous, and somewhat hooked and pointed. The male and female differ greatly in colour: in the male, the head and neck are grey, variegated with transverse lines of brown: the throat is black, and the under part of the head has a white streak on each side: the wings are of a dusky ferrugineous colour, variegated with spots of black; the back is of a deep grey, spotted also in the same manner with black; the breast and belly are white, with large cordated spots of a ferrugineous brown.

In the female, the back is of a whitish-grey, with longitudinal spots of black, and the rest of the body has, for it's ground colour, a very pale grey, variegated with spots of brown and black: the wings are not long; the tail is composed of sixteen feathers, transversely streaked with black and grey, and all of them, except the two middle ones, marked with a black fascia toward the extremity.

The legs are robust and short, and are feathered on the anterior part above half way down to the feet.

It is frequent in many parts of Europe, but not in England; it lives in thick woods, and feeds on the catkins of the hazel, and other trees, and on other vegetable substances. All the writers on birds have described it, and they have have all called it Gallina Corylorum.

Tetrao cauda cinerea superiore medietate rufo variegata.
The Tetrao, with a grey tail variegated in the upper part with brown.

The red-legged Partridge.

This is of the bigness of a large tame pigeon: the head is small; the ears are patulous; the eyes are large, and very bright, and their iris is red: the beak also is of a bright red, and is of the shape of that of the common hen; it is robust, hooked, and sharp.

The head, as also the back, rump, and neck are grey, only that there is a tinge of a deep claret colour toward the bottom of the neck: the lower part of the head is white, and there is some white also about the basis of the lower chap of the beak; over this white there runs a kind of ring of black: the breast is of a tawny colour, with a tinge of reddish; the sides are elegantly variegated with a changeable admixture of black, grey, yellowish, and red: the wings are not long or large, in proportion to the bulk of the bird; the long feathers are twenty-five in each: the tail is about a hand's breadth long, and the two middle feathers are often entirely grey; the others are variegated toward their upper part with brown: the legs are of a fine bright red.

 This

This is frequent in the cornfields of many parts of Europe, but not in England; we have it from France made into pies, and eſteem it greatly. All the writers on birds have mentioned it, and they have all called it Perdix rufa, perdix major, or perdix rufa major. The female in this ſpecies differs from the male in her colouring; the back is paler, and the breaſt browner.

Tetrao macula nuda coccinea pone oculos.
The Tetrao, with a naked ſcarlet mark behind the eyes.

The common Partridge.

This is ſomewhat ſmaller than the red-legged partridge of France laſt deſcribed, but in other reſpects greatly like it: the head is ſmall and rounded; the eyes are large and bright, and their iris is yellow; the beak is ſhort, robuſt, a little hooked, and ſomewhat gibbous on the back; there are ſome naked, red excreſcences about the eyes; the breaſt in the male has a reddiſh-mark, of the ſhape of a horſeſhoe on it: the ſides of the head and under the beak are yellowiſh, and thence they become of a bluiſh-grey, ſpotted and ſtreaked with black down to the ſpot on the breaſt; under that the colour becomes paler, and has ſome admixture of yellow with the grey.

The back, ſhoulders, ſides, and wings are all variegated with brown, grey, and black, very regularly and elegantly diſpoſed, only thoſe long feathers which hang from the ſhoulders, and the larger ones which cover the wings, have a yellowiſh-white about their middle: the wings are not long or large; the principal feathers of each are twenty-three, and theſe are brown and variegated with a yellowiſh colour: the tail is ſhort, and conſiſts of twelve feathers; the four middle ones of the colour of the body, the reſt of a duſky yellowiſh, with grey tips.

The legs are robuſt, but not long; they are of a greeniſh colour in the young birds, afterwards white; there is no mark of a ſpur in the males.

This is frequent in our fields, and all the writers on birds have deſcribed it. They all call it Perdix vulgaris, and Perdix cinerea, or Perdix cinerea minor.

Tetrao linea ſuperciliorum alba.
The Tetrao, with the line of the eye-brows white.

The Quail.

This is the ſmalleſt bird of this genus; it does not exceed the fieldfare in ſize, but it is, in all reſpects, like the reſt of it's brethren; the head is ſmall, but the eyes are large, and their iris is hazel: the beak is ſhort, but robuſt; it is not gibbous, but rather depreſſed on the upper part, but it is very ſharp; the lower chap is blackiſh, the upper of a whitiſh-brown.

The breaſt and belly are of a pale and foul yellowiſh colour; the throat is of the ſame tinge, but with ſome admixture of reddiſh, and immediately under the baſe of the lower chap there begins a black, oblong ſpot, which runs a conſiderable way down; the head is blackiſh, but the edges of the feathers are variegated with grey and brown, and there runs an elegant white line on each ſide, above the eyes: the back is variegated with brown, black, grey, and a yellowiſh colour.

The wings are not large; their principal feathers are brown: the tail is very ſhort, and conſiſts of twelve feathers; it is black, but faſciated with tranſverſe lines of a yellowiſh-brown.

The legs are ſhort, moderately robuſt, and of a pale colour; the ſkin which covers them is ſquammoſe, rather than annulated.

This ſpecies is frequent with us, and is eſteemed at our tables. The authors who have written on birds have all deſcribed it, and all under the ſame name Coturnix, or Coturnix Latinorum.

BIRDS.

BIRDS.

Class the Sixth.

PASSERES.

THE beak of the Passeres is of a conic and much attenuated figure.

COLUMBA.

THE beak of the Columba is straight, and it is furfuraceous at the base: the nostrils are oblong and membranaceous, and are in part covered: the tongue is entire or undivided.

Columba cærulescens collo nitido, macula alarum duplici nigra.
The bluish Columba, with a double blackish spot on the wing.

The Wood-pigeon.

This is somewhat larger than the common pigeon kept in our dove-houses, and is a more beautiful bird, though much like it in the general colouring, as well as in form: the head is small; the eyes are small, but bright: the beak is moderately long, straight, of a kind of conic figure, pointed, and of a pale red colour: the head is of a dusky bluish-grey; the neck is covered with feathers of a changeable colour, which in different lights are either purple or green, both colours bright and glossy: the breast, on it's upper part, is of a purplish colour, and so are the shoulders, and the tops of the wings: the wings are long; and their principal feathers are variegated with black and grey: the tail is moderately long, it is composed of twelve feathers; they are black for about one third of their length from the extremity, and the rest is grey: the legs are slender and red.

This species is frequently wild in our woods, and has been described by all the authors who have written on this subject, under the names of Œnas and Vinago.

The tame Pigeon.

From this have been propagated the common dove-house pigeon, and all the tame species, as all our varieties of the fowl-kind from the original yellowish wild cock. The Pouter, the Tumbler, the Barbary Pigeon, the Jacobin, and the rest, have all their origin from this, and are varieties, not distinct species of birds.

Columba collo utrinque albo, pone macula fusca.
The Columba, with the neck white on each side, and a brown spot behind.

The Ringdove.

This is larger than the common wild pigeon, but greatly resembles it in form: the head is small; the eyes are small and very bright, and their iris is of a pale yellow: the beak is moderately long, slender, and yellow, and is covered at the base with a reddish membrane; and above the nostrils, which are situated in this membrane, with a furfuraceous matter: the head, the back, and the wings are of a dusky bluish-grey colour; the hinder part of the neck is ornamented with a white semi-circle or ring, reaching half way round, and resembling a kind of collar: the part of the neck below, as well as above, this collar, is of an elegant changeable colour, appearing green, purple, or blue, according to the direction in which the light falls upon it: the front of of the neck, as alſo the upper part of the breaſt, are of a purpliſh colour, but with ſome admixture of grey: the belly is ſimply of a whitiſh-grey: the wings are moderately long; the principal feathers are twenty-four in each; theſe are moſt of them entirely black, but in ſome the edges are greyiſh or whitiſh: the tail is longer than in the common pigeon, and is compoſed of twelve feathers; it is black at the extremity, but grey every-where elſe: the feet are naked and red; the claws are black; the legs feathered.

We have this in England, but it is not frequent; in other parts of Europe it is more common. All the writers on birds have deſcribed it. Aldrovand calls it Palumbus major torquatus; others, ſimply, Palumbus torquatus.

The turtle is the ſame ſpecies kept tame. **The Turtle.**

Columba griſeo et nigro variegata pectore albente.
The grey and black Columba, with a white breaſt. **The Picaſpinima.**

This is a regular and beautiful pigeon, of the ſize of a common ſparrow: the head is ſmall, and flatted on the crown; the eyes are ſmall, but very bright; the iris is of a gold yellow, the pupil black: the beak is ſhort, and of a dark brown: the head, neck, back, ſhoulders, and wings are covered with feathers, the body of which is grey, but the edges black; the breaſt is white, the belly of a pale grey.

This is frequent in the woods in South America. Marcgrave has deſcribed it.

· *Columba vireſcens pedibus flavis.*
The green Columba, with yellow legs. **The St Thomas's Pigeon.**

This is of the ſize of our common pigeon, and of the ſame form: the head is ſmall and flatted; the eyes are ſmall; their iris is yellow, and the pupil black: the beak is long, and ſomewhat gibboſe on the upper part; the anterior half is blue, that near the head purple; the whole bird is green, like the common parrot, only the long feathers in the wings, and thoſe of the tail, are browniſh with the green, and there is ſome yellowneſs under the baſe of the tail: the legs are yellow.

It is a native of South America. It has it's name from an iſland called St Thomas's, where alſo it is frequent. Marcgrave has deſcribed it, and we ſometimes receive ſpecimens of it.

TURDUS.

THE beak of the Turdus is of ſubulato-conic form; it is ſtraight, ſomewhat convex on the upper part, and has no membrane at the baſe: the tongue is lacerated and emarginated.

Turdus ater, roſtro palpebriſque fulvis.
The black Turdus, with the beak and eye-lids yellowiſh. **The Blackbird.**

This is of about the ſize of the thruſh; the weight is nearly four ounces: the head is moderately large; the eyes are large and bright; there is a yellowneſs about the eye-lids: the beak is half an inch long, and ſharp-pointed; it is yellow: the male is black in all parts; the female is more brown, as are alſo the young birds of both ſexes: the wings are long and large; the principal feathers are eighteen in each, and of theſe the fourth is the longeſt: the tail is three inches, or more, in length; it is compoſed of twelve feathers, all of the ſame length, except the outer two, which are a little ſhorter than the reſt: the legs are ſlender and black; the whole bird has been ſeen white.

It

BIRDS Series 6.
Plate 22 – Page 493
The Great Plover
The Cassiowary
The Ostrich
The Cock of the Wood
The Quail
The Grouse
The Hazel Hen
The Red legd Partridge
The Gross Beake
The Wood Pigeon
The Turtle Dove
The Garrulus Bohemicus

It is frequent with us in woods and hedges, and has been described by all the writers on these subjects. They call it Merula and Merula nigra. The males of this species, when kept in cages, are eternal singers.

Turdus nigricans torque albo.
The black Turdus, with a white ring.

The ring Ouzel.

This species is a little larger than the common black-bird; it's weight is near five ounces; the head is large and flatted; the eyes are large, and their iris is hazel: the beak is half an inch long, and of a brownish colour; the inside of the mouth is yellow, and the tongue is hairy: the head, back, and wings are of a very dusky brownish colour; the breast and belly are covered with long feathers, spotted in the middle, and fringed about the edges with white: the lower part of the throat is elegantly variegated with a semi-lunar ring, or streak of white, of a finger's breadth in the middle; the narrower parts or corners terminating at the sides, and no part of the mark being continued round.

The wings are moderately long, and the long feathers in each are eighteen: the tail is composed of twelve; they are nearly of a length, only the exterior two are somewhat shorter, and they also are blacker than the rest: the legs are brown, slender, and moderately long.

This is a native of most parts of Europe; it is found in woods, and particularly on those in hilly places. All the writers on birds have described it under the name of Merula torquata.

Turdus nigricans torque fusco.
The black Turdus, with a brown ring.

The Amsel.

This is somewhat larger than the common black-bird: the head is large, flatted, and compressed; the eyes are large, their iris is of a dusky hazel: the beak is oblong, slender, sharp, and yellow: the whole upper upper part of the bird is of a dusky brown, variegated with spots and streaks of black: the throat is of a reddish-brown, variegated with black spots; the ring at the bottom of the throat is brown, and is sometimes so pale, as to be scarce distinguishable: the breast and belly are grey, and are variegated with spots of black; the legs are of a dusky colour.

This is common in many parts of Europe, but we have it not in England. Most of the writers on birds have named it. Willughby calls it Merula montana.

Turdus dorso griseo, plumis pinnatis.
The grey Turdus, with pinnated plumes.

The Stone Black-bird.

This is of the size of the common black-bird: the head is small, and very little, if at all, flatted at the crown: the eyes are large and bright; the beak is a finger's breadth long, sharp at the point, somewhat flatted on the upper part, and throughout black: the mouth is yellow within, and the tongue is divided: the head and back are of a deep brown, approaching to black, but variegated in a very singular and elegant manner with a pale grey, in so large a proportion, that it seems the principal colour; the feathers are elegantly pinnated, and the edges grey: the throat is white; the tips of the feathers which cover the breast are grey; there runs also a black line across the breast: The legs are slender, and moderately long, and are of a lead colour.

This species is frequent in many parts of Europe, but we have it not in England. Most of the authors who have written on birds have described it. They call it Merula saxatilis; and Ray in particular, Merula saxatilis Sturni genus.

Turdus variegatus capite cano.
The variegated Turdus, with a hoary head.

The Fieldfare.

This is larger than the common black-bird: the head is small, oblong, and flatted; the eyes are large, and the eye-lids and a little space about them are yellow: the ears are patulous; the beak is nearly half an inch long, and slender; it is yellow, except at the point, where it is black: the head is of a pale grey; the neck also and the rump are grey: the back and covering feathers of the wing are of a yellowish colour, variegated with spots of black; the throat is spotted with black and yellow; the lower part of the breast and the whole belly are of a whitish colour, and without spots: the legs are black, and the claws also black.

This is not a native with us, but it comes over in great abundance in winter. Authors who have written on birds have all described it. Charleton calls it Turdus Trichas; others, Turdus Pilaris.

Turdus linea supra oculos albicante.
The Turdus, with a white line over the eyes.

The common Thrush.

This is smaller than the fieldfare; it's weight is about three ounces: the head is small and flatted; the eyes are bright, their iris is hazel; the ears are patulous; the beak is half an inch long, brown, and pointed: the head is of an olive brown, spotted with a darker colour; the white line runs over the eyes: the back is of the same olive brown, variegated in the same manner; the wings have a tinge of the ferrugineous: the breast is yellow, and the belly whitish: the legs are brown.

This is frequent in our woods and hedges; it sings very agreeably. All the writers on birds have described it. They call it Turdus Iliacus, and Turdus viscivorus minor.

Turdus cinereo-flavescens pectore maculato.
The greyish-yellow Turdus, with a spotted breast.

The Missel Thrush.

This is the largest of the Turdus-kind; it's weight is near five ounces: the head is small; the eyes are bright, and their iris is hazel; the beak is half an inch long and brown: the head is of a greyish colour, with some variegation of black in the middle of the feathers: the back and the wings are of the same greyish colour, with a tinge of yellowish: the breast and belly are spotted with black; the ground colour of the breast is yellow, that of the belly whitish.

All the writers on birds have mentioned this. They call it Turdus viscivorus major.

Turdus dorso cinereo-aurantio, pectore albo.
The orange-grey Turdus, with a white breast.

The Redwing.

This is smaller than the common thrush, but it has much of the appearance of the fieldfare: the head is small and flatted; they eyes are bright, their iris is of a deep hazel; the ears are patulous, and the beak is brown, with some admixture of yellow: the head, neck, and back are of a dusky greyish, with a tinge, but a faint one, of an orange yellow: the sides under the wings, and the under part of the wings themselves, are of a yet stronger orange colour, approaching to red: the breast is white, so is also the belly, but the throat is yellowish; and over the eyes there runs a line of yellow: the legs are of a paler colour; the toes long.

All the writers on birds have mentioned this. They call it Turdus Iliacus and Tylas; the Germans, the Wyntrostell; and we, improperly thence, the Wind Thrush.

Turdus

Turdus totus cæruleſcens.
The wholly blue Turdus.

The Indian Mock-bird.

This is of the ſize of our common black-bird: the head is larger, in proportion; the eyes are bright and large; the ears patulous, and the beak half an inch long, and of a duſky colour: the whole bird is of a deep and beautiful blue colour. It has a faculty of imitating ſounds, whence it obtained the name of the Mock-bird. Authors, who have deſcribed it, have called it Cæruleus Indicus.

Turdus coccineus alis et cauda nigris.
The ſcarlet Turdus, with the wings and tail black.

This is ſomewhat of the figure of our black-bird, but thicker and ſhorter in the body; the ſize is nearly the ſame: the head is ſmall; the eyes are large, and their iris is of a bright hazel; the ears are patulous: the whole back, breaſt, and belly are of a beautiful ſcarlet colour; the wings and the tail are black: the beak is red, but black at the baſe: the legs are of an orange colour.

This is frequent in ſome parts of Europe, and in America. Aldrovand calls it Merula roſea.

Turdus niger pectore ſanguineo.
The black Turdus, with a blood-red breaſt.

The Jacapu.

This is of the ſize and ſhape of our black-bird: the head is ſmall; the ears are patulous; the iris of the eyes hazel, and the beak yellowiſh: the head, neck, ſhoulders, and back are black; the rump is grey: the breaſt is of a fine ſcarlet, and the feet are yellow.

This is a native of the Braſils. Marcgrave calls it Jacapu; and Willughby, Merula Indica pectore ſanguineo.

Turdus torquatus occipite ſanguineo.
The torquated Turdus, with the hinder part of the head red.

This is of the ſize of the common black-bird, and much reſembles in figure that ſpecies of Turdus, which we call the Redwing: the head is ſmall, and conſiderably depreſſed; the iris of the eyes is of a deep hazel: the beak is half an inch long, ſlender, and pointed; the noſtrils are oval and large, and the ears are patulous: the whole body is of a duſky colour, with few variegations; the remarkable one is a ring at the lower part of the neck, and the redneſs of the whole hinder part of the head is a very ſtriking diſtinction; the legs are ſlender and yellow.

It is a native of Italy, but few of the writers on birds have been acquainted with it.

STURNUS.

THE beak of the Sturnus is of a ſubulated figure, depreſſed in an angulated manner, and obtuſe at the extremity: the tongue is emarginated and acute. Of this genus there is but one known ſpecies; others have been referred to it, but erroneouſly.

STURNUS.

The Starling.

This is of the ſize of the common black-bird, but it ſtands more erect, and the body is ſlenderer: the head is ſmall and depreſſed; the eyes have a hazel iris; the beak is is more than half an inch long, and is yellow in the male, and brown in the female: the general colour is black, but it is variegated with spots of grey, and the tips of the feathers of the neck and back are yellowish; there is a changeable cast of bluish or purplish, as it is viewed in different lights, seen all over the back, and that particularly in the male, and the rump has a tinge of greenish or greyish: the wings are moderately long, and their principal feathers are brown: the tail is moderately long, and composed of twelve feathers; they are brown, and have some yellow at their edges.

This is frequent with us, and, when taught, will imitate the human voice. All the writers on birds have named it, and all have called it by the same name Sturnus.

ALAUDA.

THE beak of the Alauda is straight, and of a subulated form, and both the chaps are equal in length: the tongue is membranaceous, acute, and emarginated, and the hinder toe is longest.

Alauda rectricibus albo et ferrugineo partim variegatis.
The Alauda, with the long wing feathers variegated with white and brown.

The Sky-lark.

This is somewhat larger than the common sparrow, and the body is longer, in proportion to it's thickness; it's weight is about an ounce and half: the head is small, and the beak slender; the nostrils are small and round: there is a grey corona surrounding the hinder part of the head, and reaching to the eyes; and the bird has a way of erecting the feathers along the top of the head, so as to form a kind of crest; the general colour of the head is a brownish-grey, with some blackness in the middle of the feathers: the back is of the same general colouring, and has the same variegations: there is some white under the throat toward it's top, and lower down it is yellowish, with spots of brown: the wings are long and well-feathered; the tail is moderately long, and is formed of twelve feathers: the legs are brown; the toes are slender, and the hinder one longest.

This is very common with us, and is described by all the writers on birds. They call it Alauda vulgaris; and some of them Alauda cristata; others, Alauda non cristata.

Alauda lineola superciliorum alba.
The Alauda, with the line over the eyes white.

The Titlark.

This is considerably smaller than the common lark; it's weight is hardly an ounce: the head is small and depressed; the iris of the eyes is hazel; the beak is fully half an inch long, and very slender: the head, the shoulders, and the back are mottled with yellow, black, and greenish: the rump is, simply, of a yellowish-green; and the neck, on it's upper part, has all the variegations of the back, but they are obscured by a considerable admixture of grey.

The throat, the breast, and sides are of a yellowish-white, variegated with spots of black; and the lower part of the belly is, simply, of a whitish-yellow.

It is frequent with us in woods, and sits on trees, which the former species never does. All the writers on birds have described it. They call it by the general name of Alauda Pratorum.

Alauda

Alauda alis oblique albo variegatis.
The Alauda, with the wings obliquely variegated with white.

The Wood-lark.

This is of a middle ſize, between the common lark and the tit-lark; it's weight is about an ounce and a quarter, and it's body is ſhorter, in proportion to it's thickneſs, than in the common lark: the head is ſmall; the beak is brown, ſlender, and half an inch long: the noſtrils are round; the iris of the eyes is brown: the legs are of a pale ſtraw colour, and ſometimes reddiſh.

The breaſt and belly are of a yellowiſh-white, and the throat is whiter ſtill, but there are black ſpots on the middle of all the feathers: the head and back are mottled with black, brown, and yellowiſh; the neck has a great deal of grey diffuſed over it: there runs a white line backwards from each eye, which in a manner ſurrounds the head; the rump is of a reddiſh-yellow, and the long feathers in each wing; and ſome them are elegantly and obliquely divided between white and brown, eſpecially the ſecond in each wing, which is the characteriſtic of the ſpecies.

This is frequent with us; we often ſee flights of them librating themſelves in the air, and ſinging at the ſame time with a note not unlike that of the black-bird. All the writers on theſe ſubjects have deſcribed it. Ray, Willughby, and others diſtinguiſh it by the name of Alauda Arborea.

Alauda gula pectoreque flaveſcente.
The yellow-breaſted Alauda.

This is of the ſize of the common ſparrow, and more reſembles that bird in ſhape than any other ſpecies of this genus: the head is ſmall; the beak ſlender, and the iris of the eyes hazel: the ground colour of the head, neck, and back is a ruſty grey, and it is variegated with ſpots of black and brown, diſpoſed in the manner of thoſe on the wood-cock: the breaſt and the throat are yellow; there is alſo a line of yellow on each ſide of the head over the eyes, continued to the hinder part; the belly is white: the covering feathers of the wings are of a blackiſh-brown, ſpotted with white: the tail is moderately long, and not forked.

We have this in plenty in England, and in moſt other parts of Europe. Ray, Willughby, and others have called it Alauda minor campeſtris.

Alauda rectricibus nigris, lateralibus tribus albis.
The Alauda, with the tail feathers black, except the three lateral ones, which are white.

The pied Chaffinch.

The Engliſh name of this ſpecies is a very ſtrange one, for one of this genus; but it muſt be acknowledged that the bird has leſs of the general appearance of the lark-kind than any of the others; the generality of writers have not indeed known whither to refer it. It is about the ſize of the wood-lark: the head is ſmall; the beak is ſhort, and not very ſlender; and the eyes bright, and their iris of a beautiful hazel: the head, neck, and breaſt are of a whitiſh-brown; the whole under part of the body is whitiſh; the back is naturally black, but it varies with grey and brown: the wings are black at the tips, and whitiſh elſewhere: the tail is black, excepting for the three outer feathers on each ſide, which are white.

This is not a native of England, but in the more northern parts of Europe it is frequent. Authors have deſcribed it under the ſeveral names of Paſſer Lapponico-Alpinus, and Avis nivalis; and the bird deſcribed as different ſpecies, under the names of Fringilla montana, and Montifringilla calcaribus Alaudæ, ſeems to be the ſame in all reſpects with it.

FRINGILLA.

THE beak of the Fringilla is of a conic figure, and acute; the two chaps mutually receiving one another.

Fringilla alis flavo nigro et albo variegatis.
The Fringilla, with the wings variegated with black, yellow, and white.

The Gold-finch.

This is smaller than the common sparrow, and of a more elegant form: the head is large and round; the neck short and thick; the eyes are bright, and their iris is of a dark hazel: the beak is robust, short, and white, sometimes black just at the tip.

The bird is very gaily and elegantly coloured; there is a beautiful spot of red at the base of the beak: the top of the head is black, and it's hinder part white: the neck and back are of a mixed colour, composed of grey, and a reddish-brown: the belly is white; the ground colour of the wings and tail is black, but they are elegantly variegated with yellow and white: the legs are short, and the hinder toe is longer than any of the others.

This species is very frequent with us, especially on heaths, and by way-sides, where there are thistles, on the seeds of which it feeds. All the writers on birds have have named it; they all call it Carduelis.

Fringilla capite nigricante maculato.
The Fringilla, with a black-spotted head.

The Greenland Gold-finch.

This is of the size of our common linnet. The head is large, rounded, and well-feathered; the eyes are large; their iris is of a deep hazel; the beak is robust, short, and yellow, but black at the tip: the head is black, but there is an elegant spot of white on each side of it behind the eyes, and some other variegations: the hinder or upper part of the neck is of a reddish-brown; the back, the wings, and the tail are of a greyish colour, with an admixture of brown: the breast is white, but there runs all down it a longitudinal streak of black: the tail is short and forked.

This is a native of Sweden and Denmark. Rudbeck and others call it Carduelis Laponica.

Fringilla alarum basi subtus flavissima.
The Fringilla, with the base of the wings a gold yellow underneath.

The Brambling.

This is a very beautiful bird; it is of the size of the common lark: the head is moderately large and rounded; the beak is a quarter of an inch, or somewhat more, in length, very robust, and of a conic figure; the base is usually yellow, and the extremity black, but sometimes it is black all over.

The head, neck, and upper part of the back, in the male, are of a fine shining black, with a changeable tinge of purple: the hinder part toward the rump is whitish; the throat is of a reddish-brown, with an admixture of yellow: the breast is white, and the feathers behind the anus are reddish: the principal feathers of the wings are black, but they are more or less variegated with white and brown; the tail is black, but is is sometimes in part tipped or edged with brownish or whitish. In the female, the head is of a greyish-brown; the neck is grey, and the back is variegated with a great deal of brownish-grey at the edges of the feathers.

We

We have this species in England but rare; it is very frequent in other parts of Europe. All the writers on birds have named it. Ray calls it Fringilla Montana; others, Montifringilla.

Fringilla pectore ferrugineo, alis nigris albo maculatis.
The Fringilla, with a ferrugineous breast, and with the wings black and spotted with white. **The Chaffinch.**

This is somewhat smaller than the common sparrow; it's weight is less than an ounce. The head is large and rounded; the iris of the eyes is hazel; the beak is short, but very robust and strong: the head in the male is bluish, and almost black, just at the anterior part: the back is of a reddish-brown, with some admixture of greenish and greyish; the breast is of a reddish hue, and the lower part of the belly is white: the wings are black, variegated with white, and the tail is also black and white: the legs are short, robust, and brown.

It is very common with us, and all the writers on birds have described it. They call it Fringilla.

Fringilla virescens alis et cauda luteo variegatis.
The greenish Fringilla, with the wings and tail variegated with yellow. **The Green-finch.**

This is a little larger than the chaffinch: the head is large and round, the ears are patulous; the iris of the eyes hazel, and the beak conic and robust: the whole upper part of the body in the male, and especially in the summer months, has a strong tinge of green diffused over it at other times, and in general in the female; it is of a dusky brown, with only a faint tendency toward the olive: the breast and belly are yellowish, but this is also more strongly in the male than in the female, and most conspicuous in the summer months: the wings and tail are black, but they are both beautifully variegated with a bright yellow.

This is very frequent with us. All the writers on birds have described it, and all under the same name Chloris.

Fringilla dorso cinereo-lutescente, pectore maculato.
The yellowish-grey Fringilla, with a spotted breast. **The Siskin.**

This is of the bigness of the green-finch: the head is large and round; the iris of the eyes is hazel; the beak short, conic, and robust: the head is black; the back has a greenish tinge, but, when nearly viewed, the colour is found to be composed of a mixture of grey and yellow; the upper part of the neck is blackish; the rump is of a greenish-yellow: the throat and breast are spotted, and of a greenish-yellow; the belly is white; the wings are elegantly variegated with a transverse streak of yellow, and the principal feathers are brownish, but green at the edges.

We have not this species wild in England; it is frequent in Germany, and sings very sweetly. The writers on birds have almost all mentioned it. Gesner calls it Acanthus avicula; others, Spinus sive Ligurinus. It is fond of places where there are juniper-trees, in it's wild state.

Fringilla capite flavo, dorso griseo-flavicante.
The Fringilla, with a yellow head, and a greyish-yellow body. **The Yellow-hammer.**

This is somewhat larger than the common sparrow, and is an extremely beautiful bird: the head is large; the eyes have a hazel-coloured iris: the ears are patulous; the

the beak is robuſt and conic, and the ſides of the under chap of it are compreſſed, and of a ſingular form: the throat is of a fine bright yellow, as is alſo the belly; the breaſt has ſomewhat of a reddiſh tinge, mixed with that colour: the head is yellow, but there is a greeniſh tinge in it, and an admixture of brown: the ſhoulders are of a mixed green and grey; the feathers which cover the back are black in their middle, but their edges have a tinge of greeniſh and greyiſh, and ſomewhat of brown: the female is, in general, of a paler colour than the male, and has leſs yellow about her: the male, in the ſummer months, when viewed in front, appears almoſt entirely yellow.

This is very frequent with us. All the writers on birds have deſcribed it. Geſner calls it Emberiza flava; Bellonius, Hortulanus; Aldrovand, Latere alterum genus.

The Canary-bird.

Fringilla corpore albicante, alis et cauda vireſcentibus.
The whitiſh Fringilla, with the wings and tail greeniſh.

This is about the ſize of the chaffinch, and much reſembles it in ſhape: the head is large and rounded; the eyes are ſmall, but bright: the beak is robuſt and ſharp-pointed, but it is ſhort, and of a conic figure. The whole bird, in it's natural ſtate, is of a cream colour, or yellowiſh-white, except that the principal feathers of the wings and tail are greeniſh; but with us, as bred in cages, and under various accidents, it differs greatly in colour.

It is a native of the Canary Iſlands, whence it has it's name; but the melody of it's voice has occaſioned it to be brought into all parts of Europe, and it will be eaſily made to breed with us in cages. All the writers on birds have mentioned it. Geſner calls it Canaria; the others, with one voice, Paſſer Canarienſis. We are not to ſuppoſe all the accidental differences we ſee in it are any thing more than varieties.

The Berluccio.

Fringilla nigro et fuſco variegata, collo vireſcente.
The brown and black variegated Fringilla, with a green neck.

This is of the bigneſs of the yellow-hammer, but the body is ſlenderer: the head is large, and of an oval form; the eyes are ſmall, and their iris hazel; the beak is ſhort, thick, very robuſt, and in the male red all over; but in the female the upper chap is black, and the under one blue: the head is of a greyiſh-brown colour, variegated with black; the back is alſo of the ſame colours, but there is leſs of the black in it: the breaſt and the throat are grey in the male; indeed there is a redneſs diffuſed among the grey of the breaſt; the rump is of a darker brown, and has ſome red alſo in it: the wings are moderately long, and are elegantly variegated with white, greeniſh, and brown: the tail is compoſed of twelve feathers, and is principally brown, but there is ſome black, and ſome white in it; it is about an inch and half in length.

This is frequent in moſt parts of Europe, but we have it not in England. All the writers on birds have deſcribed it, and they have all called it Hortulanus; the Italians call it Berluccio.

The greater Linnet.

Fringilla capite cinereo, dorſo griſeo, pectore rubente.
The Fringilla, with a grey head and back, and a reddiſh breaſt.

This is of the bigneſs of the common ſparrow; the head is ſmall and depreſſed; the eyes are ſmall, and their iris hazel; the beak is ſmall, conic, and acute; the head is of a pale aſh colour, often with a tinge of red; the back is of a duſky browniſh-grey: the breaſt and belly in the female are of a pale whitiſh-grey, but in the male there is a great bluſh of red on the breaſt.

The wings are moderately long; the principal feathers are black, variegated with white; the tail is forked; the female never has the redneſs at the top of the head; the male always has it in ſome degree, but it is more conſpicuous in ſummer than in winter.

This is frequent with us, and has been deſcribed by all the writers on birds. They call it Linaria rubra major.

Fringilla dorſo fuſco et albo variegato, pectore ſanguineo.
The Fringilla, with the back brown and white, the breaſt red.

The common Linnet.

This is of the ſize of the gold-finch, or ſomewhat larger: the head is ſmall and flatted; the eyes are ſmall and hazel; the beak is conic, very ſmall, and ſharp at the point: the head, neck, and back are variegated with a bright brown and white; the breaſt and belly are pale; but in the male, and eſpecially in the ſummer months, there is a redneſs on the top of the head, and all over the breaſt; the tail is brown, with the tips and edges of many of the feathers pale: the long feathers of the wings are alſo brown, with the edges pale; and there is a white ſpot on the upper part of each wing, formed by the covering feathers.

This is very frequent with us. Geſner and others call it Linaria rubra; Ray and Willughby, Linaria rubra minor.

Fringilla capite nigro, maxillis rufis, torque albo.
The Fringilla, with a black head, brown at the ſides and with a white ring.

The Reed-ſparrow.

This is a very beautiful and a very ſingular bird; it is of the ſize of the common linnet: the head is ſmall and depreſſed; the beak ſhort and black, and the eyes hazel: the head is black, only that there is ſome reddiſh-brown about the angles of the mouth; the throat and breaſt are alſo black; there is a whiteneſs ſurrounding the black of the head, in form of a ring: the outer part of the neck is grey; the back is brown, and both are ſpotted very freely with black: the long feathers of the wings are black, but their exterior edge is of a ferrugineous colour: the tail-feathers are alſo moſtly black, but ſome of them are brown, and others grey at the edges: the rump is of a duſky grey, and the belly white.

It is frequent in watery places, and has a very ſweet and ſingular note. Geſner calls it Paſſer Aquaticus ſive Schœniclus; Aldrovand, Junco ſive Paſſer arundinaceus; Ray and Willughby, and from them ſeveral others, Paſſer torquatus in arundinetis nidificans.

Fringilla fuſca gula nigra, temporibus ferrugineis.
The brown Fringilla, with a black throat, and brown temples.

The common Sparrow.

This is larger than the linnet, and the male is an erect and handſome bird: the head is large; the eyes ſmall, and the beak ſhort: the general colour on the back is a duſky brown, and on the breaſt and belly whitiſh: the head in the male is grey, ornamented with a black ſpot, and the ſides browniſh: the throat is ornamented with a large black ſpot: the wings are ſhort, and their principal feathers brown; the tail is ſhort and forked.

This is common in all parts of Europe, and is deſcribed by all the writers on birds, under the name of Paſſer domeſticus.

TROCHILUS.

THE beak of the Trochilus is of a fubulated figure, but fine as a thread; it is not perfectly ftraight, and is longer than the head: the whole bird is very minute.

Trochilus aureus purpure variegatus.
The gold and purple Trochilus.

The yellow humming-bird.

This is the fmalleft of all the known birds; our largeft fpecies of the humble-bee is very nearly equal to it in fize: the head is fmall and rounded; the beak is long, very flender, and turns upward toward the extremity; the eyes are minute, but very bright; the legs are extreamly fmall: the colour in the female is perfectly, and without variegation, yellow; in the male there is a beautiful tinge of purple thrown in the way of a changeable colour over the head and neck, and in fome degree over the breaft; and, in the male, the yellow is alfo more bright and glittering than in the female.

This is a native of many parts of America, and feeds on the honey juices of flowers, which it fucks in the fame manner as the bee, and often is on the wing all the time it is feeding. The late authors on thefe fubjects have all mentioned it. They call it Mellivora and Tominein.

Trochilus viridefcens.
The green Trochilus.

The green humming-bird.

This, though very minute, is a little larger than the other, and is yet more beautiful: the head is fmall and round; the beak is long, and extreamly flender, and turns up toward the point: the eyes are minute, and their iris of a gold yellow: the whole bird is of the fame uniform colour, which is the moft elegant green imaginable, with a fhade of a deep changeable blue on many parts, and particularly about the head and neck: the wings are long, and of a brighter green than any other part; the legs and feet are wonderfully fmall; the diminutive fize and fingular colouring of thefe birds are incidents fo ftriking, that there requires little more to be faid to characterife and diftinguifh them.

This is a native of the fame parts of America with the former, and the fame writers have defcribed it.

Trochilus purpurafcens.
The purple Trochilus.

The purple humming-bird.

This is the moft beautiful of the three fpecies, and is of a middle fize between the other two: the head is very fmall, and almoft perfectly round; the eyes are minute, and their iris of a gold yellow: the beak is of nearly twice the length of the head, and is flender as a thread of filk; of a deep gloffy brown, and turned upwards at the extremity: the legs are very minute, and the feet of an elegant and beautiful ftructure: the head is of a deep blood colour, with a fhade of a changeable hue, that appears at firft fight to be black, but is in reality a very deep blue; the neck is fomewhat paler than the head; the whole upper furface is of a bright and beautiful purple, and the breaft and belly of the fame colour, only of a paler tinge; there is fome of the changeable blue diffufed alfo over the back, but none of it on the breaft.

This is a native of the fame parts of the world with the former, and is frequently brought over by way of curiofity, among the other fpecies. They are fo fmall, and fo elegantly coloured, that the women of fome parts of America wear them in their ears by way of pendants.

SITTA.

SITTA.

THE beak of the Sitta is of a conic, and somewhat cultrated, form: the tongue is lacerated and emarginated, and the feathers of the tail are rigid.

Of this genus there is but one known species, and that has been usually confounded with the picus.

SITTA. The Nuthatch.

This is of the size of our common gold-finch: the head is small and depressed; the beak is short, and is black on the upper part, and whitish on the lower, toward the throat: the nostrils are round, and are ornamented about their edges with short rigid hairs.

The head, neck, and back are grey; the sides under the covers of the wings are reddish: the throat and the breast are of a paler yellow, with a tinge of brownish or chesnut red in it; the lower part of the belly has some red feathers, with white tips: there runs on each side of the head, from the angles of the beak quite to the neck, a long black spot.

The wings are grey; the long feathers are eighteen in each, and each wing is variegated on the under part with two white spots: the tail is very short, and is composed of twelve feathers, which are some of them totally grey, others partly black; there is a transverse white streak under it, and the whole is rigid, in the manner of that of the wood-pecker-kind, and serves to the same purpose, assisting the bird in supporting itself on the trunks of trees.

It is frequent with us; it's food is sometimes insects, but, in defect of them, vegetables, as nuts and the like. All the writers on birds have described it. They call it Picus cinereus.

AMPELIS.

THE beak of the Ampelis is conic, straight, and convex: the upper chap is longer than the under, and the point is emarginated both ways: the tongue is crooked, cartilaginous, acute, and bifid.

Ampelis grisea capite nigro variegato. *The grey Ampelis, with the head variegated with black.* The Roller.

This is of the size of the common black-bird; the head is large, and somewhat depressed; the beak is black: the head is of a bright brown, variegated with black; the feathers toward the hinder part are longer than elsewhere: at the angles of the mouth there is also a line of white on each side: the whole body is grey, only that toward the head there is some admixture of a chesnut brown: the throat, the temples, and the space about the eyes are black: the feathers which cover the nostrils are also black, and so are the long feathers of the wings, and those of the tail; the legs and feet also are of the same colour: the long feathers on the hinder part of the head are moveable at the creature's pleasure, and, when they are erected, it appears crested: the rib of the long feathers of the tail is of a reddish hue, and their tips are yellowish; and the long feathers of the wings are some of them terminated in a very singular manner, by a fine purple membrane: the covering feathers of the wings are also black, but the principal of them have the tips white, which occasions some variegation.

This

This is a native of many parts of Europe, but we have it not in England; it lives principally in woods. Geſner calls it Garrulus Bohemicus; Aldrovand, Ampelis; Jonſton, and many others, Picæ glandariæ ſpecies.

Ampelis dorſo griſeo, macula ad oculos longitudinali.
The grey-backed Ampelis, with a longitudinal ſpot at the eyes.

This is a bird ſtrangely miſplaced, as to it's genus, by the generality of the naturaliſts, moſt of them making it a kind of hawk; the bird is about the bigneſs of a lark: the head is moderately large and depreſſed; the eyes are bright and piercing, and their iris hazel; the beak is black, long, ſtrong, and hooked at the end, and near the curvature there are two appendages, one on each ſide; the mouth is yellow within, and the fiſſure of the palate hairy.

There are a number of ſhort, black briſtles about the noſtrils and the angles of the mouth; the back is of a ferruginous brown, and ſo are the ſmaller covering feathers of the wings; the head is grey, and ſo is the rump; from the beak, on each ſide, there runs a longitudinal black ſtreak along by the eyes, to the back part of the head: at the extremity of this ſpot there is a white one, ſerving to divide it from the grey; the belly is white, and the throat and breaſt are alſo white, but they have a tinge of red diffuſed over them: the wings are moderately long; the tail is compoſed of twelve feathers, the middle ones longeſt; theſe are totally black, the others are more or leſs variegated with white: the legs are black, or of a very deep blue.

It is a native of England, but is not very common with us; it feeds on beetles, flies, and other inſects. All the writers on birds have deſcribed it. Willughby calls it Lanius tertius; others, Lanius minor rufus.

Ampelis cæruleſcens alis caudaque nigricantibus.
The bluiſh Lanius, with the wings and tail black.

This is of the ſize of the common black-bird; the head is ſmall and flatted; the eyes are bright and piercing, and their iris yellow; the beak is oblong, and ſomewhat hooked at the extremity: the head is of a very dark blue grey, approaching to black, and there is a line of abſolute black paſſing from the angles of the beak, to the hinder part of the head: the back alſo is of the ſame dark blue grey, and the breaſt and belly are of the ſame colour, only paler: the long feathers of the wings have their tips white; the tail is black, and the legs and feet are black.

The bird is frequent in ſome parts of England. All the writers on theſe ſubjects have mentioned it. Willughby calls it Lanius cinereus major; others, Collurio major.

L O X I A.

THE beak of the Loxia is large, thick, and ſhort, crooked, convex each way, and the upper chap of it is moveable, as well as the under: the tongue is entire and undivided.

Loxia linea alarum duplici alba. **The Groſs-beak.**
The Loxia, with a double line of white on the wings.

This is larger than the common ſparrow; the head is large and round; the beak is remarkably large, thick at the baſe, and pointed at the extremity; of a conic figure, and of a reddiſh or whitiſh colour: the eyes are large, and their iris is grey; the front of the head is of an orange colour, and between that and the eyes it is black: the reſt of the head is of a reddiſh-brown, with ſome tinge of yellow; the neck is grey; the back is of a reddiſh-brown, but the middle part of all the feathers is whitiſh: the breaſt

BIRDS Series 7.
The Ring Ouzel
The Crested Lark
The Pied Finch
The Gold Finch
The Brambling
The Bunting
The Canary Bird

breast is of a reddish-grey, and so are the sides, but the sides have more of the red than the breast: the belly, especially toward the tail, is white: the long feathers in each wing are eighteen; the tail is short, and is composed of twelve feathers, all of equal length.

This is a native of Germany, France, and Italy; it sometimes comes over to us in winter. Authors call it Coccothraustes and Coccothraustes vulgaris. We the Haw-finch or Gross-beak.

Loxia rostro forficato.
The Loxia, with a forficated beak.

The Cross-beak.

This is much of the form and shape of our bull-finch, but a little larger: the head is large and rounded; the eyes are large, and their iris is of a greyish hazel: the beak is very long, in proportion to the bird, and is of a very singular form; it is thick at the base, and both the chaps are arched or bent so, that the points cross one another, and stand at a distance: the head, neck, back, and covering feathers of the wings are of a deep blackish colour, but the edges of all the feathers have a greenish cast; there is also somewhat of grey in the head, along with the general colour: the rump is greenish; the upper part of the throat is grey, and the breast is green; the belly is white, and the feathers immediately under the tail are black. This is the general colouring, but the bird differs extremely both in the different sexes, and in ages, and even in seasons of the year; it has often more of the green, sometimes a great deal of yellow in it, and at some times is brownish.

It is frequent in Germany, and we have it also in the West of England; with us it eats apples, cherries, and other fruits, and is very ravenous. In other parts of Europe it feeds principally on the kernels in the cones of the pine and fir-kind, and it's beak seems formed on purpose for the getting at them. All the authors on birds have called it Loxia.

Loxia capite nigro, pectore rubente.
The Loxia, with a black head, and a red breast.

The Bull-finch.

This is about the size of the common sparrow: the head is large and rounded; the eyes are large, and their iris is hazel; the beak is very thick, and very short, and both chaps are convex: the front of the head is black, but in the male there is some redness at the sides: the back is of a dusky bluish-grey, approaching to black, as are also the neck, shoulders, and tops of the wings: the breast is red, as is also the throat; the belly is white, and there is also some white in the feathers, which cover the base of the tail, and the larger ones of the wings: the wings are elegantly variegated with black and red; the tail is short, black, and composed of twelve feathers.

This is common with us. Authors call it Rubicilla and Pyrrhula; we, the Alp, or Nope, and the bull-finch.

MOTACILLA.

THE beak of the Motacilla is of a subulated figure, and is straight; the tongue is lacerated.

Motacilla pectore nigro.
The Motacilla, with a black breast.

The common Wagtail.

This is a very beautiful bird; it is about equal to the gold-finch in size, but the body is longer, in proportion, and is much better covered with feathers, so that it looks much larger than it is: the head is large and rounded; the eyes are large, and their iris hazel; the beak is straight, slender, moderately long, and black: the beak and the eyes are surrounded with a space of white, which is continued in a broad line, down almost to the wings: the crown of the head, both sides of the neck, and the back are black; the breast and belly are white: the wings, when expanded, are of a semicircular figure; the long feathers in each are eighteen: the tail is long, and both that and the wings are variegated with black and white.

This is frequent with us about waters. Authors call it Motacilla alba.

Motacilla pectore abdomineque flavo.
The yellow-breasted Motacilla. **The yellow Wagtail.**

This is a more beautiful bird than the former; it is of the same figure with the former, but somewhat smaller: the head is rounded; the eyes are large, and their iris of a greyish hazel: the head and neck are of a dusky olive colour, approaching to black; the middle of the back is greenish; the breast and belly are yellow; the sides of the head are variegated also with some streaks of yellow, and the wings have some variegation of white.

This is frequent enough about waters. Authors call it Motacilla flava.

Motacilla pectore albo, corpore nigro.
The Motacilla, with a black body and white breast. **The Water Ouzel.**

This is of the bigness of the common black-bird: the head is large and round; the eyes are large, and their iris is hazel, with a tinge of grey: the beak is slender and black; the head and the upper part of the neck are of a deep rusty brown, approaching to black: the back, wings, and shoulders are variegated with a deep grey, and a bright black: the throat and the anterior part of the breast are of a snow-white; the upper part of the belly is reddish, but the lower part, toward the tail, is black.

This is frequent in many parts of England. Authors call it Merula aquatica.

Motacilla dorso cano, fronte alba.
The Motacilla, with a grey back and white forehead. **The Wheat-ear.**

This is of the bigness of the sparrow: the head is large and flatted; the eyes are small, and their iris grey: the beak is slender, oblong, and black; the head is white at the front, but there is a great deal of black about the eyes: the hinder part of the head is grey, with a blush of red; and the back also is grey, with more of the red in it; and in the female there is an admixture of green about the edges of the feathers; the rump is white; the throat and the breast are white, with a tinge of red diffused over them, and the belly is perfectly white in the female, but in the male it is yellowish: the wings are entirely black; both the long feathers and the investient ones are of that colour: the tail is moderately long; it is composed of twelve feathers, and is variegated with black and white.

This is frequent in many parts of England, and is much esteemed at table. Authors call it Oenanthe sive Vitiflora; we the Fallow-finch and the Wheat-ear.

Motacilla nigricans gula flavescente.
The black Motacilla, with a yellow throat. **The Stone-chatter.**

This is of the bigness of the linnet: the head is large, the eyes large, and their iris is hazel, and the beak black: the head in the male is coal-black; in the female it is brownish: the neck is black on the upper or outer part, but there is on each side a white spot: the back is black, but the edges of the feathers have a tinge of yellowish:

lowish: the throat is yellow; the breast is yellow, with a tinge of red, and the belly is white.

This is frequent in most parts of Europe; it lives principally on heaths, and is very noisy. All the writers on birds have described it. Bellonius calls it Rubetra; others, Muscicapa tertia.

Motacilla pectore caeruleo, macula flavescente, albedine cincta.
The blue-breasted Motacilla.

This is of the bigness of the chaffinch: the head is large; the eyes are hazel, with a tinge of grey; the beak is slender, sharp, and black: the head, the back, and the tail are grey: there is a small white spot on each side of the head, near the corner of the eye, and from the base of the beak to the breast there runs a deep blue streak: at the beginning of the breast it is terminated by a spot of yellow, and in the midst of it there is also a yellowish spot, surrounded with whitish; the rest of the breast is white; the belly also is white, but under the tail it is yellowish. The female is throughout of a paler colour than the male, and she has not that spot of white at the eye.

This is frequent in many parts of Europe, but I have not seen it in England. Authors call it Ruticilla tertia and Ulig stecklin; others, Avis Carolina.

Motacilla russo cinerea, genuum annulis cinereis.
The brownish-grey Motacilla, with the annulets of the knees grey.

The Nightingale.

This is more eminent for the sweetness of it's note, than for it's beauty; it is of the size of the linnet, but in shape it more resembles the red-breast: the head is small; the eyes are large, and their iris is pale: the beak is dusky, slender, and moderately long; the head, neck, and back are of a greyish-brown; the upper parts of the wings have a tinge of reddish mixt with this, and there is yet more of this colour about the the tail: the throat, the breast, and the belly are of a pale whitish-grey; the tail is white underneath, and the thighs are also covered with white feathers, but the knees are surrounded, as it were, with rings of grey.

Authors have all described it under the name of Luscinia.

Motacilla gula nigra, abdomine russo.
The Motacilla, with a black throat and reddish belly.

The Red-start.

This is of the size of the chaffinch, but slenderer, in proportion to it's thickness: the head is small, and somewhat depressed; the eyes are large, and their iris grey, with a tinge of hazel: the beak is slender, oblong, and of a dark colour.

The head, the neck, and the back are are of a bright grey; the anterior part of the head is white: the throat and the sides of the head under the eyes are black: the breast is of a reddish colour, and so also are the rump and the tail, only that two of the feathers in the middle are brown: the female is much less beautiful than the male; her head and neck are grey, and her breast is pale.

We have this frequently in our orchards and woods; it is a very beautiful bird. Authors call it Ruticilla and Phoenicurus; we, the Fire-tail, the Fire-finch, and the Red-start.

Motacilla

Motacilla fusca gula pectoreque subrubentibus.
The brown Motacilla, with the throat and breast reddish.

The Red-breast.

This is of the size of the nightingale: the head is moderately large and rounded; the eyes are bright and small, the beak slender and brown: the head, neck, and back are of a pale olive brown, with a tinge of grey: the throat and breast are throughout of a tawny colour, approaching to reddish: the belly is white; the wings and tail of the same brownish colour, but the covering feathers of the hinder part of the wing have a few spots of a dusky ferruginous colour: the legs and feet are brown; the hinder toe long.

This is frequent in our gardens, and sings finely. Authors call it Rubecula and Erithacus.

Motacilla testacea, subtus cinerea, pileo obscuro.
The brown Motacilla, with a black cap, and grey breast.

The Black-cap.

This is a very small bird, scarce larger than the wren; the top of the head is entirely black; the neck is grey, and the back is of a brown colour, with a very strong tinge of green in it: the throat, breast, and belly are of a pale whitish-grey; the lower part of the belly only is a little yellowish; the long feathers of the wings and tail are brown, but they have a tinge of greenish about their edges.

This is very frequent in Italy, and the warmer parts of Europe, and we have it also in England. All the writers on birds have described it. They call it Atricapilla and Ficedula.

Motacilla fusca, subtus alba, pectore maculato.
The brown Motacilla, white below, and with the breast spotted.

The Cretan Ficedula.

This is a little larger than our wren: the head is small and rounded; the eyes are hazel, and the beak is brown: the head, neck, and back are of a dusky brown: the wings are somewhat darker than the back, as is also the tail: the belly is white, and the breast is grey, spotted with black: the legs are very slender and brown.

This is frequent in the East, and in some parts of Europe; we have it in England but very rarely. Most of the writers on birds have described it. Some call it Ficedula Cannabina; others, after Aldrovand and Willoghby, Ficedula quarta.

Motacilla castanea alis nigro cinereoque variegatis.
The chesnut-coloured Motacilla, with the wings variegated with white and grey.

The Wren.

This is a very minute bird; we have not in Europe any that is smaller: the head is large and round; the eyes are dark, and the beak slender and brown: the head, neck, and back are of a dusky chesnut-brown; the rump is of a brighter brown, as is also the tail, and all this part is variegated with obscure and transverse lines of black; the throat is of a pale yellowish colour; the middle of the breast is still whiter, and the lower part of it is variegated with transverse lines of black: the sides also are variegated in the same manner, and the lower part of the belly is of a reddish-brown; there are a few round white spots on the wings, as also on the base of the tail: the tail is short, and is generally carried erect.

This is every-where frequent with us, creeping about hedges, and taking short flights. All the writers on birds have named it, and they all call it Passer troglodytes.

Motacilla

Motacilla fuſca, ſubtus albida, macula poſt oculos griſea.
The brown Motacilla, white underneath, and with a grey ſpot behind the eyes.

The Hedge-ſparrow.

This is of the bigneſs of the red-breaſt: the head is large and rounded; the eyes are ſmall, and their iris is hazel: the beak is ſlender and brown; the ears are large and patulous.

The head, neck, and back are of a mixed brown, compoſed of a reddiſh-tawny and black; the middles of all the feathers being black, and their edges ferrugineous: the head has alſo ſomewhat of a greyiſh tinge, and is ornamented on each ſide with a ſpot; the lower part of the back, toward the rump, has ſomewhat of a greeniſh tinge: the large feathers of the wings are brown, and their edges a reddiſh-brown; there are alſo ſome white ſpots on the others, one on each: the breaſt and throat are grey; the belly and thighs are white; the legs yellowiſh.

This is frequent in our hedges, and ſings very ſweetly. All the writers on birds have deſcribed it. Geſner calls it Curruca; others, Hippolais and Curruca Eliote.

Motacilla vireſcens ſubtus cinerea.
The greeniſh Motacilla, with a grey breaſt.

This is of the ſize of the hedge-ſparrow, and has been by many miſtaken for the female of that ſpecies; the head is large and rounded; the eyes are ſmall, and their iris is hazel: the beak is ſlender and pointed; the head, neck, and back are of a duſky olive colour, with a ſtrong tinge of green: the wings are of a ferrugineous brown; the throat is white, and the breaſt is of a pale grey; the belly whiter: the tail is brown, but toward the edges it has a little variegation of white.

Authors call this Ficedula ſepiaria. We have it not in England.

Motacilla alis nigro et flavo variegatis.
The Motacilla, with variegated wings.

The creſted Regulus.

This is of the ſize of the common wren, but, it's feathers lying more cloſe and even, it looks ſmaller: the head is a little depreſſed; the eyes are of a deep hazel: the beak is ſlender and brown: the head, neck, and back are of a mixt colour of greeniſh and grey; the breaſt and belly are of a pale grey, rendered duſky by an admixture of the olive brown, and the throat is paler than any other part: the wings are variegated with black and yellow, and on ſome parts with white; the head in the male is ornamented with a yellow ſpot in the middle, and has a creſt of feathers erected at pleaſure.

We have this in England, but it is leſs common than in many other parts of Europe. The authors, who have written on birds, call it Regulus criſtatus, and Parus ſylvaticus.

Motacilla cinereo-virens ſubtus flaveſcens.
The yellow-waſted, greyiſh-green Motacilla.

The Regulus, without a creſt.

This is as ſmall as the preceding ſpecies, and is a very elegant little bird: the head is ſmall, and the eyes of a bright hazel; the beak is very ſlender and ſharp, and of a bright brown: the head, neck, and back are of a very ſingular colour, which ſeems compoſed of grey, green, and brown; upon the whole, it forms a very ſingular kind of olive: the ſides of the head are ornamented with an oblong, yellow line, which runs from the eyes to the hinder part of the head: the throat is of a very pale yellowish, approaching to lemon colour; the breast and belly are of the same colour, only yellower: the long feathers of the wings, and those of the tail, are of a dusky brown, but they are somewhat brown at the edges.

This is frequent in most parts of Europe; we have it in England, but less common than elsewhere. Authors call it Regulus alius, and Regulus alius non cristatus.

PARUS.

THE beak of the Parus is of a subulated form; the point of the tongue is truncated, or, as it were, cut off abruptly, and is terminated by four bristles.

Parus capite nigro, temporibus albis, nucha lutea. **The great**
The Parus, with a black and white head, and a yellow neck. **Titmouse.**

This is the largest of the whole family of the Pari; the size is nearly that of the gold-finch: the head is depressed; the eyes are hazel; the beak is brown and bright; the back is of a shining black; the top of the head is also of a deep glossy black, but the sides of it are white: the upper part of the throat is black, and the spot is continued till it joins the blackness on the head; there runs also an oblong spot of black from that of the neck, quite down to the breast: the upper part of the neck is yellow, and the shoulders, from an admixture of that yellow, and the black of the back, appear greenish: the breast and the belly are yellow; the wings and tail are greyish, and their principal feathers are black, with some variegation.

This is frequent in the woods in some parts of England, especially about watery places. All the writers on birds have described it. They call it Fringillago and Parus major.

Parus capite cristato.
The Parus, with a crested head. **The crested Titmouse.**

This is a very pretty bird, but it is considerably smaller than the former: the head is small; the eyes are hazel; the beak is short and brown: the back is of a glossy greyish-brown; the wings and tail are of a greyish-black; the belly is white, but the feathers have a blackness on their inside: the head is ornamented on the crown with a series of feathers longer than the others, which naturally stand in an erect, or nearly erect, posture: the head itself is of a mixed black and white colour; the feathers are all black toward the base, and white at the tip; there is a black space behind the eyes, and a black collar about the neck: the tail is short, and the legs are bluish. Authors call it Parus cristatus.

Parus vertice cæruleo, alis albo variegatis. **The blue**
The blue-headed Parus, with black and white wings. **Titmouse.**

This is a very small bird, it is little larger than our common wren: the head is small and depressed; the eyes are hazel, and the beak is black: the anterior part of the head is white; the crown of it is blue, and the sides or temples white: there runs a blue line from each side of the head near the eyes, and joins the blue on the crown; this is continued to the throat, and there spreads into a large and elegant black spot: the breast is yellow; the sides also have their share of yellow: the belly is white toward the hinder part: the long feathers of the wings are black, but blue at the edges; and those which immediately cover their bases are blue, spotted with white: the back in it's upper part is greenish, in it's lower darker; the upper part of the neck is white; the legs are black.

This is frequent in most parts of Europe. Authors call it Parus cæruleus; we, the blue Titmouse.

Parus

Parus dorso cinereo, capite nigro, vertice albo. **The black**
The black and white-headed Parus, with a grey back. **Titmouse.**

This is of the bigness of the common wren: the head is large and rounded, and the beak sharp; the eyes are hazel and large; the ears very patulous: the head is white on the top, and toward the front the rest of it is of a fine deep glossy black: the back is grey, and so is the rump; the breast and belly are white: the long feathers of the wings are brown, and the legs are blue.

This is common in all parts of Europe. Authors call it Parus ater and Carbonarius.

Parus capite nigro, temporibus albis, dorso cinereo. **The Marsh**
The grey Parus, with a black head and white temples. **Titmouse.**

This is a very pretty little bird: the head is large and flatted; the eyes are of a dark hazel, and the beak is brown: the whole head is of a deep black, except that the temples are white: the back is of a dusky greenish-grey; the belly and breast are of a whitish-grey, but it is only the tips of the feathers that come in sight, and form this colour, for the body of them is black: the long feathers of the wings are black; the covering feathers, and those of the tail, are of a greenish-grey: the legs are blue; the toes long, and the claws black and sharp.

This is common in our damp woods. Authors call it Parus palustris, from it's frequenting wet places.

Parus cauda corpore longiore. **The long-tailed**
The Parus, with the tail longer than the body. **Titmouse.**

This is a very beautiful little bird; it is larger than the common Parus: the head is large and rounded; the eyes are of a dark hazel, and the beak is black: the head is white, as is also the breast; the belly is flesh-coloured: the back is brown, and is variegated in a very beautiful manner with spots of black, and the wings are variegated with blue, black, and white: the tail is very long and black.

This is a native of England, and is common in gardens and orchards in many parts of the kingdom. The authors on birds have all described it; they call it Parus caudatus.

Parus caeruleo, albo, et nigro variegatis. **The Indian**
The variegated black, blue, and white Parus. **Titmouse.**

This is a very beautiful bird, and greatly resembles the European kinds, as well in form as in colouring: the head is large; the eyes hazel; the ears very patulous, and the beak black; the whole head and upper part of the neck are of a pale blue colour: the breast, the belly, and a part of the throat are white: the wings are of a pale blue toward their top, but their long feathers, as also those of the tail, are of a deeper blue: the back is black; all the colours are very bright and glossy, and the bird makes a very beautiful appearance, even dried, as we see it in specimens.

It is a native both of the East and West Indies. Authors have described it under the name of Parus Indicus.

Those who are not aware of the different colouring of the males and females of birds, may extend the family of the Pari to a much larger number.

HIRUNDO.

THE beak of the Hirundo is very ſmall, of a ſubulated figure, crooked, and depreſſed at the baſe: the opening of the mouth is enormouſly wide.

Hirundo rectricibus macula alba notatis.
The Hirundo, with the tail feathers ſpotted with white.

The common houſe-ſwallow.

This, though little regarded for it's beauty, is a very elegant bird; it is of the ſize of the linnet, but of a very different form of body, and diſpoſition of plumage: the head is large and depreſſed; the eyes are ſmall, but very bright: the beak is very inconſiderable; it is ſharp at the point, and flatted at the baſe, but it's angles run a great way up the head, ſo that the opening of the mouth is ſurpriſingly wide: the whole upper part of the body is of an extreamly elegant bluiſh-black, bright, gloſſy, and in ſome degree changeable with the light: the breaſt and belly are of a ſnow-white, and the covering feathers of the under part of the wings are alſo white: the anterior part of the head, and the upper part of the throat, are brown; the long feathers of the wings are black, as are alſo thoſe of the tail, which is forked: the legs are extreamly ſhort, and they are naked, not feathered, as in ſome other ſpecies.

This is frequent in ſummer about our houſes. All the authors have deſcribed it under the names of Hirundo, Hirundo vulgaris, and Hirundo domeſtica.

Hirundo dorſo cæruleſcente, rectricibus immaculatis.
The Hirundo, with a blue back, and no ſpots on the tail feathers.

The Field-ſwallow.

This is ſomewhat larger than the houſe-ſwallow: the head is large and depreſſed; the beak is ſmall and black; the eyes are very bright, and the noſtrils naked: the head and back are black, but there is a very ſtrong ſhade of deep blue diffuſed over them in ſuch a manner, that in ſome lights they appear entirely and ſolely blue: the front of the head, between the eyes and the baſe of the beak, is of a pure black, without any tinge of the blue: the long feathers both of the wings and tail are brown; the throat, the breaſt, and the belly are white, as are alſo the rump, and the feathers which cover the under parts of the wings: the legs are ſhort, and they are covered all the way down with a kind of white wool.

This ſpecies is very frequent with us about waters, and among hedges. All the authors on birds have deſcribed it. They call it Hirundo agreſtis, and Hirundo ſylveſtris.

Hirundo tota nigra gula albicante.
The black Hirundo, with only the throat white.

The Martin.

This is larger than the common ſwallow, but it's body is ſomewhat ſhorter and flatter: the head is large and flatted; the eyes are large, and extreamly bright; their iris is hazel: the beak is very ſhort and inconſiderable, but the opening of the mouth is enormouſly wide: the beak is pointed and black, very weak, and depreſſed toward the noſtrils; the whole bird, as well breaſt as back, is of the ſame colour, which is properly a brown ſo very deep, that it approaches to black, and has a tinge of greeniſh over it; only on the upper part of the throat there is a moſt remarkable large ſpot of a greyiſh-white: the wings are very long; the tail is alſo very long and forked: the legs are very ſhort, and the feet ſmall.

It is very frequent with us, and all the writers on birds have deſcribed it. They call it Hirundo apus, and Cypſelus minor.

Hirundo

Hirundo cinerea gula abdomineque albis.
The grey Hirundo, with the throat and belly white.

This is the smallest of the swallow-kind, but in form it perfectly resembles the common species: the head is large and depressed; the beak is very small, black, and weak; the eyes are very large, and their iris of a hazel colour: the head, back, wings, breast, and tail are all of the same colour, which is a deep rusty greyish-brown, approaching to black, but with a very strong tinge of green diffused over it: the throat is ornamented with a large spot of white, and the belly is white; the long feathers in the wings are eighteen in each; the tail is forked, and is composed of only ten feathers: the legs are very short, and the feet small.

It is frequent with us about the banks of rivers. All the writers on birds have described it. Gesner calls it Hirundo riparia sive Drepanis; and others use the same name.

Hirundo cauda integra, ore cetis ornato.
The Hirundo, with an undivided tail and bristles at the mouth.

The Goatsucker.

This is a very singular species, and has, by the generality of authors, been separated from it's true and proper genus, and put among the owls; nay, in English, we call it the Churn-owl: it is of the size of the cuckow, and very much resembles it in shape; the back is of a very beautifully variegated hue; the colours are black, white, and brown; these are blended together in very fine undulations, and among these there are many separate and larger spots of black: the belly is of a pale brown, and there are on it some lines of black, broader than those on the back; the breast is of the same pale colour with undulations of the same kind, only smaller.

The head is large, in proportion to the body; the ears large and patulous; the beak is inconsiderable, depressed at the base, somewhat hooked at the point, black and soft: the feet are small, and the middle toe twice as long as the others: the legs are short and hairy; the tail is long and undivided, and is composed of ten feathers.

The feathers of the whole body lie loose, and are soft and beautiful; it is very singular, that there stand a kind of whiskers like those of the beasts of prey about the mouth of this bird; they are composed each of eight hairs or bristles: the opening of the mouth in all the swallow-kind is commonly large, but in this it is more so than in any: the nostrils are prominent, and of a cylindric figure.

In the male there is a little variegation of white in some of the long feathers of the wings, which distinguishes it from the female, in which sex there is no such variegation.

We have this species in many parts of England. Many of the writers on birds have described it. Bellonius calls it Strix Caprimulgus, Fur nocturnus; others have condensed the name Caprimulgus. It feeds on insects, and flies chiefly in the evening.

PROCELLARIA.

THE beak of the Procellaria is of a compressed figure; the upper and under chap are equal in length; the upper is hooked at the point: the nostrils are of a cylindric form; they run parallel to the length of the beak, and grow to it: the feet are palmated.

Of this singular genus there is but one known species.

PROCELLARIA. **The Petrel.**

This is an extremely ſingular bird; the bigneſs of it is about that of our common water wag-tail: the head is large and depreſſed, the ears are broad and open; the eyes are very large; their iris is of a deep hazel, and they have a very piercing aſpect: the beak is ſmall, ſlender, and compreſſed; the noſtrils ſtand very prominent on it, and their openings are ſeparated by a membrane: the beak is of a duſky colour, and black at the tip.

The head is of a deep and very gloſſy black on the crown, and of a ſomewhat duſkier black on the temples: the neck and back are of the ſame deep, gloſſy, and beautiful black with the head, but the rump is leſs gloſſy: the breaſt, throat, and belly are black alſo, but it is a leſs full colour, and leſs gloſſy than on the back.

The tail feathers are twelve, all of an equal length, and black; and the tail, though not ſhort, does not exceed the tips of the wings, when they are cloſed, they being very long: the lower feathers of the rump, which cover the baſes of the long feathers of the tail, are white, with black tips, ſo that the lower part of the rump appears in a great meaſure white.

The wings are very long; the principal feathers of them are totally black, but the covering feathers of the ſecond ſeries have ſome white toward their tips.

The legs are ſhort and ſlender; the thigh is naked, at leaſt half the way up: the feet are large; the toes are but ſlender, but they are connected by a fine thin membrane; there is no hinder toe, but the claw in that part is ſmall, and is connected immediately to the back of the foot: the whole foot is indeed of a very ſingular ſtructure: the membrane which connects the toes, and forms the web, is black, and very thin: the middle toe is ſhorteſt, and has only two joints; the outer is longer, and has four: the interior has three, and is longer than either: the claws are blackiſh, and nearly equal in length.

This ſingular bird has not been known till of late years; it is an inhabitant of the Northern ſeas, where it ſkims along the ſurface of the waves with great rapidity. It has the knowledge of an approaching ſtorm, and always, in that caſe, gets under covert of the ſhips that are near: the ſailors, when they ſee no other ſign of a ſtorm, prepare for one on this ſignal, and they are not diſappointed. Dampier and ſome of our other voyagers have deſcribed it under the name of Petrel; other writers have been little acquainted with it. It was firſt mentioned in the Stockholm Tranſactions, under the name of Procellaria, or the Storm-bird; the Swedes call it, ſtormwaders fogleis.

THE

BIRDS
The White Wagtail
The Rock Robin

THE

HISTORY OF ANIMALS.

PART VI.

Of QUADRUPEDS.

QUADRUPEDS are animals which have the body covered with hairs, which walk on four legs, and the females of which bring forth their young alive, not in the egg state, and nourish them with milk from their teats.

This series of animals will be considerably lessened in number, by the throwing out of it the frog, lizard, and other amphibious animals, which have already been described in their place, but which those who confound these classes have placed under this head. It will also be enlarged by the admission of the bat, which, from it's having the forefeet webbed with a membrane, and using them, as birds do their wings, in flying, has been generally ranked among that order of animals.

QUADRUPEDS.

Class the First.

GLIRES.

THE Glires are distinguished by having the fore-teeth only two in number, and those prominent.

MUS.

THE fore-teeth in the Mus are acute, and there are no canine teeth at all; the feet are divided, and the ears are naked.

Mus cauda brevi, corpore fulvo nigroque variegato.
The short-tailed Mus, with a body variegated with black and tawny. — **The Leming.**

This creature resembles the common rat in shape, but it's tail is shorter; it's length is about five inches; it's fur is very fine, and is tinged with a great variation of colours; the

the upper part of the extremity of the head is black, the lower part is paler; the crown of the head is yellowish, and the neck and shoulders are black: the body is brown, but it is variegated with a great many spots of black; these are small, of different figures, and are equally disposed over every part, quite to the tail, which is not more than half an inch in length, and is hairy and yellowish, with an admixture of black.

The head is of an acuminated figure, large at the base, and very sharp at the point; there are a great number of rigid hairs about the mouth; these form a kind of whiskers, and twelve of them, six on each side, are longer than the rest.

The mouth is small, and the upper lip is divided, as in the squirrel, and from this there stand forth two large and somewhat crooked teeth, and on the lower jaw there stand two others to answer them: the grinders stand at a considerable distance from these; they are three, the anterior of them the broadest: the tongue is large, and so long, that it reaches to the roots of the fore-teeth.

The eyes are small and black; the ears are obtuse, and are laid backwards on the neck: the legs are very small, and the anterior pair are very short; the feet are hairy, and each is armed with five sharp, strong, and hooked claws: the hinder legs are short too, but they are somewhat longer than the others: the belly is white.

It is very singular, that a creature of a tolerably bulky body, and with such very short legs, should be able to run swiftly, yet there are few creatures more nimble.

It is frequent in Norway, and some other of the northern parts of Europe; it breeds in the mountainous places, but at times comes down into the low country, where the vast troops of it destroy all the vegetable produce, and, often dying afterwards upon the place, leave a stench, which occasions pestilential fevers. Olaus Magnus first mentioned this creature in his history of the northern nations, and from him all the succeeding authors borrowed what they have said about it. Scheffer calls it Mus montanus; others, Lemmus, Mus Norwegicus; the Swedes, Fialmus and Sablemus; the Laplanders, Lemmar.

Mus cauda brevi, dorso nigro-fusco, ventre albo.
The short-tailed Mus, with a brownish-black back, and white belly.

This is somewhat larger than the common mouse, and it's body is longer, in proportion to the thickness; the head is remarkably large, and the snout short and obtuse: the eyes are small, and not prominent, as in many kinds; the ears are short, broad, of a figure approaching to round, and they are in a manner buried in the fur, which is longer than in the common mouse.

The tail is very short, and is covered with scattered hairs, more numerous than in the common rat, but not at all thick set: the legs are very short, and the toes slender; the colour is duskier than in the common mouse, and, when closely examined, has some resemblance to that of the rat, but it is darker, and has a tinge of yellowish: the belly is whitish, with an admixture of a lead colour grey.

This is frequent with us under hedges, and in pastures; and it is singular that it has escaped the notice of so many of the writers on these subjects. Ray describes it under the name of Mus agrestis capite grandi, cauda brevi; and Agricola seems to have meant it, when he talks of a wild mouse with a large head and short tail; others are silent about it.

Mus.

Mus cauda longa ſubnuda, corpore fuſco cineraſcente.
The browniſh-grey Mus, with a long and almoſt naked tail.

The Rat.

This is a quadruped too well known to need a long deſcription; it greatly reſembles the common mouſe in form, but is at leaſt five times as large; it's colour is a duſky brown, with a greater or leſſer admixture of grey: the head is large at the baſe, conic, and ſharp at the ſnout; the eyes are ſmall and bright; the legs moderately long: the tail is very long and ſlender; it is divided into more than a hundred and fifty annular joints, as it were; it is of a bluer colour than the reſt of the body, and is almoſt naked; the hairs on it being very few, and ſet at diſtances.

This is too common in all parts of the world. The writers on animals have all deſcribed it. Geſner calls it Mus domeſticus major; Ray, Mus domeſticus major ſive Rattus; Jonſton, ſimply and improperly, Glis.

Mus cauda longa villoſa, auribus brevibus.
The ſhort-eared Mus, with a ſhort hairy tail.

The Ground-rat.

This is nearly of the ſize of the common rat, and much reſembles it in it's general form: the head is large; the ſnout ſharp; the eyes black and prominent, and the ears roundiſh and naked, compoſed of a very thin bluiſh membrane, and ſo ſhort, that they are entirely buried under the down or fur of the head: the back is of a very deep duſky brown, approaching to black; the ſides are ſomewhat paler; the belly is of a pale brown, with an admixture of grey: the tail is not more than of half the length of the body, and is hairy as the body; the teeth are yellowiſh.

I ſaw two or three of theſe ſome years ſince in Yorkſhire, but they are not common there; in ſome of the northern parts of Europe they are frequent, and very miſchievous. Few of the authors who have written on quadrupeds have mentioned it. Linnæus ſays it is common in Sweden.

Mus cauda longa corpore nigro flaveſcente.
The yellowiſh-black Mus, with a long tail.

This has more of the appearance of the common mouſe than of the rat, but it is conſiderably larger than that, though much ſmaller than the other: the head is oblong, ſharp at the ſnout, and large at the baſe; the eyes are ſmall, but prominent; the ears are ſhort, broad, and naked: the back is of a duſky colour, approaching to black, but it has ſome tinge of reddiſh-yellow in it; the belly is white: the tail is long, and has ſome hairs on it; they are black on the upper ſurface, and white on the under: the legs are longer, and more ſlender than in the common kinds.

We have this in houſes, and in the fields. Ray calls it Mus domeſticus medius; the other writers have not named it.

Mus cauda longa mediuſcula, ventre ſubalbido.
The Mus, with a long and almoſt naked tail, and a white belly.

The Mouſe.

This is much ſmaller than the preceding kinds: the head is ſmall and ſhort; the eyes are large, and very prominent: the ears are ſhort and naked; the back is of a deep bluiſh-brown, and the belly is whitiſh.

It is common every-where in houſes and in fields. Authors call it Mus domeſticus vulgaris, and Mus domeſticus minor.

Mus cauda longa piloſa, gula albicante.
The Mus, with a long hairy tail, and a white throat.

The Dormouſe.

This is a very pretty creature; it is of the bigneſs of the common mouſe: the head is ſmall, and not ſo ſharp at the ſnout as in many ſpecies; the ears are broad and ſhort; the eyes are large, bluiſh-bright, and very prominent: the head is of a reddiſh-brown, very bright and ſhining; the back is of a duſky brown, with a tinge of orange colour; the belly is of the ſame colour, but ſtill paler, and the throat is white: the tail is moderately long, and is hairy, as in the ſquirrel.

We have it in our fields and gardens. Moſt of the writers on animals have deſcribed it. Ray calls it Mus avellanarum minor; we, the Dormouſe or Sleeper, from it's naturally ſleeping all the winter part of the year.

Mus cauda longa piloſa, ventre albo.
The white-bellied Mus, with a long hairy tail.

The Italian Dormouſe.

This is ſomewhat larger than the common dormouſe, and is a much handſomer creature: the head is large, and of an elegant colour, a mixt orange and brown: the ears are ſmall, naked, and obtuſe; the eyes are black, large, very bright and protuberant: the back is of a beautiful ferruginous brown, with ſome tinge of yellow: the belly is white; the tail is long and hairy, in the manner of that of the ſquirrel; it is of a fine orange brown all the way to the tip, where it is white.

This is frequent in Italy, and ſleeps all the winter; we ſometimes have it kept here.

Mus cauda abrupta, digitis connexis.
The abrupt-tailed Mus, with connected toes.

The Water Rat.

This is conſiderably larger than the common rat, and of a different colour: the head is large, and ſharp at the extremity; the teeth are long, and of a yellowiſh colour: the eyes are large, black, and prominent: the ears are ſhort and naked.

The head is of a browniſh colour, with a ferruginous tinge; the back is alſo of a ruſty brown, with an admixture of grey: the belly is pale; the tail is not ſo long as in the common rat, nor does it grow taper from the baſe to the extremity, as in that ſpecies, but is all the way of the ſame thickneſs, and is abrupt at the extremity: the legs are ſhorter than in the common rat; and the feet are longer, and the toes connected by membranes.

This is frequent about waters, and in ſhips; the banks of the Thames, at low water, often diſcover millions of them. Ray calls it Mus major aquaticus ſive Rattus aquaticus.

Mus capite parvo, cauda lata.
The broad-tailed Mus, with a ſmall head.

The Muſk-rat.

This is larger than the common rat, and very different from it in ſhape: the head is ſmall; the opening of the mouth narrow; the eyes not prominent, and very ſmall; the ears ſhort, broad, and naked, and the teeth yellowiſh.

It is about ſeven inches in length, excluſively of the tail, and it is very bulky, in proportion: the whole body is covered very thick with long and ſoft hair; this is blackiſh on the back, and grey on the belly; the head is of a paler colour, and is covered with much ſhorter hair, and the extremity of the ſnout is ſmall, and formed for burrowing in the ground, in the manner of that of the mole: the tail is longer

than

than the whole body; it is of a very ſingular form, being flatted, and is covered with a few ſcattered hairs: the fore-legs are two inches long, and each is terminated by a broad foot, the toes of which are armed with very ſharp claws: the hinder legs are larger and longer; the toes are connected by a membrane, as in the common water rat, for the advantage in ſwimming.

This is frequent in Tartary and Ruſſia, and in ſome other parts of the world, and has a very ſtrong perfumed ſmell like muſk, which is preſerved in the dried ſkins brought over to us, and called Muſk-waſh ſkins. Ray calls it Mus aquaticus exoticus, after Cluſius.

Mus cauda hirſuta, pedibus rubentibus.
The hairy-tailed Mus, with red feet. **The great Sleeper.**

This is of the ſize of the rat, but more corpulent: the head is ſhort and thick; the opening of the mouth ſmall; the noſtrils fleſh-coloured; the eyes large, black, and prominent, and the ears large and naked.

The head is of a deep lead colour, with a tinge of brown; the back is of a dark brown, with ſome admixture of grey, but with very little of the blue tinge in it: the belly is white, as are alſo the inſide of the thighs, and the under part of the tail; there is a blackneſs all about the noſtrils, as alſo round the eyes, and the ſame colour is viſible on the extremity of the tail.

This is frequent in many parts of Europe; it retires into caverns in the ground in winter, but it carries into them a very conſiderable ſtore of nuts, and other fruits. All the writers on quadrupeds deſcribe it. Pliny calls it Sorex; Geſner, Mus avellanarum; Ray, Mus avellanarum major.

Mus corpore longiore, cauda brevi.
The Mus, with a long body, and ſhort tail. **The Citille.**

This is very different in ſhape from all other of the rat-kind: the head is ſhort and ſmall; the ſnout is ſharp; the opening of the mouth is narrow, and the teeth long and ſharp; the eyes are very ſmall, and not at all prominent; the ears are ſcarce perceptible, but the apertures are very viſible, as in the mole: the whole head is of a ſomewhat depreſſed form.

The body is long and ſlender, and reſembles that of the weaſel-kind, rather than the others of this genus: the head and back are of a duſky brown, with no admixture of a ſilvery grey; the belly is paler; the tail is very ſhort.

This is a native of many parts of Europe, but not, that I know, of England. Geſner calls it Mus noricus vel Citillus; and Ray continues the ſame name. The fur is uſed, in many parts of the North, in cloathing. The creatures retire in companies into caverns for the winter.

Mus dorſo rutilo, ventre nigricante.
The Mus, with a reddiſh-brown back, and black belly. **The Cricetus.**

This is conſiderably larger than the common rat, and more bulky in the body: the head is ſmall, and pointed at the extremity; the mouth is moderately large; the eyes are bright, black, and prominent, and the ears naked, ſhort, and broad.

The head is of a bright reddiſh-brown on the top, but black on the temples; the back is of the ſame colour with the head, unleſs that it is a little more tawny; the ſides are variegated with a few ſpots of white; the belly is black, but the throat is white: the tail is three inches in length, and thick covered with hairs of a tawny colour,

lour, with a blush of reddish in it: the legs are very short, and the feet are large; the toes long, and armed with sharp claws.

It is frequent in the East, and is very voracious; it has a custom of sitting upon it's haunches, and, when in this position it is seen in a front view, it looks like a bear in miniature. All the writers have described it. Gelenius calls it Arctomys Palæstinorum; Gesner and others, Cricetus.

Mus cauda elongata nuda, corpore rufo.
The Mus, with a long naked tail, and tawny body. **The Marmotte.**

This is the largest of the rat-kind; it is bigger than a rabbit, little less than a hare: the head is large, and broad at the base, but narrow at the extremity; the mouth is small; the teeth are long and sharp; the ears are short, and, as it were, cut off: the eyes are large and prominent, and they are very bright and black: the body is corpulent, but the legs are very short: the tail is long and naked, and in form greatly resembles that of the common rat.

The head is of a reddish-brown, or tawny colour, with the admixture of something of the orange; the nose is blackish, and there are some black whiskers about it, like those of the cat: the back is of the same tawny or orange brown; sometimes it is deeper, sometimes paler, and it often degenerates into a kind of black: the back is covered within the skin by a kind of fatty matter, even when the rest of the body is ever so bare; this seems to have given it, by nature, a defence against that intense cold which it is to endure for a very great part of the year.

The teeth are yellow; the legs are well covered with hair, and it is longer, and of a paler colour on these, and on the belly, than on the back: the feet are formed somewhat like those of the bear, and the toes are armed with very sharp and black claws; it uses the hinder legs with great familiarity, and often walks on them alone.

It is a native of the mountains in Switzerland and other places, and breeds only on their tops: they sleep much in the winter, and prepare warm and comfortable lodgings for that period, which they line with fur and dried vegetables, and store with provisions. All the authors who have have written on animals have described this. Pliny calls it Mus Alpinus, a name copied by most others.

Mus corpore variegato, cauda nulla.
The Mus, with a variegated body, and without a tail. **The Guinea-pig.**

This is considerably larger than the rat, but less than the rabbit, to which it has some degree of resemblance: the head is large and thick; the extremity of it is formed like that of the hare, but more obtuse, and the teeth are disposed in the same manner: the eyes are black and bright, but not very prominent; the ears low, broad, thin, and pellucid; the hair is more like that of a young pig, than of any thing of this kind; and the voice, in some degree also, resembles that of the hog-kind, whence, and from the form and shape not greatly disagreeing, we are not to wonder that it has obtained the name of the Guinea-pig.

The legs are robust, and not very long, and it has no tail; the colouring is very variable; we most usually see it variegated with large blotches of a tawny colour, and white; sometimes it is all over white, and sometimes entirely tawny: the fore-feet have four toes; the hinder ones have only three, and the middle one is longer than the other two. The creature is very frequently rubbing it's head with it's fore-feet, in the manner of the rabbit, and it will oftener sit upon the hinder ones; from this custom, part of the hinder legs is usually bare and callous. It does not hop in the manner of the rabbit, but walks by a regular and even motion of all four legs, in the manner of the hog; and, when it fights, it strikes with the head, in the manner of the hog.

It is a great breeder, and often will produce eight young ones at a time; it feeds on vegetables of all kinds, and it's flesh is well-tasted, and has much the flavour of pork. The authors who have written on quadrupeds have all named it. Ray calls it Mus sive Cuniculus Guineensis et Americanus Porcelli pilo et voce Cobaya Brasiliensibus dictus. It is common to the East and West Indies, and lives very comfortably with us.

Mus pilis rigidis, cauda brevi.
The Mus, with rigid hairs, and a short tail. **The Aguti.**

This is of the bigness of the Guinea-pig, and in many respects greatly resembles it: the head is large; the eyes are prominent, very bright and black; the ears are broad, and very short: the head is, in the general form, like that of the rabbit, but the snout is narrower and sharper, and the upper jaw is somewhat longer than the under, as is the case in the hog-kind; the upper lip, however, is divided, as in the hare, and through this division are seen the teeth, which are long and sharp: the tail is very short and smooth; the legs also are short, and almost naked, and the anterior ones are shorter than the hinder.

The fur is very thick and harsh; the hairs of which it is composed are stiff, rigid, and shining, like those of the hog-kind; the colours are a mixture of brown and reddish, with some tinge of black.

This is a native of South America; it is a very voracious animal, and holds it's food in the fore-feet, as the squirrel, swallowing it with great quickness. When it is angry, it raises the bristles on the back, and strikes the earth forcibly with it's hinder legs; it is very swift in running. Marcgrave calls it Aguti sive Acuti Brasiliensibus; Ray, Mus sylvestris Americanus Cuniculi magnitudine, Porcelli pilis et voce.

Mus fuscus cinereo maculatus ventre albente.
The brown Mus, spotted with grey, and with a white belly. **The Paca.**

This is the largest of all the Mus-kind; it is as big as a pig of a week old; the head is short, and the extremity or snout sharp, in the manner of the former species, and the upper jaw is a little longer than the under; the upper lip is divided, and the mouth is furnished with a kind of beard: the teeth are sharp; the eyes are black and prominent; the nostrils are large and patulous; the ears are short, naked, and somewhat obtuse.

The body is thick and fleshy; the legs are short, and the fore-ones most so: the fur is short, but the hairs of which it is composed, are thick and harsh to the touch: the back is of a dusky brown, or amber colour; the sides are very beautifully spotted with grey; the belly is white.

It lives in the East Indies and South America, and in both is very frequent; the noise it makes is like that of a hog, and it strikes with the head in the manner of that animal, and raises the bristles on the back, when angry. Marcgrave calls it Paca; Ray, Mus Brasiliensis magnus Porcelli pilis et voce.

Mus dorso fusco-nigricante, ventre albo, corpore oblongo.
The white-bellied Mus, with a blackish back, and long body. **The Bell-mouse.**

This is not so thick in the body as the common rat, but it is longer, and approaches, in some degree, to the form of the weasel; the head is oblong, large at the upper part, but very slender at the snout; both the jaws are equal in length, and the upper lip is split as in the hare: the teeth are long, slender, and sharp; the eyes are black and prominent; the ears are short, naked, obtuse, and rounded: the tail is short and hairy; the legs are short, especially the anterior pair.

The back is rounded and fleshy; the fur is very thick, smooth, and fine; the hair not rigid; the back and sides are of a very deep brown, approaching to black; the belly is white.

This is a native of the woods of Germany, but is a very nimble and cunning animal, for that it is rarely taken; when seized, it bites in a violent manner. Gesner calls it Glis; and other authors have described it, but it is as if they had never seen it.

Mus corpore brevi, lateribus striatis.
The short-bodied Mus, with striated sides.

This is of the size of the common rat, and it's body is as thick, in proportion to the length: the head is short; the snout obtuse, and the upper lip divided; the ears are short, but somewhat pointed at the top; the eyes are small, black, and prominent; the teeth are sharp: the legs are short, and the toes armed with very sharp claws; the tail is short and hairy.

The back is of a deep brown; the sides are elegantly variegated with undulated streaks of brown and grey, and the belly is almost white.

This is a native of the East Indies, and lives principally on trees, in the manner of the squirrel. Ray calls it Mus Indicus striatus.

SOREX.

THE upper fore-teeth of the Sorex are bifid, and the lower ones are serrated: the upper canine teeth are very small, and are four in number. Of this genus there is only one known species.

SOREX. **The Shrew-mouse.**

This is an extremely singular little creature; it is smaller than the common mouse, and much resembles it in figure: the head is large, and broad at the base, but it is terminated by a long, slender snout, in which are the nostrils very conspicuous; the lower lip is smaller than the upper, and the whole snout is guarded or ornamented with a kind of bristles: the teeth are very singular in their structure, and as singularly disposed; the four grinders of the under jaw are acute, and each divided into four parts: the fore-teeth of the same jaw are short and serrated; the grinders of the upper jaw are also four, and they are bifid; the canine are four, and very small, and the fore-teeth of this jaw are bifid and crooked: the ears are short and broad; the legs are short, and the tail is exactly that of the common mouse.

The head, neck, and back are all of an uniform colour, a dusky brown, with a tinge of the ferrugineous: the belly is white; the whole creature has a very disagreeable smell.

It is very frequent in our corn-fields, and under dry hedges; and in the country sometimes comes into houses. The cats will kill it, as the other reptiles of it's kind, but they will not eat it. All the writers on quadrupeds have described it. Ray calls it Mus Araneus, and the rest in general preserve the same name; we call it the Hardy-shrew, or Shrew-mouse.

SCIURUS.

SCIURUS.

THE fore-teeth of the Sciurus are prominent, and there are no canine teeth: the legs are formed for climbing and for leaping.

Sciurus fusco-subrubens ventre albido.
The reddish-brown Sciurus, with a white belly.

The common Squirrel.

This is a small-bodied animal, but it is so well covered with fur, that it appears much larger than it truly is; the head is small, and of a figure approaching to oval, but pointed at the extremity; the eyes are large and bright; the fore-teeth sharp and prominent: the tail is very long, and covered with a long and fine hair, in such manner as to equal the body in apparent thickness.

The back and sides are of a brown colour, with an admixture of a tawny red; the belly is perfectly white; the colour, however, is not invariable.

There are squirrels of this species in Poland that are black; and in Russia they are, in general, of a darker or lighter grey. All the writers on quadrupeds have described it under the name of Sciurus and Sciurus vulgaris. We have instances of it's becoming perfectly white.

Sciurus grifeus cauda minore.
The grey Sciurus, with a smaller tail.

The American grey Squirrel.

This is twice as large as the common squirrel: the head is large, oval, and sharp at the extremity or snout; the eyes are very large, bright, and prominent; the ears are short; the mouth is small, and the teeth sharp and prominent: the body is more corpulent than in the common squirrel, and the tail, though large and bushy, is not so long, in proportion, as that of the common kind: the legs are very robust; the whole back and sides are of a dark iron grey, variegated with black in clouds, streaks, and blotches, and very beautiful: the belly is paler; the legs are pale toward the body, but the feet are black.

This is frequent in the woods in North America, and throws itself with great force and rapidity from tree to tree. Ray calls it Sciurus Americanus cinereus major.

Sciurus cauda maxima grifeo-nigrescens.
The blackish Sciurus, with a very large tail.

The Ceylon Squirrel.

This is of the size of the common squirrel, but still slenderer in the body; it's weight is almost nothing: the head is small, oval, but more sharp at the snout than in any other species: the eyes are black, large, and prominent; the mouth is small; the whole upper part of the body is of a grey colour, approaching to black, and down the ridge of the back there runs a streak of absolute black, formed of hairs much more stiff than those of any other part, and in some degree resembling bristles: the belly is very pale, but is not absolutely white: the legs are very slender, and of a dusky grey; the feet are blackish, and the toes long: the tail is very long, and is very thick, covered with a long and fine down, variegated with white and black.

This species is peculiar to the East, and is no where so frequent as in the forests of the Island of Ceylon. Ray calls it Sciurus Zeylanicus pilis in dorso nigricantibus.

Sciurus caudo teretiuſcula, auribus latis nudis.
The round-tailed Sciurus, with ſhort, broad ears.

The brown American Squirrel.

This, on a curſory view, has more of the appearance of ſome of the weaſel-kind, than of the ſquirrel: the head is ſmall and oblong; the mouth narrow, the eyes black, not very large, but prominent, and the ears naked, membranaceous, and roundiſh; the creature is larger than our ſquirrel: the back is of a deep brown, approaching to black; the ſides have a tinge of reddiſh among the brown; the belly is ferrugineous: the tail is long, and of a rounded form; the legs are very ſlender, and nearly black.

It is a native of many parts of America. Some authors call it, ſimply, Sciurus Americanus.

Sciurus hypochondriis prolixis volitans.
The Sciurus, with the ſides extended.

The flying Squirrel.

This is a very beautiful, as well as a very ſingular creature, though truly a quadruped, having the advantage, in ſome degree, of flying; it is much ſmaller than our common ſquirrel, and it's head is ſo ſlender, that it's weight is leſs than could be imagined: the head is oval, ſhort, and obtuſe; the mouth is ſmall, and the teeth ſlender, but prominent and ſharp; the eyes are very black and prominent; the legs are very ſlender, and the tail is of a ſingular figure, and different from that of any other of the ſquirrel-kind; it is long and broad, or flatted, and the creature does not, on any occaſion, elevate it along the back, as the common ſquirrel.

The back and ſides are of a deep and duſky grey; the belly is white, and, at the commiſſure or joining of the two colours at the ſides, there runs a long line of black: the ſkin of the ſides is extended and lax, and is continued both along the fore and hinder legs, ſo that the creature can at pleaſure extend it. This is it's prolix; when it leaps from tree to tree, it is very light and nimble, and can take ſhort leaps without any aſſiſtance from this ſtructure; but when it would go to a tree at conſiderable diſtance, and to which the mere elaſticity of the hinder legs could not carry it, this membrane is extended, and ſupports it in the air, nor does it want a limited power of vibration in it, which carries it on ſtill further, and makes it anſwer many of the purpoſes of wings.

Sciurus rufo-nigreſcens lateribus variegatis.
The darker-coloured Sciurus, with variegated ſides.

The Barbary Squirrel.

This is of the ſize of the common ſquirrel, and greatly reſembles it in form: the head is ſmall and oval; the eyes are large, bright, and prominent; the ears are ſhort, broad, and obtuſe: the mouth is ſmall, and the teeth ſharp, but not large: the back is of a deep but very beautiful colour; a tawny brown, with a conſiderable admixture of black: the ſides are variegated with oblique lines of tawny and white; the belly is of a pearly colour or white, with ſome admixture of blue: the legs are ſlender, and the tail is long, large, covered with fine thick and deep fur, and variegated with white, and a paler tawny than that of the back or ſides, and often the white is blended entirely with that colour: the toes are long and ſlender, the claws black.

This is a native of Africa, and of ſome parts of the Eaſt Indies; it's fur is very fine. Geſner calls it Sciurus Getulus; Cluſius, Muſtela Africana. Thoſe who depend on the colour only, for diſtinction of ſpecies, may enlarge this family very greatly, for the ſame ſpecies is often ſubject to many variations in this reſpect; but theſe are all that are at preſent known to be truly and certainly diſtinct.

LEPUS.

LEPUS.

THE upper fore-teeth of the *Lepus* are duplicated, the under ones are ſimple; there are no canine teeth: the ears are long, and the legs are formed for running.

Lepus cauda abrupta, pupillis atris.
The Lepus, with an abrupt tail, and black eyes.

The common Hare.

The hare greatly reſembles the rabbit in form, but it is larger, and ſomewhat longer, in proportion to it's thickneſs: the head is large, oval, and not very acute; the mouth is ſmall; the upper lip divided; the eyes large, black, and prominent; the ears remarkably long: the fore-legs are ſhort, but the hinder ones are very long, furniſhed with muſcles of a ſurpriſing ſtrength; by this ſtructure of the legs, it is qualified for running at a ſurpriſing rate: the prominence of the eyes is ſo ſingular, that the creature ſeems to have a power of ſeeing every way at once with them; and it is ſaid, that they are not cloſed, even when the creature ſleeps. As an attention to danger is the great means of this creature's eſcaping it, the ears are calculated in the ſame manner for anſwering that purpoſe: they are always in a poſition to receive the leaſt ſounds, and are moveable with a ſurpriſing eaſe.

The hare is a native of almoſt all parts of the world; it lives ſolitary, and remains quiet all the day, only feeding in the night. In the northern countries it becomes white in winter, only that the ears continue grey, and their tips black. In ſome places they are grey all the year, in others white; with us, and, in moſt parts of the South of Europe, the colour is a tawny, with a ferruginous tinge, but this is not determinate: the tail is ſhort, downy, and abrupt. All the writers on quadrupeds have deſcribed it, and all under the ſame name Lepus.

Lepus cauda breviſſima, pupillis rubris.
The red-eyed Lepus, with a very ſhort tail.

The Rabbit.

This is, though a ſmaller, a handſomer creature than the hare, and not only in different countries is of different colours, as that is, but has a great variety even in the ſame: the head is ſmall and ſhort; the eyes are large, prominent, and red; the mouth is ſomewhat larger, in proportion, than that of the hare: the ears are very long, and the legs are, in the ſame manner, as in that ſpecies, of different length; the hinder pair being much longer than the fore ones. The general colour of the rabbit, in this country, is a pale browniſh-grey on the back, and white on the belly, but we have it darker, of a ſilvery grey, and altogether white.

It is common to moſt parts of the world; and all the authors who have written on quadrupeds have deſcribed it, and all under the ſame name Cuniculus.

Lepus cauda elongata.
The long-tailed Lepus.

The Siberian Rabbit.

This is of the ſize of our common rabbit, but the body is ſomewhat thinner, in proportion to it's length: the head is ſmall, oval, and obtuſe; the eyes are very large and prominent, and of a ferruginous colour, with a changeable tinge of green: the ears are very long, and the mouth is ſmall, but the teeth are large: the fore-legs are remarkably ſhort and ſlender, but the hinder pair are proportionably longer, and more robuſt than in our hare: the tail is very remarkable; it is conſiderably long, and takes off greatly from the reſemblance to the hare or rabbit-kind: the colour is very various; they are formed of all the degrees of grey, from the very paleſt to what is nearly black: the belly in all is pale and whitiſh, and in winter they are all over white. Many of them, during the summer months, are very beautifully variegated with oblique and transverse streaks of black and grey.

It is a native of Russia and Tartary, and the fur, which is very fine and long, is much valued. Some few of the naturalists have described it; they have called it Cuniculus Sibericus.

Lepus cauda nulla.
The Lepus, without any tail. **The Tapeti.**

This is somewhat smaller than our rabbit, but it perfectly resembles it in shape: the head is small and rounded; the eyes are large, black, and very prominent; the mouth is small, and the teeth short; the ears are very long: the hinder pair of legs are greatly longer than the fore ones, and there is no tail, not so much as the rudiments of one; the want of this gives a great singularity to the appearance of the creature: the back is of a deep tawny colour; the belly is white: the head is of a paler colour than the rest, and there is a redness in the front of it; the throat is white: there is sometimes more, sometimes less, of the whiteness in this part; and even when the belly is not white, as is sometimes the case, this part still is so.

It is a native of South America. Maregrave calls it Tapeti; Ray, Cuniculus Brasiliensis Tapeti dictus.

Lepus auribus brevioribus obtusis.
The Lepus, with short and obtuse ears. **The Aperea.**

This is a very singular species; it seems to be of a middle kind, between the hare and the rat, but this is only in appearance: it's legs, eyes, &c. are of the true form of those of the hare, and the ears have the appearance of those of the rat; but, though short and obtuse, they are not rounded nor naked, as in that genus. It is smaller than our rabbit, and the body is slenderer; the head is oval; the eyes are large, black, and prominent; the lip is divided, and the mouth is small: the fore-legs are small and short, the hinder pair very long and robust.

The back is of a dusky tawny colour, with an admixture of a ferruginous brown; the sides are paler: the belly is white; there is no appearance of a tail, not even the least rudiments of such a part; the rump is paler than the rest of the body.

This is a native of the Brasils. Maregrave has described it, and from him others. The natives call it Aperea; Ray, and the rest of the zoologists, Cuniculus Brasiliensis Aperea dictus. It lives in the manner of the rabbit, and burrows holes in the ground for it's retreat.

CASTOR.

THE upper fore-teeth of the Castor are excavated at right angles; there are no canine teeth: the grinders are complicated; the feet are palmated, and formed for swimming.

Castor cauda ovata plana.
The Castor, with a flat oval tail. **The Beaver.**

This is a very large animal; the length of a full-grown one, measuring from the end of the snout to the tip of the tail, is near four feet; and it's breadth, in the lower part of the body, a foot: it's weight is between thirty and forty pounds.

The head is large, and, from the very extremity of the nose to the hinder part, it is compressed, and is nearly as broad as it is long: the eyes are small and black; the ears are also small, short, obtuse, and in some degree rounded, and they are hairy on the outside, and smooth on the inner: the fore-teeth are only two in each jaw; they are large, prominent, broad, and obliquely cut off, as it were: they are white on the inside,

inſide, but on the outſide they are of a beautiful yellow, with a tinge of red: the neck is thick and ſhort.

The fore-legs are robuſt, and the feet have each five diſtinct toes: the claws are obtuſe and rounded; they reſemble in ſhape the nails of a human hand, and are of a dun colour: theſe legs ſerve in the purpoſe of arms, and the creature holds up it's food to it's mouth in them, as the ſquirrel: the palm or inner part of the foot, and the different lengths and diſpoſition of the fingers, are juſt as in the human hand.

The hinder feet are very different from theſe; the toes are connected together by a black, thick, and ſtrong membrane, in the manner of thoſe of a gooſe, and they are terminated by oblong, obtuſe, and black claws or nails: the body is covered with two kinds of fur; they are different in length, colour, and every other circumſtance, and are both arranged in the ſame ſituation in all parts: the ſhort fur immediately covers the body, the long hangs farther from it; the long fur is compoſed of hairs of the thickneſs of thoſe of a man's head, of about an inch and a half in length, and of a deep blackiſh-brown colour, and a very gloſſy ſurface; theſe are very rigid and ſtrong, and are ſo ſolid, that, when cut aſunder, there can be no cavity diſtinguiſhed in them, even by the aſſiſtance of the microſcope: the ſhort fur is compoſed of hairs of not more than half an inch in length, extremely thin, and as ſoft as the moſt downy plumage of a bird; this is of a paler colour than the other. The otter has this double covering of different fur as well as the beaver, and ſo have ſome other animals, eſpecially thoſe which roll in the mud, or dive under water; the uſe of the inner coat ſeems to be the keeping the water from immediate contact with the body of the creature.

The tail of the beaver is of a very ſingular kind; it is of an oval figure, broad, flat, black, and has a ſcaly appearance, like the ſkin of fiſhes; it is near a foot in length, and is about three inches broad at the baſe, near five inches acroſs the middle, and a little ſmaller at the very extremity, where it terminates in an oval figure.

The penis is of a bony ſtructure, and is hid within the belly of the animal; under it are ſituated the bladders, containing the medicinal ſubſtance, called Caſtoreum, which the vulgar have ſuppoſed to be the teſticles of the animal. The teſticles in the male are ſituated near the follicle of caſtor, but, diſtinct from them, their place is under the oſſa pubis.

The beaver is a native of many parts of the world; it was once frequent in England, but the ſpecies is, at this time, like that of the wolves in Ireland, utterly extinct here. All the writers on quadrupeds have named it; they call it Caſtor and Fiber.

Caſtor cauda, longa, lanceolata plana.
The Caſtor, with a long, flat, lanceolated tail.

The flat-tailed, ſmall Beaver.

This is a very little animal, in compariſon of the former, and has ſcarce any thing of the appearance of the genus, but, when examined, is found to poſſeſs all it's characteriſtics. It is of the ſize of the common mole, and very much reſembles it in the ſhape of the body, but the length of the legs and tail afford a very obvious diſtinction. The head is ſhort and acute; the eyes are ſmall, as in the mole; the ears are alſo ſmall, the body is corpulent: the tail is equal to the body in length, and is of a flatted figure, and has a few ſcattered hairs on it: the fore-legs are ſhort, and the hinder ones longer: the back is of a deep ferrugineous brown, with an admixture of tawny and of black; the ſides are tawny; the belly is white: the toes of the hinder feet are connected by a membrane, for the uſe of ſwimming.

This is a native of many of the northern parts of the world; it generally lives about the banks of rivers, and burrows itſelf holes in the ground for it's ſecurity and defence. Authors have confounded it with the Mus Exoticus of Cluſius, but without reaſon. The fur is much eſteemed in Ruſſia; it wants the perfumed ſmell of the ſkin of the Mus Exoticus, and Cluſius has been idly cenſured for denying it to this.

HYSTRIX.

HYSTRIX.

THE fore-teeth of the Hyſtrix are obliquely truncated; there are no canine teeth; the ears are of a figure approaching to round; the body is covered with prickles or ſpines.

Hyſtrix manibus tetradactylis, plantis pentadactylis, capite criſtato.
The criſted Hyſtrix, with four toes on the fore-feet, and five on the hinder.

The Porcupine.

This is a conſiderably large animal, but, with the ſingular covering of it's ſpecies, it appears much more ſo than it truly is; it ſomewhat reſembles the badger in ſhape: it's length, from the noſe to the tail, is about two feet, and the whole body is covered with a kind of briſtles: theſe on the ſhoulders, legs, ſides, and belly are of a deep and gloſſy black; on the middle of the back, the hips and the loins, and on the tail where the ſpecies have their origin, they are variegated with black and white: theſe in figure and conſiſtence perfectly reſemble the briſtles of a hog, and are quite diſtinct from the ſpines or quills.

The neck of the creature is ſhort; the head is thick and obtuſe; the noſtrils are very large; they are not, however, open and patulous, as in ſome animals, but formed of a tranſverſe ſlit: the upper lip is ſplit in the middle, in the manner of that of a hare, and the front of the head is ornamented, in the manner of that of the cat, with long whiſkers; theſe are black, and have their origin about the noſtrils: the opening of the mouth is not very large; the fore-teeth are two in each jaw; they are large, ſtrong, bent inwards, and obliquely truncated: the grinders are eight in each jaw; the eyes are very ſmall, and their iris is blue.

The ears are very like that of the human ſpecies; about theſe there is a quantity of ſoft hair, wholly unlike that of the reſt of the body, and under the lower jaw there is alſo a quantity of ſoft and fine down.

The top of the head is decorated with a very ſingular creſt, formed of briſtles of a great length, not leſs than eight inches; theſe ſtand erect, and the creſt is continued along the neck to the ſhoulders: theſe briſtles are ſome of them black, ſome white, and ſome variegated, in a very beautiful manner, with both theſe colours.

The legs are ſhort, the feet large, and the claws not very ſharp; there are five toes on each of the hinder feet, and only four on each of the others, and, of theſe, one which is the exterior is larger than the reſt, and reſembles a thumb: the briſtles which cover the whole body are thick, and compreſſed at the baſe; they ſtand each on a ſhort and ſlender pedicle, and terminate in a fine point: the form of theſe is ſingular, but that of the ſpecies is much more ſo.

This ſpines or quills, as they are commonly called, are of two kinds; ſome are ſhorter, thicker, ſtronger, and more ſharply pointed, having a kind of double edge, like the ſhoemaker's awl; the others are longer, weaker, and more flexible; theſe are a foot long, and are compreſſed at the point: the ſpines of the firſt kind are white at the baſe, and of a duſky cheſnut colour on the upper part; the others are white at each extremity, and are variegated with black and white in the middle.

The tail is about four inches long; it is armed with numerous ſpines, diſpoſed in ſeveral annular ſeries, and at the extremity of it, inſtead of theſe ſpines, there are ten or twelve tubular bodies of the ſame thickneſs with them, but only of about half their length; theſe ſtand on ſlender pedicles, and are open at the end, and all the way tranſparent.

The

The porcupine is a native of the warmer parts of Europe, and many other places. All the writers on quadrupeds have deſcribed it. They call it Hyſtrix and Hyſtrix vulgaris.

Hyſtrix pedibus tetradactylis, cauda exerta ſeminuda.
The Hyſtrix, with four toes on the hinder feet, and an exerted, almoſt naked tail.

The American Porcupine.

This is ſmaller than the European porcupine, and it is armed alſo with ſhorter ſpines, but they are more numerous, and more rigid; it's length, from the extremity of the noſe to the tail, is about ſixteen inches; it's breadth, over the middle of the back, ten: the head is ſmall; the eyes are round, prominent, and very bright: the ears are ſmall, but rounded, and are in a manner buried under the ſpines; the mouth is ſmall; the upper lip is divided, as in the hare: the teeth are large, bent inwards, and cut off obliquely at the ſorepoints: the anterior part of the head is decorated with whiſkers, in the manner of the cat, but they are very rigid and black: the legs are ſhort, and have each four toes to the foot; the tail is a foot long, and the extremity of it is covered, but that not very thickly, with briſtles like thoſe of the hog; the reſt is naked.

The whole body is covered with very robuſt and rigid ſpines, of three inches in length, and under theſe there is a ſofter down, of more than an inch in length; two thirds of each ſpine from the baſe are brown, the upper part is black, and the very tip is white: they are all hollow, in the manner of the barrels of a quill, and adhere but looſely to the ſkin.

This is a native of many parts of America. Marcgrave calls it Cuandu, and Hernandez, Hoitztlacuatzin.

Hyſtrix pedibus pentadactylis, cauda exerta.
The five-toed Hyſtrix, with an exerted tail.

The East Indian Porcupine.

This is a large and unwieldy animal; it is larger than the European porcupine, and much more corpulent, in proportion to it's length: the head is large, oval, and obtuſe; the eyes are not very large, but they ſtand prominent, and are black; the ears are broad, obtuſe, and low: the upper lip is divided, and the teeth are two in the front of each jaw, long, thick, bent inwards, and obliquely truncated at the extremity: the legs are ſhort, and the feet have each five toes, which are conſiderably long, and armed with large but obtuſe claws: the tail is oblong, and the ſpines are large, ſharp, and robuſt; they are variegated with a cheſnut brown, white, and black, and make a very beautiful, as well as a very formidable, appearance.

This is a native of the East Indies, but few of the writers on beaſts have met with it.

Hyſtrix pedibus pentadactylis, cauda truncata.
The five toed Hyſtrix, with a truncated tail.

The Malacca Porcupine.

This is of the ſize of a ſucking pig, and is not unlike that animal in form: the head is larger, in proportion, than in any other ſpecies; the eyes alſo are large, but they are not ſo protuberant as in many ſpecies, in which they are ſmaller: the ears are ſhort, broad, and rounded; the upper lip is divided, and the teeth are long, crooked, obliquely truncated, and very white: the legs are ſhort, and the feet have each five toes; the tail is ſhort and truncated: the whole body is covered with a harſh but gloſſy fur, compoſed of tawny briſtles, and among this there ſtand a vaſt number of ſpines, variegated with tortoiſeſhell colour, and yellow, and with a little white: theſe are of two kinds, as in the common porcupine; ſome large and long, ſome ſhort, but more robuſt than the others, and more looſely fixed in the ſkin. It is theſe ſhorter quills, not the long ones, that the world ſuppoſes, and too many who ought to have known better, aſſert, that the creature throws off, by way of darts, when it is attacked. There is no truth in the account; the creature has no power of diſcharging any of them, nor do they ſerve for any other purpoſe than the ſpines of our common hedge-hog, that is, to defend it againſt other animals, which attempt to ſeize on it with their mouths.

This ſpecies is a native of many parts of the Eaſt, but few of the writers on quadrupeds have been acquainted with it. The Pedro de Porco, or famous Porcupine Bezoar, is taken from the gall-bladder of this ſpecies, and is not properly a Bezoar, though honoured with that name.

D I D E L P H I S.

THE fore-teeth of the upper jaw of the Didelphis are two obtuſe, and four conic, on each ſide: Thoſe of the lower jaw are eight, and very ſmall; the canine teeth on each ſide are three.

Didelphis mammis intra abdomen.
The Didelphis, with the paps within the abdomen. **The Opoſſum.**

This is a conſiderably large, and an extremely ſingular, animal; it's length, from the extremity of the noſe to the rump, is about fifteen inches; it's tail is equal in length to the whole body; the head is long, narrow, and acute, and has much of the appearance of that of the hog's; the ears are an inch and a half long, an inch in breadth, and obtuſe: the eyes are large, but not prominent; the mouth opens to a conſiderable width; the noſe is, as it were, truncated and obtuſe; the noſtrils are large, and there are a number of long, black whiſkers, both about the noſe, and on the ſide of the head, beyond and under the eyes: the legs are long and robuſt; the feet are large; the toes long and ſlender, and the claws ſharp, long, and crooked.

The creature is covered with a double fur; a longer, which is thin, and compoſed of elegant dark brown gloſſy hairs, conſiderably thick and rigid; a ſhorter, which is very thick-ſet, and is of a pale brown colour, and very ſoft and downy: the tail is ſlender, rounded, and naked, like that of the rat, and is tuberculated on the ſurface in the ſame manner, as in that animal; it is dark-coloured toward the baſe, but of a whitiſh-brown toward the extremity.

The ſkin of the belly in the female is looſe and large, and there is an aperture in it, at which it occaſionally takes in the young. It was at one time ſuppoſed that the young, in this ſpecies, were received into the uterus, or at leaſt into the cavity of the creature's body, but that is without all foundation. They are only taken into a cavity formed by nature between theſe two membranes, and preſerved from all ſort of dangers there.

This is a native of many parts of North America. Authors call it the Poſſum and Opoſſum.

Didelphis mammis extra abdomen.
The Didelphis, with the teats without the abdomen. **The African Rat.**

This is ſmaller than the Opoſſum, and it's legs ſhorter, by which advantages, together with the likeneſs of the tail to the rat-kind, it has miſled the few authors, who have been acquainted with it, to rank it among the rat-kind: the head is of an oval figure, but a little depreſſed, and acute at the ſnout; the eyes are large, and they ſtand very prominent: the ears are ſhort and obtuſe; the opening of the mouth is large, and the teeth are moderately long and ſharp: the back is fleſhy; the legs are robuſt, and the creature runs and climbs in an uncommon manner with them.

This

QUADRUPEDS Series 1.
Plate 25. Page 53
The Rat
The Ground Rat
The Citille
The Guinea Pig
The Squirrell
The Hare
The Marmotte
The Rabbit
The Beaver
The Porcupine

This is a native of many parts of Africa, and has been described by Seba, under the name of the Mus Africanus Hayopolin dictus. Most of the zoologists have omitted it.

QUADRUPEDS.

Class the Second.

AGRIÆ.

THE Agriæ have no teeth: the tongue is very long and cylindric. Of this singular class there are only two known genera. The Myrmecophaga and Manis.

MYRMECOPHAGA.

THE body of the Myrmecophaga is covered with hair; the ears are of a roundish figure.

Myrmecophaga manibus tridactylis, plantis pentadactylis.
The Myrmecophaga, with three toes on the fore-feet, and five on the hinder feet.

The great Ant-bear.

This is a very singular and a very beautiful creature; it is of the size of a mastiff dog: the head is large, and of a singular figure; the snout being at least a foot in length, and of a cylindric figure: the opening of the mouth is small, and there are no teeth in it: the tongue is rounded, and pointed at the end, and, when fully exerted, is a foot and half in length. This is, in the usual state, carried doubled in the mouth; but, when the creature is to eat, it thrusts out upon an ant-hill, and when it is covered with those creatures, draws it in again.

The eyes are small and black; the ears are short and rounded; the tail is very long, very large, and covered with a thick fur; the creature has a great facility in moving it, and wipes every part of the body with it at pleasure: the neck is short; the body is robust and fleshy: the legs are robust, and about a foot in length; the fore-ones are about an inch longer than the hinder.

The fur on the back is black, but with an admixture of a hoary white: the head and neck are covered with hair of the same colour, but shorter; this stands forward: that on the back and tail is six inches long, and lies backward; that on the sides is shorter, and has it's direction downward: the fore-legs are of a greyer or whitish-hoary colour, only each of them toward the foot has a spot of black; the breast is covered with one great black spot, which reaches each way to the sides, and half way down the body, and is terminated at top by a white line: the hinder legs are black; the tail is of a flatted or broad figure, and is covered with hairs long, rigid, and thick like horse-hairs.

It is a native of many parts of America; and from some resemblance of it's hinder legs to those of the bear, and it's feeding on ants, it has been called Ursus formicarius by Cardan and others. Marcgrave calls it by it's Brasilian name, Tamandua Guacu.

Myrmecophaga manibus didactylis, plantis tetradactylis.
The Myrmecophaga, with only two toes on the fore-feet, and four on the hinder feet.

This is smaller than the other, and the legs also are shorter, in proportion; otherwise it greatly resembles it in shape: the head is small, oval, and continued at the front into a very long and slender snout of a rounded figure, and conic, or growing gradually smaller all the way, from the top of the head to the extremity: the ears are short, naked, and rounded, their apertures very large; the eyes are moderately large, and very bright and sparkling: the body is rounded and fleshy; the legs are short and robust, and the tail is very long, and of a flatted figure, and very bushy, or covered thick with long hairs.

The back is of a very dark colour, and there run all along it a sort of bristles pointing backward; the rest of the body is covered with shorter, but very rigid and stiff, hairs: the creature is not swift of foot; either this or the other, indeed, may be overtaken, at their best speed, by a man who runs but a moderate pace: this has a long fleshy tongue like the other, and feeds in the same manner, by thrusting it out, where there are plenty of insects, which gather about it, and are devoured, when it is taken in again.

It is a native of the Brasils, and some other parts of South America, but has been described by none of the writers on this subject. We have skins of it sent over to us preserved.

Myrmecophaga manibus tetradactylis, plantis pentadactylis.
The Myrmecophaga, with four toes on the fore-feet, and five on the hinder.

The lesser Ant-bear.

This is a very beautiful animal; it's size is that of the fox, or smaller, and it's figure not greatly unlike that of that animal: the head is moderately long, the neck short, and the body about a foot in length, and very fleshy: the tail is like that of the other species, and is nearly equal to the body in length: the legs are short, and are not so thick and strong as in the others.

The head is altogether of a conic form, but not entirely straight; it bends a little downwards; the base towards the ears is thick, and from thence it gradually grows smaller to the other end, where it is very sharp; the mouth is small, and there are no teeth in it: the ears stand erect, they are short and open; the eyes are very small and black; the tongue is eight inches long, rounded, and tapering gradually from the base to the point.

The animal is, in general, of a pale whitish-yellow colour; the fur is deep, and the hairs of which it is composed are rigid and stiff, only the back and belly have in the male a tinge of blackness, and the hairs are longer there than in any other part: the tail is well covered with hair all over, except at the extremity, where it is naked: there are two broad lines of black on the sides of the neck, one on each side; they are continued down to the shoulders, and thence run to the extremity of the back, where they unite.

This creature lays very fast hold of any thing with it's fore-feet; it has also a way of hanging itself to the boughs of trees, by twisting the naked extremity of the tail round them. It sleeps all the day with the head between the fore-legs; in the evening it goes in search of food.

It is common in many parts of South America. Marcgrave calls it Tamandua I.

MANIS.

MANIS.

THE body of the Manis is covered with a kind of ſcales; there are no ears. Of this ſingular genus there is only one known ſpecies, which has been confounded with the lizards.

MANIS. The ſcaly Lizard.

This is a creature of great beauty, and perhaps one of the moſt ſingular in the world; it's aſpect has a great ſhew of terror, but it is the moſt inoffenſive creature imaginable, having neither inclination, nor the leaſt power, of doing hurt, and feeding only on the ſmalleſt inſects.

It is in form ſomewhat like the lizard, but not all ſo in any other reſpect, and it's covering is of the moſt extraordinary kind imaginable. It is about four feet in length, and it's body in the broadeſt part, which is toward the hinder legs, is about ten inches in breadth; it is of a rounded figure on the back: the legs are ſhort, they ſtand at about a foot diſtance; the reſt of the creature, from the hinder part to the extremity, is a tail broad, thin, and between two and three feet in length; it is not connected to the hinder part of the body, as in the generality of quadrupeds, but is continuous with it.

The head is ſmall, of a conic figure, about three inches in diameter at the baſe, and thence gradually growing ſmaller to the ſnout, which is ſharp and naked; the reſt of the head is covered with the ſame ſort of ſcales with thoſe of the body, only they are ſmaller: the eyes are moderately large; there are no ears, nor are there any teeth in the mouth, but the tongue is ten inches, or more, in length, fleſhy, a little flatted, of the thickneſs of a child's finger, and pointed at the end.

The whole upper ſurface of the creature, the back, and the outſides of the legs are covered with an armature of ſcales: the belly and the inſides of the legs are naked; the ſcales are of a firm ſubſtance, and have very much the appearance of tortoiſe-ſhell; they are, on the body, two inches in length, more than an inch in breadth, and of an oval figure, and each terminates in a kind of ſpine: the baſe is deeply ſtriated, and theſe lines are continued three fourths of the length of the ſcale, but the reſt is ſmooth, and of a natural poliſh. The whole creature is of a brown colour; the ſtriated part of the ſcales is of a red duſky brown, the ſmooth poliſhed part has an admixture of yellow: the ſides of the body, and yet more thoſe of the tail, are of a ſerrated form, the ſcales terminating one over another at ſome diſtance: the legs are ſhort, but very robuſt, and the claws are very remarkably ſtrong and thick.

This is a native both of the Eaſt Indies and South America, but it is leſs frequent in the latter. It lives in the woods, and as it's legs raiſe it very little from the ground, and it's colour is that of decayed leaves, and the ſcales with which it is covered have alſo the ſhape of them; it has very little to diſtinguiſh it, among thoſe remains with which the ground is ſtrewed in foreſts, unleſs when in motion. It feeds on inſects as the ant-bear does, and takes them in the ſame manner, thruſting out it's tongue, till covered with them, and then drawing it in loaded with the food. Bontius, Cluſius, and others call it Lacertus ſquammoſus. I purchaſed two very perfect ſpecimens of it, the one in ſpirits, the other dried at the late Duke of Richmond's ſale, and from theſe the figure in the adjoining table has been made.

QUADRUPEDS.

Class the Third.

SYLVIÆ.

THE Sylviæ have the fore-teeth, both in the upper and under jaw, four in number; and the teats are situated, not on the belly, but on the breast.

Linnæus has distinguished this class by the name of Anthropomorphæ, beasts having the form of the human species. It is an assertion of that author, that he could find no distinction in characters between man and the monkey. I am apt to believe few would join with him in this opinion, but still fewer, in the putting the ignavus or sloth in the same class; since, whatever unlucky likeness there might be between the monkey-kind and ourselves in form, this ugliest of the creation can have no claim to such a resemblance. The genera of this class are only the two, the Bradypus and Simia.

BRADYPUS.

THE face of the Bradypus is covered with hair: the claws are of a subulated form; there are no ears, nor are there any middle teeth.

Bradypus manibus tridactylis, cauda brevi. **The Sloth.**
The Bradypus, with three toes to the fore-feet, and a short tail.

This is a very extraordinary animal, both in figure and qualities; it is hard to say of what other it is equal to in size, since it is like none in shape; the length of the body is about a foot, and it is corpulent, unless when it has suffered by long hunger, as is indeed often the case. When well fed, it's thickness is equal to it's length: the neck is short and thick; the tail is extreamly short: the legs are robust, and about five inches long; the fore-pair are a little longer than the hinder.

The feet are flatted or plain, in the manner of those of the bear or monkey, but they are extreamly narrow, and ill-calculated for walking: the claws are very long and sharp; they are hollowed on the under side, and pointed at the extremity, and they are of a pale yellowish colour, and of a horny structure.

The head is small and round; the mouth is also small, and of a turbinated form: the face, in some degree, resembles that of the monkey-kind, but it is covered with short hair, not naked, as in that creature: the nose only is naked, smooth, glare, and black: the teeth are small; the eyes are small and black, and have, at all times, a sleepy aspect: there are no ears; the tail is obtuse.

The colour of the whole animal is a pale greyish-brown; the fur is very thick and deep, and is composed of extreamly soft hairs, which move about every way with the least breath of wind: there runs a brown blackish line down the middle of the back, but, excepting for this, the hair there is of a paler colour than elsewhere: the hairs of the neck are longer than those of any other part; they hang on each side in manner of a mane.

It is the slowest mover of all the quadrupeds; it is generally seen on trees, climbing in a very deliberate manner from one part of them to another; on the ground it's motion is still slower; the traversing a space of fifty yards is the labour of a day for it: it is usually seen on the top of tall trees, by way of security: it feeds on their leaves and bark, back, and never drinking. It is extremely timorous, and the head is continually thrown from one ſide to another, as if in alarms at every noiſe: it cries like a young kitten.

It is frequent in many parts of America. Marcgrave calls it Ai ſive Ignavus; Cluſius, Ignavus and Agilis; we call it the Sloth.

Bradypus manibus didactylis, cauda nulla.
The Bradypus, with only two toes on the fore-feet, and without any tail.

The Ceylon-ſloth.

This is a very ſingular animal, and is a ſtill ſlower mover than the American ſpecies, it's body being proportionately more bulky, and the feet narrower, and worſe formed for walking, than in that ſpecies: the head is ſhort; the whole face hairy, excepting only the tip of the noſe, which is naked and blue: the eyes are ſmall, and appear half ſhut; the back has ſome breadth, and the hips in particular are large; the back is of a very dark brown, with an admixture of greyiſh and of olive; the ſides are paler, and have more of the olive: the belly is duſky; the teeth are ſmall; the noſtrils wide and elate, it has no tail.

This ſpecies has been yet ſeen no where, except in the iſland of Ceylon. It is frequent in the woods there, and is ſo ſlow in it's motions, that it is eaſily taken or deſtroyed.

S I M I A.

THE face of the Simia is naked; the claws are rounded and flattiſh, in ſome degree like the nails on the human hand, and there is an eye-lid each way.

Simia ecaudata ſubtus glabra.
The Simia without a tail, with a ſmooth belly.

The Satyr.

This ſpecies has an unlucky reſemblance to the human form, but it is the moſt miſchievous, bold, and vicious of the whole monkey-kind; it ſeems highly probable, that the ſatyrs, in the antient poets, had their origin from imperfect accounts of this creature.

It is one of the larger kind; it uſually walks erect on the two hinder legs, and is in that poſition more than four feet in height: the face has no hair on it, and carries too ſtriking a reſemblance of ſome of the leſs beautiful of our own ſpecies; the eyes are large, and have an upper and under eye-lid, exactly as in our own ſpecies: the ears are broad, flat, and open in the manner of ours: the noſtrils have much the ſame ſituation, though the whole of the noſe is not ſo determinate in figure, or ſo elevated; the teeth alſo have the form and ſituation of our own, and even the eye-laſhes are like ours. Theſe ſingularities are not found in any degree in the other quadrupeds; the obſervation of them in this is as old Ariſtotle, and the modern obſervations perfectly confirm it.

The whole back, ſides, and hinder part of the legs are hairy; the whole anterior part, in the erect poſture of the creature, is ſmooth; the breaſt is naked, and has the nipples ſituated on it, juſt as in the human ſpecies; and the whole, when viewed in this direction, has very little reſemblance to any thing of the quadruped-kind: the fore-legs have very much the appearance of arms, and the creature bends and directs them in the ſame manner, and the hinder ones reſemble legs; they are more robuſt than the others: the paws of the fore-feet have the form of the human hand, but in a rude and coarſe way; and the claws are large, flat, and rounded in the manner of our nails: the paws of the hinder feet are yet more like hands; in this the ſimilarity is, in ſome degree, loſt, for the toes are not ſhort, as on our feet, but longer, and more like fingers than thoſe of the fore-ones: the anterior toe on theſe reſembles the thumb of of the human hand, being removed to a great distance from the next, and being also slender and long; whereas that on the fore-paw is short, and immediately joined to the other.

The hair on the back of this species is of a deep tawny, with an admixture of olive, and is long, thick, soft, and very glossy; the anterior, or properly the under part of the body, for such it is, when the creature stands on all four, is of a dull lead colour; the breast is paler than any other part, and the nipples purple: the buttocks are naked and of the same colour with the belly; it has no tail, not even the rudiment of one.

This is a native of Africa; I once saw a full-grown one shewn as a curiosity in London: it was a male, and when it stood erect, as it had been practised to do with a stick in it's hand, it had a very affecting, general resemblance of a decrepid old man. It was very quarrelsome, and would throw any thing that lay in it's way at those who came to see it, if the eye of the showman were not continually upon it.

Simia acauda unguibus indicis subulatis.
The tail-less Simia, with the claws of the fore-toes subulated.

This is very frequently seen in an erect posture, and then is about three feet in height: the face has much the appearance of some ill-formed one of the human species: the nostrils are very wide: the nose much depressed; the eyes large, and the ears very broad and low: the whole body in this species is hairy; the breast and belly, as well as the back: the fur is deep and glossy; is of a tawny colour, with an admixture of olive on the back, and is yellower on the breast and belly: the legs are very long and slender, and it has not the least appearance of a tail.

It is a native of many parts of the East Indies, and is very fierce in it's wild state, and will never be tamed so as to make it safe to keep it among human creatures.

Simia acauda clunibus tuberosis.
The tail-less Simia, with large buttocks.

This is more apt to walk on the hinder legs only, than any other species, and nature seems to have provided for it's keeping firm in that posture, by making it's lower part of so remarkable a bulk and weight: the face is perfectly naked, and of a blackish colour; the ears are very large, and they lie almost flat upon the head; the eyes are remarkably large, prominent, bright, and flaring; the nostrils turn up, and are very wide; the nose is depressed, and the mouth is large, and well furnished with teeth: the whole body is covered with an olive-coloured fur, which is deeper, thicker, and harsher on the back than on the belly: the legs are long, but they are not so slender as in some other species: the buttocks are very large, protuberant, fleshy, and in part naked; the naked portion is of the same colour with the face: the feet are formed in the manner of the human hand; and the claws or nails are broad, rounded, and of a bluish colour.

This is a native of many parts of the East, and is easily tamed, and less mischievous than many of the others. Prosper Alpinus has mentioned it, as have also some others.

Simia acauda capite majore.
The great-headed Simia, with no tail.

This is more frequently seen on all fours than erect: the head is remarkably large, though the monkey is but of a moderate size; when erect, it scarce measures three feet: the face is naked, and very ugly; the eyes are large; the nose is strangely depressed and flat; the ears are large and low, and the nostrils open and turned up: the forehead is flat and very broad: the whole body is covered with hair, but it is thin, and short on the breast and belly, though very thick and long on the back; it is of a tawny olive, with an admixture of black on the back, and paler on the belly: the legs are long and slender, and the head in walking hangs down in a remarkable manner.

It is a native of many parts of the East, and is a sullen, sulky animal. Professor Alpinus calls it Simia acaudis rufo-nigricans, but this is not exactly the colour of that which I saw.

Simia cauda brevi, ore vibrissato.
The whiskered Simia, with a very short tail. **The Baboon.**

This is considerably different from all the preceding species: the head is large, and hairy on the hinder part, but the face is naked; the eyes are small and depressed; the nose is flat, and the nostrils wider; it is, upon the whole, remarkably ugly in it's aspect: the body is bulky, and the legs are strong, but not so long as in the several former species: the buttocks are large and fleshy, and there is at their base the rudiment of a tail, but it is very short and inconsiderable: the back is of a dark olive; the belly is paler. The creature usually walks on all fours, but it can occasionally stand on the hinder feet only, with great facility.

It is a native both of the East and West Indies, and is often brought alive into Europe: it is a large kind, and, when erect, makes a very awkward appearance; the mouth is, in a very particular manner, furnished with whiskers, which seem to impede the eating, rather than to be of any real use to the creature. Many of the writers on quadrupeds have described it under the name of Papio, and Papio major.

Simia caudata imberbis auribus comosis.
The tailed Simia, with hairy ears.

This is a singular species, and is less ugly than the preceding, though very far from a beauty; it frequently walks on the hinder legs, and, when erect, is about two feet and a half high: the face is oblong, flatted, naked up to the forehead, and of a blackish colour, but with some tinge of yellow: the eyes are large, prominent, and hazel; the nostrils are wide; the mouth is well furnished with teeth, and the ears are broad, low, and hairy: the legs are long and slender, and the tail is longer than they, and, when the creature stands erect, draggles on the ground: the fur is short, but soft and glossy; it is of a dusky tawny-yellow on the back, and brown, but with some tinge of yellow, on the belly; the feet are black.

This is a native of the East, and is frequent in the woods there. Alpinus and some other writers have described it, but it has seldom been seen alive in Europe.

Simia caudata, imberbis, fusca, unguibus pollicum subrotundis.
The long-tailed, brown Simia, with the claws rounded.

This is often seen erect, and is, in that posture, considerably more than two feet high: the head is short, rounded, and considerably large; the forehead is prominent; the eyes are large and black; the nose is depressed, and the nostrils are large: the colour of the naked skin of the face is a deep colour, between blue and purple, but so dark, that the whole appears at a small distance black: the ears are short and broad; the back is of a deep brown colour; the fur long and glossy: the legs are long and slender; the feet, and particularly the hinder ones, have very much the appearance of hands, and the claw of the shorter toe, which resembles the thumb, is very broad and rounded.

This is frequent in the woods of China, and many other parts of the East; it is easily tamed, and becomes very familiar and tractable, but full of tricks. Some of the writers

writers on quadrupeds seem to have been acquainted with it, but their accounts of the several species of this large genus are so very confused, that it is hard to ascertain the absolute species, which they have meant, under any of them.

Simia subflavescens pectore albente, cauda floccosa.
The yellowish Simia, with a whitish breast, and a woolly tail.

This is smaller than any of the former, but more beautiful than either of them; when erect, it measures very little more than two feet in height: the head is small, oblong, and flatted; the eyes are large, and very piercing; the nose is less depressed than in most of the others; the nostrils are elevated and open; the ears are low, and not so broad as in many of the other species: the legs are long and slender; the tail is long, and covered with a thick, woolly down, especially on it's lower part: the back and sides are of a brown colour, with a very considerable tinge of yellow; the back part of the head and neck are still more yellow: the breast is white, as is also the throat, and the naked part of the face is of a dusky dun colour.

This is frequent in South America, as also in many parts of Africa, and the East Indies; but, though seen in considerable plenty in the woods, it is so light, so nimble, and so shy, that it is rarely taken: when caught, it is easily familiarised, and becomes very tractable.

Simia caudata naribus bifidis.
The long-tailed Simia, with bifid nostrils.

This is a larger species than the preceding, when erect, as it frequently stands in it's wild state, it is two feet and a half high: the head is long but slender, and not much depressed: the forehead is oval; the eyes are prominent and grey; the nose is depressed toward it's base, but it turns up at the extremity, and the nostrils are close and bifid: the ears are low, and considerably broad; the teeth are large and sharp the body is not thick, and the legs are proportionately small; the feet are formed v. like the human hand; the tail is long and small; the back is of a deep olive-brown; there are no whiskers or beard about the mouth; the belly is tawny, but the breast is of the same olive colour, with the sides only paler.

This is a native of the Brasils. Marcgrave and Piso have described it, and from them others.

Simia caudata collo pectoreque jubatis.
The tailed Simia, with a mane on the neck and breast.

This is an extremely singular species: the head looks small, in proportion to the body; but that is rather owing to the bushy mane which is under it, than to any real deficiency in size: the forehead is depressed; the eyes are large and prominent; the nose is short and broken, and the tip of it turns up; the nostrils are very wide, and of a rounded figure: the ears are short, broad, and woolly. The creature, when it stands erect, is about two feet four inches in height; the legs are slender; the feet formed perfectly like hands: the body is covered with a moderately deep fur of a glossy brown, but the neck and breast are ornamented with a very deep hairiness, in manner of a mane.

This is a native of South America. Marcgrave has described it.

Simia

Simia caudata auribus genubusque barbatis.
The tailed Simia, with tufts of hair at the ears and knees.

This, when erect, is two feet six inches high: the head is large, oblong, and depressed; the forehead is somewhat prominent, but less so than in most species; the nose is flatted; and the tip of it turns very little up; the nostrils are wide and oval; the eyes are large and black; the ears are low and broad, but they have each a tuft of hair or fur, longer than that of the body, and of a pale greyish-brown colour, which gives a very singular appearance to the face: the body is slender; the legs are long and slender; the hinder feet are formed very much in the manner of the human hand, and the fore-ones have somewhat, but less, of that appearance; the toes being shorter, and the inner one, which represents the thumb, is very short and inconsiderable, and not placed at a distance, as the thumb of the human hand.

The fur is of a very bright and beautiful brown, with a tinge of a tawny yellow: there are tufts of hair at the knees of all the legs, as well the hinder ones as the others, like those of the ears, and of the same greyish-brown colour; the tail is long.

This is a native of the Brasils, and is sometimes brought into Europe. Marcgrave and Piso have described it. It is common in the woods there, and it's tufted knees distinguish it at sight from all the other species.

Simia capite comoso.
The hairy-headed Simia.

This is sufficiently ugly, but the resemblance which the long hair of the head gives it to the human face is very striking, and indeed shocking. It is of the larger kind: when erect, as it very frequently walks by choice, it is near three feet high: the head appears large, by means of the hair, but it is not, in reality, any bigger than usual in the other species: the face is oval and naked; it is of a tawny colour, much like that of some of the darkest of the Moors: the hair falls down the sides of it, as that of the human head, and is of an extreamly dark colour; the eyes are black and prominent; the forehead is rounded; the base of the nose is depressed, and it's end turned up, as in most of the other species.

The body is not bulky; the hinder legs are robust, but the fore-ones are slender: the feet are formed more like the human hand, than those of any other species: the back is of a deep olive brown; the breast has somewhat of a tawny hue, mixed with the olive colour; the belly has some little admixture of a whitish or greyish, but the dusky brownish olive is the reigning colour throughout; the tail is moderately long.

This is a native of the East; it is frequent also in some parts of Africa. Prosper Alpinus and some others have described it; but the generality of those who have done so, seem not to have seen it. We had one in great perfection shewn in London, in the year 1751, under the name of a Man-Tyger, from which this description was taken.

Simia caudata, barbata cauda seminuda.
The bearded Simia, with the tail half naked.

This is a singular and an ugly species; it seldom stands erect, but, when it does, the height is about two feet three inches; the head is short and oval; the eyes are small, and have a kind of a sleepy aspect, as in the Ignavus or sloth: the ears are low, broad, and naked; the nose is broken down, as it were, at the base, and turns up at the end: the nostrils are frightfully patulous; the mouth is wide, and the teeth are numerous and large: the beard is composed of moderately long, rigid, and thick-set hairs, and gives a very singular aspect to the creature: the body is more bulky or corpulent paler than in most others; the legs are not so long as in the generality of the monkey-kind, and they are very robust: the feet are made less in the form of hands, than in most others: the whole body is of a deep, ferrugineous brown, but the breast and belly are paler than the back: the tail is moderately long, and is in part hairy, and part naked; the naked part looking as if meant to be taken hold of.

It is a native of the Brasils. Marcgrave has described it.

Simia caudata barbata, cauda floccosa.
The bearded Simia, with a woolly tail.

This is one of the smallest kind; the height, when it stands erect, is not so much as two feet: the head is long, but remarkably slender; the forehead is low; the nose depressed, and turning up at the head; the eyes are black and depressed; the mouth is surrounded with a moderate beard: the body is slender, and the whole creature has a lighter appearance than the others: the legs are long and slender; the tail long, thick, and floccose.

The back is of a deep but glossy colour, seeming a mixt hue of brown, ferrugineous, and olive; the belly is paler: the breast, the legs, and the tail have some tinge of yellowish.

This is a native of the East and West Indies. Marcgrave has described it, and we sometimes see it brought over alive in our East India ships.

Simia barba cana, cauda simplici.
The Simia, with a hoary beard, and naked tail.

This is about two feet four inches high, when it stands erect: the head is remarkably small, and of an oval form; the eyes are large and prominent; the ears are very broad, low, and of a corrugated surface, greatly resembling those of the human head: the beard is grey, and has much the resemblance of that of an old man.

The body is bulky, and the legs are rather robust than long; they are covered with a short, smooth down, but their feet are not so like the human hand, as in many of the species: the back is of a deep olive colour; the face is naked and black; the breast is of a pale brown, and the belly still paler; the tail is long and naked.

This is a native of Africa, and we have also had it from some parts of the East. Clusius and others have described it. It seems to have been, indeed, one of the first known species.

Simia cruribus longissimis, facie nuda.
The long-legged Simia, with no beard.

This is an extreamly singular species, and has an awkwardness about it, that distinguishes it from all the others: the height, when it stands erect, is two feet nine inches; the body is very slender, and the head long, but narrow: the forehead is oval; the base of the nose is depressed very deeply; it's end turns up, and the nostrils are wide. The eyes are large, black, and prominent; the ears very low, and considerably broad: the legs are remarkably long and slender, more so, indeed, than in any other species: the tail is also very long and slender, and all the way naked: the colour on the back is a deep and beautiful olive, with a very prevailing tinge of the green; the breast is paler, and the belly whitish: the face is black, and so are the feet; the claws are very broad, round at the ends, and of a deep blue.

I never saw any more than one of this species; that was sent over, as a present to the late Duke of Richmond, by Sir Thomas Robinson, at that time governor of Barbadoes.

Simia caudata fusco-ferruginea.
The ferruginous brown Simia.

The Rat-ape.

This is one of the smallest of the monkey-kind; it is not larger than the common rat, but, though of a dusky colour, it has a peculiar pleasantness or chearfulness of aspect: the head is small; the forehead is low; the eyes are large and bright, and their iris is of a light hazel: the nose is depressed toward the top, and is somewhat turned up, but obtuse at the extremity: the nostrils are large; the teeth are small and white, and the creature is almost continually shewing them by grinning: the neck is short; the body is tolerably fleshy; the legs are slender, robust, not remarkably long; the tail is very long and naked, and has greatly the appearance of that of the rat. The colour is a mixt ferruginous brown and grey, with somewhat of an under colour of yellowish; the belly is darker than the back or sides.

It is a native both of the East and West Indies, but few of the writers on quadrupeds have described it. Marcgrave mentions one like it, but it is not certain it is the same.

QUADRUPEDS.

Class the Fourth.

FERÆ.

THE fore-teeth of the Feræ are six each way; the canine teeth are longer than the others.

URSUS.

THE fore-teeth of the Ursus are of a conic figure, and are emarginated on the inner part; the inferior ones are lobated: the canine teeth are placed at a distance from the grinders, and are on each side two; the hinder teeth is very small: the feet are formed for walking, and, in their usual way of being placed down, the creature treads upon the heel. The penis is bony.

Ursus cauda abrupta.
The Ursus, with an abrupt tail.

The common Bear.

This is a large but an uncouth and unsightly animal; it grows to different sizes in different places, from that of a mastiff-dog to the bigness of a small heifer: the whole creature is covered with a thick and deep fur in such manner, that it appears rather a shapeless lump, than an animal: the head is large and long; the neck is short, and very thick; the thighs are long, but the under part of the legs short, and it has a knee-pan or patella at that joint.

The fur is deep both on the back and the belly, but, when stripped, the skin is remarkably thick on the former part, and very thin on the latter; the hairs which compose the fur are not rigid and harsh, as in the lion and tyger, but rather soft and woolly; they are somewhat curled also, much more so than in the goat, but not nearly so much as in the sheep-kind: the brain is in vast quantity; the skull is thin, but firm, and the eyes are very small.

The feet are each divided into five toes, the hinder ones as well as the fore-ones; that which answers to the thumb on the human hand, or great toe on our feet, is situated on the contrary side, in the place of the least toe, or little finger; they are all short, thick, and in a manner shapeless: the claws are very firmly united to the extremities of the toes; it has no tail, but yet it has a lengthened os coccygis: the body is, in reality, slender, though it appears enormously bulky, under that vast load of fur with which it is covered.

The bear is a native of America, and of many of the northern parts of Europe. Toward the pole it is very large and white; in other places it is smaller, and of a black or rusty brown. All the writers on animals have named it, and all under the common name Ursus.

Ursus cauda elongata.
The Ursus, with an elongated tail.

This creature grows to the size of a large mastiff-dog, or more, and is bulky, in proportion: the head is long, of an oval form, and narrow; the eyes are small; the nostrils are large, and the mouth is furnished in a very formidable manner with teeth; the legs are short, but they are remarkably thick; the feet are large, and are spread upon the ground in walking; the creature treading upon the heel as the human species, not in a more raised manner, on the anterior part of the foot, as the rest of the quadrupeds.

This is frequent in the northern parts of Europe. Wormius has described it.

FELIS.

THE fore-teeth of the Felis are small, obtuse, and equal: the tongue is armed with a kind of spiculæ, which have their points bent backwards: the feet are formed for climbing, and the claws may be drawn back at the creature's pleasure.

Felis cauda elongata floccosa, collo piliato.
The Felis, with an elongated floccose tail, and a mane on the neck. — The Lion.

This is the strongest and the fiercest of all quadrupeds: it is taller than the largest mastiff, and more bulky, but the fore-part is over big, in proportion to the hinder, where it is lank, and wants that majesty of appearance it has in front.

The head is large, and the breast broad; the bigness of the head is owing to the great quantity of muscular flesh which covers the skull, and to the breadth of the jawbones: the width of the breast is only apparent; the breast is, in reality, narrower and more compressed than in the common dog, but the quantity of long hair that hangs over this, gives it the external breadth: the neck is very thick, stiff, and rigid: the antients supposed, that it consisted in the skeleton of one continued bone, but that is not truth, it is formed of vertebræ or joints as regularly as in other animals; only the spinose apophyses of those joints are longer than in other animals, and the ligaments by which they are fastened to one another remarkably firm. It has been a vulgar opinion also, that the bones are solid, and have no cavity, but this is also an error; the bones of the legs are indeed remarkably thick, but they have the natural cavity, and have marrow in it, as in other animals.

The tongue is very long and large, and is all the way covered with a kind of spiculæ; these are hollow at their base, sharp at the ends, and all point backwards or toward the throat: the heart of many creatures of the fiercer kinds is small, but in the lion it is remarkably large; the whole structure of the creature is formed for strength: the legs are vastly thick and strong; the muscles of the whole body vast and tough; the claws are of a surprising length and thickness, and, when they are not exerted for service, are drawn back in a very surprising manner, so that they scarce appear: the eyes are large and fierce, and the teeth are very long and terrible: the fur of

QUADRUPEDS. Series 2.

Manis

the Manis call'd the Scaly Lizard

Sylviæ

the long tail'd brown Ape with rounded Claws

the hairy Ear'd Ape

the long leged Ape with no Beard

the Coqui

the hairy headed Simia

the Bear

the Lion

the Tyger

the Lioness

of the whole body is of a tawny yellowish colour; it is not very long, except on the mane, where it is composed of hairs of a very different length, form, and structure; the tail is long, thick, and floccose.

The lioness is, in all respects, like the lion, except in that she has no mane; but this makes so great a difference in her appearance, that she seems a creature of another genus.

The lion is a native of many parts of Asia and Africa. It will live to a great age in Europe, and sometimes breeds with us, but is no where wild in this part of the world. All the writers on quadrupeds describe it under the name of Leo, and the female under that of Leæna.

Felis cauda elongata, maculis virgatis.
The Felis, with an elongated tail, and virgated spots. **The Tyger.**

The tyger is a very large and terrible animal; it's fierceness is greater than that of the lion, nor does any living creature escape it's attacks. At it's full growth it is of the bigness of a small heifer, and, though the body is less bulky, the legs are greatly thicker than in that animal: the head is large, the eyes are very large, and of a fierce and terrible aspect; the mouth is very large, and the teeth enormously long: the neck is very thick, the shoulders large, and the legs of a monstrous thickness; yet this is principally owing to the vastness of the bones, for they are covered with very little flesh: the loins are weak, in proportion to the rest of the body, and the hinder legs have less appearance of strength than the fore-ones: the tail is thick and long, but is not floccose as the lion's.

The ground colour of the tyger is a pale tawny, with an admixture of brown, but is all over variegated with streaks of black; these are long, and considerably broad, and go the cross-way of the body.

The tyger is a native of Asia, and also of many parts of America. The antients have spoke greatly of it's swiftness, but this is not to be understood of it's running, for in that it is inferior in speed to most of the beasts of prey, but of it's leaping. It seldom pursues any thing in a fair chace. It's custom is to lie in wait, and it will throw itself forward out of these lurking-places with a surprising violence and rapidity; if it misses the prey in the leap, it very often does not turn against it, but walks away. All the writers on quadrupeds have described this under the name of Tigris.

Felis cauda elongata, maculis superioribus orbiculatis, inferioribus virgatis.
The long-tailed Felis, with the upper spots round, and the lower virgated. **The Leopard.**

This is a very beautiful animal, and, though less terrible in appearance than the lion or tyger, is not less fierce or voracious; it is considerably smaller than the tyger, and it's body is less bulky; but the limbs are surprisingly strong, and the creature is very active.

The head is large and fleshy; the eyes are fierce and large; the mouth wide, and is furnished with a terrible apparatus of teeth: the neck is longer, and less thick than in the lion or tyger, and the body is also smaller, but there is an appearance of great strength about the shoulders: the claws are of a terrible length, and are formed exactly as those of the cat: the tail is long, thick, and variegated in the same manner as the body.

The ground colour of the animal is a deep fallow or tawny brown; the back and upper part of the sides are variegated with round spots of a deep and glossy black; the lower

lower part of the sides with streaks of the same colour with those spots, which are continued quite to the belly.

It is a very nimble as well as a very fierce animal, scarce any thing escapes it. All the writers on quadrupeds have named it, but there is some confusion in the accounts of the earlier authors, and some of the later ones seem to have made two species of the male and female. They call the male *Pardus*, and the female *Panthera*. It is a native of Asia and America.

Felis cauda elongata maculis superioribus virgatis, inferioribus punctatis.
The Felis, with an elongated tail, with the upper spots virgated, the lower punctated.

The Cat-a-Mountain.

This is of the bigness of one of our mastiffs, but not of those of the very largest kind: the head is large; the eyes are prominent and very fierce; the ears short and patulous, and the mouth large, and furnished with a terrible apparatus of teeth: the neck is short and thick; the body is short, rounded, and bulky: the legs are robust, and the claws very sharp and very long.

The ground colour of the back and sides is of a dusky brown; the belly and part of the legs, as also part of the face, is white: the variegations are black, and they run along the back and upper part of the sides, in a longitudinal, not in a transverse, direction: the lower part of the sides, and the belly, instead of these streaks, have only small spots of black; the tail is long, and variegated.

This is a native of Asia and Africa, and some parts also of America. Authors have described it under the name of Catus Pardus; and some have confounded it with the leopard.

Felis cauda elongata, auribus penicilli-formibus.
The long-tailed Felis, with pencilled ears.

This is a singular and a very beautiful animal; it is of the size of a large dog, but the limbs are much stronger, and more bulky: the head is large; the ears are very broad at their base, and very open, but they terminate in a point at the top, and are each ornamented there with a pencil of hairs of a moderate length, thick, firm, and coal black: the neck is short and thick; the breast is broad; the body is corpulent and short; the tail is long and variegated: the legs are robust, but not so remarkably thick as in some of the other beasts of prey; the claws are very long, sharp, and hooked.

The ground colour is a pale brown, with a mixture of tawny; the variegations are black: they are disposed in irregular spots, large and oblong on the upper part of the animal, and smaller and rounded on the under.

This is a native of some parts of Europe, and has been, by some writers, confounded with the Cat-a-mountain; but the pencilled ears, in which it resembles the lynx, sufficiently distinguish it.

Felis cauda truncata, corpore rufescente maculato.
The truncated-tailed Felis, with the body brown and spotted with black.

The Lynx.

This is a large and robust animal, but it is less than the lion: the head is large, but not very long; the forehead is flat; the eyes are large and fierce; the ears are very large and open, but they terminate in a point at the top, and are there ornamented

ornamented with a pencil of fine black hairs: the mouth is furnished with very terrible teeth, and there are whiskers about it, as also over the eyes: the neck is long, and moderately thick: the breast is large and broad; the body is moderately corpulent; the legs are very robust, and the claws terrible; the tail is short and abrupt, but thick, and well covered with hair.

The fur upon the whole animal is very long and deep, but is not so thick as in many of the preceding species; it is of a very singular colour, a kind of a very pale reddish-brown: there is an admixture of white in it in some parts, and the whole body and legs are spotted with black: the spots are of an irregular figure, and stand at distances, and are not large.

This is a native of some parts of Europe; it is even found so far north as Denmark and Sweden, where it lives in the midst of thick woods, and will climb trees with great facility. All the writers on quadrupeds have described it. They call it Lynx and Lupus Cervarius; we, the Ounce. It is a very fierce creature, and remarkable for the quickness of it's sight.

Felis cauda truncata, corpore albo, maculato.
The short-tailed, white Felis, with black spots.

The White Ounce.

This is of the bigness of our bull-dog, and is a very beautiful, but very fierce and mischievous, animal: the head is large and short; the forehead is depressed; the eyes are large and blue, and the nostrils are wide; there are whiskers about the mouth, and over the eyes, and the ears end in a pencil of long hairs of a grey colour: the breast is broad; the body short and thick; the tail short, thick, and truncated, and the legs strong, but not so remarkably thick as in the other: the claws are very sharp and robust.

The whole body is white, only that there are a few spots of black of an irregular figure scattered over it in different places; these are largest toward the back, and small upon the legs: the tail is pretty equally variegated with black and white.

This is a native of many of the northern parts of Europe, but it is rarely seen; it lives in the midst of the thickest forests, and climbs trees with great alacrity. The Swedes call it Callo.

Felis cauda elongata, auribus æqualibus.
The long-tailed Felis, with even ears.

The Cat.

This is the smallest animal of all this genus: the head is short and rounded; the neck short and thick; the body long and slender; the legs slender and short; the tail very long, and well covered with hair. The natural colour is a brown tawny, variegated with very pale, and almost white, streaks, disposed cross-wise, and running from the ridge of the back to the lower part of the sides: the belly and the throat are white.

This is the natural colouring of the cat, as we have it wild in our woods. When it is kept in the house, the breed is so mixed with various-coloured males and females, that the diversification is almost infinite. It is singular, that this species is smaller when kept in the house, than while wild. I have seen many of them in Rockingham forest, and other woods in Northamptonshire, and elsewhere; they feed on rabbits, leverets, and the like, and do a vast deal of mischief to the game. In France the cats are all of a bluish lead colour; in the North of Europe they are all white. All the writers on quadrupeds have described the species, under the name of Felis, Felis domestica, and Felis vulgaris.

MUSTELA.

THE upper foreteeth of the Mustela are straight, distinct, and acute: the foreteeth of the lower jaw are obtuse and clustered; two of them stand inward: the feet are made for climbing.

Mustela rufo-fusca medio dorso nigra.
The reddish-brown Mustela, with the middle of the back black.

The Gulo.

This is of the size of our common cat, but the body, unless when distended with food, as is very frequently the case, is much slenderer, in proportion to it's length: the head is small, and of a kind of oval figure, slender at the snout, and rounded on the crown; the eyes are prominent, though not very large; their iris is of a deep hazel; the ears are short and pointless; the nostrils are large, and the mouth is wide, and well furnished with teeth.

The neck is short and thick; the shoulders are high; the breast narrow, and the back sharp: the legs are short, and not very robust, but the claws are extremely sharp: the whole body is of a dusky brownish colour, with a tinge of ferrugineous red, only that the belly is pale and whitish, and the ridge of the back, on the contrary, is black.

It is a native of Germany, and many of the northern parts of Europe. It is very shy; it lives in woods, and takes it's post usually on trees, from whence it precipitates itself down on any thing that is fit for it's prey. It is the most voracious of all animals. It will feed on carcasses, though it's more usual food is birds, and the smaller quadrupeds. We had one of them shewn among other wild beasts, as a sight in London, about four years ago. It would eat in such a manner of any offal that was given it, as not to be able to stir for many hours, but would lie distended, panting, and, as it were, dying. All the authors who have written on quadrupeds have mentioned it. They call it Gulo, and the Glutton, and relate many odd and idle stories of the effects of it's voracious appetite.

Mustela fulvo-nigricans gula pallida.
The blackish-brown Mustela, with a pale throat.

The Martin.

This is equal to the common cat in length, but the body is much slenderer, and the legs are shorter: the head is small, short, and pointed at the snout: the eyes are long, and their iris is of a deep hazel; the ears are short: the mouth is large, and is well furnished with teeth, and there are whiskers about it as in the cat: the breast is narrow; the back is flatted, and somewhat broad; the legs are slender: the tail is equal to the whole body in length, and is very bushy, or covered thick with long hair.

The whole body is covered with a deep and fine fur of a very dark brown, with an admixture of black, but very bright and glossy; the throat is whitish.

It is frequent with us about houses, and is also abundant in the woods in many parts of Europe, where they distinguish two kinds, those of the fir-woods, and those of the beech-woods, under the names of Martes Abietum, and Martes Fagorum. Those called the fir-wood kind are of a paler colour, with somewhat yellowish about them. It feeds on birds, and the smaller quadrupeds. All the authors who have written on beasts have named it; they call it Martes and Troyna.

Mustela flavescente-nigricans ore albo, collari flavo.
The yellowish-black Mustela, with a white mouth, and a yellow collar.

The Pole-cat.

This is somewhat smaller than the Martin, and it's legs are shorter: the head is small, oblong, pointed at the extremity, and rounded on the forehead: the ears are short, broad, pendulous, and white at the edges; the eyes are moderately large, black, and piercing in their aspect; the mouth is wide, and well furnished with teeth; the neck is short and thick; the body is long and slender, and the legs are short, and not very robust.

The head is of a dark brown, approaching to black, but with a tinge of orange colour in it: the mouth and the part about the nostrils are white; the upper part of the neck is black, the under yellowish; the whole body is of a very deep colour, mixed of a kind of dusky yellow, and black: the back is darkest, and the belly paler than the sides: the tail is long and black; the throat is also black, and the legs are black: the yellow mark on the under part of the neck begins a little beyond the angles of the mouth: the upper jaw is longer than the under.

This is frequent with us; it lives in warm hilly fields, under thick hedges, and in woods, and feeds on birds, and the smaller quadrupeds; but it chiefly is found in the neighbourhood of villages, or of farm-houses, where it commits great ravages among the poultry. All the writers on quadrupeds have described it, and all under the same name Putorius. It has a most insufferable scent, which is owing to a fœtid matter, formed in two glands, situated near the anus.

Mustela caudæ apice atro.
The Mustela, with the tip of the tail black.

The Weasel.

This is a smaller animal than the pole-cat; the head is small, of an ovated form, and sharp at the snout; the ears are small, short, and patulous; the eyes are of a fierce aspect, moderately large and prominent; the nostrils are large, and the mouth is well furnished with teeth: the upper jaw is longer than the under, and the anterior part of the head, in some degree, resembles that of a dog: the mouth and about the nostrils and eyes are ornamented with long and thick whiskers; the ears are covered with a short, but very dense, fur, and have a singular kind of a duplicature at their base; the inner aperture of the ear, which is remarkably large, in proportion to the outer, is full of a kind of fleshy protuberances or granulations of a reddish colour.

The neck is short and slender; the body is about eight inches long, and slender; the tail is little more than a third of the length of the body: the legs are very short and slender, and the feet have each five toes, armed with little but very sharp claws.

The whole body is covered with a fine and tolerably long fur; the back is of a dark colour, with some tinge of a ferrugineous or reddish-brown; the sides have more of the red, and the belly is white; the whiteness begins at the chin, and is continued over the whole under part of the creature: the legs are darker than the sides.

This is frequent in all parts of Europe. We have it in woods, and under warm hedges, and it is often found about the farmer's barns and out-houses, where it does great damage among the poultry. All the writers on animals have described it. They all call it by the same name Mustela; we, the Weasel, the Foumart, or the Fitchet.

Mustela albido-flavescens capite depresso.
The yellowish-white Mustela, with a flatted head. **The Ferret.**

This is of the middle size, between the pole-cat and weasel; it is, indeed, but little inferior to the former in length, though less corpulent: the head is small and depressed, the snout sharp; the eyes look very fierce and red; the ears are short, patulous, and erect; they are considerably wide, especially toward the base: the mouth is large, and the teeth very sharp.

The neck is short and slender; the body is long and thin: the legs are short, and are each divided at the foot into five toes, and there armed with very sharp and moderately long claws: the whole animal is of a yellowish-white colour, but the under part is darker than the upper, which is against the usual course in these animals.

It is kept very frequent with us, but is a native of Africa. Our people use it in taking of rabbits; they place nets at the mouths of the burrows, and turn in the ferret, which forces the creature out of it's retreat, and it is taken. Before they do this, they are obliged to sew up the mouth of the ferret, otherwise it would destroy the victim, and mangle it.

Mustela nivea auribus angustis, caudæ apice nigro.
The white Mustela, with narrow ears, and the tip of the tail black. **The Ermin.**

This is an extremely beautiful creature; the bigness is that of the common weasel, and there have been those who have supposed it the same species, only differing in colour, but they cannot have seen perfect specimens of the animal.

The head is very small, depressed, broad at the crown, and sharp at the snout: the anterior part, in a great degree, resembles that of the head of a greyhound: the ears are short, erect, and narrow; the eyes are small, but somewhat prominent; the mouth is large, and the teeth small but sharp: the neck is short, and somewhat thick; the body is about eight inches in length, but it is very slender: the legs are short, and the feet are each divided into five slender and weak toes, which are armed with sharp white claws.

The whole body of the animal is white, the purest and brightest colour that can be expressed by that word: the tip of the tail is of a deep black; about the eyes it is of a dusky colour between grey and yellow, and there is a spot of the same colour in the middle of the head, another between the shoulders, and, finally, a third just above the insertion of the tail.

This is a native of the northern parts of Europe; it lives in places remote from towns or houses, and feeds on moles and mice, and other small animals. The fur is extremely valued. All the writers on beasts have described it. Ray calls it Mustela candida sive Animal Ermineum; and most of the others Erminea, or Animal Ermineum.

Mustela fulva auribus cinerascentibus.
The brown Mustela, with grey ears. **The Sable.**

This is very like the common weasel in form, but it is equal to the pole-cat in size: the head is small, and of an ovated form, depressed on the forehead, and sharp at the snout: the upper jaw is longer than the under; the eyes are large, prominent, and dark; the ears are short, broad, erect, and patulous; the mouth is wide, and the teeth are sharp; the neck is very short, and tolerably thick; the body is long and slender: the tail is moderately long, and is well covered with hair, and the legs are very

very short and slender; the feet have each five toes, and those are armed with very sharp, though not with very strong, claws: the whole creature has greatly the appearance of our weasel or pole-cat, but it walks lower or nearer to the ground; though this is rather owing to the manner of it's using it's legs, than to their being in reality shorter than in those species.

The fur of this creature is very thick and deep, and is remarkably fine and glossy: the colour is a deep, but a very beautiful, brown; there are long hairs, in manner of whiskers, about the mouth and nostrils, as also over the eyes: the nose and the ears are grey, and the throat is of a yet paler grey or whitish; the rest is all of the same elegant brown.

It is a native of many of the northern parts of Europe, and it's fur is valued at a very high rate. We have a way of dying and preparing fox-skins, so that they, in some degree, imitate it; but, when seen with the true sable, the counterfeit makes but a very paltry appearance. All the writers on animals have described the creature. Ray and others call it Mustela Zibellina.

Mustela cauda annulis nigris albidisque cincta.
The Mustela, with the tail annulated with white and black. **The Genet.**

This is a very beautiful little animal, but it differs considerably in form from those which have been hitherto described; they are all slender-bodied and low, their legs being very short; this is more corpulent, and the legs are longer; it has much the appearance of a young fox: the head is large, oblong, broad, and depressed at the crown, and pointed at the nose; the ears are short, broad, and erect; the eyes are large, and they stand prominent: the nostrils are wide, and the mouth is large, and very well furnished with teeth: the neck is short and thick; the body is long, and tolerably large; the tail is long and very beautiful, and the legs are proportioned more like that of the cat, than the short ones of the sable or weasel.

The head is of a dusky tawny, with an admixture of black, but there is also some appearance of grey about the nose and ears; and all about the mouth, and over the eyes, there are black and thick whiskers: the back is of a very deep tawny, approaching to black; the sides are of a paler tawny, or yellowish colour, and are variegated with large spots of black, disposed in a very regular and beautiful manner: the legs are nearly black, and the tail is beautifully variegated with the same deep tawny, approaching to black, and with white or regular annular spots.

It is a native of many parts of the East; it lives in woods, and especially where there are springs. It feeds on birds, and the smaller quadrupeds. It is easily tamed, and becomes familiar with mankind, and in the Turkish dominions is kept in the houses to destroy vermin, as cats are with us. All the writers of beasts have described it; they call it Genetta.

Mustela grisea et albido variegata.
The grey and white variegated Mustela. **The tabbied Mungo.**

This is an extreamly beautiful animal, it is the largest of all the Mustelæ; the size is that of our largest wild cat, but the body is less corpulent, and the tail is vastly longer, and more beautiful; it is the lightest of all creatures, in proportion to it's apparent size: the head is small, oblong, depressed at the crown, and narrow and sharp at the snout: the eyes are prominent; the ears are covered with a fine down; the mouth is large, and well furnished with teeth: the neck is slender, and moderately long: the body is tolerably large and long, the legs are slender, and the tail is of a very great length, and covered with such a long fur, that it appears very thick; it is, indeed, always naturally in that state, in which the tail of the common cat appears, when that creature is provoked: the feet are each terminated by five toes, and these are armed with long, sharp, hooked claws, of a deep colour.

The whole body is variegated with a pale grey and whitish colour, disposed in transverse streaks, in the manner of the variegations of our tabby cats, and reaching from the ridge of the back to the lower part of the sides; the throat and belly are whitish.

This is a native of many parts of the East, where it lives wild in the woods, and is frequently also kept tame in houses, for the destroying of vermin. It is the most inoffensive creature imaginable, and so indefatigable in search of mice, and other reptiles, that it is impossible to open a box or a drawer almost in any part of the house, but the creature is in it on the instant. We have it brought over to Europe, and it lives very comfortably with us. I remember one at the late Duke of Richmond's, which used to run tame about the house, and afforded great diversion, in it's escapes from an Italian greyhound, that was eternally pursuing it. It is the nimblest, in it's motions, of all creatures, and the distance to which it will throw itself at one leap is surprising; it is so extremely light, and so deeply covered with fur, that it seems to float upon the air, when the motion is once given, and to be carried, without any new impulse, to a surprising distance. Ray and some others have described it; they call it Viverra Indica quæ Mungo Lusitanis, Mungathia Ceylanensibus; we call it the Mungo or Mungose.

Mustela ferruginea cruribus longioribus.
The ferruginous Mustela, with longer legs.

The brown Mungo.

This is smaller than the preceding species, but is otherwise like it in all but colour: the head is large and oval; the forehead somewhat depressed; the nose sharp, and the upper jaw somewhat, though not much, longer than the under: the eyes are large and prominent: the ears are short, erect, and patulous; there are whiskers about the mouth, and over the eyes, and the teeth are not large, but very sharp.

The neck is short and thick; the body is compressed; the legs are moderately long and slender, and the tail is very long, and covered with a deep fur; the claws are sharp, but not strong: the head is of a pale ferrugineous colour, very bright and glossy, and there is also some greyness about the eyes, and at the mouth the whiskers are black: the back is of a very deep ferrugineous brown, on the very ridge it almost approaches to black; the sides are paler, and the throat and belly are yet paler than either: the legs are dark, and the tail is irregularly, but somewhat in the annulated manner, variegated with a very deep brown, and a paler or yellowish ferrugineous.

This is a native both of the East and West Indies, and is kept in the houses there, as we do cats to destroy vermin. We have it sometimes brought alive to England, and it will live very comfortably with us. Authors have called it Mungo and Mungathia fusca. It is not so handsome as the former.

L U T R A.

THE fore-teeth in the upper jaw in the Lutra are straight, distinct, and acute; those of the under jaw are obtuse, and stand in a clustered manner with two inward: the ears are situated lower than the eyes, and the feet are palmated and formed for swimming.

Lutra digitis omnibus æqualibus.
The Lutra, with all the toes of equal length.

The Otter.

This is a large and a fierce animal; it's length, including the tail, is more than three feet; the body is long and bulky, but the legs are short, so that it walks very low: the head is short and thick; the forehead depressed; the nose thick and rounded; the mouth is large; the nostrils are small, and there are on each side, between the nostrils and mouth, a number of whiskers: the eyes are large, and of a fierce aspect; the ears are short and round, and they are situated very oddly, they do not stand on the top

top of the forehead as in most other animals; but at the very hinder part of the head, just at the neck, and much lower than the eyes, as well as a great way behind them.

The neck is very thick, and not very short: the body is rounded, and very fleshy; the tail is extreamly thick and long, and is all the way covered with a thick fur: the legs are short, but very robust; the feet are broad, and they are webbed; the toes are each armed with a sharp claw.

The whole body is covered with a fine and tolerably deep fur, but less deep as well as less glossy than that of the beaver, which animal it in general greatly resembles; the back is of a deep chesnut brown, but with some faint admixture of grey in it: the throat, the breast, and the belly are paler, or have more of the grey; the legs are of a very dark chesnut, and so is the tail, and both are covered with a much shorter fur than that of the body.

This is very frequent in most parts of Europe: it frequents waters, and feeds on fish; it will keep under water a great while, and move under it with a surprising swiftness. We have it in too great abondance in the fens in Lincolnshire, where it destroys the pike in great quantities, and will seize on the largest. In the river Nen, near Peterborough, at about half a mile above the town, there is a part where the water is very clear and deep; I have seen them there, as I have been fishing from a boat, passing with an amazing rapidity, at twenty feet under the surface. All the writers on quadrupeds have described it; they call it Lutra.

Lutra pollice digitis breviore.
The Lutra, with the inner toe shorter than the other.

The Brasilian Otter.

This is a large and a very beautiful animal; it's length, including the tail, is near four feet: the head is large, thick, rounded at the crown, but somewhat sharp at the snout: the eyes are large and black; the ears are very broad and open; they stand extreamly backward, almost at the neck, and on the sides, not on the top of the head, but are but little more than on a level with the angles of the mouth in height: the nostrils are large, and there are long whiskers about the mouth: the teeth are sharp, and the opening of the mouth is wide and terrible.

The body is very corpulent; the back broad, and the belly rounded and swelling: the legs are robust, but short: the toes are long, except the inner one, which answers to the thumb on the human hand, and is very short and inconsiderable: the others are connected by a membrane, in the manner of those of the water fowls: the tail is very long, and very thick, and is carried in a straight direction.

The whole body is covered with a deep fur, very glossy and soft to the touch, and of a fine and elegant black colour: the legs are of a less deep black than the body, and the head is of a tawny brown, but the most singular variegation is a spot of yellow on the throat.

This is a native both of the East Indies, and of South America. Marcgrave and Piso have described it, and we have of late had skins of it from the East. The Brasilians call it Jiya, a name devised in imitation of a yelping noise it makes, which is not unlike that of a puppy; they also call it Carigueibeju.

CANIS.

THE fore-teeth in the upper jaw of the Canis are acute, and there are four intermediate ones of a trilobated figure: the canine teeth of the upper jaw are remote from the fore-ones: the top of the cranium is carinated. Though all dogs, however various in name and figure, are truly but of one species, this genus comprehends so many other animals agreeing with them in character, that it is extensive.

Canis

Canis cauda recurva.
The Canis, with the tail bent upwards. **The Dog.**

The dog, in it's wild ſtate, is of a middle proportion, between the maſtiff and greyhound: the head is long, and the noſe obtuſe; the neck is long and fleſhy, and the body is moderately large; the legs are long, and the tail is very long, and naturally turns up: the colour is tawny, and the fur about as long as that on the maſtiff-kind with us; it has a ſtrong and quick ſmell, and follows beaſts of a ſmaller ſize than itſelf, as well by the noſe, as by ſight.

The dog, in this ſtate, is a native of many parts of the Eaſt, and lives comfortably in the woods. It does not attack a man, if it meets him, but neither has it any thing of that kindneſs and familiarity which we find in it, unleſs bred with us. Many animals may be made as tame as the dog, by the ſame kind of treatment. I have known it tried on the otter with perfect ſucceſs.

Authors have mentioned this ſpecies by the ſimple name of Canis, giving a variety of others to the ſeveral varieties which we have produced, and are continually producing by mixed breeds. The maſtiff, the wolf-dog, the greyhound, the hound and it's ſubdiviſions, the ſpaniel, the water-ſpaniel, the bull-dog, the lap-dog, and the reſt, are only varieties of this original ſpecies. To thoſe who would ſuppoſe them diſtinct, their forms, manners, and other particularities would render them ſubjects for a volume.

Canis cauda incurva.
The Canis, with the tail bending inward. **The Wolf.**

As ſome have been for making the lap-dog, the greyhound, and the reſt diſtinct ſpecies, others have erred in the contrary extream, and have been for confounding the wolf and the dog as of the ſame, though in reality they are perfectly different. The wolf is a very large and a very fierce animal; it is equal to the biggeſt maſtiff in ſize, and has much of the general appearance of that creature; the head is large and fleſhy; the eyes have a very bold and fierce aſpect; they are large and prominent, and their iris is hazel: the ears are ſhort, pointed, and erect; the teeth are very large, and the creature has a way of ſhewing them in a frightful manner by grinning: the neck is robuſt and thick; the antients ſuppoſed the creature could not move it, but this is an error; the wolf turns it about more readily than any of the dog-kind, and, though very ſtrong, it is not at all rigid: the body is large, and the back broad, unleſs when the creature is ſtarved: the legs are moderately long and very robuſt, and the tail is long and buſhy, like that of the fox, and naturally turns inward.

The natural colour of the wolf is black, but there are ſome tawny, and in many places they are, in winter, perfectly white as ſnow: the voice of this creature is very like the howling of the dog, but it does not bark in the manner of that animal.

It is a native of almoſt all parts of Europe, and is very miſchievous wherever it comes. Cattle are a continual ſacrifice to it; and in hard winters, when the woods afford no food, they will come down in troops, and attack houſes and villages, deſtroying every thing they can get at. All the writers on quadrupeds have deſcribed it, and all the under the ſame name Lupus.

Canis roſtro anguſtiore.
The Canis, with a ſlender ſnout. **The Jackall.**

This is a very beautiful creature, and ſo greatly approaches to the dog-kind, that it would be very natural for a perſon at firſt ſight to miſtake it for ſome mungrel breed of that animal; it is of the ſize of a ſmall hound: the head is long and narrow, eſpecially

pecially toward the nose; the ears are short and erect; the eyes are large, and of a remarkably brisk and piercing aspect: the mouth is wide, and is very well furnished with teeth: the neck is short and thick; the body is large; the legs are about in the same proportion, as to length and thickness, as in the hound: the tail is large, and covered with a deep fur.

The colour of the whole body is that sort of tan or liver colour, which we see in spots on some of the hounds, and spaniels, only that it is brighter, and has more of the yellow in it: the legs are darker than the body, and there is something dark also about the face.

This species is a native of the East in vast abundance: they hunt, and that in natural packs, it being frequent to hear the noses of more than two hundred of them together; in these companies they will seize upon animals, wh ch they would never dare to attack alone, but there is often bloody fighting about the division of the spoil. It is not impossible but lions, and other larger beasts of prey, may be alarmed by the cries of these little creatures in their chace, and may fall in and rob them of their prey, but the general opinion of their attendance on the lion is fabulous. All the writers on quadrupeds have mentioned the Jackall; they call it Lupus aureus.

Canis pilis cervicis erectis longioribus.
The Canis, with the hairs of the neck long and erect.

The Hyena.

This is a very singular and a very ugly animal; it is of the bigness of a bull-dog: the head is large and short; the nose is obtuse; the mouth is wide, and furnished with a terrible armature of teeth; the eyes are large, black, and of a very fierce aspect; the ears are short, broad, and erect: the neck is very thick, and is covered with a kind of bristles instead of hairs, which naturally stand erect, and give a very formidable appearance to the creature: the body is bulky and rounded; the shape of it is not unlike that of a pig: the legs are moderately long, and very robust; the general colour is a very dusky olive, approaching to black: the legs are darker, and the face paler than the rest.

It is a native of many parts of the East, and is an extreamly fierce and voracious animal. It is not very swift of foot, but it is continually lying in wait for other creatures, and scarce any thing that comes in it's way escapes it. It's voice is shrill, and has a mournful sound. Some have placed this creature among the monkies, others have made it of the badger-kind; and it is probable that the stories of the Man-tyger and Lamia have been founded on the misrepresentations which authors have given of this animal.

Canis cauda recta extremitate alba.
The Canis, with the tail straight and white at the tip.

The Fox.

The fox much resembles the dog in form, and is of the size of a common spaniel: the head is large, oblong, and, though considerably broad at the crown, is very narrow at the nose: the eyes are large and prominent; the ears moderately large and erect; the mouth is wide, and well furnished with teeth: the neck is short and robust; the body is large, and the legs are very strong, though not remarkably thick; the tail is long and bushy. The colour of the fox with us is a reddish-tawny, but in some parts of Germany it is naturally black, in others grey, or of a dark and glossy brown, and there are some places where it is white, either during the winter, or the whole year.

It is a native of most parts of Europe, and is a very mischievous creature about farm-houses, destroying poultry and other things in a most rapacious manner. All the authors who have written on quadrupeds have described it; they have called it Vulpes,

Canis cauda recta extremitate nigra.
The Canis, with a straight tail black at the tip.

The larger Fox.

This is somewhat larger than the common fox, and every way more robust, in proportion: the head is large, and less attenuated at the snout than in the common fox; the nose is obtuse; the nostrils are large; the eyes are large, black, and prominent; the ears are erect and patulous; the mouth is wide, and well furnished with teeth; the neck is thick and short, the body is also short; the legs are very thick, and the tail is long, straight, bushy, and it's tip black: the colour of the whole animal is a deep brown, somewhat like that of the sable, and the fur is very deep and glossy. We have the skins sometimes imposed on us, when cut up and wrought, for true sable; and indeed they come nearer it than those of any other animal.

It is a native of Tartary; few authors have been acquainted with it.

Canis cauda recta unicolore.
The Canis, with a straight tail all of one colour.

The Siberian Fox.

This is of the size of the common fox: the head is larger, and is not so much attenuated at the snout; the nostrils are large; the mouth is wide; the ears are short, erect, and patulous: the eyes are small, but they are prominent and black; the neck is short and robust; the legs are shorter and thicker than in the common fox, and the tail is longer, and more bushy, and is neither black nor white at the extremity, but of the same colour with the rest of it, which is the same with that of the body, and is a pale brown, with a tinge of grey.

This is a native of Siberia, and is frequent in the woods, and about villages, preying on poultry, and weak animals of every kind, as the common kind with us. Few authors have seen it.

PHOCA.

THE fore-teeth in the upper jaw of the Phoca are six; those in the under jaw are only four: the feet have each five toes, and are palmated and made for swimming: there are no ears.

Phoca dentibus caninis tectis.
The Phoca, with the canine teeth covered.

The Sea-calf.

This is a very singular and extraordinary animal; it seems, indeed, in some degree, to connect the quadruped and the fish-kind. It grows to five feet, or more, in length, and will become very corpulent, but those which are usually caught are smaller: the head is very large, and irregularly shaped; it is oblong, depressed on the crown, and somewhat flatted at the sides, and has a kind of protuberance on the under part, so that in the whole it appears very singular: the eyes are large, and are not prominent, but sunk in the substance of the head, as it were; the ears are only two small apertures; in this it is extremely singular, being the only creature among the viviparous quadrupeds, that has no external ear: the snout is obtuse; the nostrils are small; the mouth is wide, and over it there are placed a number of whiskers: the neck is very short and thick; the body is long, and very large: the back flatted and depressed; the belly round and prominent: the fore-legs are rounded, and very short; they are indeed buried under the skin of the body, almost to the feet: the hinder legs are broad, flat, and joined together, so as to resemble the tail of a fish; between these there is a very short tail: all the feet are webbed, or have the toes, which are five on each, connected by a membrane: from this structure of the feet the creature must be very ill qualified for walking: the hairs or bristles which compose the whiskers are of an odd figure, square and compressed; there are none of them over the eyes, but only about the mouth.

The

The whole body is covered with a short fur, of a mixed greyish and yellowish colour; it is often spotted with a darker colour toward the back, and is always paler on the breast and belly than elsewhere: the tongue is large, like that of a calf, but it is bifid, or divided into two parts at the end: the creature is contrived for living a great part of it's time under water, and the foramen ovale of the heart is to this purpose continued open in it, as it is in fœtus's, which are to live without the assistance of breathing.

It is very frequent about the sea-coasts of many parts of Europe. We have it in Cornwal, and some other places; they come out of the sea to sleep, and the people take this occasion of attacking them. They will throw stones from about them, by the assistance of their hinder legs, to a great distance, and with a dangerous violence. All the writers on quadrupeds have described this species, they call it Phoca and Vitulus marinus; we, the Sea-calf.

Phoca dentibus caninis exertis.
The Phoca, with the canine teeth exerted. **The Walrus.**

This is a much more singular and extraordinary creature than the former, and is much larger; it grows to the size of the largest ox: the head is very large, and is thick, short, and of an almost rounded figure: the eyes are very large and prominent; there are no ears, but only an aperture on each side of the head, of an oblong form, and not very large: the nose is obtuse; the nostrils are large, and the creature contracts and dilates them at pleasure; the mouth is very large, and the upper part of it is furnished with some very large, thick, and cartilaginous whiskers: the tongue is short; the canine teeth of the upper jaw are of an enormous length and size, and they hang downwards and forwards toward the breast; the creature uses these strange weapons to climb upon the ice, and to hang itself to the rocks, in it's getting on shore to sleep.

The neck is short and robust; the body is corpulent to a great degree: the fore-legs are buried a great way under the skin of the body; the hinder legs are connected together, and they form a kind of tail, like that of a fish; there is no tail.

The skin is thick and firm, and is covered with short, and not very thick-set, hairs; they are of a grey colour, and tolerably firm.

This strange animal is a native of the Northern Seas; it is frequent about Greenland, and in other places where there are whale-fisheries, and sometimes comes into warmer climates. Most of the writers on animals have described it. They call it the Rosmarus, the Wallross or Walrus, and the *Morse*; some, but very improperly, the Sea-horse. We use the teeth in the manner of ivory.

M E L E S.

THE fore-teeth of the Meles are all obtuse, and those of the upper jaw are striated. There is also, in this genus, always a bag of *secreted* matter, situated near the anus.

Meles unguibus anticis longissimis.
The Meles, with the anterior claws very long. **The Badger.**

This is a very singular animal, and is not without it's beauty; it is of the size of a small dog: the body is thick and short; the neck is extremely short: the fur, with which it is covered, is composed of hairs so rigid and thick, that they have the appearance of hogs bristles: those on the back, which are the largest and fairest, are yellow toward their base, of a blackish-brown in the middle, and of a deeper or stronger yellow at the tips, so that the colour of the creature is an odd mixture of deep

brown

brown and pale yellow, which together form a kind of grey; the animal has thence obtained one of it's English names, the people of many places with us calling it the grey: the sides and belly are in some [illegible] of a yellowish colour, but in others the whole under part of the body is black; this has a very singular appearance, as in all other creatures almost the belly is paler than the back: the throat is always black, as are also the shoulders, and all the legs, as well the hinder as the fore-ones; the head is long, and there runs all the way down it, from the crown to the extremity of the snout, a broad line of white, equal to two fingers in breadth; and below this there runs on each side a broad line of black, which is continued beyond the region of the eyes, almost to the neck; and below this, on each side, the fur is again whitish. This gives a great variation of colouring to the head, and it appears very beautiful.

The tail is thick and short, and is covered with short and rigid black hairs: the ears are short and round, but have something of the appearance of those of a rat: the eyes are small; the snout is formed very much like that of a dog: the legs are short and robust, and the toes of the fore-feet are armed with extremely sharp and long claws: the general figure of the head resembles that of the fox; it is broad across the top, and thence becomes narrow all the way to the nose, so that it is of a kind of conic figure; the cheeks are tumid.

Just under the tail, above the orifice of the excrements, there is an aperture which discharges itself into a small bag, the contents of which are thick, white, and soft, and in some degree resemble the brains of a calf, when boiled. This substance has no particular smell; but, beside this, there is a very fetid matter, secreted by two glands, situated near the anus, and having no communication with the bladder.

This creature is a native of most parts of *Europe*; we have it in England, but it is not common: with us it makes itself holes in the earth, as the rabbit does. It's food is insects, and the smaller animals, but it will sometimes prey on larger. The strength of the muscles about the mouth, to which the swelling out of the cheeks is owing, gives it a power of biting in a violent manner. All the writers on these subjects have described it. Gesner and many others call it Meles; Jonston and Aldrovand, and after them also many others, Taxus; Charleton, Taxus sive Daxus. We call it the Badger, the Pate, and the Grey.

Meles cinerea unguibus uniformibus.
The grey Meles, with uniform claws.

The Civet-Cat.

It was long before the form of the creature, to which we owe the civet, was known, and long after this, before it could be determined to what genus of quadrupeds it belonged; it was first supposed of the cat, and afterwards of the dog kind, but it is truly one of the badger species.

It is a large and fierce animal; it's size is that of the common badger, but it's body is not so bulky: the head is large, oblong, and considerably thick all the way, not conic, as in the fox: the forehead is depressed; the whole snout is rounded and thick; the nose turns up a little, and the nostrils are moderately large: the mouth is wide, and is furnished in a very formidable manner with teeth, and there are a few rigid, but not very long, whiskers placed about it; the eyes are small, and they do not stand very prominent; the ears are large, obtuse, erect, and patulous.

The neck is short, rigid, and thick; the body is long, and considerably thick; the tail is long, and resembles that of the common cat; it is covered with long hair, and there runs a ridge of the same hair also along the top of the back.

The whole animal is of a light silvery colour, variegated in a beautiful manner, with large spots of black: the legs are moderately long and robust, and they are almost entirely black: the feet are armed with claws very long and sharp, and equally so throughout, not longer on the anterior, than on the hinder, pair, as in the badger.

Under

Under the tail is situated the bag, in which is contained the perfume, which we call civet. It's situation is exactly the same with that of the bag, which contains the white sebaceous matter in the badger.

It is singular that the throat, and part of the under surface of the common badger, is black; the whole under surface of the civet animal, the throat, breast, and belly are all black, and this is contradictory to the custom of nature, in the generality of animals, which are darker-coloured on the back, and paler on the belly, that has a strange appearance; in this, however, as well as in particulars which are more essential, the badger and this creature agree in a surprising manner.

The civet animal is a native of South America, and of some other of the warmer quarters of the world; it will live with us, and even produce the perfume, but in small quantities, and of an inferior form to that which is obtained in warmer countries. All the late writers on beasts have mentioned this. They call it Felis Zibethicus, and Animal Zibethicum.

Meles unguibus uniformibus leucophaea.
The bluish Meles, with uniform claws.

The Ichneumon.

This is also an animal, which, though long known by name, has been put among very wrong genera by authors; it has been generally ranged among the mus-kind, and called a rat; but it is truly of the badger genus, and indeed extremely resembles the common badger in almost every particular. It is of the bigness of a large cat; the head is long, and the nose tolerably sharp; the eyes are large, but not prominent; the ears are short, erect, and pendulous; the nostrils are large, and the mouth is wide, very well furnished with teeth, and ornamented with whiskers; the tongue is broad, and rounded at the end: the body is long and thick; the tail is long and thick, and is well covered with hair; the legs are robust, but they are not very long; the toes are armed with sharp claws: the head is of a dark colour, approaching to black, especially about the nose; the whole body, beside, is of a grey colour, like that of our common badger, and has, in the same manner, a tinge of yellowish in it, but there is more of the black among the darker shade; so that it has a bluish tinge with the rest, which is not in our badger: the under part of the body is darker than the back or sides, and the legs also are dark or blackish.

This is a native of Egypt, and some other places; it feeds on smaller animals, and does not spare serpents, which are very numerous in that part of the world, and which it destroys in great quantities. The people at Alexandria bring the young to market for sale; those who buy them breed them up tame in their houses, as we do cats, for the destruction of vermin. It is an extremely nimble and bold animal; it will stand a battle with the largest dog, and will strangle any creature in a few minutes, if it can get hold of it's throat, which is what it always attempts. It will stand upon the hinder legs to look about for prey, and, when it has discovered any creature that is proper for it's purposes, it creeps very slowly on the ground till within a due distance, and then throws itself on the unexpecting animal with a surprising violence and rapidity.

Most of the authors who have written on animals have named it; they call it Mus Ichneumon, Ichneumon, and Mus Pharaonis.

ERINACEUS.

THE lateral fore-teeth of the Erinaceus are shorter than the others: the nostrils are cristated, and the body, instead of hairs, is cloathed, in the manner of that of the porcupine, with spines.

Erinaceus auriculatus.
The Erinaceus, with larger ears. **The hedge-hog.**

There have not been wanting authors, who have afferted, that the porcupine and our common hedge-hog were the fame animal, differing only in fize; on the contrary, they are creatures not only of a different genus, but even of diftinct claffes, nor is there indeed any thing in which they agree but the covering of their bodies with this kind of armature, inftead of hair. The hedge-hog is a little animal, confiderably thick, in proportion to it's length, and, when it draws itfelf together at the approach of danger, appears of a kind of an oval figure.

The length of the creature is about feven inches, and it is confiderably broad and thick, in proportion: the head is fmall and oblong; it is broad toward the upper part, and grows fmaller to the nofe: the mouth is formed very much like that of the badger, or in general like the dog-kind: the eyes are fmall, they are black and protuberant; the ears are fhort and broad, and very much refemble thofe of the common rat: the neck is fhort; the back is broad and prominent; the legs are fhort and robuft: the feet are formed like thofe of a dog; there are five toes on each, and the inner one is fhorter than the others, in manner of a thumb.

The whole neck, back, fhoulders, and upper part of the head are covered with ftrong and fharp fpines: the throat, breaft, and belly, as alfo the legs, are covered with a harfh and fhort fur, of a whitifh colour.

This is frequent with us, and in almoft all other parts of Europe. It lives under hedges, and in dry woods. All the writers on quadrupeds have defcribed it. They call it Echinus terreftris, Erinaceus, and Erinaceus vulgaris.

Erinaceus fubauriculatus.
The Erinaceus, with very fmall ears. **The white hedge-hog.**

This is larger than the common hedge-hog, but very like it in form; the head is fmall, and of an oblong figure, broad and depreffed on the crown, and gradually fmaller to the nofe, where, though fmall, it is ftill obtufe: the noftrils are large; the mouth is not very wide, but it is well furnifhed with teeth; the eyes are fmall, bright, black, and prominent: the ears are very low and inconfiderable, they fcarce indeed deferve the name, but the aperture is fufficiently large; the body is broad and thick, but not very long; the back is prominent; the belly flat; the legs are fhort and robuft; the claws fharp, but not very long or ftrong.

The whole upper furface of the body is covered with elegant white fpines, in the place of hair; they are longer, but not thicker, than the fpines of the common hedge-hog: the belly is covered with a pale-coloured, harfh kind of hair, and the legs are alfo covered with a hairinefs of the fame kind, fhort, harfh, and not very thick fet.

It is a native of the American iflands; Ray and others call it Erinaceus Indicus albus.

DASYPUS.

THE body of the Dafypus is covered with a kind of bony or horny coat of mail.

Dafypus

Dafypus cingulo fimplici.
The Dafypus, with a fingle belt.

The Weafel-headed Armadillo.

This is a fingular and beautiful creature: the body is about nine inches in length, and five in breadth, and is covered by nature with a kind of coat of mail, or pointed armour, the fubftance of which refembles tortoife-fhell in fome degree, but is much more firm and hard: the head is fmall, and of an oblong form; it is broad and flatted on the crown, and thence becomes gradually narrower to the extremity, where it is very fmall and fharp; the eyes are fmall; the ears are about half an inch long, patulous, and diftant about an inch and a quarter from one another; the noftrils are wide, the mouth is fmall: the neck is fhort; the body is of a flatted figure, but fomewhat prominent on the back; the tail is about four inches long, and is flender; the legs are very fhort, and the feet large; the toes remarkably long.

The head, the back, the fides, and the thighs are all covered with a cruftaceous and firm armature, which extends itfelf alfo over the tail: the coverings of the head and of the legs are compofed of fquammæ or fcales of a roundifh figure, and of about a quarter of an inch in diameter: the neck is covered with a fingle lamina, compofed of many fmall fruftules of a fquare figure: the fhoulders are covered with a compages of vaft multitudes of fquare fegments of fcales; the covering of the back, from the fhoulders to the tail, is compofed of eighteen moveable laminæ or plates, connected by intermediate cartileges or membranes; the anterior laminæ are compofed only of fquare pieces or fcales; the hinder ones of mixt fquare and round ones: the farther part of the covering at the tail is of a parabolic figure; the anterior part of the tail is covered with fix annular laminæ, compofed of fmall, fquare fcales; the hinder part of it is covered only with fcales not arranged with laminæ at all: the breaft and belly are naked.

This is a native of many parts of America, and is often brought over dried to us. All the late authors have defcribed it; they call it Tatu Muftelinum, the Weafel Armadillo.

Dafypus cingulis fex.
The Dafypus, with fix belts.

This is a larger fpecies than the former, and not lefs fingular or beautiful; it is of the bignefs of a young pig, and is not unlike one in form: the head is exactly of the hog-kind; the forehead is large and depreffed; the fnout is flender and fmall, but obtufe; the eyes are fmall, and funk deep in the head: the ears are fhort, of a brown colour, patulous at their bafe, and naked; the mouth is fmall; the tongue is narrow and pointed, and the teeth are about eighteen in each jaw; the neck is fhort and thick; the body is corpulent, thick, and elevated on the back: the tail is fhort; the legs alfo are fhort, and the feet have fomewhat of the appearance of the human hand.

The head, the whole body, and the tail are covered with a kind of coat of mail; it is of a bony ftructure, and is compofed in an elegant manner of fcales: behind the head there are two joints, thefe ferve for the motions of the neck; on the back there are feven of thefe divifions, at which the fcales are joined by a brown fkin, that throws itfelf between them; thefe ferve for the giving way to the motions of the back, as the others do to thofe of the neck; the reft of the covering is entire: the breaft, belly, and throat are without any of this armature; the fkin of thefe parts is naked, only that there grow certain hairs at a fmall diftance from one another; thefe are of a finger's breadth long, and of a white colour: there are alfo fome hairs of this kind difpofed in manner of whifkers about the mouth and the fkin, between the feveral fcales of the armature, alfo has many of the fame kind growing from it.

The legs are covered with a flighter armature; the tail is very thick at it's origin, but is grown gradually fmaller to the extremity.

This

This is a native of many parts of South America; it feeds on vegetables, and will make it's way into the ground to get at their roots, in the manner of a hog, by burrowing with it's snout. It eats also fruits, and is particularly fond of melons; it will also feed on insects. It usually is found on hilly grounds, but sometimes about waters; it's flesh is esculent, and of an agreeable flavour. The writers on the American animals have all described it. Marcgrave and others have called it Tatu vulgaris, and Armadillo vulgaris.

Dasypus cingulis septem.
The Dasypus, with seven belts.

The Tatuete.

This is of the bigness of a large cat; the head is oblong and small; it is broad at the forehead, and narrower all the way to the snout, which is very small, but obtuse; the ears are considerably long, and are carried erect; the eyes are small, black, and depressed; the nostrils are wide: the body is thick and oblong; the back is elevated, and the belly is somewhat prominent; the tail is short, but it is proportionably longer than in either of the former species: the legs are also short, though less so than in the others; and they have this singularity, that the fore-feet have only four toes, and the hinder ones have five; of these the three middle ones are large and long, the two lateral ones short: the top of the head, and the whole body are covered with a scaly armature, of the hardness of bone; and this is so large, that the creature can with ease draw back it's legs and head, so as to remain wholly covered with it: the belts, or transverse pieces of which this covering of the back is composed, are seven; they are separated by naked places, covered with a tough and firm brown skin: the scaly matter is of a deep brown colour, and very firm structure; the tail is covered with the same kind of armature, and has as many joints as there are in the body.

The throat, the breast, and the belly, as also the inside of the legs, and the under part of the tail are all covered with a tough, but not hard, skin; this is of a brown colour, and has a few short and white hairs scattered over it.

This, as well as the preceding species, is frequent in South America; they feed, in the same manner as the others, indiscriminately on animals of the smaller kind, and on vegetables, and their flesh is esteemed at table. Marcgrave and the other Brasilian writers have described this under the Brasilian name of Tatuete.

Dasypus cingulis tribus.
The Dasypus, with three belts.

The oriental Armadillo.

This is equal to the last described species in size: the head is small, in proportion to the body, and is of an oblong and somewhat conic figure; it is large, and rounded at the crown, and thence gradually becomes smaller to the snout, where it is very slender, but obtuse; the eyes are small and black; they do not stand prominent, but appear rather sunk in the head: the ears are about three quarters of an inch long, erect, patulous, and brown; the nostrils are wide; the mouth is not very large, and there are some whiskers of a grey colour about it.

The neck is short and thick; the body is oblong, rounded, and thick; the tail is very short; the legs also are very short, but the feet are large; they are formed in the manner of hands, and the toes are very long.

The whole back, the neck, the head, and the tail, and also the upper and outer part of the legs, are covered with large scales, forming a kind of bony armature for every part, where the creature could be liable to hurt: those on the back are placed transversely, and are so broad, that three of them reach the whole length; these are connected by a bluish-tough skin, on which there grow a few white hairs: the head, tail, and legs are covered with smaller and more numerous scales of the same kind; the breast and belly are covered with a very firm and tough skin, like that between the scaly belts of the back; it is of a livid colour, granulated in some degree, and

has

QUADRUPEDS Series 3.
Plate 27. Page 560
The Cat o Mountain
The Lynx
The Gulo
The Marten
The Otter
The Jackall
The Cat
The Wolf
The Siberian Fox
The Sea Calf
The Armadillo
The Civet Animal
The Hedgehog

has a few short grey hairs growing on it: the whole upper surface is of a ferruginous brown, the under one of this dusky blue.

It is a native of the East Indies. It is frequent in the woods, where it feeds not only on beetles and other insects, but also on fruits. It's flesh is not eaten, but it is probably as worthy a place on the table, as that of the several American kinds, which are all esteemed delicacies by the natives. Few authors have been acquainted with this species; those who have, call it Tatu orientalis.

Dasypus cingulis quatuor.
The Dasypus, with four belts.

The Water Armadillo.

This is larger than many of the other species: it's head is large, and shorter than in any other; it is broad and depressed across the forehead, and thence becomes smaller to the snout, but it is there of some considerable breadth, and obtuse: the eyes are large, black, and somewhat prominent; the ears are short, erect, and patulous: the nostrils are narrow; the mouth is small, and has about it a few hairs by way of whiskers; the neck is short, thick, and of a kind of conic form, largest at the joining with the shoulders, and smaller to the head: the back is broad and elevated; the belly is somewhat prominent; the tail is short, and largest at the base, whence it tapers very swiftly to the extremity: the legs are short, and not very robust, and the feet which are large have the toes very long, and disposed in form of fingers.

The head, neck, back, and tail are covered with a bony armature, of a yellowish-brown colour, disposed in the manner of scales: the head and tail have small ones; the back is covered by only four transverse pieces, which are broad, and are connected by a tough skin of a dusky colour, with a few hairs on it, standing up between the crevices: the breast, belly, and throat, as also the under part of the tail, and the insides of the legs, are all covered with the same firm and tough skin, and this has also in all those places a few of the same short brown hairs.

This is a native of America, and it is said to be also found in some parts of Europe. some of the old writers have mentioned it under the name of Chelonifcus.

Dasypus tegmine tripartito.
The Dasypus, with the covering tripartitely divided.

The African Armadillo.

The head of this species is long and slender; at the crown it has some breadth, and is depressed; thence it gradually becomes smaller to the snout, where it is small, but obtuse, and rounded: the eyes are little, and sunk in the head; the ears are short, but erect and patulous; the mouth is small, and is encompassed with a kind of whiskers.

The neck is moderately long and slender; the body is broad, oblong, of an ovated form, and elevated on the back: the tail is very short; from a very large base it tapers suddenly to a small point: the legs are short and weak; the feet are large; the toes long and slender, and the claws also long and weak.

The back is covered with a stiff and strong armature, of a bony rather than a scaly nature, and of a brown colour; this is divided beautifully in a tripartite order: the head and tail are also covered with the same sort of mail: the belly and breast are naked, and of a yellowish-brown colour, and there are a few loose hairs scattered at distances all over these parts.

This is a native of Africa, and is found about the borders of rivers; it feeds on roots, which it grubs out of the ground, as the hog does, with it's nose: it will also eat insects, and such small animals as it can destroy. The few writers who have been acquainted with this species, have called it Tatu sive Armadillo Africanus.

Daſypus cingulis novem.
The Daſypus, with nine belts.

The three-toed Armadillo.

The head of this ſpecies is oblong and ſlender; it is rounded at the forehead, where it is largeſt, and thence gradually tapers to the ſnout, where it is narrower and ſharper than in any other ſpecies: the eyes are black and large, but they do not ſtand prominent, but are rather ſunk in the head: the ears are ſhort, but they are erect and patulous; the noſtrils are ſmall, and the mouth alſo is ſmall, and is furniſhed with whiſkers: the neck is ſhort and thick; the body is oblong, conſiderably broad, and elevated a great deal on the back: the tail is ſhort, and very thick at it's baſe, but it tapers to a point: the legs are ſhort, but they are robuſt; the forefeet have only three toes each, the hinder ones have five; they are long and ſlender, and are armed with very long and ſharp claws.

This is a native of many parts of South America; it is alſo common to the Eaſt Indies, and to ſome parts of Africa. It burrows holes for itſelf in the ground, and lives in them in the manner of the badger. Few of the writers on beaſts have diſtinguiſhed it with any degree of accuracy; they have probably confounded it with the common armadillo, for it is ſcarce poſſible it ſhould have wholly eſcaped them.

TALPA.

THE feet of the Talpa are conſtructed in an elegant and ſingular manner; they have the form of hands, and are calculated for digging. It has no external ears.

Talpa caudata.
The tailed Talpa.

The Mole.

This is a little animal, and very ſeldom comes in ſight, it's natural habitation being under-ground; it is nearly of the bigneſs of the common rat in the body, but ſomewhat longer, and leſs corpulent: and the legs are extremely ſhort, neither has it the long tail of that creature, ſo that it does not in the leaſt reſemble it in form: the head is large at the forehead, but it thence becomes ſmall at the ſnout, where it is formed ſomewhat like that of the hog: the mouth is moderately large; the teeth are ſharp, and ſome of them are divided or terminated by ſeveral points, as in the mus araneus: the neck is extremely ſhort; there ſcarce ſeems, indeed, to be any, but the head may be ſaid to be affixed between the ſhoulders: the body is long and rounded; the tail is very ſhort and hairy, there are no external ears; it has been ſaid alſo that the creature has no eyes, but this is a vulgar error; the eyes are indeed extremely ſmall, and are not eaſily ſeen, unleſs the hair about the top of the head be blown open, ſo as to ſeparate it; but then we ſee, in the proper place, two very bright and black eyes, though not larger than the heads of ſmall pins: the legs are very ſhort, and the feet broad; the toes are five on each: the fore-legs have the vola of the foot extended into a kind of palm, like that of the human hand; and the claws on the toes are larger, broader, and more robuſt than on any other animal of a proportionate ſize: the ſkin is thick, and the fur which covers it is a very fine and ſoft down, of a bluiſh-black colour. We have ſeen ſome quite white, but this is an accident that happens to birds as well as quadrupeds.

It is frequent in our paſtures, burrowing at a ſmall depth under-ground: it feeds on worms, and on the roots of graſs, and is very miſchievous, by throwing up the ground.

Talpa ecauda.
The tail-leſs Talpa.

The painted Mole.

This is larger than the common mole: the head is large and oblong, broad at the top, and narrow, but obtuſe at the ſnout; the eyes are very minute, and ſcarce viſible, without ſome pains in the ſeeking after them; there are no ears: the mouth is ſmall, but the teeth are ſharp, and the creature uſes them very boldly, biting any one that touches it in a very violent manner; the body is long, thick, and rounded; there is no tail, not indeed ſo much as the leaſt rudiments of one; the legs are ſhort, and the feet are broad, and formed like hands: the fur with which the creature is covered is ſhort and fine, but very thick; it is of a mixed colour, in which a purpliſh and a yellowiſh tinge ſeem the prevailing ones, but neither of theſe is ſeen diſtinct in any light: the mixture, when the eye is thrown full upon it, forms a kind of brown, and in ſide lights has many varied tinges, in the manner of the changeable ſilks.

This is a native of the Eaſt; it lives under-ground, in the manner of our mole, and feeds on inſects, and on the roots of plants.

VESPERTILIO.

THE Veſpertilio, from it's power of flying, has been generally ranged among the birds, but with the utmoſt impropriety. That it is, in all reſpects, a quadruped, will appear by it's characters.

The fore-teeth of the upper jaw are ſix in number, they are acute and diſtant; the fore-teeth of the lower jaw are alſo ſix, and they are acute and contiguous: the canine teeth are two above, and two below, on each ſide: the feet have each five toes, as well the hinder ones as the fore ones; the fore-feet have the toes connected by a membrane, and expanded into a ſort of wings.

Veſpertilio caudatus, ore naſoque ſimplici.
The Veſpertilio, with a tail, and with the mouth and noſe ſimple.

The Bat.

This creature's having the power of flying has made it be ranked by many as obſerved among the birds, but it has the mouth of a quadruped, not the beak of a bird; it has hair on the body, not feathers, and it produces it's young alive, not under the form of eggs. It has indeed no one thing in common with that ſeries of animals but it's flying, and the flying lizard, or the flying ſquirrel, may as juſtly be numbered among them.

The common bat is of the bigneſs of the mouſe, and very much reſembles it in ſhape and colour: the head is large and oblong; the forehead is prominent, and the noſe ſmall but obtuſe: the ears are ſmall, erect, and patulous; the eyes are moderately large and prominent; the noſtrils are open, and the mouth is large, and the teeth ſharp: the neck is ſhort, the body is oblong and bulky; the tail is moderately long, and the hinder legs ſlender, but not ſhort: the toes are long, and are armed with ſharp claws.

The fore-feet are expanded with a kind of leather wings; the toes are very long, and ſerve as ribs to ſupport and move the tough membrane that is expanded between them. Theſe wings, as they are called, ſurround the body, and are continuous one to the other, and at the bending of them there is a kind of hook, by which the creature fixes itſelf to trees, walls, or other parts of buildings, or even to rocks: the colour of the fur is that of the common mouſe, only with a tinge of the olive: the membranes, forming what are called the wings, are of a duſky hue. The female has two

two teats, and the young are produced two at a birth; the parent, while they are unable to take care of themselves, flies about with them, adhering to her body.

This is common in all parts of Europe; we have it about churches, and other old buildings, in great abundance. It hides itself, during the day, and flutters about in the evening, preying on moths and other insects. All the writers on animals have described it; they call it Vespertilio, and Vespertilio vulgaris; we the Bat and the Flitter-mouse.

Vespertilio cauda nulla.
The Vespertilio, without a tail.

The Eagle Bat.

The size and fierceness of this species have obtained it the addition of eagle to it's name; it is a very singular and formidable animal; the body is twenty inches long, and considerably thick, in proportion; and, when the wings are expanded, it is not smaller in appearance than the bird from which it is named: the head is large, of an oblong figure, and obtuse; the ears are short and pointed, erect, and covered with a very fine down; the eyes are large, black, and prominent; the mouth is large, and well furnished with teeth; the neck is so short, that the head seems set on immediately to the shoulders; the body is thick and fleshy, there is no tail; the hinder legs are moderately long and fleshy, and the fore-ones are expanded into a kind of leathery wings meeting over the back, and, when open, of a most surprising expanse: the back is covered with a tolerable long fur; the belly with a shorter: that of the back is of a dusky and disagreeable dun colour, that of the belly is paler.

This is a native of the East and West Indies; it lives in caverns of the mountains, and often terrifies people by it's fluttering in a strange manner. Many authors have mentioned a bat of this enormous size, but few have described, or even named it, in such a manner, as to convey an idea of it's distinctions.

Vespertilio caudatus labio superiore bifido.
The tailed Vespertilio, with the upper lip bifid.

This is a very large species, but not nearly equal to the preceding in that respect: the head is of an oval figure, large at the crown, and smaller at the nose, but obtuse there: the ears are large and erect, but not very high, in proportion to their expanse: the eyes are large, prominent, and black; the mouth is wide, and furnished with strong and sharp teeth; the upper lip is divided in the manner of that of the hare; the neck is very short, the shoulders are large; the back is broad and flat; the whole body is corpulent, but the tail is moderately long, and the hinder legs are robust; the wings formed of the fore-pair are very broad, and of a bluish colour; the whole body is covered with a soft and fine down, of a deep lead colour; sometimes there have been seen some of them white, but that is accidental and uncommon.

This is a native of South America, and is frequent in caves, and other dark and retired places. Some of the writers on animals have called it Vespertillo cato similis. The same writers have called the preceding species also Canis volans, but these are arbitrary names, and expressive of nothing, except faintly of the size.

Vespertilio caudatus naso foliato acuminato.
The tailed Vespertilio, with a foliated, acuminated nose.

This is larger than the common bat, but it does not approach to either of the two former in size; the head is of a singular figure, oblong, depressed, and acuminated at the nose; the crown is large and flatted; the front is less depressed, and the extremity at the mouth is sharp and foliated; the ears are short, erect, and patulous; the eyes are small, but they are prominent; the mouth is wide, and well furnished with teeth: the neck is very short and thick; the body is broad and flatted; the tail is short,

ſhort, and the hinder legs are ſlender, and not very long: the wings formed of the fore-feet are large; they meet over the back, and extend to a vaſt width, when opened: the membranes which compoſe them are thin, but very tough, and of a dun colour.

The whole body is covered with a ſoft and fine down; the back is of a very deep olive colour, with an admixture of a ſooty black: the belly is paler, but of the ſame tinge; both the olive and the ſmoaky and dingy black are eaſily diſtinguiſhable in it; the front of the head is paler than any other part, and the ears are yet paler than that.

This is common in the caverns of rocks, and in buildings of all parts of America; and flies about in an evening, juſt in the manner of our bat.

Veſpertilio caudatus naſo foliato obverſe cordato.
The tailed Veſpertilio, with a foliated and obverſely cordated noſe. **The flying Mouſe.**

This is of a middle ſize, between our common bat and the laſt deſcribed, and is a very ſingular creature: the head is large, in proportion to the body; the ears are ſhort, erect, and patulous; the eyes are ſmall, black, and prominent: the mouth is large, and well furniſhed with teeth: the noſe is of a very odd figure, ſomewhat divided at the end, and foliated, as in the former, but not pointed, but obverſely cordated; the heart part or emarginated end being the extremity: the neck is ſhort and very thick; the ſhoulders are large; the body is fleſhy; the back rounded, but the belly ſomewhat flat: the tail is ſhort and ſmall; the hinder legs very ſlender and long, eſpecially in the lower joint; the wings formed of the expanded fore-feet are broad and brown: the fur of the whole body is ſhort and thick; it is of a duſky ferruginous colour, and ſomewhat paler on the breaſt than elſewhere; for on the belly it is as dark as on the back.

This is frequent in the Dutch Spice-Iſlands, and I don't know that it is met with elſewhere. The Dutch call it by a name expreſſing the Flying-mouſe; and many authors, Glis volans.

QUADRUPEDS.

Claſs the Fifth.

JUMENTA.

THE teeth of the Jumenta are few in number, and diſpoſed in an irregular manner; often they differ from one another extremely in ſize and figure.

ELEPHAS.

THERE are no fore-teeth in the mouth of the Elephas: the upper canine teeth are very long; the anterior part of the head is alſo furniſhed with a very long and flexible proboſcis: the teats are two, and they are ſituated on the breaſt. Of this ſingular genus there is only one known ſpecies.

ELEPHAS. **The Elephant.**

The ſize of the elephant alone, had it no other diſtinctive character, would be ſufficient to make it known from all the other animals in the world; but it has ſingularities of the moſt ſtriking and obvious kind, beſide this. When at full growth, it meaſures from ſeventeen to twenty feet in height, from the ground to the higheſt part of of the back, and it's body is withal so enormously bulky, that the belly reaches nearer the ground than could be easily conceived of that of a creature of such height: the head is large, and of a kind of ovated figure, large and broad at the temples, and smaller to the mouth: the extremity is continued to a great length, in form of a proboscis or trunk, which it uses, in the place of a hand, to reach it's food to it's mouth; and, on other occasions, to save the trouble of much motion in it's unwieldy body: from the sides of the upper jaw there grow two teeth of so enormous a length and size, that, to those who have not seen the animal, it appears scarce to be conceived, that any creature can carry them; these are what furnishes us with ivory, and we see them of between three and four hundred pounds weight the pair: the mouth is small, in proportion to the bigness of the creature; but, besides these enormous teeth which hang out of it, there are in each jaw four grinders confined and hid within it, which are of a scarce less surprizing magnitude and structure; they are composed, as it were, each of many other smaller teeth joined together, and they are oblong, high, flatted, and level at the top, excepting for some transverse, undulated ridges, by means of which they grind their food.

The eyes are very small, in proportion to the bulk of the head: the ears are large and membranaceous; the legs are of an amazing thickness, and the feet are very broad; they are not covered each with a single hoof, in the manner of the base, but is divided into toes; these, however, do not stand separate, as in the generality of quadrupeds, but adhere one to the other, and are all covered by the same common skin, only the extremities of them appear at the verge of the foot, and they are there armed with broad and obtuse claws or nails: the tail is small, short, and inconsiderable.

The whole body is covered with an extreamly thick and strong skin, of a kind of mouse colour, with an admixture of tawny; it is not naked, as some have asserted, but has a few long and very rigid hairs growing at distances on it: it is all over covered also with a kind of tubercles and excrescences resembling warts, which are split deeply in several parts, and in different directions, down from the surface; these are of a somewhat darker colour than the rest of the skin, and, when they are cut, have a great resemblance of whale-bone: these tubercles, though divided and cracked in this manner, adhere, in a very firm degree, to the skin, so that it is impossible to tear off either a whole one, or even but a part of one, without separating a portion of the skin with it.

The proboscis or trunk of the elephant is, properly speaking, nothing but the nose continued to a greater length; the substance of it is firm, but fleshy, and it is composed of three series or orders of fibres: the creature has, among the variety of motions which it can give, this singular organ or power of retracting or protruding it forward with great violence; and the difference of it's length, when thus contracted or extended, is not less than from one foot to five, or more. The aspera arteria or windpipe is large, and has no epiglottis to defend it from the entrance of any thing; it is quite distinct from the oesophagus, so that there is no danger of any thing getting down it in the creature's swallowing it's food, or drink; but it would be very easy for any little animal, that made it's way in at the aperture of the trunk, to creep down into the lungs; and, to defend itself against this, it always sleeps with it's trunk, applied close to the ground at the extremity, so that nothing but air can get in.

The ears, which are very large, are dentated round the edges, and they hang in an irregular pendulous manner, nor are ever erected, except when the creature is provoked; at the same time that these are contracted, and in some degree erected, for it is but in an imperfect manner that the elephant has this in it's power: the snout is also contracted to it's shortest dimensions, so as to be held in readiness to dart out on the enemy with full violence: the soals of the feet are not covered with any thing horny or firm, but with a mere skin, and this indeed thinner than that of the rest of the body, and easily cut through with a knife. The eminences in the circumference of the foot are five; they are short, and answer to so many toes; and under the skin which covers the toes, down to the division of these, there is lodged a great quantity of fat for the keeping them soft, and preventing their rubbing against one another: the

large

large teeth are hollow, and have a medullary substance running up a great way on them; the grinders are solid, and their weight is very great.

The elephant is a native of Africa, and some other of the warmer countries, and even of some colder climates. We have seen them occasionally preserved alive with us, and shewn as curiosities. All the writers on animals have described it, under the name of Elephas.

RHINOCEROS.

THE Rhinoceros has in each jaw eleven fore-teeth; there are no canine teeth: the nose is ornamented with a single or double horn, which is permanent.

Rhinoceros cornu unico. **The common**
The Rhinoceros, with a single horn. **Rhinoceros.**

This, of all quadrupeds, approaches nearest to the elephant in size, but it is not equal to it in that respect: the body is nearly as bulky, but the legs are much shorter. A full-grown Rhinoceros measures fourteen feet from the ground to the highest part of the back, and the legs are so remarkably short, that, with all this height, the belly comes near the ground: the head is very large and oblong, of an irregular figure, broad at the top, and narrower and depressed toward the snout: the ears are very large and long; they in some degree resemble those of a hog, and are soft, and covered with a tender skin: the eyes are very small, and there is something extreamly singular in their situation; they do not stand on the upper part of the head, as in other animals, but at a small distance from the extremity of the snout: on the upper part of the snout, near the extremity of it, there stands a horn of a conic figure, and very strong; it grows to about two feet and a half in length, and is a little bent backwards; it's colour is black, and it's substance very firm and hard.

The neck is short and very thick; the body rounded, and enormously big; the legs are very thick and clumsy to appearance, but all that strength is necessary to their being able to support so immense a bulk of body: the feet are broad, and divided into toes; and the tail is short, and furnished with some long and extreamly thick black hairs.

The colour of the creature is a dirty tawny; the skin is remarkably thick and hard; it is, indeed, so hard, that the creature could not easily turn itself in any directions, but that nature has formed a kind of joints and folds in it; by means of these it moves it's body, though in an unwieldy and awkward manner.

It is a native of some parts of Asia and Africa; it generally frequents the parts of the country which are far from the resort of men: it feeds on vegetables. We sometimes have it brought into Europe, and shewn as a curiosity. The skin of this creature, like that of the elephant, is covered at little distances with a kind of low protuberances, resembling warts; these have all hairs growing out of them, but they are but few, and are very thick and black. It is not easy to say to what length they would grow, were they left to themselves; for the creature is subject to itchings of the skin, and rubs them all off at but a little height above the skin. The horn also often shares the same fate. There is one now kept at a shew in London, in which the horn is not more than three inches high, and obtuse, which is owing to the creature's continually rubbing it down against the walls and boards of the place where it is kept.

Rhinoceros cornu gemino.
The double-horned Rhinoceros.

This is a large, unwieldy animal, in most respects greatly resembling the former: the head is enormously large and long; it is so bulky, that the creature seems to find pain

pain in holding it up, and is always seen in the wild state with it in a depending posture: the ears are large; the eyes small, and placed near the extremity of the snout: at the upper part of the nose, and near it's tip, there grows a large horn, as in the other species; and, just behind it, another of the same form and colour, but smaller; they are both of a firm and hard texture, and pointed at the extremities; but the larger never grows to so great a length in this species, as in the other, which has it single.

The neck is moderately long and thick, especially toward the shoulders; the body is of the same enormous bulk as in the other; the legs also are very thick and short, and the feet in the same manner divided into toes: the tail is about a foot and a half long, not very thick, naked for three fourths of it's length, but at the tip furnished with hairs not very numerous, but thick, and considerably long.

The skin is thick and hard, as in the preceding species, and in the same manner is covered with little protuberances, in form of warts; the hairs are few, but they are very thick and strong; they arise from these, and are black, and of some length.

This is a native of some parts of the world with the former; travellers have seen it living, but we never had one brought over to Europe. We have sometimes met with the horns preserved, with a part of the skin of the head to which they grow; some of our museums afford specimens of these, by which we are assured of their situation, a circumstance concerning which many have erred.

HIPPOPOTAMUS.

THE fore-teeth of the upper jaw of the Hippopotamus are four, they are placed in pairs, and at some distance; those of the lower jaw are prominent, and the intermediate ones are protruded forward: the canine teeth are single and obliquely truncated; the teats are only two, and they are placed near the groin.

Hippopotamus cauda brevi.
The Hippopotamus, with a short tail. **The Sea-horse.**

This is a very large and unwieldy animal; it is equal to a common ox in size; the head is very large, oblong, and somewhat depressed, and is obtuse and thick toward the rostrum or extremity: the ears are about three inches in length, so that they appear small, in proportion to the bulk of the head: the eyes are small, and stand not very high in the head; the mouth is very wide, and the teeth very large and strong: the canine ones are not exerted as in the boar, but, when the mouth is opened, they make a very formidable figure; they are six inches long, and nearly as much in circumference, and are not rounded, but of a somewhat flatted and trigonal figure; they are harder than the teeth of any other known animal, very few tools will make the least impression on them, and they will give fire in great plenty on being struck against a steel, in the manner of flint.

The neck is moderately long, and very thick; the body is extreamly bulky, and rounded: the legs are very thick, and are between three and four feet in length: the foot is twelve inches in breadth, and is divided into four toes each, armed with a claw of near three inches in breadth, rounded and obtuse.

The tail is very singular in it's form, it more resembles that of the tortoise than of any other animal; it is not more than six inches in length, and is of a conic figure, very thick at the base, and thence tapering very swiftly to a point: the creature moves it up or down at pleasure, but it has no power of bending it.

The skin is very thick and black, it is perfectly naked; there is not the least appearance of hair on it, only that about the mouth there are a kind of whiskers, as cats have.

This

This is a native of Egypt, and some other warm climates. It frequents rivers, and passes a great part of it's time under water, but it comes out of it to sleep and to breed. The Nile and the Niger, and some other large rivers, abound with it. It feeds on vegetables, and is particularly fond of the roots of some of the water plants. It does not swim under water, as most of the amphibious animals do, nor are it's feet at all formed for that purpose; it's weight and bulk would also render it very unfit for such motions. It walks on the bottom, and feeds on the vegetables which are produced there. All the writers on animals have named it. Ray and others call it Hippopotamus; Bochart supposes it to be the Behemoth, mentioned in the book of Job.

Hippopotamus acaudatus capite graciliore.
The slender-headed Hippopotamus, without a tail. **The Tapijerete.**

This is of the bigness of an ass, only that it's body is much more bulky and unwieldy: the head is large and oblong, it somewhat resembles that of a hog in shape, and is slenderer, in proportion to it's length, than that of any other of this genus: the ears are short, but they are very broad, and approach to a roundish figure; they usually hang pendulous in some degree, but the creature at pleasure renders them rigid, and they are in that state not erected, but only protended forward: the eyes are small and black; the mouth is large and formidable, and the teeth are thick and obtuse.

The neck is short, and very thick; the body is thick, and in some degree resembles that of a hog in figure; the skin is very tough and thick, and is not naked, as in the former species, but covered with short and glossy hair: it is of a dusky umber colour, when the creature is full-grown; when younger, it is of a paler brown, and spotted with black: there is no tail, nor any the smallest rudiments of one: the legs are short, but robust.

This is a native of South America, and lives principally on land, though it will occasionally frequent the waters for it's food. It eats only vegetables, and it's flesh is much esteemed at the tables of the Portuguese, who have settlements in South America. It is frequent in the woods of the Brasils. Maregrave and others who have written on the animals of that part of the world, have described it. They call it, by it's Brasilian name, Tapijerete; and some, by it's Portuguese one, Anta.

Hippopotamus acaudatus capite crasso.
The thick-headed Hippopotamus, with no tail. **The Copy-Bara.**

This is the smallest animal of this genus; the bigness is not more than that of a small hog, and it considerably resembles that animal in the shape of it's body: the head is enormously large and thick, it appears greatly disproportioned to the body: the ears are short, broad, and roundish; the eyes are large and black; the mouth is very large, and it's armature of teeth is very formidable: the neck is short, and very thick; the body is about two feet in length, and a foot and a half in diameter, and is covered with a very tough and firm skin, which is not naked, as in the great Hippopotamus, but has a fine glossy, though not thick, fur: this is composed of thick and moderately long hairs, and is of a dusky tawny, with a strong tinge of brown.

The legs are short, and the feet divided into toes, each of which has a broad and obtuse nail: the creature has no tail.

This is a native of South America, and spends it's time principally in the woods, though it will descend into the rivers occasionally. It feeds on vegetables, and it's flesh is well tasted, and esteemed at table. Maregrave and others have described it under it's Brasilian name Copy-Bara.

EQUUS.

THE fore-teeth of the Equus are six; the upper ones are incurvated, and the inferior are prominent: the canine teeth are not exerted; they are on each side separated by a space from the other: the hoof is undivided, and the teats are two, and are situated in the groin.

Equus cauda undique setosa.
The Equus, with the tail hairy all over. **The Horse.**

The horse is one of the noblest animals of the creation; he is in strength and natural fierceness equal to any, and is yet easily tamed, and made fit for our purposes: scarce any creature excels him in swiftness, any more than in strength, and hardly any in beauty.

The head is long and large; the eyes large and prominent; the ears erect and beautiful: the neck is long and thick, elegantly formed and decorated with a mane of long hair, like that of the lion: the body is rounded, and beautifully turned; the legs are strong without being bulky; the tail is long and hairy all the way; the hairs on it resemble those of the mane, but that they are longer, thicker, and more beautiful.

The horse is a native of some parts of Europe, and in it's wild state has a great majesty of appearance; it is kept tame for the purposes of draught and carriage, in all the known countries of the world, where it is either native, or to be had by means of commerce.

Equus cauda extrema setosa.
The Equus, with the tail hairy at the end. **The Ass.**

The ass considerably resembles the horse in many respects, but it gives but a faint and mean copy of that noble animal; it is smaller, and wants the symmetry as well as the dignity of appearance so conspicuous in that generous animal.

The head of the ass is long; the ears are very long and narrow: they seem much over proportioned in length to the head, and have no very elegant effect in the appearance they give to the whole head: the eyes are large, but they have nothing bright or striking in their aspect: the neck is moderately long, but it is lank, and not finely turned: the body is rounded, and the back somewhat depressed; the legs are long and slender, and the tail is very long, but it is not hairy all the way, as in the horse, but only at the end.

The ass is covered with a short and coarse fur, of a pale dun colour, and has a streak of black running down it's back, and across the shoulders; it's neck does not wholly want the beauties of a mane, but it is shorter and less regular than in the horse.

The ass is wild in many of the warmer countries; and authors, who are fond of multiplying species, suppose it different in this wild state from the tame ass, and call it Onager. Many have described it under this name as a distinct species, not observing that their descriptions gave no specific mark of distinction.

The horse and ass are so nearly allied, that they will copulate together, and the produce is a mule; a creature of a middle nature between it's two parents, but incapable of propagating it's species; so careful is nature to avoid filling the world with monsters.

Equus lineis transversis versicolor.
The transversly streaked Equus.

The Zebra, or wild Ass.

This is an extreamly beautiful animal, and, though in colouring so extreamly different from all the other kinds and varieties of the Equus, agrees with it in all other respects; it is about equal to the common ass in size, but of a much more elegant figure: the head is small and short; the ears are long; the eyes are large and bright, and the mouth is considerably large: the neck is long, slender, but elegantly turned; the body is rounded and small, in comparison of that of the ass: the legs are long and slender, but very strong; they seem all bone, only just covered with the skin: the hoof is undivided, and of a deep brown: the tail is long and beautiful, but it is only hairy at the end.

The whole animal is particoloured, or beautifully striped in a transverse direction, with long and broad streaks, alternately of a deep glossy and shining brown and whitish, with some absolutely black: those of the body have their origin from the ridge of the back, and are carried down to the belly, surrounding the whole body.

This is a native of many parts of the East; there are usually seen great numbers of them together, in manner of flocks of sheep, and they are extreamly swift of foot. All the writers on animals have described it; they call it Zebra and Asinus Africanus.

SUS.

THE upper fore-teeth of the Sus are four in number, and are convergent: those of the lower jaw are eight, and are patulous: the canine teeth of the upper jaw are two, and they are short; those of the under jaw are single, and they are exerted: the crown of the head is carinated, and the hoof is divided.

Sus dorso antice setoso, cauda pilosa.
The Sus, with a bristly back, and a hairy tail.

The common hog.

Descriptions of animals so common as the hog, the horse, and the like, may seem impertinent and superfluous; but the utter disregard to certain objects, because universally known at the time, has been the source of one of the great deficiencies in the natural history of the antients. There is also this to be said farther, in defence of a succinct mention of the more remarkable and characteristical parts, that the creature, in it's native wildness, is often unlike, in many respects, to those of the same species, which are kept about our houses, and that the marks of such, while they preserve the characteristics of the animal, may also keep up a knowledge of it's genuine condition. Many authors, in regard to this particular species, have distinguished the wild boar and the tame one as if of distinct species, but they are the same.

The head of the hog is large and long; the snout is terminated by a rounded plane, in which are situated the nostrils; the eyes are small, and not prominent: the ears are shorter in the wild state, than as the creature is kept tame, and the rostrum appears longer, but this is principally from the tame hog being fatter: the neck is short and very thick; the body is extreamly bulky, the belly often hanging down almost to the ground: the tail is short and hairy all the way along; the legs are robust, but not very thick or long: the back is furnished with a broad series of bristles, and the whole body is covered with two kinds of hair in the wild state, a longer and a shorter, but this is less distinguishable in the tame hog.

The hog is a native of Germany, Italy, and many other parts of Europe; it is fond of mountainous places, and generally lives in thick woods. When fed with us, it grows fat in a degree beyond that of any other animal. No creature is more worth attention in regard to profit from keeping, for it is fed at small expence, and will

produce

produce as far as twenty pigs at a litter. All the writers on animals have deſcribed it. They call it Aper, Sus, and Sus domeſticus. It is frequent to ſee hogs with undivided hoofs; theſe have been deſcribed as a diſtinct ſpecies, under the name of Sus ungulis ſimplicibus, but this is only an accidental variety; the naked African hog, and the Chineſe hog, though deſcribed alſo as diſtinct ſpecies, are of the ſame kind, only varieties of the common hog.

Sus dorſo pone ſetoſo, cauda nuda.
The Sus, with the back briſtly behind, and with a naked tail.

The American Hog.

This is of the ſize of our common hog: the head is very long; the ſnout truncated and flat; the eyes are ſmall, and the ears are remarkably long, and pointed at the extremity: the neck is ſhort and thick; the body alſo is very thick and clumſy; the legs are ſhort, but remarkably bony and ſtrong; the tail is very long, and has no hairs upon it.

The whole creature is of a duſky and diſagreeable browniſh-red colour; the back is not all the way furniſhed with briſtles, as in the European hog, but only has them on the hinder part, toward the inſertion of the tail.

This is a native of many parts of South America; it is not only wild in the woods, but is kept tame about houſes, for the ſake of it's fleſh, which is equal to that of our hog. The writers on the Braſilian animals have deſcribed it, and from them others. Ray and moſt of the moderns have called it Porcus Guinienſis, the Guinea Hog.

Sus dorſo cyſtifero, cauda nulla.
The Sus, with a cyſt on the back, and with no tail.

The Musk-hog.

This is ſmaller than the common hog, but in moſt reſpects it greatly reſembles it in form; it meaſures about two feet in length, from the crown of the head to the rump, and it is conſiderably thick, in proportion: the head is long and ſlender, as in our hog; the eyes are ſmall, and ſeem in a manner buried in the head: the ears are two inches and a half long, erect, and pointed at the extremity: the neck is ſhort and thick, the body corpulent; the legs ſhort and robuſt, and formed perfectly like thoſe of our hog, and the feet cloven in the ſame manner, but it is ſingular that it has no tail: the whole body is of a duſky iron-grey colour, and the hair is ſtiffer, and more rigid than in our hog, the whole deſerving the name of briſtles, and every hair being variegated with annular marks of a blackiſh and a deep grey, in ſome with annules of a duſky white; it is from the combination of theſe that the general grey colour ariſes. The briſtles are ſhorteſt on the lower part of the ſides, and they thence become gradually longer to the top of the back, where there is a ridge of them of about five inches long, and in all reſpects perfectly reſembling the common hog's briſtles.

On the middle of the head there ariſes a kind of creſt, compoſed of a large cluſter of briſtles; theſe are very rigid, and are in general black, and on the middle of the back there is a kind of cyſtic gland, with an opening at the upper part, in which there is ſecreted a perfumed fluid matter of a mixt ſmell, between that of muſk and civet.

This ſingular creature is a native of Mexico; it's fleſh is eſteemed at table, but it is a cuſtom to cut out the perfumed gland on the back, as ſoon as the creature is killed, otherwiſe it is ſaid to communicate it's ſcent in a very diſagreeable manner to the fleſh. Moſt of the writers on quadrupeds have named it. Marcgrave and others call it Tajacu; Ray after Tyſon, and others after him, Aper Mexicanus moſchiferus, or the perfumed Mexican Boar.

Sus dentibus duabus fronti innatis.
The Sus, with two teeth growing on the forehead.

The Babyrouſſa.

This is a very extraordinary animal; it is of the bigneſs of our largeſt hogs, but is leſs corpulent, in proportion to it's height: the head is oblong, and narrow toward the ſnout, at the extremity of which it is truncated and plain; the neck is thick, and moderately long; the body is rounded and fleſhy, but not ſo immoderately as in our hog: the legs are moderately long, and very robuſt; the hoofs are cloven, and the tail is ſhort and hairy all the way, as in the common hog.

The eyes are ſmall, and do not ſtand at all prominent; the ears are long, erect, and acute: the mouth is large, and well furniſhed with teeth, but there is ſomething ſo ſingular in the ſituation and growth of one of the pairs, that it is ſcarce neceſſary to mention any other diſtinction or character of the animal: there are a pair of exerted teeth in the lower jaw, not unlike those of many other animals; but there are alſo two in the upper jaw, which perforate the fleſh of the head, and ſtand forward in the manner of horns: they have been called horns by many, but their inſertion by Gomphoſis in the jaw ſhews them to be in reality teeth.

This ſingular ſpecies is a native of ſome parts of the Eaſt Indies. Authors call it Porcus Indicus and Babyrouſſa.

QUADRUPEDS.

Claſs the Sixth.

PECORA.

THE Pecora have no fore-teeth in the upper jaw; thoſe in the lower are ſix or eight: the feet are covered with divided hoofs; and the teats are two, and are ſituated in the groin.

CAMELUS.

THE Camelus has no horns; the lip is divided: the fore-teeth of the lower jaw are ſix, and they are broad, and ſtand prominent: the canine teeth of the upper jaw are three; thoſe of the under jaw are only two, and they are ſituated at ſome diſtance from one another.

d

Camelus topho dorſi unico.
The Camelus, with a ſingle bunch on the back.

The Dromedary.

This is a large and tolerably beautiful animal; it is higher in the back than the horſe, and, when the head is held erect, it is much more ſo, as the neck is longer; but the neck is ſlenderer, as well as longer, than in the horſe, and what it has in height it wants in dignity; the body alſo is proportionably ſmaller; the head of the dromedary is ſmall, and the upper lip is divided in the manner of that of the hare; the ears are ſhort; the eyes are large, and the opening of the mouth but ſmall: the neck is ſlender and rounded; the body is alſo rounded, and on the back there is a ſingle and large callous protuberance: the legs are long and tolerably robuſt, but leſs ſo than in the horſe-kind: the feet are large and broad, and each has two obtuſe nails at the front: the ſoal of the foot is very broad; it is flat and fleſhy, and is covered only with a ſoft ſkin.

There are ſix calloſities on and about the knees, which nature has provided to be of uſe to it, in that frequent bending and reſting on theſe parts, which is neceſſary for it's own purposes, as well as those of mankind who employ it; and there is a seventh callosity much larger than those on the breast; this is eight inches long, six broad, and two thick.

The dromedary is a native of many parts of the East, but it is less frequent than the camel: it is swifter than that creature, but less strong, and is used for riding on, more than for carrying heavy loads.

The Bactrian Camel.

Camelus tophis dorsi duobus.
The Camelus, with two bunches on the back.

There has been great confusion of names between this and the former species. As most of the more accurate writers have called this the camel, and that the dromedary, there have been many who have called that the camel; and, when we apply the term dromedary to express a creature that moves slowly, we seem to have understood this species of the camelus originally by that name. The vulgar seem to continue their opinions; but the more accurate all use the words in this sense, in which they stand here.

The camel is larger than the dromedary, and, when it holds up it's head, is indeed of an immoderate size, in regard to height: the head is small and short; the snout is obtuse, and the upper lip is split or divided in the manner of that of a hare: the eyes are not large, but they stand in a bold and prominent manner; the ears are short, and the upper part of the head is somewhat depressed: the neck is very long and slender, though not so much so as in the dromedary; it is of a figure approaching to rounded, but somewhat flatted; the breast is broad; the body is rounded and bulky, and on the top of the back there stand two large bunches or protuberances, which seem as if formed by nature for the fixing of burthens: the legs are robust, but slender, and considerably long, and the feet are broad and tender, and divided just in the manner of those of the dromedary: the tail is somewhat shorter than in that species; from the base to the extream hairs it is not more than two feet in length.

The covering of the whole body is a soft and fine fur; the hairs are shorter than those of our ox-kind, and greatly softer to the touch: those on the head, at the throat, and on the top of the neck, are much longer than the others, and the longest of all grow about the bunches on the back: there are some in that part which are nearly a foot in length, and they still are soft and flexible, not rigid and harsh, as those of the mane and tail of the horse.

The camel is a native of many parts of Asia, particularly of Bactria, the name of which country makes a general part of it's denomination. It is of great use in the East in carrying burthens. Almost all the writers on this subject have described it under the name of Camelus and Bactrianus.

The Glama.

Camelus dorso laevi, pectore gibboso.
The Camelus, with the back even, and the breast gibbose.

This is an extreamly singular animal; it has been supposed by most a species of sheep, though very absurdly; it is truly a camel, though in many respects different from the two generally known species just described; it is about four feet in height from the ground to the top of the back, but if measured to the head, when that is held erect, it is greatly taller: the length of the creature, from the head to the rump, is about six feet: the head is small; the rostrum is obtuse, and the upper lip is divided in the manner of that of the hare, or exactly as in the common camel: the eyes are large and protuberant, and the ears are short; the neck is very remarkably long, even for an animal of the camel-kind: the breast is broad, and has on it's middle a very large protuberance or bunch, exactly resembling that of the camel's back; there is continually a kind of fluid secreted from this part: the back is rounded, and the body tolerably fleshy: the tail is short; the legs are moderately long: the whole body is

of

QUADRUPEDS. Series 4
The Elephant
The Rhinoceros
The Horse
The Boar
The Dromedary
The Camel
The Elk
The Ibex

of a pale dun colour, and the down is soft and fine: the feet are broad and fleshy; the sole flat and soft, covered only with a skin, as in the camel, and the front of the foot is divided into toes in the same manner.

This is a native of Peru; it is a tame and inoffensive animal, and serves the natives as a beast of burthen. Ray calls it Camelus Peruvianus Glama dictus; others, Ovis Glama dictus, Matthiolus Elapho-camelus.

Camelus gibbis nullis.
The Camelus, without any gibbosity. **The Pacos.**

This has also been reckoned by many a species of sheep, though it be truly and properly a camel; it is about three feet and a half high, from the ground to the top of the back, but the neck is very long, and, when the head is carried erect, it is very tall: the head is short; the ears are also short and broad; the eyes are moderately large and prominent, and the upper lip is divided or split as in the hare: the neck is long and slender; the body is plump and rounded; the tail is a foot and a half in length, and the legs are long and slender, but strong, and are covered with a very short and fine down.

The head has some considerably long hair about it, and the neck is covered with a surprising quantity of the same kind, but of a greater length: this is used as wool in the country where the creature is produced, and it is in consequence of this that the animal which produces it has been called a sheep, for there is not any one thing beside that can have led to the ranking it with that animal.

It is a native of Peru, and is sometimes employed, as the Glama, in carrying burthens; but it is less fit for this; it is valued for the hair of the neck, and for it's flesh, which is well tasted. Authors call it Pacos, and Ovis Peruviana.

MOSCHUS.

THE Moschus has no horns; the canine teeth of the upper jaw are exerted. Of this genus there is only one known species, which is the animal that produces the perfume, from which it is named.

MOSCHUS. **The Musk Animal.**

This creature has not only been misrepresented by those who have written on animals in general, but it has been referred to genera, with the species of which it has not the least affinity. Some authors have made it a kind of goat; others have called it a stag, but this is extremely improper, and it is very strange that those who thus arranged it, should not have attended to it's having no horns, and to it's having exerted tusks.

The creature, when full-grown, is three feet in length, from the tip of the nose to the rump: the head is oblong; it's length is about six inches, and it is narrow; it is not more than three inches in diameter across the forehead, and it thence gradually tapers to the nose, where it is very small: the anterior part of the head is, indeed, much like that of the greyhound: the ears are long and erect; they resemble those of the rabbit, and are equal in length to the diameter of the forehead: the tail is very short; it is not more than two inches in length, and the creature always carries it erect: the body is tolerably fleshy and rounded; the legs are moderately long, they are more than a foot in length, and are tolerably robust: the feet are deeply divided each into two claws in the anterior part, and as many heels behind: the heels are as long as the anterior divisions of the foot, and are therefore remarkably conspicuous.

The fur on the head, and that on the legs, is about half an inch long; that on the belly is an inch and half in length, and that which grows on the back is three inches: these

theſe hairs are thicker than in any other known animal, and are variegated from the baſe to the extremity with diſtinct ſpaces of brown and white; thoſe on the head and legs, indeed, are ſolely and uniformly brown, and thoſe on the belly, and under the tail, white; and theſe, as well as thoſe on the back, are not ſtraight, but curled or waved, with two or three undulations.

On each ſide of the lower jaw, near the angle of the mouth, there ſtands a tuft of ſhort and rigid hairs; theſe are all of the ſame length, and theſe tufts of them have a very ſingular appearance. The veſſel or bag in which the perfume, called muſk, is contained, is three inches long, and two broad, and hangs under the belly, protuberating near three quarters of an inch beyond the ſurface.

This is a native of the Eaſt. All the writers on quadrupeds have deſcribed it under the name of Moſchus, Animal Moſchiferum, Capra Moſchifera, and Cervus Moſchiferus.

CERVUS.

THE horns of the Cervus are ſolid, divaricated, and deciduous; they are hairy at firſt, but afterwards ſmooth: the canine teeth of the upper jaw are ſingle and remote.

Cervus cornubus ſimpliciſſimis, pedibus anticis longiſſimis.
The Cervus, with the ſimple horns, and the fore-legs very long.

The Camelopardalis.

This is one of the moſt extraordinary animals in the world. Authors have been ſtrangely puzzled among what others to arrange it; ſome have made it a camel, others a ſheep, and the generality have called it neither the one nor the other, but an animal ſui generis: the horns differ extreamly in form from thoſe of the generality of the ſtag-kind, but it is evidently of that genus, nor is there any neceſſity to eſtabliſh the palmated figure of the horns, as one of the characters of it.

It is at once one of the moſt beautiful and aſtoniſhing creatures in the world; when it ſtands erect, it meaſures to the head not leſs than fifteen feet from the ground, and from the front of the noſe to the tail is eighteen feet; this vaſt height and length, however, are, in a great meaſure, owing to the length of the neck, which is enormous, beyond that of all other animals, for the body is very ſmall, in proportion, and eſpecially at the hinder part is but low.

The head is of the form of that of the common deer, but it is much ſmaller, in proportion: the eyes are large and prominent; the ears very large and patulous, and the horns ſimple, very ſhort, and obtuſe: the neck is ſlender; though ſo monſtrouſly long, it carries it uſually erect and ſtraight, and the hinder part of it is ornamented with a very elegant mane, which flows gracefully in every direction about the neck.

The body is not very bulky, but it is rounded and ſhort: the breaſt is broad; the hinder part ſmall, and as it were diſproportioned: the tail hangs down to the hams, and is beſet with hairs as ſtiff and thick as thoſe of a horſe's tail.

The proportion of the legs is another extream ſingularity in this creature: the anterior pair are very long, the hinder ones ſo ſhort, that the body is carried in ſo oblique a direction, that the creature ſeems riſing into an erect poſture. Nature has not formed it for eating graſs, for it is with great dificulty that it can get it's head to the ground, as it ſtands; it naturally feeds on the leaves and young ſhoots of trees; the whole body is covered with a fur, compoſed of thick hairs of a moderate length, and it is ſpotted with black, and a very deep brown on a light ground, in the manner of the panther.

It

It is a native of some parts of the East, and is extremely tame and inoffensive. The Arabians call it Zarnapa and Giraffa; most of our writers, Camelopardalis. The figure is so extremely singular, that many people thought it a creature of imagination, till Bellonius described it from the life.

Cervus cornubus acaulibus palmatis.
The Cervus, with the horns palmated, and without a stem. **The Elk.**

The elk is a very large and strong animal; it is equal to a well-sized horse in magnitude, and not inferior in strength, but is a much less beautiful creature: the head is large and oblong, very broad at the forehead, and tapering to the nose, where it is small but obtuse: the mouth is large, the lips thick, and the teeth strong: the eyes are moderately large, but they do not stand prominent; the ears are very remarkably large and long; they in this respect resemble those of the ass, and indeed the female elk is not greatly unlike that creature in her general figure; the ears are nine inches long, and four in breadth, and are erect and patulous: the horns are not tall and ramose, in the manner of those of the generality of the deer-kind, but they arise with a thick and short trunk, which almost immediately spreads into a vast breadth; and this expansion, which resembles in some degree an open hand, is ornamented at the extremity with some denticulations resembling fingers; the neck is short and thick; the body is large, and the back broad and flatted; the legs are moderately long, and extreamly robust: the horns in the male, for the female has none, refer the creature at sight to the stag-class, but the bulk of the body, and general proportions, if the female alone were examined, would not lead any one to think it such.

The whole body is covered with a thick and tolerably deep fur; the hairs are rigid, and very firm: the colour of the whole creature is a dusky brownish-grey, only the belly is somewhat paler, and the legs darker than the rest: the hoofs are divided exactly in the manner of those of the stag, and are supposed to have great virtue in nervous disorders, but 'tis probably no more than that of the horns and hoofs of any other species.

This is a native of many of the northern parts of Europe; it bears the frozen countries very well, and runs upon the ice in a strange manner. All the writers on quadrupeds have described it, and all under the names of Alce and Elen.

Cervus cornubus ramosis, teretibus, incurvis.
The Cervus, with ramose, cylindric, and crooked horns. **The Stag.**

This is a very stately and a very beautiful animal: people are apt to confound it with the common fallow deer, but with great impropriety; it is of twice the size, and different in many other respects: the head is remarkably large, and the neck strong and thick, as is indeed necessary, were it only for the supporting it's vast load of horns.

The head is oblong, very large and broad on the forehead, and gradually smaller, but obtuse at the end of the snout: the eyes are full and large; the ears are long and patulous; the horns are tall, almost erect, and of a beautiful form; they rise each with a single and rounded stem, which continues it's form to the top, only sending off branches and divarications: they are hairy, when once formed, but afterwards they become very strong, and lose all that downy appearance.

The neck is short, and extreamly robust; the body is oblong, rounded, and plump; the back somewhat flatted, and the belly prominent: the legs are long, and very robust; the hoofs cloven: the fur is deep, and the hairs of which it is composed are remarkably thick: the colour is a tawny reddish, with a mixture of brown.

This is a native of England, and of most other parts of Europe, but the sportsmen will not suffer it to be very frequent. The male is called a Stag or Hart, the female a Hind; we call both sexes by the general name of the Red Deer.

Cervus cornubus ramosis teretibus, summitatibus palmatis.
The Stag, with the horns ramose and cylindric, and their tops palmated.

The Rein-deer.

This is a large and very beautiful species; it is not inferior to the elk in size or strength, but it greatly exceeds it in form, and has all the appearance of the deer-kind; it is of the size of a small horse, but it's shape is exactly that of the red deer: the head is oblong, large, and gradually smaller from the region of the ears to the snout, where it is not so obtuse as most of the deer-kind: the eyes are not large; the ears are moderately long, and resemble those of the common or fallow deer in shape: the horns are very large, and very beautiful; they are of a middle kind, between those of the stag and of the elk: they rise with a very large cylindric and strong trunk, which is continued to a great height, as in the stag, and in the same manner as in that species sends off it's ramifications, but toward the top it spreads into breadth, and becomes palmated so, as in the whole to exceed all the other horns of the deer-kind in beauty.

The neck is short, but very strong; the breast and shoulders are very large, and the whole body seems calculated for extraordinary strength: the legs are long and very robust; the tail is longer, in proportion, than in the stag; the hoofs are divided in the usual manner: the colour is a tawny brown, with an admixture of a ferruginous red.

This is a native of the frozen regions; there is no country so far north as not to afford it. It is calculated for living in regions where the common herbage cannot grow, for it will feed on the moss of mountains, and branches of trees. It is of vast use to the inhabitants of the northern countries, as a beast of draught. Authors call it Rangifer.

Cervus cornubus ramosis compressis, summitatibus palmatis.
The Cervus, with ramose, compressed horns, palmated at top.

The Fallow Deer.

This is a very beautiful species; it perfectly resembles the stag or red deer in form, but is much smaller, and exceeds not only that, but every species, in beauty: the head is oblong, rounded, and obtuse; the eyes are large and prominent; the ears are moderately long and patulous: the neck is robust, and moderately long; the body is tolerably bulky and rounded; the tail is short; the legs are long, slender, and yet strong: the horns approach more in figure to those of the rein-deer, than of any other species: they are large at the base, and run up a great way with a bulky stem; but this is not rounded, as in the rein-deer, but compressed at the top; they are divaricated and palmated.

This is frequent in our parks, where we have a great many varieties of it in regard to the colouring, all which are, however, from the same original species. All the writers on quadrupeds have described it; they call it Cervus Platyceros and Dama.

Cervus cornubus ramosis, teretibus, erectis.
The Cervus, with ramose, cylindric, and erect horns.

The Roe-deer.

This is the smallest of the Cervus-kind; it is considerably less than the fallow deer: the head is obtuse and oblong; the eyes are small, but prominent; the ears are moderately long, and very patulous: the horns are like those of the common deer, except that they are not at all palmated at the extremities, but are all the way taper and divided; they also send off a number of shoots on the sides, and are rougher on the surface, than those of any other species: the neck is short and thick; the body is less corpulent than in any other species, and the legs are slender, and very beautiful.

The

The ground colour is a deep brown, but it is variegated in an elegant manner with black, and a very deep grey; it is ſometimes quite plain, and ſometimes has either of theſe marks in a beautiful wildneſs.

We have this in our parks. Authors call it Capreolus and Caprea, but it has no reſemblance of the goat-kind.

CAPRA.

THE horns of the Capra are hollow, and are turned upwards, and are erect, and ſcabrous: the fore-teeth are eight, and the exterior ones are ſhorter than the others, and acute: there are no canine teeth.

Capra cornubus carinatis arcuatis.
The Capra, with carinated, arcuated horns. **The Goat.**

The common goat is nearly of the ſize of the ſheep, but the wool of the latter makes it appear much the larger: the head is oblong, conſiderably broad at the top, and thence gradually taper to the extremity of the noſe, where it is ſmall and ſharp: the eyes are large and bright; the noſtrils wide, and the mouth large: the neck is ſhort and thick; the body bulky, and the legs ſhort, but robuſt, and very ſtiff; the hoof is divided and brown.

The fur of the goat is deep, and the hairs rigid; they are waved, but not curled as in the ſheep, ſo that the longeſt of them have not at all the appearance of wool: the beard is long, and hangs down from the chin; the horns are moderately long, ſtraight, or but little contorted, and of a deep brown: the general colour of the goat, in it's wild ſtate, is a pale dun, but, as kept tame, the varieties in this reſpect are endleſs.

It is a native of moſt parts of Europe. It is fond of mountainous and rocky places, and runs up precipices, and climbs rocks in a ſurpriſing manner, though it's feet ſeem by no means formed for it. All the writers on beaſts have deſcribed it. They call it Hircus and Caper vulgaris.

Capra pedibus digito humano anguſtioribus.
The Capra, with feet ſmaller than a man's finger. **The Guinea-deer.**

This elegant little creature is univerſally referred to the deer-kind, but improperly: it is truly and in all reſpects a goat: it is not ſo tall as a common cat, and extremely fine limbed: the head is ſmall and oblong; the eyes are large, in proportion, and are bright and piercing; the ears are pendulous: the horns are very ſhort; they ſtand almoſt erect, and are furrowed on the ſurface, and of a deep brown: the body is moderately bulky; the legs are very ſlender, the hoofs are divided, as in the common goat, and of a pale brown: the whole animal is covered with a fine coat, of a bright ſhining yellowiſh-brown, and very gloſſy.

This is a native of Guinea, and ſome other warm countries. Authors call it Cervus puſillus Guineenſis.

Capra cornubus erectis uncinatis.
The Capra, with erect, uncinated horns. **The Rupicapra, or Chamoiſe.**

This is a very beautiful animal; it's horns refer it evidently to the goat-kind, otherwiſe it's whole form has more of the appearance of the deer: the head is long and narrow, rounded at the top, and obtuſe, but very ſmall at the extremity of the noſe: the eyes are large, bright, and tolerably prominent; the ears are pendulous; the horns are of a ſingular make; they ſtand nearly erect, and are ſeven inches long; both the male

male and female have them; they are ftraight to very near the top, where they are bent back in the fhape of a hook, and are fharp at the ends; they are of a very dark brown, nearly approaching to black, and are annulated on the under part, and longitudinally ftriated on the upper; they are hollow, and the cavity is filled up by a bony matter growing from the fkull.

The neck is moderately long and flender; the breaft full, broad, and well formed; the body is not very bulky: the legs are flender and long; the whole body is covered with a deep fur, waved, and fomewhat curled at the inner part of the ears; the forehead, the throat, and the belly are white; the upper part of the head, over the eyes, has on each fide an oblong fpot of yellow: the whole body befide is of a blackifh colour, not bright and gloffy, but obfcure; the tail is blacker than the reft, and is is black on both fides, not white underneath, as in many animals of this genus.

This is a native of many of the warm climates, and has been defcribed by moft of the writers on animals. They call it Rupicapra; the French, Chamoife; and the Germans, Gempe, or Gemfe. The hooked form of the tips of the horns, in this fpecies, has given occafion to the error of fuppofing it hangs by them to the rocks, and to many other tales equally abfurd and ridiculous.

Capra cornubus nodofis in dorfum reclinatis.
The Capra, with nodofe horns bending towards the back. **The Ibex.**

This is a very beautiful animal, but in nothing fo fingular as in the length of it's horns: the head is fmall, and beautifully formed; the eyes are large and bright; the ears are pendulous; the nofe is obtufe, and the noftrils wide; the horns are of a furprifing length, they hang backward, and often extend confiderably beyond the rump, being more than equal to the neck and body in length: they are of a blackifh colour, and annulated on the furface.

The body is not fo large, in proportion to the height, as in the common goat, but more refembles the deer-kind; the legs are alfo perfectly like thofe of the deer, ftraight, flender, and elegant; the hoofs are blackifh: the whole body is of a dark dufky colour, and the male has a very long and bufhy black beard.

This is frequent in many parts of Europe; it lives in mountainous places, and, notwithftanding the incumbrance of that vaft length of horns, it runs and leaps with a furprifing force and rapidity. Moft of the writers on quadrupeds have defcribed it. They call it the Ibex and Rupicapra; the Germans, Steinbock.

Capra cornubus teretibus dimidiato arcuatis, annulatis.
The Capra, with the horns cylindric, and half way arched. **The Antelope.**

Many of the goat-kind approach to the deer in fhape, but none fo much as this; it is a very beautiful creature: the head is fmall, oblong, and obtufe; the eyes are large; the ears are pendulous, and the noftrils wide; the horns are of an extremely fingular figure; they are not of any great length; they ftand in an erect pofition, but are not ftraight, for toward the middle they are turned or bent outwards, and from thence to the top they bend inwards, fo that they in fome degree reprefent the lyre of the antients: they are of a coal-black colour, and feem ornamented with a fpiral twift, but the lines are in reality circular: the neck is flender, and elegantly turned, and is longer than in the generality of the goat-kind; the body is rounded, and not very thick, but the breaft is broad and beautiful: the legs are very flender, and the hoofs divided in the manner of thofe of the goat, and black.

This is a native of Africa, and is a very fwift and bold animal; it will attack almoft any thing in it's defence. Moft of the writers on animals have named it. They call it Capricerva, Dorcas Lybica, and Strepficeros Plinii. We have it fometimes kept in parks, but it is apt to be mifchievous.

Capra

Capra cornubus teretibus, rectiſſimis, longiſſimis, baſi annulatis.
The Capra, with cylindric, ſtraight, and very long horns, annulated at the baſe.

The Indian Antelope, or Bezoar-goat.

This is an extreamly beautiful animal, and the horns are different from thoſe of all others of this genus; the creature is of the bigneſs of the common fallow deer: the head is ſmall, oblong, and obtuſe at the noſtrils; the ears are patulous; the eyes are large and prominent: the horns are not leſs than three feet in length; when full-grown, they ſtand erect, and are perfectly ſtraight; they are annulated or marked with diviſions ſtanding circularly above the reſt of the ſurface, toward the bottom, but all the way up from the part they are ſmooth, and of a ſhining ſurface, and black colour: the neck is ſhort and thick; the body is of an elegant form, rounded, and not very thick: the legs are long and ſlender; the tail is a foot in length, and is hairy all the way down: the whole body is ſometimes of a pale and bright brown colour; ſometimes it is darker, and often is variegated in a very elegant manner, and has the ſpots diſpoſed like thoſe in the panther: the hoofs are always black, and the lower part of the legs alſo is uſually ſo.

This is a native of the Eaſt. Moſt authors have deſcribed it, though many of them very inaccurately. They have called it Gazella Indica, and Capra Bezoartica. The Bezoar-ſtone, a drug of more price than virtue, though not than fame, is found in the ſtomach of this animal.

Capra cornubus teretibus, perfecte annulatis, arcuatis.
The Capra, with cylindric, arched, and perfectly annulated horns.

The African Antelope.

This is a ſmall but a very beautiful ſpecies; it is not equal to the common deer in ſize, but it greatly reſembles it in ſhape, and, were it not for the horns, would be more aptly referred to that than to the goat-kind: the head is larger than in the others, but it is not broad but rounded; the noſe is obtuſe, and the noſtrils are wide: the eyes are large and prominent; the ears are large and patulous; the horns ariſe from the middle of the forehead, juſt between the eyes; they are not remarkable for their length, but they are of a beautiful black colour, and are annulated all the way from the baſe to the very tips; they riſe up ſtraight from the head, but they are bent near the middle: the neck is ſhort, and ſomewhat thick; the body is long and ſlender; the legs are ſlender, and the tail is moderately long and black.

The back is of a duſky grey; the belly white, and the ſides are of a mixed colour, between grey and white; there alſo runs down a broad line of white from the forehead to the noſe; the inſide of the ears alſo, and the under part of the tail, are white.

This is a native of Africa, but it is ſometimes brought over to us alive, and kept in our parks. Many of the writers on animals have named it, but few have accurately deſcribed it. They call it Capra Africana.

Capra capite faſciculo tophoſo, cavitate infra oculos.
The Capra, with a tophoſe bunch on the head, and a hollow under the eyes.

This is a very beautiful ſpecies; it is conſiderably ſmaller than the common goat: the head is large, and is ornamented with a very ſingular protuberance in the middle; this is of a fleſhy ſubſtance, and covered with a deep fur: the ears are moderately large and patulous; the eyes are large and protuberant, and, what is very ſingular, there is a cavity on each ſide of the head, between theſe and the noſe, in which there is ſecreted a yellowiſh fœtid liquor, which ſoon coagulates, and becomes black: the preſent quantity being wiped away, there is immediately a freſh ſupply ſecreted: theſe cavities have no communication with the eyes; the smell of the secreted matter is of a middle kind, between that of musk and castor, but the latter is prevalent.

The neck is short and thick; the body is thick, and the belly often hangs down a little: the tail is short, and the legs are robust, but not long.

The colour of the whole animal is a deep grey; it is almost black on the back, paler at the sides, and whitish at the belly: the hoofs are black, as are also the bottoms of the legs. Ray has described this species under the name of Capra sylvestris Africana Grimmii.

Capra auribus pendulis longissimis.
The Capra, with very long, pendulous ears. **The long-eared, Syrian Goat.**

This is of the size of the common goat, and greatly resembles it in form: the head is oblong, small, and obtuse, though slender at the snout: the nostrils are wide; the ears are remarkably large; they are not only very long, but very broad, and do not stand erect, but are pendulous, in the manner of those of our hounds: the eyes are large and prominent; the horns are short, and have but one incortion; they are of a deep black, and sharp at the points: the body is thick and short; the legs are robust, and not long: the colour is a reddish-tawny, not unlike that of the fox.

The creature is a native of Syria, but it is sometimes brought over alive to us, and shewn among other beasts as a curiosity. Gesner first described it. He calls it Capra Mambrina sive Syriaca; Ray continues the same name.

O V I S.

THE horns of the Ovis are hollow, bent backward, twisted, and rugose: the fore-teeth are eight; the slender ones are narrower than the others, there are no canine teeth.

Ovis cornubus compressis lunatis.
The Ovis, with compressed and lunated horns. **The Sheep.**

The common sheep is an animal more remarkable for it's use than beauty, though, when perfectly clean, it is not without it's share of comeliness: the head is small, oblong, and narrow at the snout; the eyes are large and prominent; the ears are patulous, and the horns of a twisted figure turning backwards, and they are not rounded but flat, and annulated on the surface: the face is covered only with a short kind of hair, as are also the legs, which are rather long than robust, but the whole body beside is covered with a thick and deep wool, curled and twisted, and obscuring the shape of the body, and making it seem much thicker and clumsier than it really is.

The tail is short with us, in comparison of that extent to which it grows in some parts of Arabia, where it spreads into a vast breadth as well as length, and trails after the animal. Some have distinguished the sheep under this variety, by the name of Ovis laticauda, as if a species different from the common kind; but this is an error, for the difference is no more than a variety.

Ovis cornubus erectis spiralibus.
The Ovis, with erect and spiral horns. **The Cretic Sheep.**

This is of the bigness of the common sheep, and resembles it in form: the head is oblong, broad at the top, and very small at the nose; the eyes are large and prominent; the ears are patulous; the horns are not at all like those of our sheep; they are short, erect, and straight, large at the base, very sharp at the point, and elegantly cochleated

cochleated or marked with a spiral twist all the way up: the neck is short and thick; the body appears bulky, but that is rather owing to the thick covering of wool which it carries, than to it's real size: the legs are very long, in comparison of those of our sheep; they and the face are covered with short and rigid greyish-fur; the body with a soft white curling wool.

This is frequent in Crete, and other parts of the *Levant*. Bellonius calls it Strepsiceros Creticus; and others have continued the name, only adding to it the generical term Ovis.

Ovis auribus pendulis, palearibus laxis, occipite prominente.
The Ovis, with pendulous ears, a lax dewlap, and with the back of the head prominent.

The Angolan Sheep.

This is a very singular species, and is extremely from the common sheep in many respects; it is somewhat larger than the common kind; the head is shorter, and more obtuse, and on the hinder part swells out in a singular and very conspicuous manner: the nose is obtuse; the nostrils are large; the eyes are large and prominent: the ears are very long and broad, and are not carried erect, but hang down; the horns are small, but they are of the form of those of our ram, and turn round till their point comes near the eyes: the neck is short, and the flesh of the under part of it is loose, and is ornamented with a kind of mane of long hairs; the body is bulky, and the legs are long and robust; the tail is long and hairy.

The body is not covered with wool, in the manner of our sheep, but with short, rigid, and undulated hair, like that of the goat-kind.

This is a native of the coast of Guinea, and some other parts of Africa. Ray and others call it Ovis Guineensis sive Angolensis.

BOS.

THE horns of the Bos are hollow, and are turned forward; they are of a lunated figure, and a smooth surface: the fore-teeth are eight in number, and there are no canine teeth.

Bos cornubus teretibus flexis.
The Bos, with cylindric, crooked horns.

The Bull.

The bull is a very heavy, but a stately and fierce-looking, animal: the head is large, oblong, and very broad; the nose is obtuse; the nostrils are wide; the eyes are large, and have a very fierce aspect; the ears are long and patulous: the horns are short, and but little; bent smooth on the surface, and sharp at the point.

The forehead of the bull is decorated with short curled hair; the skin hangs loose under his throat: the neck is very thick and robust; the body is very large, and the legs strong, and moderately long; the tail is long: the colour is generally a deep reddish-brown, but it varies greatly.

This is wild in many parts of Europe, we have it bred for the sake of it's flesh; when castrated, it becomes what we call the ox. There are some parts of Europe where the bull, in it's wild state, grows to the full bigness of our ox, and is very fierce. Authors have called it in this state Urus, as if of a different species. We call it, in the tame state, Taurus, and the female Vacca, the Cow.

Bos juba longissima, cornubus inflexis.
The Bos, with a very long mane, and bent horns.

The Bonasus.

This is a very bulky and unwieldy animal; it is larger than our bull: the head is short and broad, and the forehead flat; the nose is obtuse; the nostrils are wide, and the

the ears are long and broad; the horns are very ſhort, they do not exceed eight inches in length, and are turned in ſuch a manner, as to be quite unfit for wounding: the neck is ſhort and thick, and is ornamented with a mane like that of a horſe, which hangs down quite to the breaſt: the body is very bulky, much more ſo than in the common bull; the legs are very thick and robuſt, but ſhort; the tail is long: the colour of the whole animal is a deep tawny, only the forehead and the breaſt are white: the mane is of a darker and browner colour than any other part.

This is a native of the Eaſt, and of many other of the warmer climates. When purſued, it does not attempt to defend itſelf with it's horns, but kicks and diſcharges it's dung to a great diſtance againſt the purſuers. Ariſtotle, and all the old authors, have deſcribed it, they call it Bonaſus.

Bos juba longiſſima, dorſo gibboſo.
The Bos, with a very long mane, and a gibboſe back. **The Biſon.**

This is a robuſt and fierce animal; it is equal to the common bull in ſize: the head is very large, broad, and ſhort; the noſe is obtuſe; the noſtrils are large, and the eyes are large and fierce in their aſpect: the horns are longer than in the common bull, and point more forwards: the neck is very robuſt and thick; there hangs all down it a mane conſiderably long and ſhaggy quite to the breaſt; the body is very bulky and unwieldy: the legs are ſhort and thick; the tail reaches to the ground; there is a large bunch on the ridge of the back, like that of a camel.

This is a native of America, and is a very fine creature. Authors call it Bos Camelita and Biſon.

Bos cornubus vaſtis, intortis, reſupinatis.
The Bos, with very large, crooked, and reſupinated horns **The Buffaloe.**

This is a very large ſpecies; it is equal in ſize to our biggeſt oxen: the head is very large; the forehead remarkably broad, and the aſpect very fierce and terrible: the eyes are large and prominent; the ears are large, long, and pendulous: the horns are vaſtly large, of a coal-black colour, and intorted; they are very thick at the baſe, but ſharp at the point: the neck is thick, and remarkably ſhort; the fleſh hangs very looſe under the throat; the body is more bulky, in proportion, than in our bull, and the legs are thicker, but about equal in length.

The colour is uſually a blackiſh-grey, but in this there is great variation: there is uſually ſome white under the belly, and about the forehead.

This is a native of the Eaſt, but it has been introduced into Italy, and ſome other parts of Europe, where it is kept for ſervice. It is a beaſt of burthen and draught in the pope's territories; and the milk of the female is uſed as that of our cows. But it is apt to be miſchievous, and is leſs fit to be truſted looſe in fields where people walk. All the writers on quadrupeds have mentioned it. There has been ſome diſpute whether the name of the Bubalus or Buffaloe belonged to this ſpecies, or to ſome other. It has been, by ſome authors, applied to the other ſpecies, but this moſt agrees with the oldeſt deſcriptions.

F I N I S.

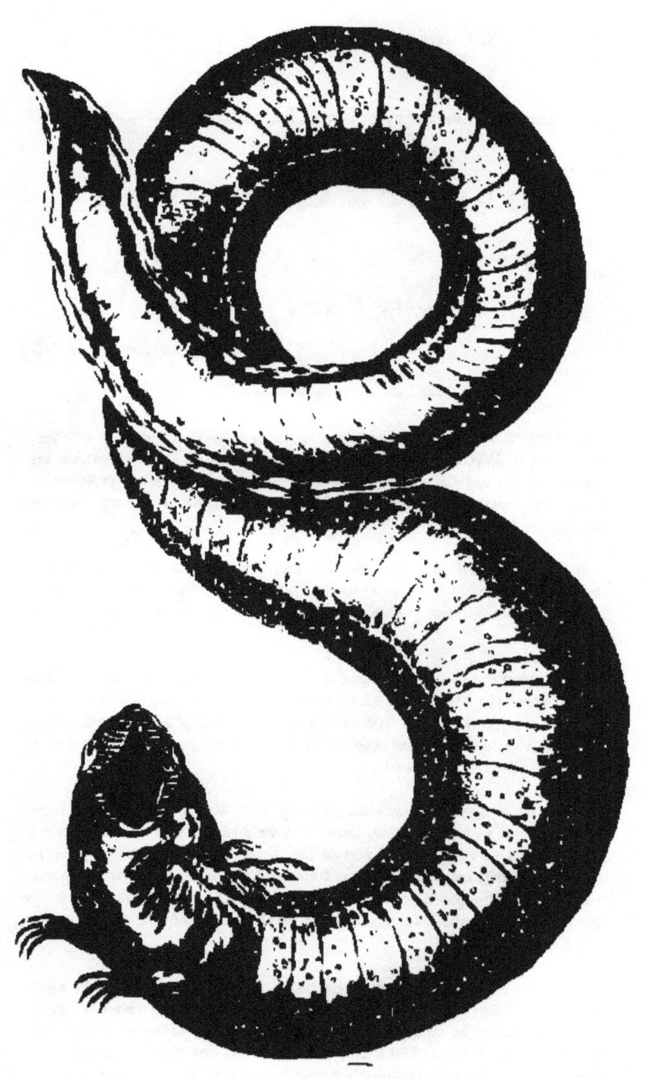

Siren lacertina — The Iguana Siren

See page. 38

ADDITIONAL ANIMAL.

SIREN LACERTINA.

THE IGUANA SIREN.

See Frontispiece of Vol. III.

ABOUT twelve years since there came into England, amongst a Collection of Insects and Reptiles of America, two small creatures, which the possessor of the collection called Two-leg'd Lizards. They were preserved in spirits: they were above six inches long; and all their parts were tolerably perfect.

It was the general opinion at that time that they were Larvæ of some Newt; not perfect Animals. To this it may be remembered, that I objected; because there were palpable and perfect claws upon the toes; and that those toes were four.

Thus rested the matter in uncertainty till the year 1765, when the ingenious and excellent Dr. Garden, of Charles Town, in South Carolina, sent over to England a larger and more perfect specimen. This was laid before the Royal Society; and some anatomical observations were afterwards published about it, in the Transactions of that body.

These observations not perfectly agreeing with some general opinions I had established within my own mind, concerning the unalterable nature of the principal organs in animal bodies, I sent to Dr. Garden, requesting him to send me over one of these creatures, the largest and most perfect he could find, with cautions that no violence were used in catching it: and to this application I owe the specimen, of which an engraving is prefixed as frontispiece to the present volume.

The same ingenious Gentleman who had given the Society those observations, saw and dissected this creature, at my request, with me; in company with a physician, whom he brought for that purpose. We found the parts, as I had expected we should find them; and that what had before appeared out of the course of Nature, was probably owing to the Animal having been hurt.

So

So fair an opportunity offering of ſeeing the creature in its extreme perfection, I could not omit adding an account of it to the preſent Work.

The Animal is harmleſs, tho' its aſpect be formidable; its ſize, in length and thickneſs, was exactly what is given in the figure: it was taken from under the hollow bank of a muddy pond; and when laid on the graſs, ſent forth a very ſingular ſound; 'twas not unlike the chirping of ſome large bird: but in no manner reſembling the hiſſing of ſnakes.

Its colour is a dull lead-like grey; it has no ſcales, but is covered with innumerable pale ſpots: the Body is rounded, bulky, and divided by remote lines, in number about forty: its Head is flatted: its Eyes are ſmall, and covered as they are in eels: the Mouth is ſmall, the under Jaw is ſhorter than the upper, and it has many rows of teeth.

Its Gills ſtand out, and are formed like a kind of fins; there are three on each ſide; and no other perfect creature whatſoever has ſuch. There are, under theſe, three openings on each ſide.

Its Legs are only two; and they are placed very near the head; they are ſmall, in proportion to the bulk of the Animal, and each has four toes, armed with ſharp claws. There neither are any hinder Legs, nor any appearance, where there ſhould be ſuch. Its Tail is flatted; and has an evident fin.

Such is the Iguana Siren, a palpable *Animal bipes et implume*, of all others the moſt ſingular: and in nothing more ſo than this, that to thoſe who arrange creatures in an artificial method there requires not only a new Genus to be erected for it; nay, not alone a new Claſs; but that much greater diviſion, a new Order of Animal Nature. To me, who have laboured always after a natural method, theſe difficulties, of an artificial, are not ſtrange: I know, for I ſee it every where, that Nature proceeds from Species to Species by an exact and equidiſtant ſcale; and that this Animal ſtands as a frontier inſtance, between the artificial orders; connecting two of them: equally allied to both, and properly belonging to neither. And ſo perfect is Nature in her gradations, ſo abſolute in her laws, that this is not the ſingle inſtance of a frontier Animal among the amphibious, or of a two-leg'd creature without feathers. There is a ſnake with two legs; they are two hinder ones, as theſe are fore ones; and 'tis true, they are very poor ones, but ſtill they are legs: and they have toes as theſe have, the number of which diſtinguiſhes them from all other creatures of their kind, for they are only two.

Theſe are the inſtances which perplex artificial methods; but they are the great connecting links in the natural.

F I N I S.

INDEX.

INDEX.

Lophius

INDEX.

Sear.

INDEX.

FINIS.

www.ingramcontent.com/pod-product-compliance
Lightning Source LLC
Chambersburg PA
CBHW022123020426
42334CB00015B/729

* 9 7 8 3 7 4 2 8 0 1 2 5 8 *